VOLUME ONE HUNDRED AND FIFTEEN

CURRENT TOPICS IN DEVELOPMENTAL BIOLOGY

Craniofacial Development

CURRENT TOPICS IN DEVELOPMENTAL BIOLOGY

"A meeting-ground for critical review and discussion of developmental processes"
A.A. Moscona and Alberto Monroy (Volume 1, 1966)

VOLUME ONE HUNDRED AND FIFTEEN

CURRENT TOPICS IN DEVELOPMENTAL BIOLOGY

Craniofacial Development

Edited by

YANG CHAI

Professor
George and MaryLou Boone Chair in Craniofacial Biology
Director
Center for Craniofacial Molecular Biology
Associate Dean of Research
Herman Ostrow School of Dentistry
University of Southern California
Los Angeles, California, USA

ELSEVIER

AMSTERDAM • BOSTON • HEIDELBERG • LONDON
NEW YORK • OXFORD • PARIS • SAN DIEGO
SAN FRANCISCO • SINGAPORE • SYDNEY • TOKYO
Academic Press is an imprint of Elsevier

Academic Press is an imprint of Elsevier
225 Wyman Street, Waltham, MA 02451, USA
525 B Street, Suite 1800, San Diego, CA 92101-4495, USA
The Boulevard, Langford Lane, Kidlington, Oxford OX5 1GB, UK
125 London Wall, London, EC2Y 5AS, UK

First edition 2015

Notices

Knowledge and best practice in this field are constantly changing. As new research and experience broaden our understanding, changes in research methods, professional practices, or medical treatment may become necessary.

Practitioners and researchers must always rely on their own experience and knowledge in evaluating and using any information, methods, compounds, or experiments described herein. In using such information or methods they should be mindful of their own safety and the safety of others, including parties for whom they have a professional responsibility.

To the fullest extent of the law, neither the Publisher nor the authors, contributors, or editors, assume any liability for any injury and/or damage to persons or property as a matter of products liability, negligence or otherwise, or from any use or operation of any methods, products, instructions, or ideas contained in the material herein.

ISBN: 978-0-12-408141-3
ISSN: 0070-2153

For information on all Academic Press publications visit our website at http://store.elsevier.com/

CONTENTS

CONTRIBUTORS

Annita Achilleos
Stowers Institute for Medical Research, Kansas City, Missouri, USA

Zhengwen An
Craniofacial Development and Stem Cell Biology, Dental Institute, Kings College London, London, United Kingdom

Kristin B. Artinger
Department of Craniofacial Biology, University of Colorado Anschutz Medical Campus, Aurora, Colorado, USA

Rebecca Babb
Craniofacial Development and Stem Cell Biology, Dental Institute, Kings College London, London, United Kingdom

Anamaria Balic
Research Program in Developmental Biology, Institute of Biotechnology, University of Helsinki, Helsinki, Finland

Ryan F. Bloomquist
School of Biology, Petit Institute for Bioengineering and Bioscience, Georgia Institute of Technology, Atlanta, Georgia, USA

Jeffrey O. Bush
Program in Craniofacial Biology; Department of Cell and Tissue Biology, and Institute for Human Genetics, University of California San Francisco, San Francisco, California, USA

Yang Chai
Center for Craniofacial Molecular Biology, University of Southern California, Los Angeles, California, USA

David E. Clouthier
Department of Craniofacial Biology, University of Colorado Anschutz Medical Campus, Aurora, Colorado, USA

Gage Crump
Eli and Edythe Broad CIRM Center for Regenerative Medicine and Stem Cell Research at the University of Southern California, Los Angeles, California, USA

Tamar Eigler
Department of Biological Regulation, Weizmann Institute of Science, Rehovot, Israel

Jian Q. Feng
Department of Biomedical Sciences, Texas A&M Baylor College of Dentistry, Dallas, Texas, USA

Teresa E. Fowler
School of Biology, Petit Institute for Bioengineering and Bioscience, Georgia Institute of Technology, Atlanta, Georgia, USA

Scott E. Fraser
Molecular and Computational Biology, University of Southern California, Los Angeles, California, USA

Jennifer C. Fuchs
Department of Craniofacial Development & Stem Cell Biology, King's College London, London, United Kingdom

Alice F. Goodwin
Program in Craniofacial Biology, and Department of Orofacial Sciences, University of California San Francisco, San Francisco, California, USA

Yongchao Gou
State Key Laboratory of Oral Diseases, Department of Orthodontics, West China Hospital of Stomatology, Sichuan University, Chengdu, China, and Center for Craniofacial Molecular Biology, School of Dentistry, University of Southern California, Los Angeles, USA

Rebecca Green
Department of Cell Biology and Anatomy, Alberta Children's Hospital Research Institute, and McCaig Bone and Joint Institute, University of Calgary, Calgary, Alberta, Canada

Benedikt Hallgrimsson
Department of Cell Biology and Anatomy, Alberta Children's Hospital Research Institute, and McCaig Bone and Joint Institute, University of Calgary, Calgary, Alberta, Canada

Belinda R. Hauser
Matrix and Morphogenesis Section, Laboratory of Cell and Developmental Biology, National Institute of Dental and Craniofacial Research, National Institutes of Health, Bethesda, Maryland, USA

Robert J. Hinton
Department of Biomedical Sciences, Texas A&M Baylor College of Dentistry, Dallas, Texas, USA

Matthew P. Hoffman
Matrix and Morphogenesis Section, Laboratory of Cell and Developmental Biology, National Institute of Dental and Craniofacial Research, National Institutes of Health, Bethesda, Maryland, USA

Mamoru Ishii
Department of Biochemistry and Molecular Biology, USC Norris Comprehensive Cancer Center, Keck School of Medicine, University of Southern California, Los Angeles, California, USA

Rulang Jiang
Division of Developmental Biology, and Division of Plastic Surgery, Cincinnati Children's Hospital Medical Center, Cincinnati, Ohio, USA

Junjun Jing
State Key Laboratory of Oral Diseases, West China Hospital of Stomatology, Sichuan University, Sichuan, PR China

Rebecca Kim
Program in Craniofacial Biology, and Department of Orofacial Sciences, University of California San Francisco, San Francisco, California, USA

Ophir D. Klein
Program in Craniofacial Biology; Department of Orofacial Sciences; Department of Pediatrics, and Institute for Human Genetics, University of California San Francisco, San Francisco, California, USA

Yu Lan
Division of Developmental Biology, and Division of Plastic Surgery, Cincinnati Children's Hospital Medical Center, Cincinnati, Ohio, USA

Ralph Marcucio
Department of Orthopaedic Surgery, San Francisco General Hospital, Orthopaedic Trauma Institute, University of California San Francisco, San Francisco, California, USA

Robert E. Maxson
Department of Biochemistry and Molecular Biology, USC Norris Comprehensive Cancer Center, Keck School of Medicine, University of Southern California, Los Angeles, California, USA

Amy E. Merrill
Department of Biochemistry and Molecular Biology, Keck School of Medicine, and Center for Craniofacial Molecular Biology, Ostrow School of Dentistry, University of Southern California, Los Angeles, California, USA

Inbal Michailovici
Department of Biological Regulation, Weizmann Institute of Science, Rehovot, Israel

Washington Mio
Department of Mathematics, Florida State University, Tallahassee, Florida, USA

Lindsey Mork
Eli and Edythe Broad CIRM Center for Regenerative Medicine and Stem Cell Research at the University of Southern California, Los Angeles, California, USA

Cynthia L. Neben
Department of Biochemistry and Molecular Biology, Keck School of Medicine, University of Southern California, Los Angeles, California, USA

Carolina Parada
Center for Craniofacial Molecular Biology, University of Southern California, Los Angeles, California, USA

Christopher J. Percival
Department of Cell Biology and Anatomy, Alberta Children's Hospital Research Institute, and McCaig Bone and Joint Institute, University of Calgary, Calgary, Alberta, Canada

Pedro A. Sanchez-Lara
Director of Craniofacial Genetics, Children's Hospital Los Angeles; Assistant Professor of Clinical Pediatrics; Assistant Professor of Clinical Pathology, and Research Instructor Center for Craniofacial Molecular Biology, Herman Ostrow School of Dentistry, University of Southern California, Los Angeles, California, USA

Richard A. Schneider
Department of Orthopaedic Surgery, University of California at San Francisco, San Francisco, California, USA

Paul T. Sharpe
Craniofacial Development and Stem Cell Biology, Dental Institute, Kings College London, London, United Kingdom

Jingjing Sun
Department of Biochemistry and Molecular Biology, USC Norris Comprehensive Cancer Center, Keck School of Medicine, University of Southern California, Los Angeles, California, USA

Andre L.P. Tavares
Department of Craniofacial Biology, University of Colorado Anschutz Medical Campus, Aurora, Colorado, USA

Irma Thesleff
Research Program in Developmental Biology, Institute of Biotechnology, University of Helsinki, Helsinki, Finland

Man-Chun Ting
Department of Biochemistry and Molecular Biology, USC Norris Comprehensive Cancer Center, Keck School of Medicine, University of Southern California, Los Angeles, California, USA

J. Todd Streelman
School of Biology, Petit Institute for Bioengineering and Bioscience, Georgia Institute of Technology, Atlanta, Georgia, USA

Paul A. Trainor
Stowers Institute for Medical Research, Kansas City, Missouri, and Department of Anatomy & Cell Biology, University of Kansas Medical Center, Kansas City, Kansas, USA

Le A. Trinh
Molecular and Computational Biology, University of Southern California, Los Angeles, California, USA

Abigail S. Tucker
Department of Craniofacial Development & Stem Cell Biology, King's College London, London, United Kingdom

Eldad Tzahor
Department of Biological Regulation, Weizmann Institute of Science, Rehovot, Israel

Ana Angelova Volponi
Craniofacial Development and Stem Cell Biology, Dental Institute, Kings College London, London, United Kingdom

Jian Xu
Center for Craniofacial Molecular Biology, School of Dentistry, University of Southern California, Los Angeles, USA

Jingyue Xu
Division of Developmental Biology, Cincinnati Children's Hospital Medical Center, Cincinnati, Ohio, USA

Nathan M. Young
Department of Orthopaedic Surgery, San Francisco General Hospital, Orthopaedic Trauma Institute, University of California San Francisco, San Francisco, California, USA

Tian Yu
Craniofacial Development and Stem Cell Biology, Dental Institute, Kings College London, London, United Kingdom

Tingwei Zhang
Center for Craniofacial Molecular Biology, School of Dentistry, University of Southern California, Los Angeles, USA, and State Key Laboratory of Oral Diseases, Department of Prosthodontics, West China Hospital of Stomatology, Sichuan University, Chengdu, China

PREFACE

INTRODUCTION TO *CRANIOFACIAL DEVELOPMENT*

The development of the craniofacial complex lies at the core of what makes us human. The face houses our sensory organs and the cranium cradles the brain. Any aberrations in craniofacial development can therefore jeopardize fundamental processes that allow us to survive, thrive, and enjoy our world: our senses of sight, hearing, smell, and taste; our ability to eat, drink, and breathe; our intellectual capacity. Moreover, the uniquely individual face represents one of the most important and salient parts of the identity we present to our peers. For all of these reasons, craniofacial development has long been a source of fascination to scientists.

Beginning in the mid-twentieth century, it became clear that the unique challenges of understanding craniofacial biology are best suited to an interdisciplinary team approach, due in large part to luminaries of the field such as Hans Spemann, Hilde Mangold, and Samuel Pruzansky. The contributions to the present volume continue this tradition, building on scientific and technological advances in developmental biology, molecular genetics, and medicine to bring a multitude of perspectives to bear on issues pertaining to craniofacial development. The authors hail from the fields of developmental patterning, cell and molecular biology, stem cell biology, evolutionary biology, clinical genetics, and everything in between. As a result, this book provides a comprehensive overview of the current status of our understanding of craniofacial development.

This book is organized into three sections. The first section discusses aspects of craniofacial morphogenesis and regeneration on multiple scales, from cells to tissues to organs. Many of our unique organs are housed in the craniofacial region, yet the regulatory mechanisms of cell–cell interactions, tissue development, and organogenesis share common features with those that control the development of the rest of the body. The major families of growth factors (BMP/TGF-β, FGF, Wnt, and Hh) are all involved in the regulation of craniofacial development. The signaling specificity of these growth factors lies with the cells, which have unique transcription factors and other genetic and epigenetic regulators that collectively determine the outcome of growth factor signaling during craniofacial morphogenesis. The chapters in this section offer evolutionary and embryonic perspectives on the regulation of the facial muscles, mandible, tongue, palate, temporomandibular joint, cranial sutures, tooth, and ear development. These studies provide an in-depth overview of cell fate determination and how continued tissue–tissue interactions control craniofacial development and regeneration.

The second section focuses on craniofacial patterning and signaling mechanisms. The principal goal of developmental biology is to understand how tissues are induced and patterned to generate different organs at the proper location and time. Evidence suggests that signals from the anterior visceral endoderm, anterior neural ridge, and mesoderm are required for the development of craniofacial structures. The formation of the vertebrate "New Head" depends upon the presence of cranial neural crest cells and additional sensory placodes. Multiple molecules have been identified as crucial regulators for tissue–tissue interactions during craniofacial development. Using fish, chick, mouse, and other animal models, the chapters in this section provide a summary of our current understanding of how early craniofacial patterning is established and how signaling networks control craniofacial morphogenesis.

The third section of the volume is centered around disease models, human genetics, genomics, and new approaches in dynamic imaging. Animal models have clearly played an important role in our understanding of normal and abnormal craniofacial development. Yet despite the vast amount of knowledge that we have gained from animal studies, shortcomings remain: for example, apparently only one third of mouse mutations associated with craniofacial defects have been linked to mutations of the orthologous human genes with similar dysmorphologies. It is not known what defects might arise from mutations of the other two thirds of these genes, as some of these mutations result in early embryonic lethality. It is of paramount importance that there are constant interactions between researchers involved in human and animal studies, as this will ensure a rapid translation of fundamental discoveries into treatment and prevention of craniofacial malformations in humans. Craniofacial malformations represent complex multifactorial diseases, amplified or attenuated by gene–gene interactions, gene–environmental interactions, environmental influences, and mechanical factors. The rapid technological advancement and decreased cost of sequencing the human genome will undoubtedly facilitate the discovery of genetic mutations involved in human craniofacial malformations. Furthermore, this new information will facilitate the design and implementation of precision medicine. Finally, the improved spatial and temporal resolution of dynamic and live imaging will also elevate our study of craniofacial development to the next level of insight and excellence.

It has been an extremely rewarding experience working with so many talented and accomplished scientists and clinicians in putting together this book. I would like to express my tremendous gratitude to them for the time and effort they devoted to this project. It is my hope that our authors' diverse perspectives will provide a well-rounded view of the current state of affairs in craniofacial development and a springboard for further advances in the field.

YANG CHAI, DDS, PhD Professor
George and MaryLou Boone Chair in Craniofacial Biology
University of Southern California

ACKNOWLEDGMENTS

First of all, I would like to acknowledge the tremendous debt I owe to my wife Lihua and our two beautiful daughters, Audrey and Stephanie. Their continued and unconditional love and support have made everything possible in my life. I dedicate this book to them. I also want to thank each contributor individually and all of them collectively. With their collective wisdom, we have put together a remarkable book on the current knowledge of craniofacial development and tissue regeneration. Finally, special thanks to Bridget Samuels and Julie Mayo for their help in editing this book.

PART 1

Craniofacial Morphogenesis and Regeneration: From Cells to Tissues to Organs

CHAPTER ONE

Craniofacial Muscle Development

Inbal Michailovici, Tamar Eigler, Eldad Tzahor[1]
Department of Biological Regulation, Weizmann Institute of Science, Rehovot, Israel
[1]Corresponding author: e-mail address: eldad.tzahor@weizmann.ac.il

Contents

Abstract

The developmental mechanisms that control head muscle formation are distinct from those that operate in the trunk. Head and neck muscles derive from various mesoderm populations in the embryo and are regulated by distinct transcription factors and signaling molecules. Throughout the last decade, developmental, and lineage studies in vertebrates and invertebrates have revealed the peculiar nature of the pharyngeal mesoderm that forms certain head muscles and parts of the heart. Studies in chordates, the ancestors of vertebrates, revealed an evolutionarily conserved cardiopharyngeal field that progressively facilitates the development of both heart and craniofacial structures during vertebrate evolution. This ancient regulatory circuitry preceded and facilitated the emergence of myogenic cell types and hierarchies that exist in vertebrates. This chapter summarizes studies related to the origins, signaling circuits, genetics, and evolution of the head musculature, highlighting its heterogeneous characteristics in all these aspects, with a special focus on the FGF–ERK pathway. Additionally, we address the processes of head muscle regeneration, and the development of stem cell-based therapies for treatment of muscle disorders.

Current Topics in Developmental Biology, Volume 115
ISSN 0070-2153
http://dx.doi.org/10.1016/bs.ctdb.2015.07.022

GLOSSARY BOX AND ABBREVIATIONS

CPM (cranial paraxial mesoderm): a subset of the PM located on both sides of the neural tube.
EOM (extraocular muscles): a group of muscles that control eye movement and mainly originate from the prechordal mesoderm.
ERK extracellular signal-regulated kinase
FGF fibroblast growth factor
Head muscles group of muscles that develop from the unsegmented (anterior) head mesoderm.
IFT (inflow tract): the venous pole of the heart.
LPM lateral plate mesoderm
NCC neural crest cell
Neck muscle a group of skeletal muscles that originate from both somitic and nonsomitic mesoderm that facilitate the free movement of the vertebrate head.
OFT (outflow tract): the arterial pole of the heart.
PA (pharyngeal arches, also termed branchial arches): the templates of the adult craniofacial structures. Bilateral bulges on either side of the pharynx comprising outer (ectodermal) and inner (endodermal) epithelia, neural crest-derived mesenchyme, and a mesoderm core.
PM (pharyngeal mesoderm): the anterior mesoderm of the embryo located around the pharynx and which gives rise to both head and cardiac-derived tissues.
Prechordal mesoderm mesodermal cells located anterior to the notochord that contribute primarily to the extraocular muscles.
SpM (splanchnic mesoderm): mesodermal cells located on the ventral side beneath the floor of the pharynx that contribute both to head and heart muscles.

1. HEAD MUSCLE FORMATION AND HETEROGENEITY: EXTRAOCULAR MUSCLES, PHARYNGEAL MESODERM, AND NECK MUSCLES

Muscles in the head are highly heterogeneous with respect to both origin and regulatory mechanisms. This section summarizes the different mesoderm tissues that contribute to the formation of the head muscles, presented here in the order from caudal to rostral. The human head contains more than 60 muscles, which control eye movements, food uptake, respiration, and facial expression (Bothe, Ahmed, Winterbottom, Scheven von, & Dietrich, 2007; Diogo et al., 2015; Grifone & Kelly, 2007; Noden & Francis-West, 2006; Sambasivan, Kuratani, & Tajbakhsh, 2011; Tzahor, 2009; Tzahor & Evans, 2011; Wachtler & Jacob, 1986).

Strikingly, the head includes at least six different groups of muscles with distinct developmental origins and evolutionary histories (Fig. 1; Diogo et al., 2015). Progenitor cells from the occipital somites move into the head

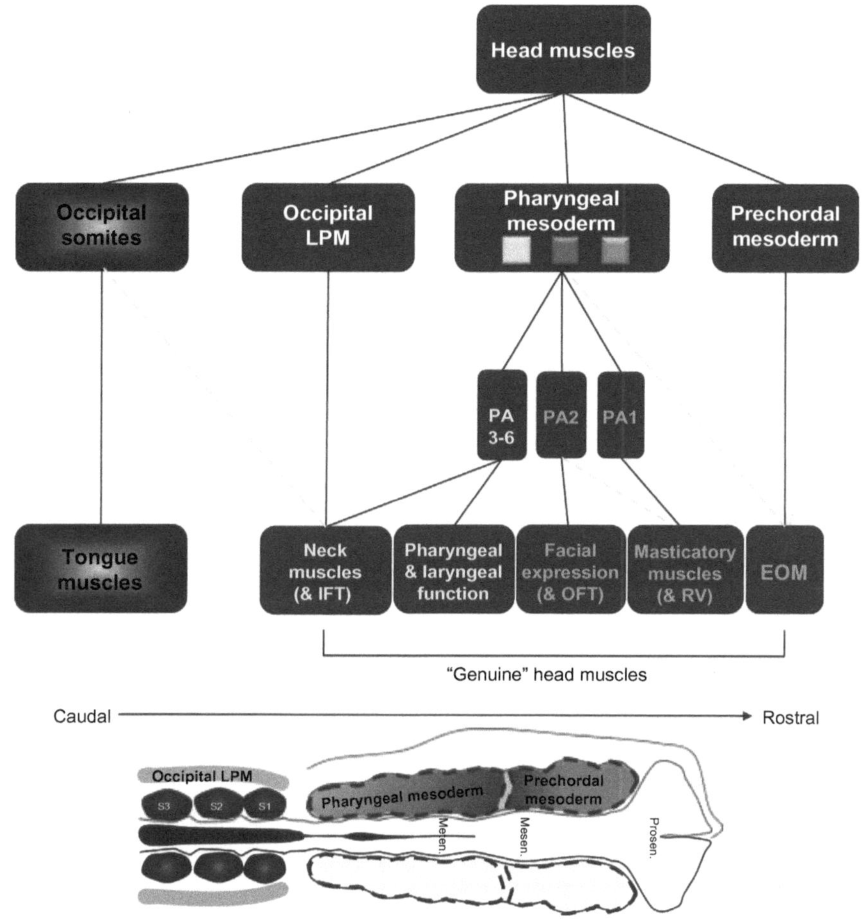

Figure 1 The mesodermal origins of the head muscles. A diagram explaining the contribution of each embryonic tissue to the mature head muscles in caudal to rostral (left to right) order. The summary map shows the mesoderm origin sites of muscles in the chick embryo: prechordal mesoderm, pharyngeal mesoderm, occipital lateral plate mesoderm (LPM), and the occipital somites (S1–S3). Pharyngeal arch (PA), inflow tract (IFT), outflow tract (OFT), right ventricle (RV), and extraocular muscles (EOM). (A color version of this figure is located at the end of this chapter.)

and give rise to the tongue muscle. Despite their localization in the head, myogenic precursors from the occipital somites follow the trunk myogenesis program (Mootoosamy & Dietrich, 2002). The origin of the tongue muscles is still controversial. Based on lineage tracing, the Birchmeier group utilized two different mouse Cre lines, $Lbx1^{Cre}$ and $Mesp1^{Cre}$, to differentially mark somitic and cranial mesoderm, respectively (Czajkowski, Rassek, Lenhard,

Bröhl, & Birchmeier, 2014). They suggest that tongue muscle is derived not only from the somitic but also from the cranial ($Mesp1^{+}$) mesoderm. Notably, this Cre-meditated cell lineage encompasses the anterior part of the mouse including the heart and the occipital somites (Harel et al., 2009; Saga et al., 1996). Evidence for the somitic origin of these muscles came from lineage studies in mice showing that the tongue muscles are $Pax3^{+}$, $Mesp1^{+}$, and $Isl1^{-}$ in terms of their lineages (Harel et al., 2009).

The development of the neck muscles enables the free movement of the head. This group of muscles is highly conserved in jawed vertebrates (Kuratani, 2008). The origin of the neck muscles has long been controversial (Allis, 1897; Couly, Coltey, & Le Douarin, 1993; Huang, Zhi, Ordahl, & Christ, 1997; Huang, Zhi, Patel, Wilting, & Christ, 2000; Noden, 1983b; Noden & Francis-West, 2006; Piekarski & Olsson, 2007). Neck muscles, located in the transition zone between head and trunk, have both somitic and nonsomitic origins. The trapezius and sternocleidomastoid neck muscle (also termed the cucullaris muscle in jawed vertebrates) progenitor cells express markers of the pharyngeal mesoderm (PM), indicating their branchiomeric origin, whereas other muscles in the neck are derived from $Pax3^{+}$ somitic cells (Lescroart et al., 2015; Theis et al., 2010). Likewise, homotopic and heterotopic transplantation experiments in the chick revealed that the occipital lateral plate mesoderm (LPM) (adjacent to somites 1–3) gives rise to neck muscles (Theis et al., 2010). Recent retrospective clonal analysis in the mouse embryo shows that a distinct subpopulation within the PM contributes to the formation of the trapezius muscle as well as to the myocardium at the venous pole of the heart (IFT, inflow tract) (Lescroart et al., 2015). This study demonstrates that while there is a clonal relationship between the trapezius and sternocleidomastoid muscles and the laryngeal muscles (both muscles that derived from the posterior pharyngeal arches (PAs) 3–6), there is no clonal relation with other head or somite-derived muscles. Hence, these branchiomeric neck muscles constitute an independent head muscle lineage that is distinct from those which give rise to first and second arch-derived head muscles (marked in yellow, Fig. 1).

The PM is a subset of head mesoderm that surrounds the developing pharynx (Tzahor & Evans, 2011). The PM consists of two regions located anterior to the somitic series: the cranial paraxial mesoderm (CPM), located on both sides of the neural tube, and the splanchnic mesoderm (SpM), located on the ventral side beneath the floor of the pharynx. The PM precursors stream into the neighboring PAs (also known as branchial arches), which serve as templates for the adult craniofacial structures. These head muscle precursors fill the myogenic core within the PAs (Tzahor & Evans, 2011), where cranial neural crest cells (cNCC) surround the muscle *anlagen*, separating the myoblasts from the overlying surface ectoderm (Noden, 1983a; Trainor, Tan, & Tam, 1994). Later on in development, muscle progenitors within the PAs differentiate and form the muscles responsible for operating the jaw (PA1 and PA2), facial expression (PA2), and pharyngeal and laryngeal functions (PA3) (Fig. 1).

The prechordal mesoderm, located anteriorly to the notochord, provides myogenic precursor cells that contribute to skeletal muscles around the eye (termed extraocular muscles, EOM) (Couly, Coltey, & Le Douarin, 1992; Hacker & Guthrie, 1998; Mootoosamy & Dietrich, 2002; Wachtler & Jacob, 1986). The PM together with the prechordal mesoderm and the occipital LPM provide the precursors for more than 60 distinct skeletal muscles in the vertebrate head and neck that control eating and facial expression, as well as eye and head movement.

1.1 Head Versus Trunk Myogenesis

Head muscles differ from all other skeletal muscles. Whereas head muscles originate from the head mesoderm (outlined above), trunk and limb muscles originate from the somites (Bailey, Holowacz, & Lassar, 2001; Christ & Ordahl, 1995; Noden & Francis-West, 2006; Sambasivan et al., 2011; Theis et al., 2010). Consistent with the distinct embryological origins of head and trunk muscles, their regulatory signals and genetic program are also diverse.

1.1.1 Network Governing Head Muscle Development

In comparison to the trunk, the molecular mechanisms underlying head myogenesis are in general less well characterized. While BMP and FGF–ERK block both cranial and trunk myogenic differentiation programs, Wnt/β-catenin signaling pathway plays an opposite role in head and trunk myogenesis (Michailovici et al., 2014; Tzahor et al., 2003) (see Section 3 for extended discussion on the role of FGF–ERK signaling during myogenesis). Myogenesis in the trunk is induced by Wnt signals from the dorsal neural tube (Münsterberg et al., 1995; Stern, Brown, & Hauschka, 1995; Tajbakhsh et al., 1998), whereas myogenesis in the head is blocked by these same signals. Not only is head muscle formation locally repressed by BMP and Wnt signals, but it is also induced by antagonists of these signaling pathways (Noggin, Gremlin, and Frzb, respectively) which are secreted by adjacent tissues like the cNCC (Tzahor et al., 2003).

Another pathway that acts differentially in head and trunk myogenesis is Notch, particularly with respect to the Notch ligand, Delta 1 Dll1. While Notch signaling inhibits both head and trunk myogenic differentiation by repression of MyoD (Bröhl et al., 2012; Czajkowski et al., 2014; Delfini & Duprez, 2000), other functions of Notch signaling differ between the two sites. In contrast to somite-derived muscle progenitors, Dll1 signaling is neither required for satellite cell population nor for Pax7 expression in cranial myogenic progenitors (Czajkowski et al., 2014). Although several studies demonstrate that Notch activation promotes a progenitor phenotype while inhibiting terminal differentiation (Czajkowski et al., 2014; Mayeuf-Louchart et al., 2014; Mourikis, Gopalakrishnan, Sambasivan, & Tajbakhsh, 2012; Wilson-Rawls, Molkentin, Black, & Olson, 1999), there remains disagreement in the literature as to the inhibitory role of Notch (Pascoal, de Lima, Leslie, Hughes, & Saúde, 2013; Rios, Serralbo, Salgado, & Marcelle, 2011).

1.1.2 Intrinsic Genetic Program Directing Head Muscle Development

Skeletal muscle development is controlled by myogenic regulatory factors (MRFs), a family of muscle-specific basic helix-loop-helix (bHLH) transcription factors. Skeletal muscle differentiation takes place when MRF genes of the MyoD family (e.g., Myf5, MyoD, Myogenin, and MRF4) are activated in muscle progenitors, and this genetic program is operative in both the trunk and head regions (Buckingham, 2006). MRFs are responsible, in concert with additional cofactors, for cell cycle withdrawal that eventually triggers the expression of muscle structural proteins such as myosin and actin and the assembly of functional myofibers (Molkentin & Olson, 1996).

Upstream regulators of the MRFs in head muscle progenitors differ from those required for trunk and limb muscles. Tcf21 (also known as Capsulin), Msc (MyoR), Tbx1, Pitx2, and Lhx2 are required for MRF gene activation during head myogenesis but are nonessential for trunk myogenesis (Bothe et al., 2007; Bryson-Richardson & Currie, 2008; Grifone & Kelly, 2007; Sambasivan et al., 2011; Tzahor, 2009, 2015; Tzahor & Evans, 2011).

Indeed, MyoR/Capsulin double knockout mouse embryos lacked a specific type of muscles within the first PA. However, trunk muscles of the double mutants were unaffected by elimination of these two genes (Lu et al., 2002). Loss of Tbx1 results in sporadic activation of Myf5 and MyoD in the mandibular arch (PA1) and the presence of hypoplastic PA1-derived muscles at later developmental stages, while no effect was seen on trunk muscles (Kelly, Jerome-Majewska, & Papaioannou, 2004). Tbx1 cooperates with Myf5 in branchiomeric muscle formation; thus only in their combined absence (Tbx1:Myf5 double null mutants) MyoD is not activated, and these muscles are missing (Sambasivan et al., 2009). Additionally, Pax3, encoding a regulator of MRF activation in the trunk, is neither expressed nor required during head muscle formation (Hacker & Guthrie, 1998; Harel et al., 2009; Mootoosamy & Dietrich, 2002; Tajbakhsh, Rocancourt, Cossu, & Buckingham, 1997). Furthermore, no muscle defects were found in the head of Pax3/Myf5 double knockout embryos (Tajbakhsh et al., 1997; Tremblay et al., 1998).

In addition, there is also a distinct genetic program within the head muscle groups that differentially governs EOM and branchiomeric muscle cell fates. While EOM progenitor fate is determined by Myf5 or Mrf4, which are both downstream of Pitx2, pharyngeal muscle progenitor fate is determined by Tbx1 and Myf5 whereas Mrf4 is dispensable for these muscles (Sambasivan et al., 2009).

A recent study from our lab (Harel et al., 2012) identified Lhx2 as a novel player during pharyngeal muscle and cardiac development. Loss of this gene resulted in a pharyngeal muscle specification defect, as well as phenotypes resembling DiGeorge syndrome (22q11.2 deletion). A series of knockout mice experiments revealed epistatic relationships between Tbx1, Lhx2, and Myf5 within the pharyngeal muscle network, affecting early pharyngeal muscle specification and patterning. In addition, these experiments together with ChIP-seq analysis also identified a genetic link between Tcf21, Tbx1, and Lhx2 in the pharyngeal muscle transcriptional circuit.

The development of the neck muscles (the cucullaris muscle in jawed vertebrates and the trapezius and the sternocleidomastoid muscles in mammals) recapitulates the head muscle program in terms of the genetic program and the late differentiation onset, although a considerable part of these muscles lies within the trunk territory. Head myogenesis markers like Capsulin, MyoR, Tbx1, and Pitx2 are also expressed during cucullaris muscle development (Sambasivan et al., 2009; Theis et al., 2010). Likewise, the lack of the early trunk muscle specification markers such as Pax3 and Pax7 rules out the possibility that the trunk myogenic program is involved. The cucullaris mammalian homologues, the trapezius and the sternocleidomastoid muscles, were absent in the $Tbx1^{-/-}$ mutant mice (Huynh, Chen, Terrell, & Baldini, 2007; Kelly et al., 2004; Lescroart et al., 2015; Theis et al., 2010), whereas in the Pax3:Myf5 double mutants, which lack all trunk and limb muscles, these neck muscles remained intact (Tajbakhsh et al., 1997).

Similar to head muscle differentiation, the onset of cucullaris muscle differentiation is also delayed compared to the trunk (Lescroart et al., 2015; Theis et al., 2010). There seems to be a conserved myogenic mechanism for the cucullaris muscle in chick, mouse, and turtle. Where the differentiation time window of the cucullaris muscle takes place after those of the trunk, limb, and head muscles. An interesting question remains: why do the neck muscles specifically start to differentiate after all other muscles?

2. OVERLAP BETWEEN CARDIAC AND PHARYNGEAL MUSCLE PROGENITORS

A strong link between the evolution of head muscles and the heart has been revealed in recent years. Both muscle cell types are formed by a common pool of mesoderm progenitor cells within the cardiopharyngeal field (CPF) of vertebrate embryos. The origin of this field could be traced back in evolution to chordates, the sister group to the vertebrates (Diogo et al., 2015).

PM cells give rise to both pharyngeal muscles and cardiac progenitors depending upon signals from adjacent pharyngeal endoderm, surface ectoderm, and cNCC (Grifone & Kelly, 2007; Kelly, 2010; Sambasivan et al., 2011; Tzahor, 2009, 2015; Tzahor & Evans, 2011). Heart progenitors in the cardiac crescent are derived from the lateral SpM, known as the first heart field (FHF), which is contiguous with the PM, differentiates earlier, and eventually populates the left ventricle (reviewed by Buckingham, Meilhac, & Zaffran, 2005; Dyer & Kirby, 2009; Evans, Yelon, Conlon, & Kirby, 2010; Tzahor & Evans, 2011; Tzahor & Lassar, 2001; Vincent & Buckingham, 2010). The secondary heart field (SHF) is situated slightly caudal and medial to these cells, and it gives rise to the myocardium and smooth muscle of the distal outflow tract (OFT) (Kelly, 2012). Hence, cells added at the arterial and venous poles of the heart derive from both overlapping and distinct regions of the PM (Evans et al., 2010; Moorman, Christoffels, Anderson, & van den Hoff, 2007).

Studies in both chick and mouse embryos have established that PM-derived cardiac progenitor cells, collectively referred to as the anterior heart field, populate the cardiac OFT, right ventricle, and atria during heart looping stages (Kelly, Brown, & Buckingham, 2001; Mjaatvedt et al., 2001;

Waldo et al., 2001). The anterior heart field is a subset of the SHF, and the SHF is the cardiogenic part of the PM.

Genetic lineage analyses have shown that some head skeletal muscle and cardiac progenitors of the SHF derive initially from a contiguous pool of PM cells that express overlapping sets of marker genes (Capsulin, MyoR, Tbx1, and Pitx2 with Isl1 and Nkx2.5, markers of the SHF) (Bryson-Richardson & Currie, 2008; Diogo et al., 2015; Grifone & Kelly, 2007; Nathan et al., 2008; Tirosh-Finkel, Elhanany, Rinon, & Tzahor, 2006; Tzahor, 2007; Tzahor & Evans, 2011).

PM progenitors harbor cardiogenic potential (Tzahor & Lassar, 2001). These cells migrate ventrally and enter the OFT where they contribute to both endocardial and myocardial lineages (Tirosh-Finkel et al., 2006). The PM cells also form skeletal muscle progenitors. Within the PM, CPM cells contribute to the proximal region of the myogenic core, which gives rise to the mandibular muscle (PA1-derived), while the SpM that lies ventral and lateral to the CPM contributes to the myogenic cells in the distal region of the PA that later form the intermandibular and digastric muscles (Nathan et al., 2008).

Confirmation that PM progenitors can be considered multipotent cells that give rise to branchiomeric skeletal muscles and SHF-derived regions of the heart comes from retrospective clonal analyses in mice, a method for analyzing cell lineage in intact embryos (Lescroart et al., 2010). It has been shown that head muscles and SHF derivatives originate from multipotent PM progenitors. The first lineage (marked in blue, Fig. 1) contributes to branchiomeric muscles (temporalis and masseter) as well as to the EOMs and to the right ventricular (RV) myocardium. The second lineage (marked in pink, Fig. 1) contributes to facial expression muscles, and to OFT myocardium. Recently, a third PM cell lineage (marked in yellow, Fig. 1) was identified as contributing to both neck and myocardial cells derived from the posterior SHF (Lescroart et al., 2015). Specifically, the trapezius muscle group showed a clonal relationship to myocardium at the venous pole of the heart and the pulmonary trunk at the arterial pole of the heart, yet no clonal relation with head or somite-derived muscles was observed (Lescroart et al., 2015). This study supports the notion that neck muscles can be considered branchiomeric muscles.

Although head muscles and SHF derivatives originate from multipotent PM progenitors, they can respond differently to certain regulatory signals. For example, FGF–ERK signaling blocks the premature differentiation of both cardiac and head muscle progenitors (Michailovici et al., 2014;

Tirosh-Finkel et al., 2010). On the other hand, BMP signaling plays opposing roles in these muscle populations. While it induces cardiac differentiation of PM progenitors, it inhibits their differentiation to skeletal muscles (Tirosh-Finkel et al., 2010; Tzahor et al., 2003).

The head mesoderm gastrulates from the primitive streak before the trunk mesoderm (Psychoyos & Stern, 1996), yet myogenic differentiation is delayed in the head, relative to the trunk (Hacker & Guthrie, 1998; Noden, Marcucio, & Borycki, 1999). A common view regarding the delay in cranial muscle development stems from the need to initially execute the embryonic heart program and later allow the proper development of head muscles (Bothe & Dietrich, 2006; Grifone & Kelly, 2007). This could also explain why a single molecule, BMP, acts differently on adjacent cell populations derived from multipotent PM progenitors.

2.1 The Cardiopharyngeal Field

A new evolutionary paradigm for the coemergence of heart and head muscles has been recently reviewed (Diogo et al., 2015). The basic evolutionary unit is the CPF, which is a developmental domain that gives rise to the heart and branchiomeric (pharyngeal) muscles. The CPF includes the anterior lateral mesoderm of the FHF plus PM that gives rise to regions of the heart and branchiomeric muscles (Diogo et al., 2015). The idea is based on our increased understanding that developmental programs of the heart and branchiomeric muscles are tightly linked, suggesting that these tissues share common evolutionary origins. Indeed cellular and molecular parallels between cardiac and head muscles are probably more ancient in evolution and could go back to 600–700 million years (Diogo et al., 2015; Grifone & Kelly, 2007; Olson, 2006; Sambasivan et al., 2011; Tzahor, 2009; Tzahor & Evans, 2011; Wang, Razy-Krajka, Siu, Ketcham, & Christiaen, 2013; Fig. 2).

The new head hypothesis about the origin of vertebrates (Gans & Northcutt, 1983) proposed that the emergence of branchiomeric muscles was associated with a shift from "passive" filtration to more active feeding modes and the emergence of neural crest and placode-derived sensory organs. However, recent studies have suggested that these structures and the CPF can be found in the ascidian *Ciona intestinalis*, a model urochordate (Diogo et al., 2015). Several molecular and genetic parallels between the CPF of ascidians and vertebrates in the regulatory circuitry underlying the formation of specialized muscle cells were identified (e.g., Tbx1,

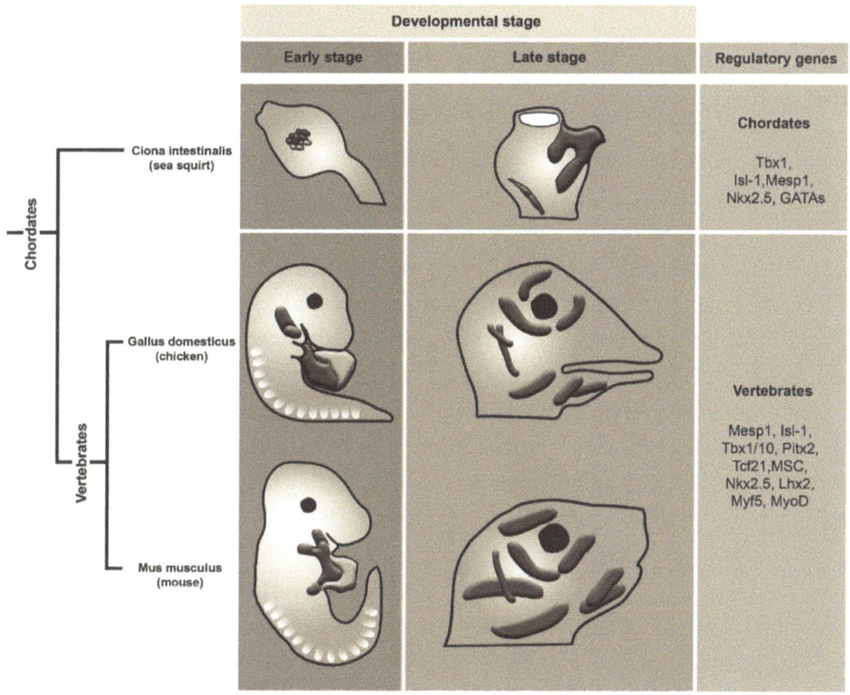

Figure 2 Evolutionarily conserved cardiopharyngeal muscle progenitors. Schematic of early- and late-stage embryonic head muscle development, indicating conservation of cardiopharyngeal precursors. The chordate *Ciona intestinalis* possesses first heart precursors (FHP, red), and second heart precursors (SHP, blue), which contribute to the development of the heart, whereas oral siphon muscles (OSM, purple) are derived from a population of trunk lateral cells. The vertebrate mouse and chick embryos exhibit similar first heart field precursors (FHF, red) and second heart field precursors (SHF, blue), as well as branchiomeric progenitors contributing to craniofacial skeletal muscle (purple). Genes involved in the regulation of craniofacial muscle development and differentiation programs are indicated. *Adapted from Diogo et al. (2015).* (A color version of this figure is located at the end of this chapter.)

Nkx2.5, Isl1; see Fig. 2) (Stolfi et al., 2010; Wang, Razy-Krajka, et al., 2013). We argue that the presence of a CPF, with dual cardiac and skeletal myogenic capacity, is probably a synapomorphy of olfactores (i.e., a derived feature shared by urochordates and vertebrates). We propose that this ancient regulatory circuitry preceded and facilitated the emergence of myogenic cell types and hierarchies that exist in vertebrates (Diogo et al., 2015).

3. FGF–ERK SIGNALING DURING MYOGENESIS

3.1 Fibroblast Growth Factor Signaling

Fibroblast growth factors (FGFs) are polypeptide growth factors with diverse biological activities. The FGF family is comprised of at least 22 members, which act predominantly via a set of tyrosine kinase receptors (FGFR 1–4). Binding of FGF to its receptor induces receptor dimerization, with subsequent activation of the cytoplasmic tyrosine kinase domain followed by intermolecular transphosphorylation and activation of downstream signaling events. This activation induces different cellular responses including multiple developmental processes such as differentiation, proliferation, and migration. There are five main pathways by which the FGFs can transduce their signals: RAS/MAPK pathway, RAS/June pathway, PI3 kinase/Akt pathway, RAS/P38 pathway, and PLCγ/PKC pathway (Itoh, 2007; Mason, 2007). FGF signaling affects skeletal muscle progenitors in several ways, promoting both progenitor cell proliferation and differentiation depending on the cellular and spatiotemporal context.

A growing number of studies investigating the role of FGF signaling in myogenesis have emerged, although our knowledge regarding this issue remains unclear. Myoblasts grown in culture start to differentiate when the amount of growth factors in the media is reduced. The key growth factor repressing myogenic differentiation in these cultures was found to be FGF (Clegg, Linkhart, Olwin, & Hauschka, 1987; Olwin & Rapraeger, 1992). However, myoblasts were found to respond differently to FGF treatment: In some, FGF delayed myogenic differentiation, while in other cases, FGF was found to be required for the differentiation (Seed & Hauschka, 1988).

More than two decades ago, Olson and colleagues described the inhibitory influence of FGF on myogenic differentiation in transfected cells (Spizz, Hu, & Olson, 1987). Other *in vitro* studies demonstrated that FGF receptor availability also regulates skeletal myogenesis (Scata, 1999). FGF null embryos ($FGF8^{-/-}$) die during gastrulation (Sun, Meyers, Lewandoski, & Martin, 1999). Studies on FGF8 mutant mice revealed that FGF8 has a role in PA development and in cNCC survival (Abu-Issa, Smyth, Smoak, Yamamura, & Meyers, 2002; Frank et al., 2002; Trumpp, Depew, Rubenstein, Bishop, & Martin, 1999). The $FGF8^{neo/-}$ mutant embryos, which had reduced (but not completely null) *FGF8* transcript levels, had smaller and fused PAs due to massive cNCC death along with loss of mesoderm expression of *Tbx1* in the arches (Abu-Issa et al., 2002). This implies that FGF signaling is upstream of *Tbx1* in the mouse.

Studies on early head muscle development using implantation of various beads into the head mesoderm of chick embryos extended our understanding on how FGF signaling regulates head muscle development. Implantation of FGF8 loaded beads into the head mesoderm resulted in downregulation of *Myf5, Pitx2,* and upregulation of *MyoR* and *Tbx1*. Inhibition of FGF signaling via SU5402-loaded beads mildly upregulated *Pitx2* and suppressed *MyoR* and *Tbx1* expression. Likewise, inhibition of FGF signaling using retinoic acid (RA) downregulated *MyoR* and *Tbx1*. Thus, low levels of RA promote FGF signaling and consequently *MyoR* and *Tbx1* activation (Bothe, Tenin, Oseni, & Dietrich, 2011; Scheven von, Alvares, Mootoosamy, & Dietrich, 2006).

FGF signaling has been found to be an important regulator of myogenesis in the limb. In the chick embryo, FGFR4 is expressed in $Myf5^+$ $MyoD^+$ myogenic cells in the limb. Furthermore, inhibition of FGFR4 in the chick limb buds suppressed muscle differentiation (MyoD and myosin heavy chain) (Marcelle, Wolf, & Bronner-Fraser, 1995; Marics, Padilla, Guillemot, Scaal, & Marcelle, 2002). A screen for genes regulated by Pax3 in mouse embryos identified FGFR4 as a direct target of Pax3 in the somites (Lagha et al., 2008). Interestingly, the electroporation and expression of *FGF8* in chick somites upregulated *FGFR4* expression and enhanced myogenic differentiation (Marics et al., 2002). Yet, the inhibition of FGF signaling by the overexpression of its negative regulator Sprouty 2 (Spry2), prevented myogenic progenitor cells ($Pax3^+$ $Pax7^+$) from entering the skeletal muscle differentiation program (Lagha et al., 2008). Together, these *in vivo* studies suggest that FGF signaling is required for trunk myogenesis.

3.2 The Cross Talk Between Neural Crest Cells and FGF Signaling

Several studies performed by our lab and others (outlined below) elucidated the importance of neural crest cells (NCC) in regulation of FGF signaling. These studies used different NCC ablation techniques in the chick to gain insight into the role of NCC during both head and heart development. Cranial NCC ablation involves the ablation of the dorsal half of the neural plate of the stage 8 chick embryo bilaterally from the mid-diencephalon down through rhombomere 2, which yields mostly Hox-negative NCC ablation (Fig. 3A). This procedure was shown to inhibit the expression of *FGF8* and cause improper development of the PAs. Beads soaked with FGF8 rescued

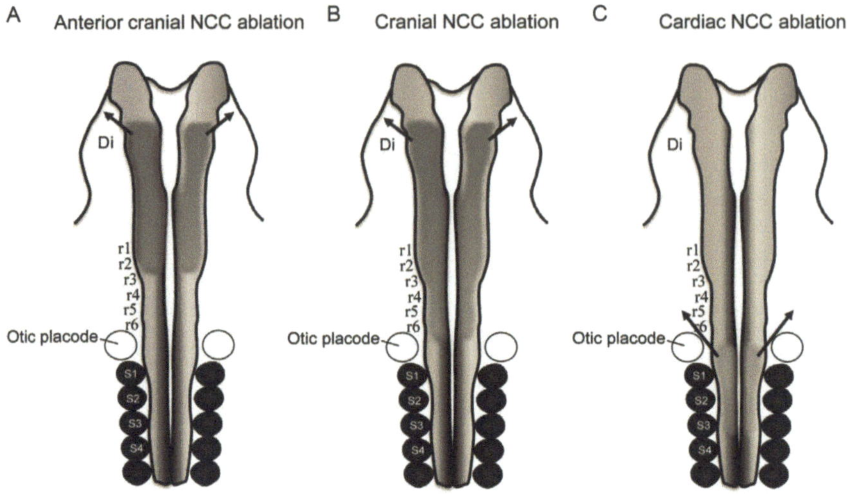

Figure 3 Neural crest cell ablation procedures. (A) Anterior cranial NCC ablation is the ablation of the dorsal half of the neural plate, bilaterally from the mid-diencephalon (Di) down through rhombomere 2 (r2). (B) Cranial NCC ablation is carried out from the diencephalic level to the level of rhombomere 5/6 (anterior to somite 1, S1). (C) Cardiac NCC ablation includes the ablation of the neural folds, bilaterally from the level of the mid-otic placode through the posterior limit of somite 3. (A color version of this figure is located at the end of this chapter.)

this phenotype (Creuzet, Schuler, Couly, & Le Douarin, 2004). RNAi-mediated knockdown of *Fgf8* in normal chick embryos resulted in almost no migration of cNCC into PA1 on the transfected side, suggesting a role of the FGF8 protein as an attractant for NCC migration (Creuzet et al., 2004).

Cranial NCC ablation of the dorsal half of the neural plate of stage 8 chick embryos from the diencephalon to the level of rhombomere 5/6 (anterior to somite 1, see Fig. 3B) has been widely used (Rinon et al., 2007; Tzahor et al., 2003). Alternatively, we used a molecular ablation technique that was established by the Marcelle lab (Rios et al., 2011) to ablate NCC by electroporation of a DTA construct under the control of a NCC-specific promoter (U2-DTA: the U2 is an evolutionary conserved *Sox10* enhancer sequence, cloned upstream to the diphtheria toxin gene). Ablation of the cNCC perturbed head myogenic differentiation and resulted in a strong upregulation of *FGF8* expression in the ectoderm of the PAs (Michailovici et al., 2014; Rinon et al., 2007). Hence, in the absence of cranial cNCC, myoblasts are kept in a proliferative state, most likely due to the elevated *FGF8* levels. Taken together, the data imply that cNCC downregulate *FGF8* gene expression and thus promote myogenic differentiation;

however, the precise mechanism by which this phenomenon occurs remains unclear. We further asked whether ectopic inhibition of FGF signaling, using SU5402, would induce myogenic differentiation in the absence of NCC. SU5402 treatment strongly induced differentiation in PM explants taken from both control and the NCC-ablated embryos (Michailovici et al., 2014). This experiment indicates that inhibition of FGF signaling is mostly required within the muscle progenitors in a cell-autonomous manner.

Cardiac NCC ablation is also a commonly used technique and involves the ablation of the neural folds of the stages 8–10 chick embryo, from the level of the mid-otic placode through the posterior limit of somite 3 (Hutson et al., 2006; Kirby, Kumiski, Myers, Cerjan, & Mishima, 1993; Fig. 3C). This ablation resulted in abnormal heart looping with increased expression of *FGF8* and FGF target genes in the caudal pharynx of the ablated embryos. Anti-FGF8b antibodies or SU5402 treatments rescued the abnormal heart looping phenotype. The normal SHF myocardium was extremely sensitive to both high and low levels of FGF signaling (Hutson et al., 2010, 2006; Vincent & Buckingham, 2010). FGF8 signaling is not only chemotactic for cranial NCC but also for cardiac NCC (Sato et al., 2011). Blocking FGF8 signaling with SU5402 or FGFR3 antibodies delayed cardiac NCC migration *in vivo*. This chemotactic effect of FGF8 is likely executed through FGFR1–3 expressed in the migrating cardiac NCC (Sarkar, Petiot, Copp, Ferretti, & Thorogood, 2001; Sato et al., 2011).

As already mentioned, anterior cranial NCC ablation resulted in inhibition of *FGF8* expression, suggesting that these NCC endogenously upregulate *FGF8* expression. On the other hand, both cranial and cardiac NCC ablation resulted in the opposite outcome, an elevation of *FGF8* expression. Therefore, it has been concluded that the posterior Hox-positive NCC endogenously downregulate *FGF8* expression. However, the precise mechanism by which NCC regulates *FGF8* expression is not clear and should be further examined.

3.3 The Extracellular Signal-Regulated Kinase

Extracellular signal-regulated kinase 1/2 (ERK, also known as p42/44 mitogen-activated protein kinase (MAPK)) can be activated by a variety of growth factors/mitogens (such as FGF) and has many substrates. ERK contains two phosphorylation sites. The first is the TEY sequence, which is phosphorylated by MEK and is important for the release of ERK from its cytoplasmic anchors. The second phosphorylation site is the SPS sequence that upon phosphorylation by ERK and CK2, induces nuclear translocation of ERK2, through binding to importin7 (Chuderland, Konson, & Seger, 2008; Plotnikov, Chuderland, Karamansha, Livnah, & Seger, 2011; Plotnikov, Zehorai, Procaccia, & Seger, 2011). Phosphorylated SPS amino acid stretch seems to act as a general nuclear translocation signal. It has been shown that a phosphomimetic EPE peptide, composed of 18-mer myristylated peptide based on the KID (kinase insert domain) of ERK2, replacing the serine residues of the SPS sequence to glutamic acid, delays the nuclear translocation of ERK, probably by competitive inhibition of binding to importin7 (Chuderland et al., 2008; Plotnikov, Chuderland, et al., 2011; Plotnikov, Zehorai, et al., 2011).

The majority of studies regarding the role of ERK during myogenesis have been done in cultured myoblasts. These studies have shown that ERK is critical for growth factor-induced cellular proliferation of myoblasts and subsequently for myoblast fusion (Jones, Fedorov, Rosenthal, & Olwin, 2001; Knight & Kothary, 2011), making ERK a key regulator of both myoblast proliferation and differentiation. During proliferation, ERK activity prevents cell cycle exit during G1 (Heller, Gredinger, & Bengal, 2001). On the other hand, it has been shown that ERK2 is required for efficient terminal differentiation of skeletal myoblasts (Li & Johnson, 2006). In addition, overexpression of constitutively active MEK1 enhances the transcriptional activity of MyoD (Gredinger, Gerber, Tamir, Tapscott, & Bengal, 1998; Zetser, Frank, & Bengal, 2001). A recent study showed that MEK1 stabilizes MyoD through direct phosphorylation at tyrosine 156 during myogenic differentiation (Jo, Cho, & Jo, 2011). In contrast, treating myoblasts with insulin-like growth factor-1, basic fibroblast growth factor, or leukemia inhibitory factor inhibited myogenic differentiation through activation of ERK (Jo et al., 2005; Tortorella, Milasincic, & Pilch, 2001; Weyman & Wolfman, 1998). Ectopic expression of constitutively active RAF or MEK1, which act upstream to ERK, also perturbed the differentiation of myoblasts (Dorman & Johnson, 1999; Perry, Parker, & Rudnicki, 2001).

Nevertheless, it is still unclear how a single molecule (e.g., ERK) can regulate two opposing cellular outcomes (proliferation vs. differentiation). Using both chick and mouse models, corroborated by mathematical models, a recent study from the Tzahor group shed light on the mechanism responsible for the switch from proliferation to differentiation of muscle progenitors (Michailovici et al., 2014). This study implicated ERK subcellular

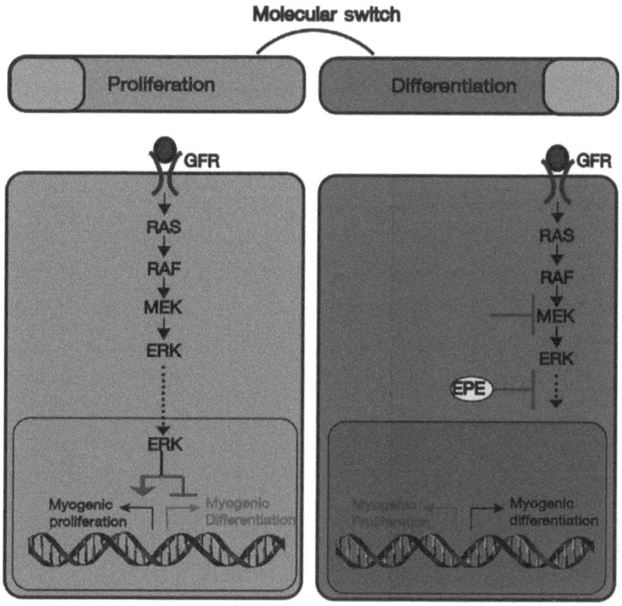

Figure 4 ERK-mediated molecular switch determines myogenic differentiation. As demonstrated by Michailovici et al. (2014), muscle differentiation is determined by the cellular localization of the signaling molecule ERK. Nuclear localization of ERK promotes a progenitor/proliferative phenotype (left side), while cytoplasmic ERK is associated with myogenic differentiation (right side). (A color version of this figure is located at the end of this chapter.)

localization as the major component controlling this switch. Inhibition of ERK nuclear translocation induced robust myogenic differentiation of muscle progenitor/stem cells in both head and trunk regions. These findings suggest that ERK shuttling between the nucleus and the cytoplasm provides a switch-like transition between proliferation and differentiation of muscle progenitors (Michailovici et al., 2014; Fig. 4).

4. HEAD MUSCLE REGENERATION

To date, investigation of muscle regeneration has focused largely on trunk/limb skeletal muscles only. There have been several clinical trials aimed at testing safety and efficacy of stem cell-based therapies to treat skeletal muscle injury, muscular dystrophy, and myocardial infarction. Yet, advances in the field of craniofacial skeletal muscle regeneration remain behind the curve. Approaches to skeletal muscle regenerative medicine have concentrated on the use of growth factors, satellite cell transplants, various scaffolds, or a combination of these to injured muscle with varying results. As of yet, there are no clinical reports of head muscle regeneration.

Although limb and head muscles share many common properties, head muscles have distinct embryonic origins and exhibit variations in their satellite cell populations, likely contributing to differences in their regenerative capacity in response to injury (McLoon, Thorstenson, Solomon, & Lewis, 2007; Pavlath et al., 1998). Adult skeletal muscle contains resident myogenic progenitor cells, called satellite cells, which are essential for maintenance of muscle mass and muscle regeneration (Dumont, Wang, & Rudnicki, 2015; Relaix & Zammit, 2012; Yin, Price, & Rudnicki, 2013). Satellite cells can proliferate, self-renew both *in vitro* and *in vivo*, and differentiate into mononucleated myoblasts which ultimately differentiate into multinucleated functional myofibers (Abou-Khalil et al., 2015; Dumont et al., 2015; Murphy, Lawson, Mathew, Hutcheson, & Kardon, 2011; Relaix & Zammit, 2012). Satellite cells derived from trunk/limb muscle show regenerative potential both *in vitro* and *in vivo* (Beauchamp, Morgan, Pagel, & Partridge, 1999; Collins et al., 2005; Montarras et al., 2005; Rossi et al., 2010).

Head and trunk myogenic programs vary considerably (reviewed by Bothe et al., 2007; Grifone & Kelly, 2007; Noden & Francis-West, 2006; Tzahor, 2009, 2015; Tzahor & Evans, 2011). Head muscle satellite cells are derived from a separate genetic lineage and follow different genetic programs compared to trunk satellite cells (Harel et al., 2009; Sambasivan et al., 2009). Lineage tracing analysis indicates that satellite cells are associated with their own muscles with respect to their embryonic origin and gene expression pattern. For example, Pax3 lineage-derived satellite cells exist in limb muscles but are not found in head muscles (Brown, Engleka, Wenning, Min Lu, & Epstein, 2005; Harel et al., 2009). In contrast, Isl1 lineage solely marks PA-derived head muscles and their associated satellite cells (Harel et al., 2009). Differences between satellite cell populations are retained in their myoblast progeny and into adulthood (Ono, Boldrin, Knopp, Morgan, & Zammit, 2010; Porter et al., 2006). Interestingly, myogenic differentiation of head muscles is delayed compared to myogenesis derived from the somites (Hacker & Guthrie, 1998; Noden et al., 1999). Furthermore, head satellite cells are more proliferative and display delayed differentiation compared to trunk satellite cells (Ono et al., 2010). Another example of these differences is the cardiogenic potential of head- but not trunk-derived satellite cells (Harel et al., 2009). These disparities may partially explain the unequal occurrence of myopathies in head and limb muscles (Benveniste, Jacobson, Farrugia, Clover, & Vincent, 2005; Emery, 2002;

Karpati, Carpenter, & Opitz, 1986; Muller et al., 2001; Thomas, Joseph, Adkins, Andrade, & Stemple, 2008) and may suggest that head muscles exhibit properties that provide resistance to some muscular dystrophies but more susceptibility to others.

In a recent study, canine craniofacial muscle satellite cells exhibited more efficient and faster differentiation and a stronger stem cell phenotype compared to somite-derived satellite cells (La Rovere et al., 2014). In contrast, satellite cells derived from mouse masseter muscle were able to regenerate muscle in a trunk muscle injury model as well as trunk-derived satellite cells (Harel et al., 2009; Ono et al., 2010).

Despite our current knowledge of the embryonic and genetic origin of craniofacial skeletal muscle as well as our growing understanding of mechanisms involved in regulation of satellite cell proliferation and differentiation, functional craniofacial muscle regeneration remains an elusive goal. The discovery of mechanisms that contribute to craniofacial muscle regeneration would have significant clinical impact: muscle regeneration is an essential component of craniofacial structure repair such as that of cleft palate, as well as reconstruction after acquired injuries, congenital deformities, and muscle wasting associated with aging and Bell's palsy. To date, the most effective therapies require tissue transfer from donor sites, which incurs additional risks of infection, pain, and loss of functionality.

The use of stem cell transplants for craniofacial muscle regeneration therapy shows strong promise, but still remains in the early discovery phases. Bone marrow-derived mesenchymal stem cells have been shown to have myogenic potential, can be induced into myogenic progenitors both *in vivo* and *in vitro*, and may be able to contribute to muscle regeneration (Bianco, Robey, & Simmons, 2008; Torrente et al., 2007). Interestingly, the satellite cell is not the only cell type within the muscle compartment with myogenic potential; mesangioblasts, for example, can be induced into myoblasts and can proliferate in culture without tumorigenic potential (Cossu & Bianco, 2003). Similarly, the pericyte has been shown to be myogenic both *in vitro* and *in vivo* (Quattrocelli et al., 2011). Ecto-mesenchymal stem cells originating from the cranial neural crest exhibit myogenic and regenerative potential both *in vitro* and *in vivo* (Nie et al., 2014).

Craniofacial-derived skeletal muscle satellite cells would be the ideal candidates for craniofacial muscle regeneration therapy, however, and have been shown to exhibit regeneration potential *in vivo* (Harel et al., 2009). Craniofacial skeletal muscles, specifically EOM and laryngeal muscles, are rich in activated satellite cells with strong myogenic potential, which may explain their ability to resist certain muscle disorders such as dystrophin deficiencies including Duchenne muscular dystrophy (Kallestad et al., 2011; McLoon, 2003; Marques, Ferretti, Vomero, Minatel, & Neto, 2007; Wang, Razy-Krajka, et al., 2013; Wang, Fu, et al., 2013). Recently, it was shown that satellite cells derived from the EOM possess greater proliferative, renewal, and myogenic potential than those derived from limb muscle. Moreover, EOM satellite cells derived from dystrophin-null (mdx) mice exhibit greater growth, renewal, and regenerative properties compared to limb muscles, which are severely affected by the dystrophin deficiency (Stuelsatz et al., 2015). Taken together, the data suggests that craniofacial muscle may be a reliable source for obtaining muscle progenitor cells for treatment of skeletal muscle defects.

Additionally, head-derived satellite cells may be a viable source for autologous cell transplantation for the purpose of heart regeneration as they are derived from similar embryonic origin and express markers of the cardiac lineage (Harel et al., 2009). The adult mammalian heart, unlike other tissues, does not regenerate following injury, and to date the current approaches to cell-autologous heart regeneration have not been impressively successful (Behfar, Crespo-Diaz, Terzic, & Gersh, 2014; Buikema, Van Der Meer, Sluijter, & Domian, 2013; Garbern & Lee, 2013; Pearl Quijada, 2014; Ptaszek, Mansour, Ruskin, & Chien, 2012). Therefore, the potential of using head muscle satellite cells for therapeutic application for heart regeneration should be further examined.

The development of novel approaches for craniofacial muscle regeneration has been largely ignored, yet remains a crucial component of successful craniofacial reconstruction postinjury and treatment of congenital disorders, as well as the treatment of age-associated muscle atrophy (wasting). Not only will further discovery contribute to our clinical approach to facial reconstruction, but it may also elucidate mechanisms related to trunk and heart muscle regeneration.

REFERENCES

Abou-Khalil, R., Yang, F., Lieu, S., Julien, A., Perry, J., Pereira, C., et al. (2015). Role of muscle stem cells during skeletal regeneration. *Stem Cells*, *33*, 1501–1511.

Abu-Issa, R., Smyth, G., Smoak, I., Yamamura, K.-I., & Meyers, E. N. (2002). Fgf8 is required for pharyngeal arch and cardiovascular development in the mouse. *Development*, *129*, 4613–4625.

Allis, E. P. (1897). The cranial muscles and cranial and first spinal nerves in amia calva. *Journal of Morphology*, *12*, 487–808.

Bailey, P., Holowacz, T., & Lassar, A. B. (2001). The origin of skeletal muscle stem cells in the embryo and the adult. *Current Opinion in Cell Biology*, *13*, 679–689.

Beauchamp, J. R., Morgan, J. E., Pagel, C. N., & Partridge, T. A. (1999). Dynamics of myoblast transplantation reveal a discrete minority of precursors with stem cell-like properties as the myogenic source. *The Journal of Cell Biology*, *144*, 1113–1122.
Behfar, A., Crespo-Diaz, R., Terzic, A., & Gersh, B. J. (2014). Cell therapy for cardiac repair—Lessons from clinical trials. *Nature Reviews. Cardiology*, *11*, 232–246.
Benveniste, O., Jacobson, L., Farrugia, M. E., Clover, L., & Vincent, A. (2005). MuSK antibody positive myasthenia gravis plasma modifies MURF-1 expression in C2C12 cultures and mouse muscle in vivo. *Journal of Neuroimmunology*, *170*, 41–48.
Bianco, P., Robey, P. G., & Simmons, P. J. (2008). Mesenchymal stem cells: Revisiting history, concepts, and assays. *Cell Stem Cell*, *2*, 313–319.
Bothe, I., Ahmed, M. U., Winterbottom, F. L., von Scheven, G., & Dietrich, S. (2007). Extrinsic versus intrinsic cues in avian paraxial mesoderm patterning and differentiation. *Developmental Dynamics*, *236*, 2397–2409.
Bothe, I., & Dietrich, S. (2006). The molecular setup of the avian head mesoderm and its implication for craniofacial myogenesis. *Developmental Dynamics*, *235*, 2845–2860.
Bothe, I., Tenin, G., Oseni, A., & Dietrich, S. (2011). Dynamic control of head mesoderm patterning. *Development*, *138*, 2807–2821.
Bröhl, D., Vasyutina, E., Czajkowski, M. T., Griger, J., Rassek, C., Rahn, H.-P., et al. (2012). Colonization of the satellite cell niche by skeletal muscle progenitor cells depends on notch signals. *Developmental Cell*, *23*, 469–481.
Brown, C. B., Engleka, K. A., Wenning, J., Min Lu, M., & Epstein, J. A. (2005). Identification of a hypaxial somite enhancer element regulating Pax3 expression in migrating myoblasts and characterization of hypaxial muscle Cre transgenic mice. *Genesis*, *41*, 202–209.
Bryson-Richardson, R. J., & Currie, P. D. (2008). The genetics of vertebrate myogenesis. *Nature Reviews. Genetics*, *9*, 632–646.
Buckingham, M. (2006). Myogenic progenitor cells and skeletal myogenesis in vertebrates. *Current Opinion in Genetics & Development*, *16*, 525–532.
Buckingham, M., Meilhac, S., & Zaffran, S. (2005). Building the mammalian heart from two sources of myocardial cells. *Nature Reviews. Genetics*, *6*, 826–835.
Buikema, J. W., Van Der Meer, P., Sluijter, J. P. G., & Domian, I. J. (2013). Concise review: Engineering myocardial tissue: The convergence of stem cells biology and tissue engineering technology. *Stem Cells*, *31*, 2587–2598.
Christ, B., & Ordahl, C. P. (1995). Early stages of chick somite development. *Anatomy and Embryology*, *191*, 381–396.
Chuderland, D., Konson, A., & Seger, R. (2008). Identification and characterization of a general nuclear translocation signal in signaling proteins. *Molecular Cell*, *31*, 850–861.
Clegg, C. H., Linkhart, T. A., Olwin, B. B., & Hauschka, S. D. (1987). Growth factor control of skeletal muscle differentiation: Commitment to terminal differentiation occurs in G1 phase and is repressed by fibroblast growth factor. *The Journal of Cell Biology*, *105*, 949–956.
Collins, C. A., Olsen, I., Zammit, P. S., Heslop, L., Petrie, A., Partridge, T. A., et al. (2005). Stem cell function, self-renewal, and behavioral heterogeneity of cells from the adult muscle satellite cell niche. *Cell*, *122*, 289–301.
Cossu, G., & Bianco, P. (2003). Mesoangioblasts—Vascular progenitors for extravascular mesodermal tissues. *Current Opinion in Genetics & Development*, *13*, 537–542.
Couly, G. F., Coltey, P. M., & Le Douarin, N. M. (1992). The developmental fate of the cephalic mesoderm in quail-chick chimeras. *Development*, *114*, 1–15.
Couly, G. F., Coltey, P. M., & Le Douarin, N. M. (1993). The triple origin of skull in higher vertebrates: A study in quail-chick chimeras. *Development*, *117*, 409–429.
Creuzet, S., Schuler, B., Couly, G., & Le Douarin, N. M. (2004). Reciprocal relationships between Fgf8 and neural crest cells in facial and forebrain development. *Proceedings of the National Academy of Sciences*, *101*, 4843–4847.

Czajkowski, M. T., Rassek, C., Lenhard, D. C., Bröhl, D., & Birchmeier, C. (2014). Divergent and conserved roles of Dll1 signaling in development of craniofacial and trunk muscle. *Developmental Biology, 395*, 307–316.
Delfini, M.-C., & Duprez, D. (2000). Paraxis is expressed in myoblasts during their migration and proliferation in the chick limb bud. *Mechanisms of Development, 96*, 247–251.
Diogo, R., Kelly, R. G., Christiaen, L., Levine, M., Ziermann, J. M., Molnar, J. L., et al. (2015). A new heart for a new head in vertebrate cardiopharyngeal evolution. *Nature, 520*, 466–473.
Dorman, C. M., & Johnson, S. E. (1999). Activated Raf inhibits avian myogenesis through a MAPK-dependent mechanism. *Oncogene, 18*, 5167–5176.
Dumont, N. A., Wang, Y. X., & Rudnicki, M. A. (2015). Intrinsic and extrinsic mechanisms regulating satellite cell function. *Development, 142*, 1572–1581.
Dyer, L. A., & Kirby, M. L. (2009). The role of secondary heart field in cardiac development. *Developmental Biology, 336*, 137–144.
Emery, A. (2002). The muscular dystrophies. *The Lancet, 359*, 687–695.
Evans, S. M., Yelon, D., Conlon, F. L., & Kirby, M. L. (2010). Myocardial lineage development. *Circulation Research, 107*, 1428–1444.
Frank, D. U., Fotheringham, L. K., Brewer, J. A., Muglia, L. J., Tristani-Firouzi, M., Capecchi, M. R., et al. (2002). An Fgf8 mouse mutant phenocopies human 22q11 deletion syndrome. *Development, 129*, 4591–4603.
Gans, C., & Northcutt, R. G. (1983). Neural crest and the origin of vertebrates: A new head. *Science, 220*, 268–273.
Garbern, J. C., & Lee, R. T. (2013). Cardiac stem cell therapy and the promise of heart regeneration. *Cell Stem Cell, 12*, 689–698.
Gredinger, E., Gerber, A. N., Tamir, Y., Tapscott, S. J., & Bengal, E. (1998). Mitogen-activated protein kinase pathway is involved in the differentiation of muscle cells. *Journal of Biological Chemistry, 273*, 10436–10444.
Grifone, R., & Kelly, R. G. (2007). Heartening news for head muscle development. *Trends in Genetics, 23*, 365–369.
Hacker, A., & Guthrie, S. (1998). A distinct developmental programme for the cranial paraxial mesoderm in the chick embryo. *Development, 125*, 3461–3472.
Harel, I., Maezawa, Y., Avraham, R., Rinon, A., Ma, H.-Y., Cross, J. W., et al. (2012). Pharyngeal mesoderm regulatory network controls cardiac and head muscle morphogenesis. *Proceedings of the National academy of Sciences of the United States of America, 109*, 18839–18844.
Harel, I., Nathan, E., Tirosh-Finkel, L., Zigdon, H., GuimarAes-Camboa, N., Evans, S. M., et al. (2009). Distinct origins and genetic programs of head muscle satellite cells. *Developmental Cell, 16*, 822–832.
Heller, H., Gredinger, E., & Bengal, E. (2001). Rac1 inhibits myogenic differentiation by preventing the complete withdrawal of myoblasts from the cell cycle. *Journal of Biological Chemistry, 276*, 37307–37316.
Huang, R., Zhi, Q., Ordahl, P., & Christ, B. (1997). The fate of the first avian somite. *Anatomy and Embryology, 195*, 435–449.
Huang, R., Zhi, Q., Patel, K., Wilting, J., & Christ, B. (2000). Contribution of single somites to the skeleton and muscles of the occipital and cervical regions in avian embryos. *Anatomy and Embryology, 202*, 375–383.
Hutson, M. R., Zeng, X. L., Kim, A. J., Antoon, E., Harward, S., & Kirby, M. L. (2010). Arterial pole progenitors interpret opposing FGF/BMP signals to proliferate or differentiate. *Development (Cambridge, England), 137*, 3001–3011.
Hutson, M. R., Zhang, P., Stadt, H. A., Sato, A. K., Li, Y.-X., Burch, J., et al. (2006). Cardiac arterial pole alignment is sensitive to FGF8 signaling in the pharynx. *Developmental Biology, 295*, 486–497.

Huynh, T., Chen, L., Terrell, P., & Baldini, A. (2007). A fate map of Tbx1 expressing cells reveals heterogeneity in the second cardiac field. *Genesis, 45*, 470–475.
Itoh, N. (2007). The Fgf families in humans, mice, and zebrafish: Their evolutional processes and roles in development, metabolism, and disease. *Biological & Pharmaceutical Bulletin, 30*, 1819–1825.
Jo, C., Cho, S.-J., & Jo, S. A. (2011). Mitogen-activated protein kinase kinase 1 (MEK1) stabilizes MyoD through direct phosphorylation at tyrosine 156 during myogenic differentiation. *The Journal of Biological Chemistry, 286*, 18903–18913.
Jo, C., Kim, H., Jo, I., Choi, I., Jung, S.-C., Kim, J., et al. (2005). Leukemia inhibitory factor blocks early differentiation of skeletal muscle cells by activating ERK. *Biochimica et Biophysica Acta, 1743*, 187–197.
Jones, N. C., Fedorov, Y. V., Rosenthal, R. S., & Olwin, B. B. (2001). ERK1/2 is required for myoblast proliferation but is dispensable for muscle gene expression and cell fusion. *Journal of Cellular Physiology, 186*, 104–115.
Kallestad, K. M., Hebert, S. L., McDonald, A. A., Daniel, M. L., Cu, S. R., & McLoon, L. K. (2011). Sparing of extraocular muscle in aging and muscular dystrophies: A myogenic precursor cell hypothesis☆. *Experimental Cell Research, 317*, 873–885.
Karpati, G., Carpenter, S., & Opitz, J. M. (1986). Small-caliber skeletal muscle fibers do not suffer deleterious consequences of dystrophic gene expression. *American Journal of Medical Genetics, 25*, 653–658.
Kelly, R. G. (2010). Core issues in craniofacial myogenesis. *Experimental Cell Research, 316*, 3034–3041.
Kelly, R. G. (2012). The second heart field. *Current Topics in Developmental Biology, 100*, 33–65.
Kelly, R. G., Brown, N. A., & Buckingham, M. E. (2001). The arterial pole of the mouse heart forms from Fgf10-expressing cells in pharyngeal mesoderm. *Developmental Cell, 1*, 435–440.
Kelly, R. G., Jerome-Majewska, L. A., & Papaioannou, V. E. (2004). The del22q11.2 candidate gene Tbx1 regulates branchiomeric myogenesis. *Human Molecular Genetics, 13*, 2829–2840.
Kirby, M. L., Kumiski, D. H., Myers, T., Cerjan, C., & Mishima, N. (1993). Backtransplantation of chick cardiac neural crest cells cultured in LIF rescues heart development. *Developmental Dynamics, 198*, 296–311.
Knight, J. D., & Kothary, R. (2011). The myogenic kinome: Protein kinases critical to mammalian skeletal myogenesis. *Skeletal Muscle, 1*, 29.
Kuratani, S. (2008). Evolutionary developmental studies of cyclostomes and the origin of the vertebrate neck. *Development, Growth & Differentiation, 50*, S189–S194.
La Rovere, R. M. L., Quattrocelli, M., Pietrangelo, T., Di Filippo, E. S., Maccatrozzo, L., Cassano, M., et al. (2014). Myogenic potential of canine craniofacial satellite cells. *Frontiers in Aging Neuroscience, 6*, 90.
Lagha, M., Sato, T., Bajard, L., Daubas, P., Esner, M., Montarras, D., et al. (2008). Regulation of skeletal muscle stem cell behavior by Pax3 and Pax7. *Cold Spring Harbor Symposia on Quantitative Biology, 73*, 307–315.
Lescroart, F., Hamou, W., Francou, A., Théveniau-Ruissy, M., Kelly, R. G., & Buckingham, M. (2015). Clonal analysis reveals a common origin between nonsomite-derived neck muscles and heart myocardium. *Proceedings of the National Academy of Sciences, 112*, 1446–1451.
Lescroart, F., Kelly, R. G., Le Garrec, J.-F., Nicolas, J.-F., Meilhac, S. M., & Buckingham, M. (2010). Clonal analysis reveals common lineage relationships between head muscles and second heart field derivatives in the mouse embryo. *Development, 137*, 3269–3279.
Li, J., & Johnson, S. E. (2006). ERK2 is required for efficient terminal differentiation of skeletal myoblasts. *Biochemical and Biophysical Research Communications, 345*, 1425–1433.

Lu, J.-R., Bassel-Duby, R., Hawkins, A., Chang, P., Valdez, R., Wu, H., et al. (2002). Control of facial muscle development by MyoR and capsulin. *Science*, *298*, 2378–2381.
Marcelle, C., Wolf, J., & Bronner-Fraser, M. (1995). The in vivo expression of the FGF receptor FREK mRNA in avian myoblasts suggests a role in muscle growth and differentiation. *Developmental Biology*, *172*, 100–114.
Marics, I., Padilla, F., Guillemot, J.-F., Scaal, M., & Marcelle, C. (2002). FGFR4 signaling is a necessary step in limb muscle differentiation. *Development*, *129*, 4559–4569.
Marques, M. J., Ferretti, R., Vomero, V. U., Minatel, E., & Neto, H. S. (2007). Intrinsic laryngeal muscles are spared from myonecrosis in the mdx mouse model of Duchenne muscular dystrophy. *Muscle & Nerve*, *35*, 349–353.
Mason, I. (2007). Initiation to end point: The multiple roles of fibroblast growth factors in neural development. *Nature Reviews. Neuroscience*, *8*, 583–596.
Mayeuf-Louchart, A., Lagha, M., Danckaert, A., Rocancourt, D., Relaix, F., Vincent, S. D., et al. (2014). Notch regulation of myogenic versus endothelial fates of cells that migrate from the somite to the limb. *Proceedings of the National academy of Sciences of the United States of America*, *111*, 8844–8849.
McLoon, L. K. (2003). Activated satellite cells in extraocular muscles of normal adult monkeys and humans. *Investigative Ophthalmology & Visual Science*, *44*, 1927–1932.
McLoon, L. K., Thorstenson, K. M., Solomon, A., & Lewis, M. P. (2007). Myogenic precursor cells in craniofacial muscles. *Oral Diseases*, *13*, 134–140.
Michailovici, I., Harrington, H. A., Azogui, H. H., Yahalom-Ronen, Y., Plotnikov, A., Ching, S., et al. (2014). Nuclear to cytoplasmic shuttling of ERK promotes differentiation of muscle stem/progenitor cells. *Development*, *141*, 2611–2620.
Mjaatvedt, C. H., Nakaoka, T., Moreno-Rodriguez, R., Norris, R. A., Kern, M. J., Eisenberg, C. A., et al. (2001). The outflow tract of the heart is recruited from a novel heart-forming field. *Developmental Biology*, *238*, 97–109.
Molkentin, J. D., & Olson, E. N. (1996). Defining the regulatory networks for muscle development. *Current Opinion in Genetics & Development*, *6*, 445–453.
Montarras, D., Morgan, J., Collins, C., Relaix, F., Zaffran, S., Cumano, A., et al. (2005). Direct isolation of satellite cells for skeletal muscle regeneration. *Science*, *309*, 2064–2067.
Moorman, A. F. M., Christoffels, V. M., Anderson, R. H., & van den Hoff, M. J. B. (2007). The heart-forming fields: One or multiple? *Philosophical Transactions of the Royal Society of London. Series B, Biological Sciences*, *362*, 1257–1265.
Mootoosamy, R. C., & Dietrich, S. (2002). Distinct regulatory cascades for head and trunk myogenesis. *Development*, *129*, 573–583.
Mourikis, P., Gopalakrishnan, S., Sambasivan, R., & Tajbakhsh, S. (2012). Cell-autonomous Notch activity maintains the temporal specification potential of skeletal muscle stem cells. *Development*, *139*, 4536–4548.
Muller, J., Vayssiere, N., Royuela, M., Leger, M. E., Muller, A., Bacou, F., et al. (2001). Comparative evolution of muscular dystrophy in diaphragm, gastrocnemius and masseter muscles from old male mdx mice. *Journal of Muscle Research and Cell Motility*, *22*, 133–139.
Münsterberg, A. E., Kitajewski, J., Bumcrot, D. A., McMahon, A. P., & Lassar, A. B. (1995). Combinatorial signaling by Sonic hedgehog and Wnt family members induces myogenic bHLH gene expression in the somite. *Genes & Development*, *9*, 2911–2922.
Murphy, M. M., Lawson, J. A., Mathew, S. J., Hutcheson, D. A., & Kardon, G. (2011). Satellite cells, connective tissue fibroblasts and their interactions are crucial for muscle regeneration. *Development*, *138*, 3625–3637.
Nathan, E., Monovich, A., Tirosh-Finkel, L., Harrelson, Z., Rousso, T., Rinon, A., et al. (2008). The contribution of Islet1-expressing splanchnic mesoderm cells to distinct branchiomeric muscles reveals significant heterogeneity in head muscle development. *Development*, *135*, 647–657.

Nie, X., Xing, Y., Deng, M., Gang, L., Liu, R., Zhang, Y., et al. (2014). Ecto-mesenchymal stem cells from facial process: Potential for muscle regeneration. *Cell Biochemistry and Biophysics*, *70*, 615–622.
Noden, D. M. (1983a). The role of the neural crest in patterning of avian cranial skeletal, connective, and muscle tissues. *Developmental Biology*, *96*, 144–165.
Noden, D. M. (1983b). The embryonic origins of avian cephalic and cervical muscles and associated connective tissues. *The American Journal of Anatomy*, *168*, 257–276.
Noden, D. M., & Francis-West, P. (2006). The differentiation and morphogenesis of craniofacial muscles. *Developmental Dynamics*, *235*, 1194–1218.
Noden, D. M., Marcucio, R., & Borycki, A. G. (1999). Differentiation of avian craniofacial muscles: I. Patterns of early regulatory gene expression and myosin heavy chain synthesis. *Developmental Dynamics*, *216*, 96–112.
Olson, E. N. (2006). Gene regulatory networks in the evolution and development of the heart. *Science*, *313*, 1922–1927.
Olwin, B. B., & Rapraeger, A. (1992). Repression of myogenic differentiation by aFGF, bFGF, and K-FGF is dependent on cellular heparan sulfate. *The Journal of Cell Biology*, *118*, 631–639.
Ono, Y., Boldrin, L., Knopp, P., Morgan, J. E., & Zammit, P. S. (2010). Muscle satellite cells are a functionally heterogeneous population in both somite-derived and branchiomeric muscles. *Developmental Biology*, *337*, 29–41.
Pascoal, S., de Lima, J. E., Leslie, J. D., Hughes, S. M., & Saúde, L. (2013). Notch signalling is required for the formation of structurally stable muscle fibres in zebrafish. *PloS One*, *8*, e68021.
Pavlath, G. K., Thaloor, D., Rando, T. A., Cheong, M., English, A. W., & Zheng, B. (1998). Heterogeneity among muscle precursor cells in adult skeletal muscles with differing regenerative capacities. *Developmental Dynamics*, *212*, 495–508.
Pearl Quijada, M. A. S. (2014). Making it stick: Chasing the optimal stem cells for cardiac regeneration. *Expert Review of Cardiovascular Therapy*, *12*, 1275–1288.
Perry, R. L. S., Parker, M. H., & Rudnicki, M. A. (2001). Activated MEK1 binds the nuclear MyoD transcriptional complex to repress transactivation. *Molecular Cell*, *8*, 291–301.
Piekarski, N., & Olsson, L. (2007). Muscular derivatives of the cranial most somites revealed by long-term fate mapping in the Mexican axolotl (Ambystoma mexicanum). *Evolution & Development*, *9*, 566–578.
Plotnikov, A., Chuderland, D., Karamansha, Y., Livnah, O., & Seger, R. (2011). Nuclear extracellular signal-regulated kinase 1 and 2 translocation is mediated by casein kinase 2 and accelerated by autophosphorylation. *Molecular and Cellular Biology*, *31*, 3515–3530.
Plotnikov, A., Zehorai, E., Procaccia, S., & Seger, R. (2011). The MAPK cascades: Signaling components, nuclear roles and mechanisms of nuclear translocation. *Biochimica et Biophysica Acta*, *1813*, 1619–1633.
Porter, J. D., Israel, S., Gong, B., Merriam, A. P., Feuerman, J., Khanna, S., et al. (2006). Distinctive morphological and gene/protein expression signatures during myogenesis in novel cell lines from extraocular and hindlimb muscle. *Physiological Genomics*, *24*, 264–275.
Psychoyos, D., & Stern, C. D. (1996). Fates and migratory routes of primitive streak cells in the chick embryo. *Development*, *122*, 1523–1534.
Ptaszek, L. M., Mansour, M., Ruskin, J. N., & Chien, K. R. (2012). Towards regenerative therapy for cardiac disease. *The Lancet*, *379*, 933–942.
Quattrocelli, M., Palazzolo, G., Floris, G., Schöffski, P., Anastasia, L., Orlacchio, A., et al. (2011). Intrinsic cell memory reinforces myogenic commitment of pericyte-derived iPSCs. *The Journal of Pathology*, *223*, 593–603.
Relaix, F., & Zammit, P. S. (2012). Satellite cells are essential for skeletal muscle regeneration: The cell on the edge returns centre stage. *Development*, *139*, 2845–2856.

Rinon, A., Lazar, S., Marshall, H., Büchmann-Møller, S., Neufeld, A., Elhanany-Tamir, H., et al. (2007). Cranial neural crest cells regulate head muscle patterning and differentiation during vertebrate embryogenesis. *Development, 134*, 3065–3075.

Rios, A. C., Serralbo, O., Salgado, D., & Marcelle, C. (2011). Neural crest regulates myogenesis through the transient activation of NOTCH. *Nature, 473*, 532–535.

Rossi, C. A., Pozzobon, M., Ditadi, A., Archacka, K., Gastaldello, A., Sanna, M., et al. (2010). Clonal characterization of rat muscle satellite cells: Proliferation, metabolism and differentiation define an intrinsic heterogeneity. *PloS One, 5*, e8523.

Saga, Y., Hata, N., Kobayashi, S., Magnuson, T., Seldin, M. F., & Taketo, M. M. (1996). MesP1: A novel basic helix-loop-helix protein expressed in the nascent mesodermal cells during mouse gastrulation. *Development, 122*, 2769–2778.

Sambasivan, R., Gayraud-Morel, B., Dumas, G., Cimper, C., Paisant, S., Kelly, R. G., et al. (2009). Distinct regulatory cascades govern extraocular and pharyngeal arch muscle progenitor cell fates. *Developmental Cell, 16*, 810–821.

Sambasivan, R., Kuratani, S., & Tajbakhsh, S. (2011). An eye on the head: The development and evolution of craniofacial muscles. *Development, 138*, 2401–2415.

Sarkar, S., Petiot, A., Copp, A., Ferretti, P., & Thorogood, P. (2001). FGF2 promotes skeletogenic differentiation of cranial neural crest cells. *Development, 128*, 2143–2152.

Sato, A., Scholl, A. M., Kuhn, E. B., Stadt, H. A., Decker, J. R., Pegram, K., et al. (2011). FGF8 signaling is chemotactic for cardiac neural crest cells. *Developmental Biology, 354*, 18–30.

Scata, K. (1999). FGF receptor availability regulates skeletal myogenesis. *Experimental Cell Research, 250*, 10–21.

Seed, J., & Hauschka, S. D. (1988). Clonal analysis of vertebrate myogenesis. *Developmental Biology, 128*, 40–49.

Spizz, G., Hu, J.-S., & Olson, E. N. (1987). Inhibition of myogenic differentiation by fibroblast growth factor or type β transforming growth factor does not require persistent c-myc expression. *Developmental Biology, 123*, 500–507.

Stern, H. M., Brown, A. M., & Hauschka, S. D. (1995). Myogenesis in paraxial mesoderm: Preferential induction by dorsal neural tube and by cells expressing Wnt-1. *Development, 121*, 3675–3686.

Stolfi, A., Gainous, T. B., Young, J. J., Mori, A., Levine, M., & Christiaen, L. (2010). Early chordate origins of the vertebrate second heart field. *Science, 329*, 565–568.

Stuelsatz, P., Shearer, A., Li, Y., Muir, L. A., Ieronimakis, N., Shen, Q. W., et al. (2015). Extraocular muscle satellite cells are high performance myo-engines retaining efficient regenerative capacity in dystrophin deficiency. - PubMed - NCBI. *Developmental Biology, 397*, 31–44.

Sun, X., Meyers, E. N., Lewandoski, M., & Martin, G. R. (1999). Targeted disruption of Fgf8 causes failure of cell migration in the gastrulating mouse embryo. *Genes & Development, 13*, 1834–1846.

Tajbakhsh, S., Borello, U., Vivarelli, E., Kelly, R., Papkoff, J., Duprez, D., et al. (1998). Differential activation of Myf5 and MyoD by different Wnts in explants of mouse paraxial mesoderm and the later activation of myogenesis in the absence of Myf5. *Development, 125*, 4155–4162.

Tajbakhsh, S., Rocancourt, D., Cossu, G., & Buckingham, M. (1997). Redefining the genetic hierarchies controlling skeletal myogenesis: Pax-3 and Myf-5 act upstream of MyoD. *Cell, 89*, 127–138.

Theis, S., Patel, K., Valasek, P., Otto, A., Pu, Q., Harel, I., et al. (2010). The occipital lateral plate mesoderm is a novel source for vertebrate neck musculature. *Development, 137*, 2961–2971.

Thomas, L. B., Joseph, G. L., Adkins, T. D., Andrade, F. H., & Stemple, J. C. (2008). Laryngeal muscles are spared in the dystrophin deficient mdx mouse. *Journal of Speech, Language, and Hearing Research, 51*, 586–595.

Tirosh-Finkel, L., Elhanany, H., Rinon, A., & Tzahor, E. (2006). Mesoderm progenitor cells of common origin contribute to the head musculature and the cardiac outflow tract. *Development*, *133*, 1943–1953.
Tirosh-Finkel, L., Zeisel, A., Brodt-Ivenshitz, M., Shamai, A., Yao, Z., Seger, R., et al. (2010). BMP-mediated inhibition of FGF signaling promotes cardiomyocyte differentiation of anterior heart field progenitors. *Development*, *137*, 2989–3000.
Torrente, Y., Belicchi, M., Marchesi, C., D'Antona, G., Cogiamanian, F., Pisati, F., et al. (2007). Autologous transplantation of muscle-derived $CD133^+$ stem cells in duchenne muscle patients. *Cell Transplantation*, *16*, 563–577.
Tortorella, L. L., Milasincic, D. J., & Pilch, P. F. (2001). Critical proliferation-independent window for basic fibroblast growth factor repression of myogenesis via the p42/p44 MAPK signaling pathway. *Journal of Biological Chemistry*, *276*, 13709–13717.
Trainor, P. A., Tan, S. S., & Tam, P. P. (1994). Cranial paraxial mesoderm: Regionalisation of cell fate and impact on craniofacial development in mouse embryos. *Development*, *120*, 2397–2408.
Tremblay, P., Dietrich, S., Mericskay, M., Schubert, F. R., Li, Z., & Paulin, D. (1998). A crucial role for Pax3 in the development of the hypaxial musculature and the long-range migration of muscle precursors. *Developmental Biology*, *203*, 49–61.
Trumpp, A., Depew, M. J., Rubenstein, J. L. R., Bishop, J. M., & Martin, G. R. (1999). Cre-mediated gene inactivation demonstrates that FGF8 is required for cell survival and patterning of the first branchial arch. *Genes & Development*, *13*, 3136–3148.
Tzahor, E. (2007). Wnt/β-catenin signaling and cardiogenesis: Timing does matter. *Developmental Cell*, *13*, 10–13.
Tzahor, E. (2009). Heart and craniofacial muscle development: A new developmental theme of distinct myogenic fields. *Developmental Biology*, *327*, 273–279.
Tzahor, E. (2015). Head muscle development. *Results and Problems in Cell Differentiation*, *56*, 123–142.
Tzahor, E., & Evans, S. M. (2011). Pharyngeal mesoderm development during embryogenesis: Implications for both heart and head myogenesis. *Cardiovascular Research*, *91*, 196–202.
Tzahor, E., Kempf, H., Mootoosamy, R. C., Poon, A. C., Abzhanov, A., Tabin, C. J., et al. (2003). Antagonists of Wnt and BMP signaling promote the formation of vertebrate head muscle. *Genes & Development*, *17*, 3087–3099.
Tzahor, E., & Lassar, A. B. (2001). Wnt signals from the neural tube block ectopic cardiogenesis. *Genes & Development*, *15*, 255–260.
Vincent, S. D., & Buckingham, M. E. (2010). How to make a heart: The origin and regulation of cardiac progenitor cells. *Current Topics in Developmental Biology*, *90*, 1–41.
von Scheven, G., Alvares, L. E., Mootoosamy, R. C., & Dietrich, S. (2006). Neural tube derived signals and Fgf8 act antagonistically to specify eye versus mandibular arch muscles. *Development*, *133*, 2731–2745.
Wachtler, F., & Jacob, M. (1986). Origin and development of the cranial skeletal muscles. *Bibliotheca Anatomica*, *29*, 24–46.
Waldo, K. L., Kumiski, D. H., Wallis, K. T., Stadt, H. A., Hutson, M. R., Platt, D. H., et al. (2001). Conotruncal myocardium arises from a secondary heart field. *Development*, *128*, 3179–3188.
Wang, Z., Fu, M., Wang, L., Liu, J., Li, Y., Brakebusch, C., et al. (2013). p21-activated kinase 1 (PAK1) can promote ERK activation in a kinase-independent manner. *Journal of Biological Chemistry*, *288*, 20093–20099.
Wang, W., Razy-Krajka, F., Siu, E., Ketcham, A., & Christiaen, L. (2013). NK4 antagonizes Tbx1/10 to promote cardiac versus pharyngeal muscle fate in the ascidian second heart field. *PLoS Biology*, *11*, e1001725.
Weyman, C. M., & Wolfman, A. (1998). Mitogen-activated protein kinase kinase (MEK) activity is required for inhibition of skeletal muscle differentiation by insulin-like growth factor 1 or fibroblast growth factor 2. *Endocrinology*, *139*, 1794–1800.

Wilson-Rawls, J., Molkentin, J. D., Black, B. L., & Olson, E. N. (1999). Activated notch inhibits myogenic activity of the MADS-Box transcription factor myocyte enhancer factor 2C. *Molecular and Cellular Biology, 19*, 2853–2862.

Yin, H., Price, F., & Rudnicki, M. A. (2013). Satellite cells and the muscle stem cell niche. *Physiological Reviews, 93*, 23–67.

Zetser, A., Frank, D., & Bengal, E. (2001). MAP kinase converts MyoD into an instructive muscle differentiation factor in Xenopus. *Developmental Biology, 240*, 168–181.

CHAPTER TWO

Mandible and Tongue Development

Carolina Parada[1], Yang Chai[1]
Center for Craniofacial Molecular Biology, University of Southern California, Los Angeles, California, USA
[1]Corresponding authors: e-mail address: paradabo@usc.edu; ychai@usc.edu

Contents

Abstract

The tongue and mandible have common origins. They arise simultaneously from the mandibular arch and are coordinated in their development and growth, which is evident from several clinical conditions such as Pierre Robin sequence. Here, we review in detail the molecular networks controlling both mandible and tongue development. We also discuss their mechanical relationship and evolution as well as the potential for stem cell-based therapies for disorders affecting these organs.

Current Topics in Developmental Biology, Volume 115
ISSN 0070-2153
http://dx.doi.org/10.1016/bs.ctdb.2015.07.023

1. EARLY DEVELOPMENT OF THE MANDIBULAR ARCH

1.1 Induction, Delamination, and Migration of Neural Crest Cells

The vertebrate neural crest is a pluripotent cell population that gives rise to a wide array of cell types (Le Douarin, Creuzet, Couly, & Dupin, 2004). Although this differentiation ability has been assumed for decades, only recently has *in vivo* evidence based on mouse genetic tools confirmed the multipotent nature of neural crest cells (NCCs) (Baggiolini et al., 2015). The neural crest is induced at the dorsal region of the neural folds between the surface ectoderm and the neural plate via molecular interactions involving BMP, FGF, and WNT proteins (Trainor & Krumlauf, 2000). Simultaneously with their induction, NCCs undergo epithelial-to-mesenchymal transformation, which leads to their delamination and consequent migration from the neural tube to precise destinations. The neural crest can be subdivided into four distinct axial populations, namely the cranial, cardiac, vagal, and trunk NCC, each of which contributes to a distinctive set of specific cell and tissue types. Cranial neural crest cells (CNCCs) can be further subdivided into forebrain-, midbrain-, and hindbrain-derived populations. This subdivision is achieved through the action of neuroepithelial organizing centers and gradients of FGF, retinoic acid, and WNT signals that specify the character of cells located in these three cephalic vesicles (Gavalas, Trainor, Ariza-McNaughton, & Krumlauf, 2001).

Following this regionalization, two different domains are defined by the expression of Hox genes. Hox genes are expressed along the cranial–caudal axis that defines the posterior hindbrain neuroepithelium (r4 to r8) and it's NCC (Trainor & Krumlauf, 2000, 2001). In contrast, NCC from the forebrain, midbrain, and anterior hindbrain (r1 to r2) do not express any Hox gene (Couly, Grapin-Botton, Coltey, Ruhin, & Le Douarin, 2002). This division produces the specific derivatives of the NCC: a rostral, Hox-negative domain that originates the entire facial skeleton, and a caudal, Hox-positive domain. This pattern of expression highlights the influence of Hox genes in craniofacial evolution (see below; Couly et al., 2002). Interestingly, the skeletogenic capacities of the Hox-negative and Hox-positive NCC domains are different: both are able to generate cartilage whereas only the anterior region yields intramembranous bones (Creuzet, Couly, & Le Douarin, 2005).

Segmental streaming and migration of CNCC are controlled locally by a combination of intrinsic factors and paraxial exclusion zones in the ectoderm and mesoderm, which restrict the migration of CNCC through the territory adjacent to the odd-numbered rhombomeres (r3 and r5) (Kulesa, Bailey, Kasemeier-Kulesa, & McLennan, 2010; Trainor, Sobieszczuk, Wilkinson, & Krumlauf, 2002). This constraint on the streams avoids fusion of the cranial ganglia and mixing of CNCC with distinct anterior-posterior (AP) genetic identities (Trainor & Krumlauf, 2000). The analysis of *Wnt1-Cre;R26R* transgenic embryos has provided valuable information regarding the contribution of postmigratory CNCC to the head in mammals (Chai & Maxson, 2006; Chai et al., 2000). Upon arrival of CNCC at the ventral region of the embryo, they have contact with both ectoderm and endoderm, and their proliferative activity produces distinct swellings known as branchial arches (BAs) as well as the frontonasal prominence. The first BA and the frontonasal prominence give rise to most of the structures in the mammalian face.

1.2 Interarch Patterning

Patterning of the BAs that are Hox-negative is mainly controlled by the specific expression of distal-less (Dlx) genes. A Dlx code provides CNCC with patterning information and intra-arch polarity along the dorsal–ventral (DV)/proximal–distal axis. In each BA, Dlx1/2, Dlx5/6, and Dlx3/4 transcripts overlap distally but exhibit offset proximal expression boundaries. In the first BA, *Dlx1* and *Dlx2* are expressed in both the maxillary and mandibular processes. *Dlx5* and *Dlx6* are expressed only in the mandibular process; their expression extends close to the position of the future hinge region between the maxilla and mandible (Depew, Simpson, Morasso, & Rubenstein, 2005). *Dlx3* and *Dlx4* expression domains are further restricted to the distal-most end of the mandibular process (Depew, Lufkin, & Rubenstein, 2002; Depew et al., 1999; Jeong et al., 2008). In posterior BAs, the nested DV expression domains of *Dlx* genes intersect with the AP Hox code in NCC (Santagati & Rijli, 2003). Because Dlx5/6 control Dlx3/4 expression (Depew et al., 2002; Jeong et al., 2008), the subdivision of the first BA into maxilla and mandible is mainly achieved with two Dlx combinations: Dlx1/2 for the maxillary and Dlx1/2/5/6 for the mandibular process. Thus, a Dlx combinatorial code in CNCC establishes intra-arch polarity (Depew et al., 2005; Minoux & Rijli, 2010). Loss of *Dlx5* and

Dlx6 results in a homeotic transformation of mandible to maxilla, supporting a model of patterning within the BA that relies on a nested pattern of *Dlx* gene expression (Depew et al., 2002). This observation suggests that at one point during development, the upper and lower jaws are very similar in terms of gene expression and that *Dlx5/6* genes constitute the main difference. In addition, jaw development is sensitive to the dosage of *Dlx* genes as haploinsufficiency of single or multiple *Dlx* genes has a gradient effect on mandibular development (Depew et al., 2005). This observation clearly suggests that the expression of *Dlx* genes must be tightly regulated. Although FGF8-soaked beads in the first arch epithelium are able to induce *Dlx2* and *Dlx5* expression in the mandibular mesenchyme, loss of *Fgf8* in the ectoderm results in unaltered *Dlx2* and *Dlx5* expression in the first BA (Trumpp, Depew, Rubenstein, Bishop, & Martin, 1999). Similarly, BMP-soaked beads are able to induce *Dlx* gene expression, but the endogenous BMP and *Dlx* expression patterns do not suggest a direct regulatory relationship. Overall, *Dlx* genes are clearly important for intrabranchial arch patterning (for more information on Dlx genes, see Section 4). Once specified, maxillary and mandibular prominences are established through local migration and regionalized proliferation of CNCC. The maxillary prominence gives rise to part of the upper lip, the maxillary bone, and the secondary palate, whereas the mandibular prominence forms the mandible and part of the tongue.

1.3 Intra-Arch Patterning: Proximal–Distal and Oral–Aboral Axis

CNCC-derived mesenchyme in the mandibular prominence is patterned along the proximal–distal and oral–aboral (rostral–caudal) axis in a complex manner, with the activity of the epithelium being essential (Chai & Maxson, 2006). The mandibular ectoderm can be generally divided into the proximal domain, which expresses *Fgf8*, and the distal domain, which expresses *Bmp4*. FGF and BMP act antagonistically to restrict *Barx1* and *Dlx2* expression to the proximal domain of the first BA mesenchyme and *Msx1*, *Msx2*, and *Alx4* expression to the distal domain (Chai & Maxson, 2006). The biological significance of such a regional molecular specification has been demonstrated by inhibiting BMP signaling, resulting in ectopic *Barx-1* expression in the distal mesenchyme, and producing a morphological change of tooth shape from incisor to molar configuration (Tucker, Matthews, & Sharpe, 1998). The function of *Barx1* is not restricted to the dentition but is also involved in the development of other structures of the digestive system, which are essential for mammalian evolution (Miletich, Buchner, & Sharpe, 2005). Overall, these studies show that FGF and BMP growth factor gradients are critical determinants for specifying the proximal–distal pattern.

Furthermore, FGF signals contribute to the establishment of the oral–aboral (rostral–caudal) axis. Postmigratory CNCC distribution significantly overlaps with the expression of two specific Lim-homeobox domain genes, *Lhx6* and *Lhx7*, within the oral mesenchyme. In the aboral region, the homeobox gene *Goosecoid* (*Gsc*) is expressed in an *Lhx6/7*-negative area. Thus, the mandibular arch mesenchyme can be roughly divided into the *Lhx*-positive rostral (oral) domain and the *Gsc*-positive caudal (aboral) domain (Tucker, Yamada, Grigoriou, Pachnis, & Sharpe, 1999). Apparently, FGF8 is responsible for the regulation of *Lhx* and *Gsc* expression and thereby controls the establishment of the oral–aboral axis in the mandible and the maxilla.

2. MANDIBULAR BONE DEVELOPMENT

2.1 Meckel's Cartilage

After the mandibular arch is formed and patterned, condensation of the CNCC-derived mesenchyme occurs at the level of the presumptive first molar (Fig. 1). This condensation is the first step in the formation of the Meckel's cartilage (MC). Mesenchymal cells then differentiate into chondrocytes and organize into a characteristic bilateral rod-shaped cartilage prior to any sign of ossification in the mandible. The MC is surrounded by a perichondrium composed of fibrous mesenchymal cells, which separates it from neighboring nonchondrogenic osteogenic cells (Fig. 1). The MC first lengthens ventromedially and dorsolaterally on both sides of the mandibular arch and fuses at the most distal tip. The distal ends of the MC develop into the mandibular symphysis, whereas the proximal ends curve to develop the primordium of the malleus and incus bones of the middle ear (Fig. 1E; Richany, Bast, & Anson, 1956). The intermediate portion of the MC degrades in mammals. In the most posterior part of this intermediate region, the chondrocytes transform into fibroblasts to form the sphenomandibular ligament after resorption of the cartilaginous matrix.

Histologically, the MC is a typical hyaline cartilage. Chondrocytes are embedded in extracellular matrix enriched in type 2 collagen and proliferate in the cartilage lacuna. In addition, the MC undergoes additional growth resulting from activity of the chondrocytes located in the perichondrium (Amano et al., 2010). The molecular mechanism underlying the formation

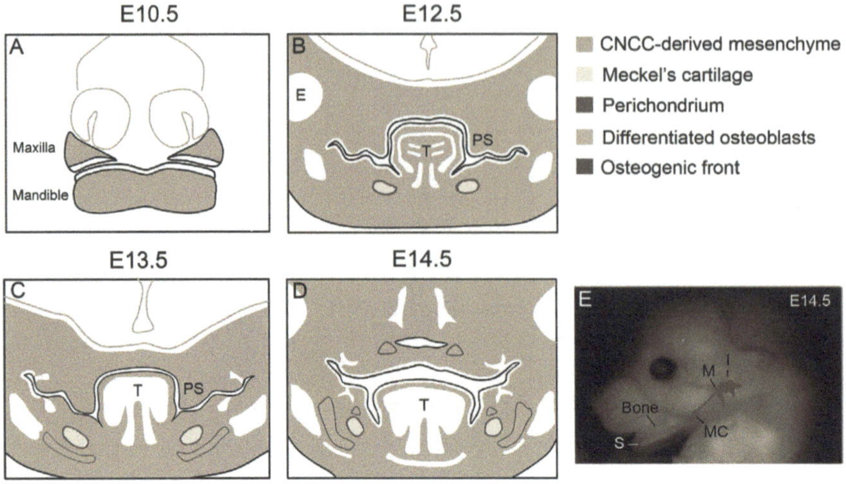

Figure 1 Scheme of mandible development. (A) The first BA subdivides into maxillary and mandibular prominences at E10.5. At this stage, both prominences are solely constituted by CNCC-derived mesenchyme, which is covered by epithelium. (B) At E12.5, the MC develops in the mandibular primordium and mesenchymal cells condense close to it. (C) Mesenchymal cells start differentiating into osteoblasts. An osteogenic front composed of undifferentiated cells surrounds the differentiated osteoblasts. (D) Differentiation progresses in the mandibular primordium at E14.5. (E) The MC gives rise to the symphysis in the distal region and to the malleus and incus in the proximal region of the mandible. *Abbreviations*: E, eye; MC, Meckel's cartilage; I, incus; M, malleus; PS, palatal shelf; S, symphysis; T, tongue. (A color version of this figure is located at the end of this chapter.)

and degeneration of the MC is still unknown. *Sox9*, a well-known chondrogenic master gene, is expressed in all chondroblasts of the MC. Some recent evidence suggests that *Sox9* expression in the mandible is necessary for MC initiation and development. In the absence of *Sox9* in the CNCC-derived mesenchyme, the MC is completely absent (Mori-Akiyama, Akiyama, Rowitch, & de Crombrugghe, 2003). Although the mandible is smaller than in control mice, mandibular gross morphology and bone formation are not severely affected in *Sox9*-mutant mice. This study suggests that the MC may primarily control the size of the mandible but not its initial development including patterning (Mori-Akiyama et al., 2003).

Another factor potentially involved in the formation of the MC is connective tissue growth factor (CTGF). CTGF is expressed along the entire length of the MC from E12.5, when the cartilaginous condensation is first detectable, to E15.5. At this stage, differentiation into mature chondrocytes is accompanied by a significant reduction in CTGF expression although strong expression is still detectable in the peripheral chondrocytes and the perichondrium (Parada, Li, Iwata, Suzuki, & Chai, 2013; Shimo et al., 2004). Ablation of CTGF leads to severe alteration of the MC morphology, including folding of the proximal ends. This phenotype is associated with micrognathia (Ivkovic et al., 2003). The cause of the small mandible in *Ctgf*$^{-/-}$ mutant mice is not known but researchers speculate that it is due to the disruption of the mechanical properties of the MC, similar to the defects in other cartilages in these mice (Ivkovic et al., 2003). The function of CTGF in MC development has also been addressed using *Wnt1-Cre; Tgfbr2*$^{fl/fl}$ mice, which exhibit micrognathia and a defect in chondroblast proliferation in the MC. In these mice, *Ctgf* expression is downregulated in the MC and, interestingly, exogenous CTGF rescues the proliferation defect *in vitro* (Oka et al., 2007). Deletion of Alk5 (Type 1 receptor of TGFβ) causes a similar phenotype including a smaller mandible and the defective MC morphology. *Ctgf, Tgfbr2*, and *Alk5* mutant mice display a common defect in which the MC is severely disrupted in the proximal region, whereas the distal region is comparable to controls. Ossification is also normal in the distal region.

FGF signaling is also involved in MC development. Injection of a dominant negative form of FGFR3 in the developing mandible of chick embryos at early stages leads to truncation of the rostral end, shortening of the MC, and the absence of five of the mandibular bones (Havens et al., 2008). Conversely, overexpression of Fgf10 causes deformation of the MC and a significant increase in size. Fgf10 overexpression also induces upregulation of cartilage-specific genes such as *Col2a1* and *Sox9 in vitro* (Terao et al., 2011). The function of the WNT pathway in the development of the MC is not clear. It has been suggested from studies on limb chondrogenesis that WNT is an inhibitor of this process. Accordingly, chick MC and mandibles treated with WNT5a are severely malformed *in vivo* and *in vitro*. Both the cartilage and mandibular bones are affected (Hosseini-Farahabadi et al., 2013). However, in *Fuz*$^{-/-}$ mice, the MC is expanded due to increased cell proliferation linked to the upregulation of Wnt canonical target genes (Zhang et al., 2011). Overall, these findings suggest that Sox9 might control initiation of MC development, whereas CTGF, FGF, and TGFβ signaling might be involved in regulating later events such as the establishment of cell fate.

Degeneration of the MC appears to involve autophagia. At around E16.5, chondrocytes of the central portion of the MC become hypertrophic and express Beclin1 and LC3. Moreover, caspase 3 is expressed in the lateral population of terminal hypertrophic chondrocytes along the degeneration area of the MC, which indicates that autophagy occurs in hypertrophic chondrocytes during the degradation of the MC and happens prior to chondrocyte cell death (Yang, Zhang, Liu, Zhou, & Li, 2012). During the degeneration process, hypertrophic chondrocytes also become p53-positive (Trichilis & Wroblewski, 1997). Recent studies have suggested that BMP signaling participates in the control of the degeneration process. In the absence of Noggin, the MC is significantly thickened due to elevated cell proliferation, enhanced phosphorylated Smad1/5/8 expression, and lack of degeneration (Wang, Zheng, Chen, & Chen, 2013). The size of the mandibular bone in *Noggin*$^{-/-}$ mice appears to be increased. The MC in these mice does not degrade; instead, it differentiates into bone, mimicking the process of mandibular bone development in other species (Wang et al., 2013). Although the MC was previously proposed to serve as a template for bone deposition and to control endochondral and intramembranous bone formation during mammalian mandibular development, the phenotypes of the mutant mice described above do not directly support this hypothesis.

2.2 Mandibular Osteogenesis

Formation of the mandibular bone occurs in different ways along the proximal–distal axis. As mentioned before, the distal region undergoes endochondral-like ossification to form the symphysis. The middle (and largest) part undergoes intramembranous ossification. The proximal region of the mandible is classified as secondary cartilage and is formed by endochondral ossification. Intramembranous ossification is characterized by an initial condensation of mesenchymal cells followed by differentiation of those cells into osteoblasts. Next, osteoblasts secrete a collagen–proteoglycan (osteoid) matrix that is able to bind calcium salts (Amano et al., 2010). Osteoblast differentiation occurs through a multistep molecular pathway regulated by different transcription factors and signaling proteins. *Dlx5*, which is expressed from the early stages, induces the expression of *Runx2*. *Runx2* (also known as cbfa1) is one of the first genes expressed by mesenchymal cells committed to the osteogenic lineage and is required for differentiation of mesenchymal cells into preosteoblasts (Baek, Kim, de Crombrugghe, & Kim, 2013). Ablation of *Runx2* causes generalized malformations of the skeleton. In the mandible, *Runx2*-null mice display ectopic cartilaginous processes in the MC and lack the condylar cartilage and the mandibular bone (Shibata et al.,

2004). In addition to Dlx5, Hand2 also controls *Runx2* expression but in an inhibitory manner, resulting in the negative regulation of osteoblast differentiation. In the mandibular primordium, downregulation of *Hand2* precedes Runx2-driven osteoblast differentiation (Funato et al., 2009). *Osterix* (*Osx* or *SP7*), a *Runx2* downstream gene, is required for the differentiation of preosteoblasts into mature osteoblasts and is specifically expressed in all osteoblasts (Zhang, 2010). Mutation of *Osx* in CNCC derivatives leads to a tiny and rudimentary mandible although the MC is unaffected (Baek et al., 2013). This finding suggests again that the development of the mandibular bone and MC are not interdependent.

Endochondral ossification occurs when bone is formed after the establishment of a cartilaginous template (Wagner & Karsenty, 2001). At the onset of this process, mesenchymal cells aggregate into condensations that define the future skeletal elements. FGF and BMP are essential for the formation of chondrogenic condensations. Following mesenchymal condensation, cells in the core of the condensations differentiate into chondrocytes that secrete collagen types II, IX, and XI and aggrecan. Cells at the periphery of the condensation form the perichondrium. As mentioned above, Sox9 is the earliest identified factor required for chondrogenesis. Although it is dispensable for condensation, it is necessary for the subsequent steps toward chondrocyte differentiation (Barna & Niswander, 2007). BMP signaling also stimulates chondrocyte differentiation (Yoon et al., 2005), whereas retinoic acid, WNT, and NOTCH pathways inhibit late stages of chondrogenesis. Chondrocytes within the cartilage progressively undergo maturation from proliferating to prehypertrophic to hypertrophic chondrocytes (Wagner & Karsenty, 2001). Prehypertrophic chondrocytes are characterized by the expression of Indian hedgehog (Ihh) (Vortkamp et al., 1996) and hypertrophic chondrocytes by the expression of type X collagen. Coupled with chondrocyte hypertrophy, osteoblast differentiation first occurs within the perichondrium and continues within the marrow cavity. The differentiation of osteoblasts from their mesenchymal progenitors is regulated by signaling pathways such as IHH, NOTCH, WNT, and BMP and also controlled by the activity of transcription factors including Dlx5, Runx2, and Osx, among others (Long & Ornitz, 2013).

In the most distal region of the mandible, Tgfβ-1/Runx2-expressing mesenchymal cells condense at the prospective symphysis. From embryonic day 16.5 to 18.5, chondrocytes expressing Sox9 form in this region and organize into a structure that resembles a growth plate with distinct Ihh, collagen X, and osteopontin expression patterns. Ihh signaling appears to be essential for symphyseal cartilage development. In *Ihh*$^{-/-}$ mice, the development of the symphysis is defective due to enhanced chondrocyte maturation and reduced proliferation of the chondroblast progenitors. This phenotype is rescued upon ablation of Gli3, which thus acts as a negative regulator of symphyseal development. Gli3 function is specifically related to the control of chondroblast proliferation (Sugito et al., 2011). In postnatal life, Ihh also plays an important role. Mesenchymal cells expressing the Ihh receptor Patched1 are present anterior to the Ihh-expressing secondary cartilage, proliferate, differentiate into chondrocytes, and contribute to anterior growth of alveolar bone (Sugito et al., 2011).

In the proximal region, development of the condyle and mandibular angle starts at around E14.5. Both structures are first observed as Sox9-positive, highly condensed mesenchymal cell areas in the posterior osteogenic front of the mandibular bone. These cells in the proximal part of the mandible have the potential to become either osteoblasts or chondrocytes; thus, they are called osteochondroprogenitor cells (Shibata, Suda, Suzuki, Fukuoka, & Yamashita, 2006; Silbermann et al., 1987). Reduced expression of Osterix in combination with Sox9–Sox5 expression appears to be important for the specification of the chondrogenic fate of those cells and consequently the onset of condylar cartilage formation (Shibata et al., 2006). By E15.5, cartilage matrix is clearly detectable in both the condylar and angular processes. TGFβ signaling controls the fate of the osteochondrogenic progenitors in the proximal region of the mandible. This pathway is crucial for cell lineage determination during endochondral ossification, as a positive regulator for chondrocytes and an inhibitor of osteoblasts, via regulation of the expression of critical transcriptional factors. In *Wnt1-Cre;Tgfbr2*$^{fl/fl}$ mice, *Sox9* expression is reduced in osteochondroprogenitor cells in the proximal region of the mandible, whereas osteoblast markers such as *Runx2* and *Dlx5* expression are enhanced. Changes in the expression levels of these three genes result in bone formation without a cartilaginous intermediate in these mice (Oka et al., 2007). Ihh is also involved in the development of proximal structures of the mandible. Ihh expression is detectable in the condylar cartilage by E15.5, and expression of its receptors and effector genes, such as *Gli1*, *Gli2*, *Gli3*, and *PTHrP*, suggest that its range of activity extends to apical condylar tissue layers, including the polymorphic chondroprogenitor layer. In *Ihh*$^{-/-}$ mice, cartilage growth, cell proliferation, and PTHrP expression are reduced as well as chondrocyte gene expression. As in the symphyseal region, these severe alterations are partially rescued in double *Ihh*$^{-/-}$;*Gli3*$^{-/-}$ mice, confirming that

Gli3 modulates the action of Ihh (Shibukawa et al., 2007). Details on the molecular regulation of the condyle and the temporomandibular junction can be found elsewhere in this volume (Hinton, Jing, & Feng, 2015).

3. TONGUE DEVELOPMENT

The development of the tongue and mandible is tightly connected. The tongue begins with the formation of a medial triangular elevation on top of the mandibular arch called the median lingual swelling. If the early mandibular arch is abnormal, tongue development is disrupted in most cases (see below). Next, lateral lingual swellings form on each side of the median tongue bud at E10.5. At this stage, the tongue primordium is exclusively composed of CNCC-derived mesenchyme and covered by epithelium (Han et al., 2012). The lateral swellings grow, fuse with each other, and overgrow the medial lingual swelling. Myoblasts from the occipital somites start invading the tongue primordium at E11.5. The merged lateral lingual swellings form the anterior two-thirds of the tongue. The fibrous CNCC-derived lingual septum is the fusion site of those lateral swellings. Two outgrowths from the third BA, the copula and the hypopharyngeal eminence, compose the posterior third of the tongue. As development advances, the copula is progressively overgrown by the hypopharyngeal eminence and disappears (Moore & Persaud, 2008). Finally, the tongue primordium undergoes rapid enlargement and the muscular component occupies most of it (Huang, Zhi, Izpisua-Belmonte, Christ, & Patel, 1999). Tongue connective tissue and vasculature are derived from CNCC, whereas the skeletal muscles originate from myoblasts (Noden & Francis-West, 2006). Reciprocal interactions between CNCC and myogenic cells play an important role in regulating tongue development.

3.1 Development of the Muscular Component of the Tongue

Myoblasts from the occipital somites migrate to the tongue primordium along the hypoglossal cord (Huang et al., 1999). This process is tightly controlled by at least three main molecules: c-met, Gab1, and Lbx1 (Amano et al., 2002; Gross et al., 2000; Sachs et al., 2000). Pax3 and Pax7 regulate proliferation and differentiation of myogenic progenitors shortly after their arrival in the limb or the tongue primordium (Tajbakhsh & Buckingham, 2000; reviewed in Parada, Han, & Chai, 2012). Pax3 is essential for muscle development because it controls the gene hierarchy that activates myogenic regulatory factors (MRFs), including myogenic factor 5 (Myf5), muscle-specific regulatory factor 4 (MRF4), myoblast determination protein (MyoD), and myogenin (Tajbakhsh & Cossu, 1997). Together, these MRFs direct the determination and differentiation of myoblasts in the limbs (Berkes & Tapscott, 2005; Nabeshima et al., 1993; Rudnicki et al., 1993). In the developing tongue, MRFs are expressed in different subpopulations of myoblasts throughout development (Han et al., 2012; Hosokawa et al., 2010; Zhong, Zhao, Mayo, & Chai, 2015). Recent studies have shown that ablation of the *Myf5*-expressing progenitors in the tongue does not significantly affect muscle pattern or size (Zhong et al., 2015). This finding suggests that a *Myf5*-independent myogenic lineage can compensate for loss of *Myf5*-expressing myogenic cells, as suggested by other studies of limb muscles (Gensch, Borchardt, Schneider, Riethmacher, & Braun, 2008; Haldar, Karan, Tvrdik, & Capecchi, 2008). In contrast, ablation of the *MyoD*-positive population in the tongue leads to microglossia that is associated with a striking reduction in muscle fiber formation in the tongue (Zhong et al., 2015). Myogenic differentiation is followed by myoblast fusion. Individual myoblasts fuse with one another to generate myotubes. Afterward, additional differentiated myoblasts incorporate into the forming myotubes, leading to the further maturation of the myofibers (Rochlin, Yu, Roy, & Baylies, 2010). Myoblast fusion requires extracellular calcium and changes in cell membrane topography and cytoskeletal organization. Recently, studies have identified several cell adhesion proteins, transmembrane lipids, and intracellular domain-associated signaling or adaptor proteins that accumulate at sites of contact between two myogenic cells (Hindi, Tajrishi, & Kumar, 2013).

Major signaling pathways such as WNT, TGFβ, and FGF are involved in the regulation of myogenesis at different key developmental stages. Canonical Wnt signaling plays important roles in dermomyotome and myotome formation (Linker, Lesbros, Stark, & Marcelle, 2003; Otto, Schmidt, & Patel, 2006). In the somites, Wnt1 ligands preferentially activate Myf5, whereas Wnt7a ligands activate MyoD in myogenic progenitors (Tajbakhsh & Cossu, 1997). Wnt signaling is activated in the same region of the hypoglossal cord as the *Myf5*-derived population. In the tongue primordium, canonical Wnt signaling is activated in the myogenic region right after migration of the myoblasts, and it shifts to the CNCC-derived mesenchyme at later stages. The Wnt signaling pathway first controls myoblast migration and later regulates differentiation of the myogenic progenitors in the tongue, based on the analysis of *Myf5-Cre;β-catenin*$^{fl/fl}$ mice, which exhibit microglossia (Zhong et al., 2015).

Recent studies have shown that TGFβ pathway members are also involved in the regulation of proliferation, differentiation, and myoblast fusion during tongue myogenesis. Interestingly, both inhibition and promotion of myogenesis by TGFβ have been reported (McPherron, Lawler, & Lee, 1997; Wang, Noulet, Edom-Vovard, Le Grand, & Duprez, 2010). Ablation of *Smad4*, a mediator of TGFβ signaling, in myogenic progenitors disrupts myogenic terminal differentiation and myoblast fusion in the tongue primordium but does not interfere with early myogenic determination. The disruption in these processes leads to reduced myotube length, decreased average number of myonuclei per myotube, and numerous centrally located nuclei (Han et al., 2012).

FGF pathway members act as downstream mediators of TGFβ signaling that control myogenic differentiation and fusion. *Fgf6* and its receptor, *Fgfr4*, are expressed from early stages in the myogenic component of the mouse developing tongue, although their expression patterns do not overlap entirely (Han et al., 2012). Both members of the FGF family act downstream of Smad4-mediated TGFβ signaling during tongue myogenic differentiation and myoblast fusion. Accordingly, in *Myf5-Cre;Smad4*$^{fl/fl}$ mice, exogenous FGF6 partially rescues the compromised tongue myoblast differentiation and fusion (Han et al., 2012). *Fgf6*$^{-/-}$ mice have a severe regeneration defect with fibrosis and myotube degeneration, consistent with a function during tongue myogenic differentiation (Armand, Laziz, & Chanoine, 2006). Although *Fgfr4*$^{-/-}$ mice show no phenotype at birth, muscle regeneration in *Fgfr4*-null mice becomes highly abnormal at the time point when *Fgfr4* is normally expressed (Zhao et al., 2006). Previous studies suggest that TGFβ2 may function upstream of Smad4 during myogenesis. TGFβ2 promotes fusion of myoblasts *in vitro* and is specifically expressed in the region of myoblast-to-myotube fusion (Olson, Sternberg, Hu, Spizz, & Wilcox, 1986). This expression pattern suggests that myoblasts may induce TGFβ2 to stimulate the fusion of adjacent myoblasts. Taken together, these findings indicate that a genetic hierarchy involving TGFβ and FGF plays a critical role in the regulation of specific myogenic factors during tongue muscle development.

3.2 CNCC-Derived Mesenchyme in the Developing Tongue

CNCCs initiate and potentially direct tongue development (Han et al., 2012). From the few studies published addressing the function of the CNCC-derived mesenchyme in tongue development, it is tempting to speculate that CNCCs in the tongue act as a scaffold for the organization of migrating myoblasts into the myogenic core and also may operate as a niche that releases molecular instructions to direct survival, proliferation, and differentiation of myogenic progenitors as well as patterning of the muscles. Early mandibular development is characterized by the differential expression of specific Dlx genes. *Dlx5/6* expression also regulates the development of the anterior (and largest) region of the tongue. Craniofacial myogenesis depends on normal *Dlx5/6* expression by CNCC, because ablation of both genes leads to loss of masticatory muscles and disrupted tongue development (Heude et al., 2010). Because *Dlx5/6* are only expressed by the CNCC-derived mesenchyme in the craniofacial region, these genes might be playing an early instructive function to control muscle formation. In *Dlx5/6*$^{-/-}$ mice, the intrinsic muscles of the tongue and sublingual muscles are severely affected, although they express both determination and differentiation markers (Heude et al., 2010). *Dlx5/6* work together with the transcription factor *Hand2*, which is also expressed early in the mandibular arch, to establish the dorsoventral/proximodistal pattern of the mandibular arch and, consequently, that of the tongue. *Hand2* contributes to the establishment of the proximal–distal patterning through a negative-feedback loop in which it represses *Dlx5/6* expression in the distal arch mesenchyme following *Dlx5*- and *Dlx6*-mediated induction of *Hand2* expression in the same region. Failure to inhibit distal *Dlx5/6* expression leads to the absence of lateral lingual swelling expansion, resulting in aglossia. Therefore, *Hand2* seems to determine a distal mandibular arch domain that is favorable for lower jaw development, including the induction of tongue morphogenesis (Barron et al., 2011).

Hh signaling in the CNCC is also crucial for early tongue development. *Ptch1*, a *Shh* receptor, is expressed in the mesenchyme of the first BA and in the tongue CNCC-derived mesenchyme, suggesting that this tissue is responsive to Shh signals coming from the tongue epithelium. Interestingly, *Wnt1-Cre;Smo*$^{n/c}$ embryos, in which Shh signaling is disrupted in the CNCC component, exhibit a vestigial tongue. The tongue defect in these mice is detectable early and is associated with the absence of *Myf5*-expressing myogenic progenitors. The CNCC component is also reduced. Thus, Hh signaling in the CNCC-derived mesenchyme might be involved in transmitting information from the epithelium to the myogenic progenitors to coordinate tongue formation (Jeong, Mao, Tenzen, Kottmann, & McMahon, 2004). The relevance of Shh signaling in tongue development is confirmed by the conditional inactivation of *β-catenin* in the lingual epithelium, which causes downregulation of *Shh* expression in the tongue epithelium and reduction of *Ptch1* and *Gli1* expression in the underlying mesenchyme. In *β-catenin*-mutant mice, the two lateral swellings of the primitive tongue are smaller and remain separated. In addition, the merger between the tuberculum impar and two lingual swellings is defective. At late stages, the tongue is much smaller, deformed, and completely lacks taste papillae. The number of CNCC-derived mesenchymal cells is severely reduced, which appears to be the cause of the microglossia (Lin et al., 2011). These findings are also consistent with the phenotype of $Fuz^{-/-}$ mice that exhibit severe craniofacial deformities including agenesis of the tongue. Malformations in these mice are associated with downregulation of Hedgehog signaling (Zhang et al., 2011).

TGFβ family members expressed in the CNCC also regulate skeletal muscle development via tissue–tissue interactions. Ablation of TGFβ pathway members in CNCC lead to severe tongue defects. For instance, deletion of *Tgfbr2* in CNCC-derived mesenchyme results in defects in tongue myogenesis and a smaller overall tongue size, which are due to a significant decrease in proliferation of myogenic cells (Hosokawa et al., 2010). The reduction in myogenic cell proliferation is associated with downregulation of *Fgf10* expression, which is only detectable in the CNCC component. Interestingly, exogenous FGF10 reverses the reduction of tongue myogenic cell numbers in *Tgfbr2*-mutant mice, which suggests noncell autonomous activity (Hosokawa et al., 2010). More recently, another study suggested that Fgf10 overexpression in the CNCC-derived tongue mesenchyme in *Wnt1-Cre;Tak1*$^{fl/fl}$ mice also causes tongue malformations (Song et al., 2013). These results are consistent with previous studies showing that FGF signaling is required for skeletal muscle formation and suggest that its function is dose dependent.

Tak1 is an element of one of the noncanonical pathways mediating TGFβ signals. In the developing tongue of *Wnt1-Cre;Tgfbr2*$^{fl/fl}$ mice, noncanonical TGFβ signaling via a TβRI/TβRIII complex results in increased phosphorylation of p38 (Iwata, Suzuki, Pelikan, Ho, & Chai, 2013), which is accompanied by overactivation of ABL1 and PKC. The activation of these pathways in *Tgbr2*-mutant mice is associated with upregulation of the expression of *follistatin* and downregulation of *Fgf4*. Additionally, elastic and collagen fibers are poorly organized and immature and *tenascin C* expression is compromised in *Wnt1-Cre;Tgfbr2*$^{fl/fl}$ tongues (Iwata et al., 2013). Moreover, reduction of altered noncanonical TGFβ signaling rescues microglossia in *Wnt1-Cre;Tgfbr2*$^{fl/fl}$ mice by restoring proliferation of the myogenic progenitors. However, muscle disorganization persists, suggesting that the noncanonical TGFβ pathway affects myoblast proliferation, whereas canonical TGFβ signaling might be required for muscle organization (Iwata et al., 2013).

3.3 Tendon Development and Connection to Bone

The initial relationship between CNCC and myogenic progenitors is maintained throughout myogenesis and helps establish the attachment of tongue and head muscles to the skeleton during craniofacial development. Both muscle and tendon cells are involved in the formation of this attachment and regulate each other at different stages of development throughout the body. For example, the loss of stripe, which is involved in early steps of tendon differentiation, results in the disruption of the entire somatic muscle pattern (Frommer, Vorbrüggen, Pasca, Jackle, & Volk, 1996). Similarly, ablation of Periostin, an adhesion molecule, in the myoseptum causes a differentiation defect in myoblasts (Kudo, Amizuka, Araki, Inohaya, & Kudo, 2004). On the other hand, although muscles are not necessary for the initiation of tendon formation in the branchiomeric and extraocular regions, they are required for further tendon development (Grenier, Teillet, Grifone, Kelly, & Duprez, 2009).

During tendon development, collagen fibrillogenesis and assembly occur to form a structurally and functionally mature tissue. *In vivo* evidence suggests that the condensation, determination, and differentiation of tendon progenitors are dependent on the activity of *Scleraxis* (*Scx*) (Murchison et al., 2007). All of the forming tendons associated with the tongue are of CNCC origin and express *Scx* (Hosokawa et al., 2010 and our unpublished data). One of the relevant functions of *Scx* during tendon formation is the control of collagen expression (Léjard et al., 2007; Terraz, Brideau, Ronco, & Rossert, 2002). *Type 1 collagen* is expressed in the tongue from early stages of development. It is detectable in the CNCC-derived mesenchyme adjacent to the tongue epithelium (Hosokawa et al., 2010) and in the tendons of the extrinsic muscles, which connect the tongue to the mandible (our unpublished results), where *Scx* is also expressed.

TGFβ and FGF also act at later stages in the induction of tendon formation and maintenance of tendon progenitor cells (Pryce et al., 2009). TGFβ may mediate the recruitment of new tendon cells by promoting the differentiation of muscles and cartilage to establish the connections between tendon primordia and their respective musculoskeletal counterparts (Pryce et al., 2009). TGFβ signaling is a potent inducer of *Scx* throughout the organism. Disruption of TGFβ signaling in *Tgfb2*$^{-/-}$;*Tgfb3*$^{-/-}$embryos or through inactivation of *Tgfbr2* results in the loss of most tendons and ligaments in the limbs, trunk, tail, and head (Pryce et al., 2009). In *Wnt1-Cre; Tgfbr2*$^{fl/fl}$ mice, *Scx* expression is diminished in the tendons of the tongue muscles (Hosokawa et al., 2010). Bead implantation experiments indicate that TGFβ signaling, particularly TGFβ2, induces *Scx* expression in CNCC during tongue development. Thus, in the tongue primordium, TGFβ signaling is required for the cell-autonomous induction of *Scx*, *type I collagen* expression, and the regulation of the fate of CNCC (Hosokawa et al., 2010). Other TGFβ ligands, such as growth/differentiation factor 8 (GDF-8, encoded by *MSTN*), could also be involved, as *Mstn*$^{-/-}$ mice have hypocellular tendons, in addition to hypertrophic muscles (Mendias, Bakhurin, & Faulkner, 2008). However, the function of GDF-8 in tendon formation in the craniofacial region is still unclear.

FGFs have both early and late functions in tendon development. During somite development, FGF signals secreted from the myotome induce the formation of a *Scx*-expressing tendon progenitor population in the sclerotome, at the juncture between the future muscle and cartilage lineages. The activation of Pea3 and Erm in response to FGF signaling is both necessary and sufficient for *Scx* expression in the somite, which helps restrict the domain of somitic tendon progenitors (Brent & Tabin, 2004). Later, *Pea3* and *Sprouty1* and *2* are expressed in muscles and tendons, and their expression is enhanced at the myotendinous junctions in limbs. Analysis of *Pea3* and *Sprouty* gene expression in muscleless limbs of *Pax3*-mutant mice indicates strong expression in muscles and expression in tendons that depends on muscles (Eloy-Trinquet, Wang, Edom-Vovard, & Duprez, 2009), demonstrating again the importance of interactions between connective tissues and muscle during development and adult stages alike.

4. EVOLUTION OF THE MANDIBLE AND TONGUE

The mandible is a recent evolutionary novelty characteristic of gnathostomes. The anatomical configuration of the BAs is different in agnathans and gnathostomes. This different configuration is related to changes in the CNCC-derived mesenchyme that is distributed in an embryonic domain known as the trigeminal region, the rostral part of the embryonic head corresponding to the peripheral innervation of the trigeminal nerve (Kuratani, 2012). CNCC populating the first BA do not express any Hox genes. The absence of *Hox* genes from the first BA is thought to be a permissive and indispensable condition favoring maxilla and mandible development. Interestingly, in agnathans like lamprey and hagfish, the homologous structures do express Hox genes. The most anteriorly expressed *Hox* gene in gnathostomes is *Hoxa2*, which is only expressed in the more posterior NCC that migrate into the second BA. Accordingly, ectopic expression of *Hoxa2* in the first BA in several animal models impedes jaw development and ablation of Hoxa2 in mice leads to duplications of some first BA skeletal components in the second BA, but not of the jaw (Creuzet, Couly, Vincent, & Le Douarin, 2002). In agnathans, the *HoxL6* gene is expressed in the first BA, and this is also the case for the homologous gene in amphioxi, which do not exhibit jaws or BAs (Cohn, 2002). It is still not clear how this Hox-negative zone in gnathostomes is controlled from a molecular point of view. Researchers have suggested that it might be due to the action of the midbrain–hindbrain isthmus, a signaling center that expresses and releases FGF8, which in turns prevents the expression of *Hoxa2* in its rostral domain (Trainor et al., 2002).

The development and evolution of the mandible are highly dependent on the influence of other tissues on the CNCC-derived mesenchyme. For instance, the foregut endoderm provides patterning information to the mesenchyme in the first BA. Interestingly, this tissue can only pattern *Hox*-negative neural crest-derived mesenchyme but not *Hox*-positive cells (Creuzet et al., 2002; Couly et al., 2002). In lamprey, *HoxL6*-positive NCC migrating into the first BA are unresponsive to the endoderm and consequently incapable of developing jaws (Cohn, 2002). Another remarkable feature of lamprey is that they show no Dlx code, meaning that there is no restricted expression of any *Dlx* genes along the proximal–distal axis of the BAs, unlike in gnathostomes (Cohn, 2002; Kuratani, Nobusada, Horigome, & Shigetani, 2001). The *Bapx1* gene is also absent in lamprey, which suggests that it might contribute to the acquisition of jaws and the establishment of maxilla in gnathostomes that express it focally in the first BA (Cerny et al., 2010). At later stages, the absence of *Hox* expression might allow the cartilage to develop and then be sculpted by the differential expression of *Dlx* genes. Even at these late developmental stages, *Hoxa2* represses endochondral ossification in the second BA (Kanzler, Kuschert, Liu, & Mallo, 1998), whereas *Dlx5* activates chondrocyte differentiation throughout the body (Ferrari & Kosher, 2002). A fine balance between the opposing activities of these genes could then be the basis for differentiation of the jaw and other skeletal elements of the vertebrate face.

The feeding mechanism is probably the most important element determining the success of adaptation of vertebrates to their environment. In feeding, both the jaws and tongue play a crucial function. The tongue has a characteristic form in tetrapods. Fish have a slight elevation of the mucosa on the floor of the mouth but this structure does not contain any voluntary muscles, unlike the tongues of land vertebrates (Iwasaki, 2002), one exception being the *Xenopus laevis* (Toyoshima & Shimamura, 1982). Most adult amphibians and all known reptiles, birds, and mammals have a tongue. Therefore, it has been suggested that the tongue appeared with the establishment of tetrapods and is related to the terrestrial lifestyle. In agnathans, particularly in lamprey, there is a tongue-like structure with the shape of a piston, but this organ is not homologous to the tongues of gnathostomes. The lamprey tongue and the tongues of tetrapods originated independently during evolution. The musculature of the gnathostome tongue originates in the hypobranchial system. In agnathans and gnathostome fishes, the hypobranchial apparatus forms musculature that is related to the gills and reforms during embryonic development to produce the lingual musculature. With the reduction of the gills in tetrapods, the mobile tongue appeared. Thus, the tongue musculature is derived from the hypobranchial musculature, which is anchored to the hyoid apparatus. Most likely, one of the main roles of the tongue is to facilitate eating on land, in collaboration with other anatomical structures within and near the oral cavity. It has been proposed that, during adaptation from a wet to a dry habitat in the evolution of vertebrates, stratification and keratinization were the most important changes in the lingual epithelium (Iwasaki, 2002).

5. MECHANICAL RELATIONSHIP BETWEEN THE MANDIBLE AND TONGUE

The tongue and mandible have a common origin. They arise simultaneously from the mandibular arch and their development is coordinated (Fig. 1). Mutations affecting early mandibular development have deleterious effects on tongue formation. In wild-type mice, the mandibular primordium grows down and lengthens from E12.5 to newborn stage (Ramaesh & Bard, 2003). This growth pattern seems to provide the tongue with physical space to move downward, and this movement closely coincides with another important event in craniofacial morphogenesis, the reorientation of the palatal shelves from a vertical to a horizontal position (Ferguson, 1977). In humans, morphometric analyses have shown that the growth of the tongue and mandible is intrinsically linked in size and shape between 20 gestational weeks and 24 months postnatally (Hutchinson, Kieser, & Kramer, 2014). The mechanical relationship between the mandible and tongue is also supported by certain clinical conditions in which alteration of one affects the other. One of the best examples is Pierre Robin Sequence (PRS; Fig. 2). PRS was originally described as a form of respiratory obstruction accompanied by malposition of the tongue and caused by dysmorphic atresia of the mandible. The current definition of PRS includes respiratory obstruction, glossoptosis, mandibular hypoplasia, and cleft palate (Fig. 2; Tan, Kilpatrick, & Farlie, 2011). Several previous studies have reported cleft palate due to physical obstruction by the tongue in mice. However, only mutation of *Prdm16* causes failed palate elevation associated with a highly positioned tongue and smaller mandibular bone, mimicking the clefting of human PRS (Bjork, Turbe-Doan, Prysak, Herron, & Beier, 2010). Yet, no direct evidence has been presented showing that the tongue malposition and cleft palate are due to the mandibular malformation. Further investigation is required to clarify the influence of mandibular growth on the size and position of the tongue.

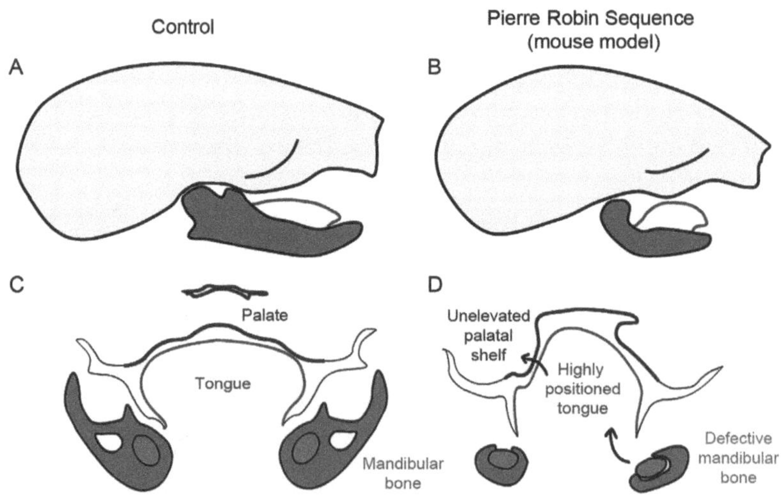

Figure 2 Mechanical relationship between the mandible and tongue. (A) and (B) show schemes of a sagittal view of the skull of control and a PRS mouse model at newborn stage, respectively. (C) and (D) are schemes of corresponding coronal sections. (A color version of this figure is located at the end of this chapter.)

6. STEM CELLS IN MANDIBLE AND TONGUE REGENERATION

Mandibular distraction osteogenesis (MDO) is the conventional treatment for infants with compromised airways due to micrognathia with glossoptosis and other congenital anomalies resulting in unilateral or bilateral mandibular hypoplasia. Typically in MDO, a corticotomy is used to fracture the bone into two segments, and the ends of the bone are gradually moved apart during the distraction phase, allowing new bone to form in the gap. Studies in multiple animal models have demonstrated diverse methods to increase bone-healing processes with the addition of stem cells, growth factors, experimental medications, and laser and ultrasound treatment. Mesenchymal stem cells (MSCs) from the bone marrow also enhance bone growth at distraction sites. Recently, it has been shown that stromal cell-derived factor-1 facilitates migration of endogenous MSCs *in vitro* and *in vivo*, which makes it a potential target that can be manipulated to upregulate an ongoing endogenous process. Moreover, the use of BMP and FGF along with transplantation of MSCs dramatically improves outcomes in osteogenic distraction compared to the cell transplant alone (Earley & Butts, 2014).

Stem cell-mediated tissue regeneration also has the potential to restore tongue tissue lost to cancer. Tongue squamous cell carcinoma is one of the most prevalent malignant cancers of the head and neck region. Surgical management remains the basis of treatment. After excision of the lesion, reconstruction is required to maintain function. However, almost half of patients have difficulties in eating and drinking after treatment. Stem cell-mediated tissue regeneration would restore the normal tongue volume, shape, and function, which is generally not possible using surgical procedures. The presence of *Pax7*-positive cells in the adult tongue provides the opportunity to develop a stem cell-based regeneration approach. Pax7 is crucial for the specification and survival of satellite cells, which are quiescent myogenic cells (Morgan & Partridge, 2003). In the adult muscle, they act as a reserve population, able to proliferate in response to injury, and give rise to regenerated muscle and additional satellite cells (Oustanina, Hause, & Braun, 2004). Interestingly, $Pax7^{-/-}$ mice do not show muscle defects during embryogenesis but display severely compromised muscle regeneration capability. Overexpression of *Pax7* results in downregulation of *MyoD* and promotes cell-cycle withdrawal from the proliferating state; therefore, Pax7 plays a critical role in the maintenance of the satellite cell pool

(Olguin & Olwin, 2004). In the tongue of adult mice, the number of *Pax7*-positive cells is low but increases after injury, consistent with their role in forming new muscles (our unpublished data). In order to use *Pax7*-positive satellite cells in regenerative medicine, it is necessary to gain a comprehensive understanding of the regulatory mechanisms that control tongue morphogenesis and myogenesis. New approaches will likely be based on stem cell and tissue engineering technologies, using isolated CNCC-derived stem cells and skeletal muscle satellite *Pax7*-positive cells under the precise control of the proper molecular regulatory mechanisms to generate a functional tongue.

7. CONCLUSION

The mandible and tongue share a common origin in the CNCC populating the first BA and develop simultaneously and coordinately. Their relationship is not only based on their CNCC origins, but is also mechanical in nature. Growth of the tongue appears to be sensitive to that of the mandible. In humans, conditions in which one of these two structures is affected usually disrupt the development of the other. Understanding normal development of the mandible and tongue will facilitate the treatment of both developmental syndromes and cancer affecting this region using regenerative medicine and tissue engineering approaches.

REFERENCES

Amano, O., Doi, T., Yamada, T., Sasaki, A., Sakiyama, K., Kanegae, H., et al. (2010). Meckel's cartilage: Discovery, embryology and evolution—Overview of the specificity of Meckel's cartilage. *Journal of Oral Biosciences, 52*, 125–135.

Amano, O., Yamane, A., Shimada, M., Koshimizu, U., Nakamura, T., & Iseki, S. (2002). Hepatocyte growth factor is essential for migration of myogenic cells and promotes their proliferation during the early periods of tongue morphogenesis in mouse embryos. *Developmental Dynamics, 223*, 169–179.

Armand, A. S., Laziz, I., & Chanoine, C. (2006). FGF6 in myogenesis. *Biochimica et Biophysica Acta, 1763*, 773–778.

Baek, W. Y., Kim, Y. J., de Crombrugghe, B., & Kim, J. E. (2013). Osterix is required for cranial neural crest-derived craniofacial bone formation. *Biochemical and Biophysical Research Communications, 432*, 188–192.

Baggiolini, A., Varum, S., Mateos, J. M., Bettosini, D., John, N., Bonalli, M., et al. (2015). Premiratory and migratory neural crest cells are multipotent *in vivo*. *Cell Stem Cell, 16*, 314–322.

Barna, M., & Niswander, L. (2007). Visualization of cartilage formation: Insight into cellular properties of skeletal progenitors and chondrodysplasia syndromes. *Developmental Cell, 12*, 931–941.

Barron, F., Woods, C., Kuhn, K., Bishop, J., Howard, M. J., & Clouthier, D. E. (2011). Downregulation of Dlx5 and Dlx6 expression by Hand2 is essential for initiation of tongue morphogenesis. *Development*, *138*, 2249–2259.

Berkes, C. A., & Tapscott, S. J. (2005). MyoD and the transcriptional control of myogenesis. *Seminars in Cell and Developmental Biology*, *16*, 585–595.

Bjork, B. C., Turbe-Doan, A., Prysak, M., Herron, B. J., & Beier, D. R. (2010). Prdm16 is required for normal palatogenesis in mice. *Human Molecular Genetics*, *19*, 774–789.

Brent, A. E., & Tabin, C. J. (2004). FGF acts directly on the somitic tendon progenitors through the Ets transcription factors Pea3 and Erm to regulate scleraxis expression. *Development*, *131*, 3885–3896.

Cerny, R., Cattell, M., Sauka-Spengler, T., Bronner-Fraser, M., Yu, F., & Medeiros, D. M. (2010). Evidence for the prepattern/cooption model of vertebrate jaw evolution. *Proceedings of the National Academy of Sciences of the United States of America*, *107*, 17262–17267.

Chai, Y., Jiang, X., Ito, Y., Bringas, P., Han, J., Rowitch, D. H., et al. (2000). Fate of the mammalian cranial neural crest during tooth and mandibular morphogenesis. *Development*, *127*, 1671–1679.

Chai, Y., & Maxson, R. E. (2006). Recent advances in craniofacial morphogenesis. *Developmental Dynamics*, *235*, 2353–2375.

Cohn, M. J. (2002). Evolutionary biology: Lamprey Hox genes and the origin of jaws. *Nature*, *416*, 386–387.

Couly, G., Grapin-Botton, A., Coltey, P., Ruhin, B., & Le Douarin, N. M. (2002). Determination of the identity of the derivatives of the cephalic neural crest: Incompatibility between Hox gene expression and lower jaw development. *Development*, *125*, 3445–3459.

Creuzet, S., Couly, G., & Le Douarin, N. M. (2005). Patterning the neural crest derivatives during development of the vertebrate head: Insights from avian studies. *Journal of Anatomy*, *207*, 447–459.

Creuzet, S., Couly, G., Vincent, C., & Le Douarin, N. M. (2002). Negative effect of Hox gene expression on the development of the neural crest-derived facial skeleton. *Development*, *129*, 4301–4313.

Depew, M. J., Liu, J. K., Long, J. E., Presley, R., Meneses, J. J., Pedersen, R. A., et al. (1999). Dlx5 regulates regional development of the branchial arches and sensory capsules. *Development*, *126*, 3831–3846.

Depew, M. J., Lufkin, T., & Rubenstein, J. L. (2002). Specification of jaw subdivisions by Dlx genes. *Science*, *298*, 381–385.

Depew, M. J., Simpson, C. A., Morasso, M., & Rubenstein, J. L. R. (2005). Reassessing the Dlx code: The genetic regulation of branchial arch skeletal pattern and development. *Journal of Anatomy*, *207*, 501–561.

Earley, M., & Butts, S. C. (2014). Update on mandibular distraction osteogenesis. *Current Opinion in Otolaryngology & Head and Neck Surgery*, *22*, 276–283.

Eloy-Trinquet, S., Wang, H., Edom-Vovard, F., & Duprez, D. (2009). Fgf signaling components are associated with muscles and tendons during limb development. *Developmental Dynamics*, *238*, 1195–1206.

Ferguson, M. W. (1977). The mechanism of palatal shelf elevation and the pathogenesis of cleft palate. *Virchows Archiv A, Pathological Anatomy and Histology*, *375*, 97–113.

Ferrari, D., & Kosher, R. A. (2002). Dlx5 is a positive regulator of chondrocyte differentiation during endochondral ossification. *Developmental Biology*, *252*, 257–270.

Frommer, G., Vorbrüggen, G., Pasca, G., Jackle, H., & Volk, T. (1996). Epidermal egr-like zinc finger protein of Drosophila participates in myotube guidance. *EMBO Journal*, *15*, 1642–1649.

Funato, N., Chapman, S. L., McKee, M. D., Funato, H., Morris, J. A., Shelton, J. M., et al. (2009). Hand2 controls osteoblast differentiation in the branchial arch by inhibiting DNA binding of Runx2. *Development, 136*, 615–625.
Gavalas, A., Trainor, P., Ariza-McNaughton, L., & Krumlauf, R. (2001). Synergy between Hoxa1 and Hoxb1: The relationship between arch patterning and the generation of cranial neural crest. *Development, 128*, 3017–3027.
Gensch, N., Borchardt, T., Schneider, A., Riethmacher, D., & Braun, T. (2008). Different autonomous myogenic cell populations revealed by ablation of Myf5-expressing cells during mouse embryogenesis. *Development, 135*, 1597–1604.
Grenier, J. K., Teillet, M. A., Grifone, R., Kelly, R. G., & Duprez, D. (2009). Relationship between neural crest cells and cranial mesoderm during head muscle development. *PLoS One, 4*, e4381.
Gross, M. K., Moran-Rivard, L., Velasquez, T., Nakatsu, M. N., Jagla, K., & Goulding, M. (2000). Lbx1 is required for muscle precursor migration along a lateral pathway into the limb. *Development, 127*, 413–424.
Haldar, M., Karan, G., Tvrdik, P., & Capecchi, M. R. (2008). Two cell lineages, myf5 and myf5-independent, participate in mouse skeletal myogenesis. *Developmental Cell, 14*, 437–445.
Han, D., Zhao, H., Parada, C., Hacia, J. G., Bringas, P., Jr., & Chai, Y. (2012). A TGF-beta-Smad4-FGF6 signaling cascade controls myogenic differentiation and myoblast fusion during tongue development. *Development, 139*, 1640–1650.
Havens, B. A., Velonis, D., Kronenberg, M. S., Lichtler, A. C., Oliver, B., & Mina, M. (2008). Roles of FGFR3 during morphogenesis of Meckel's cartilage and mandibular bones. *Developmental Biology, 316*, 336–349.
Heude, E., Bouhali, K., Kurihara, Y., Kurihara, H., Couly, G., Janvier, P., et al. (2010). Jaw muscularization requires Dlx expression by cranial neural crest cells. *Proceedings of the National Academy of Sciences of the United States of America, 107*, 11441–11446.
Hindi, S. M., Tajrishi, M. M., & Kumar, A. (2013). Signaling mechanisms in mammalian myoblast fusion. *Science Signaling, 6*, re2.
Hinton, R. J., Jing, J., & Feng, J. Q. (2015). Genetic influences on temporomandibular joint development and growth. *Current Topics in Developmental Biology, 115*, 85–109.
Hosokawa, R., Oka, K., Yamaza, T., Iwata, J., Urata, M., Xu, X., et al. (2010). TGF-[beta] mediated FGF10 signaling in cranial neural crest cells controls development of myogenic progenitor cells through tissue–tissue interactions during tongue morphogenesis. *Developmental Biology, 341*, 186–195.
Hosseini-Farahabadi, S., Geetha-Loganathan, P., Fu, K., Nimmagadda, S., Yang, H. J., & Richman, J. M. (2013). Dual functions for WNT5A during cartilage development and in disease. *Matrix Biology, 32*, 252–264.
Huang, R., Zhi, Q., Izpisua-Belmonte, J. C., Christ, B., & Patel, K. (1999). Origin and development of the avian tongue muscles. *Anatomy and Embryology (Berlin), 200*, 137–152.
Hutchinson, E. F., Kieser, J. A., & Kramer, B. (2014). Morphometric growth relationships of the immature human mandible and tongue. *European Journal of Oral Sciences, 122*, 181–189.
Ivkovic, S., Yoon, B. S., Popoff, S. N., Safadi, F. F., Libuda, D. E., Stephenson, R. C., et al. (2003). Connective tissue growth factor coordinates chondrogenesis and angiogenesis during skeletal development. *Development, 130*, 2779–2791.
Iwasaki, S. (2002). Evolution of the structure and function of the vertebrate tongue. *Journal of Anatomy, 201*, 1–13.
Iwata, J. I., Suzuki, A., Pelikan, R. C., Ho, T. V., & Chai, Y. (2013). Noncanonical transforming growth factor β (TGFβ) signaling in cranial neural crest cells causes tongue muscle developmental defects. *Journal of Biological Chemistry, 288*, 29760–29770.
Jeong, J., Li, X., McEvilly, R. J., Rosenfeld, M. G., Lufkin, T., & Rubenstein, J. L. R. (2008). Dlx genes pattern mammalian jaw primordium by regulating both lower jaw-specific and upper jaw-specific genetic programs. *Development, 135*, 2905–2916.

Jeong, J., Mao, J., Tenzen, T., Kottmann, A. H., & McMahon, A. P. (2004). Hedgehog signaling in the neural crest cells regulates the patterning and growth of facial primordia. *Genes and Development*, *18*, 937–951.
Kanzler, B., Kuschert, S. J., Liu, Y. H., & Mallo, M. (1998). Hoxa-2 restricts the chondrogenic domain and inhibits bone formation during development of the branchial area. *Development*, *125*, 2587–2597.
Kudo, H., Amizuka, N., Araki, K., Inohaya, K., & Kudo, A. (2004). Zebrafish periostin is required for the adhesion of muscle fiber bundles to the myoseptum and for the differentiation of muscle fibers. *Developmental Biology*, *267*, 473–487.
Kulesa, P. M., Bailey, C. M., Kasemeier-Kulesa, J. C., & McLennan, R. (2010). Cranial neural crest migration: New rules for an old road. *Developmental Biology*, *344*, 543–554.
Kuratani, S. (2012). Evolution of the vertebrate jaw from developmental perspectives. *Evolution & Development*, *14*, 76–92.
Kuratani, S., Nobusada, Y., Horigome, N., & Shigetani, Y. (2001). Embryology of the lamprey and evolution of the vertebrate jaw: Insights from molecular and developmental perspectives. *Philosophical Transactions of the Royal Society of London. Series B, Biological Sciences*, *356*, 1615–1632.
Le Douarin, N. M., Creuzet, S., Couly, G., & Dupin, E. (2004). Neural crest cell plasticity and its limits. *Development*, *131*, 4637–4650.
Léjard, V., Brideau, G., Blais, F., Salingcarnboriboon, R., Wagner, G., Roehrl, M. H., et al. (2007). Scleraxis and NFATc regulate the expression of the pro-alpha1(I) collagen gene in tendon fibroblasts. *Journal of Biological Chemistry*, *282*, 17665–17675.
Lin, C., Fisher, A. V., Yin, Y., Maruyama, T., Veith, G. M., Dhandha, M., et al. (2011). The inductive role of Wnt-[beta]-Catenin signaling in the formation of oral apparatus. *Developmental Biology*, *356*, 40–50.
Linker, C., Lesbros, C., Stark, M. R., & Marcelle, C. (2003). Intrinsic signals regulate the initial steps of myogenesis in vertabretes. *Development*, *130*, 4797–4807.
Long, F., & Ornitz, D. M. (2013). Development of the endochondral skeleton. *Cold Spring Harbor Perspectives in Biology*, *5*, a008334.
McPherron, A. C., Lawler, A. M., & Lee, S.-. J. (1997). Regulation of skeletal muscle mass in mice by a new TGF-beta superfamily member. *Nature*, *387*, 83–90.
Mendias, C. L., Bakhurin, K. I., & Faulkner, J. A. (2008). Tendons of myostatin-deficient mice are small, brittle, and hypocellular. *Proceedings of the National Academy of Sciences of the United States of America*, *105*, 388–393.
Miletich, I., Buchner, G., & Sharpe, P. T. (2005). Barx1 and evolutionary changes in feeding. *Journal of Anatomy*, *207*, 619–622.
Minoux, M., & Rijli, F. M. (2010). Molecular mechanisms of cranial neural crest cell migration and patterning in craniofacial development. *Development*, *137*, 2605–2621.
Moore, K. L., & Persaud, T. V. (2008). *The developing human*. Philadelphia: Saunders.
Morgan, J. E., & Partridge, T. A. (2003). Muscle satellite cells. *International Journal of Biochemistry & Cell Biology*, *35*, 1151–1156.
Mori-Akiyama, Y., Akiyama, H., Rowitch, D. H., & de Crombrugghe, B. (2003). Sox9 is required for determination of the chondrogenic cell lineage in the cranial neural crest. *Proceedings of the National Academy of Sciences of the United States of America*, *100*, 9360–9365.
Murchison, N. D., Price, B. A., Conner, D. A., Keene, D. R., Olson, E. N., Tabin, C. J., et al. (2007). Regulation of tendon differentiation by scleraxis distinguishes force-transmitting tendons from muscle-anchoring tendons. *Development*, *134*, 2697–2708.
Nabeshima, Y., Hanaoka, K., Hayasaka, M., Esuml, E., Li, S., Nonaka, I., et al. (1993). Myogenin gene disruption results in perinatal lethality because of severe muscle defect. *Nature*, *364*, 532–535.
Noden, D. M., & Francis-West, P. (2006). The differentiation and morphogenesis of craniofacial muscles. *Developmental Dynamics*, *235*, 1194–1218.

Oka, K., Oka, S., Sasaki, T., Ito, Y., Bringas, P., Jr., Nonaka, K., et al. (2007). The role of TGF-beta signaling in regulating chondrogenesis and osteogenesis during mandibular development. *Developmental Biology, 303*, 391–404.

Olguin, H. C., & Olwin, B. B. (2004). Pax-7 up-regulation inhibits myogenesis and cell cycle progression in satellite cells: A potential mechanism for selfrenewal. *Developmental Biology, 275*, 375–388.

Olson, E. N., Sternberg, E., Hu, J. S., Spizz, G., & Wilcox, C. (1986). Regulation of myogenic differentiation by type beta transforming growth factor. *Journal of Cell Biology, 103*, 1799–1805.

Otto, A., Schmidt, C., & Patel, K. (2006). Pax3 and Pax7 expression and regulation in the avian embryo. *Anatomy and Embryology (Berlin), 211*, 293–310.

Oustanina, S., Hause, G., & Braun, T. (2004). Pax7 directs postnatal renewal and propagation of myogenic satellite cells but not their specification. *EMBO Journal, 23*, 3430–3439.

Parada, C., Han, D., & Chai, Y. (2012). Molecular and cellular regulatory mechanisms of tongue myogenesis. *Journal of Dental Research, 91*, 528–535.

Parada, C., Li, J., Iwata, J., Suzuki, A., & Chai, Y. (2013). CTGF mediates Smad-dependent transforming growth factor β signaling to regulate mesenchymal cell proliferation during palate development. *Molecular and Cellular Biology, 33*, 3482–3493.

Pryce, B. A., Watson, S. S., Murchison, N. D., Staverosky, J. A., Dünker, N., & Schweitzer, R. (2009). Recruitment and maintenance of tendon progenitors by TGFbeta signaling are essential for tendon formation. *Development, 136*, 1351–1361.

Ramaesh, T., & Bard, J. B. (2003). The growth and morphogenesis of the early mouse mandible: A quantitative analysis. *Journal of Anatomy, 203*, 213–222.

Richany, S. F., Bast, T. H., & Anson, B. J. (1956). The development of the first branchial arch in man and the fate of Meckel's cartilage. *Quarterly Bulletin of Northwestern University Medical School, 30*, 331–355.

Rochlin, K., Yu, S., Roy, S., & Baylies, M. K. (2010). Myoblast fusion: When it takes more to make one. *Developmental Biology, 341*, 66–83.

Rudnicki, M. A., Schnegelsberg, P. N., Stead, R. H., Braun, T., Arnold, H. H., & Jaenisch, R. (1993). MyoD or Myf-5 is required for the formation of skeletal muscle. *Cell, 75*, 1351–1359.

Sachs, M., Brohmann, H., Zechner, D., Mueller, T., Huelsken, J., Walther, I., et al. (2000). Essential role of Gab1 for signaling by the C-Met receptor *in vivo*. *Journal of Cell Biology, 150*, 1375–1384.

Santagati, F., & Rijli, F. M. (2003). Cranial neural crest and the building of the vertebrate head. *Nature Reviews. Neuroscience, 4*, 806–818.

Shibata, S., Suda, N., Suzuki, S., Fukuoka, H., & Yamashita, Y. (2006). An in situ hybridization study of Runx2, Osterix, and Sox9 at the onset of condylar cartilage formation in fetal mouse mandible. *Journal of Anatomy, 208*, 169–177.

Shibata, S., Suda, N., Yoda, S., Fukuoka, H., Ohyama, K., Yamashita, Y., et al. (2004). Runx2-deficient mice lack mandibular condylar cartilage and have deformed Meckel's cartilage. *Anatomy and Embryology (Berlin), 208*, 273–280.

Shibukawa, Y., Young, B., Wu, C., Yamada, S., Long, F., Pacifici, M., et al. (2007). Temporomandibular joint formation and condyle growth require Indian hedgehog signaling. *Developmental Dynamics, 236*, 426–434.

Shimo, T., Kanyama, M., Wu, C., Sugito, H., Billings, P. C., Abrams, W. R., et al. (2004). Expression and roles of connective tissue growth factor in Meckel's cartilage development. *Developmental Dynamics, 231*, 136–147.

Silbermann, M., Reddi, A. H., Hand, A. R., Leapman, R., von der Mark, K., & Franzen, A. (1987). Chondroid bone arises from mesenchymal stem cells in organ culture of mandibular condyles. *Journal of Craniofacial Genetics and Developmental Biology, 7*, 59–79.

Song, Z., Liu, C., Iwata, J., Gu, S., Suzuki, A., Sun, C., et al. (2013). Mice with Tak1 deficiency in neural crest lineage exhibit cleft palate associated with abnormal tongue development. *Journal of Biological Chemistry, 288*, 10440–10450.

Sugito, H., Shibukawa, Y., Kinumatsu, T., Yasuda, T., Nagayama, M., Yamada, S., et al. (2011). Ihh signaling regulates mandibular symphysis development and growth. *Journal of Dental Research, 90*, 625–631.

Tajbakhsh, S., & Buckingham, M. (2000). The birth of muscle progenitor cells in the mouse: Spatiotemporal considerations. *Current Topics in Developmental Biology, 48*, 225–268.

Tajbakhsh, S., & Cossu, G. (1997). Establishing myogenic identity during somitogenesis. *Current Opinion in Genetics & Development, 7*, 634–641.

Tan, T. Y., Kilpatrick, N., & Farlie, P. G. (2011). Developmental and genetic perspectives on Pierre Robin sequence. *American Journal of Medical Genetics. Part C, Seminars in Medical Genetics, 163C*, 295–305.

Terao, F., Takahashi, I., Mitani, H., Haruyama, N., Sasano, Y., Suzuki, O., et al. (2011). Fibroblast growth factor 10 regulates Meckel's cartilage formation during early mandibular morphogenesis in rats. *Developmental Biology, 350*, 337–347.

Terraz, C., Brideau, G., Ronco, P., & Rossert, J. (2002). A combination of cis-acting elements is required to activate the pro-alpha 1(I) collagen promoter in tendon fibroblasts of transgenic mice. *Journal of Biological Chemistry, 277*, 19019–19026.

Toyoshima, K., & Shimamura, A. (1982). Comparative study of ultrastructures of the lateral-line organs and the palatal taste organs in the African clawed toad, Xenopus laevis. *Anatomical Record, 4*, 371–381.

Trainor, P. A., & Krumlauf, R. (2000). Patterning the cranial neural crest: Hindbrain segmentation and Hox gene plasticity. *Nature Reviews. Neuroscience, 1*, 116–124.

Trainor, P. A., & Krumlauf, R. (2001). Hox genes, neural crest cells and branchial arch patterning. *Current Opinion in Cell Biology, 13*, 698–705.

Trainor, P. A., Sobieszczuk, D., Wilkinson, D., & Krumlauf, R. (2002). Signalling between the hindbrain and paraxial tissues dictates neural crest migration pathways. *Development, 129*, 433–442.

Trichilis, A., & Wroblewski, J. (1997). Expression of p53 and hsp70 in relation to apoptosis during Meckel's cartilage development in the mouse. *Anatomy and Embryology (Berlin), 196*, 107–113.

Trumpp, A., Depew, M. J., Rubenstein, J. L., Bishop, J. M., & Martin, G. R. (1999). Cre-mediated gene inactivation demonstrates that FGF8 is required for cell survival and patterning of the first branchial arch. *Genes and Development, 13*, 3136–3148.

Tucker, A. S., Matthews, K. L., & Sharpe, P. T. (1998). Transformation of tooth type induced by inhibition of BMP signaling. *Science, 282*, 1136–1138.

Tucker, A. S., Yamada, G., Grigoriou, M., Pachnis, V., & Sharpe, P. T. (1999). Fgf-8 determines rostral-caudal polarity in the first branchial arch. *Development, 126*, 51–61.

Vortkamp, A., Lee, K., Lanske, B., Segre, G. V., Kronenberg, H. M., & Tabin, C. J. (1996). Regulation of rate of cartilage differentiation by Indian hedgehog and PTH-related protein. *Science, 273*, 613–622.

Wagner, E. F., & Karsenty, G. (2001). Genetic control of skeletal development. *Current Opinion in Genetics & Development, 11*, 527–532.

Wang, H., Noulet, F., Edom-Vovard, F., Le Grand, F., & Duprez, D. (2010). Bmp signaling at the tips of skeletal muscles regulates the number of fetal muscle progenitors and satellite cells during development. *Developmental Cell, 18*, 643–654.

Wang, Y., Zheng, Y., Chen, D., & Chen, Y. (2013). Enhanced BMP signaling prevents degeneration and leads to endochondral ossification of Meckel's cartilage in mice. *Developmental Biology, 381*, 301–311.

Yang, R. T., Zhang, C., Liu, Y., Zhou, H. H., & Li, Z. B. (2012). Autophagy prior to chondrocyte cell death during the degeneration of Meckel's cartilage. *Anatomical Record (Hoboken), 295*, 734–741.

Yoon, B. S., Ovchinnikov, D. A., Yoshii, I., Mishina, Y., Behringer, R. R., & Lyons, K. M. (2005). Bmpr1a and Bmpr1b have overlapping functions and are essential for chondrogenesis *in vivo*. *Proceedings of the National Academy of Sciences of the United States of America, 102*, 5062–5067.

Zhang, C. (2010). Transcriptional regulation of bone formation by the osteoblast-specific transcription factor Osx. *Journal of Orthopaedic Surgery and Research, 5*, 37.

Zhang, Z., Wlodarczyk, B. J., Niederreither, K., Venugopalan, S., Florez, S., Finnell, R. H., et al. (2011). Fuz regulates craniofacial development through tissue specific responses to signaling factors. *PLoS One, 6*, e24608.

Zhao, P., Caretti, G., Mitchell, S., McKeehan, W. L., Boskey, A. L., Pachman, L. M., et al. (2006). Fgfr4 is required for effective muscle regeneration *in vivo*. Delineation of a MyoD-Tead2-Fgfr4 transcriptional pathway. *Journal of Biological Chemistry, 281*, 429–438.

Zhong, Z., Zhao, H., Mayo, J., & Chai, Y. (2015). Different requirements for Wnt signaling in tongue myogenic subpopulations. *Journal of Dental Research, 94*, 421–429.

CHAPTER THREE

Cellular and Molecular Mechanisms of Palatogenesis

Yu Lan[*,†,1], **Jingyue Xu**[*], **Rulang Jiang**[*,†,1]
[*]Division of Developmental Biology, Cincinnati Children's Hospital Medical Center, Cincinnati, Ohio, USA
[†]Division of Plastic Surgery, Cincinnati Children's Hospital Medical Center, Cincinnati, Ohio, USA
[1]Corresponding authors: e-mail address: yu.lan@cchmc.org; rulang.jiang@cchmc.org

Contents

Abstract

Palatogenesis involves the initiation, growth, morphogenesis, and fusion of the primary and secondary palatal shelves from initially separate facial prominences during embryogenesis to form the intact palate separating the oral cavity from the nostrils. The palatal shelves consist mainly of cranial neural crest-derived mesenchymal cells covered by a simple embryonic epithelium. The growth and patterning of the palatal shelves are controlled by reciprocal epithelial–mesenchymal interactions regulated by multiple signaling pathways and transcription factors. During palatal shelf outgrowth, the embryonic epithelium develops a "teflon" coat consisting of a single, continuous layer of periderm cells that prevents the facial prominences and palatal shelves from forming aberrant interepithelial adhesions. Palatal fusion involves not only spatiotemporally regulated disruption of the periderm but also dynamic cellular and molecular processes that result in adhesion and intercalation of the palatal medial edge epithelia to form an intershelf epithelial seam, and subsequent dissolution of the epithelial seam to form the intact roof of the oral cavity. The complexity of regulation of these morphogenetic processes is reflected by the common occurrence of cleft palate in humans. This review will summarize major recent advances and discuss major remaining gaps in the understanding of cellular and molecular mechanisms controlling palatogenesis.

Current Topics in Developmental Biology, Volume 115
ISSN 0070-2153
http://dx.doi.org/10.1016/bs.ctdb.2015.07.002

1. INTRODUCTION

Cleft lip and/or cleft palate are among the most common birth defects in humans, occurring at a frequency of about 1 in 500–2500 live births (Dixon, Marazita, Beaty, & Murray, 2011; Gorlin, Cohen, & Hennekam, 2001; Mangold, Ludwig, & Nothen, 2011; Schutte & Murray, 1999; Vanderas, 1987). Clinically, cleft lip with or without cleft palate (CL/P) is a unilateral or bilateral gap between the philtrum and the lateral upper lip, often extending through the upper jaw into the nostril and accompanied by cleft of the secondary palate—the roof of the oral cavity. Cleft palate (CP) also frequently occurs without cleft lip. Both CL/P and CP appear in syndromic and nonsyndromic forms. The Online Mendelian Inheritance in Man (OMIM) database (www.ncbi.nlm.nih.gov/omim/) lists over 500 syndromic disorders in which CL/P or CP is a feature. The genetic basis of some of these syndromes has been identified through positional or candidate gene cloning (reviewed by Dixon et al., 2011). Approximately 70% of CL/P and 50% of CP cases are nonsyndromic, for which the etiology and pathogenesis are complex and poorly understood (Dixon et al., 2011; Mangold et al., 2011; Rahimov, Jugessur, & Murray, 2012).

The major reason why CL/P and CP are common birth defects is because development of the midface involves highly coordinated growth and morphogenesis of initially separate primordia. In humans, development of the face begins in the fourth week of embryogenesis with migrating neural crest cells that combine with mesodermal and ectodermal cells to establish the facial primordia. The facial primordia initially consist of five separate parts surrounding the primitive mouth, known as the stomodeum (reviewed by Dixon et al., 2011; Jiang, Bush, & Lidral, 2006; Levi et al., 2011). At the rostral midline is the frontonasal prominence, which is populated by mesenchymal cells derived from the forebrain and midbrain neural crest cells. The stomodeum is bound laterally by a pair of maxillary processes and caudally by a pair of mandibular processes, which are populated by neural crest cells derived from the rostral region of the hindbrain. By the end of the fourth week, nasal pits form bilaterally in the lower part of the frontonasal prominence and extend into the stomodeum such that the frontonasal prominence is divided distally into paired medial and lateral nasal processes. Subsequently, rapid growth of the maxillary processes results in the medial nasal processes merging with each other, forming the intermaxillary segment. The lateral nasal processes fuse with the maxillary processes to form the lateral nose and cheek region. The intact upper lip forms by the maxillary processes fusing medially with the medial nasal processes (Jiang et al., 2006).

The palate begins to develop during the fifth week of human embryogenesis and is divided into two regions: the primary and secondary palates (reviewed by Bush & Jiang, 2012; Dixon et al., 2011; Levi et al., 2011). The primary palate arises from the intermaxillary segment and contains the philtrum, the incisors, and a small part of the hard palate situated anterior to the incisive foramen (located immediately behind the incisor gum). The secondary palate includes both the hard and soft palate posterior to the incisive foramen. Development of the secondary palate begins during the sixth week of human embryogenesis with the initiation of two outgrowths, termed palatal shelves, from the oral side of the maxillary processes. The palatal shelves initially grow downward on both sides of the developing tongue. During the eighth week, as the head expands and the tongue moves downward, the bilateral palatal shelves elevate to a horizontal position above the dorsum of the tongue and merge with each other at the midline. In addition, the palatal shelves also fuse anteriorly with the primary palate and dorsally with the nasal septum, both of which are derived from the medial nasal processes. These fusion processes are complete by the 12th week of gestation.

The dynamic morphogenetic processes giving rise to the midface are often disturbed by genetic or environmental insults. For example, CL/P may result from disturbances in proliferation, migration, and survival of the neural crest cells or by fusion between the medial nasal and maxillary processes, whereas CP without cleft lip may result from disruption of palatal shelf growth, elevation, or fusion. Moreover, since palate development occurs concurrently with significant growth and expansion of the whole craniofacial complex, malformation of structures in the vicinity of the palatal shelves, such as the tongue and mandible, sometimes hinders palatal shelf elevation or contact, resulting in CP (reviewed by Bush & Jiang, 2012; Chai & Maxson, 2006; Dotto, 2012). In addition to extensive human genetic studies that have led to identification of the gene mutations responsible for many CP syndromes (reviewed by Dixon et al., 2011), genetic and developmental biology studies in model animal systems, particularly laboratory mice in which the palatal growth and morphogenetic processes are remarkably similar to that in humans (Bush & Jiang, 2012; Fig. 1), in the last 25 years have provided significant insights into the cellular and molecular processes underlying palatogenesis. Since studies of the roles of various signaling pathways in controlling palate development have been extensively reviewed recently (Bush & Jiang, 2012; Cobourne & Green, 2012;

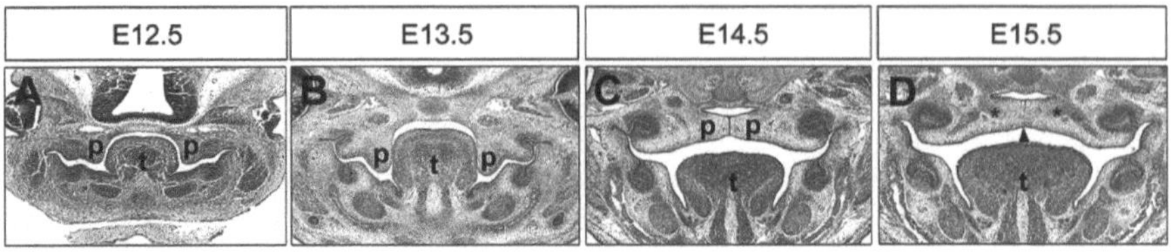

Figure 1 Palate development in mice. (A–D) Representative HE-stained coronal sections through the middle of the anterior–posterior axis of the developing palatal shelves at E12.5 (A), E13.5 (B), E14.5 (C), and E15.5 (D). The palatal shelves grow vertically from the oral side of the maxillary processes from E11.5 to past E13.5 (A and B). By E14.5, the palatal shelves have elevated to a horizontal position above the tongue, adhered to each other at the midline and formed the midline epithelial seam (MES) (C). At around E15.5, the MES is disintegrated to allow confluence of the palatal mesenchyme (D). Arrowhead in (D) points to the disintegrating MES. Asterisks in (D) mark the sites of ossification that will form the palatal bone. p, palatal shelf; t, tongue. (A color version of this figure is located at the end of this chapter.)

He & Chen, 2012; Lane & Kaartinen, 2014; Parada & Chai, 2012; Smith, Lozanoff, Iyyanar, & Nazarali, 2012; Stanier & Pauws, 2012), we focus this chapter on integrating recent studies of the molecular pathways in palatal growth and patterning as well as on highlighting new advances in understanding the cellular and molecular mechanisms regulating epithelial adhesion and palate fusion.

2. REGULATION OF PALATAL SHELF GROWTH AND PATTERNING

The central player in the regulation of palatal shelf growth appears to be sonic hedgehog (Shh) signaling. Shh is expressed throughout the early oral epithelium prior to palatal shelf outgrowth (Rice, Connor, & Rice, 2006). Mice with tissue-specific inactivation of *Smoothened* (*Smo*), which encodes the obligate transducer of hedgehog signaling, in the cranial neural crest cells showed complete agenesis of the secondary palate (Jeong, Mao, Tenzen, Kottmann, & McMahon, 2004). Studies of mice with tissue-specific inactivation of *Smo* in the early palatal mesenchyme or of *Shh* in the epithelium confirmed that Shh signaling is critical for palatal shelf growth (Lan & Jiang, 2009; Rice et al., 2004). Shh signaling is required for activation and/or maintenance of expression of several transcription factors, including Foxf1, Foxf2, and Osr2, in the developing palatal mesenchyme (Lan & Jiang, 2009). Although whether Foxf1 and Foxf2 play primary roles in palatal shelf growth remains to be elucidated, Osr2 has been shown to function as an intrinsic regulator of palatal mesenchyme cell proliferation (Lan et al.,

2004). In addition, Shh signaling regulates expression of cell cycle regulators Cyclin D1 and Cyclin D2 in the palatal mesenchyme (Lan & Jiang, 2009).

The developing palatal shelves exhibit both morphological and molecular heterogeneity along the anteroposterior and oronasal axes (reviewed by Bush & Jiang, 2012; Hilliard, Yu, Gu, Zhang, & Chen, 2005; Smith et al., 2012). Whereas *Shh* is initially expressed throughout the palatal epithelium at the onset of palatal outgrowth (Rice et al., 2004), its expression becomes highly restricted to the palatal rugae, the periodic epithelial ridges that form in species-specific patterns on the oral side of the palatal epithelium, in the anterior region and in the sensory papilla in the posterior region of the palate (Baek et al., 2011; Pantalacci et al., 2008; Welsh & O'Brien, 2009; Fig. 2). Whereas the periodic formation of rugae during palate development appears to be regulated by a Turing-type reaction–diffusion mechanism (Economou et al., 2012), the restriction of *Shh* expression to the oral side of the palatal epithelium correlates with differential expression of *Fgf10* and *Fgf7* along the oronasal axis of the palatal mesenchyme, with *Fgf10* preferentially expressed on the oral side and *Fgf7* preferentially on the nasal side (Rice et al., 2004; Veistinen, Aberg, & Rice, 2009). In palatal explant culture assays, Fgf10 induced, whereas Fgf7 repressed, *Shh* expression in the palatal epithelium (Han et al., 2009; Rice et al., 2004). Moreover, application of function-neutralizing antibody against Fgf7 caused increased *Shh* expression in the palatal epithelium, suggesting that Fgf7 normally competes with Fgf10 and represses *Shh* expression in the nasal side of the palatal epithelium (Han et al., 2009). Expression of *Fgf7* in the palatal mesenchyme is maintained by the Dlx5 transcription factor, whereas Fgf10 and Shh act in a positive feedback loop to regulate each other's expression (Han et al., 2009; Lan & Jiang, 2009; Rice et al., 2004; Fig. 3). Interestingly, exogenous Shh protein inhibited *Fgf7* expression in palatal mesenchyme explants (Han et al., 2009), suggesting that Shh signaling plays an active role in maintaining the oronasal asymmetry of the palatal shelves through differential regulation of expression of *Fgf10* and *Fgf7* in the mesenchyme (Fig. 3). In addition, maintenance of *Shh* expression in the most anterior rugae and in the posterior palatal epithelium requires the functions of the Msx1 and Pax9 transcription factors, respectively, in the palatal mesenchyme (Zhang et al., 2002; Zhou, Gao, Lan, Jia, & Jiang, 2013). *Msx1* expression is restricted to the anterior region of the developing palatal mesenchyme and regulates anterior palatal mesenchyme proliferation through activation of expression of *Bmp4*, which in turn signals to the epithelium to maintain *Shh* expression (Zhang et al., 2002). Whereas *Bmp4* expression in the anterior palate is

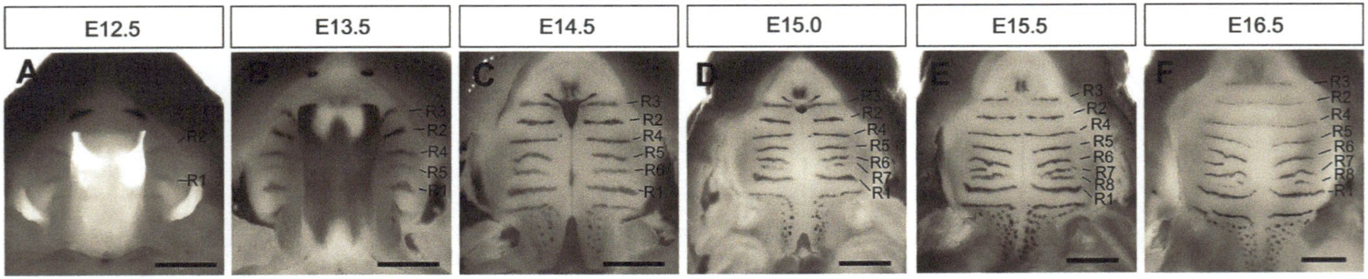

Figure 2 Whole-mount oral view of growth and patterning of the developing palatal shelves in mice. Patterns of *Shh* mRNA expression were detected by whole-mount *in situ* hybridization. R1–R8 mark the Shh-expressing palatal rugae in the sequence of their formation from E12.5 (A) to E16.5 (F). R1 marks the anterior–posterior boundary (all panels are oriented with the tip of the snout pointing up). The patchy *Shh* expression in the posterior palate marks the sensory papilla buds. Scale bar, 500 μm. (A color version of this figure is located at the end of this chapter.)

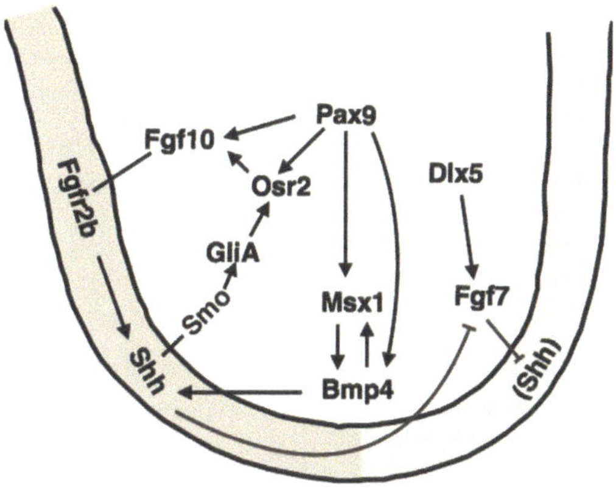

Figure 3 Molecular regulation of palatal shelf growth and oronasal patterning. Schematic diagram of a coronal section of the developing palatal shelf, with the oral side to the left. Black arrows indicate positive regulation whereas red lines indicate repression. Shh is expressed in the palatal epithelium and positively regulates Fgf10 expression through the Osr2 transcription factor in the palatal mesenchyme. Fgf10 signals through the Fgfr2b receptor to maintain Shh expression in the palatal epithelium. Dlx5 activates Fgf7 expression in the palatal mesenchyme. Fgf7 and Shh repress each other's expression such that Fgf7 expression is restricted to the nasal side of the palatal mesenchyme and Shh to the oral side of the palatal epithelium. Msx1 maintains Shh expression through Bmp4 in the anterior palate whereas Pax9 acts upstream of Bmp4, Fgf10, Msx1, and Osr2 in the palatal mesenchyme. (A color version of this figure is located at the end of this chapter.)

dependent on Msx1 function, *Bmp4* is also expressed in the posterior region of the developing palatal shelves where *Msx1* is normally not expressed (Zhou et al., 2013). The expression of *Bmp4* in the posterior palate depends on Pax9 function for its maintenance (Zhou et al., 2013). Mice lacking *Pax9* function exhibit complete penetrance of CP and palatal shelf growth defects (Peters, Neubuser, Kratochwil, & Balling, 1998; Zhou et al., 2011, 2013). In $Pax9^{-/-}$ mutant mouse embryos, expression of *Fgf10* and *Osr2* in the developing palatal mesenchyme as well as expression of *Shh* in the palatal epithelium were also reduced in comparison with wild-type littermates. Zhou et al. (2013) showed that *Fgf10* expression was also reduced in the $Osr2^{-/-}$ palatal mesenchyme and that expression of *Osr2* from a $Pax9^{Osr2KI}$ knockin allele restored *Fgf10* expression in the palatal mesenchyme and partly rescued posterior palate morphogenesis in the *Pax9* mutant embryos; although the reductions in *Shh* expression in the palatal epithelium and in *Bmp4* expression in the posterior palatal mesenchyme were not rescued (Zhou et al., 2013). Together, these studies suggest that Pax9 regulates the Osr2–Fgf10 and the

Bmp4–Shh pathways independently during palate development, with Pax9 and Shh signaling converging on the regulation of the Osr2 transcription factor (Fig. 3).

In addition to *Msx1* and *Pax9*, several other transcription factors, including short stature homeobox-2 (Shox2), BarH-like homeobox-1 (Barx1), meningioma-1 (Mn1), and T-box factor 22 (Tbx22), exhibit differential expression along the anteroposterior axis (Fig. 4). *Shox2* is expressed throughout the anterior half of the palatal mesenchyme but completely absent from the posterior half, with the anterior–posterior boundary corresponding exactly to the first formed palatal rugae (Li & Ding, 2007; Pantalacci et al., 2008; Welsh & O'Brien, 2009; Yu et al., 2005; Fig. 4). Expression of *Tbx22* is largely restricted to the posterior palatal mesenchyme, whereas expression of *Barx1* and *Mn1* shows preference in the posterior palatal mesenchyme, with expression expanded into the anterior palate domain during palatal shelf growth (Liu et al., 2008; Welsh & O'Brien, 2009). These patterns of expression during palatal development correlate with palatal developmental defects in several mutant mice, with *Shox2* specifically required for anterior palatal growth whereas *Mn1* acts upstream of *Tbx22* to regulate posterior palatal growth (Liu et al., 2008; Pauws et al., 2009; Yu et al., 2005).

Whereas the differential gene expression patterns and mutant mouse phenotypes clearly demonstrate regionalized control of palatal growth along the anteroposterior axis, the molecular mechanism that sets up the anteroposterior patterns of the developing palatal shelves remains to be elucidated. Hilliard et al. (2005) showed that exogenous Bmp4 induced *Msx1* expression in the anterior but not posterior palatal mesenchyme in explant culture assays, which is probably due to the fact that *Bmpr1a* is preferentially expressed in the anterior palate (Baek et al., 2011). Yu et al. (2005) showed that anterior palatal epithelium was able to induce *Shox2* expression in posterior palatal mesenchyme in recombinant explant culture assays (Yu et al., 2005). Together, these results indicate that both the palatal epithelium and mesenchyme are patterned along the anteroposterior axis.

3. PROPER PERIDERM DIFFERENTIATION IS CRITICAL FOR NORMAL PALATOGENESIS

During early embryogenesis, the ectoderm consists initially of a single layer of undifferentiated cuboidal epithelial cells covering the entire embryo, including lining the primitive oral cavity (Richardson et al., 2014).

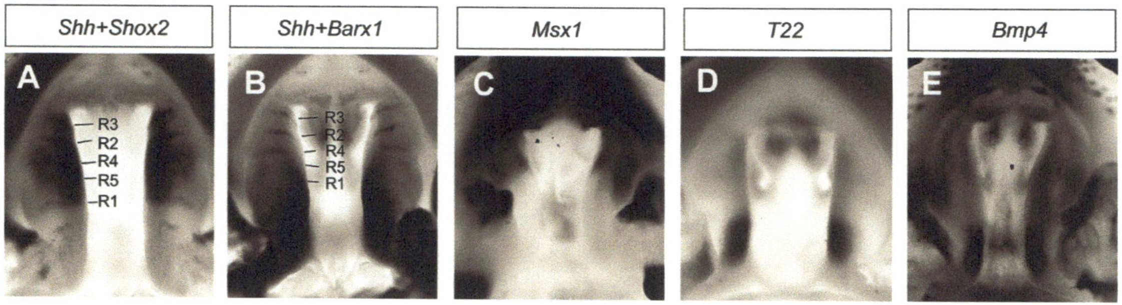

Figure 4 Differential patterns of gene expression along the anterior–posterior axis of the developing palatal shelves at E13.5. (A) Dual color *in situ* hybridization detection of *Shh* (brown) and *Shox2* (blue) mRNA expression. *Shh* is expressed in the palatal rugae (marked as R1–R5 in the sequence of their formation), whereas *Shox2* mRNAs are restricted to the anterior half of the developing palatal shelves. R1 coincides with the posterior boundary of *Shox2* mRNA expression. (B) Dual color *in situ* hybridization detection of *Shh* (brown) and *Barx1* (blue) mRNA expression. *Barx1* is mainly expressed in the posterior halves of the developing palatal shelves. (C) Whole-mount *in situ* hybridization detection of *Msx1* (purple) mRNA expression. *Msx1* mRNAs were detected in the anterior-most quarter of the developing palatal shelves. (D) *Tbx22* mRNA expression is highly restricted in the posterior half of the developing palatal shelves. (E) *Bmp4* mRNA expression was detected in the anterior and posterior subregions of the developing palatal shelves. (A color version of this figure is located at the end of this chapter.)

As development proceeds, the embryonic skin undergoes a series of defined stratification and differentiation events to generate the mature epidermis. The first stratification event produces the periderm, a single layer of flattened epithelial cells covering the embryonic epithelia during subsequent developmental stages until shortly before birth (Holbrook & Odland, 1975; M'Boneko & Merker, 1988; Polakowska, Piacentini, Bartlett, Goldsmith, & Haake, 1994; Richardson et al., 2014; Sanes, Rubenstein, & Nicolas, 1986). Several functions have been proposed for the periderm, including protection from the environment (Hayward, 1983), regulation of underlying mesenchyme (Scott, Nau, Wittfoht, & Merker, 1987), and contribution to cornified envelope formation (Akiyama et al., 1999). Whereas all of these functions have yet to be proven, a series of mouse genetic studies and human disease phenotypes have shown a critical role for the periderm in craniofacial and palate development, as discussed below.

Through characterization of the cellular mechanisms underlying CP pathogenesis in the *Jag2*$^{-/-}$ mutant mice, we first reported a critical role for the oral periderm in preventing aberrant oral epithelial adhesions that interfered with palatal shelf elevation (Casey et al., 2006; Jiang, Lan, Chapman, et al., 1998; Richardson, Dixon, Jiang, & Dixon, 2009). We showed that *Jag2* is expressed in the developing oral epithelium and activates the Notch1 receptor during periderm differentiation. In the *Jag2*$^{-/-}$ mutant embryos, although oral epithelial stratification occurred, the resultant suprabasal cells did not have the flattened morphology characteristic of normal periderm cells and the mutant oral epithelia was disorganized and subsequently showed multiple intraoral interepithelial adhesions, including adhesion between the maxillary and mandibular oral epithelia as well as adhesion of the palatal shelves to the developing tongue (Casey et al., 2006; Jiang, Lan, Chapman, et al., 1998; Fig. 5A). The correlation of aberrant oral epithelial adhesions with the disruption in oral periderm differentiation in the *Jag2*$^{-/-}$ mutant mice suggests that periderm normally provides a nonsticking barrier for the formation of the oral cavity. The specific activation and nuclear localization of the intracellular domain of Notch1 in the periderm cells in wild-type embryos and the dramatic reduction of nuclear accumulation of Notch1 in the suprabasal oral epithelium in *Jag2*$^{-/-}$ mutant embryos suggest that Jag2-Notch1 signaling plays a critical role in periderm cell differentiation.

Aberrant oral epithelial adhesions have been associated with several congenital ectodermal disorders, including Van der Woude syndrome (VWS) and popliteal pterygium syndrome (PPS), which are also characterized by

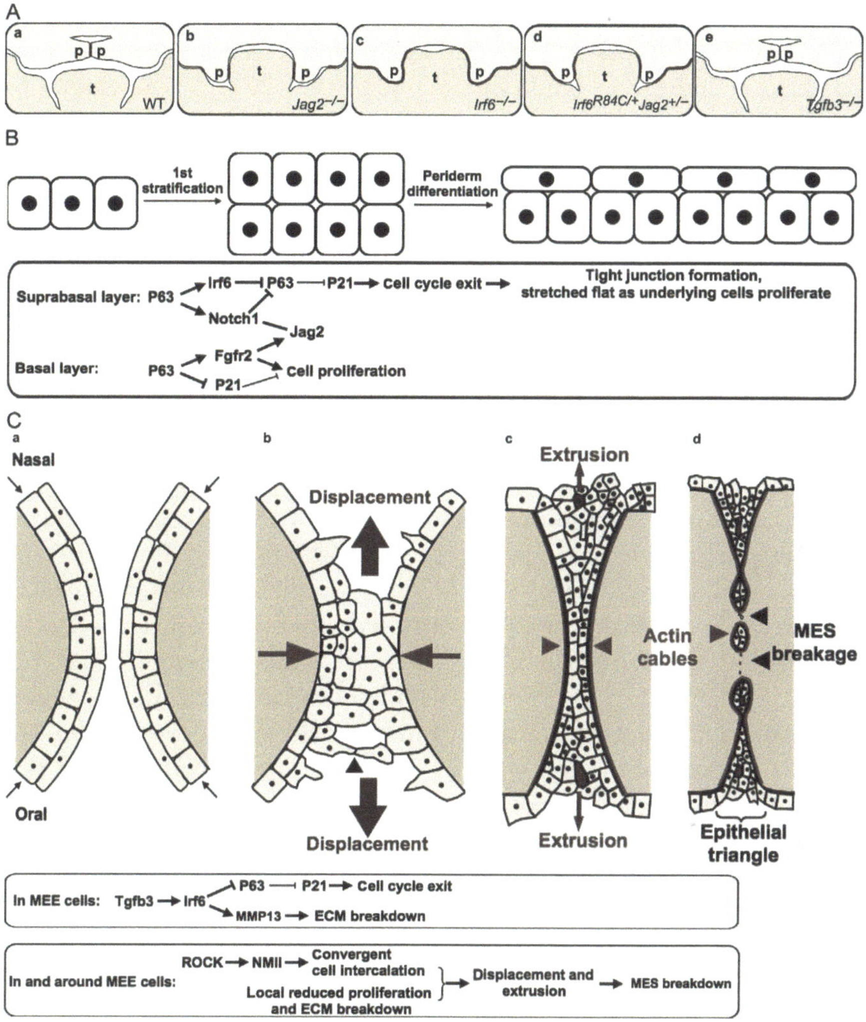

Figure 5 Periderm differentiation plays a critical role in palatogenesis. (A) Schematic diagrams depicting palatal shelf adhesion at the onset of palatal fusion in wild-type mouse embryos (a); palate–tongue and maxillary–mandibular adhesions in the *Jag2*$^{-/-}$ mutant embryos (b); extensive intraoral epithelial adhesions in *Irf6*$^{-/-}$ mutant embryos (c); palate–tongue, palate–mandible, and maxillary–mandibular adhesions in the *Irf6*$^{R84C/+}$*Jag2*$^{+/-}$ compound heterozygous mouse embryos (d); and failure of palatal adhesion in *Tgfb3*$^{-/-}$ mutant mouse embryos (e). Red line marks sites of interepithelial adhesion. p, palatal shelf; t, tongue. (B) Schematic diagram of 1st epithelial stratification and periderm differentiation, with the molecular and cellular pathways in the suprabasal and basal cells outlined in the rectangular box below. (C) Schematic diagram of coronal sections of palatal shelves depicting cell convergence and extrusion during

(Continued)

varying degrees of cleft lip and CP (Kondo et al., 2002; Richardson et al., 2006; Van Der Woude, 1954). Mutations in the *IRF6* gene have been shown to underlie both syndromes (Kondo et al., 2002). Gene expression studies in mice showed that *Irf6* is highly specifically expressed during periderm cell differentiation (Knight, Schutte, Jiang, & Dixon, 2006; Richardson et al., 2009, 2014). Mouse embryos deficient in *Irf6* as well as mouse embryos homozygous for the PPS mutation $Irf6^{R84C}$ showed severe intraoral epithelial adhesions, including maxillary–mandibular and palatal–tongue adhesions that caused CP by blocking palatal shelf elevation (Ingraham et al., 2006; Richardson et al., 2006; Fig. 5A). More recently, Richardson et al. (2014) carefully examined periderm development, using multiple molecular markers, in wild-type and $Irf6^{R84C/R84C}$ mutant mouse embryos and demonstrated that Irf6 function is required for normal periderm differentiation. Furthermore, two other mutant mouse strains, $Ikka^{-/-}$ and $Sfn^{Er/Er}$, which exhibit remarkably similar phenotypes to the $Irf6^{-/-}$ mutant mice, also lack periderm differentiation (Richardson et al., 2014). Furthermore, genetic ablation of periderm cells during embryogenesis caused aberrant intraoral epithelial adhesions, which led to CP in the severely affected embryos (Richardson et al., 2014). Together, these results indicate that proper periderm formation prevents pathological interepithelial adhesions and is critical for orofacial and palatal morphogenesis.

How do Irf6 and the Jag2-Notch1 pathways regulate periderm differentiation? Recent studies suggest that both Irf6 and Notch signaling are activated by and feedback to negatively regulate the transcription factor p63 to drive periderm differentiation. *P63* is expressed in the earliest embryonic epithelium and continues to be expressed in basal undifferentiated epithelial

Figure 5—Cont'd palatal adhesion and fusion, with the molecular and cellular processes regulating MES breakdown outlined in the rectangular box below. Palatal epithelial cells are in light blue color and the palatal mesenchyme is in gray. The paired palatal shelves are each covered with a single layer of periderm cells prior to interpalatal adhesion (a). At the onset of palatal fusion, periderm cells and basal MEE cells extend protrusions and converge toward (green arrows) the midline (b), with some cells in the multicellular seam displaced toward the oral or nasal side. As the MES forms, cellular rosettes form and cells at the center of the rosettes are squeezed out by extrusion (marked in red) (c). The actomyosin cables pull the MES apart and surround the epithelial islands, continuing to extrude cells from cellular rosettes (d). Palatal MES breakdown is driven by both the nonmyosin II-mediated cell convergence and extrusion as well as by Tgfb3-Irf6 regulated local cell cycle arrest and ECM breakdown. *Panel (C): The cell convergence and extrusion diagrams are modified from Kim et al. (2015).* (A color version of this figure is located at the end of this chapter.)

cells but is downregulated upon periderm differentiation (Richardson et al., 2009, 2014). The *p63* gene is transcribed using two alternative promoters, producing two major classes of protein isoforms that differ in the N-terminal amino acid sequences (Crum & McKeon, 2010; King & Weinberg, 2007; Romano et al., 2012). Mice deficient in all p63 isoforms and mice with specific inactivation of ΔNp63, the predominant isoform in epidermal cells, exhibit a thin, undifferentiated epidermis with a significantly reduced level of *Irf6* expression in oral epithelial cells (Mills et al., 1999; Moretti et al., 2010; Thomason et al., 2010; Yang et al., 1999). Moreover, the p63 protein binds to enhancer sequences upstream of the *Irf6* gene and can activate reporter gene expression driven by the *Irf6* gene sequences, suggesting that p63 directly activates *Irf6* expression during early epithelial development. However, upon periderm differentiation, *Irf6* is strongly expressed but p63 is downregulated in the periderm cells (Richardson et al., 2009, 2014). Moretti et al. (2010) showed that IRF6 promotes proteasome-mediated degradation of the ΔNp63 protein in human keratinocytes, suggesting that Irf6 negatively regulates p63 during periderm differentiation. Moreover, whereas p63 has previously been shown to positively regulate Jag1 and Jag2 expression (Candi et al., 2007; Sasaki et al., 2002), a recent study suggests that active Notch signaling represses p63 expression in the developing surface ectoderm (Tadeu & Horsley, 2013). Furthermore, although Irf6 and Jag2 are not required for each other's expression in the developing oral epithelium, $Irf6^{R84C/+};Jag2^{+/-}$ compound heterozygous mice exhibited similar persistent expression of p63 in stratified and aberrantly adhering oral epithelium (Richardson et al., 2009; Fig. 5A), suggesting that Irf6 and Jag2-Notch signaling act synergistically to downregulate p63 during periderm differentiation. It has been shown that p63 represses transcription of *p21 in vitro* and *in vivo* (Laurikkala et al., 2006; Welsh & O'Brien, 2009; Westfall, Mays, Sniezek, & Pietenpol, 2003). Thus, the Irf6 and Jag2-Notch signaling pathways drive periderm differentiation through indirect activation of p21-mediated cell cycle exit, with continued proliferation of the basal epithelial cells and new tight junctions that form between the suprabasal cells (Yoshida et al., 2012) contributing to the stretching and flattening of the periderm cells (Fig. 5B).

The functions of many other genes, including *Grhl3*, *Ikka*, and *Sfn*, are required for periderm differentiation. Mutations in *GRHL3* have been associated with a subset of VWS patients with no linkage to *IRF6* (Peyrard-Janvid et al., 2014). $Grhl3^{-/-}$ mice also exhibit aberrant interepithelial adhesions (Peyrard-Janvid et al., 2014). Studies in zebrafish showed that

Grhl3 is specifically expressed in periderm cells and its expression is dependent on Irf6 (de la Garza et al., 2013). *Ikka*$^{-/-}$ and *Sfn*$^{Er/Er}$ mice exhibit failure of periderm differentiation similar to *Irf6*$^{-/-}$ mice, but these genes do not appear to be regulated by Irf6 and it is not clear yet how they regulate periderm differentiation (Richardson et al., 2014). Mice lacking Fgf10 or its epithelial receptor Fgfr2b also exhibit aberrant palate–mandible or palate–tongue adhesions (Alappat et al., 2005; Rice et al., 2004). Alappat et al. (2005) showed that *Fgf10*$^{-/-}$ embryos exhibit loss of Jag2 expression in the oral and palatal epithelium, suggesting that Fgf10 signals through Fgfr2b to maintain *Jag2* expression in the oral epithelium. Recently, Ferone et al. (2012) showed that *Fgfr2* is a direct target of p63 in the developing epithelium (Ferone et al., 2012). Together with previous studies showing upregulation of *Jag2* expression by p63 and the loss of periderm integrity in *Jag2*$^{-/-}$ mice (Candi et al., 2007; Casey et al., 2006; Jiang, Lan, Chapman, et al., 1998; Sasaki et al., 2002), these studies suggest that p63 indirectly regulates *Jag2* expression in the basal epithelial cells through Fgfr2b and that Jag2 signals to the suprabasal cells to maintain the integrity of the periderm (Dotto, 2012; Fig. 5B).

4. PALATAL FUSION: FORMATION AND DISSOLUTION OF THE INTERSHELF EPITHELIAL SEAM BY CELL CONVERGENCE AND EXTRUSION

In order for the paired palatal shelves to initiate adhesion and fusion after they have elevated to the horizontal position above the dorsum of the developing tongue, the periderm covering the medial edge of the shelves has to be disrupted (Yoshida et al., 2012). Classic embryological as well as transmission and scanning electron microscopy studies have shown that, just prior to intershelf adhesion, the periderm cells at the medial edge change shape dramatically by elongating and contracting, and their nuclei become pyknotic (Fitchett & Hay, 1989; Griffith & Hay, 1992; Waterman & Meller, 1974; Yano, Ohtsuru, Ito, Fuji, & Yamashita, 1996). Similar apoptotic events occur in the periderm cells of the edge epithelia of medial nasal and maxillary prominences prior to lip fusion (Sun, Baur, & Hay, 2000). The desquamation of the periderm cells enables the apposing pair of palatal shelves to adhere to each other at the midline, with the adhered medial edge epithelial layers (MEE) subsequently undergoing intercalation to form the midline epithelial seam (MES) (Yoshida et al., 2012).

Although the cellular processes of palatal fusion have been extensively studied, the exact mechanisms regulating palatal periderm desquamation and MES dissolution are still incompletely understood (Bush & Jiang, 2012; Cuervo & Covarrubias, 2004; Hu et al., 2015; Kim et al., 2015; Sun et al., 2000). Historically, three distinct but nonexclusive cellular mechanisms driving MES dissolution, including epithelial–mesenchymal transformation, apoptotic cell death, and migration of MES cells into the oral and nasal sides of the palatal epithelia were proposed and supported by experimental evidence from palatal culture studies (Cuervo & Covarrubias, 2004; Sun et al., 2000; reviewed by Bush & Jiang, 2012; Dudas, Li, Kim, Yang, & Kaartinen, 2007; Gritli-Linde, 2007; Nawshad, 2008). With the development of Cre/loxP-mediated genetic lineage labeling methods, the fate of the MES cells during palatal fusion in mouse embryos has been analyzed by several laboratories using different epithelial Cre-driver transgenic mice (Jin & Ding, 2006a; Vaziri Sani et al., 2005; Xu et al., 2006). Since no genetically labeled derivatives of the MES are found in the postfusion palatal mesenchyme, results from these studies mostly favored the apoptosis hypothesis for MES dissolution although MES migration was also detected (reviewed by Bush & Jiang, 2012). However, direct analysis of the requirement for important apoptosis regulators in palatal fusion has been inconclusive (Cecconi, Alvarez-Bolado, Meyer, Roth, & Gruss, 1998; Cuervo, Valencia, Chandraratna, & Covarrubias, 2002; Jin & Ding, 2006b; Takahara, Takigawa, & Shiota, 2004). Most recently, Kim et al. (2015) examined cell behavior during palatal fusion using a combination of genetic lineage labeling, tissue-specific gene inactivation, and live imaging. These studies revealed an essential role for actomyosin contractility-driven convergence and cell intercalation in the formation of MES and subsequent cell displacement and extrusion during MES breakdown (Fig. 5C). Whereas previously an argument against cell death as the major mechanism for MES dissolution suggested that extensive cell death of the MES cells would weaken the fusion site and lead to the bilateral palatal shelves being pulled apart (Sun et al., 2000), Kim et al. (2015) demonstrated by using live imaging the process of MES cell extrusion, during which converging MES cells form rosettes and the cells in the center of these rosettes are squeezed out by multicellular actin cables. Immunodetection of active caspase-3 and E-cadherin of E14.5 mouse embryonic palatal tissues showed that about half of the apoptotic cells in the MES were part of the cellular rosettes. Thus, apoptosis in the MES, at least during early stages of MES breakdown, does not involve simultaneous death of neighboring cells that might weaken the fusion site but rather involves extrusion of the apoptotic cell by its neighbors. Previous studies showed that an epithelial cell destined for apoptosis initiates signaling to its neighbors to form a contractile ring of actin and myosin prior to procaspase activation (Rosenblatt, Raff, & Cramer, 2001). Furthermore, pharmacological inhibition of stretch-activated ion channels blocked palatal fusion in explant culture, indicating that live cell extrusion due to crowding, in addition to apoptosis, also plays a key role in MES dissolution. Together with a series of genetic and pharmacological experiments demonstrating the requirement for Rho kinase- and myosin light-chain kinase-mediated activation of nonmuscle myosin II in MES cell intercalation and displacement during palatal fusion, these studies reveal the mechanistic connections of all previously observed cellular behaviors during palatal fusion, including extensive epithelial filapodia and cell shape changes at the beginning of palatal shelve adhesion, cell displacement, and collective cell migration in the oronasal directions during MES formation and breakdown, and apoptotic cell death, to actomyosin-driven MES cell convergence (Fig. 5C).

What are the molecular mechanisms underlying the spatiotemporal regulation of MEE fusion competence and the fusion process? Embryological and genetic studies have demonstrated an essential role for Tgfb3 signaling regulating palatal fusion. During mouse palate development, *Tgfb3* expression is specifically activated in the MEE by E13 (Fitzpatrick, Denhez, Kondaiah, & Akhurst, 1990; Pelton, Hogan, Miller, & Moses, 1990). Mice lacking Tgfb3 function and mice with epithelium-specific deletion of either *Tgfbr1* or *Tgfbr2* exhibit CP due to failure of palatal fusion (Dudas et al., 2006; Kaartinen et al., 1995; Proetzel et al., 1995; Xu et al., 2006). In *Tgfb3*$^{-/-}$ embryos, although the palatal shelves grow and elevate normally, approach and contact each other at the midline, appropriate adhesion does not occur. Even when placed in direct contact with each other in explant culture, the *Tgfb3*$^{-/-}$ mutant palatal shelves adhere poorly or not at all (Kaartinen, Cui, Heisterkamp, Groffen, & Shuler, 1997). Recent studies demonstrated that periderm cells persisted as an intact single layer covering the palatal shelves in *Tgfb3*$^{-/-}$ mutant embryos (Hu et al., 2015; Wu et al., 2013). Remarkably, whereas initial differentiation of the periderm cells requires Irf6 function, recent studies suggest that Irf6 is also required to mediate Tgfb3-induced palatal fusion. This second important function of Irf6 in palate development is brought about by Tgfb3 activation of its expression in the basal layer of the MEE cells just prior to palatal fusion (Fakhouri et al., 2012; Iwata et al., 2013). *Tgfb3*$^{-/-}$ as well as *Tgfbr2*$^{fl/fl}$; *K14-Cre* mutant embryos show significantly reduced *Irf6* expression in

MEE cells (Iwata et al., 2013; Knight et al., 2006). Transgenic expression of *Irf6* in the basal epithelial layer driven by the *keratin-14* gene promoter was able to rescue palatal fusion in *Tgfbr2*$^{fl/fl}$;*K14-Cre* mutant mice (Iwata et al., 2013). Iwata et al. (2013) suggested that Irf6 might employ the same downstream molecular cascade, involving repression of p63 and indirectly activating p21 as in periderm differentiation, to regulate palatal MES degeneration (Iwata et al., 2013; Moretti et al., 2010). Whereas the upregulation of p21 likely contributes to MEE cell cycle exit, whether Irf6 contributes to Tgfb3 regulation of periderm disruption remains to be investigated. On the other hand, Irf6 is required downstream of Tgfb3 for MES breakdown since *Irf6*$^{R84C/R84C}$ mutant embryos showed persistence of the aberrantly adhered MEE and tongue epithelial cells whereas the aberrant palate–tongue epithelial seam was disintegrated in *Jag2*$^{-/-}$ mutant embryos (Casey et al., 2006; Richardson et al., 2009). Like Tgfb3, Irf6 function is also required for the expression of *Mmp13* in MEE cells (Blavier et al., 2001; Richardson et al., 2009). Thus, Irf6 might also mediate Tgfb3-induced MES disintegration through the upregulation of *Mmp13* and other factors involved in degradation of the basal lamina (Fig. 5C).

Upon activation of the receptor complexes, Tgfβ signaling activates receptor-activated Smads, including Smad2 and Smad3, which then partner with the common mediator Smad4 to regulate expression of downstream target genes (Shi & Massague, 2003). Receptor Smads can also bind to Trim33, a chromatin reader (He et al., 2006). In addition, Tgfβ signaling can activate Smad-independent pathways, including the Tgfβ-activated kinase-1 (Tak1) and p38 MAPK kinase cascade (Derynck & Zhang, 2003). Studies of mice with epithelium-specific inactivation of *Smad4*, *Tak1*, and Trim33, show that all three pathways act partly redundantly to mediate Tgfβ3 signaling in palatal MES breakdown (Lane et al., 2015; Xu et al., 2008).

Mice with *K14-Cre*-mediated deletion of *Tgfbr1*, *Tgfbr2,* or *Tgfb3* exhibit milder palatal fusion defects than the *Tgfb3*$^{-/-}$ germ line knockout mice, such that interpalatal shelf adhesion occurred initially but the MES failed to disintegrate in the *K14-Cre* conditional mutant mouse strains (Dudas et al., 2006; Kaartinen et al., 1995; Lane et al., 2014; Proetzel et al., 1995; Xu et al., 2006). This has been attributed to lack of Cre activity in the palatal periderm cells in *K14-Cre* embryos (Lane et al., 2014). The milder phenotypes in the *K14-Cre* conditional mutants than in the *Tgfb3* null mice indicate that direct effects of Tgfb3 signaling on the periderm are critical for disruption of periderm integrity and initiation of intershelf adhesion. Whereas a recent report suggested that Tgfb3 regulates periderm removal through repression of ΔNp63 (Hu et al., 2015), it is likely that Tgfb3 affects another pathway in the periderm cells for their desquamation since ΔNp63 is already downregulated during periderm differentiation, as discussed above. On the other hand, some other transcription factors expressed in the MEE, in particular Runx1 and Snail1, have been shown to be required for palatal fusion (Charoenchaikorn et al., 2009; Murray, Oram, & Gridley, 2007). Charoenchaikorn et al. (2009) showed that *Runx1*$^{-/-}$ mutant embryos with a transgenic rescue of hematopoietic defects exhibit aberrant MEE cell behavior and cleft anterior palate. Whether *Runx1* deficiency affected expression of *Tgfb3* in the MEE was not characterized, however. Whereas *Snail1*$^{-/-}$ embryos die during early embryogenesis and *Snail2*$^{-/-}$ mice exhibit an incomplete penetrance of CP, all *Snail1*$^{+/-}$;*Snail2*$^{-/-}$ compound mutant mice have CP (Carver, Jiang, Lan, Oram, & Gridley, 2001; Jiang, Lan, Norton, Sundberg, & Gridley, 1998; Murray et al., 2007). Although the palatal shelves grow and elevate normally and make contact at the midline, they remain covered with periderm cells and fail to form the MES in *Snail1*$^{+/-}$;*Snail2*$^{-/-}$ compound mutant embryos. Expression of *Tgfb3* is not affected in the *Snail1*$^{+/-}$; *Snail2*$^{-/-}$ mutant MEE, indicating that these transcription factors regulate interpalatal shelf adhesion downstream of, or in parallel to, Tgfb3 signaling. Since the Snail family transcription factors have been implicated in epithelial–mesenchymal transition by directly repressing the expression of components of epithelial cell adhesive complexes (Nieto, 2002), it is possible that Snail1 and Snail2 act downstream of or in conjunction with Tgfb3 signaling to loosen adhesion between the MEE periderm cells at the onset of palatal fusion. Further studies are necessary to delineate these molecular relationships and to elucidate how they relate to activation of the actomyosin contractility-driven cell convergence behavior during MES formation.

5. CONCLUDING REMARKS

Nearly 30 years ago, the introduction of the gene-targeting technology for manipulating the mouse genome via homologous recombination revolutionized mouse genetics and propelled the laboratory mouse to the forefront of biomedical research (reviewed by Capecchi, 2005). Since then, mice carrying each of thousands of constitutive or conditional alleles have been generated and analyzed, which have provided much of the current understanding of mammalian biology and mechanisms of human diseases.

Much of the progress in our understanding of the cellular and molecular mechanisms of palatogenesis is also due to the widespread application of sophisticated genetic manipulations in mice and detailed morphological and molecular analysis of mutant mouse models (Bush & Jiang, 2012). Whereas the major signaling pathways governing palatogenesis have been well studied genetically, the underlying biochemical mechanisms remain largely uncharacterized. Moreover, as whole exome and whole-genome sequencing approaches are increasingly effectively applied for human disease gene discovery (Newman & Black, 2014; Tabor et al., 2014; The Deciphering Developmental Disorders Study, 2015), it can be expected that the number of genes associated with craniofacial and palate defects will increase rapidly. The major challenge is to integrate new discoveries into the understanding of the gene regulatory networks controlling craniofacial and palate development. Combining high-throughput molecular and biochemical approaches, such as chromatin immunoprecipitation and high-throughput sequencing (ChIP-seq), whole transcriptome RNA-seq-based gene expression profiling, and mass spectrometry with bioinformatics analysis will help generate gene regulatory and protein–protein interaction network models that can be further tested genetically and biochemically to provide new insights into the mechanisms of palate development. Moreover, the recent adaptation of the versatile CRISPR-Cas9 system for genome editing has once again revolutionized genetics and developmental biology research and enabled unprecedented access to investigating the detailed functions of proteins, genes, RNAs, and other elements as well as cellular and molecular processes in the endogenous context (Doudna & Charpentier, 2014; Harrison, Jenkins, O'Connor-Giles, & Wildonger, 2014). Together, these new technological tools will help rapidly advance our understanding of the mechanisms of palate development and lead to the development of new and better treatment or prevention strategies for CP and other craniofacial birth defects.

ACKNOWLEDGMENTS

We apologize to many colleagues whose work we could not discuss in this chapter due to space limitations. We thank Yang Gao for contributing data used in Fig. 4A and B. We thank Jeff Bush for sharing the PLoS Biology publication on cell convergence and extrusion during palate fusion prior to online access. Palate development research in the Lan and Jiang laboratories was supported by the National Institutes of Health National Institute of Dental and Craniofacial Research (NIH/NIDCR) grants R03DE023864 (Y. L.) and R01DE013681 (R. J.).

REFERENCES

Akiyama, M., Smith, L. T., Yoneda, K., Holbrook, K. A., Hohl, D., & Shimizu, H. (1999). Periderm cells form cornified cell envelope in their regression process during human epidermal development. *Journal of Investigative Dermatology, 112*, 903–909.

Alappat, S. R., Zhang, Z., Suzuki, K., Zhang, X., Liu, H., Jiang, R., et al. (2005). The cellular and molecular etiology of the cleft secondary palate in Fgf10 mutant mice. *Developmental Biology, 277*, 102–113.

Baek, J. A., Lan, Y., Liu, H., Maltby, K. M., Mishina, Y., & Jiang, R. (2011). Bmpr1a signaling plays critical roles in palatal shelf growth and palatal bone formation. *Developmental Biology, 350*, 520–531.

Blavier, L., Lazaryev, A., Groffen, J., Heisterkamp, N., DeClerck, Y. A., & Kaartinen, V. (2001). TGF-beta3-induced palatogenesis requires matrix metalloproteinases. *Molecular Biology of the Cell, 12*, 1457–1466.

Bush, J. O., & Jiang, R. (2012). Palatogenesis: Morphogenetic and molecular mechanisms of secondary palate development. *Development, 139*, 231–243.

Candi, E., Rufini, A., Terrinoni, A., Giamboi-Miraglia, A., Lena, A. M., Mantovani, R., et al. (2007). DeltaNp63 regulates thymic development through enhanced expression of FgfR2 and Jag2. *Proceedings of the National Academy of Sciences of the United States of America, 104*, 11999–12004.

Capecchi, M. R. (2005). Gene targeting in mice: Functional analysis of the mammalian genome for the twenty-first century. *Nature Reviews Genetics, 6*, 507–512.

Carver, E. A., Jiang, R., Lan, Y., Oram, K. F., & Gridley, T. (2001). The mouse snail gene encodes a key regulator of the epithelial-mesenchymal transition. *Molecular and Cellular Biology, 21*, 8184–8188.

Casey, L. M., Lan, Y., Cho, E. S., Maltby, K. M., Gridley, T., & Jiang, R. (2006). Jag2-Notch1 signaling regulates oral epithelial differentiation and palate development. *Developmental Dynamics, 235*, 1830–1844.

Cecconi, F., Alvarez-Bolado, G., Meyer, B. I., Roth, K. A., & Gruss, P. (1998). Apaf1 (CED-4 homolog) regulates programmed cell death in mammalian development. *Cell, 94*, 727–737.

Chai, Y., & Maxson, R. E., Jr. (2006). Recent advances in craniofacial morphogenesis. *Developmental Dynamics, 235*, 2353–2375.

Charoenchaikorn, K., Yokomizo, T., Rice, D. P., Honjo, T., Matsuzaki, K., Shintaku, Y., et al. (2009). Runx1 is involved in the fusion of the primary and the secondary palatal shelves. *Developmental Biology, 326*, 392–402.

Cobourne, M. T., & Green, J. B. (2012). Hedgehog signalling in development of the secondary palate. *Frontiers of Oral Biology, 16*, 52–59.

Crum, C. P., & McKeon, F. D. (2010). p63 in epithelial survival, germ cell surveillance, and neoplasia. *Annual Review of Pathology, 5*, 349–371.

Cuervo, R., & Covarrubias, L. (2004). Death is the major fate of medial edge epithelial cells and the cause of basal lamina degradation during palatogenesis. *Development, 131*, 15–24.

Cuervo, R., Valencia, C., Chandraratna, R. A., & Covarrubias, L. (2002). Programmed cell death is required for palate shelf fusion and is regulated by retinoic acid. *Developmental Biology, 245*, 145–156.

de la Garza, G., Schleiffarth, J. R., Dunnwald, M., Mankad, A., Weirather, J. L., Bonde, G., et al. (2013). Interferon regulatory factor 6 promotes differentiation of the periderm by activating expression of Grainyhead-like 3. *Journal of Investigative Dermatology, 133*, 68–77.

Derynck, R., & Zhang, Y. E. (2003). Smad-dependent and Smad-independent pathways in TGF-beta family signalling. *Nature, 425*, 577–584.

Dixon, M. J., Marazita, M. L., Beaty, T. H., & Murray, J. C. (2011). Cleft lip and palate: Understanding genetic and environmental influences. *Nature Reviews Genetics*, *12*, 167–178.
Dotto, G. P. (2012). p63 and FGFR: When development meets proliferation. *EMBO Molecular Medicine*, *4*, 165–167.
Doudna, J. A., & Charpentier, E. (2014). Genome editing. The new frontier of genome engineering with CRISPR-Cas9. *Science*, *346*, 1258096.
Dudas, M., Kim, J., Li, W. Y., Nagy, A., Larsson, J., Karlsson, S., et al. (2006). Epithelial and ectomesenchymal role of the type I TGF-beta receptor ALK5 during facial morphogenesis and palatal fusion. *Developmental Biology*, *296*, 298–314.
Dudas, M., Li, W. Y., Kim, J., Yang, A., & Kaartinen, V. (2007). Palatal fusion—Where do the midline cells go? A review on cleft palate, a major human birth defect. *Acta Histochemica*, *109*, 1–14.
Economou, A. D., Ohazama, A., Porntaveetus, T., Sharpe, P. T., Kondo, S., Basson, M. A., et al. (2012). Periodic stripe formation by a Turing mechanism operating at growth zones in the mammalian palate. *Nature Genetics*, *44*, 348–351.
Fakhouri, W. D., Rhea, L., Du, T., Sweezer, E., Morrison, H., Fitzpatrick, D., et al. (2012). MCS9.7 enhancer activity is highly, but not completely, associated with expression of Irf6 and p63. *Developmental Dynamics*, *241*, 340–349.
Ferone, G., Thomason, H. A., Antonini, D., De Rosa, L., Hu, B., Gemei, M., et al. (2012). Mutant p63 causes defective expansion of ectodermal progenitor cells and impaired FGF signalling in AEC syndrome. *EMBO Molecular Medicine*, *4*, 192–205.
Fitchett, J. E., & Hay, E. D. (1989). Medial edge epithelium transforms to mesenchyme after embryonic palatal shelves fuse. *Developmental Biology*, *131*, 455–474.
Fitzpatrick, D. R., Denhez, F., Kondaiah, P., & Akhurst, R. J. (1990). Differential expression of TGF beta isoforms in murine palatogenesis. *Development*, *109*, 585–595.
Gorlin, R. J., Cohen, M. M., & Hennekam, R. C. M. (2001). *Syndromes of the head and neck*. Oxford, New York: Oxford University Press.
Griffith, C. M., & Hay, E. D. (1992). Epithelial-mesenchymal transformation during palatal fusion: Carboxyfluorescein traces cells at light and electron microscopic levels. *Development*, *116*, 1087–1099.
Gritli-Linde, A. (2007). Molecular control of secondary palate development. *Developmental Biology*, *301*, 309–326.
Han, J., Mayo, J., Xu, X., Li, J., Bringas, P., Jr., Maas, R. L., et al. (2009). Indirect modulation of Shh signaling by Dlx5 affects the oral-nasal patterning of palate and rescues cleft palate in Msx1-null mice. *Development*, *136*, 4225–4233.
Harrison, M. M., Jenkins, B. V., O'Connor-Giles, K. M., & Wildonger, J. (2014). A CRISPR view of development. *Genes and Development*, *28*, 1859–1872.
Hayward, A. F. (1983). The permeability of the epithelium of the skin of fetal rats demonstrated with a lanthanum-containing solution. *Journal of Anatomy*, *136*, 379–388.
He, F., & Chen, Y. (2012). Wnt signaling in lip and palate development. *Frontiers of Oral Biology*, *16*, 81–90.
He, W., Dorn, D. C., Erdjument-Bromage, H., Tempst, P., Moore, M. A., & Massague, J. (2006). Hematopoiesis controlled by distinct TIF1gamma and Smad4 branches of the TGFbeta pathway. *Cell*, *125*, 929–941.
Hilliard, S. A., Yu, L., Gu, S., Zhang, Z., & Chen, Y. P. (2005). Regional regulation of palatal growth and patterning along the anterior-posterior axis in mice. *Journal of Anatomy*, *207*, 655–667.
Holbrook, K. A., & Odland, G. F. (1975). The fine structure of developing human epidermis: Light, scanning, and transmission electron microscopy of the periderm. *Journal of Investigative Dermatology*, *65*, 16–38.

Hu, L., Liu, J., Li, Z., Ozturk, F., Gurumurthy, C., Romano, R. A., et al. (2015). TGFbeta3 regulates periderm removal through DeltaNp63 in the developing palate. *Journal of Cellular Physiology*, *230*, 1212–1225.

Ingraham, C. R., Kinoshita, A., Kondo, S., Yang, B., Sajan, S., Trout, K. J., et al. (2006). Abnormal skin, limb and craniofacial morphogenesis in mice deficient for interferon regulatory factor 6 (Irf6). *Nature Genetics*, *38*, 1335–1340.

Iwata, J., Suzuki, A., Pelikan, R. C., Ho, T. V., Sanchez-Lara, P. A., Urata, M., et al. (2013). Smad4-Irf6 genetic interaction and TGFbeta-mediated IRF6 signaling cascade are crucial for palatal fusion in mice. *Development*, *140*, 1220–1230.

Jeong, J., Mao, J., Tenzen, T., Kottmann, A. H., & McMahon, A. P. (2004). Hedgehog signaling in the neural crest cells regulates the patterning and growth of facial primordia. *Genes and Development*, *18*, 937–951.

Jiang, R., Bush, J. O., & Lidral, A. C. (2006). Development of the upper lip: Morphogenetic and molecular mechanisms. *Developmental Dynamics*, *235*, 1152–1166.

Jiang, R., Lan, Y., Chapman, H. D., Shawber, C., Norton, C. R., Serreze, D. V., et al. (1998). Defects in limb, craniofacial, and thymic development in Jagged2 mutant mice. *Genes and Development*, *12*, 1046–1057.

Jiang, R., Lan, Y., Norton, C. R., Sundberg, J. P., & Gridley, T. (1998). The Slug gene is not essential for mesoderm or neural crest development in mice. *Developmental Biology*, *198*, 277–285.

Jin, J. Z., & Ding, J. (2006a). Analysis of cell migration, transdifferentiation and apoptosis during mouse secondary palate fusion. *Development*, *133*, 3341–3347.

Jin, J. Z., & Ding, J. (2006b). Analysis of Meox-2 mutant mice reveals a novel postfusion-based cleft palate. *Developmental Dynamics*, *235*, 539–546.

Kaartinen, V., Cui, X. M., Heisterkamp, N., Groffen, J., & Shuler, C. F. (1997). Transforming growth factor-beta3 regulates transdifferentiation of medial edge epithelium during palatal fusion and associated degradation of the basement membrane. *Developmental Dynamics*, *209*, 255–260.

Kaartinen, V., Voncken, J. W., Shuler, C., Warburton, D., Bu, D., Heisterkamp, N., et al. (1995). Abnormal lung development and cleft palate in mice lacking TGF-beta 3 indicates defects of epithelial-mesenchymal interaction. *Nature Genetics*, *11*, 415–421.

Kim, S., Lewis, A. E., Singh, V., Ma, X., Adelstein, R., & Bush, J. O. (2015). Convergence and extrusion are required for normal fusion of the mammalian secondary palate. *PLoS Biology*, *13*, e1002122.

King, K. E., & Weinberg, W. C. (2007). p63: Defining roles in morphogenesis, homeostasis, and neoplasia of the epidermis. *Molecular Carcinogenesis*, *46*, 716–724.

Knight, A. S., Schutte, B. C., Jiang, R., & Dixon, M. J. (2006). Developmental expression analysis of the mouse and chick orthologues of IRF6: The gene mutated in Van der Woude syndrome. *Developmental Dynamics*, *235*, 1441–1447.

Kondo, S., Schutte, B. C., Richardson, R. J., Bjork, B. C., Knight, A. S., Watanabe, Y., et al. (2002). Mutations in IRF6 cause Van der Woude and popliteal pterygium syndromes. *Nature Genetics*, *32*, 285–289.

Lan, Y., & Jiang, R. (2009). Sonic hedgehog signaling regulates reciprocal epithelial-mesenchymal interactions controlling palatal outgrowth. *Development*, *136*, 1387–1396.

Lan, Y., Ovitt, C. E., Cho, E. S., Maltby, K. M., Wang, Q., & Jiang, R. (2004). Odd-skipped related 2 (Osr2) encodes a key intrinsic regulator of secondary palate growth and morphogenesis. *Development*, *131*, 3207–3216.

Lane, J., & Kaartinen, V. (2014). Signaling networks in palate development. *Wiley Interdisciplinary Reviews: Systems Biology and Medicine*, *6*, 271–278.

Lane, J., Yumoto, K., Azhar, M., Ninomiya-Tsuji, J., Inagaki, M., Hu, Y., et al. (2015). Tak1, Smad4 and Trim33 redundantly mediate TGF-beta3 signaling during palate development. *Developmental Biology*, *398*, 231–241.

Lane, J., Yumoto, K., Pisano, J., Azhar, M., Thomas, P. S., & Kaartinen, V. (2014). Control elements targeting Tgfb3 expression to the palatal epithelium are located intergenically and in introns of the upstream Ift43 gene. *Frontiers in Physiology*, *5*, 258.

Laurikkala, J., Mikkola, M. L., James, M., Tummers, M., Mills, A. A., & Thesleff, I. (2006). p63 regulates multiple signalling pathways required for ectodermal organogenesis and differentiation. *Development*, *133*, 1553–1563.

Levi, B., Brugman, S., Wong, V. W., Grova, M., Longaker, M. T., & Wan, D. C. (2011). Palatogenesis: Engineering, pathways and pathologies. *Organogenesis*, *7*, 242–254.

Li, Q., & Ding, J. (2007). Gene expression analysis reveals that formation of the mouse anterior secondary palate involves recruitment of cells from the posterior side. *International Journal of Developmental Biology*, *51*, 167–172.

Liu, W., Lan, Y., Pauws, E., Meester-Smoor, M. A., Stanier, P., Zwarthoff, E. C., et al. (2008). The Mn1 transcription factor acts upstream of Tbx22 and preferentially regulates posterior palate growth in mice. *Development*, *135*, 3959–3968.

Mangold, E., Ludwig, K. U., & Nothen, M. M. (2011). Breakthroughs in the genetics of orofacial clefting. *Trends in Molecular Medicine*, *17*, 725–733.

M'Boneko, V., & Merker, H. J. (1988). Development and morphology of the periderm of mouse embryos (days 9–12 of gestation). *Acta Anatomica (Basel)*, *133*, 325–336.

Mills, A. A., Zheng, B., Wang, X. J., Vogel, H., Roop, D. R., & Bradley, A. (1999). p63 is a p53 homologue required for limb and epidermal morphogenesis. *Nature*, *398*, 708–713.

Moretti, F., Marinari, B., Lo Iacono, N., Botti, E., Giunta, A., Spallone, G., et al. (2010). A regulatory feedback loop involving p63 and IRF6 links the pathogenesis of 2 genetically different human ectodermal dysplasias. *Journal of Clinical Investigation*, *120*, 1570–1577.

Murray, S. A., Oram, K. F., & Gridley, T. (2007). Multiple functions of Snail family genes during palate development in mice. *Development*, *134*, 1789–1797.

Nawshad, A. (2008). Palatal seam disintegration: To die or not to die? That is no longer the question. *Developmental Dynamics*, *237*, 2643–2656.

Newman, W. G., & Black, G. C. (2014). Delivery of a clinical genomics service. *Genes (Basel)*, *5*, 1001–1017.

Nieto, M. A. (2002). The snail superfamily of zinc-finger transcription factors. *Nature Reviews Molecular Cell Biology*, *3*, 155–166.

Pantalacci, S., Prochazka, J., Martin, A., Rothova, M., Lambert, A., Bernard, L., et al. (2008). Patterning of palatal rugae through sequential addition reveals an anterior/posterior boundary in palatal development. *BMC Developmental Biology*, *8*, 116.

Parada, C., & Chai, Y. (2012). Roles of BMP signaling pathway in lip and palate development. *Frontiers of Oral Biology*, *16*, 60–70.

Pauws, E., Hoshino, A., Bentley, L., Prajapati, S., Keller, C., Hammond, P., et al. (2009). Tbx22null mice have a submucous cleft palate due to reduced palatal bone formation and also display ankyloglossia and choanal atresia phenotypes. *Human Molecular Genetics*, *18*, 4171–4179.

Pelton, R. W., Hogan, B. L., Miller, D. A., & Moses, H. L. (1990). Differential expression of genes encoding TGFs beta 1, beta 2, and beta 3 during murine palate formation. *Developmental Biology*, *141*, 456–460.

Peters, H., Neubuser, A., Kratochwil, K., & Balling, R. (1998). Pax9-deficient mice lack pharyngeal pouch derivatives and teeth and exhibit craniofacial and limb abnormalities. *Genes and Development*, *12*, 2735–2747.

Peyrard-Janvid, M., Leslie, E. J., Kousa, Y. A., Smith, T. L., Dunnwald, M., Magnusson, M., et al. (2014). Dominant mutations in GRHL3 cause Van der Woude Syndrome and disrupt oral periderm development. *The American Journal of Human Genetics*, *94*, 23–32.

Polakowska, R. R., Piacentini, M., Bartlett, R., Goldsmith, L. A., & Haake, A. R. (1994). Apoptosis in human skin development: Morphogenesis, periderm, and stem cells. *Developmental Dynamics*, *199*, 176–188.

Proetzel, G., Pawlowski, S. A., Wiles, M. V., Yin, M., Boivin, G. P., Howles, P. N., et al. (1995). Transforming growth factor-beta 3 is required for secondary palate fusion. *Nature Genetics, 11*, 409–414.

Rahimov, F., Jugessur, A., & Murray, J. C. (2012). Genetics of nonsyndromic orofacial clefts. *Cleft Palate-Craniofacial Journal, 49*, 73–91.

Rice, R., Connor, E., & Rice, D. P. (2006). Expression patterns of Hedgehog signalling pathway members during mouse palate development. *Gene Expression Patterns, 6*, 206–212.

Rice, R., Spencer-Dene, B., Connor, E. C., Gritli-Linde, A., McMahon, A. P., Dickson, C., et al. (2004). Disruption of Fgf10/Fgfr2b-coordinated epithelial-mesenchymal interactions causes cleft palate. *Journal of Clinical Investigation, 113*, 1692–1700.

Richardson, R. J., Dixon, J., Jiang, R., & Dixon, M. J. (2009). Integration of IRF6 and Jagged2 signalling is essential for controlling palatal adhesion and fusion competence. *Human Molecular Genetics, 18*, 2632–2642.

Richardson, R. J., Dixon, J., Malhotra, S., Hardman, M. J., Knowles, L., Boot-Handford, R. P., et al. (2006). Irf6 is a key determinant of the keratinocyte proliferation-differentiation switch. *Nature Genetics, 38*, 1329–1334.

Richardson, R. J., Hammond, N. L., Coulombe, P. A., Saloranta, C., Nousiainen, H. O., Salonen, R., et al. (2014). Periderm prevents pathological epithelial adhesions during embryogenesis. *Journal of Clinical Investigation, 124*, 3891–3900.

Romano, R. A., Smalley, K., Magraw, C., Serna, V. A., Kurita, T., Raghavan, S., et al. (2012). DeltaNp63 knockout mice reveal its indispensable role as a master regulator of epithelial development and differentiation. *Development, 139*, 772–782.

Rosenblatt, J., Raff, M. C., & Cramer, L. P. (2001). An epithelial cell destined for apoptosis signals its neighbors to extrude it by an actin- and myosin-dependent mechanism. *Current Biology, 11*, 1847–1857.

Sanes, J. R., Rubenstein, J. L., & Nicolas, J. F. (1986). Use of a recombinant retrovirus to study post-implantation cell lineage in mouse embryos. *EMBO Journal, 5*, 3133–3142.

Sasaki, Y., Ishida, S., Morimoto, I., Yamashita, T., Kojima, T., Kihara, C., et al. (2002). The p53 family member genes are involved in the Notch signal pathway. *Journal of Biological Chemistry, 277*, 719–724.

Schutte, B. C., & Murray, J. C. (1999). The many faces and factors of orofacial clefts. *Human Molecular Genetics, 8*, 1853–1859.

Scott, W. J., Jr., Nau, H., Wittfoht, W., & Merker, H. J. (1987). Ventral duplication of the autopod: Chemical induction by methoxyacetic acid in rat embryos. *Development, 99*, 127–136.

Shi, Y., & Massague, J. (2003). Mechanisms of TGF-beta signaling from cell membrane to the nucleus. *Cell, 113*, 685–700.

Smith, T. M., Lozanoff, S., Iyyanar, P. P., & Nazarali, A. J. (2012). Molecular signaling along the anterior-posterior axis of early palate development. *Frontiers in Physiology, 3*, 488.

Stanier, P., & Pauws, E. (2012). Development of the lip and palate: FGF signalling. *Frontiers of Oral Biology, 16*, 71–80.

Sun, D., Baur, S., & Hay, E. D. (2000). Epithelial-mesenchymal transformation is the mechanism for fusion of the craniofacial primordia involved in morphogenesis of the chicken lip. *Developmental Biology, 228*, 337–349.

Tabor, H. K., Auer, P. L., Jamal, S. M., Chong, J. X., Yu, J. H., Gordon, A. S., et al. (2014). Pathogenic variants for Mendelian and complex traits in exomes of 6,517 European and African Americans: Implications for the return of incidental results. *The American Journal of Human Genetics, 95*, 183–193.

Tadeu, A. M., & Horsley, V. (2013). Notch signaling represses p63 expression in the developing surface ectoderm. *Development, 140*, 3777–3786.

Takahara, S., Takigawa, T., & Shiota, K. (2004). Programmed cell death is not a necessary prerequisite for fusion of the fetal mouse palate. *International Journal of Developmental Biology*, *48*, 39–46.

The Deciphering Developmental Disorders Study. (2015). Large-scale discovery of novel genetic causes of developmental disorders. *Nature*, *519*, 223–228.

Thomason, H. A., Zhou, H., Kouwenhoven, E. N., Dotto, G. P., Restivo, G., Nguyen, B. C., et al. (2010). Cooperation between the transcription factors p63 and IRF6 is essential to prevent cleft palate in mice. *Journal of Clinical Investigation*, *120*, 1561–1569.

Van Der Woude, A. (1954). Fistula labii inferioris congenita and its association with cleft lip and palate. *The American Journal of Human Genetics*, *6*, 244–256.

Vanderas, A. P. (1987). Prevalence of craniomandibular dysfunction in children and adolescents: A review. *Pediatric Dentistry*, *9*, 312–316.

Vaziri Sani, F., Hallberg, K., Harfe, B. D., McMahon, A. P., Linde, A., & Gritli-Linde, A. (2005). Fate-mapping of the epithelial seam during palatal fusion rules out epithelial-mesenchymal transformation. *Developmental Biology*, *285*, 490–495.

Veistinen, L., Aberg, T., & Rice, D. P. (2009). Convergent signalling through Fgfr2 regulates divergent craniofacial morphogenesis. *Journal of Experimental Zoology Part B: Molecular and Developmental Evolution*, *312B*, 351–360.

Waterman, R. E., & Meller, S. M. (1974). Alterations in the epithelial surface of human palatal shelves prior to and during fusion: A scanning electron microscopic study. *Anatomical Record*, *180*, 111–135.

Welsh, I. C., & O'Brien, T. P. (2009). Signaling integration in the rugae growth zone directs sequential SHH signaling center formation during the rostral outgrowth of the palate. *Developmental Biology*, *336*, 53–67.

Westfall, M. D., Mays, D. J., Sniezek, J. C., & Pietenpol, J. A. (2003). The Delta Np63 alpha phosphoprotein binds the p21 and 14-3-3 sigma promoters in vivo and has transcriptional repressor activity that is reduced by Hay-Wells syndrome-derived mutations. *Molecular and Cellular Biology*, *23*, 2264–2276.

Wu, C., Endo, M., Yang, B. H., Radecki, M. A., Davis, P. F., Zoltick, P. W., et al. (2013). Intra-amniotic transient transduction of the periderm with a viral vector encoding TGFbeta3 prevents cleft palate in Tgfbeta3(−/−) mouse embryos. *Molecular Therapy*, *21*, 8–17.

Xu, X., Han, J., Ito, Y., Bringas, P., Jr., Deng, C., & Chai, Y. (2008). Ectodermal Smad4 and p38 MAPK are functionally redundant in mediating TGF-beta/BMP signaling during tooth and palate development. *Developmental Cell*, *15*, 322–329.

Xu, X., Han, J., Ito, Y., Bringas, P., Jr., Urata, M. M., & Chai, Y. (2006). Cell autonomous requirement for Tgfbr2 in the disappearance of medial edge epithelium during palatal fusion. *Developmental Biology*, *297*, 238–248.

Yang, A., Schweitzer, R., Sun, D., Kaghad, M., Walker, N., Bronson, R. T., et al. (1999). p63 is essential for regenerative proliferation in limb, craniofacial and epithelial development. *Nature*, *398*, 714–718.

Yano, H., Ohtsuru, A., Ito, M., Fuji, T., & Yamashita, S. (1996). Involvement of c-Fos proto-oncogene during palatal fusion and interdigital space formation in the rat. *Development, Growth & Differentiation*, *38*, 351–357.

Yoshida, M., Shimono, Y., Togashi, H., Matsuzaki, K., Miyoshi, J., Mizoguchi, A., et al. (2012). Periderm cells covering palatal shelves have tight junctions and their desquamation reduces the polarity of palatal shelf epithelial cells in palatogenesis. *Genes to Cells*, *17*, 455–472.

Yu, L., Gu, S., Alappat, S., Song, Y., Yan, M., Zhang, X., et al. (2005). Shox2-deficient mice exhibit a rare type of incomplete clefting of the secondary palate. *Development*, *132*, 4397–4406.

Zhang, Z., Song, Y., Zhao, X., Zhang, X., Fermin, C., & Chen, Y. (2002). Rescue of cleft palate in Msx1-deficient mice by transgenic Bmp4 reveals a network of BMP and Shh signaling in the regulation of mammalian palatogenesis. *Development, 129*, 4135–4146.

Zhou, J., Gao, Y., Lan, Y., Jia, S., & Jiang, R. (2013). Pax9 regulates a molecular network involving Bmp4, Fgf10, Shh signaling and the Osr2 transcription factor to control palate morphogenesis. *Development, 140*, 4709–4718.

Zhou, J., Gao, Y., Zhang, Z., Zhang, Y., Maltby, K. M., Liu, Z., et al. (2011). Osr2 acts downstream of Pax9 and interacts with both Msx1 and Pax9 to pattern the tooth developmental field. *Developmental Biology, 353*, 344–353.

CHAPTER FOUR

Genetic Influences on Temporomandibular Joint Development and Growth

Robert J. Hinton*[,1], Junjun Jing[†], Jian Q. Feng*

*Department of Biomedical Sciences, Texas A&M Baylor College of Dentistry, Dallas, Texas, USA
[†]State Key Laboratory of Oral Diseases, West China Hospital of Stomatology, Sichuan University, Sichuan, PR China
[1]Corresponding author: e-mail address: bhinton@bcd.tamhsc.edu

Contents

Abstract

The temporomandibular joint (TMJ) is a small synovial joint at which the mandible articulates with the skull during movements involved in speaking and mastication. However, the secondary cartilage lining its joint surfaces is indicative of a very different developmental history than limb cartilages. This review summarizes our current knowledge of genes that regulate the formation of primary components of the TMJ, as well as genes that regulate postnatal growth of the TMJ. Although the TMJ is regulated by some of the same genes that are important in limb joints, others appear unique to the TMJ or have different actions. *Runx2*, *Sox9*, and members of the *TGF-β/BMP* family are critical drivers of chondrogenesis during condylar cartilage morphogenesis, and *Indian hedgehog (Ihh)*

ISSN 0070-2153
http://dx.doi.org/10.1016/bs.ctdb.2015.07.008

is important for formation of the articular disc and cavitation. *Osterix (Osx)* is a critical regulator of endochondral bone formation during postnatal TMJ growth.

1. INTRODUCTION

For many years, the temporomandibular joint (TMJ) has been a misunderstood joint. Thought by many in the 1920 to the 1940s to be a nonloaded joint (Robinson, 1946; Wilson, 1920), it is now generally agreed that significant loads are delivered to the TMJ as a consequence of mastication and incision (Hylander, 1986; Iwasaki et al., 2010). As a synovial joint with an articular disc, the TMJ was long presumed to be a smaller version of the knee or hip joint by cartilage researchers in the orthopedic community and therefore mostly ignored. It is only relatively recently that this small joint has become of interest to those outside dental and craniofacial research. Over the last 20–30 years, the implications of its unusual morphogenesis have driven investigations into the molecular and cell biology of the TMJ. However, it is only in the last decade that specific genes that are important for TMJ development and growth have been identified using transgenic mouse models.

Building on the early work of Hall (1972), the TMJ has long been understood to be a joint comprised of *secondary* cartilage (Beresford, 1981), a tissue that differs in several ways from the primary cartilages that are present in the developing limb joints and cranial base. It is designated as secondary because the cartilage capping the mandibular condyle (mandibular condylar cartilage or MCC) develops after the bone of the mandibular ramus, the opposite of the "cartilage first, then bone" sequence that characterizes primary cartilages. As a result of this developmental hierarchy, the primary cartilaginous joints of the limbs are often mostly formed by the time that the MCC and other intraarticular joint structures of the TMJ are beginning to form. The MCC develops adjacent to the intramembranous bone of the mandible, distinct from Meckel's cartilage (Radlanski, Renz, & Klarkowski, 2003; Vinkka, 1982), arising either from alkaline phosphatase (AP)-positive cells of likely periosteal origin (Shibata, Fukada, Suzuki, Ogawa, & Yamashita, 2002) or as a condensation separate from the developing bone (Anthwal, Chai, & Tucker, 2008; Vinkka, 1982). The MCC is only one of many secondary cartilages that are found, transiently in some instances, at many locations in the developing craniofacial complex (Vinkka, 1982). Secondary cartilages continue postnatally in regions such as the mandibular condyle, angular process of the mandible, and intermaxillary suture (Hinton, 1988; Petrovic, 1972). Because of its persistence and role in the growth of the mandible in height and length, the MCC is one of the most important secondary cartilages. Another oddity of the TMJ is that it is a purely mammalian joint; it does not exist in reptiles, but develops in mammals in conjunction with the loads produced by a specialized dentition. Thus, the periosteum of formerly intramembranous bones was transformed into a perichondrium in force-bearing areas where the "new" TMJ would develop (Crompton, 1985). This evolutionary transition is reflected in the formation of the TMJ by three separate mesenchymal condensations (condylar, temporal, and disc) that grow toward each other during development (Sperber, 2001), as compared to the single block of mesenchyme within which the joints involving primary cartilaginous articulations develop.

At least three important differences delineate the secondary cartilage at the MCC from primary cartilages of the limbs and cranial base. First, the cells that proliferate in the MCC are not differentiated chondrocytes, but rather only partially differentiated along the chondrogenic pathway—i.e., "prechondroblasts" that are polymorphic in shape rather that rounded like chondrocytes and that secrete type I collagen rather than type II collagen that is characteristic of the underlying chondrocytes (Mizoguchi, Nakamura, Takahashi, Kagayama, & Mitani, 1990; Silbermann et al., 1987). A relatively hypocellular fibroblastic layer (the articular layer) overlies the prechondroblastic layer, and together they comprise the perichondrium that surmounts the cartilage proper of the MCC (Fig. 1). In keeping with its origins adjacent to the periosteum of the ramus of the intramembranous mandible, the articular and prechondroblastic layers of the MCC perichondrium are contiguous with the fibrous and osteogenic layers of the ramus periosteum. Second, the MCC, like other secondary cartilages (Fig. 2) but unlike primary cartilages, can be transient in its expression, requiring conditions of low oxygen tension and/or function to maintain its cartilaginous phenotype. Because the MCC developed in mammals via the localized transformation of a periosteum to a perichondrium in a region (the "new" TMJ) subjected to biomechanical forces, the MCC can diminish in size or be replaced by bone if the evoking stimulus is reduced (Bouvier & Hylander, 1984; Hinton, 1988) or eliminated by lack of movement (Carlson, McNamara, Graber, & Hoffman, 1980; Glineberg, Laskin, & Blaustein, 1982) or placement in a nonfunctional environment (Duterloo & Wolters, 1971; Engelsma, Jansen, & Duterloo, 1980; Petrovic, Stutzmann, & Oudet, 1975). Last, the MCC acts as a site of growth for the mandible to increase

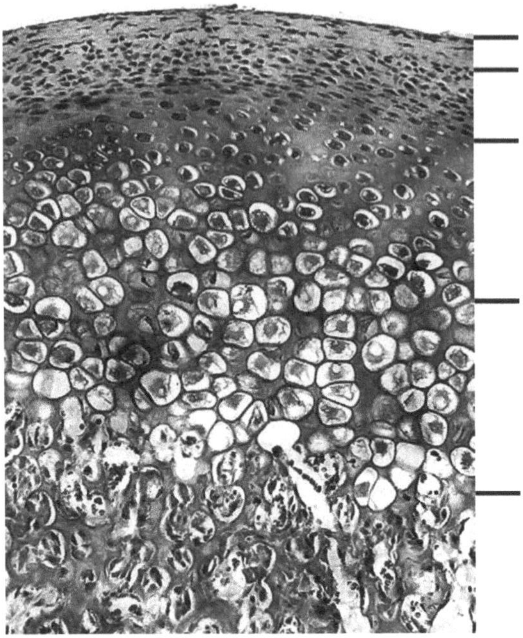

Figure 1 Histology of the growing (postnatal) rat mandibular condylar cartilage. Coronal section, Attwood's stain. (A color version of this figure is located at the end of this chapter.)

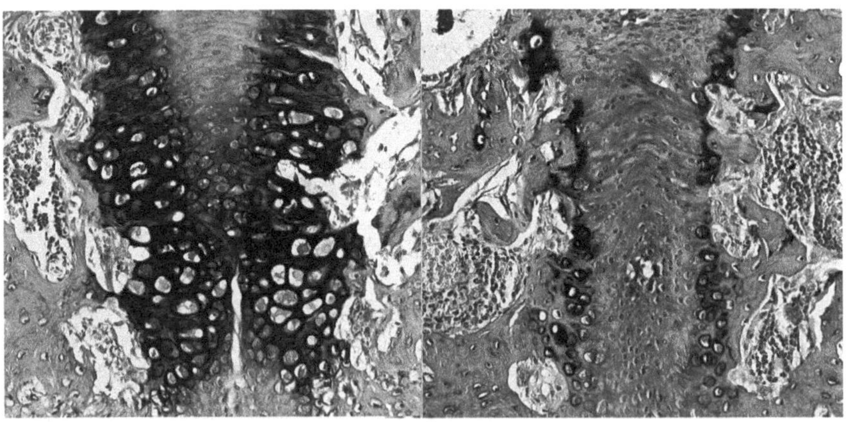

Figure 2 Secondary cartilage at intermaxillary suture in rats fed a normal hard diet (left) or soft, mushy diet (right) for 2.5 weeks. Coronal sections, alcian blue-Kernechtrot aluminum sulfate. Original magnification ×40. Darkest densities correspond to alcian blue staining. Note marked reduction in extent of alcian blue staining and loss of hypertrophic chondrocytes in soft diet animals. (A color version of this figure is located at the end of this chapter.)

in length and height during prenatal and early postnatal life, and also as an articular cartilage beginning with the ingestion of hard food. Thus, it subsumes the functions of both the growth plate and articular cartilage, but in a single tissue that never develops an epiphysis.

Given its unusual manner of development and the distinctive characteristics of its dividing cells, it would not be surprising if the molecular determinants of development and growth of the MCC and TMJ were somewhat different from those in primary cartilaginous joints. The goal of this chapter is to summarize what is known to date of the genetic regulation of TMJ development and growth. It is a "snapshot" to be sure, but appreciation of the state of our knowledge can facilitate comparisons with limb joints and identify areas for future study.

2. NORMAL MORPHOGENESIS AND CELL BIOLOGY OF THE TMJ

In order to evaluate the TMJ phenotypes in transgenic mice models, an appreciation of the normal development of the murine TMJ (Fig. 3) is necessary. Joint formation commences with the appearance of mesenchymal condensations that will become the condyle and mandibular fossa, with a third condensation between them that will become the articular disc (E13.5). Next, chondrocytes differentiate in the deeper layers of the MCC (E15.0), and bone formation becomes evident in the mandibular fossa. At E17.5, the articular disc and the upper and lower joint cavities can be distinguished (Gu, Wei, Yu, Fei, & Chen, 2008; Shibata, Suzuki, Tengan, Ishii, & Kuroda, 1996), completing the elements of a recognizable TMJ. Since the TMJ is not formed when reflex jaw movements occur at or before E14 (Tsunekawa, Arata, & Obata, 2005), some have suggested that these early movements likely take place at the malleus/incus joint at the rostral end of Meckel's cartilage (Merida-Velasco et al., 1999). Since paralysis *in ovo* or *in vitro* has been shown to prevent joint cavitation (Fell & Canti, 1934; Mitrovic, 1982), it is reasonable to suspect that cavitation, or even chondrogenesis, in the TMJ is stimulated by these early jaw movements. This supposition is supported by the demonstration that *ex utero* limitation of jaw movements in mouse fetuses reduces growth at the condylar cartilage (Habib et al., 2005).

The cells in the MCC that divide and subsequently differentiate to effect growth are not differentiated chondrocytes. As noted above, these prechondrocytes are highly dependent on jaw movement and functional

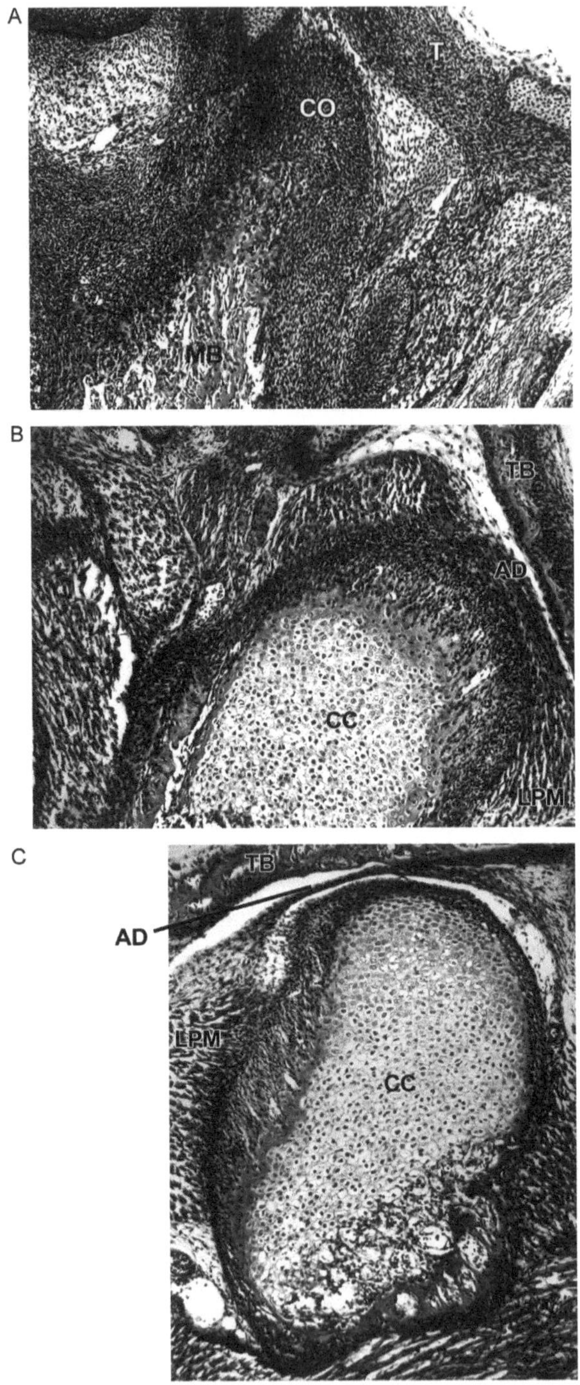

Figure 3 See legend on opposite page.

loading to maintain the chondrogenic phenotype. Thus, dividing cells in the prechondroblastic layer of the condylar cartilage exhibit a dual potential. This phenotypic lability is underscored by the expression of both *Sox 9* and *Runx2* RNA (markers for the chondrogenic and osteogenic pathways, respectively) in the mesenchymal cell condensation and later in the prechondroblastic cells of the developing MCC (Shibata, Suda, Suzuki, Fukoka, & Yamashita, 2006) and by the preferential localization of the cell fate mediators *Notch* and *Twist* to these cells (Serrano, So, & Hinton, 2014; Serrano, So, Svoboda, & Hinton, 2011). In light of these characteristics of MCC perichondrial cells, morphogenesis of the condylar cartilage should be influenced by genes that encourage differentiation along the chondrogenic pathway. There exists evidence that some of the above genes, and likely others, are critical for formation of the condylar cartilage. Still other genes affect the postnatal growth of the MCC.

3. GENES THAT AFFECT MANDIBULAR CONDYLAR CARTILAGE

3.1 Agenesis of Mandibular Condylar Cartilage

Inactivation of either *Runx2* or *Sox 9*, which are both expressed in the mesenchymal condensation that initiates the formation of the mandibular condyle and later expressed in the prechondroblastic zone, has been shown to result in agenesis of the condylar cartilage. Because Runx2 is necessary for bone formation, mice-lacking *Runx2* do not form bone at all. Curiously, while *Runx2(−/−)* mice exhibited no disruption of primary cartilage formation (e.g., in the limbs and Meckel's cartilage), the MCC and other secondary cartilages, such as the angular and coronoid, were completely absent

Figure 3 (A) Temporomandibular joint at *blastematic* stage showing condylar (CO) and temporal (T) blastemas separated by less dense mesenchyme that will form the future disc. Mandibular bone, MB. Mouse, embryonic day 15. Hematoxylin and eosin, sagittal section. (B) Temporomandibular joint at *cavitation* stage illustrating chondro-differentiation of condylar cartilage and formation of joint spaces in progress. Superior joint space is evident (upper right), but inferior joint space is not yet formed. CC, condylar cartilage; AD, articular disc; LPM, lateral pterygoid muscle; TB, temporal bone. Mouse*, embryonic day 17. Hematoxylin and eosin, sagittal section. *The superior joint space forms earlier in a mouse, opposite the human condition. (C) Temporomandibular joint at *maturation* stage. Both inferior and superior joint spaces are fully formed, lateral pterygoid muscle is attached to condylar neck and articular disc begins to resemble postnatal form. CC, condylar cartilage; AD, articular disc; LPM, lateral pterygoid muscle; TB, temporal bone. Mouse, embryonic day 19. Hematoxylin and eosin, sagittal section. (A color version of this figure is located at the end of this chapter.)

(Shibata et al., 2004). A condylar condensation that was AP-positive and exhibited weak *Col 1* expression was present, but reduced in size at E18.5, compared to the wild-type condensation at E14. However, no genetic markers of secondary cartilage formation (e.g., *Col2* or *aggrecan*) were evident. A subsequent study showed that *Runx2*(−/−) condylar explants could be induced to undergo chondrogenic differentiation (but not hypertrophy) by the addition of exogenous BMP-2 (Fukuoka, Shibata, Suda, Yamashita, & Komori, 2007), suggesting that a partial rescue was possible. Complete agenesis of the condylar cartilage also occurs following inactivation of *Sox9* in cranial neural crest cells (Mori-Akiyama, Akiyama, Rowitch, & de Crombrugghe, 2003; Wang et al., 2011). Meckel's cartilage was absent, as were any bones formed by endochondral ossification. Interestingly, *Runx2* expression was still present in the mandibular ramus and early on in the mandibular fossa condensation of the *Sox9*-null mice.

Interference with signaling by members of the TGF-β family also results in severe inhibition of condylar cartilage morphogenesis. Conditional inactivation in cranial neural crest cells of the *Tgfbr2* receptor, which is specifically required for proliferation in both osteoprogenitor and chondroprogenitor cells, resulted in a vestigial condylar cartilage (Oka et al., 2007). Although a mesenchymal condensation for the condyle was present in *Tgfbr2* mutant mice at E14.5, metachromatic staining of the cartilage was greatly diminished in the mutant condyles by E16.5. By E18.5, when joint formation was complete in wild-type animals, chondrogenesis of the condylar process remained greatly reduced compared to mutant condyles, and the hypertrophic zone was entirely absent. In a subsequent study, this group demonstrated a greatly reduced proliferation and diminished expression of *Sox 9* in the condylar cartilage of mutant animals (Oka et al., 2008). However, *Runx2* expression was increased in mutant condyles, as was the expression of *Dlx5*, a transcription factor important for osteoblast differentiation. The authors conclude that TGF-β signaling influences cell fate in the dividing cells of the MCC by regulating *Sox9* expression, which in turn represses *Runx2* expression. Still another study indicates that TGF-β is necessary for *initiation* of secondary cartilages in the mandible. Explant cultures of half mandibles from E13.5 and E14.5 mice exposed to an inhibitor of the downstream signaling of type I TGF-β receptors that partner with Tgfbr2 showed a complete loss of secondary cartilage compared to vehicle-treated explants, and preexisting secondary cartilages in E15.5 explants disappeared after 4 days in culture (Anthwal et al., 2008). Mice lacking the BMP receptor Alk2 were completely lacking a TMJ (Dudas,

Sridurongrit, Nagy, Okazaki, & Kaartinen, 2004). The condylar cartilage failed to form and, although the squamous part of the temporal bone developed in attenuated form, the mandibular fossa was missing. The secondary cartilages at the coronoid and angular processes were either rudimentary or missing altogether.

3.2 Disrupted Condylar Cartilage Formation

Significant effects on the condylar cartilage, albeit less drastic than agenesis, were seen in mice deficient in other genes. In mice-lacking Indian hedgehog (*ihh*(−/−)), proliferation in prechondroblastic zone cells was inhibited, cell organization disrupted, and chondrocyte gene expression altered (Shibukawa et al., 2007). *Ihh* is a well-known regulator of long bone development and growth, including intramembranous bone collar formation, chondrocyte proliferation and maturation, and endochondral ossification (Koyama et al., 2007). A TMJ phenotype is also evident in mice deficient in *Trps,* a gene coding for a transcription factor mutated in human Tricho–Rhino–Phalangeal syndrome. *Trps1* is expressed in the perichondrium and chondrocytes of the developing MCC at E15.5 (Michikami et al., 2012) and, similar to *ihh*, regulates various aspects of growth plate proliferation and maturation, as well as the intramembranous ossification of the perichondrium (Napierala et al., 2008). Moreover, *Trps1* interacts with *ihh/Gli3* signaling in perichondrial ossification (Wuelling et al., 2009) and is a strong repressor of *Runx2* (Napierala et al., 2008). *Trps*(−/−) mice exhibited a significant delay in condylar cartilage formation compared to wild-type mice. By E18.5, the normal layers were disrupted, consisting of a thickened perichondrium overlying smaller than normal hypertrophic chondrocytes (Fig. 4).

A similar hypoplastic MCC phenotype, with a considerable reduction in *Col 10* expression, was present in mice in which BMPr1a was inactivated in cranial neural crest cells (Gu et al., 2014). Notably, *ihh* was strongly downregulated in *BMPr1a*(−/−) mice, suggesting that BMPr1a likely regulates TMJ morphogenesis via its action on *ihh*. In mice with constitutively active *BMPr1a*, Gu and colleagues (2014) found mostly an opposite phenotype: the condylar cartilage became considerably larger, with a noticeable increase in hypertrophic cells and a downregulation of *Runx2*. Most interestingly, the condyle and mandibular fossa degenerated by E18.5.

Patterning genes that affect the formation of first and second branchial structures also affect MCC formation. *Foxc1*-null mice, which have a

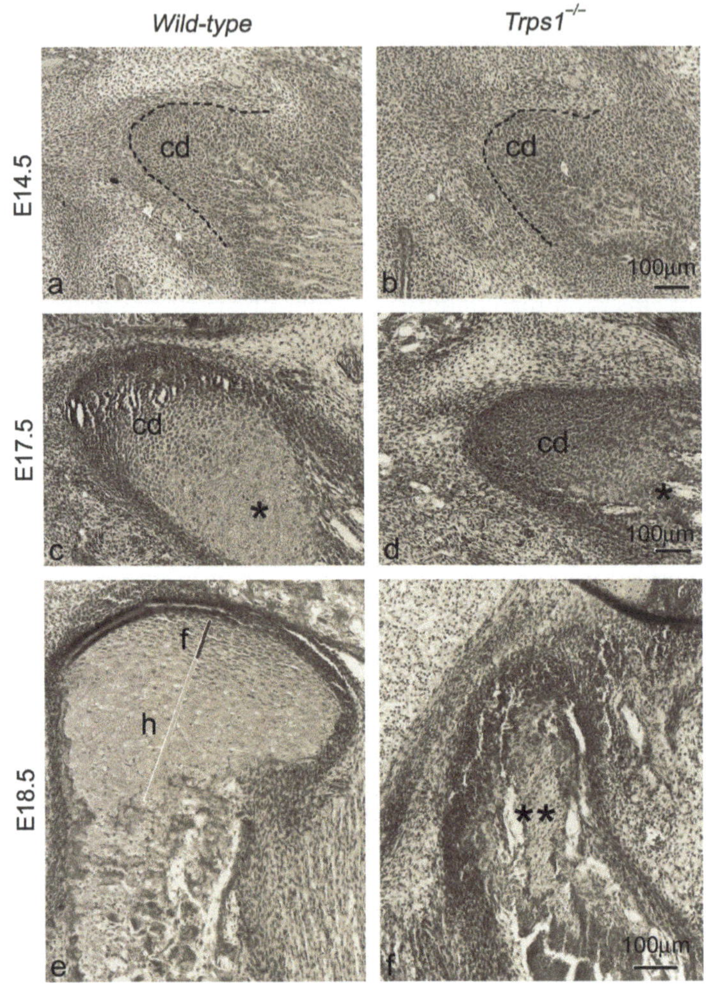

Figure 4 Development of the TMJ in *Trps1* knockout mice at E14.5 (A, B), E17.5 (C, D), and E18.5 (E, F). H–E staining of sagittal sections of the wild-type (A, C, E) and of mutants (B, D, F). Right is anterior in each section. (A, B) The dotted line shows the margin of the developing condyle. (C, D) The star indicates the region of maturating chondrocytes in the condyle. (E, F) Flattened (f) and hypertrophic (h) cell layers are indicated by the red line and green line, respectively (double stars in f indicate abnormally matured chondrocytes in the mutant condyle, cd condyle). *Used with permission of Michikami et al. (2012).* (A color version of this figure is located at the end of this chapter.)

bilateral fusion of the maxillary complex to the mandible, display a hypoplastic condyle and nonfunctional TMJ lacking a disc or joint cavities (Inman, Purcell, Kume, & Trainor, 2013). Although marker genes such as *Sox9*, *aggrecan*, and *ihh* were similar in expression between control and mutant condyles, *Col 10*-expressing chondrocytes constituted a greater proportion of the cartilage in mutant mice. Likewise, conditional inactivation of the homeobox gene *Shox2* in neural crest cells results in a delayed condensation of cells forming the future condyle, accompanied by downregulation of *Sox9, Runx2, Osterix,* and *ihh* in the condensation (Gu et al., 2008).

3.3 Postnatal Effects

Conditional inactivation of genes has been used to study the effects on postnatal growth and maturation of the MCC. Mice in which *ihh* knockout was conditionally induced with tamoxifen (Fig. 5) displayed disruptions in the zonal architecture of the MCC, with greatly reduced zones of prechondroblastic and early chondrocytic cells, as well as reduced *Sox9* and *Col2* expression and proliferation (Ochiai et al., 2010), suggesting that *ihh* may be very important for the postnatal growth of the TMJ. In limb cartilage, parathyroid hormone-related protein (*PTHrP*) forms a negative feedback loop with *ihh*, with PTHrP maintaining the chondrocytes in a proliferative, less differentiated state. Mice with a constitutively active *PTHrP* receptor in cartilage also show postnatal TMJ abnormalities. This feedback loop appears to work similarly in the prechondroblasts of the MCC as it does in proliferative chondrocytes of the growth plate (Shibukawa et al., 2007), but in this instance, the condylar cartilage is mostly composed of immature chondrocytes and fibroblastic cells, with only a few islands of hypertrophic cartilage (Tsutsui, Riminucci, Holmbeck, Bianco, & Robey, 2008). Inactivation of *Ndst1*, which catalyzes the constituent sulfation of heparan sulfate proteoglycans (HS-PGs) resulted in TMJ agenesis in severely affected mice, but mildly affected mutants at E18.5 exhibited a thicker prechondroblastic cell layer with increased proliferation compared to wild-type animals (Yasuda et al., 2010). Although the *ihh* expression was unaffected in mutants, the topography of *ihh* signaling activity, as indicated by its receptor *Ptch,* was much broader than in wild-type animals. The authors suggest that this finding reflects a more extensive field of *ihh* diffusion allowed by the lack of *ihh* binding to the defective HS-PGs in the mutant condyles.

Like disruption of *BMPr1a* in developing mice, conditional deletion of *BMPr1a* in cartilage of early postnatal mice has pronounced effects. In 4-week-old mice, overall mandibular length was significantly reduced and the extent of the condylar cartilage was drastically attenuated along with cell proliferation and Sox 9 immunoreactivity (Jing, Hinton, Mishina, et al., 2014). In addition, a virtual cessation of endochondral bone formation was noted deep to the MCC, as reflected by a complete lack of

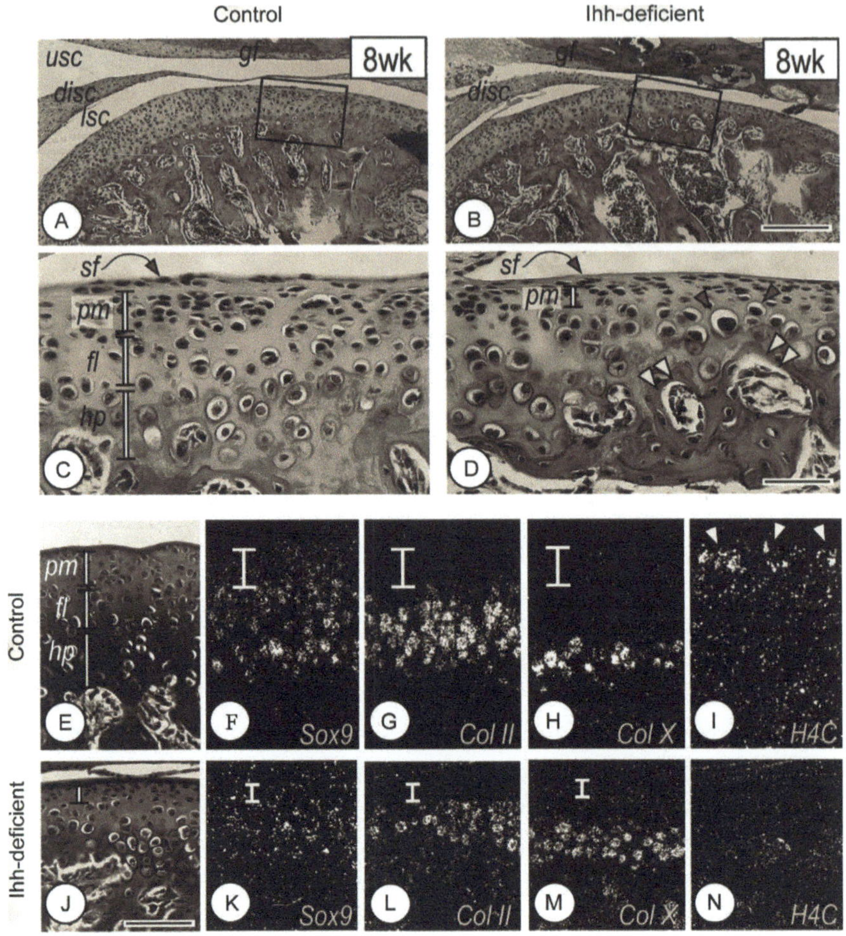

Figure 5 Mandibular condylar cartilage organization is abnormal in *Ihh*-deficient mice. H&E-stained parasagittal sections from 8-week-old control (A, C) and *Ihh*-deficient (B, D) condyles. Control (E) and ihh-deficient (J) condylar cartilage stained with safranin O/fast green show the reduced extent in cKO mice. Serial parasagittal sections from control (F–I) and *Ihh*-deficient condyles (K–N) processed for *in situ* hybridization of: *Sox9* (F, K), *collagen II* (*Col II*) (G, L), collagen X (H, M), and *histone H4C* (*H4C*) (I, N) demonstrate zonal disruption of gene expression and proliferation in cKO mice. Vertical bars in C and E indicate zonal organization in control condyles. *sf*, superficial layer; *pm*, polymorphic zone; *fc*, flattened chondrocyte zone; *hc*, hypertrophic zone. *Used with permission of Ochiai et al. (2010)*. (A color version of this figure is located at the end of this chapter.)

safranin-O-stained "cartilage remnants" in the subchondral bone and reduced VEGF immunoreactivity. Interestingly, immunoreactivity for Osterix (Osx), a transcription factor that is essential for osteoblast differentiation and bone mineralization (Nakashima et al., 2002) and is also found in cartilage, was greatly reduced in *BMPr1a*-null mice. When the same investigators examined MCC morphology in mice with conditional knockout of *Osx* in cartilage using Aggrecan-Cre, they found some similarities to and some differences from the *BMPr1a*-null mice. In 4-week-old mutant mice, the MCC was greatly thickened, due to a threefold greater thickness of hypertrophic chondrocytes (Jing, Hinton, Jing, et al., 2014). Moreover, the subchondral bone phenotype was very similar to the *BMPr1a*-null mice, with reduced TUNEL staining and almost no immunoreactivity for VEGF, TRAP, or MMP-13. In 5-month-old *Osx*-mutant mice, radiographic and micro-CT images of the cKO condyles demonstrate a pronounced radiopaque region separated by a radiolucent gap from the bone of the ramus (Fig. 6). This tissue comprising the "gap" contained cells that resembled chondrocytes and was positive for Col 2, but was also positive for von Kossa and osteopontin staining, suggesting that it is calcified cartilage. The authors suggest that Osx may be important for coupling terminal chondrogenic differentiation to endochondral ossification.

A variety of cellular processes, including cartilage growth and endochondral bone formation, are regulated by fibroblast growth factors (FGFs) that signal via four receptors (Ornitz, 2005). Three of the FGF receptors (*fgfr1*, *fgfr2*, and *fgfr3*) are expressed in different layers of the condylar cartilage (Molteni, Modrowski, Hott, & Marie, 1999; Purcell et al., 2012). Mice with a knock-in mutation in *Fgfr3* that greatly enhances the receptor's affinity to its ligands display a set of alterations in the TMJ structure and growth beginning at postnatal day 21 (Yasuda et al., 2012). Condyle size, as well as total cartilage thickness, was reduced in the mutant mice, with diminished proliferation, *Sox 9* and *Col 10* expression, and subchondral bone trabeculation. Reduced proliferation and downregulation of *Ptch*, the receptor for *ihh*, and other components of the *ihh* feedback system was noted following treatment of mutant explants with ligands for *Fgfr3*. Thus, the authors suggest that *Fgfr3* and *ihh* may be antagonists in the regulation of proliferation and chondrocyte maturation in the MCC. The Sprouty genes *Spry1* and *Spry2* also encode intracellular signaling inhibitors that affect FGF signaling. In the *Spry1(−/−)/Spry2(−/−)* mice, the condylar cartilage was 50% smaller in length and 25% smaller in width compared to wild-type animals, albeit with a normal histological structure (Purcell et al., 2012).

3.4 Degenerative Changes

Several recent studies have focused on genes that, if disrupted, may trigger degenerative changes at the MCC. Ishizuka and colleagues (2014) have

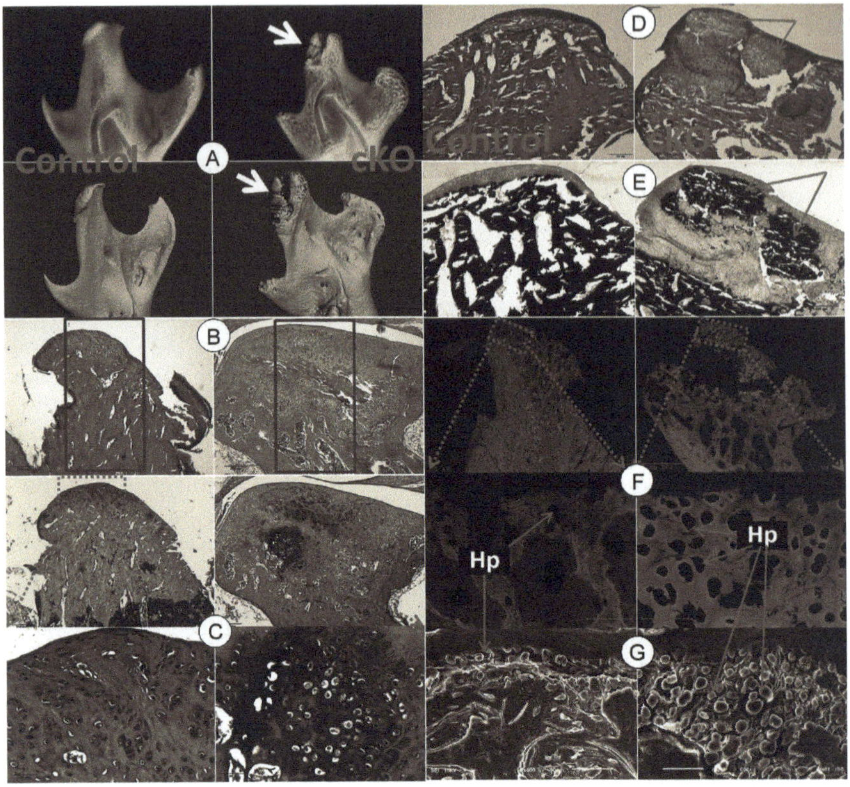

Figure 6 Large calcified cartilage masses are present in condylar cartilage from 5-month-old Osx-cKO mice. (A) Radiographic and μ-CT images showed that the surface of the calcified region in the mandibular condyle of cKO mice was separated from the underlying bone of the condylar neck by a radiolucent gap. (B, C) (upper images) Discrete areas of safranin-O-stained cartilage were present in the MCC of cKO mice, separated by stained regions similar to those of the underlying bone; (lower images) hypertrophic chondrocytes were more numerous in cKO condyles and sometimes clustered. Goldner (D) and von Kossa (E) staining showed large calcified areas within the MCC in cKO mice. (F, G) Back-scattered SEM and acid-etch SEM indicated that the calcified areas in the mutant mice MCC were composed of hypertrophic chondrocytes. Hp, hypertrophic chondrocyte. *Used with permission of Jing, Hinton, Jing, et al. (2014).* (A color version of this figure is located at the end of this chapter.)

demonstrated that SAMP8 mice, which develop early-onset osteoarthritic changes, display sharp reductions in the expression of *ihh* and its associated signaling molecules. This is accompanied by a significant decrease in the number of articular zone cells and attenuation of the expression of *lubrican*, a protein produced by articular zone or synovial cells and secreted into the joint space. SAMP8 mice with an induced malocclusion had more rapid manifestation of these changes. Another study showed that mice deficient in the *Proteoglycan 4 (Prg4)* gene, which is responsible for *lubrican* production, display significant alterations in all components of the TMJ (Koyama et al., 2014). By 3 months of age, *Prg4*(−/−) mice displayed a greatly thickened MCC, and by 6 months of age the articular zone had lost adherence to the underlying cartilage and the subchondral bone had developed an increased marrow space volume. These changes were mirrored by a decrease in *Col 1* and an increase in apoptosis in articular zone cells of the MCC, coupled with expanded zones of cells expressing *Col 2* and *Col 10*. An increase in *Col 2*-expressing cells was noted in the articular layer, and the synovial membrane cells became hyperplastic. Still another investigation has shown that mice with constitutive activation of beta catenin signaling develop TMJ changes that resemble those seen in osteoarthritis (Wang et al., 2014). From 1 to 6 months of age, mice-overexpressing *β-catenin* showed MCC alterations of increasing severity: decreasing cellularity, reduced Alcian blue and safranin O staining (suggestive of reduced proteoglycan content), and a progressive decrease in overall cartilage thickness. *Aggrecan* and *Col 2* expression were downregulated, and expression of cartilage-degrading proteinases like *MMP-13* and *Adamts5* were upregulated.

4. GENES AFFECTING THE MANDIBULAR FOSSA

4.1 Agenesis or Abbreviation

Less is known of the genetic effects on development of the mandibular fossa and articular disc because the focus is often on the condylar cartilage and the other structures are either unreported or not imaged. However, in studies where mandibular fossa status is reported, condylar cartilage agenesis is typically associated with agenesis or significant reduction in the mandibular fossa. Disrupted disc morphogenesis, often accompanied by dysfunctional development of joint cavities, is another corollary of condylar cartilage agenesis. In the *Sox9*-null mice described above (Wang et al., 2011), mandibular fossa formation was greatly abbreviated, and the articular disc appeared more diffuse than in the controls, with incomplete joint cavity formation by E18.5. Similarly, mandibular fossa formation was completely abolished in the Alk2-null mice (Dudas et al., 2004); while it was not imaged, it is highly likely that articular disc formation was absent also.

Since the mandibular fossa is derived from neural crest cells and does not express *Sox9*, Wang and coworkers (2011) were able to evaluate the effect of the missing condylar cartilage on fossa development. Although the mesenchymal condensation for the fossa formed normally at E14.5, its ossification (expression of *Runx2*) was delayed and it began to gradually regress, so that only a small bit of bone remained laterally at day E18.5. Thus, the mandibular fossa, although not congenitally absent, was unable to sustain its development in the absence of the condyle and actually regressed over time. Two mouse models served as "natural experiments" to further test this relationship. In one model, the developing condyle became dislocated from the developing mandibular fossa, mimicking the absence of the condyle, resulting in the arrested development of the fossa. In the second model, a greatly enlarged Meckel's cartilage displaced the condyle from its normal position relative to the mandibular fossa but the presence of Meckel's cartilage in articulation with the fossa permitted normal fossa development to continue (Wang et al., 2011). These two experiments suggest that normal fossa development depends on normal condyle development.

Another genetic study that describes an alteration in mandibular fossa morphogenesis is for mice deficient in both the genes *Sprouty 1* and *Sprouty 2* (*Spry1(−/−)*; *Spry2(−/−)*), which encode intracellular inhibitors of the receptor tyrosine kinase signaling pathways, including those triggered by FGFs (Purcell et al., 2012). In these mice, the mandibular fossa fails to develop (Purcell et al., 2012). *Spry1* and *Spry2* were expressed in the lateral pterygoid and temporalis muscles adjacent to the TMJ structures, while the FGF receptors *Fgfr1*, *Fgfr2*, and *Fgfr3* were expressed differentially in the periosteum, perichondrium, and MCC. The overall effects of the *Spry* deletion were specific to muscle: the temporalis and lateral pterygoid muscles were 48% and 69% larger, respectively, than in the control embryos. Interestingly, the articular disc formed normally, and the condyle was histologically normal but somewhat smaller than wild-type condyles. However, no condensation was observed at E14.5 for the mandibular fossa, and the fossa was absent except for a small lateral portion, similar to that seen in the study by Wang and colleagues (2011). The effect on fossa development was only observed when both *Spry1* or *Spry2* were inactivated, and inactivation of *Spry1* and/or *Spry2* in bone or cartilage had no TMJ phenotype. The authors suggested that the mandibular fossa phenotype may be a mechanical one due to an enlarged temporalis muscle hindering the condensations of the temporal bone from fusing to form the mandibular fossa in the mutant animals. A somewhat similar situation was observed in mice-lacking *Foxc1*: the MCC was smaller than usual with an increased proportion of hypertrophic chondrocytes while the fossa/eminence never formed. Interestingly, the masseter and temporalis were *reduced* in size in mutant animals (Inman et al., 2013).

When the condylar cartilage is present but defective in cellular morphology or smaller than normal, the mandibular fossa is sometimes affected in a parallel manner. In mice-lacking *BMPr1a*, the condyle is hypoplastic by E18.5 and the fibrocartilage that is normally present in the fossa is absent (Gu et al., 2014). Augmented BMPr1a signaling, on the other hand, results in lack of osteogenesis (*Runx2* expression) in the fossa/eminence area at E15.5 and E17.5, and by E18.5 the area degenerates along with the condyle (Gu et al., 2014).

4.2 Postnatal Effects

Although there is no true articular eminence in the mouse TMJ, the postnatal morphology of the mandibular fossa in growing mice resembles the tissue lining the eminence in species that possess one: a fibrocartilage that is an attenuated mirror image of the layers of the MCC. An articular layer overlies a thin prechondroblastic layer (Fig. 7), deep to which is a layer of Col 2-positive secondary chondrocytes (but not hypertrophic chondrocytes), and finally the bone of the articular eminence deep to that (Hinton & Carlson, 1983; Hinton & McNamara, 1984). However, most reports of genetic regulation of TMJ postnatal growth do not describe changes in this area. When they do, the effect on the eminence is often similar to that seen in the tissue layers of the MCC. *FgfR3*^{P244R} mutant mice, which displayed reduced thickness of the MCC with decreased *Col 2* and *ihh*

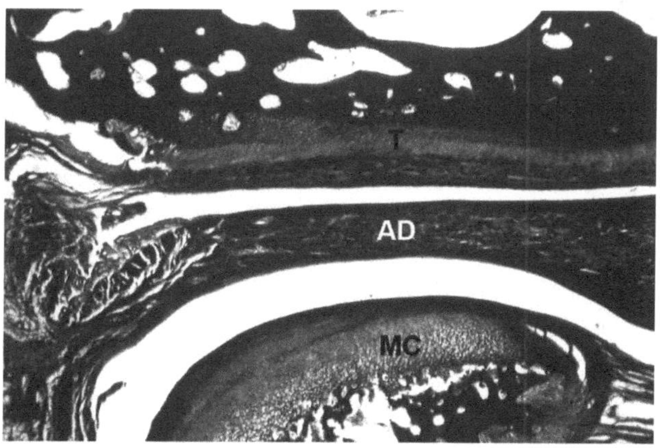

Figure 7 Sagittal section through temporomandibular joint of growing monkey showing mandibular condyle (MC), temporal bone (T), and articular disc (AD). Note that the bluish stain characterizing cartilage is present deep to the perichondrium on the temporal bone as well as the condyle. Masson trichrome. (A color version of this figure is located at the end of this chapter.)

expression, also showed a reduction in *Col2* in the eminence chondrocytes (Yasuda et al., 2012). Similarly, *Prg4(−/−)* mice exhibited hyperplasia of the articular eminence secondary cartilage, in similarity to the increased thickness and expansion of the *Col2* and *Col 10*-positive regions of the MCC (Koyama et al., 2014).

5. GENES AFFECTING THE ARTICULAR DISC AND JOINT CAVITATION

5.1 Disc Agenesis and Lack of Joint Cavity Formation

The articular disc develops from a third mesenchymal condensation interposed between the condensations that will become the MCC and the mandibular fossa. A condensation of cells in this intermediate region is discernible by E16.5, and gradually assumes the shape of a disc. Simultaneously, small clefts in the mesenchyme appear between the forming disc and the temporal bone. The initial coalescence of these clefts forms the superior joint cavity (space); the inferior joint space between the disc and condyle is typically present by E17.5 (Gu et al., 2008). The earliest descriptions of disruption in articular disc formation were in studies of mice lacking the Indian hedgehog (*ihh*) gene (Shibukawa et al., 2007) or one of its downstream effectors such as Smoothened (*Smo*) or the *Gli* family of transcription factors (Purcell et al., 2009). In *Ihh*-null mice (Shibukawa et al., 2007), the disc failed entirely to develop, and joint cavitation did not occur, leaving a thin layer of mesenchyme separating the condyle and mandibular fossa. The disc did form in mice having disrupted downstream effectors of *ihh* signaling, but joint cavitation was incomplete (Purcell et al., 2009). The authors concluded that *ihh* signaling is required not only to initiate disc morphogenesis but also to promote disc maturation and cavitation at a later stage of development.

Another model involving a gene that interacts with *ihh* exhibited agenesis of the articular disc. Mice deficient in *Trps1* show a nearly identical phenotype to that in *ihh*(−/−) mutants, i.e., the complete absence of a disc and joint cavities (Michikami et al., 2012). As previously noted, *Trps1* interacts with *ihh* signaling (Wuelling et al., 2009). Since *Trps1* is also expressed in the mesenchyme where the presumptive disc will form (Michikami et al., 2012), it is perhaps not surprising that in both *Trps*(−/−) and *ihh*(−/−) mice the articular disc never forms.

These models suggest that *ihh*-related signaling may be critical for disc formation. Since *ihh* is secreted by chondrocytes, it may be that the failure of the condylar cartilage to form in the *Sox9*(−/−) mice results in a lack of *ihh* that partly explains the poorly formed disc and lack of definitive joint cavities in these mice (Wang et al., 2011). Similarly, a model in which mice exhibit a wide open mouth due to skin abnormalities caused by ectopic activity of canonical Wnt signaling (Wang et al., 2011) results in a regression of the mandibular fossa and no articular disc. The authors postulated that the increased distance between the condyle and the disc condensation occasioned by the open mouth posture could inhibit diffusible factors that would normally mediate disc formation. It is possible that ihh could be one such diffusible factor. Mice with conditional inactivation of the homeobox gene *Shox2* in cranial neural crest cells display retarded appearance of TMJ structures, including the articular disc and joint cavitation (Gu et al., 2008). The *Shox2(−/−)* mice, which showed a considerably reduced expression of both *Runx2* and *Sox9*, also exhibited reduced *ihh* expression in the MCC, which may play a role in the disc/cavitation dysmorphies.

The potentially prominent role of *ihh* in regulating disc formation and joint cavitation is also supported by work of Gu and coworkers (2014) in *BMPr1a*-null mice. These mice were characterized by failure of a putative disc-like structure to separate from the MCC surface, as well as the persistence of interzone cells where cavitation was to occur. A sharp downregulation of *ihh* in the developing MCC was also present. The authors suggest that *BMPr1a* exerts its effects on TMJ morphogenesis via its regulation of *ihh*.

5.2 Postnatal Effects

Considerably less has been documented on disc morphology in postnatal mice. In *ihh*-null mice, partial adherence of the disc to the condylar surface was observed in both P4 and 8-week-old mice, suggesting that disc defects observed in embryological *ihh(−/−)* mice persist well into postnatal life (Ochiai et al., 2010). Interestingly, the authors found that *lubrican* expression was significantly upregulated in disc fibroblasts, synoviocytes, and the fibrocartilage covering the mandibular fossa/articular eminence, perhaps as a compensatory response. Two studies of possible genes regulating degenerative changes also noted articular disc abnormalities. The constitutively active Fgfr3 mice ($Fgfr3^{P244R}$) showed fusion of the articular disc to the temporal bone or condyle in adult mice (Yasuda et al., 2012). Moreover, *Prg4*-null mice (lacking *lubrican*) displayed a loss of the hourglass shape of the disc, resulting from thickening of the intermediate zone of the disc die to increased proliferation and chondrogenic differentiation (Koyama et al., 2014).

6. DISCUSSION

Although more detailed studies on existing genes and additional genetic models are needed, some broad conclusions can be drawn based on our current knowledge. Clearly, certain genes are critical for the specification of a chondrogenic path for the MCC prechondroblastic cells. In mice lacking *Runx2, Sox9, BMPr1a*, and *TGFBr2*, the condylar cartilage never forms or forms in highly abbreviated fashion. Although not examined in all instances, the temporal component of the TMJ fails to form or develops abnormally in mice lacking these genes. Still other genes likely act downstream of these "fate specifying" genes to regulate appropriate proliferation and differentiation of MCC cells as well as govern articular disc formation and the creation of the upper and lower joint cavities in the interzone between the developing condyle and mandibular fossa. Chief among these is Indian hedgehog (*ihh*), which is downregulated in developing *BMPr1a*-null mice (Gu et al., 2014) and also downregulated in postnatal mice with constitutively active *Fgfr3* signaling (Yasuda et al., 2012). Because the phenotypes in the *BMPr1a*-null and *ihh*-null mice are very similar, Gu and colleagues have suggested that *BMPr1a* regulates TMJ morphogenesis using *ihh* as a mediator. Yasuda et al. (2012) have suggested that postnatal TMJ growth may be regulated by antagonistic effects between *ihh* and *FGFr3*. Another possible regulatory hierarchy concerns the control of endochondral bone formation at the cartilage–bone interface. In mice-lacking *BMPr1a*, endochondral bone formation virtually ceases (Jing, Hinton, Mishina, et al., 2014). A very similar subchondral bone defect is seen in *Osterix*-null mice (Jing, Hinton, Jing, et al., 2014). Since Osterix is greatly reduced in *BMPr1a*(−/−) mice, it is possible that Osterix is a mediator for BMPr1a in regulating endochondral bone formation deep to the MCC. Unfortunately, the appropriate deletion and "rescue" experiments to evaluate these and other possible regulatory hierarchies have not yet been performed.

Because the regulatory mechanisms for TMJ development and growth were long assumed to be similar to those for the primary cartilages of the limb, we are only now beginning to realize that differences may in some instances exist for the secondary cartilage at the TMJ. However, there exist clear-cut differences in gene expression between the MCC and limb cartilages for which the significance is currently unknown. Examples include: the rapid chondrocyte hypertrophy that occurs in the MCC as reflected by the greatly overlapping expression of *Col1*, *Col2*, and *Col10* (Shibata, Fukada,

Suzuki, & Yamashita, 1997); the absence in the MCC of the Gdf5 and Gdf6 proteins seen in the joint cavities of the developing limb (Purcell et al., 2009); and differences in osteopontin and bone sialoprotein expression (Gu et al., 2008). In addition, qualitatively different responses to gene inactivation in MCC and limb cartilage have been documented. As an example, the phenotype for the *Osterix* conditional knockout mice in growth plate (Suppl. fig. 3 in Jing, Hinton, Jing, et al., 2014) was minimal in comparison to the striking phenotype in the TMJ (Jing, Hinton, Jing, et al., 2014), suggesting that cartilage-produced *Osx* may be of greater importance for regulating subchondral bone formation during MCC growth than in limb cartilage. Similarly, Gu and colleagues (2014) noted that the MCC in *BMPr1a*-null mice displayed *delayed* terminal hypertrophy, while in growth plate *BMPr1a*-null mice exhibit *accelerated* hypertrophy. Phenotypes unique to the TMJ were also noted for *Prg4*(−/−) mice (Koyama et al., 2014) and *Shox2*(−/−) mice (Gu et al., 2008). Clearly, there is a need for a more systematic comparison of the developmentally different cartilages in the TMJ and limbs.

Relatively few studies have contrasted the effects of genetic alteration during embryological development of the TMJ with its effects during TMJ postnatal growth. A laudable and illuminating example of this comparison is the series of studies documenting deficiency of *ihh* and its signaling partners on prenatal development (Purcell et al., 2009; Shibukawa et al., 2007) and postnatal growth (Ochiai et al., 2010). Likewise, more studies contrasting the transgenic effects of loss-of-function or gain-of-function for a single gene similar to the recent study on *BMPr1a* by Gu and colleagues (2014) would also greatly enhance our understanding of TMJ morphogenesis and growth. Finally, although our understanding to date suggests that the condylar cartilage is of primary importance in the developmental hierarchy (i.e., the fossa and disc do not form if the MCC is absent), many studies fail to describe or include images of the fossa and disc that could aid in our understanding of their genetic control.

We now appreciate that the TMJ, long considered "dentistry's joint," exhibits morphological differences with limb cartilages like articular cartilage and growth plate, differences that are the result of a very different developmental history. Recent studies of the genetic regulation of TMJ morphogenesis and growth (and lately of its degeneration) have identified a patchwork of interacting and hierarchical players that usually, but not always, resemble those that we know from limb cartilages. More systematic studies of these genes and others yet to be discovered will deepen our understanding of the causes of dentofacial deformities, and aid in the fabrication of TMJ components to replace those that are damaged or diseased.

ACKNOWLEDGMENTS

I am greatly appreciative of the willingness of Drs. Eiki Koyama, Makoto Abe, and Jerry Feng to allow me to use images from their work in this review. The research performed in Dr. Hinton's lab was supported by National Institutes of Health (NIH) grant DE015401. The work performed in Dr. Feng's lab was supported by his NIH grants DE018486 and R56DE022789.

REFERENCES

Anthwal, N., Chai, Y., & Tucker, A. S. (2008). The role of transforming growth factor-β signaling in the patterning of the proximal processes of the murine dentary. *Developmental Dynamics*, *237*, 1604–1613.

Beresford, W. A. (1981). *Chondroid bone, secondary cartilage, and metaplasia*. Baltimore: Urban and Schwartzenberg.

Bouvier, M., & Hylander, W. L. (1984). The effect of dietary consistency on gross histologic morphology in the craniofacial region of young rats. *The American Journal of Anatomy*, *170*, 117–126.

Carlson, D. S., McNamara, J. A., Jr., Graber, L. W., & Hoffman, D. H. (1980). Experimental studies of growth and adaptation of TMJ. In W. B. Irby (Ed.), *Current advances in oral surgery* (pp. 28–77). St. Louis: Mosby.

Crompton, A. W. (1985). Development and adaptation of the temporomandibular Joint. In D. S. Carlson, J. A. McNamara, Jr., & K. A. Ribbens (Eds.), *Developmental aspects of temporomandibular joint disorders* (pp. 1–18). Ann Arbor: Center for Human Growth and Development, University of Michigan.

Dudas, M., Sridurongrit, S., Nagy, A., Okazaki, K., & Kaartinen, V. (2004). Craniofacial defects in mice lacking Alk2 in neural crest cells. *Mechanisms of Development*, *121*, 173–182.

Duterloo, H. S., & Wolters, J. M. (1971). Experiments on the significance of articular function as a stimulating chondrogenic factor for the growth of secondary cartilages of the rat mandible. *Transactions. European Orthodontic Society*, 103–115.

Engelsma, S. O., Jansen, H. W. B., & Duterloo, H. S. (1980). An in vivo transplantation study of growth of the mandibular condyle in a functional position in the rat. *Archives of Oral Biology*, *25*, 305–311.

Fell, H. B., & Canti, R. G. (1934). Experiments on the development *in vitro* of the avian knee-joint. *Proceedings of the Royal Society of London B*, *116*, 316–351.

Fukuoka, H., Shibata, S., Suda, N., Yamashita, Y., & Komori, T. (2007). Bone morphogenetic protein rescues the lack of secondary cartilage in Runx2-deficient mice. *Journal of Anatomy*, *211*, 8–15.

Glineberg, R. W., Laskin, D. M., & Blaustein, D. I. (1982). The effects of immobilization on the primate temporomandibular joint. A histologic and histochemical study. *Journal of Oral and Maxillofacial Surgery*, *40*, 3–8.

Gu, S., Wei, N., Yu, L., Fei, J., & Chen, Y. P. (2008). Shox-2-deficiency leads to dysplasia and ankylosis of the temporomandibular joint in mice. *Mechanisms of Development*, *125*, 729–742.

Gu, S., Wu, W., Liu, C., Yang, L., Sun, C., Ye, W., et al. (2014). BMPR1A mediated signaling is essential for temporomandibular joint development in mice. *PLoS One*, *9*(8), e101000. http://dx.doi.org/10.1371/journal.pone.0101000.

Habib, H., Hatta, T., Udagawa, J., Zhang, L., Yoshimura, Y., & Otani, H. (2005). Fetal jaw movement affects condylar cartilage development. *Journal of Dental Research, 84*, 474–479.
Hall, B. K. (1972). Immobilization and cartilage transformation into bone in the embryonic chick. *The Anatomical Record, 173*, 391–404.
Hinton, R. J. (1988). Response of the intermaxillary suture cartilage to alterations in masticatory function. *The Anatomical Record, 220*, 376–387.
Hinton, R. J., & Carlson, D. S. (1983). Histological changes in the articular eminence and mandibular fossa during growth of the rhesus monkey (Macaca mulatta). *The American Journal of Anatomy, 166*, 99–116.
Hinton, R. J., & McNamara, J. A., Jr. (1984). Temporal bone adaptations in response to protrusive function in juvenile and young adult rhesus monkeys (Macaca mulatta). *European Journal of Orthodontics, 6*, 155–174.
Hylander, W. L. (1986). Loading patterns in the temporomandibular joint. In L. W. Graber (Ed.), *Orthodontics: State of the art, essence of the science* (pp. 48–58). St. Louis: Mosby.
Inman, K. E., Purcell, P., Kume, T., & Trainor, P. A. (2013). Interaction between Foxc1 and fgf8 durig mammalian jaw patterning and in the pathogenesis of sygnathia. *PLoS Genetics, 9*(12), e1003949. http://dx.doi.org/10.1371/journal.pgen.1003949.
Ishizuka, Y., Shibukawa, Y., Nagayama, M., Decker, R., Kinumatsu, T., Saito, A., et al. (2014). TMJ degeneration in SAMP8 mice is accompanied by deranged Ihh signaling. *Journal of Dental Research, 93*, 281–287.
Iwasaki, L. R., Crosby, M. J., Marx, D. B., Gonzalez, Y., McCall, W. D., Jr., Ohrbach, R., et al. (2010). Human temporomandibular joint eminence shape and load minimization. *Journal of Dental Research, 89*, 722–777.
Jing, J., Hinton, R. J., Jing, Y., Liu, Y., Zhou, X., & Feng, J. Q. (2014). Osterix couples chondrogenesis and osteogenesis in postnatal condyle growth. *Journal of Dental Research, 93*, 1014–1021.
Jing, J., Hinton, R. J., Mishina, Y., Liu, Y., Zhou, X., & Feng, J. Q. (2014). Critical role of Bmpr1a in mandibular condyle growth. *Connective Tissue Research, 55*(Suppl. 1), 73–78.
Koyama, E., Ochiai, T., Rountree, R. B., Kingsley, D. M., Enomoto-Iwamoto, M., Iwamoto, M., et al. (2007). Synovial joint formation during mouse limb skeletogenesis: Roles of Indian hedgehog signaling. *Annals of the New York Academy of Sciences, 1116*, 100–112.
Koyama, E., Saunders, C., Salhab, I., Decker, R. S., Chen, I., Um, H., et al. (2014). Lubricin is required fore the structural integrity and post-natal maintenance of TMJ. *Journal of Dental Research, 93*, 663–670.
Merida-Velasco, J. R., Rodriguez-Vasquez, J. F., Merida-Velasco, J. A., Sanchez-Montesinos, I., Espin-Ferra, J., & Jimenez-Collado, J. (1999). Development of the human temporomandibular joint. *The Anatomical Record, 255*, 20–33.
Michikami, I., Fukushi, T., Honma, S., Yoshioka, S., Itoh, S., Muragaki, Y., et al. (2012). Trps1 is necessary for normal temporomandibular joint development. *Cell and Tissue Research, 348*, 131–140.
Mitrovic, D. (1982). Development of the articular cavity in paralyzed chick embryos and in chick limb buds cultured on chorioallantoic membranes. *Acta Anatomica, 113*, 313–324.
Mizoguchi, I., Nakamura, M., Takahashi, I., Kagayama, M., & Mitani, H. (1990). An immunohistochemical study of localization of type I and type II collagens in mandibular condylar cartilage compared with tibial growth plate. *Histochemistry, 93*, 593–599.
Molteni, A., Modrowski, D., Hott, M., & Marie, P. J. (1999). Differential expression of fibroblast growth factor receptor-1, -2, and -3 and syndecan-1, -2, and -4 in neonatal rat mandibular condyle and calvaria during osteogenic differentiation in vitro. *Bone, 24*, 337–347.

Mori-Akiyama, Y., Akiyama, H., Rowitch, D. H., & de Crombrugghe, B. (2003). Sox9 is required for determination of the chondrogenic cell lineage in the cranial neural crest. *Proceedings of the National Academy of Sciences, 100*, 9360–9365.

Nakashima, K., Zhou, X., Kunkel, G., Zhang, Z., Deng, J. M., Behringer, R. R., et al. (2002). The novel zinc finger-containing transcription factor osterix is required for osteoblast differentiation and bone formation. *Cell, 8*, 17–29.

Napierala, D., Sam, K., Morello, R., Zheng, Q., Munivez, E., Shivdasani, R. A., et al. (2008). Uncoupling of chondrocyte differentiation and perichondrial mineralization underlies the skeletal dysplasia in tricho-rhino-phalangeal syndrome. *Human Molecular Genetics, 17*, 2244–2254.

Ochiai, T., Shibukawa, Y., Nagayama, M., Mundy, C., Yasuda, T., Okabe, T., et al. (2010). Indian hedgehog roles in post-natal TMJ development and organization. *Journal of Dental Research, 89*, 349–354.

Oka, K., Oka, S., Hosokawa, R., Hosokawa, R., Bringas, P., Brockhoff, H. C., et al. (2008). TGF-β-mediated Dlx5 signaling plays a crucial role in osteo-chondroprogenitor cell lineage determination during mandible development. *Developmental Biology, 321*, 303–309.

Oka, K., Oka, S., Sasaki, T., Ito, Y., Bringas, P., Nonaka, K., et al. (2007). The role of TGF-β signaling in regulating chondrogenesis and osteogenesis during mandibular development. *Developmental Biology, 303*, 391–414.

Ornitz, D. M. (2005). FGF signaling in the developing endochondral skeleton. *Cytokine & Growth Factor Reviews, 16*, 205–2013.

Petrovic, A. G. (1972). Mechanisms and regulation of mandibular condylar growth. *Acta Morphologica Neerlando-Scandinavica, 10*, 25–34.

Petrovic, A. G., Stutzmann, J. J., & Oudet, C. L. (1975). Control processes in the postnatal growth of the condylar cartilage of the mandible. In J. A. McNamara, Jr., (Ed.), *Determinants of mandibular form and growth* (pp. 101–154). Ann Arbor: Center for Human Growth and Development, University of Michigan.

Purcell, P., Jheon, A., Vivero, M. P., Rahimi, H., Joo, A., & Klein, O. D. (2012). *Spry1* and *Spry2* are essential for development of the temporomandibular joint. *Journal of Dental Research, 91*, 387–393.

Purcell, P., Joo, B. W., Hu, J. K., Tran, P. V., Calicchio, M. L., O'Connell, D. J., et al. (2009). Temporomandibular joint formation requires two distinct hedgehog-dependent steps. *Proceedings of the National Academy of Sciences, 106*, 18297–18302.

Radlanski, R. J., Renz, H., & Klarkowski, M. C. (2003). Prenatal development of the human mandible. *Anatomy and Embryology, 207*, 221–232.

Robinson, M. (1946). The temporomandibular joint: Theory of reflex controlled non-lever action of the mandible. *Journal of the American Dental Association (1939), 33*, 1260–1271.

Serrano, M. J., So, S., & Hinton, R. J. (2014). Roles of Notch signaling in mandibular condylar cartilage. *Archives of Oral Biology, 59*, 735–740.

Serrano, M. J., So, S., Svoboda, K. K., & Hinton, R. J. (2011). Cell fate mediators Notch and Twist in mouse mandibular condylar cartilage. *Archives of Oral Biology, 56*, 607–613.

Shibata, S., Fukada, K., Suzuki, S., Ogawa, T., & Yamashita, Y. (2002). In situ hybridization and immunohistochemistry of bone sialoprotein and secreted phosphoprotein 1 (osteopontin) in the developing mouse mandibular condylar cartilage compared with limb bud cartilage. *Journal of Anatomy, 200*, 309–320.

Shibata, S., Fukada, K., Suzuki, S., & Yamashita, Y. (1997). Immunohistochemistry of collagen types II and X, and enzyme-histochemistry of alkaline phosphatase in the developing condylar cartilage of the fetal mouse mandible. *Journal of Anatomy, 191*, 561–570.

Shibata, S., Suda, N., Suzuki, S., Fukoka, H., & Yamashita, Y. (2006). An *in situ* hybridization study of Runx2, Osterix, and Sox9 at the onset of condylar cartilage formation in fetal mouse mandible. *Journal of Anatomy, 208*, 169–177.

Shibata, S., Suda, N., Yoda, S., Fukuoka, H., Ohyama, K., Yamashita, Y., et al. (2004). Runx2-deficient mice lack mandibular condylar cartilage and have deformed Meckel's cartilage. *Anatomy and Embryology*, *208*, 273–280.

Shibata, S., Suzuki, S., Tengan, T., Ishii, M., & Kuroda, T. (1996). A histological study of the developing condylar cartilage of the fetal mouse mandible using coronal sections. *Archives of Oral Biology*, *41*, 47–54.

Shibukawa, Y., Young, B., Wu, C., Yamada, S., Long, F., Pacifici, M., et al. (2007). Temporomandibular joint formation and condyle growth require Indian hedgehog signaling. *Developmental Dynamics*, *236*, 426–434.

Silbermann, M., Reddi, A. H., Hand, A. R., Leapman, R. D., von der Mark, K., & Franzen, A. (1987). Further characterisation of the extracellular matrix in the mandibular condyle in neonatal mice. *Journal of Anatomy*, *151*, 169–188.

Sperber, G. H. (2001). *Craniofacial development*. Hamilton, Ontario: BC Decker.

Tsunekawa, N., Arata, A., & Obata, K. (2005). Development of spontaneous mouth/tongue movement and related neural activity, and their repression in fetal mice lacking glutamate decarboxylase 67. *The European Journal of Neuroscience*, *21*, 173–178.

Tsutsui, T. W., Riminucci, M., Holmbeck, K., Bianco, P., & Robey, P. G. (2008). Development of craniofacial structures in transgenic mice with constitutively active PTH/PTHrP receptor. *Bone*, *42*, 321–331.

Vinkka, H. (1982). Secondary cartilages in the facial skeleton of the rat. *Proceedings of the Finnish Dental Society*, *78*(Suppl. 7), 1–137.

Wang, M., Li, S., Xie, W., Shen, J., Im, H.-J., Holz, J. D., et al. (2014). Activation of β-catenin signaling leads to temporomandibular joint defects. *European Cells & Materials*, *28*, 223–235.

Wang, Y., Liu, C., Rohr, J., Liu, H., He, F., Yu, J., et al. (2011). Tissue interaction is required for glenoid fossa development during temporomandibular joint formation. *Developmental Dynamics*, *240*, 2466–2473.

Wilson, G. H. (1920). The anatomy and physics of the temporomandibular joint. *The Journal of the National Dental Association*, *7*, 414–420.

Wuelling, M., Kaiser, F. J., Buelens, L. A., Braunholz, D., Shivdasani, R. A., Depping, R., et al. (2009). Trps1, a regulator of chondrocytes proliferation and differentiation, interacts with the activator form of Gli3. *Developmental Biology*, *328*, 40–53.

Yasuda, T., Mundy, C., Kinumatsu, T., Shibukawa, Y., Shibutani, T., Grobe, K., et al. (2010). Sulfotransferase Ndst1 is needed for mandibular and TMJ development. *Journal of Dental Research*, *89*, 1111–1116.

Yasuda, T., Nah, H. D., Laurita, J., Kinumatsu, T., Shibukawa, Y., Shibutani, T., et al. (2012). Muenke syndrome mutation, FgfR3^{P244R}, causes TMJ defects. *Journal of Dental Research*, *91*, 683–689.

CHAPTER FIVE

Regulatory Mechanisms Driving Salivary Gland Organogenesis

Belinda R. Hauser, Matthew P. Hoffman[1]
Matrix and Morphogenesis Section, Laboratory of Cell and Developmental Biology, National Institute of Dental and Craniofacial Research, National Institutes of Health, Bethesda, Maryland, USA
[1]Corresponding author: e-mail address: mhoffman@mail.nih.gov

Contents

Abstract

Salivary glands develop as highly branched structures designed to produce and secrete saliva. Advances in mouse genetics, stem cell biology, and regenerative medicine are having a tremendous impact on our understanding of salivary gland organogenesis. Understanding how submandibular gland (SMG) initiation, branching morphogenesis, and cell differentiation occur, as well as defining the progenitor/stem cells and cell and tissue interactions that drive SMG development will help guide regenerative approaches for patients suffering from loss of salivary gland function. This review focuses on recent literature from the past 5 years investigating the regulatory mechanisms driving SMG organogenesis.

Current Topics in Developmental Biology, Volume 115
ISSN 0070-2153
http://dx.doi.org/10.1016/bs.ctdb.2015.07.029

1. INTRODUCTION

1.1 Function and Anatomy

The three pairs of major salivary glands, the submandibular (SMG), sublingual (SLG), and parotid glands (PG), produce ~90% of saliva secreted into the oral cavity; the other 10% is produced by minor glands distributed throughout the oral mucosa. The major functions of saliva include lubrication of the oral cavity to enable speaking and eating, digestion of food, antimicrobial activity, maintenance of mucosal integrity, and oral homeostasis. A reduction in salivary flow can cause clinical problems that include increased caries, xerostomia, oral infections, and difficulties with mastication, swallowing, and speech (Delli, Spijkervet, Kroese, Bootsma, & Vissink, 2014). The reader is directed to recent reviews on the physiological mechanisms of salivary secretion (Ambudkar, 2014; Catalan, Nakamoto, & Melvin, 2009) as well as the anatomy and function of the salivary glands (Holmberg & Hoffman, 2014; Tucker, 2007). Briefly, secretory acinar cells produce the isotonic primary salivary secretion. The acini are surrounded by myoepithelial cells, which are within a basement membrane. The myoepithelial cells may facilitate saliva secretion by contracting, although this has not been demonstrated experimentally. The primary salivary secretion is modified by ductal cells, which reabsorb ions resulting in hypotonic saliva, supersaturated with calcium and phosphate. The types of ducts, characterized by their morphology and histological appearance, are the intercalated, striated, granular, and excretory ducts. The epithelial compartment of the SMG is surrounded by mesenchymal stroma, which also contains immune cells, blood vessels, lymphatic vessels, fibroblasts, and nerves. Innervation is also essential for organogenesis, secretory function, and maintenance of acini, as denervation results in glandular atrophy. Both the sympathetic and parasympathetic branches of the autonomic nervous system innervate SMGs. The reader is referred to recent reviews of innervation of the developing salivary gland for details (Ferreira & Hoffman, 2013; Proctor & Carpenter, 2014).

2. REGULATION OF BRANCHING MORPHOGENESIS

2.1 SMG Initiation

SMG initiation and organogenesis involve complex interactions among multiple stem and/or progenitor cells. By definition, a stem cell is capable of both self-renewal and differentiation into all mature gland cell types. As stem cells differentiate into more committed progenitor cells they lose their ability to self-renew and become restricted to one lineage. However, no single salivary stem cell has been identified and there is currently no clear distinction between salivary stem and progenitor cells. Therefore, we will use the term stem/progenitor in this review. The reader is also directed to recent comprehensive reviews on SMG organogenesis and branching morphogenesis (Knosp, Knox, & Hoffman, 2012; Kwon & Larsen, 2015; Miletich, 2010; Patel & Hoffman, 2014).

A major research focus has been to identify epithelial stem/progenitor cells involved in murine SMG initiation. These stem/progenitors function within a niche or local microenvironment, which includes extracellular matrix (ECM) and mesenchymal, neuronal, and endothelial cells. During fetal development these other cell types may include stem/progenitor cells; therefore, different types of stem/progenitors may influence each other via multidirectional signaling networks involving both secreted factors and physical interactions. At embryonic day 11 (E11) in the mouse, the migrating neural crest cells form a loose mesenchymal aggregate beside a local thickening of the oral epithelium, which forms the SMG epithelial placode. In terms of the developmental origin of the salivary gland epithelium, lineage tracing with an endodermal-specific Sox17-cre suggests that all major salivary glands of the mouse are not endodermal, but likely ectodermal in origin (Rothova, Thompson, Lickert, & Tucker, 2012). At E12 the endothelial cells form a plexus within the surrounding mesenchyme, which begins to condense. The signals that cause mesenchymal condensation have not been identified. The neural crest also contains neuronal cell bodies that coalesce to form the parasympathetic submandibular ganglion (PSG). Gland initiation involves enlargement of the salivary epithelial placode, and a simultaneous invagination of the oral epithelium into the adjacent mesenchyme. This results in the placode forming a primary endbud on a stalk of oral epithelium. Importantly, the mesenchyme produces fibroblast growth factor 10 (Fgf10) which induces epithelial proliferation via its epithelial receptor FGFR2b (Kwon & Larsen, 2015). In the absence of Fgf10 the SMG epithelium does not develop (Fig. 1). The formation of the PSG and subsequent innervation occur in parallel with epithelial development. Recently, the developmental origin of the PSG was identified as peripheral Schwann cell precursors that migrate along the preganglionic nerves and form both the glial cells and the PSG neurons of the glands (Dyachuk et al., 2014; Espinosa-Medina et al., 2014). These neurons then

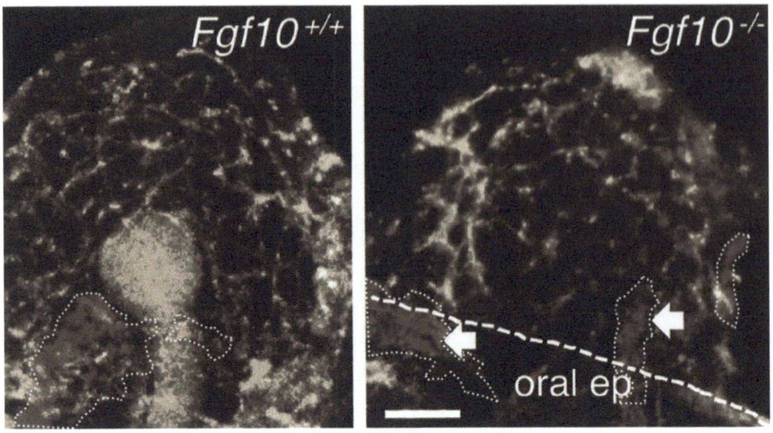

Figure 1 Fgf10 is essential for SMG epithelial development. Whole-mount immunostaining of an E12 wild-type (*Fgf10*$^{+/+}$) SMG and *Fgf10*$^{-/-}$ SMG mesenchyme. SMG development involves the formation of an epithelial endbud and primary duct (E-cadherin, green) within a condensed mesenchyme, which is surrounded by an endothelial cell plexus (PCAM, cyan blue) and the formation of a PSG (Tubb3, red). White dotted lines outline PSG, white dashed line is the oral epithelium (oral ep); arrows indicate nerves in the adjacent tongue and oral epithelium. Scale bar, 50 μm. *From Fig. 2B in Knosp et al. (2015).* (A color version of this figure is located at the end of this chapter.)

coalesce around the primary epithelial duct to form the PSG, establishing communication with the developing epithelium as it begins branching morphogenesis.

Recently four Wnt ligands were identified. They are mainly produced by the keratin (K)-5 expressing (K5+) progenitors in the primary salivary gland duct. These Wnts provide the signal to initiate gangliogenesis, in part by inducing neuronal proliferation and cell survival. This finding came unexpectedly from studying a mouse model of increased FGF signaling, in which two Sprouty (*Spry*) genes were deleted. Sproutys are key intracellular modulators of FGF signaling and act as negative-feedback antagonists (Tang, Marshall, McMahon, Metzger, & Martin, 2011). The genetic deletion of *Spry1* and *Spry2* (*Spry1/2DKO*) from the SMG epithelium resulted in a striking loss of the PSG and SMG innervation (Fig. 2), resulting in reduced branching morphogenesis (Knosp et al., 2015). During PSG formation it was shown that increasing FGF signaling reduced the expression of the Wnt signals in the epithelium. The K5+ progenitor cells mainly produced these Wnt signals that promoted neuronal survival and proliferation,

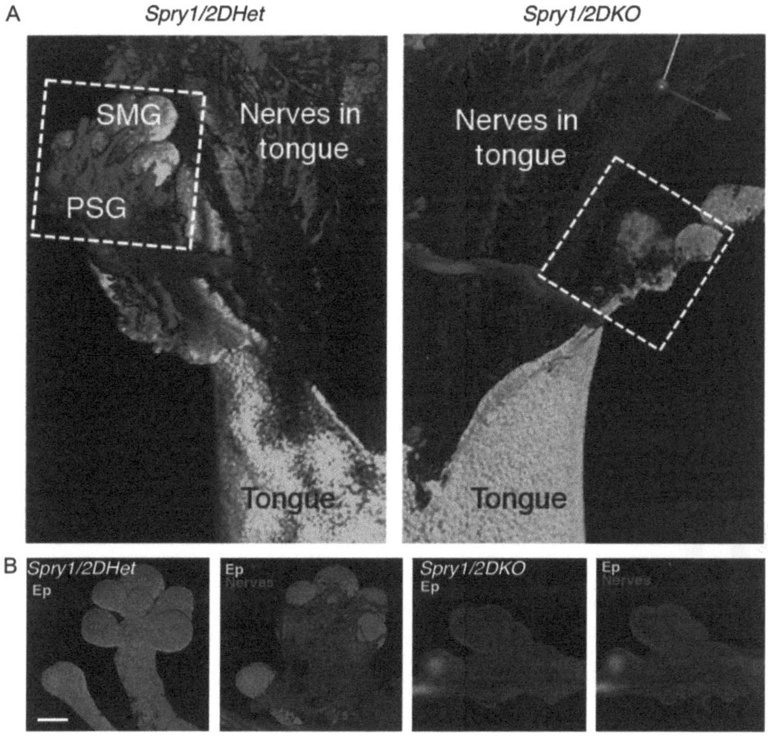

Figure 2 There is loss of the PSG and SMG innervation in the *Spry1/2DKO* mouse, although the sensory innervation of the tongue is not affected. (A) Whole-mount immunostaining for the epithelium (Ep, E-cadherin) and nerves (Tubb3) in *Spry1/2DHet* (control) and *Spry1/2DKO* tongues with SMGs attached. (B) The SMGs (green) and PSG (red) that are in white box are shown separately in the lower panels. Scale bar, 100 μm. *From Figure S1A in Knosp et al. (2015).* (A color version of this figure is located at the end of this chapter.)

PSG formation, and gland innervation. Inhibiting Wnt signaling or treatment with exogenous Fgfs disrupted PSG formation, the association of the PSG with the epithelium, and organ innervation. Wnt expression in the epithelium was inhibited by FGFR2b/MAPK signaling and promoted via neuregulin ErbB2/3/PI3K-dependent signaling. This provided a new mechanism by which the K5+ progenitor cells regulated their own innervation by integrating both FGFR2b and ErbB signaling to produce Wnts. Moreover, it is the balance of FGFR/ErbB signaling that regulates Wnts, which act on neuronal cells to promote survival and proliferation (Knosp et al., 2015).

2.2 Cleft Formation

After the formation of a primary duct, initial endbud, and PSG, the endbud begins the process of branching morphogenesis. This involves the processes of clefting, epithelial proliferation, migration, and differentiation, as well as innervation and vascular development. Endbud expansion and clefting with branch-point formation and duct elongation result in three to five endbuds forming by E13. Cleft formation requires several interrelated cellular processes, such as proliferation, migration, cell–cell adhesion, cell–ECM adhesion, ECM accumulation, and cellular contraction. The basement membrane separates the epithelium from the surrounding mesenchyme and is a specialized ECM containing laminins, collagen IV, proteoglycans, nidogen, and agrin (Hohenester & Yurchenco, 2013), as well as fibronectin. Accumulation of fibronectin during cleft initiation induces Btdb7, a factor made in the epithelium. Btdb7 induces the expression of Snail2 and suppresses E-cadherin levels (Onodera et al., 2010), which reduces columnar organization and cell-cell adhesion molecules in the cells of the outer epithelial layer, allowing cleft progression.

Contraction of the cytoskeleton also promotes progression of clefts. Rho-associated coiled-coil containing kinase (ROCK) regulates cytoskeletal contraction at discrete stages and inhibition of ROCK leads to ectopic clefting in *ex vivo* SMG culture. ROCK controls tissue organization and cell polarity via PAR-1b protein. ROCK regulates the transition of initiated clefts to a stabilized state, which is then able to undergo cleft progression, a proliferation-independent process (Daley et al., 2012). Cleft stabilization and progression occur through the stabilization of actin (Ray, Fanti, Macedo, & Larsen, 2014), where LIM-kinase (LIMK), a regulator of early and late-stage cleft formation and initiation, regulates both microfilaments and microtubules. LIMK-dependent regulation of the cytoskeleton controls fibronectin assembly and activation of β1 integrins. Furthermore, the microtubule assembly factor p25 regulates the stabilization and elongation of late-stage progressing clefts. In sum, multiple actin- and microtubule-dependent stabilization steps are controlled by LIMK and are required for cleft progression.

2.3 Migration and ECM Proteolysis

Cleft formation is coordinated with cell migration in the endbud. Using a single cell tracking technique, epithelial cell migration was shown to be highest in outer bud cells near the basement membrane, lower in the inner bud cells, and lowest in duct cells (Hsu et al., 2013). Inhibitors of integrin α6β1 and nonmuscle myosin II reduced the peripheral cell motility whereas inhibiting E-cadherin reduced inner bud motility. These findings suggest cell motility in different regions of the endbuds is dependent on different cellular mechanisms.

In addition, highly dynamic remodeling of ECM drives epithelial proliferation during branching morphogenesis. Membrane-type 2 matrix metalloproteinase (MT2-MMP)-dependent collagen IV proteolysis releases small collagen fragments called NC1 domains (Rebustini et al., 2009). These collagen fragments increase MT2-MMP expression and genes related to proliferation, by binding to β1 integrins and signaling via PI3K and AKT. The epithelium also produces heparin-binding epidermal growth factor (HBEGF), which increases MT2-MMP expression and further release of NC1 domains. Therefore, feedback from multiple protease-dependent pathways increases branching morphogenesis.

Remodeling of the basement membrane also controls local epithelial expansion. The basement membrane at the tip of the E13 endbud becomes perforated with small holes as rapid epithelial expansion occurs (Harunaga, Doyle, & Yamada, 2014). This structure appears as a mesh-like net and allows epithelial expansion while maintaining tissue integrity. The basement membrane is remodeled in a distal direction, appearing to accumulate around the secondary ducts. These local and global dynamics require both protease and myosin II activity, suggesting epithelial expansion requires proteolytic degradation and remodeling via actomyosin contractility during branching morphogenesis. Taken together, cleft initiation, stabilization, and progression are coordinated with cell migration and ECM remodeling during SMG branching morphogenesis.

2.4 Cell Differentiation

Branching morphogenesis is coordinated with the cytodifferentiation of the ductal and acinar compartments of the gland. Recently, it was reported that the transcription factor NFIB (nuclear factor IB) was involved in secretory cell differentiation during SMG development (Mellas et al., 2015). NFIB has clinical importance because a gene fusion of *NFIB* with *MYB* (*MYB-NFIB*) occurs in adenoid cystic carcinoma (ACC) of salivary glands (Stenman, 2013). ACC is the second most common salivary malignancy and more than 80% of patients with head and neck ACC die 10–15 years after diagnosis. SMG from *Nfib*$^{-/-}$ mice were hypoplastic at E18, although some branching morphogenesis did occur. The SMGs of *Nfib*$^{-/-}$ mice showed reduced apicobasal polarity, as measured by disrupted apical ZO-1 staining, had reduced ductal and acinar lumen formation, and appeared disorganized. The terminal tubules did not differentiate into secretory cells, as evidenced by a lack of staining for SMG-C, a proacinar cell secretory marker, and aquaporin 5, a water channel that is present in the apical membranes of acinar and intercalated duct cells. Thus NFIB is required for acinar cell differentiation and is important for lumen formation during mouse SMG development.

During branching morphogenesis, lumen formation occurs at the distal ends of epithelial branches and proximal microlumens coalesce in a distal direction to form a contiguous lumen. The interaction between FGF signaling and the canonical (beta-catenin dependent) and noncanonical Wnt signaling coordinates this processes. Using Axin2^{LacZ} reporter mice to show where Wnt signaling occurs, it was demonstrated that Wnt signaling appeared first in the mesenchyme at E12 and then at E14, switched to the ductal epithelium. Gain- and loss-of-function experiments showed that Wnts exert an inhibitory effect on salivary gland branching morphogenesis. Furthermore, endbuds do not have active Wnt signaling due to FGF-mediated inhibition. Thus, FGF signaling prevents lumenization of epithelial endbuds and slows down the lumenization of presumptive ducts. The mechanism involves reducing Cp2l1, a marker of duct differentiation, by repressing noncanonical Wnt5b expression and signaling via regulating sFRP1, a secreted Wnt inhibitor. Thus FGF signaling maintains undifferentiated endbud cells by inhibiting ductal fate and lumenization (Patel, Sharpe, & Miletich, 2011).

Similarly, core-binding factor beta (CBFb), a cotranscription factor that forms a heterodimer with Runx transcription factors, was shown to influence postnatal duct differentiation. CBFb is expressed in SMG ducts and conditional deletion of CBFb with a K14-cre decreases the size of the SMG and reduces saliva secretion in adult male mice (Islam et al., 2015). There was a loss of a specific ductal compartment, the granular convoluted tubules (GCT), with reduced expression of genes that are expressed in GCT, such as *Klk1*, *Ngf*, and *Egf*. GCT development is androgen-dependent, but circulating testosterone levels were not affected by CBFb deletion, suggesting that CBFb signaling regulates androgen receptor signaling pathway, not circulating testosterone levels. Therefore, Runx/CBFb-dependent transcription is required for the postnatal development of androgen-dependent GCT in the SMG. Together, these studies illustrate the critical roles transcription factors play in cell differentiation during gland development.

3. SMG INNERVATION

3.1 Parasympathetic-Epithelial Communication

Both the parasympathetic and sympathetic branches of the autonomic nervous system richly innervate the adult salivary gland and are critical for saliva secretion. The PSG releases acetylcholine (ACh), which activates the muscarinic receptors 1 and 3 (*Chrm1* and *Chrm3*) to stimulate fluid secretion. A recent review of the interactions between developing nerves and salivary glands provides a comprehensive background (Ferreira & Hoffman, 2013). Parasympathetic innervation occurs along the epithelium during SMG branching morphogenesis and sympathetic innervation occurs along the vasculature. Parasympathetic innervation is required for maintaining epithelial K5+ cells as a pool of undifferentiated progenitor cells for further development (Knox et al., 2010). Keratins are cytoskeletal proteins used as epithelial markers; K5 is used to label progenitor cells and K19 to label ductal cells. Removal of the PSG from the SMG in culture resulted in a striking decrease in the number of K5+ cells. This was dependent on ACh production by the PSG and muscarinic receptor M1 signaling in the K5+ cells. Furthermore, K5+ cell maintenance and differentiation into K19+ ductal cells were dependent on HBEGF/EGFR signaling (Knox et al., 2010). These data highlight the importance of the bidirectional communication between the PSG and the developing epithelium.

A recent advance in understanding parasympathetic-epithelial communication during organogenesis concerns the epithelial production of the neurotrophic factor neurturin (NRTN). NRTN increases PSG function and promotes neuronal survival and directional axon outgrowth. During SMG organogenesis, the PSG axons that extend along the ducts to envelop the endbuds respond to localized neurotropic cues. NRTN promotes innervation by binding to its receptor GFRα2 and the tyrosine kinase coreceptor RET, and signals via Src kinase. Gene targeting of RET, GFRα2, or NRTN in mice results in smaller PSG and reduced innervation of multiple organs. Using isolated PSG from SMGs, it was shown that NRTN reduced neuronal apoptosis and increased axon outgrowth and the expression of genes involved in parasympathetic function (Knox et al., 2013). Furthermore, antibodies blocking NRTN reduced parasympathetic nerve outgrowth and function, which reduced branching morphogenesis in intact SMG culture. The functions of NRTN on SMGs were also investigated after irradiation of the gland. In this case, NRTN reduced neuronal apoptosis and restored parasympathetic function, which helped the epithelium to regenerate. The hypothesis that NRTN may improve regeneration after irradiation in adult glands is currently under investigation.

Another report highlighting parasympathetic-epithelial crosstalk showed that parasympathetic innervation regulates ductal tubulogenesis of the epithelium (Nedvetsky et al., 2014). Vasoactive intestinal peptide (VIP) is produced by the nerves and promotes ductal growth and the formation of a contiguous ductal lumen. Isolated SMG epithelia treated with VIP in *ex vivo* culture formed lumens, which expanded via cAMP/PKA-dependent pathway. The lumen expansion was independent of apoptosis and involved the cystic fibrosis transmembrane conductance channel (CFTR), which is a cAMP-regulated chloride channel. Interestingly, ductal tubulogenesis did not require ACh/M1 signaling. In addition, analysis of the SMGs in neuregulin 1 (Nrg1) null mice, which have depleted innervation of multiple tissues, showed that they had aberrant ductal morphogenesis and reduced branching morphogenesis. A similar problem with ductal morphogenesis was also observed in the CFTR null SMGs. An emerging theme is the importance of multiple neuronal-derived factors on the branching epithelium. ACh, VIP, and Nrg1 regulate K5 + progenitor cell function and ductal tubulogenesis during SMG development. It is likely that other neuropeptides and regulatory factors produced by the nerves will influence organogenesis.

3.2 Sympathetic Innervation

The sympathetic nerves also stimulate secretion via the α and β adrenoreceptors, which increase fluid and protein-rich secretions, respectively (Patel & Hoffman, 2014). An autocrine Wnt5a-Ror signaling loop was shown to mediate sympathetic axon branching during SMG innervation (Ryu, Collins, Ho, Zhao, & Kuruvilla, 2013). Using a conditional knockout approach, the noncanonical Wnt5a was deleted with either Wnt1:cre or tyrosine hydroxylase:cre, the latter of which is specific for sympathetic nerves. Wnt5a, which is produced by the nerves, was involved in the autocrine signaling loop and target innervation. Deletion of Wnt5a reduced the extension and arborization of sympathetic fibers in the gland, but did not affect overall tissue patterning or the proliferation, migration, or differentiation of neuronal progenitors. Less is known about the function of the sympathetic nerves than the parasympathetic nerves during organogenesis. Further studies using genetic ablation or knockdown of genes required for sympathetic innervation may provide insight into the function of these nerves during development.

4. STEM/PROGENITOR CELLS

4.1 Investigating Stem/Progenitor Cells During Development

Identifying stem/progenitor cells present during fetal development and understanding how they function will impact approaches to regenerating adult glands. Epithelial progenitor cells have been isolated from adult murine SMGs using fluorescent-activated cell sorting (FACS) of the cell surface marker KIT and combinations of CD24, CD49f, CD133, and Sca1. These cells have been used to partly regenerate function in irradiated adult murine salivary glands (Lombaert et al., 2008; Nanduri et al., 2011). However, KIT labels a heterogeneous population of cells, and understanding how KIT functions during development and its role in cell fate decisions and regeneration are important research questions. KIT + cells are in epithelial endbuds during fetal development and are important for organogenesis. Epithelial *Kit* expression is upregulated by FGFR2b signaling, which is also essential for gland development (Lombaert et al., 2013). The combined effects of KIT and FGFR2b signaling via separate AKT and MAPK pathways, respectively, amplified expression of transcription factors downstream of FGFR2b signaling, including Sox10, Myc, Etv4, and Etv5. Importantly, the combined signaling increased the number of KIT + cells in the distal endbuds that also expressed K14 and Sox10. The distinct cell types within the KIT + population were characterized by their keratin expression. Whereas K5 usually pairs with K14 in other tissues, in SMG endbuds there are distinct populations of KIT+K14+ and KIT+K5+ cells (Fig. 3A). K8 was used as a pan-epithelial marker and K19 as a ductal marker. Genetic lineage tracing with a K14-cre confirmed that the K14 cells were multipotent and gave rise to progeny that included K5+ ductal cells, acinar cells, and myoepithelial cells (Fig. 3B). FACS analysis confirmed the KIT + cells were heterogeneous, and that there were distinct K14 + and K5 + populations, but a surprisingly large (56%) subset of the KIT+K8+ cells did not express K14, K5, or K19, and these cells localized between the basal layer and duct of the endbud (Fig. 3C). Furthermore, SMG with a signaling-defective form of KIT had reduced branching, demonstrating that the expansion of the distal KIT+K14+ progenitors was essential for branching morphogenesis (Lombaert et al., 2013). Surprisingly, the reduction in KIT also caused a loss of K5 + cells in the duct, due to a loss of innervation. Reduced innervation was due to reduced NRTN, which is produced by the KIT+ endbuds. Thus, the distal epithelial progenitor cells

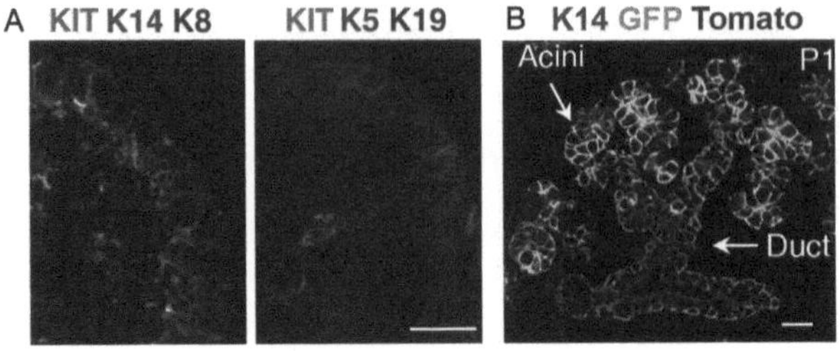

C

FACS (%) of E13 KIT+K8+ cells:

K14+	14±3
K5+	10±2
K5+K19+	7±2
K14+K5+	7±1
K19+	2±1
K14-K5-K19-	56±4

Figure 3 SMG endbuds contain distinct populations of KIT+K14+ and KIT+K5+ progenitors, and the K14+ progenitors are multipotent. (A) Immunostaining of KIT, K14, K8, and K19 in an E13 endbud. Pink cells coexpress K14 and K8 or K5 and K19, 1 μm optical sections. (B) Lineage tracing of K14 in a postnatal day 1 SMG from a K14-Cre x Rosa^{mTmG} mouse. K14 progeny cells (GFP, green) are throughout the gland and label acini, ducts, and myoepithelial cells. The image also shows the endogenous K14 (red and yellow) and non-K14 cells (Tomato, blue), 2 μm sections. Scale bar, 20 μm. (C) FACS analysis to quantitate the subpopulations of epithelial KIT+K8+ cells sorted by K5, K14, and K19 expression from E13 SMGs. Mean ± SEM; $n > 3$ biological samples. *From Figure 4 in Lombaert et al. (2013).* (A color version of this figure is located at the end of this chapter.)

secrete neurotrophic factors to stimulate neuronal innervation that maintains proximal epithelial progenitor cells.

Other markers that have been used to label progenitor cells in the developing SMG include Sox2, which is a putative stem cell marker in the sublingual gland. Lineage tracing during gland ontogenesis showed Sox2 progeny in the sublingual ducts and acinar cells of adult glands (Arnold et al., 2011). Sox2 is also expressed in some K5+ cells in the SMG duct early in development (Lombaert, Knox, & Hoffman, 2011), although further studies are required to investigate their role as stem/progenitors in the adult gland. Another progenitor population in the SMG is marked by Ascl3, a transcription factor in the ducts of SMGs. Ascl3+ cells are proliferating progenitors and lineage tracing showed they generate both ductal and acinar cells in the adult. However, genetic ablation showed that gland development occurred normally in the absence of Ascl3 progenitors, which were shown to be a separate population from the K5+ progenitor pool (Arany, Catalan, Roztocil, & Ovitt, 2011). We speculate that in the absence of one progenitor population, compensation by another can occur during development.

4.2 Postnatal Homeostasis and Label Retaining Cells

A recent report has challenged the current dogma that salivary gland homeostasis in the adult SMG is stem cell dependent (Aure, Konieczny, & Ovitt,

2015). In fact, few studies had addressed the extent to which a particular stem/progenitor cell type contributes to adult gland homeostasis. Using an inducible Mist1-Cre, which is acinar cell-specific, they genetically labeled differentiated adult acinar cells and monitored their progeny over a chase period of up to 3 months. Surprisingly, the results showed that acinar cell replacement occurs by acinar cell division, not by the differentiation of an unlabeled cell population from the ducts. Using a Mist1^{CreER} with Rosa26$^{Brainbow2.1}$ reporter mice to label individual cells and their progeny, Aure et al. (2015) beautifully showed proliferation and clonal expansion of differentiated acinar cells in all three pairs of major salivary glands (Fig. 4). Therefore, salivary gland acinar homeostasis is based on self-duplication of differentiated acinar cells. Interestingly, Mist1-cre labeling with a 6-month chase revealed that there was a significant number of single-labeled cells in the sublingual gland, and a subpopulation of these cells expresses Sox2. Since

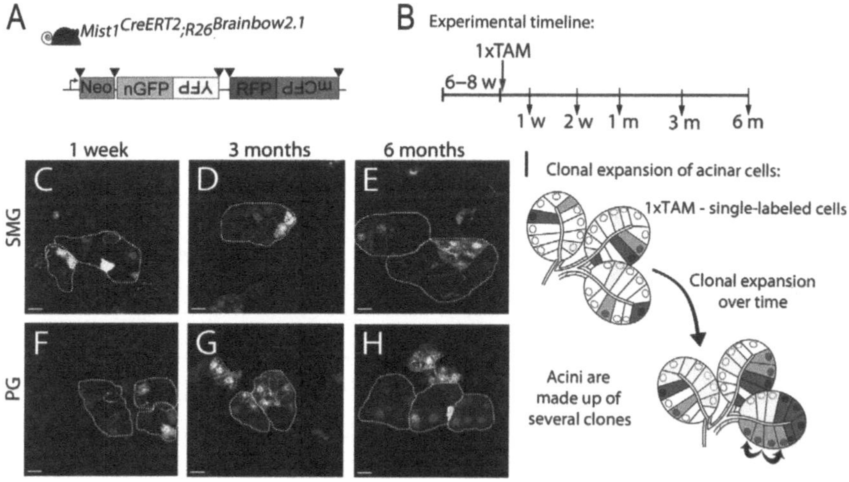

Figure 4 Acinar cell renewal occurs by self-duplication, resulting in clonal expansion of acinar cells in the SMG and PG. (A) The inducible Mist1-CreERT2 was used in combination with the R26$^{Brainbow2.1}$ reporter strain (schematic). Black triangles represent LoxP sites. (B) Adult mice (6–8 weeks old) were treated with a single dose of tamoxifen. Tissues were harvested at indicated time points. (C) Labeled cells in the SMG at 1 week following tamoxifen treatment. (D) Clonal expansion of labeled cells in the SMG at 3 months after tamoxifen treatment. (E) Acini comprised of several clones demonstrate tissue turnover 6 months after tamoxifen treatment. (F) Single-labeled cells in the PG 1 week following tamoxifen treatment. (G) Clonal expansion of labeled cells in the PG at 3 months. (H) Clonal analysis after a 6-month chase in the PG. (I) Model of acinar cell proliferation and clonal expansion in adult SMG and PG. *From Figure 2 in Aure et al. (2015).* (A color version of this figure is located at the end of this chapter.)

these Sox2+ cells appeared quiescent, it was concluded they are not likely involved in gland homeostasis but may play a role during regeneration following injury. An emerging concept is that the stem/progenitor cells involved in homeostasis may be different from those responding to damage of the gland. This is supported by studies on liver regeneration that show that hepatocytes are both maintained and regenerated independently from facultative stem cells (Schaub, Malato, Gormond, & Willenbring, 2014; Yanger et al., 2014).

In the search for a quiescent stem cell population in the adult gland, a recent study used a doxycycline-inducible histone 2B-green fluorescent protein (H2BGFP) to label K14 ductal cells (Kwak & Ghazizadeh, 2015). The K14 cells were genetically labeled with H2BGFP by doxycycline treatment throughout *in utero* gland development and for 2 weeks after birth, during postnatal gland development. The glands were then analyzed 12 weeks later to see if any H2BGFP-labeled cells remained, which would indicate they were quiescent stem cells. A quiescent adult K14 stem cell population was not detected during homeostasis, as the labeled cells had proliferated and lost their labels. A similar experiment labeled all cells in the SMG using a doxycycline-inducible ubiquitous driver, Rosa26. After a 12-week chase, again no evidence was found for a quiescent H2BGFP+ LRC population in an undifferentiated compartment. Staining for K14 (which stains the basal cells in excretory ducts), K19 (which stains granular ducts, striated ducts, and the luminal cells in excretory ducts), and KIT (which stains intercalated ducts), mapped the location of H2BGFP-LCRs to differentiated excretory, striated, and intercalated ducts. Since striated ducts have very low cellular turnover, the data suggested that there were actively dividing pools of stem/progenitor cells in the intercalated ducts and the basal layer of excretory ducts, and that these stem/progenitor populations function independently during homeostasis. The data are also consistent with the report mentioned above, which showed that acinar cells do not arise from a quiescent stem cell pool or from ductal cells during homeostasis (Aure et al., 2015).

Another study investigating the identity of LRCs in adult glands used 5-ethynyl-2′-deoxyuridine (EdU) to label proliferating cells in postnatal day 10 mice and an 8-week chase (Chibly, Querin, Harris, & Limesand, 2014). EdU+ LRCs colocalized with cells expressing the progenitor markers K5, K14, or KIT, further indicating that LRCs are a heterogeneous population of progenitors. The proliferative potential of the LRCs was shown with sphere assays, in which LRCs were stained with the proliferation marker Ki67. These data are consistent with the previous studies showing that the ductal compartment of the gland is proliferative during homeostasis. In addition, LRCs in irradiated mice did not undergo apoptosis following irradiation. This is consistent with the survival of ductal cells in patients whose salivary glands have been irradiated; however, it is the loss of acinar cells after irradiation and their inability to regenerate that presents a challenge.

Overall, these studies highlight that specific progenitor populations may be involved in homeostasis and regeneration depending on the type of damage that has occurred. It will be important in the future to identify the progenitors that are affected by, and that respond to, different types of gland damage, such as irradiation, ductal ligation, partial or full gland extirpation, and genetic ablation of specific cell populations or genetic activation of a signaling pathway

4.3 Regulation of Epithelial Progenitors: Heparan Sulfate and MicroRNAs

Fgf10 and its receptor FGFR2b are required for both human and mouse SMG development (reviewed by Patel & Hoffman, 2014). Since FGFR2b signaling is essential for progenitor survival and proliferation during organogenesis, understanding how receptor function is regulated may provide targets for progenitor expansion or gland regeneration. Importantly, FGFR2b signaling requires a heparan sulfate (HS)-containing coreceptor, which is present on the cell surface or in the ECM. HS increases the affinity of Fgf10 for FGFR2b, stabilizing the ternary signaling complex. HS is the most diverse polysaccharide due to the abundance of sulfate modifications on its sugar chains. The sulfate modifications create specific sulfated epitopes that regulate biological outcomes such as proliferation, duct elongation, and endbud expansion (Patel et al., 2008). One important mechanism that regulates epithelial morphogenesis is the variation in binding affinities that different FGFs have for HS, which affects their diffusion through the ECM, creating morphogenic gradients (Makarenkova et al., 2009). Additionally, specific types of sulfate modifications create specific epitopes that influence signaling in progenitor cells. Surprisingly, analysis of the enzymes that add sulfates to HS in KIT+ progenitor cells showed an enrichment of the HS 3-*O*-sulfotransferase (Hs3st) family. There are seven Hs3st isoforms in mice and humans; however, they produce the least abundant sulfate modification on HS. The expression of Hs3st3 enzyme isoforms, which make 3-*O*-sulfated epitopes, were rapidly upregulated in response to Fgf10/FGFR2b signaling. This resulted in an autocrine increase in FGFR2b-dependent MAPK signaling and expression of *Hs3st3* genes, KIT, and transcription factor genes that are downstream of FGFR2b signaling. The rapid response to 3-*O*-sulfated HS (3st3-HS) increased the number of KIT+ and K14+ progenitors, proliferation, and endbud morphogenesis (Fig. 5). Thus, rapidly modifying 3-*O*-sulfation provides a cellular mechanism to modulate the biologic response to FGFR2b signaling and control progenitor expansion (Patel et al., 2014). This mechanism also improved the maintenance and expansion of epithelial KIT+FGFR2b+ progenitors in salisphere culture, and biosynthetic 3-*O*-sulfated-HS may specifically expand these cells for regenerative therapies. These studies suggest that HS may be a useful tool to stimulate FGFR signaling on progenitor cells to control cellular functions.

MicroRNAs (miRNAs) have emerged as important regulators of stem/progenitor gene expression. miRNAs are small, single-stranded, noncoding RNAs that target the 3′UTR of multiple mRNAs and regulate gene expression posttranscriptionally. miRNAs regulate the activity of canonical signaling cascades such as the Wnt pathway and are implicated in branching morphogenesis. miR-21, a mesenchymal miRNA that downregulates target genes *Reck* and *Pdcd4*, enhances branching morphogenesis through the degradation of ECM by activated MMPs (Hayashi, Koyama, Azuma, & Kashimata, 2011). In contrast, the miR-200c family is highly expressed in SMG epithelial endbuds during development and influences epithelial proliferation. mir-200c targets *Zeb1*, *Hs3st1*, and *Vldlr*, thus regulating E-cadherin, HS, and very low density lipoprotein receptor function, respectively. By targeting *Vldlr* expression and its ligand reelin, miR-200c reduces expression of FGFR-dependent genes and epithelial proliferation (Rebustini et al., 2012). The appeal of using miRNAs to regulate progenitor cell gene expression and function for regenerative purposes is that a single miRNA may influence multiple genes, and thus have a broad effect on a specific biological process rather than on a single gene/signaling pathway.

5. CONCLUSION

SMG organogenesis involves interactions among multiple cell types, which add to the complexity of regulatory mechanisms that drive development. While understanding epithelial–neuronal communication has been very informative in understanding organogenesis, there is still much to learn about endothelial–epithelial interactions and endothelial–neuronal interactions. Recent studies suggest that specific stem/progenitor populations may respond to different types of damage to regenerate the gland and that these may be different from the cells that maintain homeostasis. Understanding

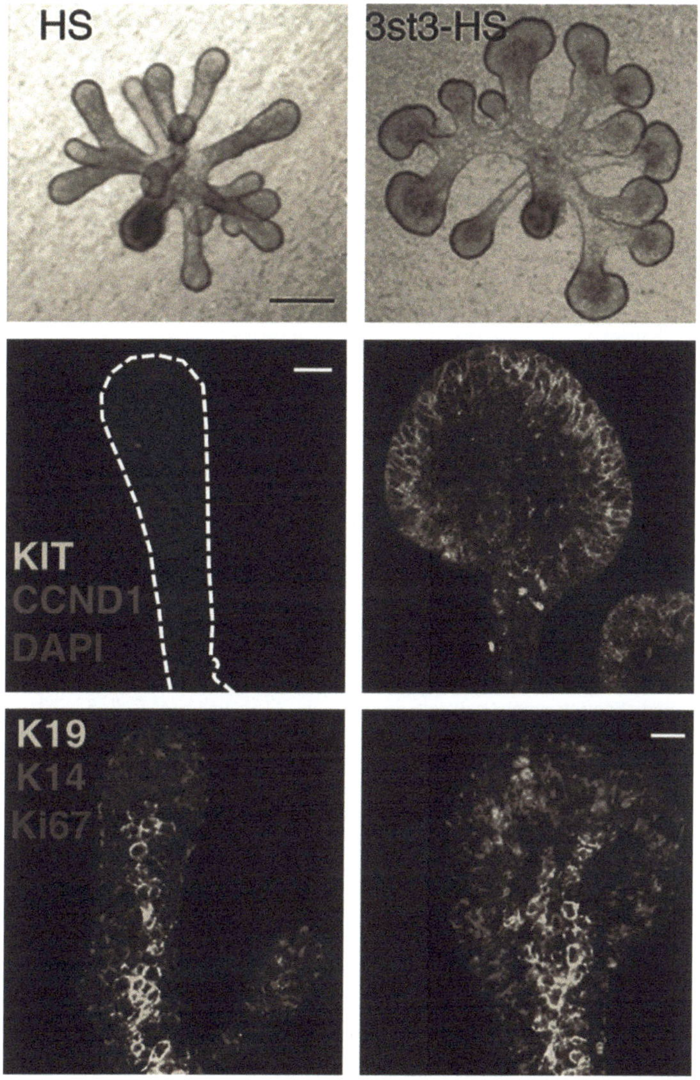

Figure 5 3-*O*-sulfated-HS (3st3-HS) increases epithelial endbud morphogenesis and proliferation as well as KIT and K14 expression. Brightfield images of E13 SMG epithelia cultured with Fgf10 and either HS or 3st3-HS for 28 h, top two panels. The epithelia were stained for proliferating cells with CCND1 (red), KIT (green), and nuclei (blue), middle two panels. Lower two panels show K14 (red), K19 (green), and proliferation (Ki67, blue). Images are 2 μm confocal sections. Scale bars, 10 μm. *From Figures 2 and 6 in Patel et al. (2014).* (A color version of this figure is located at the end of this chapter.)

how these different stem/progenitor cells respond to changes in their local microenvironments will be important for directing regeneration after damage. Little is known about the mechanisms of quiescence, maintenance, and expansion of specific stem/progenitor cells in response to specific damage. These are emerging and exciting areas of salivary gland research that remain to be explored.

ACKNOWLEDGMENTS

The authors would like to thank Drs. Vaishali Patel, Isabelle Lombaert, Jennifer Symonds, and Wendy Knosp for critical reading of this chapter. This work was supported by the intramural research program of the NIDCR at NIH.

REFERENCES

Ambudkar, I. S. (2014). Ca^{2+} signaling and regulation of fluid secretion in salivary gland acinar cells. *Cell Calcium, 55*, 297–305.

Arany, S., Catalan, M. A., Roztocil, E., & Ovitt, C. E. (2011). Ascl3 knockout and cell ablation models reveal complexity of salivary gland maintenance and regeneration. *Developmental Biology, 353*, 186–193.

Arnold, K., Sarkar, A., Yram, M. A., Polo, J. M., Bronson, R., Sengupta, S., et al. (2011). Sox2(+) adult stem and progenitor cells are important for tissue regeneration and survival of mice. *Cell Stem Cell, 9*, 317–329.

Aure, M. H., Konieczny, S. F., & Ovitt, C. E. (2015). Salivary gland homeostasis is maintained through acinar cell self-duplication. *Developmental Cell, 33*, 231–237.

Catalan, M. A., Nakamoto, T., & Melvin, J. E. (2009). The salivary gland fluid secretion mechanism. *The Journal of Medical Investigation, 56*(Suppl.), 192–196.

Chibly, A. M., Querin, L., Harris, Z., & Limesand, K. H. (2014). Label-retaining cells in the adult murine salivary glands possess characteristics of adult progenitor cells. *PLoS One, 9*, e107893.

Daley, W. P., Gervais, E. M., Centanni, S. W., Gulfo, K. M., Nelson, D. A., & Larsen, M. (2012). ROCK1-directed basement membrane positioning coordinates epithelial tissue polarity. *Development, 139*, 411–422.

Delli, K., Spijkervet, F. K., Kroese, F. G., Bootsma, H., & Vissink, A. (2014). Xerostomia. *Monographs in Oral Science, 24*, 109–125.

Dyachuk, V., Furlan, A., Shahidi, M. K., Giovenco, M., Kaukua, N., Konstantinidou, C., et al. (2014). Neurodevelopment. Parasympathetic neurons originate from nerve-associated peripheral glial progenitors. *Science, 345*, 82–87.

Espinosa-Medina, I., Outin, E., Picard, C. A., Chettouh, Z., Dymecki, S., Consalez, G. G., et al. (2014). Neurodevelopment. Parasympathetic ganglia derive from Schwann cell precursors. *Science, 345*, 87–90.

Ferreira, J. N., & Hoffman, M. P. (2013). Interactions between developing nerves and salivary glands. *Organogenesis, 9*(3), 199–205. http://dx.doi.org/10.4161/org.25224.

Harunaga, J. S., Doyle, A. D., & Yamada, K. M. (2014). Local and global dynamics of the basement membrane during branching morphogenesis require protease activity and actomyosin contractility. *Developmental Biology, 394*, 197–205.

Hayashi, T., Koyama, N., Azuma, Y., & Kashimata, M. (2011). Mesenchymal miR-21 regulates branching morphogenesis in murine submandibular gland in vitro. *Developmental Biology, 352*, 299–307.

Hohenester, E., & Yurchenco, P. D. (2013). Laminins in basement membrane assembly. *Cell Adhesion & Migration*, *7*, 56–63.
Holmberg, K. V., & Hoffman, M. P. (2014). Anatomy, biogenesis and regeneration of salivary glands. *Monographs in Oral Science*, *24*, 1–13.
Hsu, J. C., Koo, H., Harunaga, J. S., Matsumoto, K., Doyle, A. D., & Yamada, K. M. (2013). Region-specific epithelial cell dynamics during branching morphogenesis. *Developmental Dynamics*, *242*, 1066–1077.
Islam, M. N., Itoh, S., Yanagita, T., Sumiyoshi, K., Hayano, S., Kuremoto, K., et al. (2015). Runx/Cbfb signaling regulates postnatal development of granular convoluted tubule in the mouse submandibular gland. *Developmental Dynamics*, *244*, 488–496.
Knosp, W. M., Knox, S. M., & Hoffman, M. P. (2012). Salivary gland organogenesis. *Wiley Interdisciplinary Reviews. Developmental Biology*, *1*, 69–82.
Knosp, W. M., Knox, S. M., Lombaert, I. M., Haddox, C. L., Patel, V. N., & Hoffman, M. P. (2015). Submandibular parasympathetic gangliogenesis requires sprouty-dependent Wnt signals from epithelial progenitors. *Developmental Cell*, *32*, 667–677.
Knox, S. M., Lombaert, I. M., Haddox, C. L., Abrams, S. R., Cotrim, A., Wilson, A. J., et al. (2013). Parasympathetic stimulation improves epithelial organ regeneration. *Nature Communications*, *4*, 1494.
Knox, S. M., Lombaert, I. M., Reed, X., Vitale-Cross, L., Gutkind, J. S., & Hoffman, M. P. (2010). Parasympathetic innervation maintains epithelial progenitor cells during salivary organogenesis. *Science*, *329*, 1645–1647.
Kwak, M., & Ghazizadeh, S. (2015). Analysis of histone H2BGFP retention in mouse submandibular gland reveals actively dividing stem cell populations. *Stem Cells and Development*, *24*, 565–574.
Kwon, H. R., & Larsen, M. (2015). The contribution of specific cell subpopulations to submandibular salivary gland branching morphogenesis. *Current Opinion in Genetics & Development*, *32*, 47–54.
Lombaert, I. M., Abrams, S. R., Li, L., Eswarakumar, V. P., Sethi, A. J., Witt, R. L., et al. (2013). Combined KIT and FGFR2b signaling regulates epithelial progenitor expansion during organogenesis. *Stem Cell Reports*, *1*, 1–16.
Lombaert, I. M., Brunsting, J. F., Wierenga, P. K., Faber, H., Stokman, M. A., Kok, T., et al. (2008). Rescue of salivary gland function after stem cell transplantation in irradiated glands. *PLoS One*, *3*, e2063.
Lombaert, I. M., Knox, S. M., & Hoffman, M. P. (2011). Salivary gland progenitor cell biology provides a rationale for therapeutic salivary gland regeneration. *Oral Diseases*, *17*, 445–449.
Makarenkova, H. P., Hoffman, M. P., Beenken, A., Eliseenkova, A. V., Meech, R., Tsau, C., et al. (2009). Differential interactions of FGFs with heparan sulfate control gradient formation and branching morphogenesis. *Science Signaling*, *2*, ra55.
Mellas, R. E., Kim, H., Osinski, J., Sadibasic, S., Gronostajski, R. M., Cho, M., et al. (2015). NFIB regulates embryonic development of submandibular glands. *Journal of Dental Research*, *94*, 312–319.
Miletich, I. (2010). Introduction to salivary glands: Structure, function and embryonic development. *Frontiers of Oral Biology*, *14*, 1–20.
Nanduri, L. S., Maimets, M., Pringle, S. A., van der Zwaag, M., van Os, R. P., & Coppes, R. P. (2011). Regeneration of irradiated salivary glands with stem cell marker expressing cells. *Radiotherapy and Oncology*, *99*, 367–372.
Nedvetsky, P. I., Emmerson, E., Finley, J. K., Ettinger, A., Cruz-Pacheco, N., Prochazka, J., et al. (2014). Parasympathetic innervation regulates tubulogenesis in the developing salivary gland. *Developmental Cell*, *30*, 449–462.

Onodera, T., Sakai, T., Hsu, J. C., Matsumoto, K., Chiorini, J. A., & Yamada, K. M. (2010). Btbd7 regulates epithelial cell dynamics and branching morphogenesis. *Science*, *329*, 562–565.

Patel, V. N., & Hoffman, M. P. (2014). Salivary gland development: A template for regeneration. *Seminars in Cell & Developmental Biology*, *25–26*, 52–60.

Patel, V. N., Likar, K. M., Zisman-Rozen, S., Cowherd, S. N., Lassiter, K. S., Sher, I., et al. (2008). Specific heparan sulfate structures modulate FGF10-mediated submandibular gland epithelial morphogenesis and differentiation. *The Journal of Biological Chemistry*, *283*, 9308–9317.

Patel, V. N., Lombaert, I. M., Cowherd, S. N., Shworak, N. W., Xu, Y., Liu, J., et al. (2014). Hs3st3-modified heparan sulfate controls KIT+ progenitor expansion by regulating 3-O-sulfotransferases. *Developmental Cell*, *29*, 662–673.

Patel, N., Sharpe, P. T., & Miletich, I. (2011). Coordination of epithelial branching and salivary gland lumen formation by Wnt and FGF signals. *Developmental Biology*, *358*, 156–167.

Proctor, G. B., & Carpenter, G. H. (2014). Salivary secretion: Mechanism and neural regulation. *Monographs in Oral Science*, *24*, 14–29.

Ray, S., Fanti, J. A., Macedo, D. P., & Larsen, M. (2014). LIM kinase regulation of cytoskeletal dynamics is required for salivary gland branching morphogenesis. *Molecular Biology of the Cell*, *25*, 2393–2407.

Rebustini, I. T., Hayashi, T., Reynolds, A. D., Dillard, M. L., Carpenter, E. M., & Hoffman, M. P. (2012). miR-200c regulates FGFR-dependent epithelial proliferation via Vldlr during submandibular gland branching morphogenesis. *Development*, *139*, 191–202.

Rebustini, I. T., Myers, C., Lassiter, K. S., Surmak, A., Szabova, L., Holmbeck, K., et al. (2009). MT2-MMP-dependent release of collagen IV NC1 domains regulates submandibular gland branching morphogenesis. *Developmental Cell*, *17*, 482–493.

Rothova, M., Thompson, H., Lickert, H., & Tucker, A. S. (2012). Lineage tracing of the endoderm during oral development. *Developmental Dynamics*, *241*, 1183–1191.

Ryu, Y. K., Collins, S. E., Ho, H. Y., Zhao, H., & Kuruvilla, R. (2013). An autocrine Wnt5a-Ror signaling loop mediates sympathetic target innervation. *Developmental Biology*, *377*, 79–89.

Schaub, J. R., Malato, Y., Gormond, C., & Willenbring, H. (2014). Evidence against a stem cell origin of new hepatocytes in a common mouse model of chronic liver injury. *Cell Reports*, *8*, 933–939.

Stenman, G. (2013). Fusion oncogenes in salivary gland tumors: Molecular and clinical consequences. *Head and Neck Pathology*, 7(Suppl. 1), S12–S19.

Tang, N., Marshall, W. F., McMahon, M., Metzger, R. J., & Martin, G. R. (2011). Control of mitotic spindle angle by the RAS-regulated ERK1/2 pathway determines lung tube shape. *Science*, *333*, 342–345.

Tucker, A. S. (2007). Salivary gland development. *Seminars in Cell & Developmental Biology*, *18*, 237–244.

Yanger, K., Knigin, D., Zong, Y., Maggs, L., Gu, G., Akiyama, H., et al. (2014). Adult hepatocytes are generated by self-duplication rather than stem cell differentiation. *Cell Stem Cell*, *15*, 340–349.

CHAPTER SIX

The Development of the Calvarial Bones and Sutures and the Pathophysiology of Craniosynostosis

Mamoru Ishii, Jingjing Sun, Man-Chun Ting, Robert E. Maxson[1]
Department of Biochemistry and Molecular Biology, USC Norris Comprehensive Cancer Center, Keck School of Medicine, University of Southern California, Los Angeles, California, USA
[1]Corresponding author: e-mail address: maxson@usc.edu

Contents

Abstract

The skull vault is a complex, exquisitely patterned structure that plays a variety of key roles in vertebrate life, ranging from the acquisition of food to the support of the sense organs for hearing, smell, sight, and taste. During its development, it must meet the dual challenges of protecting the brain and accommodating its growth. The bones and sutures of the skull vault are derived from cranial neural crest and head mesoderm. The frontal and parietal bones develop from osteogenic rudiments in the supraorbital ridge. The coronal suture develops from a group of Shh-responsive cells in the head

ISSN 0070-2153
http://dx.doi.org/10.1016/bs.ctdb.2015.07.004

mesoderm that are collocated, with the osteogenic precursors, in the supraorbital ridge. The osteogenic rudiments and the prospective coronal suture expand apically by cell migration. A number of congenital disorders affect the skull vault. Prominent among these is craniosynostosis, the fusion of the bones at the sutures. Analysis of the pathophysiology underling craniosynostosis has identified a variety of cellular mechanisms, mediated by a range of signaling pathways and effector transcription factors. These cellular mechanisms include loss of boundary integrity, altered sutural cell specification in embryos, and loss of a suture stem cell population in adults. Future work making use of genome-wide transcriptomic approaches will address the deep structure of regulatory interactions and cellular processes that unify these seemingly diverse mechanisms.

1. INTRODUCTION

The flat bones at the top of the skull compose the skull vault, or calvaria. In mammals, these bones include the paired frontal and parietal bones, as well as the single interparietal. Between the bones are sutures, fibrous joints that serve the dual purposes of allowing the skull vault to flex during childbirth and acting as a growth center in postnatal growth and homeostasis. A number of congenital disorders affect the skull vault. Prominent among these is craniosynostosis, the fusion of the bones at the sutures. In this review, we discuss the early development of the skull vault and sutures from precursor mesenchymal populations; we consider later development and homeostasis in postnatal animals; finally, we review recent work on major disease mechanisms that underlie craniosynostosis.

2. EARLY DEVELOPMENT

2.1 Embryological Origins of the Bones of the Skull Vault

The embryological origins of the bones of the skull vault have been investigated in several vertebrates, including chick, frog (*Xenopus*), and mouse (reviewed by Chai & Maxson, 2006; Noden & Trainor, 2005). Work performed by Le Douarin and colleagues using a quail–chick transplantation approach showed that the skeletogenic tissue of the skull vault, including the frontal, parietal, and squamosal bones, is derived from cranial neural crest cells (Couly, Coltey, & Le Douarin, 1993; Le Douarin & Kalcheim, 1999; Le Douarin, Creuzet, Couly, & Dupin, 2004; Le Lievre, 1978). Noden and colleagues, using a retroviral lineage tracing approach, reached a different conclusion, demonstrating the dual neural crest–mesoderm origin of the frontal bone and an exclusively mesodermal origin for the parietal bone

(Noden & Trainor, 2005). Although these findings have not been reconciled, Noden and Trainor (2005) point out the difficulty of performing "...quail-chick transplantations centered around the ability to graft neural crest or mesoderm progenitors exclusive of contamination by the other...."

In mice, Wnt1-Cre;R26R lineage analysis showed that the frontal bones have a major contribution from neural crest, and Mesp1-Cre;R26R lineage analysis showed that the parietal bones originate from head mesoderm. The interparietal bone has contributions from both neural crest and mesoderm (Jiang, Iseki, Maxson, Sucov, & Morriss-Kay, 2002; Yoshida, Vivatbutsiri, Morriss-Kay, Saga, & Iseki, 2008). Evidence that the frontal bones also receive a small contribution from mesoderm came from Deckelbaum et al. (2012), who showed that a Gli1-expressing subpopulation of mesoderm is incorporated into the frontal bone.

Analysis of the neural crest contribution to the skull vault in zebrafish provided results consistent with findings in the mouse. Using a transgene system based on Cre recombinase to genetically label neural crest in the zebrafish, Fisher and colleagues (Kague et al., 2012) found that only the anterior portions of the frontal bones are derived from neural crest; the more posterior calvarial bones, including the parietals and the exoccipital, do not receive a neural crest contribution.

Finally, Henken's group, working with *Xenopus laevis*, labeled cranial neural crest with fluoresceinated dextran and followed the labeled cells as they migrated into the developing skull vault. This work provided evidence that cranial neural crest cells contribute to the frontoparietal bone along its entire length (Gross & Hanken, 2005). Whether mesoderm also makes a contribution to the frontoparietal bone is not clear (Gross & Hanken, 2005).

Looking across these several vertebrate groups, it seems clear that both cranial neural crest and head mesoderm make contributions to the skull vault. Although homologous relationships of the elements of the skull vault need to be investigated carefully, it appears to be the case that the boundary between head mesoderm and neural crest occupies a variable position relative to the coronal suture. In mouse, it coincides with the coronal suture, the suture itself being of mesodermal origin (Jiang et al., 2002; Yoshida et al., 2008). But in the other species, it is located in different positions—in the middle of the frontal bone in teleost fish, for example. In any case, the location of this lineage boundary at the coronal suture in mice has made it possible to investigate the potential importance of boundaries between osteogenic and nonosteogenic compartments—a topic that we consider in detail below.

2.2 Specification of the Osteogenic Rudiments

In mice, neural crest-derived and mesoderm-derived osteogenic precursor populations migrate from their sites of origin to the supraorbital ridge by E12.5 (Jiang et al., 2002; Yoshida et al., 2008). There they coalesce into the rudiments of the calvarial bones, identifiable between E12.5 and E13.5 by their expression of early osteogenic markers, including alkaline phosphatase (ALP) and Runx2.

The BMP signaling pathway plays a major role in the early specification of the calvarial bone primordia. At E11.5, *Bmp4* is expressed in the cranial mesenchyme that corresponds to the area where the bone primordium will develop within the next day (Han et al., 2007). Subsequently, at E12.5, both *Bmp2* and *Bmp4* are expressed in the primordia (Sun, Ishii, Ting, & Maxson, 2013). Loss of function of *Bmp2*, *Bmp4*, and *Bmp7* in cranial neural crest results in severe calvarial bone growth defects (Bonilla-Claudio et al., 2012) suggesting that the Bmp pathway plays an essential role in neural crest-derived calvarial bone formation.

The homeodomain transcriptional factor, *Msx2*, is an immediate early effector of the Bmp pathway (Brugger et al., 2004). In mice, a deficiency of *Msx2* results in a frontal foramen, an ossification defect in the frontal bone (Satokata et al., 2000). In humans, haploinsufficiency of *MSX2* causes familial parietal foramina (Wilkie et al., 2000). We showed that *Msx2* is required to promote the initial process of osteogenic specification during frontal bone development in the mouse (Ishii et al., 2003). Msx2, together with its congener, *Msx1*, positively controls the expression of the osteogenic gene *Runx2* in the calvarial mesenchyme at E12.5 (Han et al., 2007; Ishii et al., 2003). Thus, a Bmp–Msx axis is essential for the proper specification of the calvarial bone primordia.

Recently, we presented evidence that the transcription factor Foxc1 is part of a mechanism that sets a threshold of sensitivity of an Msx2 enhancer to Bmp signaling (Sun et al., 2013) (Fig. 1). In the Foxc1 mutant, the domain of Msx2 expression expands within the supraorbital ridge mesenchyme, coinciding with a failure of the calvarial rudiments to grow apically. These results, together with an analysis of the expression of osteogenic markers in the Foxc1 mutant, suggest that the bone primordia differentiate prematurely in the mutant, thus preventing apical expansion. A series of siRNA, cotransfection, and ChIP experiments in a newly developed cranial neural crest cell line (Ishii et al., 2012) showed that Foxc1 negatively regulates the BMP responsive element of Msx2 through direct interaction with a Foxc1 site (Sun et al., 2013). This work thus revealed a part of the molecular

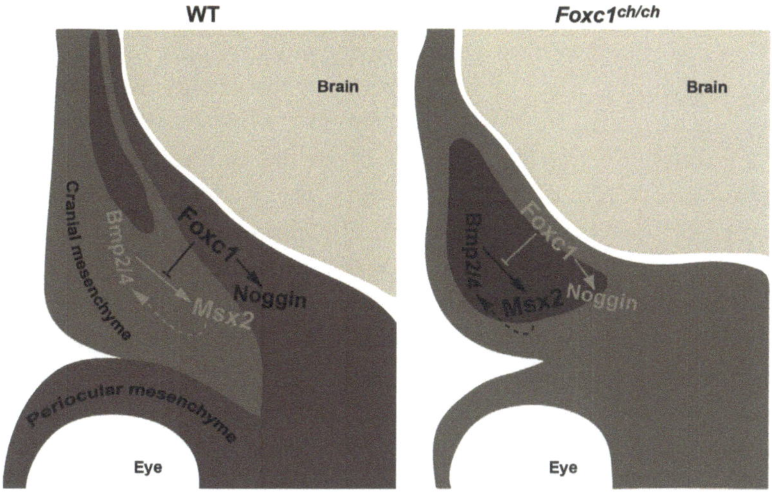

Figure 1 Schematic model showing role of Foxc1 in controlling Bmp-induced *Msx2* expression and specification of osteoprogenitors in the supraorbital ridge. At E12.5, *Foxc1* is expressed in the meninges and adjacent mesenchyme (in red), including the periocular domain. Osteogenic cells are marked in fuchsia, representing the expression of ALP. In wild-type embryos, Foxc1 negatively regulates *Msx2* and *Bmp2/4* and positively regulates *Noggin*, thereby maintaining a level of Bmp signaling appropriate for the specification of the frontal bone primordia as well as the apical migration of undifferentiated progenitor cells. In *Foxc1* mutants, loss of *Foxc1* results in upregulation of *Msx2*, which initiates a positive-feedback loop between *Msx2* and *Bmp2/4*. The result is the propagation of a wave of Bmp–Msx2 pathway from medial to lateral across the supraorbital ridge and the lateral expansion of the ALP-expressing differentiating osteoblast domain. *After Fig. 7, Sun et al. (2013).* (A color version of this figure is located at the end of this chapter.)

mechanism controlling the Bmp–Msx pathway and the size and rate of differentiation of the osteoprogenitor population that forms the calvarial rudiments.

The Wnt pathway also plays an essential role in the specification of the primordia. Atit's group showed that β-catenin promotes the early specification of the osteogenic lineage and also represses the chondrogenic lineage in the calvarial mesenchyme (Tran et al., 2010). This group also demonstrated that Twist1 mediates the chondrogenic repression exerted by β-catenin (Goodnough et al., 2012). Although these results show that a Wnt signal is indispensable for the specification of the calvarial osteogenic lineage, the source of the signal has remained elusive. Intriguingly, however, Atit's group recently found that this signal originates in the adjacent surface ectoderm (Goodnough et al., 2014). *Wntless* (*Wls*) is a Wnt ligand transporter that is necessary for Wnt protein secretion. Wls is expressed from E11.5–E12.5 in the cranial surface ectoderm and underlying mesenchyme. The authors studied the effect of *Wls* deletion in the cranial mesenchyme and the surface ectoderm by means of different Cre drivers (*Dermo1-Cre* and *Crect-Cre*, respectively). The outcome of the deletion of *Wls* was dramatic. Inactivation of Wls in the ectoderm led to loss of cranial bone and the formation of ectopic cartilage. At E12.5, the expression of early osteogenic markers including *Runx2, Bmp4, Alpl, PTHrP,* and *Osx* was significantly reduced in the mutant. Reciprocally, an increased expression of the chondrogenic marker *Sox9* was evident (Goodnough et al., 2014). The mesenchymal expression of the Wnt ligands *Wnt5a, Wnt11, Wnt4, Wnt3a,* and *Wnt16* is also controlled by Wls in ectoderm. Thus, ectoderm-derived Wnt ligands govern the specification of the osteogenic lineage and repression of the chondrogenic lineage in the cranial mesenchyme.

Mesenchymal deletion of *Wls* leads to a reduced mass of mineralized bone (Goodnough et al., 2014), despite normal early specification of the osteogenic lineage. In E13.5 mutants, the levels of *Axin2* and *Lef1* expression in the mesenchyme were reduced. Thus, mesenchyme-derived Wnt ligands may facilitate maintenance of osteoblast differentiation by transducing a Wnt signal in a tissue-autocrine manner.

Do the Wnt and Bmp pathways function synergistically in calvarial rudiment specification? Intriguingly, transcriptional activation of *Msx2* in E12.5 mesenchyme is independent of Wnt signaling (Goodnough et al., 2014), suggesting that different pathways function independently during calvarial bone specification. However, it remains possible that the Bmp and Wnt pathways interact by synergistically regulating a common effector such as Runx2. This idea is consistent with a previous finding that *Msx2* and *Twist1* genetically interact to control the size of the bone primordium during frontal bone specification (Ishii et al., 2003).

2.3 Expansion of the Osteogenic Rudiments

After the bone primordia are specified, they expand apically and laterally. The mechanism of such expansion has been a subject of active investigation. Possible mechanisms include (1) a basal to apical wave of differentiation of osteoprogenitors prepositioned in the areas of the prospective calvarial bones, (2) apical migration of osteogenic precursors and addition of such precursors to the leading edge of the growing bone, or (3) a combination of these two mechanisms.

Yoshida and colleagues (2008) addressed these possibilities with an *exo utero* DiI-labeling approach in which DiI was microinjected into the supraorbital ridge in the area of the frontal bone primordia of E13.5 embryos. Injected embryos were allowed to develop *exo utero* for 4–5 days and the distribution of the dye was assessed. The result was that 4–5 days after DiI injection, labeled cells had moved a substantial distance apically, which suggested that cell migration contributes to the expansion of the primordium. This finding was confirmed by the work of Roybal et al. (2010) and Ting et al. (2009), who showed in addition that cells labeled in the supraorbital ridge move apically within a layer of cells outside (ectocranial to) the prospective osteogenic layer before ultimately inserting at the leading edge of the growing bone (Fig. 2). EphrinA/EphA signaling in this ectocranial layer seems to have a role in the apical guidance of osteogenic precursors (Ting et al., 2009). Thus, cell migration probably contributes to the apical expansion of the calvarial rudiments. Whether this is the only mechanism of expansion remains to be determined.

Evidence for further heterogeneity in the cell populations of the developing calvaria came from an analysis of frontal bone development in *Msx1/2* conditional mutants (Roybal et al., 2010) Inactivation of *Msx1/2* in neural crest with Wnt1-Cre resulted, as expected, in a dramatic reduction in frontal bone growth. Surprisingly, this also resulted in a number of mispatterned Wormian-like bones in place of the frontal bones. DiI lineage tracing showed that these bones were derived from a population of cells that normally do not have an osteogenic fate. Wnt1-Cre;R26R lineage tracing showed that this population is composed of neural crest cells that migrate apically between E9.5 and E10.5, prior to the migration of the population that forms the osteogenic rudiments. The normal function of this population is unclear, but one possibility is that it provides an environment favorable for the growth and differentiation of the osteogenic rudiments. In *Wnt1-Cre; Msx1/2*$^{cko/cko}$ mutants—in which the frontal rudiments fail to grow—cells of this population alter their fate and become osteogenic. Whether this cell population is generally responsible for Wormian bone formation remains to be determined.

2.4 Osteogenic Cell Differentiation and Proliferation Within the Rudiments

After the calvarial rudiments expand, osteoblasts within the rudiments undergo further differentiation and lay down a matrix that becomes mineralized. Mineralization begins around E14.0 in mice as the basal area of the

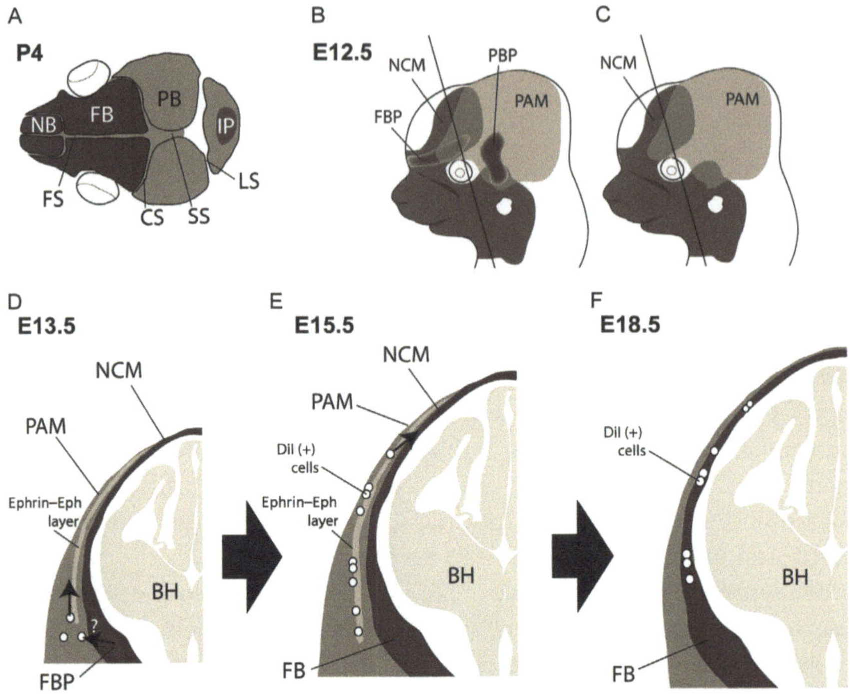

Figure 2 Development of the skull vault. (A) Origins of bones of the skull vault. Blue, neural crest; pink, mesoderm. Note that the paired frontal bones are derived from neural crest and parietal bones from mesoderm. The interparietal bone has a dual origin: the central portion is neural crest-derived; the flanking portions come from mesoderm. (B) and (C) side views of head showing cranial layers. Alp-expressing rudiments are shown in magenta. (D)–(F) Cross-sections at the level of the frontal bone showing cranial layers at different developmental stages. Note that the frontal bone rudiment elongates within layer of neural crest-derived mesenchyme. Flanking this mesenchyme laterally is a layer of ephrin-Eph expressing cells upon which migratory osteogenic cells are found. DiI labeling shows that these migratory cells move from the area of the rudiment apically, ultimately inserting into the growing bone. NB, nasal bone; FB, frontal bone; PB, parietal bone; IP, interparietal bone; FS, frontal suture; CS, coronal suture; SS, sagittal suture; LS, lambdoid suture; NCM, neural crest-derived mesenchyme; PAM, paraxial mesoderm-derived mesenchyme; FBP, frontal bone primordium; PBP, parietal bone primordium; BH, brain hemisphere. (A color version of this figure is located at the end of this chapter.)

bone primordium or "core" region of the frontal bone plate can be stained by Alizarin red (Han et al., 2007).

Wnt signaling is required in the postspecification phase of calvarial development. The mesenchyme-specific deletion of *Wls* by Dermo1-Cre results in severely reduced calvarial bone ossification (Goodnough et al., 2014). In these mutants, the expression of *Axin2* and *Lef1* in cranial mesenchyme is profoundly reduced at E13.5. Thus, Wnt ligands function to maintain the Wnt responsiveness in the calvarial mesenchyme as early as E13.5 (Goodnough et al., 2014). Early osteogenic specification is not affected in these mutants as Runx2 is normally expressed in the bone rudiment from E12.5 through E13.5. However, at E15.5, Osx expression is strikingly reduced in the mutant (Goodnough et al., 2014). These results suggest that a mesenchyme-derived Wnt signal is a prerequisite for the final phase of osteoblast differentiation and the ossification of calvarial bones.

In addition, a TGFβ signal is required to maintain continuous osteogenic lineage commitment in the bone primordium in E14.5 mouse embryos. Conditional inactivation of *Tgfbr2* in the neural crest by Wnt1-Cre results in significantly reduced expression of *Runx2, Osx, Ibsp, Alpl, Fgfr2*, and *Dlx5* in the frontal bone rudiment at the postspecification stage (E14.5) (Sasaki et al., 2006). These mutant mice have a severe defect in calvarial bone ossification (Ito et al., 2003; Sasaki et al., 2006). Tgfbr2 is also required for the proliferation of cells in the primordium at E14.5, but not during the early specification period at E12.5 (Sasaki et al., 2006). The smaller size of the bone primordium in *Msx2* mutants is accompanied by a reduced ratio of proliferative cells in E12.5 as well as E14.5 embryos (Ishii et al., 2003). However, the specification of the bone primordium and defects in osteoblast differentiation in *Wls* mutants are not linked to altered cell proliferation (Goodnough et al., 2014). Thus, we do not have simple model that can explain the relationship between osteoblast proliferation and differentiation.

3. CALVARIAL SUTURE DEVELOPMENT AND THE PATHOPHYSIOLOGY OF CRANIOSYNOSTOSIS

3.1 Early Development

The calvarial sutures include the frontal, sagittal, coronal, lambdoid, and squamosal sutures. All are derived from migratory mesenchymal cells, and all can be affected in craniosynostosis disorders. The coronal suture has been the focus of much of the recent work on suture development, and will be the focus of this review. This suture is of particular interest because it is affected in several craniosynostosis disorders, including Apert and Saethre–Chotzen syndrome (Wilkie, 1997). In mice, it is also the location of a boundary between the neural crest-derived and mesoderm-derived portions of the skull vault. It has thus served as a model for the relationship between developmental boundary formation and suture development and pathology.

In a landmark study, Deckelbaum et al. (2012) used an inducible genetic fate mapping approach to identify precursors of the coronal suture in a group of Shh-responsive cells in head mesoderm at E7.5–8.0. These cells migrate first to the supraorbital ridge at E8.5–9.0—where they transiently express En1 (E10.5–11.0)—then between E11.5 and E13.5 migrate apically, ultimately giving rise to the coronal suture and to some adjacent bone tissue.

By E13.5, sutural cells and prospective frontal and parietal bones have assumed stereotypic positions. ALP staining reveals that the frontal and parietal bones are overlapped, with the parietal on top. The coronal suture lies between them, above the frontal and below the parietal. During subsequent development, the frontal and parietal rudiments grow together in a basal to apical direction.

In the postnatal development of the calvarial bones and sutures, discrete osteogenic fronts become evident at the edges of the bones. The suture mesenchyme lies between the bones. During this phase of development, the osteogenic fronts become the major centers of osteogenic cell proliferation and bone growth. As will be discussed in more detail below, Zhao, Chai, and colleagues (Zhao et al., 2015) found recently that the suture mesenchyme is a source of stem cells for the postnatal growth and homeostasis of the calvarial bones. This population is also required for the maintenance of the calvarial sutures.

3.2 Craniosynostosis

Craniosynostosis is a common birth defect (1/2500 live births). Its most striking manifestation is abnormalities in skull shape, which are sometimes severe (Morriss-Kay & Wilkie, 2005; Wilkie, 1997). In humans, it has been shown to be caused by mutations in many genes (Cohen, 2006; Morriss-Kay & Wilkie, 2005). These include *MSX2* (Florisson et al., 2013; Jabs et al., 1993; Janssen et al., 2013), *FGFRs 1, 2, and 3* (Hajihosseini, 2008), *TWIST1* (Howard et al., 1997), *TGFBR1*, *TGFBR2* (Loeys et al., 2005), *FIBRILLIN-1* (*FBN1*) (Sood, Eldadah, Krause, McIntosh, & Dietz, 1996), *EFNB1* (Twigg et al., 2004), *EPHRINA4* (*EFNA4*) (Merrill et al., 2006), *RAB23* (Jenkins et al., 2007), the Notch ligand, *JAGGED1* (Kamath et al., 2002), *RECQL4* (Van Maldergem et al., 2006), *MASP1* (Sirmaci et al., 2010), *SH3PXD2B* (Iqbal et al., 2010), *FREM1* (Vissers et al., 2011), *ALX4* (Yagnik et al., 2012), and *SKI (Doyle et al., 2012)*. Studies in mice have identified, in addition, *Nell1* (Zhang et al., 2002), *Gdf6* (Settle et al., 2003), *Axin2 (Yu et al., 2005)*, *Dusp6*

(Li, Scott, Hatch, Tian, & Mansour, 2007), *EphA4* (Ting et al., 2009), and *Pdgfr alpha* (Moenning et al., 2009). These genes are components of several signaling pathways: *Tgfbr1, Msx2, SKI,* and *Gdf6* function in the Bmp pathway; *Jagged1* in the Notch pathway; *EphrinA4, EphrinB1, EphA4, Dusp6, Pdgfr alpha,* and *Fgfr1–3* in the RTK pathway. *RAB23* is a component of the Hedgehog pathway and *Axin2* of the Wnt pathway. *Twist1* functions to coordinate the activities of the Bmp and RTK pathways (Connerney et al., 2008; Rice et al., 2000; Ting et al., 2009). It is clear that a wide variety of genes and pathways can cause craniosynostosis. Thus, a diverse set of processes likely underlies its pathophysiology.

In the following, we describe selected recent studies that examine cellular and genetic mechanisms underlying craniosynostosis. One gene that has been of great interest is *Twist1*, which encodes a bHLH transcription factor (Bourgeois, Stoetzel, Bolcato-Bellemin, Mattei, & Perrin-Schmitt, 1996). In humans, heterozygous loss of function mutations in *Twist1* cause Saethre–Chotzen syndrome (el Ghouzzi et al., 1997; Howard et al., 1997). Affected individuals have syndactyly and coronal synostosis. Mice with a loss of function of a single Twist1 allele exhibit a similar suture phenotype. Typically synostosis is unilateral, and the suture is affected over its entire length or a portion of its length.

3.3 Twist1 Regulatory Network

Investigations of the pathophysiology of craniosynostosis in *Twist1* mutant mice have raised the possibility that a deficiency in a compartment boundary may have a role. In mammals, the coronal suture lies at the boundary between the neural crest-derived frontal bone and the mesoderm-derived parietal bone, the suture itself being mesodermal in origin (Jiang, Rowitch, Soriano, McMahon, & Sucov, 2000). ALP staining of sutures of *Twist1* mutant embryos at E13.5 suggested that loss of integrity of the boundary between osteogenic cells of the frontal and parietal bones and nonosteogenic cells of the suture could contribute to ossification of the suture.

Cre-based lineage tracing was used to test this hypothesis. Wnt1-Cre marks neural crest cells; Mesp1-Cre marks mesodermal cells (Jiang et al., 2002; Saga et al., 1999). *We used* the Wnt1-Cre;R26R and Mesp1-Cre; R26R systems to mark the neural crest–mesoderm boundary and assess the influence of *Twist1* on its integrity (Merrill et al., 2006). We found that indeed, heterozygous loss of function of Twist1 caused neural crest cells to become ectopically located in the coronal suture, and mesodermal cells to cross into the frontal bone territory. Thus, there was a loss of osteogenic–nonosteogenic boundary integrity in *Twist1* mutant mice (Fig. 3). This was accompanied by a reduction in the expression of ephrinA2, ephrinA4,

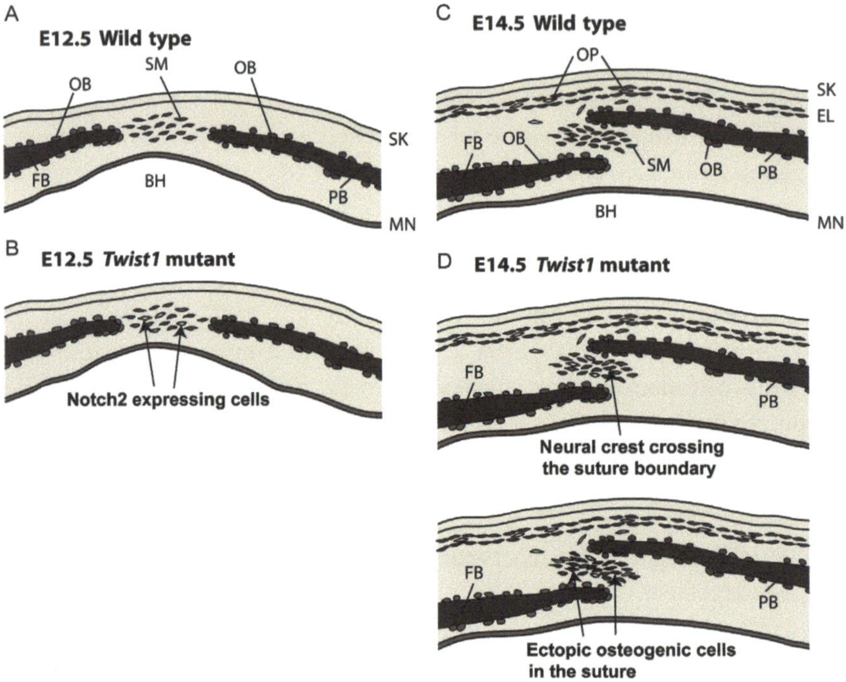

Figure 3 A model of craniosynostosis in *Twist1* mutants; a two-step process leading to ectopic osteogenesis in the suture mesenchyme. (A) and (B) Altered sutural identity. (A) At E12.5, in wild-type mice, nonosteogenic suture mesenchyme cells (SM) exclude expression of Notch2, an early marker of osteogenesis. (B) In the *Twist1* mutants, however, prominent Notch2 expression is evident in this population (shown in white), thus suggesting that the suture assumes a partial osteogenic identity. (C) and (D) Suture boundary defect. (C) At E14.5, in mice, a boundary between neural crest (blue) and mesoderm (red) lineage is formed between the frontal bone (FB, osteogenic) and the SM (nonosteogenic). (D) Top: In *Twist1* mutants, the neural crest–mesoderm boundary is disrupted by intrusion of neural crest-lineage cells (shown in blue) into the SM. Thus, there is a loss of osteogenic-nonosteogenic boundary integrity in *Twist1* mutant mice. Osteogenic precursor cells (OP) migrate through the ectocranial layer (EL), contributing to this loss of boundary integrity. Bottom: The change in sutural cell identity together with the boundary-crossing osteogenic cells induces sutural cells to become fully osteogenic (purple). BH, brain hemisphere; MN, meninges; OB, osteoblast; PB, parietal bone; SK, skin. (A color version of this figure is located at the end of this chapter.)

and EphA4 in a layer of cells ectocranial to the coronal suture. Ephrins are membrane-bound ligands that interact with Eph receptors, a large family of receptor tyrosine kinases (Klein, 2004; Kullander & Klein, 2002; Wilkinson, 2001). Ephrin–Eph signaling regulates a variety of developmental processes including the establishment of developmental boundaries (Klein, 2004; Kullander & Klein, 2002; Martinez & Soriano, 2005; Palmer & Klein, 2003; Pasquale, 2005; Poliakov, Cotrina, & Wilkinson, 2004; Surawska, Ma, & Salgia, 2004).

Intriguingly, we also found that loss of EFNA4 function in humans is associated with coronal synostosis (Merrill et al., 2006). This finding, together with the well-documented role of ephrin–Eph signaling in boundary formation, led us to propose a model in which ephrin signaling, under the control of Twist1, acts to maintain the osteogenic–nonosteogenic boundary at the coronal suture. We proposed that loss of integrity of this boundary results in osteogenic cells crossing into the suture compartment, causing the suture to take on an osteogenic identity (Merrill et al., 2006) and leading ultimately to ossification of the suture. We showed further that loss of function of the ephrinA4 receptor EphA4 causes coronal synostosis in mice, greatly strengthening the case for a relationship between craniosynostosis and loss of ephrinA signaling suggested by our earlier human genetic findings (Ting et al., 2009).

Finally, we documented a genetic interaction between Twist1 and EphA4, and showed by DiI labeling that Twist1 and EphA4 control the migration of osteogenic cells to the leading edges of the growing frontal and parietal bones (Ting et al., 2009). Twist1 and EphA4 are also required to exclude such osteogenic cells from the normally nonosteogenic territory of the coronal suture and thus to maintain the suture in an open configuration.

Analysis of Jagged/Notch signaling in the coronal suture provided further evidence of importance of an intact boundary between the suture and adjacent osteogenic territories, but also indicated that the picture is more complicated (Yen, Ting, & Maxson, 2010). Heterozygous loss of function of JAGGED1 causes Alagille syndrome in humans, characterized by impaired development of bile ducts, as well as defects in the heart, eye, kidney, face, and skull (Alagille et al., 1987; Emerick et al., 1999; Kamath et al., 2004; Krantz et al., 1998; Oda et al., 1997). Synostosis of the coronal suture is also a feature (Kamath et al., 2002). Jagged1 is expressed in a layer of sutural cells, mesodermal in origin, that are located along the osteogenic–nonosteogenic boundary between the frontal and parietal bones

(Yen et al., 2010). Conditional targeting of *Jagged1* in mesoderm, but not neural crest, results in coronal synostosis. Jagged1 expression in the suture is reduced in *Twist1* mutants, suggesting that *Twist1* is epistatic to *Jagged1*. Combination *Twist1/Jagged1* mutants exhibit a synostosis phenotype of increased severity compared to individual mutants, also consistent with an epistatic relationship between *Twist1* and *Jagged1*.

Subsequently, we found that the boundary loss was preceded by a change in sutural cell identity (Yen et al., 2010). In wild-type embryos at E12.5, Notch2, an early marker of osteogenesis, is expressed in osteogenic territories (frontal and parietal bones) but is excluded from the coronal suture. In *Twist1* mutants, Notch2 is expressed ectopically in the suture, thus suggesting that the suture takes on an osteogenic character as early as E12.5. It is important to note, however, that this change in specification state does not include expression of the committed osteoprogenitor markers ALP or Runx2. These markers are not expressed in the mutant suture until a day later at E13.5. These findings led us to propose that craniosynostosis in *Twist1* and *Jagged 1* mutants is a two-step event. First, sutural cells assume a partial osteogenic identity; second, osteogenic cells invade the suture and induce sutural cells to become fully osteogenic (Fig. 3).

A prediction of this model is that *Twist1* function would have to be reduced in both mesoderm and neural crest to produce craniosynostosis. This prediction was verified by conditional inactivation of Twist1 in mesoderm, neural crest, or both, demonstrating that reduced Twist1 function in either lineage was insufficient to cause craniosynostosis, whereas reduced Twist1 function in both lineages did cause craniosynostosis (Schafer, C., Ting, M., Brockop, M., & Maxson, R., unpublished observations).

Is the boundary mechanism unique to the coronal suture? As part of our analysis of the relationship between *Twist1* and *Jagged1*, we produced Twist1/Jagged1 double-mutant mice (Yen et al., 2010). These mice had very severe coronal synostosis, as expected, but also had synostosis of the squamosal, lambdoid, occipitointerparietal, and interfrontal sutures. These sutures do not appear to coincide with lineage boundaries, although they are boundaries between osteogenic and nonosteogenic compartments. We note that a developmental boundary may exist even in the absence of a lineage boundary. For example, if the expression of genes mediating cell–cell interaction crucial for boundary formation is not regulated by heritable mechanisms, but instead by local signaling, then a developmental boundary can be established in the absence of a lineage boundary. It is thus possible that a set of mechanisms similar to those we have documented in the coronal suture also function in other sutures to establish and maintain boundaries between osteogenic and nonosteogenic compartments.

3.4 bHLH Partners of Twist1

That Twist1 may function with a partner protein to control coronal suture development was suggested by the work of Connerney and colleagues (2006). Twist1 is a class II bHLH protein (reviewed by Massari & Murre, 2000). This class includes tissue-specific bHLH proteins such as MyoD. Class I proteins include the broadly expressed E proteins. Twist1 is capable of forming both homodimers and heterodimers with class I proteins. Twist1 is expressed throughout the developing coronal suture (Johnson, Iseki, Wilkie, & Morriss-Kay, 2000; Rice et al., 2000). The E proteins E2a and Tcf12 (Heb) are also expressed in the suture (Funato, Ohtani, Ohyama, Kuroda, & Nakamura, 2001).

Using a forced dimer approach, Connerney and colleagues (Connerney et al., 2006, 2008) tested the activity of Twist1 homodimers and Twist1/E heterodimers in cultured cells and in sutures in skull explants. They found that Twist1/E heterodimers and Twist1 homodimers differentially influence suture patency. They proposed a model in which a decrease in Twist1 expression due to Twist1 haploinsufficiency results in a higher Twist1/ Twist1 to Twist1/E ratio and thus to synostosis.

Although there is as yet no direct evidence that Twist1/Twist1 to Twist1/E ratios are important for the development of sutures *in vivo*, new human genetic findings have provided clear support for the hypothesis that Twist1 functions together with class I bHLH proteins to regulate suture development.

Wilkie's group initially identified heterozygous *TCF12* mutations in 38 unrelated families with coronal craniosynostosis (Sharma et al., 2013), Subsequent work revealed additional *TCF12* mutations, including two in individuals with clinical diagnoses of Saethre–Chotzen syndrome who did not have identified mutations in *TWIST1*. The prevalence of nonsense, frameshift, and splicing mutations strongly suggested a loss of function mechanism.

Analysis of calvarial development in Tcf12 mutant mice showed that heterozygotes do not have discernable defects in the coronal suture (Sharma et al., 2013). However, homozygous mutants, which survive for approximately 1 week after birth, have partial coronal synostosis (Brockop, M., Ting, M., & Maxson, R., unpublished observations). Thus, like Twist1, Tcf12 is required for the development of the coronal suture.

These findings suggest that the total quantity of Tcf12/Twist1 heterodimers may be a critical factor in coronal suture development. In strong support of this hypothesis, doubly heterozygous Tcf12–Twist1 mutants exhibit severe bicoronal synostosis, consistent with a strong genetic interaction between these two genes (Sharma et al., 2013). Cotransfection experiments showed a synergism between Twist1 and Tcf12 in the activation of a test promoter, further supporting a functional interaction. Genetic and biochemical data in combination thus suggest that Twist1 and Tcf12 function together to control coronal suture development. Given that Tcf12 can form heterodimers with Twist1 (Connerney et al., 2006), it is likely that the strong genetic interaction between Twist1 and Tcf12 reflects a direct protein–protein interaction that is crucial for function.

In summary, Twist1 and Tcf12 function together in a regulatory hierarchy controlling coronal suture development. It is likely that together they regulate EphA4, ephrinA2, Notch2, and Jagged1, thus controlling the targeting of osteogenic precursor cells and the specification of the coronal suture. Whether these influences involve direct effects on transcription or are indirect is not yet known.

3.5 FGF Receptors

Mutations in FGF receptors are a major cause of craniofacial malformations (Hajihosseini, 2008; Holmes, 2012; Marie, Coffin, & Hurley, 2005). Missense mutations in FGFR1–3 result in chondrodysplasia and craniosynostosis syndromes, including Apert, Crouzon, Jackson–Weiss, and Pfeiffer syndromes (Bellus et al., 1996; Ornitz, 2005). Here, we focus on recent work in which mouse models have been used to investigate the pathophysiology of Apert syndrome, which is caused by S252W or P253R mutations in FGFR2 (reviewed by Holmes & Basilico, 2012; Holmes et al., 2009). These mutations result in an increase in the affinity and reduction in the specificity of the receptor for FGF ligands (Park et al., 1995; Yu et al., 2005). Both mutations have been modeled in mice by knockin approaches, and both result in a series of malformations resembling those seen in humans with Apert syndrome (Chen, Li, Li, Engel, & Deng, 2003; Wang et al., 2005; Yin et al., 2008).

Deng and colleagues (Chen et al., 2003) produced S252W mutant mice in which the expression of the mutant receptor was dependent on Cre recombinase. Mutant mice have synostosis of the coronal suture as well as brachycephaly, distorted skulls, and widely spaced eyes. Jabs and colleagues also produced and characterized an S252W mutant mouse (Wang et al., 2005), documenting morphological phenotypes closely similar to those described by Deng and colleagues. Subsequently, the Deng group demonstrated that a small hairpin RNA targeting Fgfr2 S252W in mice prevented Apert-like craniofacial defects (Shukla, Coumoul, Wang, Kim, & Deng, 2007). Moreover, when mice were treated with the MEK1/2 inhibitor NO126, craniosynostosis was significantly inhibited, leading the authors to propose a pathogenic role for ERK activation in craniosynostosis.

Yin and colleagues (2008) produced mice with the P253R mutation. They showed that these mice had coronal synostosis, midface hypoplasia, malocclusion, and dysplasia of the skull base. This set of phenotypes is similar to that of Apert patients, with the exception of syndactyly, which is common in patients but rare in P253R mice.

Holmes and colleagues undertook a more detailed characterization of S252W mutant mice (Holmes & Basilico, 2012; Holmes et al., 2009). They used a Cre/lox approach to conditionally express the S252W mutation in the mesoderm or neural crest. They found that expression in the mesoderm, from which the coronal suture is derived, was necessary and sufficient to cause coronal synostosis (Holmes & Basilico, 2012). Expression in the neural crest had no consequences for coronal suture development. Interestingly, Cre-based lineage tracing showed that the boundary between the suture and the adjacent osteogenic territories remained intact in S252W mutants. Thus, mixing of osteogenic cells with sutural mesenchyme was not a cause of craniosynostosis. Instead, they found that the suture mesenchyme itself underwent ossification (Holmes & Basilico, 2012; Holmes et al., 2009). Thus, the pathophysiological mechanisms underlying craniosynostosis in this Apert mouse model appear distinct from those caused by mutations in Twist1, EphA4, and Jagged1.

3.6 ERF/ERK1/2

Wilkie's group found that in humans, reduced dosage of ERF causes synostosis of multiple calvarial sutures (Twigg et al., 2013). ERF encodes an ETS transcription factor that binds directly to ERK1/2. Mice with reduced Erf activity exhibit a similar multi-suture phenotype. A ChIP-seq approach in mouse embryo fibroblasts showed that Erf binding sites are often close to binding sites for Ap-1 and Runx, and transfection experiments using a DNA target with a Runx core sequence showed that Erf inhibited Runx2-mediated transactivation. Erf may thus antagonize the activity of Runx2, thereby influencing osteogenesis and suture patency.

3.7 IL11-GP130

Nieminen and colleagues showed that mutations in IL11RA, which encodes interleukin receptor alpha, cause a recessively inherited human syndrome characterized by craniosynostosis, maxillary hypoplasia, delayed tooth eruption, and supernumerary teeth (Nieminen et al., 2011). Il11ra-deficient mice exhibit similar defects, although synostosis of the calvarial sutures occurs rarely. IL11RA signals through the JAK/STAT1/3 or the SHP2/MAPK/ERK cascades (Dahmen et al., 1998; Kiessling et al., 2004).

It is interesting, given this potential link between IL11RA and the Jak/Stat1/3 pathway, that loss of function mutations in STAT3 in humans can cause hyper IgE syndrome, which has craniosynostosis as an occasional feature (Hoger, Boltshauser, & Hitzig, 1985). Moreover, we found recently that conditional inactivation of Stat3 in neural crest causes coronal synostosis in mice (Dasgupta, K., & Maxson, R., unpublished observation).

3.8 The Bmp Pathway

The Bmp pathway is fundamental to the development of the calvarial bones and sutures (Maxson & Ishii, 2008). In a recent study, Mishina and colleagues (Komatsu et al., 2013) tested the effect of enhanced Bmp signaling in cranial neural crest cells on suture development. Using transgenic mice that conditionally express a constitutively active form of Bmpr1a, they found that enhanced Bmp signaling in undifferentiated cranial neural crest cells but not in committed osteoblasts caused craniosynostosis. These results are consistent with findings of Longaker and colleagues to the effect that reduced expression of the Bmp antagonist, Noggin, can cause craniosynostosis (Warren, Brunet, Harland, Economides, & Longaker, 2003).

Finally, a recent genome-wide association study identified susceptibility loci for nonsyndromic sagittal craniosynostosis (Justice et al., 2012). One locus was near *Bmp2*, a gene with well-documented functions in calvarial growth and craniosynostosis. The other was within BBS9, which has a crucial role in ciliary function.

3.9 The Hh/Gli Pathway

Mutations in GLI3, encoding a kruppel family transcription factor, cause Grieg Cephalopolysyndactyly syndrome, which includes craniosynostosis as a feature (Vortkamp, Gessler, & Grzeschik, 1991). Gli3 is an effector of Hedgehog signaling. Rice and colleagues showed that Gli3 mutant mice exhibit craniosynostosis in the interfrontal and lambdoid sutures (Rice et al., 2010; Veistinen et al., 2012). This is associated with aberrant of Runx2 expression and reduced Twist1 expression in sutural cells. Implantation of FGF-soaked beads in the suture, a maneuver that caused Twist1 to become elevated, reduced Runx2 expression and reduced osteoprogenitor differentiation, rescuing craniosynostosis. This finding led to the expectation that genetic reduction of Runx2 function should also rescue craniosynostosis, which Rice's group subsequently verified (Tanimoto, Veistinen, Alakurtti, Takatalo, & Rice, 2012). They showed, moreover, that reduced Runx2 dosage normalized cell proliferation, which was increased in the Gli3 mutant, and also normalized p-Smad 1/5/8 phosphorylation, which was upregulated in the mutant. They interpreted these data to mean that Gli3 and Shh signaling functions in suture development through Runx2 and the Bmp–Smad pathway. Their results raise the intriguing possibility that targeting Runx2 may offer a means of mitigating craniosynostosis in patients.

Further implicating the Hh pathway in craniosynostosis, Klopocki and colleagues (Klopocki et al., 2011) made the intriguing observation that copy number variations in conserved noncoding elements are associated with syndactyly and craniosynostosis. These elements are capable of driving reporter gene expression in a manner similar to endogenous Ihh expression, suggesting that the duplications lead to misexpression of Ihh and thus cause changes in downstream gene activity.

New findings from Chai's group suggest a role for Gli transcription factors and Hh signaling in postnatal growth and homeostasis of the calvarial sutures (Zhao et al., 2015). This group identified a population of Gli1-expressing cells in adult sutures. Lineage analysis showed that these cells give rise to the osteogenic front, periosteum, dura, and the calvarial bones. These cells appear to be stem cells for calvarial bones. Intriguingly, ablation of this population resulted in craniosynostosis, suggesting that it is required for suture patency. Heterozygous loss of Twist1 function resulted in a reduction of this population, raising the possibility that Twist1 may not only control suture formation during embryonic development but also suture maintenance and bone homeostasis in adults. Future work will no doubt focus on the relationship between craniosynostosis caused by loss of the Gli1-expressing cell population in adults and craniosynostosis caused by embryonic mechanisms, such as loss of boundary integrity or changes in the specification state of sutural cells.

In conclusion, analysis of the pathophysiology underling craniosynostosis has identified a variety of cellular mechanisms, mediated by a range of signaling pathways and effector transcription factors. Cellular mechanisms include loss of boundary integrity, altered sutural cell specification in embryos, and loss of a suture stem cell population in adults. Whether there exists a deep structure of regulatory interactions and cellular processes that unify these seemingly diverse mechanisms remains to be determined and will no doubt be a major focus of research.

ACKNOWLEDGMENTS

This work was supported by National Institutes of Health grants (numbers DE012450 and DE016320) to R. E. M.

REFERENCES

Alagille, D., Estrada, A., Hadchouel, M., Gautier, M., Odievre, M., & Dommergues, J. P. (1987). Syndromic paucity of interlobular bile ducts (Alagille syndrome or arteriohepatic dysplasia): Review of 80 cases. *Journal of Pediatrics*, *110*(2), 195–200.

Bellus, G. A., Gaudenz, K., Zackai, E. H., Clarke, L. A., Szabo, J., Francomano, C. A., et al. (1996). Identical mutations in three different fibroblast growth factor receptor genes in autosomal dominant craniosynostosis syndromes. *Nature Genetics*, *14*(2), 174–176.

Bonilla-Claudio, M., Wang, J., Bai, Y., Klysik, E., Selever, J., & Martin, J. F. (2012). Bmp signaling regulates a dose-dependent transcriptional program to control facial skeletal development. *Development*, *139*(4), 709–719.

Bourgeois, P., Stoetzel, C., Bolcato-Bellemin, A. L., Mattei, M. G., & Perrin-Schmitt, F. (1996). The human H-twist gene is located at 7p21 and encodes a B-HLH protein that is 96% similar to its murine M-twist counterpart. *Mammalian Genome*, 7(12), 915–917.

Brugger, S. M., Merrill, A. E., Torres-Vazquez, J., Wu, N., Ting, M. C., Cho, J. Y., et al. (2004). A phylogenetically conserved cis-regulatory module in the Msx2 promoter is sufficient for BMP-dependent transcription in murine and Drosophila embryos. *Development*, *131*(20), 5153–5165.

Chai, Y., & Maxson, R. E., Jr. (2006). Recent advances in craniofacial morphogenesis. *Developmental Dynamics*, *235*(9), 2353–2375.

Chen, L., Li, D., Li, C., Engel, A., & Deng, C. X. (2003). A Ser252Trp [corrected] substitution in mouse fibroblast growth factor receptor 2 (Fgfr2) results in craniosynostosis. *Bone*, *33*(2), 169–178.

Cohen, M. M., Jr. (2006). The new bone biology: Pathologic, molecular, and clinical correlates. *American Journal of Medical Genetics Part A*, *140*(23), 2646–2706.

Connerney, J., Andreeva, V., Leshem, Y., Mercado, M. A., Dowell, K., Yang, X., et al. (2008). Twist1 homodimers enhance FGF responsiveness of the cranial sutures and promote suture closure. *Developmental Biology*, *318*(2), 323–334.

Connerney, J., Andreeva, V., Leshem, Y., Muentener, C., Mercado, M. A., & Spicer, D. B. (2006). Twist1 dimer selection regulates cranial suture patterning and fusion. *Developmental Dynamics*, *235*(5), 1345–1357.

Couly, G. F., Coltey, P. M., & Le Douarin, N. M. (1993). The triple origin of skull in higher vertebrates: A study in quail-chick chimeras. *Development*, *117*(2), 409–429.

Dahmen, H., Horsten, U., Kuster, A., Jacques, Y., Minvielle, S., Kerr, I. M., et al. (1998). Activation of the signal transducer gp130 by interleukin-11 and interleukin-6 is mediated by similar molecular interactions. *Biochemical Journal*, *331*(Pt. 3), 695–702.

Deckelbaum, R. A., Holmes, G., Zhao, Z., Tong, C., Basilico, C., & Loomis, C. A. (2012). Regulation of cranial morphogenesis and cell fate at the neural crest-mesoderm boundary by engrailed 1. *Development, 139*(7), 1346–1358.
Doyle, A. J., Doyle, J. J., Bessling, S. L., Maragh, S., Lindsay, M. E., Schepers, D., et al. (2012). Mutations in the TGF-beta repressor SKI cause Shprintzen-Goldberg syndrome with aortic aneurysm. *Nature Genetics, 44*(11), 1249–1254.
el Ghouzzi, V., Le Merrer, M., Perrin-Schmitt, F., Lajeunie, E., Benit, P., Renier, D., et al. (1997). Mutations of the TWIST gene in the Saethre-Chotzen syndrome. *Nature Genetics, 15*(1), 42–46.
Emerick, K. M., Rand, E. B., Goldmuntz, E., Krantz, I. D., Spinner, N. B., & Piccoli, D. A. (1999). Features of Alagille syndrome in 92 patients: Frequency and relation to prognosis. *Hepatology, 29*(3), 822–829.
Florisson, J. M., Verkerk, A. J., Huigh, D., Hoogeboom, A. J., Swagemakers, S., Kremer, A., et al. (2013). Boston type craniosynostosis: Report of a second mutation in MSX2. *American Journal of Medical Genetics Part A, 161A*(10), 2626–2633.
Funato, N., Ohtani, K., Ohyama, K., Kuroda, T., & Nakamura, M. (2001). Common regulation of growth arrest and differentiation of osteoblasts by helix-loop-helix factors. *Molecular and Cellular Biology, 21*(21), 7416–7428.
Goodnough, L. H., Chang, A. T., Treloar, C., Yang, J., Scacheri, P. C., & Atit, R. P. (2012). Twist1 mediates repression of chondrogenesis by beta-catenin to promote cranial bone progenitor specification. *Development, 139*(23), 4428–4438.
Goodnough, L. H., Dinuoscio, G. J., Ferguson, J. W., Williams, T., Lang, R. A., & Atit, R. P. (2014). Distinct requirements for cranial ectoderm and mesenchyme-derived wnts in specification and differentiation of osteoblast and dermal progenitors. *PLoS Genetics, 10*(2), e1004152.
Gross, J. B., & Hanken, J. (2005). Cranial neural crest contributes to the bony skull vault in adult Xenopus laevis: Insights from cell labeling studies. *Journal of Experimental Zoology Part B: Molecular and Developmental Evolution, 304*(2), 169–176.
Hajihosseini, M. K. (2008). Fibroblast growth factor signaling in cranial suture development and pathogenesis. *Frontiers of Oral Biology, 12*, 160–177.
Han, J., Ishii, M., Bringas, P., Jr., Maas, R. L., Maxson, R. E., Jr., & Chai, Y. (2007). Concerted action of Msx1 and Msx2 in regulating cranial neural crest cell differentiation during frontal bone development. *Mechanisms of Development, 124*(9-10), 729–745.
Hoger, P. H., Boltshauser, E., & Hitzig, W. H. (1985). Craniosynostosis in hyper-IgE-syndrome. *European Journal of Pediatrics, 144*(4), 414–417.
Holmes, G. (2012). The role of vertebrate models in understanding craniosynostosis. *Child's Nervous System, 28*(9), 1471–1481.
Holmes, G., & Basilico, C. (2012). Mesodermal expression of Fgfr2S252W is necessary and sufficient to induce craniosynostosis in a mouse model of Apert syndrome. *Developmental Biology, 368*(2), 283–293.
Holmes, G., Rothschild, G., Roy, U. B., Deng, C. X., Mansukhani, A., & Basilico, C. (2009). Early onset of craniosynostosis in an Apert mouse model reveals critical features of this pathology. *Developmental Biology, 328*(2), 273–284.
Howard, T. D., Paznekas, W. A., Green, E. D., Chiang, L. C., Ma, N., Ortiz de Luna, R. I., et al. (1997). Mutations in TWIST, a basic helix-loop-helix transcription factor, in Saethre-Chotzen syndrome. *Nature Genetics, 15*(1), 36–41.
Iqbal, Z., Cejudo-Martin, P., de Brouwer, A., van der Zwaag, B., Ruiz-Lozano, P., Scimia, M. C., et al. (2010). Disruption of the podosome adaptor protein TKS4 (SH3PXD2B) causes the skeletal dysplasia, eye, and cardiac abnormalities of Frank-Ter Haar Syndrome. *American Journal of Human Genetics, 86*(2), 254–261.
Ishii, M., Arias, A. C., Liu, L., Chen, Y. B., Bronner, M. E., & Maxson, R. E. (2012). A stable cranial neural crest cell line from mouse. *Stem Cells and Development, 21*(17), 3069–3080.

Ishii, M., Merrill, A. E., Chan, Y. S., Gitelman, I., Rice, D. P., Sucov, H. M., et al. (2003). Msx2 and Twist cooperatively control the development of the neural crest-derived skeletogenic mesenchyme of the murine skull vault. *Development*, *130*(24), 6131–6142.

Ito, Y., Yeo, J. Y., Chytil, A., Han, J., Bringas, P., Jr., Nakajima, A., et al. (2003). Conditional inactivation of Tgfbr2 in cranial neural crest causes cleft palate and calvaria defects. *Development (Cambridge, England)*, *130*(21), 5269–5280.

Jabs, E. W., Muller, U., Li, X., Ma, L., Luo, W., Haworth, I. S., et al. (1993). A mutation in the homeodomain of the human MSX2 gene in a family affected with autosomal dominant craniosynostosis. *Cell*, *75*(3), 443–450.

Janssen, A., Hosen, M. J., Jeannin, P., Coucke, P. J., De Paepe, A., & Vanakker, O. M. (2013). Second family with the Boston-type craniosynostosis syndrome: Novel mutation and expansion of the clinical spectrum. *American Journal of Medical Genetics Part A*, *161A*(9), 2352–2357.

Jenkins, D., Seelow, D., Jehee, F. S., Perlyn, C. A., Alonso, L. G., Bueno, D. F., et al. (2007). RAB23 mutations in Carpenter syndrome imply an unexpected role for hedgehog signaling in cranial-suture development and obesity. *American Journal of Human Genetics*, *80*(6), 1162–1170.

Jiang, X., Iseki, S., Maxson, R. E., Sucov, H. M., & Morriss-Kay, G. M. (2002). Tissue origins and interactions in the mammalian skull vault. *Developmental Biology*, *241*(1), 106–116.

Jiang, X., Rowitch, D. H., Soriano, P., McMahon, A. P., & Sucov, H. M. (2000). Fate of the mammalian cardiac neural crest. *Development*, *127*(8), 1607–1616.

Johnson, D., Iseki, S., Wilkie, A. O., & Morriss-Kay, G. M. (2000). Expression patterns of Twist and Fgfr1, -2 and -3 in the developing mouse coronal suture suggest a key role for twist in suture initiation and biogenesis. *Mechanisms of Development*, *91*(1-2), 341–345.

Justice, C. M., Yagnik, G., Kim, Y., Peter, I., Jabs, E. W., Erazo, M., et al. (2012). A genome-wide association study identifies susceptibility loci for nonsyndromic sagittal craniosynostosis near BMP2 and within BBS9. *Nature Genetics*, *44*(12), 1360–1364.

Kague, E., Gallagher, M., Burke, S., Parsons, M., Franz-Odendaal, T., & Fisher, S. (2012). Skeletogenic fate of zebrafish cranial and trunk neural crest. *PLoS One*, 7(11), e47394.

Kamath, B. M., Spinner, N. B., Emerick, K. M., Chudley, A. E., Booth, C., Piccoli, D. A., et al. (2004). Vascular anomalies in Alagille syndrome: A significant cause of morbidity and mortality. *Circulation*, *109*(11), 1354–1358.

Kamath, B. M., Stolle, C., Bason, L., Colliton, R. P., Piccoli, D. A., Spinner, N. B., et al. (2002). Craniosynostosis in Alagille syndrome. *American Journal of Medical Genetics*, *112*(2), 176–180.

Kiessling, S., Muller-Newen, G., Leeb, S. N., Hausmann, M., Rath, H. C., Strater, J., et al. (2004). Functional expression of the interleukin-11 receptor alpha-chain and evidence of antiapoptotic effects in human colonic epithelial cells. *Journal of Biological Chemistry*, *279*(11), 10304–10315.

Klein, R. (2004). Eph/ephrin signaling in morphogenesis, neural development and plasticity. *Current Opinion in Cell Biology*, *16*(5), 580–589.

Klopocki, E., Lohan, S., Brancati, F., Koll, R., Brehm, A., Seemann, P., et al. (2011). Copy-number variations involving the IHH locus are associated with syndactyly and craniosynostosis. *American Journal of Human Genetics*, *88*(1), 70–75.

Komatsu, Y., Yu, P. B., Kamiya, N., Pan, H., Fukuda, T., Scott, G. J., et al. (2013). Augmentation of Smad-dependent BMP signaling in neural crest cells causes craniosynostosis in mice. *Journal of Bone and Mineral Research*, *28*(6), 1422–1433.

Krantz, I. D., Colliton, R. P., Genin, A., Rand, E. B., Li, L., Piccoli, D. A., et al. (1998). Spectrum and frequency of jagged1 (JAG1) mutations in Alagille syndrome patients and their families. *American Journal of Human Genetics*, *62*(6), 1361–1369.

Kullander, K., & Klein, R. (2002). Mechanisms and functions of Eph and ephrin signalling. *Nature Reviews Molecular Cell Biology*, *3*(7), 475–486.

Le Douarin, N. M., Creuzet, S., Couly, G., & Dupin, E. (2004). Neural crest cell plasticity and its limits. *Development*, *131*(19), 4637–4650.

Le Douarin, N. M., & Kalcheim, C. (1999). *The neural crest* (2nd ed.). New York: Cambridge University Press.

Le Lievre, C. S. (1978). Participation of neural crest-derived cells in the genesis of the skull in birds. *Journal of Embryology and Experimental Morphology*, *47*, 17–37.

Li, C., Scott, D. A., Hatch, E., Tian, X., & Mansour, S. L. (2007). Dusp6 (Mkp3) is a negative feedback regulator of FGF-stimulated ERK signaling during mouse development. *Development*, *134*(1), 167–176.

Loeys, B. L., Chen, J., Neptune, E. R., Judge, D. P., Podowski, M., Holm, T., et al. (2005). A syndrome of altered cardiovascular, craniofacial, neurocognitive and skeletal development caused by mutations in TGFBR1 or TGFBR2. *Nature Genetics*, *37*(3), 275–281.

Marie, P. J., Coffin, J. D., & Hurley, M. M. (2005). FGF and FGFR signaling in chondrodysplasias and craniosynostosis. *Journal of Cellular Biochemistry*, *96*(5), 888–896.

Martinez, A., & Soriano, E. (2005). Functions of ephrin/Eph interactions in the development of the nervous system: Emphasis on the hippocampal system. *Brain Research. Brain Research Reviews*, *49*(2), 211–226.

Massari, M. E., & Murre, C. (2000). Helix-loop-helix proteins: Regulators of transcription in eucaryotic organisms. *Molecular and Cellular Biology*, *20*(2), 429–440.

Maxson, R., & Ishii, M. (2008). The Bmp pathway in skull vault development. *Frontiers of Oral Biology*, *12*, 197–208.

Merrill, A. E., Bochukova, E. G., Brugger, S. M., Ishii, M., Pilz, D. T., Wall, S. A., et al. (2006). Cell mixing at a neural crest-mesoderm boundary and deficient ephrin-Eph signaling in the pathogenesis of craniosynostosis. *Human Molecular Genetics*, *15*(8), 1319–1328.

Moenning, A., Jager, R., Egert, A., Kress, W., Wardelmann, E., & Schorle, H. (2009). Sustained platelet-derived growth factor receptor alpha signaling in osteoblasts results in craniosynostosis by overactivating the phospholipase C-gamma pathway. *Molecular and Cellular Biology*, *29*(3), 881–891.

Morriss-Kay, G. M., & Wilkie, A. O. (2005). Growth of the normal skull vault and its alteration in craniosynostosis: Insights from human genetics and experimental studies. *Journal of Anatomy*, *207*(5), 637–653.

Nieminen, P., Morgan, N. V., Fenwick, A. L., Parmanen, S., Veistinen, L., Mikkola, M. L., et al. (2011). Inactivation of IL11 signaling causes craniosynostosis, delayed tooth eruption, and supernumerary teeth. *American Journal of Human Genetics*, *89*(1), 67–81.

Noden, D. M., & Trainor, P. A. (2005). Relations and interactions between cranial mesoderm and neural crest populations. *Journal of Anatomy*, *207*(5), 575–601.

Oda, T., Elkahloun, A. G., Pike, B. L., Okajima, K., Krantz, I. D., Genin, A., et al. (1997). Mutations in the human Jagged1 gene are responsible for Alagille syndrome. *Nature Genetics*, *16*(3), 235–242.

Ornitz, D. M. (2005). FGF signaling in the developing endochondral skeleton. *Cytokine and Growth Factor Reviews*, *16*(2), 205–213.

Palmer, A., & Klein, R. (2003). Multiple roles of ephrins in morphogenesis, neuronal networking, and brain function. *Genes and Development*, *17*(12), 1429–1450.

Park, W. J., Theda, C., Maestri, N. E., Meyers, G. A., Fryburg, J. S., Dufresne, C., et al. (1995). Analysis of phenotypic features and FGFR2 mutations in Apert syndrome. *American Journal of Human Genetics*, *57*(2), 321–328.

Pasquale, E. B. (2005). Eph receptor signalling casts a wide net on cell behaviour. *Nature Reviews Molecular Cell Biology*, *6*(6), 462–475.

Poliakov, A., Cotrina, M., & Wilkinson, D. G. (2004). Diverse roles of eph receptors and ephrins in the regulation of cell migration and tissue assembly. *Developmental Cell*, *7*(4), 465–480.

Rice, D. P., Aberg, T., Chan, Y., Tang, Z., Kettunen, P. J., Pakarinen, L., et al. (2000). Integration of FGF and TWIST in calvarial bone and suture development. *Development*, *127*(9), 1845–1855.

Rice, D. P., Connor, E. C., Veltmaat, J. M., Lana-Elola, E., Veistinen, L., Tanimoto, Y., et al. (2010). Gli3Xt-J/Xt-J mice exhibit lambdoid suture craniosynostosis which results from altered osteoprogenitor proliferation and differentiation. *Human Molecular Genetics*, *19*(17), 3457–3467.

Roybal, P. G., Wu, N. L., Sun, J., Ting, M. C., Schafer, C. A., & Maxson, R. E. (2010). Inactivation of Msx1 and Msx2 in neural crest reveals an unexpected role in suppressing heterotopic bone formation in the head. *Developmental Biology*, *343*(1-2), 28–39.

Saga, Y., Miyagawa-Tomita, S., Takagi, A., Kitajima, S., Miyazaki, Ji, & Inoue, T. (1999). MesP1 is expressed in the heart precursor cells and required for the formation of a single heart tube. *Development*, *126*(15), 3437–3447.

Sasaki, T., Ito, Y., Bringas, P., Jr., Chou, S., Urata, M. M., Slavkin, H., et al. (2006). TGFbeta-mediated FGF signaling is crucial for regulating cranial neural crest cell proliferation during frontal bone development. *Development*, *133*(2), 371–381.

Satokata, I., Ma, L., Ohshima, H., Bei, M., Woo, I., Nishizawa, K., et al. (2000). Msx2 deficiency in mice causes pleiotropic defects in bone growth and ectodermal organ formation. *Nature Genetics*, *24*(4), 391–395.

Settle, S. H., Jr., Rountree, R. B., Sinha, A., Thacker, A., Higgins, K., & Kingsley, D. M. (2003). Multiple joint and skeletal patterning defects caused by single and double mutations in the mouse Gdf6 and Gdf5 genes. *Developmental Biology*, *254*(1), 116–130.

Sharma, V. P., Fenwick, A. L., Brockop, M. S., McGowan, S. J., Goos, J. A., Hoogeboom, A. J., et al. (2013). Mutations in TCF12, encoding a basic helix-loop-helix partner of TWIST1, are a frequent cause of coronal craniosynostosis. *Nature Genetics*, *45*(3), 304–307.

Shukla, V., Coumoul, X., Wang, R. H., Kim, H. S., & Deng, C. X. (2007). RNA interference and inhibition of MEK-ERK signaling prevent abnormal skeletal phenotypes in a mouse model of craniosynostosis. *Nature Genetics*, *39*(9), 1145–1150.

Sirmaci, A., Walsh, T., Akay, H., Spiliopoulos, M., Sakalar, Y. B., Hasanefendioglu-Bayrak, A., et al. (2010). MASP1 mutations in patients with facial, umbilical, coccygeal, and auditory findings of Carnevale, Malpuech, OSA, and Michels syndromes. *American Journal of Human Genetics*, *87*(5), 679–686.

Sood, S., Eldadah, Z. A., Krause, W. L., McIntosh, I., & Dietz, H. C. (1996). Mutation in fibrillin-1 and the Marfanoid-craniosynostosis (Shprintzen-Goldberg) syndrome. *Nature Genetics*, *12*(2), 209–211.

Sun, J., Ishii, M., Ting, M. C., & Maxson, R. (2013). Foxc1 controls the growth of the murine frontal bone rudiment by direct regulation of a Bmp response threshold of Msx2. *Development*, *140*(5), 1034–1044.

Surawska, H., Ma, P. C., & Salgia, R. (2004). The role of ephrins and Eph receptors in cancer. *Cytokine and Growth Factor Reviews*, *15*(6), 419–433.

Tanimoto, Y., Veistinen, L., Alakurtti, K., Takatalo, M., & Rice, D. P. (2012). Prevention of premature fusion of calvarial suture in GLI-Kruppel family member 3 (Gli3)-deficient mice by removing one allele of Runt-related transcription factor 2 (Runx2). *Journal of Biological Chemistry*, *287*(25), 21429–21438.

Ting, M. C., Wu, N. L., Roybal, P. G., Sun, J., Liu, L., Yen, Y., et al. (2009). EphA4 as an effector of Twist1 in the guidance of osteogenic precursor cells during calvarial bone growth and in craniosynostosis. *Development*, *136*(5), 855–864.

Tran, T. H., Jarrell, A., Zentner, G. E., Welsh, A., Brownell, I., Scacheri, P. C., et al. (2010). Role of canonical Wnt signaling/ss-catenin via Dermo1 in cranial dermal cell development. *Development*, *137*(23), 3973–3984.

Twigg, S. R., Kan, R., Babbs, C., Bochukova, E. G., Robertson, S. P., Wall, S. A., et al. (2004). Mutations of ephrin-B1 (EFNB1), a marker of tissue boundary formation, cause craniofrontonasal syndrome. *Proceedings of the National Academy of Sciences of the United States of America*, *101*(23), 8652–8657.

Twigg, S. R., Vorgia, E., McGowan, S. J., Peraki, I., Fenwick, A. L., Sharma, V. P., et al. (2013). Reduced dosage of ERF causes complex craniosynostosis in humans and mice and links ERK1/2 signaling to regulation of osteogenesis. *Nature Genetics*, *45*(3), 308–313.

Van Maldergem, L., Siitonen, H. A., Jalkh, N., Chouery, E., De Roy, M., Delague, V., et al. (2006). Revisiting the craniosynostosis-radial ray hypoplasia association: Baller-Gerold syndrome caused by mutations in the RECQL4 gene. *Journal of Medical Genetics*, *43*(2), 148–152.

Veistinen, L., Takatalo, M., Tanimoto, Y., Kesper, D. A., Vortkamp, A., & Rice, D. P. (2012). Loss-of-function of Gli3 in mice causes abnormal frontal bone morphology and premature synostosis of the interfrontal suture. *Frontiers in Physiology*, *3*, 121.

Vissers, L. E., Cox, T. C., Maga, A. M., Short, K. M., Wiradjaja, F., Janssen, I. M., et al. (2011). Heterozygous mutations of FREM1 are associated with an increased risk of isolated metopic craniosynostosis in humans and mice. *PLoS Genetics*, 7(9), e1002278.

Vortkamp, A., Gessler, M., & Grzeschik, K. H. (1991). GLI3 zinc-finger gene interrupted by translocations in Greig syndrome families. *Nature*, *352*(6335), 539–540.

Wang, Y., Xiao, R., Yang, F., Karim, B. O., Iacovelli, A. J., Cai, J., et al. (2005). Abnormalities in cartilage and bone development in the Apert syndrome FGFR2(+/S252W) mouse. *Development*, *132*(15), 3537–3548.

Warren, S. M., Brunet, L. J., Harland, R. M., Economides, A. N., & Longaker, M. T. (2003). The BMP antagonist noggin regulates cranial suture fusion. *Nature*, *422*(6932), 625–629. Department of Surgery, Stanford University School of Medicine, Stanford, California, USA.

Wilkie, A. O. (1997). Craniosynostosis: Genes and mechanisms. *Human Molecular Genetics*, *6*(10), 1647–1656.

Wilkie, A. O., Tang, Z., Elanko, N., Walsh, S., Twigg, S. R., Hurst, J. A., et al. (2000). Functional haploinsufficiency of the human homeobox gene MSX2 causes defects in skull ossification. *Nature Genetics*, *24*(4), 387–390.

Wilkinson, D. G. (2001). Multiple roles of EPH receptors and ephrins in neural development. *Nature Reviews Neuroscience*, *2*(3), 155–164.

Yagnik, G., Ghuman, A., Kim, S., Stevens, C. G., Kimonis, V., Stoler, J., et al. (2012). ALX4 gain-of-function mutations in nonsyndromic craniosynostosis. *Human Mutation*, *33*(12), 1626–1629.

Yen, H. Y., Ting, M. C., & Maxson, R. E. (2010). Jagged1 functions downstream of Twist1 in the specification of the coronal suture and the formation of a boundary between osteogenic and non-osteogenic cells. *Developmental Biology*, *347*(2), 258–270.

Yin, L., Du, X., Li, C., Xu, X., Chen, Z., Su, N., et al. (2008). A Pro253Arg mutation in fibroblast growth factor receptor 2 (Fgfr2) causes skeleton malformation mimicking human Apert syndrome by affecting both chondrogenesis and osteogenesis. *Bone*, *42*(4), 631–643.

Yoshida, T., Vivatbutsiri, P., Morriss-Kay, G., Saga, Y., & Iseki, S. (2008). Cell lineage in mammalian craniofacial mesenchyme. *Mechanisms of Development*, *125*(9-10), 797–808.

Yu, H. M., Jerchow, B., Sheu, T. J., Liu, B., Costantini, F., Puzas, J. E., et al. (2005). The role of Axin2 in calvarial morphogenesis and craniosynostosis. *Development*, *132*(8), 1995–2005.

Zhang, X., Kuroda, S., Carpenter, D., Nishimura, I., Soo, C., Moats, R., et al. (2002). Craniosynostosis in transgenic mice overexpressing Nell-1. *Journal of Clinical Investigation*, *110*(6), 861–870.

Zhao, H., Feng, J., Ho, T. V., Grimes, W., Urata, M., & Chai, Y. (2015). The suture provides a niche for mesenchymal stem cells of craniofacial bones. *Nature Cell Biology*, *17*(4), 386–396.

CHAPTER SEVEN

Tissue Interactions Regulating Tooth Development and Renewal

Anamaria Balic, Irma Thesleff[1]
Research Program in Developmental Biology, Institute of Biotechnology, University of Helsinki, Helsinki, Finland
[1]Corresponding author: e-mail address: Irma.Thesleff@helsinki.fi

Contents

Abstract

Reciprocal interactions between epithelial and mesenchymal tissues play a fundamental role in the morphogenesis of teeth and regulate all aspects of tooth development. Extensive studies on mouse tooth development over the past 25 years have uncovered the molecular details of the signaling networks mediating these interactions (reviewed by Jussila & Thesleff, 2012; Lan, Jia, & Jiang, 2014). Five conserved signaling pathways, namely, the Wnt, BMP, FGF, Shh, and Eda, are involved in the mediation

Current Topics in Developmental Biology, Volume 115
ISSN 0070-2153
http://dx.doi.org/10.1016/bs.ctdb.2015.07.006

of the successive reciprocal epithelial–mesenchymal cross talk which follows the general principle of morphogenetic interactions (Davidson, 1993). The pathways regulate the expression of transcription factors which confer the identity of dental epithelium and mesenchyme. The signals and transcription factors are integrated in complex signaling networks whose fine-tuning allows the generation of the variation in tooth morphologies. In this review, we describe the principles and molecular mechanisms of the epithelial–mesenchymal interactions regulating successive stages of tooth formation: (i) the initiation of tooth development, with special reference to the shift of tooth-forming potential from epithelium to mesenchyme; (ii) the morphogenesis of the tooth crown, focusing on the roles of epithelial signaling centers; (iii) the differentiation of odontoblasts and ameloblasts, which produce dentin and enamel, respectively; and (iv) the maintenance of dental stem cells, which support the continuous growth of teeth.

1. INTRODUCTION

Interactions between groups of cells and different domains of the developing embryo constitute key mechanisms of morphogenesis. Hans Spemann and Hilde Mangold were the first to experimentally demonstrate the concept of "inductive interactions" between tissues in transplantation studies using tissues derived from different newt species (Spemann & Mangold, 2001). Later, transplantation experiments revealed the importance of tissue interactions in the development of a variety of morphogenetic processes in the embryos and showed them to be "the most important mechanism regulating embryonic development," as John Gurdon stated in 1971 (Gurdon, 1971).

All morphogenetic tissue interactions in the embryo follow similar general principles and key molecular players, as presented schematically by Davidson (1993) (Fig. 1). Tissue interactions are sequential and reciprocal, and are mediated by several conserved families of signaling molecules including Wnts, BMPs, Hedgehogs, and FGFs. These molecules regulate gene expression, in particular, the expression of transcription factors that are shared between various tissues and organs. Generation of tissue and organ variety is the result of unique and exclusive combinations of transcription factors that regulate cell fate and determine the identity of the target cells during the successive stages of a given morphogenetic process (Davidson, 1993). As illustrated in our review, these same principles also apply to tooth development.

The development of all organs which form from the ectodermal and endodermal sheets lining the embryo is regulated by communication

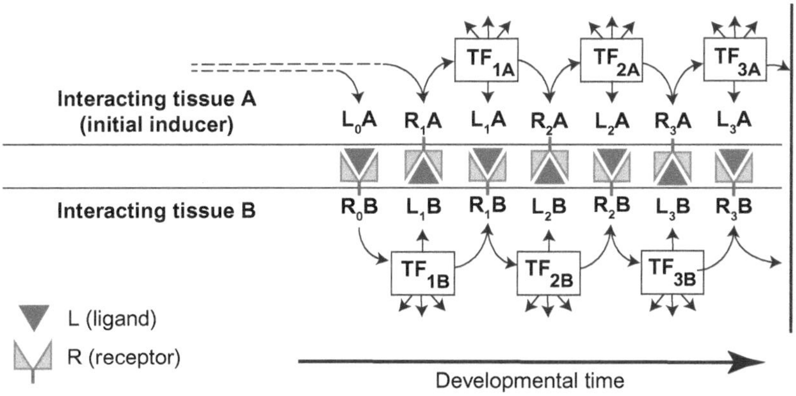

Figure 1 A chain of interactions between adjacent tissues undergoing progressive morphogenetic inductions. Release of the ligand (L) (signaling molecule) from the inductive tissue and its binding to the receptor (R) on the adjacent tissue triggers a sequence of regulatory changes in each tissue (A and B). These changes are controlled by the appearance of new transcription factors (TF) in response to signals from the adjacent tissue. New transcription factors activate a battery of genes (indicated by arrows emanating from the boxes). *Figure adapted from Davidson (1993).* (A color version of this figure is located at the end of this chapter.)

between the epithelium and underlying mesenchyme. A chain of reciprocal epithelial–mesenchymal interactions regulates all aspects of tooth development. This was first demonstrated in classic tissue recombination studies at the end of the 1960s by Edward J. Kollar, in which the epithelial and mesenchymal components of the developing mouse tooth germs were separated and combined with dental and nondental tissues, followed by intraocular engraftment in adult mice. These experiments showed that the tooth identity (molar vs. incisor) as well as the ability to reprogram nondental epithelium to tooth fate are determined by the mesenchyme (Kollar & Baird, 1969, 1970). Later tissue recombination experiments in the 1980s demonstrated that the odontogenic potential initially resides within the oral epithelium, which can induce tooth formation when recombined with neural crest-derived second branchial arch mesenchyme or with the premigratory neural crest cells (Lumsden, 1988; Mina & Kollar, 1987). These experiments indicated that the odontogenic potential shifts to the mesenchyme prior to the onset of epithelial morphogenesis at E (embryonic day) 12 (Fig. 2). In addition, tissue recombination experiments showed that epithelial–mesenchymal interactions regulate differentiation of the two major tooth-specific cell types, the odontoblasts and ameloblasts (reviewed by Hurmerinta & Thesleff, 1981; Ruch, Lesot, Karcher-Djuricic, Meyer, & Olive, 1982).

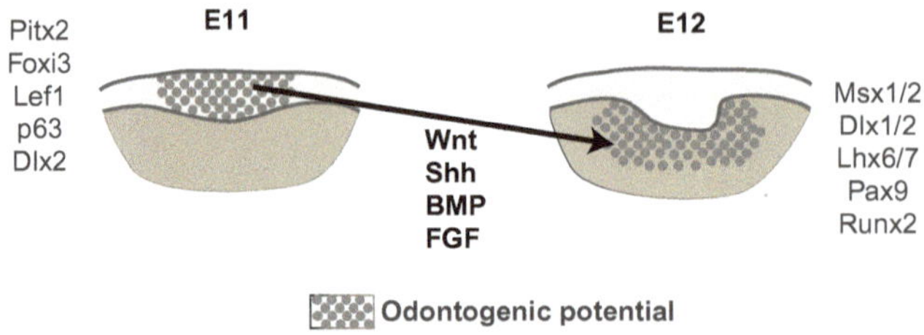

Figure 2 Shift of the odontogenic potential from epithelium to mesenchyme between the dental lamina and placode stages of tooth development. The dental lamina expresses specific transcription factors, which are associated with the dental identity and regulate the expression of signals in four conserved families. These signals mediate the shift of odontogenic potential to the mesenchyme and regulate an array of transcription factors which determine the identity and odontogenic potential of the dental mesenchyme. (A color version of this figure is located at the end of this chapter.)

In the past 25 years, great effort has been made to uncover the molecular details of the signaling networks mediating the interactions that regulate mouse tooth development (reviewed by Jussila & Thesleff, 2012; Lan et al., 2014; Thesleff, 2003). In this review, we describe the current understanding of the molecular mechanisms of the epithelial–mesenchymal interactions regulating various phases of tooth formation: (i) the initiation of tooth development, (ii) the morphogenesis of the tooth crown, (iii) the differentiation of the cells which produce mineralized dental tissues, and (iv) the maintenance of dental stem cells which support the continuous growth of teeth. The focus is on the mediation of tissue interactions by five conserved signaling pathways: the Wnt, BMP, FGF, Shh (Sonic Hedgehog), and Eda (ectodysplasin) pathways, which play fundamental roles at all stages of tooth development.

2. EPITHELIAL–MESENCHYMAL TISSUE INTERACTIONS REGULATING TOOTH INITIATION AND MORPHOGENESIS

2.1 Epithelial Signaling Centers

Teeth are initiated from the *dental lamina*, a stripe of stratified epithelium first discovered in histological sections that forms at E11 at the sites of future tooth rows (Fig. 2). Mouse embryonic dental lamina can be visualized by localized expression of several transcription factors and signaling molecules. *Pitx2* is the most specific marker of dental epithelium, and its expression persists in all epithelial cells of the developing tooth during crown morphogenesis (Mucchielli et al., 1997). Other transcription factors in the dental lamina include *Foxi3* (Shirokova et al., 2013), *Dlx2*, *Lef1*, and *p63*, which together with *Pitx2* may well be linked to the acquisition of tooth fate and odontogenic potential in the oral epithelium. The dental lamina also expresses signaling molecules, such as *Shh*, *Bmp2*, *Bmp4*, *Bmp7*, *Fgf8*, *Fgf9*, *Wnt10a*, and *Wnt10b* (Dassule & McMahon, 1998), that likely function as mediators of the odontogenic potential from the epithelium to the mesenchyme.

The morphogenesis of the first teeth in mouse is initiated from *placodes*, which appear in E12 mouse embryos as local thickenings of the dental lamina in the incisor and molar regions (reviewed by Biggs & Mikkola, 2014; Jussila & Thesleff, 2012). Thereafter, successional tooth development occurs in molars as the second and third molars (M2 and M3) develop posterior to the first (M1). The following text refers to the development of M1. The expression of many dental lamina genes becomes restricted to the placodes, including *Pitx2* and *Foxi3* as well as several signaling pathway-associated molecules in all five signaling families. Notably, the expression of signals such as *Shh*, *Wnt10*, *Bmp2*, and *Fgf20* is further restricted to a small cluster of placodal cells called the *early signaling center* (Fig. 3; Haara et al., 2012;

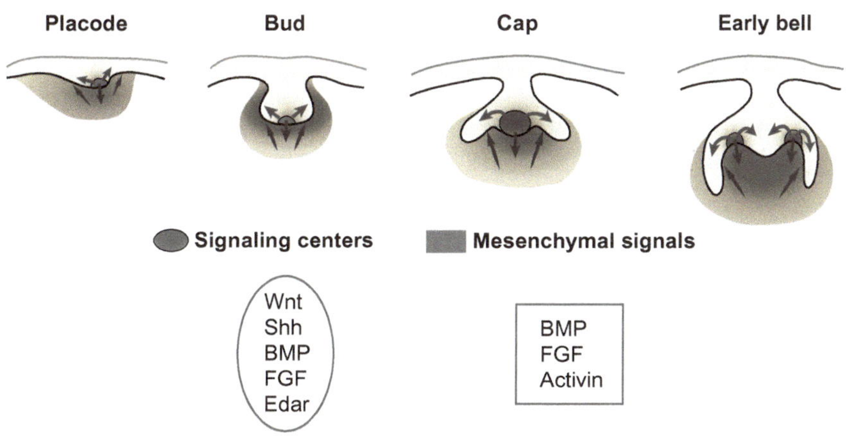

Figure 3 Epithelial–mesenchymal tissue interactions and epithelial signaling centers regulating tooth morphogenesis. Of particular significance for morphogenesis are the epithelial signaling centers which express signals of the four conserved signal families. They also express Edar, the receptor of ectodysplasin (Eda). The early signaling center in the placode regulates epithelial budding, while the primary enamel knot regulates the shift from bud to cap stage as well as the advancement of morphogenesis to bell stage. The secondary enamel knots regulate patterning of the tooth cusps in molar teeth during bell stage. The reciprocal signals expressed in mesenchyme include FGFs, BMPs, and Activin. Arrows indicate signaling across the two tissues and within the epithelium. (A color version of this figure is located at the end of this chapter.)

Jussila & Thesleff, 2012; Keranen, Kettunen, Aberg, Thesleff, & Jernvall, 1999). These signals regulate the proliferation and downgrowth of the placodal epithelium which gives rise to the tooth bud (Fig. 3).

The next signaling center, known as the (primary) *enamel knot,* forms at the tip of the bud and is characterized by the exit of cells from cell cycle (expression of *p21*) as well as localized expression of signaling molecules and signaling pathway-linked genes. The signals stimulate the growth of the flanking epithelium which gives rise to the cervical loops (CLs), and tooth morphogenesis advances from bud to cap stage. The enamel knot matures and becomes a morphologically discernible feature that is a source of many signals from the different families (reviewed by Jussila & Thesleff, 2012).

During the following bell stage, the epithelial growth and folding determine the shape and size of the tooth crown. In the molars, *secondary enamel knots* are induced in the epithelium and they act as signaling centers, expressing largely the same signals as the primary enamel knot (Fig. 3). It is the secondary enamel knots that determine the characteristic and reproducible positions and heights of tooth cusps. All five conserved signaling pathways are important for cusp patterning, as shown by the aberrant cusp patterns in mouse mutants in which signaling has been modulated (reviewed by Jernvall & Thesleff, 2012, and see below).

2.2 The Shift of Odontogenic Potential from Epithelium to Mesenchyme: Signals and Transcription Factors

The capacity to form teeth was shown to reside in the oral epithelium of E10 and E11 mouse embryos (dental lamina stage), and to shift to the underlying neural crest-derived mesenchyme by E12 (placode stage) (Fig. 2; Mina & Kollar, 1987). Signals emanating from the dental lamina are involved in transmission of the odontogenic potential to the mesenchyme and are preserved through subsequent stages of tooth development in the epithelial signaling centers, possibly contributing to the maintenance of the odontogenic potential in the mesenchyme. However, there is so far no evidence of specific signal(s) that would be sufficient to transmit the odontogenic potential. This is because its molecular basis is not known yet.

The shift of odontogenic potential coincides with the condensation of the dental mesenchymal cells, and it is therefore likely that the same epithelial signals are involved in both processes. Interestingly, epithelial *Fgf8* and *Sema3f* were shown to attract and repulse, respectively, the dental mesenchymal cells, and the mechanical stimuli associated with their compaction were sufficient to induce the expression of some dental mesenchyme-specific genes, namely, *Msx1*, *Pax9*, and *Bmp4* (Mammoto et al., 2011).

Cellular identity is determined by the unique combination of transcription factors expressed by a cell, and many different cell types can be reprogrammed by specific sets of transcription factors (Iwafuchi-Doi & Zaret, 2014). Transcription factors induced in the mesenchyme between the dental lamina and placode stages include *Msx1*, *Msx2*, *Pax9*, *Lhx6*, *Lhx7*, *Dlx1*, *Dlx2*, and *Runx2*, and it is likely that at least some of these are responsible for the acquisition of odontogenic potential in the mesenchyme. Notably, all of the listed genes are necessary for early tooth morphogenesis, as demonstrated in single and compound knockout mice (reviewed by Bei, 2009). Tooth development is arrested at the dental lamina stage in some compound mutants, including *Msx1;Msx2, Dlx1;Dlx2*, and *Lhx6; Lhx7* (only molars arrested), while arrest at the bud stage occurs when only *Pax9, Msx1,* or *Runx2* is deleted. In addition, arrested development at the lamina stage in *Gli2;Gli3* compound mutants and at the bud stage in *Lef1* knockouts indicates the necessary functions of the Shh and Wnt/β-catenin pathways. It remains to be demonstrated which transcription factors can program the acquisition of dental mesenchyme identity.

The gene regulatory networks in the mesenchyme have been elucidated particularly in association with FGF and BMP signaling. Many of the transcription factors mentioned above are regulated by epithelial BMPs and FGFs during the early steps of tooth formation. *Bmp4* was the first epithelial signal associated with the shift of odontogenic potential to the mesenchyme, while epithelial *Fgf8* and *Fgf9* were shown to be necessary for tooth initiation (Trumpp, Depew, Rubenstein, Bishop, & Martin, 1999; Vainio, Karavanova, Jowett, & Thesleff, 1993). Availability of these molecules as recombinant proteins enabled the elucidation of their function in organ culture through use of beads soaked in the proteins. Collectively, these studies have unraveled signaling cascades demonstrating that epithelial *Bmp4* regulated the expression of transcription factors *Msx1* and *Msx2* as well as *Bmp4* in the dental mesenchyme, while FGF signaling exclusively regulated *Dlx1* and *Pax9* (Vainio et al., 1993). Further studies identified common targets for epithelial BMP and FGF signaling, including *Msx1*, *Dlx2*, and *Runx2* in the dental mesenchyme (Aberg et al., 2004; Bei & Maas, 1998; Neubuser, Peters, Balling, & Martin, 1997; Thomas, Liu, Rubenstein, & Sharpe, 2000; Vainio et al., 1993). Upregulation of *Msx1* by BMP and FGF signaling represents two independent signaling cascades that result in activation of mesenchymal *Bmp4* and *Fgf3*, respectively (Bei & Maas, 1998). Therefore, the same mesenchymal transcription factors can be used to regulate different downstream targets indicating complex signaling networks formed during tooth development. Zhao, Gupta, Raj, Roussel, and Bei (2013) unraveled a complex network of transcription factors belonging to different signaling pathways (*Msx1*, *Lef1*, *Snail*, *Lhx6/8*, and *Sp3*) that regulates cell cycle. This *Msx1*-dependent signaling network was further expanded by study showing antagonistic interactions between Msx1 and Tbx2 that regulate *Bmp4* expression (Saadi et al., 2013).

2.3 Signaling Pathways and Networks in the Mediation of Epithelial–Mesenchymal Interactions During Morphogenesis

The epithelial signaling centers together with the specified dental mesenchyme are the key players in the tissue interactions which regulate tooth morphogenesis from the placode to bell stage (Fig. 3). Diagrams of proposed regulatory networks underlying epithelial–mesenchymal cross talk can be found in many publications (Aberg et al., 2004; Bei, Kratochwil, & Maas, 2000; Bei & Maas, 1998; Klein et al., 2006; Lan et al., 2014). The battery of signaling molecules expressed by the *early signaling center in the placode* and subsequently by the *primary and secondary enamel knots* regulates epithelial budding, transition to cap stage, and crown morphogenesis during bell stage. In addition to the five signaling pathways that are the focus of this review, signals in other families have significant roles in mediation of tissue interactions in the tooth, such as Activin, a TGFβ signal essential in the mesenchyme prior to bud formation (Ferguson et al., 1998), and its feedback inhibitor Follistatin, which is required in enamel knots for normal crown and cusp formation (Wang et al., 2004).

There is evidence indicating that Wnt/β-catenin may be the most upstream signal initiating the formation of signaling centers. Wnt/β-catenin signaling has been shown to be upstream of *Fgf4* and *Fgf20* as well as *Lef1* in the enamel knot (Haara et al., 2012; Kratochwil, Galceran, Tontsch, Roth, & Grosschedl, 2002). Inhibition of Wnt signaling by overexpression of the Wnt inhibitor *Dkk4* arrests tooth development at the dental lamina stage and no placodes form (Liu et al., 2008). The strongest evidence for the importance of Wnt signaling was provided by studies in which forced expression of β-catenin in the oral epithelium driven by the K14 promoter-induced continuous formation of ectopic signaling centers which gave rise to teeth (Jarvinen et al., 2006; Liu et al., 2008; Wang et al., 2009).

The ectopic signaling centers were characterized by intense focal β-catenin activity and the expression of various placode-specific molecular markers.

The morphogenesis from bud to cap stage and the formation and function of the primary enamel knot are perhaps the most actively studied topics in tooth morphogenesis research and these transitions are fully dependent on Wnt, BMP, FGF, Shh, and Eda signals. Mesenchymal *BMP4* regulates the bud-to-cap stage transition and initiates enamel knot formation through induction of cyclin-dependent kinase inhibitor *p21* in the bud stage epithelium, which probably signals the exit from the cell cycle in the forming enamel knot (Jernvall, Aberg, Kettunen, Keranen, & Thesleff, 1998). Furthermore, the arrested tooth development at bud stage in *Msx1* knockout mice was accompanied by the absence of *Bmp4* in the mesenchyme, and addition of exogenous BMP4 protein rescued the development of these teeth to cap and bell stage (Bei et al., 2000). However, conditional inactivation of *Bmp4* in dental mesenchyme in $Bmp4^{(f/f)}$*;Wnt1Cre* mice resulted in the arrest of the mandibular molar development at bud stage, while maxillary molars and incisors developed into mineralized teeth (Jia et al., 2013). The higher sensitivity of mandibular molars to inhibition of BMP signaling was explained by higher expression of the Wnt inhibitor *Dkk2* and the BMP inhibitor *Osr2* in the mandibular mesenchyme. This is an excellent example of the delicate fine-tuning of the signaling pathways during morphogenesis.

Of the five signaling families, only BMPs and FGFs transmit bidirectional signaling across the epithelial and mesenchymal tissues (Fig. 3). In the large *FGF signaling family,* the expression of individual signals is typically restricted to either epithelium or mesenchyme and they commonly mediate bidirectional signaling between the tissues. *Fgf3* and *Fgf10* are expressed in dental mesenchyme, while *Fgf4*, *Fgf9*, and *Fgf20* are expressed in the placodes and enamel knots during tooth morphogenesis. The analysis of the tooth phenotypes of *Fgf9;Fgf20* compound knockout mice indicated compensatory functions in the enamel knot, while *Fgf4* mutants had no tooth phenotype (Haara et al., 2012). The epithelial FGFs induce the expression of mesenchymal FGFs. For example, FGF4 in the enamel knot induces the expression of *Fgf3* in the dental mesenchyme. This induction is mediated by *Runx2*, a transcription factor required in the mesenchyme for bud-to-cap stage transition (Aberg et al., 2004).

A typical feature of FGF signaling is its feedback inhibition by Sprouty (Spry) molecules. Different *Sprouty* genes are expressed in the epithelium

(*Spry2*) and mesenchyme (*Spry4*), and molecular and genetic evidence from mouse models have indicated their importance in limiting FGF signaling in the enamel knot (Klein et al., 2006). Loss of *Spry2* function results in supernumerary tooth formation in the diastema region (in front of M1) due to the persistent enamel knot caused by hypersensitivity to FGF signaling.

Shh is the only member of the hedgehog signal family in teeth and its expression is restricted to dental epithelium. After being expressed in the dental lamina, *Shh* is restricted to the placodes and enamel knots, and during the bell stage, its expression spreads from the enamel knots to the differentiating epithelium. The expression patterns of the hedgehog targets, *Gli1* and *Ptc1*, during tooth morphogenesis indicate high Shh signaling activity throughout the dental mesenchyme and epithelium, excluding the enamel knots (Bitgood & McMahon, 1995). Conditional deletion of *Shh* from the bud stage onward using K14-Cre led to disrupted tooth morphogenesis (Dassule, Lewis, Bei, Maas, & McMahon, 2000). The mutant teeth were small and the growth of CLs and cusps was hampered. However, the enamel knots did form and they expressed all of the analyzed marker genes. Inhibition of epithelial Shh signaling through conditional ablation of *Smoothened* (*Smo*), a transducer of Shh signaling, using the same K14-Cre line resulted in a milder tooth morphogenesis phenotype (Gritli-Linde et al., 2002), indicating the importance of Shh signaling in the mesenchyme. In addition, the first and second molars were fused, and epithelial cell proliferation and ameloblast polarization were disturbed in the conditional *Smo* mutants. Taken together, the results indicated that Shh mediates both epithelial–mesenchymal and epithelial–epithelial tissue interactions (Gritli-Linde et al., 2002).

Restricted epithelial expression of *Wnt ligands* during tooth morphogenesis, with the exception of the noncanonical *Wnt5a* (Sarkar & Sharpe, 1999), indicates that canonical Wnts do not mediate signaling from mesenchyme to epithelium. Furthermore, Wnt/β-catenin reporter analyses have demonstrated localized activity predominantly in dental epithelium, and thus the majority of Wnt signaling is thought to be intraepithelial (Fujimori et al., 2010; Liu et al., 2008; Suomalainen & Thesleff, 2010). However, there is significant β-catenin activity in the dental mesenchyme during early tooth morphogenesis as well as during cell differentiation. The conditional deletion of β-catenin using *Osr2-Cre* leads to arrested tooth development at the bud stage, indicating epithelial–mesenchymal Wnt signaling (Chen, Lan,

Baek, Gao, & Jiang, 2009). In addition, conditional deletion of β-catenin in the incisor mesenchyme using *Prx1-Cre* resulted in a split incisor placode and duplication of the tooth (Fujimori et al., 2010). This phenotype was attributed to the downregulated *Bmp4* expression. A similar phenotype was observed after *ex vivo* addition of BMP inhibitor Noggin (Fujimori et al., 2010; Munne et al., 2010). These studies indicate the requirement for BMP signaling for placode integrity and exemplify the interactions between the Wnt and BMP pathways. A series of Wnt–BMP feedback circuits operating between the tooth epithelium and mesenchyme during tooth morphogenesis was uncovered through massive gene expression profiling of isolated dental epithelia and mesenchyme (O'Connell et al., 2012). It is obvious that these circuits are affected or interrupted in many mouse models in which either BMP or Wnt signaling has been targeted.

Studies using *Sosdc1* (also known as *Wise*, *Ectodin*, and *USAG1*)-null mutants and bead experiments have unraveled interactions of Wnt, BMP, FGF, and Shh pathways. *Sostdc1* is a Wnt inhibitor which in some contexts also inhibits BMP signaling (Lintern, Guidato, Rowe, Saldanha, & Itasaki, 2009). *Sostdc1* is regulated by BMP4 and is intensely expressed in teeth, except the enamel knot and its surrounding tissue (Laurikkala, Kassai, Pakkasjarvi, Thesleff, & Itoh, 2003). Supernumerary teeth as well as the fusion of molars observed in mice lacking *Sostdc1* were directly linked to its inhibitory effect on Wnt signaling (Ahn, Sanderson, Klein, & Krumlauf, 2010). Furthermore, this study demonstrated that FGF and Shh signaling are major downstream targets of *Sostdc1*.

Eda is a tumor necrosis family signaling molecule which specifically regulates the development of ectodermal appendages, including teeth. Eda was originally discovered as the causative gene for human X-linked hypohidrotic ectodermal dysplasia, a syndrome which is characterized by missing teeth, sparse hair, and deficient formation of nails and several exocrine glands (reviewed by Mikkola, 2009). The *Eda* expression domain in the early oral epithelium and dental epithelium is regulated by intraepithelial Wnt signaling, while mesenchymal Activin regulates the expression domain of its receptor, *Edar*, which is restricted to the dental lamina, placodes, and enamel knots. It has been demonstrated that Eda regulates all dental signaling centers in the dental epithelium and mediates interactions between epithelial compartments (Haara et al., 2012; Laurikkala et al., 2001). In addition, Eda signaling is integrated with all of the other four pathways as well as epithelial–mesenchymal interactions (reviewed by Mikkola, 2009). Eda deficiency in mice decreases the size and the number of teeth, while conditional *Eda* overexpression in dental epithelium (by the K14 promoter) increases size of tooth placodes and leads to larger and supernumerary teeth (Mustonen et al., 2003; Pispa et al., 1999). Proposed Eda targets in the placodes and enamel knots include *Fgf20*, *Wnt10a*, and *Shh* as well as the signal inhibitors *Dkk4* and *Follistatin* (Haara et al., 2012; Mikkola, 2009). Notably, many of the Eda targets are shared with Wnt signaling.

The *patterning of the secondary enamel knots in* molars illustrates the significance of the delicate fine-tuning of signaling pathways and their interactions in tooth morphogenesis. This patterning process determines the locations and heights of tooth cusps and is directed by tissue interactions mediated by signals in all conserved signal families as well as their inhibitors. The process is regulated by a Turing-type reaction–diffusion mechanism which has been implicated in patterning of the placodes of ectodermal organs and involves diffusible signaling molecules, their inhibitors, and lateral inhibition (Biggs & Mikkola, 2014; Jiang et al., 2004; Salazar-Ciudad & Jernvall, 2010). The enamel knots locally produce many signaling molecules and inhibitors and are also the targets of signals. Tinkering with signaling pathways in mutant mice such as *K14-Eda, K14-Follistatin*, *Sostdc1*$^{-/-}$, and *K14Cre-Smo*$^{-/-}$ resulted in abnormal cusp patterns (Gritli-Linde et al., 2002; Laurikkala et al., 2003; Mustonen et al., 2003; Wang et al., 2004). In addition, exposing embryonic molars to different signals and their combinations in *ex vivo* cultures can have dramatic effects in cusp numbers and patterns, and interestingly, the different patterns can be generated by computer models (Harjunmaa et al., 2012; Jernvall & Thesleff, 2012; Salazar-Ciudad & Jernvall, 2010). Fine-tuning of signaling in and around the enamel knots is considered to be the most important mechanism regulating the patterning of the tooth crowns and generating the diverse cusp patterns during evolution.

Deletion of the *Foxi3* transcription factor in the dental epithelium leads to a dramatic crown phenotype featured by supernumerary, shallow, and incorrectly patterned cusps (Jussila et al., 2015). Gene expression profiling by microarray analysis placed *Foxi3* upstream of Shh, BMP, FGF, and Wnt signaling, and *ex vivo* studies demonstrated that *Foxi3* expression is regulated by BMP4, Activin, and Eda. Thus, Foxi3 links several signaling pathways during cusp patterning. Gene expression patterns in the *Foxi3* mutant teeth indicated that the intercuspal epithelium had adopted enamel knot fate, suggesting that Foxi3 functions as an inhibitor of enamel knot formation (Jussila et al., 2015).

3. EPITHELIAL–MESENCHYMAL INTERACTIONS REGULATING DENTIN AND ENAMEL FORMATION

Dentin and enamel are mineralized tissues produced at the late bell stage of tooth development by tooth-specific cells: the mesenchyme-derived odontoblasts that produce dentin and the epithelial ameloblasts that produce enamel. Formation of these tissues occurs at the interface between the epithelium and mesenchyme and is regulated by epithelial–mesenchymal interactions and the same signals as tooth morphogenesis.

At the bell stage, the enamel organ consists of two basal cell layers: (i) the inner enamel epithelium or IEE, which encloses the dental papilla and gives rise to odontoblasts and pulp cells and (ii) the outer enamel epithelium or OEE, which faces the dental follicle and later generates the cementoblasts and periodontal ligament. Together the IEE and OEE surround the loosely arranged stellate reticulum (SR) cells and a thin layer of stratum intermedium (SI) cells, which are in direct contact with the IEE (Fig. 4A).

Differentiation of the IEE cells into ameloblasts and the directly underlying dental papilla cells into odontoblasts occurs in stages that include (i) the initial or presecretory stage, (ii) secretory stage, and (iii) maturation stage. Differentiation starts from the tips of the future cusps and is marked by polarization of preodontoblasts and preameloblasts as well as matrix secretion (Fig. 4B). Secretion of collagen type I-based predentin at the epithelial–mesenchymal interface causes preodontoblasts to retreat, leaving behind a cell extension called the odontoblast process from which the noncollagenous proteins (including the most abundant, Dentin Sialophosphoprotein or DSPP) that initiate dentin mineralization are secreted. Odontoblast processes remain embedded in dentin and thus the tubular morphology of the dentin arises.

Dental epithelium and mesenchyme are separated by a basement membrane that has been suggested to support polarization of preodontoblasts, which align themselves perpendicular to it (reviewed by Hurmerinta & Thesleff, 1981; Fig. 4B). Degradation of the basement membrane subsequent to the onset of predentin formation enables direct contacts between odontoblasts and preameloblasts suggested to mediate the induction of ameloblast differentiation (Slavkin & Bringas, 1976). This was recently supported by mice lacking *EMMPRIN* (CD147), a membrane glycoprotein expressed exclusively in the preameloblasts and SI cells. The persistence of the basement membrane in *EMMPRIN*$^{-/-}$ mice leads to delayed ameloblast

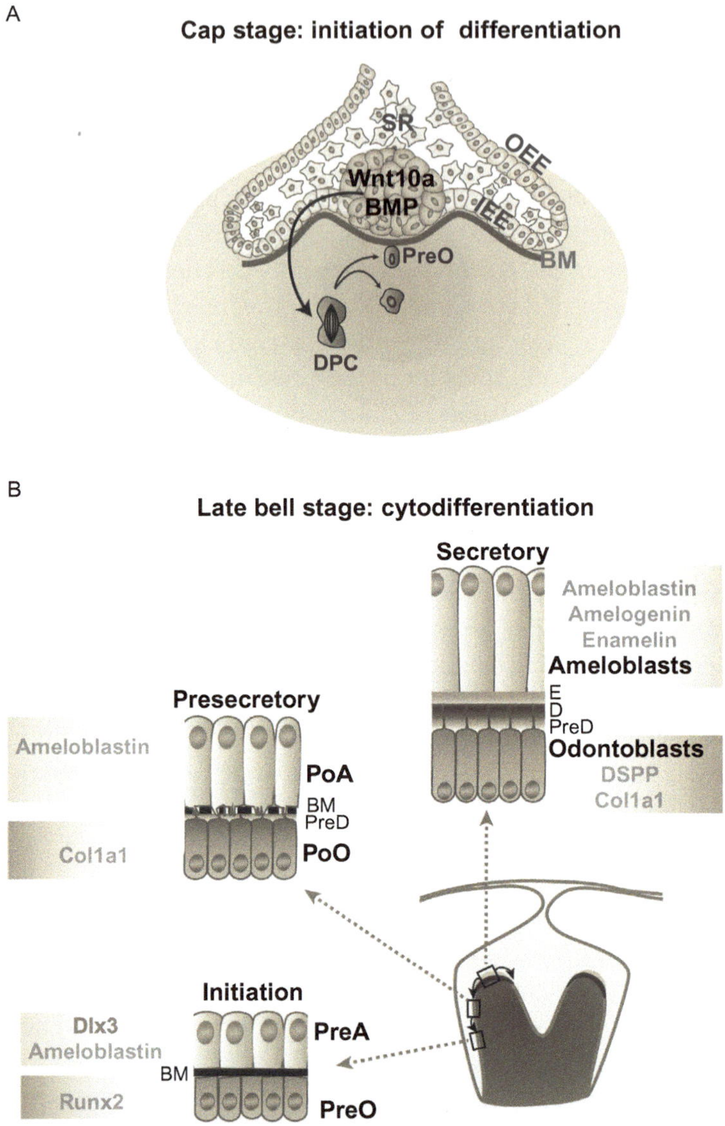

Figure 4 Epithelial–mesenchymal tissue interactions regulating differentiation of odontoblasts and ameloblasts. (A) The enamel organ during cap stage consists of inner and outer enamel epithelium (IEE and OEE, respectively) enclosing the stellate reticulum (SR) compartment. The fully developed primary enamel knot (pink cells in spherical organization) expresses Wnt10A that signals to dental papilla cells in the underlying mesenchyme (green color), which then commence odontoblast differentiation. (B) Ameloblasts and odontoblasts are generated during the bell stage of tooth development. Indicated genes represent the molecular markers that can be used to follow the progression of ameloblast (yellow boxes) and odontoblast (green boxes) lineages. *Abbreviations*: BM, basement membrane; D, dentin; E, enamel; PoA, polarized ameloblasts; PoO, polarized odontoblasts; PreA, preameloblasts; PreD, predentin; PreO, preodontoblasts; DPC, dental papilla cell. (A color version of this figure is located at the end of this chapter.)

differentiation and enamel production, and is caused by the loss of MMP-20, a matrix degrading metalloprotease that plays a role in degradation of basement membrane (Khaddam et al., 2014).

3.1 Initiation and Progression of Odontoblast Differentiation

Initiation of odontoblast differentiation starts at the tips of the future cusps where enamel knots form. The timing and the location of enamel knot formation suggest that enamel knots are the source of the first inductive signals for odontoblast differentiation (Fig. 4A; Thesleff, Keranen, & Jernvall, 2001; Yamashiro et al., 2007). *Wnt10a* is expressed in E14 mouse incisors and molars in the primary enamel knot and subsequently in the secondary enamel knots. From E14 onward the expression of *Wnt10a* shifts to the mesenchyme and becomes intense in the preodontoblasts underlying the enamel knot, which differentiate into DSPP-expressing odontoblasts (Yamashiro et al., 2007). Studies on cultured cells and tooth germs have shown that Wnt10a inhibits proliferation of dental papilla cells and initiates their differentiation by upregulating *Dspp* expression (Liu, Han, Wang, & Feng, 2013; Yamashiro et al., 2007).

BMP/TGFβ signaling has been most frequently implicated in the induction of odontoblast differentiation. *In vitro* studies and dentin regeneration experiments showed long before the development of the genetic mouse models that signals of this gene family are capable of inducing terminal differentiation of odontoblasts (Lesot et al., 2001). Mouse models lacking BMP/TGFβ signaling at earlier stages of tooth development (*Tgfβr2*$^{fl/fl}$*; Wnt1-Cre*, *Bmp4*$^{fl/fl}$*;3.6Col1a1-Cre*, and *Bmp2*$^{fl/fl}$*;Osx-Cre* null mice) display defects in dentinogenesis, namely, delayed odontoblast differentiation and reduced dentin production (Feng et al., 2011; Kim et al., 2010, 2011). Conditional ablation of BMP/TGFβ signaling at later stages of odontoblast differentiation (*DsppCre;Tgf-βr2*$^{fl/fl}$ *mice*) does not cause any noticeable dentin abnormality (Arany et al., 2014). Similarly, lack of *Smad4*, mediator of both TGFβ and BMP signaling, at an early stage of tooth development (*Smad4*$^{fl/fl}$*;Osr2-Cre* null mice) results in the generation of nonpolarized odontoblasts which deposit atubular, bone-like tissue (Bae et al., 2013). A similar phenotype was seen in transgenic mice overexpressing *Runx2* in the odontoblasts (Han et al., 2014). Lack of *Dspp* expression and production of bone matrix in teeth of both models indicate acquisition of osteoblast phenotype, suggesting that Smad4 and Runx2 are necessary for the maintenance of the odontoblast cell fate. Deletion of *Smad4* from the mesenchyme at later stages of tooth development results in a milder phenotype, which includes various degrees of altered odontoblast polarization and thin dentin (Lian et al., 2006). Together, these studies clearly indicate that BMP/TGFβ signaling regulates odontoblast differentiation in a stage- and site-specific manner.

Changes in the morphology of *Smad4* mutant teeth were attributed to increases in Wnt signaling and indicated an inhibitory effect of BMP/TGFβ signaling on Wnt pathway activity (Bae et al., 2013). However, the published data are contradictory and it is not yet determined whether BMP and Wnt signaling interactions are synergistic or antagonistic. This can perhaps be explained by the reciprocal Wnt–BMP feedback circuits involved in the epithelial–mesenchymal interactions during tooth morphogenesis (O'Connell et al., 2012).

Furthermore, there is evidence that Wnt signaling is important for terminal odontoblast differentiation. High Wnt/β-catenin reporter activity is present in odontoblasts during differentiation (Suomalainen & Thesleff, 2010). Excessive Wnt/β-catenin activity due to constitutive stabilization of β-catenin in early odontoblasts of *OC-Cre:Catnb*$^{(+/lox(ex3))}$ mutant mice induces premature differentiation of odontoblasts that deposit high amounts of insufficiently mineralized dentin, closing the pulp chamber and expressing low levels of *DSPP* (Bae et al., 2013; Kim et al., 2011). *Smad4* is significantly upregulated in these prematurely differentiating odontoblasts, suggesting that Smad4 tunes the level of Wnt activity necessary for successful odontoblast differentiation (Bae et al., 2013). Conversely, decreased Wnt signaling activity in early odontoblasts of *OC-Cre;Wls*$^{(CO/CO)}$ mutant mice caused reduction of dentin deposition resulting in a thinner dentin wall and wider pulp chamber (Bae et al., 2015). The similarities in the dentin phenotype of the *OC-Cre;Wls*$^{(CO/CO)}$ mutant mice and mice lacking BMP/TGFβ signaling indicates the synergistic activity of the two pathways. In addition, some mouse mutants, such as mice lacking *Epiprofin* (*Epfn*) *6*, demonstrate that the defects in the terminal stages of odontoblast and ameloblast differentiation are directly related to the decreases in both β-catenin and BMP signaling (Jimenez-Rojo et al., 2010).

3.2 Ameloblast Differentiation

The majority of mouse mutants with abnormal odontoblast differentiation and dentin production described above did not display significant changes in ameloblast differentiation and enamel formation. Terminal ameloblast differentiation is triggered by paracrine signals emanating from functional odontoblasts secreting predentin matrix. Odontoblast-derived BMPs are required for ameloblast differentiation, in particular, BMP4 that induces expression of ameloblast markers *p21* and *ameloblastin* (Wang et al., 2004). These BMP signals were shown to be fine-tuned by Follistatin, which acts as a BMP antagonist in the IEE, blocking ameloblast differentiation. The patterns of *Follistatin* expression during mouse incisor and molar development are closely related to inhibition of ameloblast differentiation and absence of enamel. *Follistatin* is expressed in the IEE at the tips of the mouse molar cusps and on the lingual aspect of the mouse incisors, which are both enamel-free areas (Wang et al., 2004).

Conditional deletion of either *Bmp4* or *Bmp2* significantly reduced enamel thickness, and changes in amelogenesis were similar to those caused by *amelogenin* mutations in mice and humans (Feng et al., 2011). The abnormal amelogenesis in these mutants was linked to decreased levels of *Dlx3* in the preameloblasts and a subsequent decrease in amelogenin production. In addition, *in vitro* studies on epithelial cell lines demonstrated that *Dlx3* positively regulates enamel matrix protein genes including *enamelin* and *amelogenin* (Zhang et al., 2015).

Enamel defects in mice carrying compound mutations of cell–cell adhesion molecules *Nectin-1* and *-3*, as well as mice lacking the cell membrane protein PERP, indicate the importance of the integrity of the ameloblast cell layer and its tight contacts with the SI cell layer (Jheon et al., 2011; Neupane et al., 2014; Yoshida, Miyoshi, Takai, & Thesleff, 2010). The SI consists of a few layers of epithelial cells adjacent to the ameloblasts. Ameloblasts and SI cells are tightly bound by desmosomes, recruited by interaction between Nectin-1 expressed in ameloblasts and Nectin-3 in SI cells (Yoshida et al., 2010).

The SI has traditionally been regarded as an important cell layer supporting ameloblast functions. The SI cells are characterized by high levels of expression of alkaline phosphatase as well as *Shh* (Koyama et al., 2001; Wise & Fan, 1989). Shh plays an important role in ameloblast polarization and secretion. Conditional deletion of either *Shh* or its downstream mediator *Smoothened* (*Smo*) in dental epithelium led to disrupted polarity and organization of the ameloblast layer and production of disorganized enamel matrix (Dassule et al., 2000; Gritli-Linde et al., 2002). The findings indicate that Shh function is required in dental epithelium and that Shh mediates the interaction between SI and (pre)ameloblasts during ameloblast polarization and secretion.

4. EPITHELIAL–MESENCHYMAL INTERACTIONS REGULATING STEM CELLS IN CONTINUOUSLY GROWING MOUSE INCISORS

4.1 Development of Cervical Loops

During the cap and bell stages of tooth development, the enamel organ extends in the apical direction to envelop the underlying dental mesenchyme. The leading edge of the extending epithelium consists of the CL where the IEE and OEE enclose loosely arranged SR cells. In teeth that form roots (e.g., all human teeth and mouse molars), the SR cells are depleted when crown development is completed and *root formation* starts. The remaining bilayered basal epithelium (IEE and OEE) is called *H*ertwig's *e*pithelial *r*oot *s*heath (HERS) (Fig. 5A; reviewed by Thesleff & Tummers, 2008). Downregulation of *Fgf10* in the mesenchyme surrounding the CL is necessary for the initiation of root formation, as shown by the addition of FGF10 to molar germ cultures prior to HERS formation. This preserved SR cells and prolonged extension of the CL in the apical direction (Yokohama-Tamaki et al., 2006). A similar phenotype was recently reported when *Smad4* was ablated in dental epithelium. The maintenance of the SR cells in the CL was attributed to a lack of the inhibitory function of mesenchymal BMP on epithelial *Shh* expression (Li et al., 2015, and see below). In continuously growing teeth-like rodent incisors, the SR cell compartment is maintained in the CLs by stem cells (Fig. 5A).

4.2 Epithelial Stem Cells

At the proximal end of the rodent incisor, two CLs flank the putative niche of mesenchymal stem cells. The bigger labial cervical loop (LaCL) contains a small population of slowly dividing label-retaining cells (LRCs) (Smith & Warshawsky, 1975). Harada et al. (1999) demonstrated that isolated LaCLs can generate new dental tissues when cultured *in vitro*, providing evidence for the presence of stem cells. The other CL, located on the lingual side (LiCL), is smaller with fewer stem cells, and the stem cells do not give rise to ameloblasts or enamel, hence the asymmetrical enamel deposition in the incisor (Fig. 5A; reviewed by Jussila & Thesleff, 2012; Kuang-Hsien Hu, Mushegyan, & Klein, 2014).

The growth of the rodent incisor is fine-tuned by a balance between the continuous enamel production (from stem cells) and tooth wear. Thus, enamel morphology and phenotype are used as a read-out to detect changes

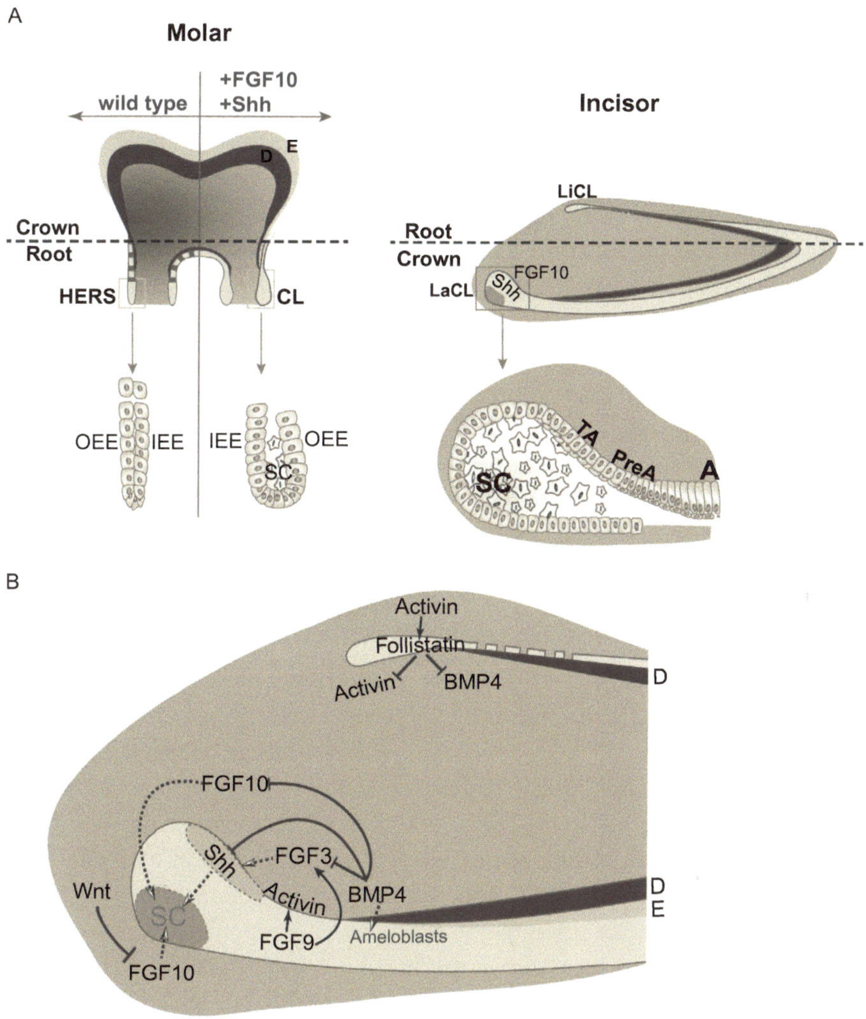

Figure 5 Cervical loops in molar root formation and continuous growth of the incisor, and the gene regulatory network that controls maintenance of epithelial stem cells. (A) In the wild-type molar tooth, disappearance of stellate reticulum (SR) cells in the cervical loop (CL) leads to formation of Hertwig's epithelial root sheath (HERS) and root formation, and is caused by downregulation of Fgf10 and Shh (illustrated in left-side root). Persistence of *Fgf10* and *Shh* expression support maintenance of the SR compartment that houses stem cells in the CL of the molar root (illustrated in right-side root) and in the LaCL of the continuously growing incisor (orange area) (Li et al., 2015; Yokohama-Tamaki et al., 2006). (B) Schematic view of the signals involved in epithelial–mesenchymal tissue interactions regulating the stem cell niche in the continuously growing incisors. Black arrows indicate positive gene regulation, red arrows indicate negative gene regulation, and green dotted arrows indicate positive effects on stem cells (orange area) and TA proliferation (blue area). *Abbreviations*: A, ameloblasts; IEE, inner enamel epithelium; OEE, outer enamel epithelium; PreA, preameloblasts; TA, transit-amplifying cells; SC, stem cells; LaCL, labial cervical loop; LiCL, lingual cervical loop. (A color version of this figure is located at the end of this chapter.)

in the stem cell niche viability, regulation, and maintenance. In addition, the labial–lingual asymmetry of the incisor can be modulated by tinkering with signaling pathways, making the incisor a valuable model to study epithelial stem cells and their regulation (reviewed by Jussila & Thesleff, 2012; Kuang-Hsien Hu et al., 2014).

Epithelial stem cells in the LaCL express molecules also found in other stem cells, such as Sox2, Lgr5, ABCG2, Bmi-1, Oct3/4, Yap, Gli1, E-cadherin, and the integrins α6 and β4 (Juuri et al., 2012; Li et al., 2011; Seidel et al., 2010; Suomalainen & Thesleff, 2010). Transgenic mouse models in which GFP expression is driven by promoters of stem cell marker genes (*Sox2-GFP, Bmi1-GFP, Lgr5-GFP*) have enabled identification and isolation of this population. Other transgenic models have provided means to genetically manipulate the stem cell niche (*Gli1-CreER, Ptch1-CreER, Bmi1-GFP*) (Biehs et al., 2013; Juuri et al., 2012; Seidel et al., 2010).

4.3 FGF Signaling and Cross Talks with BMP/TGFβ Pathways

Fgf3 and *Fgf10* are expressed intensely in the mesenchyme adjacent to the TA cells in the LaCL. In addition, *Fgf10* expression surrounds the LaCL and is also expressed at very low levels in the mesenchyme surrounding the LiCL. Fgfr1b and Fgfr2b expressing cells located within the LaCL, including the stem cell population and TA cells, respond to FGF3 and FGF10 (Harada et al., 1999, reviewed by Kuang-Hsien Hu et al., 2014).

The function of FGF10 is indispensable for stem cell maintenance in the LaCL. *Fgf10* knockout mice die at birth, but histological analysis of their incisors after development under the kidney capsule indicated that LaCL was thin and the SR cells were completely absent (Harada et al., 2002). Consistent with this data, conditional deletion or downregulation of Fgfr2b receptor led to depletion of stem cells and impaired incisor growth (Lin et al., 2009; Parsa et al., 2010). Mouse models with impaired or overactive FGF signaling have demonstrated the redundant roles of FGF3 and FGF10 signaling in the proliferation of the epithelial stem cell progeny (Klein et al., 2008; Wang et al., 2007). In addition, *Fgf3*$^{-/-}$;*Fgf10*$^{+/-}$ mutant mice display a hypoplastic LaCL and reduced enamel formation, in contrast to the lack of phenotype in *Fgf3*$^{-/-}$ mice (Wang et al., 2007). Deletion of *Tbx1*, an FGF-regulated transcription factor expressed in TA cells, led to loss of enamel formation (Caton et al., 2009).

Enhanced FGF signaling through deletion of the feedback inhibitors *Spry4* and *2* (*Spry4*$^{-/-}$;*Spry2*$^{+/-}$ mice) results in ectopic development of enamel-producing ameloblasts on the lingual, enamel-free side. Formation of enamel on both sides prevents tooth abrasion and the incisors acquire a tusk-like phenotype, indicating the important role of fine-tuning FGF signaling in regulation of incisor asymmetry (Klein et al., 2008; Fig. 5B).

A complex signaling network involving FGF3, BMP4, Activin, and Follistatin regulates the asymmetric proliferation of stem cell progeny (Fig. 5B; Wang et al., 2007). *Bmp4* and *Activin* both overlap with *Fgf3* expression in the mesenchyme under the TA cells and the balance between these signals and their antagonists controls Fgf3 and its effects on epithelial stem cell proliferation. BMP4 negatively regulates epithelial stem cells while it promotes ameloblast differentiation (Wang et al., 2004). In line with this, mice overexpressing BMP4 inhibitor *Noggin* in the epithelium (*K14-Noggin* mice) show overgrown incisors lacking enamel (Plikus et al., 2005). Complete lack of enamel in *K14-Follistatin* mice and symmetrical formation of enamel in *Follistatin*$^{-/-}$ mice can be related to changes in *Fgf3* expression (Wang et al., 2007).

The expression of *Fgf3* and *Fgf10* in dental mesenchyme also depends on TGFβ receptors type I (Alk5) and type II. The deletion of these receptors reduces the expression of *Fgf3* and *Fgf10* and decreases the number of stem cells in the LaCL (Yang et al., 2014; Zhao, Li, Han, Kaartinen, & Chai, 2011).

4.4 Hedgehog Signaling and the Importance of Nerves

In the incisor, *Shh* expression is restricted to the TA cells and maturing ameloblasts (Bitgood & McMahon, 1995; Seidel et al., 2010; Zhao et al., 2014). Gli1+ cells and Ptch1+ cells (i.e., Hh-responsive cells) are located in the incisor stem cell niches in both the epithelium (CLs) and mesenchyme and, as demonstrated by lineage tracing, are stem cells (Seidel et al., 2010; Zhao et al., 2014). Pharmacological inhibition of Shh function *in vivo* by a Smo inhibitor demonstrated that Hh signaling is required for the maintenance of the epithelial stem cell niche and the continuous generation of ameloblasts in the adult mice (Seidel et al., 2010; Zhao et al., 2014). Most recent data demonstrated that Shh/Gli1 signaling regulates the fate of the Sox2+ epithelial stem cells in the LaCL through cross talk with the BMP/Smad4 pathway (Li et al., 2015). Loss of the inhibitory effect of mesenchymal BMP on epithelial Shh resulted in the persistence of the dental epithelial stem cell population identified by Sox2-GFP (Fig. 5B).

Hedgehog signaling is also necessary for production of differentiated odontoblasts. Interestingly, severing the inferior alveolar nerve affects dentinogenesis in a pattern similar to Shh inhibition, demonstrating that nerves provide an additional source of Shh (Zhao et al., 2014). Loss of innervation and, consequently, Shh signal initially affected only mesenchyme, but over time, it also affected the epithelium, resulting in a reduced number of LRCs and decreased cell proliferation in the CLs. Reduced *Gli1* expression in both the epithelium and the mesenchyme confirmed that the most affected populations were the dental stem cells (Zhao et al., 2014). Therefore, the proper growth and patterning of the dental sensory nerve is a requirement for the establishment of the mesenchymal stem cell niche that provides *Fgf10* expressing cells and, consequently, the signal that maintains the epithelial stem cells.

The role of the nerve bundle in the biology of the incisor tooth is dual: not only does it provide a source of Shh that regulates the stem cells in the mesenchyme, it also provides a source of mesenchymal stem cells. Lineage tracing of mesenchymal cells during development, renewal, and repair of the incisor indicated that they are in part derived from peripheral nerve-associated glial cells that differentiate into pulp cells and odontoblasts (Kaukua et al., 2014).

4.5 Wnt Signaling

Wnt signaling has been implicated as the key regulator of various epithelial stem cells, for example, in the hair follicle and intestine (Sato & Clevers, 2013; Yang & Peng, 2010). However, the epithelial stem cells in the incisor seem to lack Wnt signaling activity, as evidenced by the absence of Wnt reporters, including TOP gal, BAT gal, and *Axin2* (Juuri et al., 2012; Suomalainen & Thesleff, 2010). However, a small cell population that expresses *Lgr5* exists in the SR of the CLs (Chang et al., 2013; Suomalainen & Thesleff, 2010). In contrast to the CL epithelium, Wnt/β-catenin signaling activity is present in the mesenchyme surrounding the CL, from where it regulates the apoptosis in the CL, in particular through control of the epithelial Lgr5+ stem cells. This mesenchymal–epithelial interaction was shown to be mediated by FGF10, which was downregulated by Wnt/β-catenin signaling (Yang, Balic, Michon, Juuri, & Thesleff, 2015). In addition, this study also pointed to a role of apoptosis in the regulation of stem cell homeostasis in the continuously growing incisors (Yang et al., 2015).

5. CONCLUDING REMARKS

The epithelial–mesenchymal interactions regulate all aspects of tooth development, including initiation, morphogenesis, differentiation of hard tissue producing cells, and the maintenance of stem cells. The communication between the tissues is mediated by conserved signaling molecules which act in complex gene regulatory networks with transcription factors. Detailed understanding of developmental regulation will be crucial for bioengineering of teeth, a goal that has been set for tooth development research during the past decade. So far, the only feasible protocol for tooth bioengineering is based on the classic tissue recombination experiments demonstrating that separated dental epithelium and mesenchyme can form teeth when recombined and cultured as transplants, and that depending on timing the other tissue can be substituted with nondental tissue (Kollar & Baird, 1969; Lumsden, 1988; Mina & Kollar, 1987). These classical tissue recombination methods were refined in the past decade by utilizing dental epithelium and mesenchyme dissociated into single cells prior to recombination. The recombinants formed functional teeth when transplanted to the jaws of adult mice (Nakao et al., 2007; Oshima et al., 2011), indicating that patient-derived cells could be used for the same purpose. However, unavailability of dental progenitors requires reprogramming of patient-derived cells, either induced pluripotent stem cells or adult cells for tooth bioengineering. Therefore, deciphering the "transcription factor codes" of dental epithelial and mesenchymal stem/progenitor cells will be instrumental. Several obvious candidates for transcription factors conferring the identity of dental epithelium and mesenchyme are already known (Fig. 2). However, more work is needed to characterize the progenitor cells in the dental lamina epithelium and early tooth mesenchyme and to examine their programming as well as abilities to form teeth. The continuously growing mouse incisors are also useful models for examining dental stem and progenitor cells.

ACKNOWLEDGMENTS

We thank Dr. Leah Biggs for critical reading and editing of the chapter.

REFERENCES

Aberg, T., Wang, X. P., Kim, J. H., Yamashiro, T., Bei, M., Rice, R., et al. (2004). Runx2 mediates FGF signaling from epithelium to mesenchyme during tooth morphogenesis. *Developmental Biology*, *270*(1), 76–93. http://dx.doi.org/10.1016/j.ydbio.2004.02.012.

Ahn, Y., Sanderson, B. W., Klein, O. D., & Krumlauf, R. (2010). Inhibition of Wnt signaling by Wise (Sostdc1) and negative feedback from Shh controls tooth number and patterning. *Development*, *137*(19), 3221–3231. http://dx.doi.org/10.1242/dev.054668.

Arany, P. R., Cho, A., Hunt, T. D., Sidhu, G., Shin, K., Hahm, E., et al. (2014). Photoactivation of endogenous latent transforming growth factor-beta1 directs dental stem cell differentiation for regeneration. *Science Translational Medicine*, *6*(238), 238ra269. http://dx.doi.org/10.1126/scitranslmed.3008234.

Bae, C. H., Kim, T. H., Ko, S. O., Lee, J. C., Yang, X., & Cho, E. S. (2015). Wntless regulates dentin apposition and root elongation in the mandibular molar. *Journal of Dental Research*, *94*(3), 439–445. http://dx.doi.org/10.1177/0022034514567198.

Bae, C. H., Lee, J. Y., Kim, T. H., Baek, J. A., Lee, J. C., Yang, X., et al. (2013). Excessive Wnt/beta-catenin signaling disturbs tooth-root formation. *Journal of Periodontal Research*, *48*(4), 405–410. http://dx.doi.org/10.1111/jre.12018.

Bei, M. (2009). Molecular genetics of tooth development. *Current Opinion in Genetics & Development*, *19*(5), 504–510. http://dx.doi.org/10.1016/j.gde.2009.09.002.

Bei, M., Kratochwil, K., & Maas, R. L. (2000). BMP4 rescues a non-cell-autonomous function of Msx1 in tooth development. *Development*, *127*(21), 4711–4718.

Bei, M., & Maas, R. (1998). FGFs and BMP4 induce both Msx1-independent and Msx1-dependent signaling pathways in early tooth development. *Development*, *125*(21), 4325–4333.

Biehs, B., Hu, J. K., Strauli, N. B., Sangiorgi, E., Jung, H., Heber, R. P., et al. (2013). BMI1 represses Ink4a/Arf and Hox genes to regulate stem cells in the rodent incisor. *Nature Cell Biology*, *15*(7), 846–852. http://dx.doi.org/10.1038/ncb2766.

Biggs, L. C., & Mikkola, M. L. (2014). Early inductive events in ectodermal appendage morphogenesis. *Seminars in Cell & Developmental Biology*, *25–26*, 11–21. http://dx.doi.org/10.1016/j.semcdb.2014.01.007.

Bitgood, M. J., & McMahon, A. P. (1995). Hedgehog and Bmp genes are coexpressed at many diverse sites of cell-cell interaction in the mouse embryo. *Developmental Biology*, *172*(1), 126–138. http://dx.doi.org/10.1006/dbio.1995.0010.

Caton, J., Luder, H. U., Zoupa, M., Bradman, M., Bluteau, G., Tucker, A. S., et al. (2009). Enamel-free teeth: Tbx1 deletion affects amelogenesis in rodent incisors. *Developmental Biology*, *328*(2), 493–505. http://dx.doi.org/10.1016/j.ydbio.2009.02.014.

Chang, J. Y., Wang, C., Jin, C., Yang, C., Huang, Y., Liu, J., et al. (2013). Self-renewal and multilineage differentiation of mouse dental epithelial stem cells. *Stem Cell Research*, *11*(3), 990–1002. http://dx.doi.org/10.1016/j.scr.2013.06.008.

Chen, J., Lan, Y., Baek, J. A., Gao, Y., & Jiang, R. (2009). Wnt/beta-catenin signaling plays an essential role in activation of odontogenic mesenchyme during early tooth development. *Developmental Biology*, *334*(1), 174–185. http://dx.doi.org/10.1016/j.ydbio.2009.07.015.

Dassule, H. R., Lewis, P., Bei, M., Maas, R., & McMahon, A. P. (2000). Sonic hedgehog regulates growth and morphogenesis of the tooth. *Development*, *127*(22), 4775–4785.

Dassule, H. R., & McMahon, A. P. (1998). Analysis of epithelial-mesenchymal interactions in the initial morphogenesis of the mammalian tooth. *Developmental Biology*, *202*(2), 215–227. http://dx.doi.org/10.1006/dbio.1998.8992.

Davidson, E. H. (1993). Later embryogenesis: Regulatory circuitry in morphogenetic fields. *Development*, *118*(3), 665–690.

Feng, J., Yang, G., Yuan, G., Gluhak-Heinrich, J., Yang, W., Wang, L., et al. (2011). Abnormalities in the enamel in bmp2-deficient mice. *Cells, Tissues, Organs*, *194*(2–4), 216–221. http://dx.doi.org/10.1159/000324644.

Ferguson, C. A., Tucker, A. S., Christensen, L., Lau, A. L., Matzuk, M. M., & Sharpe, P. T. (1998). Activin is an essential early mesenchymal signal in tooth development that is required for patterning of the murine dentition. *Genes & Development, 12*(16), 2636–2649.

Fujimori, S., Novak, H., Weissenbock, M., Jussila, M., Goncalves, A., Zeller, R., et al. (2010). Wnt/beta-catenin signaling in the dental mesenchyme regulates incisor development by regulating Bmp4. *Developmental Biology, 348*(1), 97–106. http://dx.doi.org/10.1016/j.ydbio.2010.09.009.

Gritli-Linde, A., Bei, M., Maas, R., Zhang, X. M., Linde, A., & McMahon, A. P. (2002). Shh signaling within the dental epithelium is necessary for cell proliferation, growth and polarization. *Development, 129*(23), 5323–5337.

Gurdon, J. B. (1971). Gene activity during embryogenesis. *Triangle, 10*(1), 23–28.

Haara, O., Harjunmaa, E., Lindfors, P. H., Huh, S. H., Fliniaux, I., Aberg, T., et al. (2012). Ectodysplasin regulates activator-inhibitor balance in murine tooth development through Fgf20 signaling. *Development, 139*(17), 3189–3199. http://dx.doi.org/10.1242/dev.079558.

Han, N., Zheng, Y., Li, R., Li, X., Zhou, M., Niu, Y., et al. (2014). beta-catenin enhances odontoblastic differentiation of dental pulp cells through activation of Runx2. *PLoS One, 9*(2), e88890. http://dx.doi.org/10.1371/journal.pone.0088890.

Harada, H., Kettunen, P., Jung, H. S., Mustonen, T., Wang, Y. A., & Thesleff, I. (1999). Localization of putative stem cells in dental epithelium and their association with Notch and FGF signaling. *The Journal of Cell Biology, 147*(1), 105–120.

Harada, H., Toyono, T., Toyoshima, K., Yamasaki, M., Itoh, N., Kato, S., et al. (2002). FGF10 maintains stem cell compartment in developing mouse incisors. *Development, 129*(6), 1533–1541.

Harjunmaa, E., Kallonen, A., Voutilainen, M., Hamalainen, K., Mikkola, M. L., & Jernvall, J. (2012). On the difficulty of increasing dental complexity. *Nature, 483*(7389), 324–327. http://dx.doi.org/10.1038/nature10876.

Hurmerinta, K., & Thesleff, I. (1981). Ultrastructure of the epithelial-mesenchymal interface in the mouse tooth germ. *Journal of Craniofacial Genetics and Developmental Biology, 1*(2), 191–202.

Iwafuchi-Doi, M., & Zaret, K. S. (2014). Pioneer transcription factors in cell reprogramming. *Genes & Development, 28*(24), 2679–2692.

Jarvinen, E., Salazar-Ciudad, I., Birchmeier, W., Taketo, M. M., Jernvall, J., & Thesleff, I. (2006). Continuous tooth generation in mouse is induced by activated epithelial Wnt/beta-catenin signaling. *Proceedings of the National Academy of Sciences of the United States of America, 103*(49), 18627–18632. http://dx.doi.org/10.1073/pnas.0607289103.

Jernvall, J., Aberg, T., Kettunen, P., Keranen, S., & Thesleff, I. (1998). The life history of an embryonic signaling center: BMP-4 induces p21 and is associated with apoptosis in the mouse tooth enamel knot. *Development, 125*(2), 161–169.

Jernvall, J., & Thesleff, I. (2012). Tooth shape formation and tooth renewal: Evolving with the same signals. *Development, 139*(19), 3487–3497. http://dx.doi.org/10.1242/dev.085084.

Jheon, A. H., Mostowfi, P., Snead, M. L., Ihrie, R. A., Sone, E., Pramparo, T., et al. (2011). PERP regulates enamel formation via effects on cell-cell adhesion and gene expression. *Journal of Cell Science, 124*(Pt. 5), 745–754. http://dx.doi.org/10.1242/jcs.078071.

Jia, S., Zhou, J., Gao, Y., Baek, J. A., Martin, J. F., Lan, Y., et al. (2013). Roles of Bmp4 during tooth morphogenesis and sequential tooth formation. *Development, 140*(2), 423–432. http://dx.doi.org/10.1242/dev.081927.

Jiang, T. X., Widelitz, R. B., Shen, W. M., Will, P., Wu, D. Y., Lin, C. M., et al. (2004). Integument pattern formation involves genetic and epigenetic controls: Feather arrays simulated by digital hormone models. *The International Journal of Developmental Biology, 48*(2–3), 117–135. http://dx.doi.org/10.1387/ijdb.041788tj.

Jimenez-Rojo, L., Ibarretxe, G., Aurrekoetxea, M., de Vega, S., Nakamura, T., Yamada, Y., et al. (2010). Epiprofin/Sp6: A new player in the regulation of tooth development. *Histology and Histopathology*, *25*(12), 1621–1630.

Jussila, M., Aalto, A., Sanz Navarro, M., Shirokova, V., Kallonen, A., Ohyama, T., et al. (2015). Suppression of epithelial differentiation by Foxi3 is essential for molar crown patterning. *Development* (in press). pii: dev.124172. [Epub ahead of print].

Jussila, M., & Thesleff, I. (2012). Signaling networks regulating tooth organogenesis and regeneration, and the specification of dental mesenchymal and epithelial cell lineages. *Cold Spring Harbor Perspectives in Biology*, *4*(4), a008425. http://dx.doi.org/10.1101/cshperspect.a008425.

Juuri, E., Saito, K., Ahtiainen, L., Seidel, K., Tummers, M., Hochedlinger, K., et al. (2012). Sox2+ stem cells contribute to all epithelial lineages of the tooth via Sfrp5+ progenitors. *Developmental Cell*, *23*(2), 317–328. http://dx.doi.org/10.1016/j.devcel.2012.05.012.

Kaukua, N., Shahidi, M. K., Konstantinidou, C., Dyachuk, V., Kaucka, M., Furlan, A., et al. (2014). Glial origin of mesenchymal stem cells in a tooth model system. *Nature*, *513*(7519), 551–554. http://dx.doi.org/10.1038/nature13536.

Keranen, S. V., Kettunen, P., Aberg, T., Thesleff, I., & Jernvall, J. (1999). Gene expression patterns associated with suppression of odontogenesis in mouse and vole diastema regions. *Development Genes and Evolution*, *209*(8), 495–506.

Khaddam, M., Huet, E., Vallee, B., Bensidhoum, M., Le Denmat, D., Filatova, A., et al. (2014). EMMPRIN/CD147 deficiency disturbs ameloblast-odontoblast cross-talk and delays enamel mineralization. *Bone*, *66*, 256–266. http://dx.doi.org/10.1016/j.bone.2014.06.019.

Kim, T. H., Lee, J. Y., Baek, J. A., Lee, J. C., Yang, X., Taketo, M. M., et al. (2011). Constitutive stabilization of ss-catenin in the dental mesenchyme leads to excessive dentin and cementum formation. *Biochemical and Biophysical Research Communications*, *412*(4), 549–555. http://dx.doi.org/10.1016/j.bbrc.2011.07.116.

Kim, Y. S., Min, K. S., Jeong, D. H., Jang, J. H., Kim, H. W., & Kim, E. C. (2010). Effects of fibroblast growth factor-2 on the expression and regulation of chemokines in human dental pulp cells. *Journal of Endodontia*, *36*(11), 1824–1830. http://dx.doi.org/10.1016/j.joen.2010.08.020.

Klein, O. D., Lyons, D. B., Balooch, G., Marshall, G. W., Basson, M. A., Peterka, M., et al. (2008). An FGF signaling loop sustains the generation of differentiated progeny from stem cells in mouse incisors. *Development*, *135*(2), 377–385. http://dx.doi.org/10.1242/dev.015081.

Klein, O. D., Minowada, G., Peterkova, R., Kangas, A., Yu, B. D., Lesot, H., et al. (2006). Sprouty genes control diastema tooth development via bidirectional antagonism of epithelial-mesenchymal FGF signaling. *Developmental Cell*, *11*(2), 181–190.

Kollar, E. J., & Baird, G. R. (1969). The influence of the dental papilla on the development of tooth shape in embryonic mouse tooth germs. *Journal of Embryology and Experimental Morphology*, *21*(1), 131–148.

Kollar, E. J., & Baird, G. R. (1970). Tissue interactions in embryonic mouse tooth germs. II. The inductive role of the dental papilla. *Journal of Embryology and Experimental Morphology*, *24*(1), 173–186.

Koyama, E., Wu, C., Shimo, T., Iwamoto, M., Ohmori, T., Kurisu, K., et al. (2001). Development of stratum intermedium and its role as a Sonic hedgehog-signaling structure during odontogenesis. *Developmental Dynamics*, *222*(2), 178–191. http://dx.doi.org/10.1002/dvdy.1186.

Kratochwil, K., Galceran, J., Tontsch, S., Roth, W., & Grosschedl, R. (2002). FGF4, a direct target of LEF1 and Wnt signaling, can rescue the arrest of tooth organogenesis in Lef1 (−/−) mice. *Genes & Development*, *16*(24), 3173–3185. http://dx.doi.org/10.1101/gad.1035602.

Kuang-Hsien Hu, J., Mushegyan, V., & Klein, O. D. (2014). On the cutting edge of organ renewal: Identification, regulation, and evolution of incisor stem cells. *Genesis*, *52*(2), 79–92. http://dx.doi.org/10.1002/dvg.22732.

Lan, Y., Jia, S., & Jiang, R. (2014). Molecular patterning of the mammalian dentition. *Seminars in Cell & Developmental Biology*, *25–26*, 61–70. http://dx.doi.org/10.1016/j.semcdb.2013.12.003.

Laurikkala, J., Kassai, Y., Pakkasjarvi, L., Thesleff, I., & Itoh, N. (2003). Identification of a secreted BMP antagonist, ectodin, integrating BMP, FGF, and SHH signals from the tooth enamel knot. *Developmental Biology*, *264*(1), 91–105.

Laurikkala, J., Mikkola, M., Mustonen, T., Aberg, T., Koppinen, P., Pispa, J., et al. (2001). TNF signaling via the ligand-receptor pair ectodysplasin and edar controls the function of epithelial signaling centers and is regulated by Wnt and activin during tooth organogenesis. *Developmental Biology*, *229*(2), 443–455. http://dx.doi.org/10.1006/dbio.2000.9955.

Lesot, H., Lisi, S., Peterkova, R., Peterka, M., Mitolo, V., & Ruch, J. V. (2001). Epigenetic signals during odontoblast differentiation. *Advances in Dental Research*, *15*, 8–13.

Li, J., Feng, J., Liu, Y., Ho, T. V., Grimes, W., Ho, H. A., et al. (2015). BMP-SHH signaling network controls epithelial stem cell fate via regulation of its niche in the developing tooth. *Developmental Cell*, *33*(2), 125–135. http://dx.doi.org/10.1016/j.devcel.2015.02.021.

Li, L., Kwon, H. J., Harada, H., Ohshima, H., Cho, S. W., & Jung, H. S. (2011). Expression patterns of ABCG2, Bmi-1, Oct-3/4, and Yap in the developing mouse incisor. *Gene Expression Patterns*, *11*(3–4), 163–170. http://dx.doi.org/10.1016/j.gep.2010.11.001.

Lian, J. B., Stein, G. S., Javed, A., van Wijnen, A. J., Stein, J. L., Montecino, M., et al. (2006). Networks and hubs for the transcriptional control of osteoblastogenesis. *Reviews in Endocrine & Metabolic Disorders*, 7(1–2), 1–16. http://dx.doi.org/10.1007/s11154-006-9001-5.

Lin, Y., Cheng, Y. S., Qin, C., Lin, C., D'Souza, R., & Wang, F. (2009). FGFR2 in the dental epithelium is essential for development and maintenance of the maxillary cervical loop, a stem cell niche in mouse incisors. *Developmental Dynamics*, *238*(2), 324–330. http://dx.doi.org/10.1002/dvdy.21778.

Lintern, K. B., Guidato, S., Rowe, A., Saldanha, J. W., & Itasaki, N. (2009). Characterization of wise protein and its molecular mechanism to interact with both Wnt and BMP signals. *The Journal of Biological Chemistry*, *284*(34), 23159–23168. http://dx.doi.org/10.1074/jbc.M109.025478.

Liu, F., Chu, E. Y., Watt, B., Zhang, Y., Gallant, N. M., Andl, T., et al. (2008). Wnt/beta-catenin signaling directs multiple stages of tooth morphogenesis. *Developmental Biology*, *313*(1), 210–224. http://dx.doi.org/10.1016/j.ydbio.2007.10.016.

Liu, Y., Han, D., Wang, L., & Feng, H. (2013). Down-regulation of Wnt10a affects odontogenesis and proliferation in mesenchymal cells. *Biochemical and Biophysical Research Communications*, *434*(4), 717–721. http://dx.doi.org/10.1016/j.bbrc.2013.03.088.

Lumsden, A. G. (1988). Spatial organization of the epithelium and the role of neural crest cells in the initiation of the mammalian tooth germ. *Development*, *103*(Suppl.), 155–169.

Mammoto, T., Mammoto, A., Torisawa, Y. S., Tat, T., Gibbs, A., Derda, R., et al. (2011). Mechanochemical control of mesenchymal condensation and embryonic tooth organ formation. *Developmental Cell*, *21*(4), 758–769. http://dx.doi.org/10.1016/j.devcel.2011.07.006.

Mikkola, M. L. (2009). Molecular aspects of hypohidrotic ectodermal dysplasia. *American Journal of Medical Genetics Part A*, *149A*(9), 2031–2036. http://dx.doi.org/10.1002/ajmg.a.32855.

Mina, M., & Kollar, E. J. (1987). The induction of odontogenesis in non-dental mesenchyme combined with early murine mandibular arch epithelium. *Archives of Oral Biology*, *32*(2), 123–127.

Mucchielli, M. L., Mitsiadis, T. A., Raffo, S., Brunet, J. F., Proust, J. P., & Goridis, C. (1997). Mouse Otlx2/RIEG expression in the odontogenic epithelium precedes tooth initiation and requires mesenchyme-derived signals for its maintenance. *Developmental Biology, 189*(2), 275–284. http://dx.doi.org/10.1006/dbio.1997.8672.

Munne, P. M., Felszeghy, S., Jussila, M., Suomalainen, M., Thesleff, I., & Jernvall, J. (2010). Splitting placodes: Effects of bone morphogenetic protein and Activin on the patterning and identity of mouse incisors. *Evolution & Development, 12*(4), 383–392. http://dx.doi.org/10.1111/j.1525-142X.2010.00425.x.

Mustonen, T., Pispa, J., Mikkola, M. L., Pummila, M., Kangas, A. T., Pakkasjarvi, L., et al. (2003). Stimulation of ectodermal organ development by Ectodysplasin-A1. *Developmental Biology, 259*(1), 123–136.

Nakao, K., Morita, R., Saji, Y., Ishida, K., Tomita, Y., Ogawa, M., et al. (2007). The development of a bioengineered organ germ method. *Nature Methods, 4*(3), 227–230. http://dx.doi.org/10.1038/nmeth1012.

Neubuser, A., Peters, H., Balling, R., & Martin, G. R. (1997). Antagonistic interactions between FGF and BMP signaling pathways: A mechanism for positioning the sites of tooth formation. *Cell, 90*(2), 247–255.

Neupane, S., Sohn, W. J., Rijal, G., Lee, Y. J., Lee, S., Yamamoto, H., et al. (2014). Developmental regulations of Perp in mice molar morphogenesis. *Cell and Tissue Research, 358*(1), 109–121. http://dx.doi.org/10.1007/s00441-014-1908-7.

O'Connell, D. J., Ho, J. W., Mammoto, T., Turbe-Doan, A., O'Connell, J. T., Haseley, P. S., et al. (2012). A Wnt-bmp feedback circuit controls intertissue signaling dynamics in tooth organogenesis. *Science Signaling, 5*(206), ra4. http://dx.doi.org/10.1126/scisignal.2002414.

Oshima, M., Mizuno, M., Imamura, A., Ogawa, M., Yasukawa, M., Yamazaki, H., et al. (2011). Functional tooth regeneration using a bioengineered tooth unit as a mature organ replacement regenerative therapy. *PLoS One, 6*(7), e21531. http://dx.doi.org/10.1371/journal.pone.0021531.

Parsa, S., Kuremoto, K., Seidel, K., Tabatabai, R., Mackenzie, B., Yamaza, T., et al. (2010). Signaling by FGFR2b controls the regenerative capacity of adult mouse incisors. *Development, 137*(22), 3743–3752. http://dx.doi.org/10.1242/dev.051672.

Pispa, J., Jung, H. S., Jernvall, J., Kettunen, P., Mustonen, T., Tabata, M. J., et al. (1999). Cusp patterning defect in Tabby mouse teeth and its partial rescue by FGF. *Developmental Biology, 216*(2), 521–534. http://dx.doi.org/10.1006/dbio.1999.9514.

Plikus, M. V., Zeichner-David, M., Mayer, J. A., Reyna, J., Bringas, P., Thewissen, J. G., et al. (2005). Morphoregulation of teeth: Modulating the number, size, shape and differentiation by tuning Bmp activity. *Evolution & Development, 7*(5), 440–457. http://dx.doi.org/10.1111/j.1525-142X.2005.05048.x.

Ruch, J. V., Lesot, H., Karcher-Djuricic, V., Meyer, J. M., & Olive, M. (1982). Facts and hypotheses concerning the control of odontoblast differentiation. *Differentiation, 21*(1), 7–12.

Saadi, I., Das, P., Zhao, M., Raj, L., Ruspita, I., Xia, Y., et al. (2013). Msx1 and Tbx2 antagonistically regulate Bmp4 expression during the bud-to-cap stage transition in tooth development. *Development, 140*(13), 2697–2702. http://dx.doi.org/10.1242/dev.088393.

Salazar-Ciudad, I., & Jernvall, J. (2010). A computational model of teeth and the developmental origins of morphological variation. *Nature, 464*(7288), 583–586. http://dx.doi.org/10.1038/nature08838.

Sarkar, L., & Sharpe, P. T. (1999). Expression of Wnt signalling pathway genes during tooth development. *Mechanisms of Development, 85*(1–2), 197–200.

Sato, T., & Clevers, H. (2013). Growing self-organizing mini-guts from a single intestinal stem cell: Mechanism and applications. *Science, 340*(6137), 1190–1194. http://dx.doi.org/10.1126/science.1234852.

Seidel, K., Ahn, C. P., Lyons, D., Nee, A., Ting, K., Brownell, I., et al. (2010). Hedgehog signaling regulates the generation of ameloblast progenitors in the continuously growing mouse incisor. *Development, 137*(22), 3753–3761. http://dx.doi.org/10.1242/dev.056358.

Shirokova, V., Jussila, M., Hytonen, M. K., Perala, N., Drogemuller, C., Leeb, T., et al. (2013). Expression of Foxi3 is regulated by ectodysplasin in skin appendage placodes. *Developmental Dynamics, 242*(6), 593–603. http://dx.doi.org/10.1002/dvdy.23952.

Slavkin, H. C., & Bringas, P., Jr. (1976). Epithelial-mesenchyme interactions during odontogenesis. IV. Morphological evidence for direct heterotypic cell-cell contacts. *Developmental Biology, 50*(2), 428–442.

Smith, C. E., & Warshawsky, H. (1975). Cellular renewal in the enamel organ and the odontoblast layer of the rat incisor as followed by radioautography using 3H-thymidine. *The Anatomical Record, 183*(4), 523–561. http://dx.doi.org/10.1002/ar.1091830405.

Spemann, H., & Mangold, H. (2001). Induction of embryonic primordia by implantation of organizers from a different species. 1923. *The International Journal of Developmental Biology, 45*(1), 13–38.

Suomalainen, M., & Thesleff, I. (2010). Patterns of Wnt pathway activity in the mouse incisor indicate absence of Wnt/beta-catenin signaling in the epithelial stem cells. *Developmental Dynamics, 239*(1), 364–372. http://dx.doi.org/10.1002/dvdy.22106.

Thesleff, I. (2003). Epithelial-mesenchymal signalling regulating tooth morphogenesis. *Journal of Cell Science, 116*(Pt. 9), 1647–1648.

Thesleff, I., Keranen, S., & Jernvall, J. (2001). Enamel knots as signaling centers linking tooth morphogenesis and odontoblast differentiation. *Advances in Dental Research, 15*, 14–18.

Thesleff, I., & Tummers, M. (2008). *Tooth organogenesis and regeneration*. Cambridge (MA): StemBook.

Thomas, B. L., Liu, J. K., Rubenstein, J. L., & Sharpe, P. T. (2000). Independent regulation of Dlx2 expression in the epithelium and mesenchyme of the first branchial arch. *Development, 127*(2), 217–224.

Trumpp, A., Depew, M. J., Rubenstein, J. L., Bishop, J. M., & Martin, G. R. (1999). Cre-mediated gene inactivation demonstrates that FGF8 is required for cell survival and patterning of the first branchial arch. *Genes & Development, 13*(23), 3136–3148.

Vainio, S., Karavanova, I., Jowett, A., & Thesleff, I. (1993). Identification of BMP-4 as a signal mediating secondary induction between epithelial and mesenchymal tissues during early tooth development. *Cell, 75*(1), 45–58.

Wang, X. P., O'Connell, D. J., Lund, J. J., Saadi, I., Kuraguchi, M., Turbe-Doan, A., et al. (2009). Apc inhibition of Wnt signaling regulates supernumerary tooth formation during embryogenesis and throughout adulthood. *Development, 136*(11), 1939–1949. http://dx.doi.org/10.1242/dev.033803.

Wang, X. P., Suomalainen, M., Felszeghy, S., Zelarayan, L. C., Alonso, M. T., Plikus, M. V., et al. (2007). An integrated gene regulatory network controls stem cell proliferation in teeth. *PLoS Biology, 5*(6), e159. http://dx.doi.org/10.1371/journal.pbio.0050159.

Wang, X. P., Suomalainen, M., Jorgez, C. J., Matzuk, M. M., Werner, S., & Thesleff, I. (2004). Follistatin regulates enamel patterning in mouse incisors by asymmetrically inhibiting BMP signaling and ameloblast differentiation. *Developmental Cell, 7*(5), 719–730. http://dx.doi.org/10.1016/j.devcel.2004.09.012.

Wise, G. E., & Fan, W. (1989). Changes in the tartrate-resistant acid phosphatase cell population in dental follicles and bony crypts of rat molars during tooth eruption. *Journal of Dental Research, 68*(2), 150–156.

Yamashiro, T., Zheng, L., Shitaku, Y., Saito, M., Tsubakimoto, T., Takada, K., et al. (2007). Wnt10a regulates dentin sialophosphoprotein mRNA expression and possibly links odontoblast differentiation and tooth morphogenesis. *Differentiation*, *75*(5), 452–462. http://dx.doi.org/10.1111/j.1432-0436.2006.00150.x.

Yang, Z., Balic, A., Michon, F., Juuri, E., & Thesleff, I. (2015). Mesenchymal Wnt/beta-catenin signaling controls epithelial stem cell homeostasis in teeth by inhibiting the anti-apoptotic effect of Fgf10. *Stem Cells*, *33*(5), 1670–1681. http://dx.doi.org/10.1002/stem.1972.

Yang, L., & Peng, R. (2010). Unveiling hair follicle stem cells. *Stem Cell Reviews*, *6*(4), 658–664. http://dx.doi.org/10.1007/s12015-010-9172-z.

Yang, G., Zhou, J., Teng, Y., Xie, J., Lin, J., Guo, X., et al. (2014). Mesenchymal TGF-beta signaling orchestrates dental epithelial stem cell homeostasis through Wnt signaling. *Stem Cells*, *32*(11), 2939–2948. http://dx.doi.org/10.1002/stem.1772.

Yokohama-Tamaki, T., Ohshima, H., Fujiwara, N., Takada, Y., Ichimori, Y., Wakisaka, S., et al. (2006). Cessation of Fgf10 signaling, resulting in a defective dental epithelial stem cell compartment, leads to the transition from crown to root formation. *Development*, *133*(7), 1359–1366. http://dx.doi.org/10.1242/dev.02307.

Yoshida, T., Miyoshi, J., Takai, Y., & Thesleff, I. (2010). Cooperation of nectin-1 and nectin-3 is required for normal ameloblast function and crown shape development in mouse teeth. *Developmental Dynamics*, *239*(10), 2558–2569. http://dx.doi.org/10.1002/dvdy.22395.

Zhang, Z., Tian, H., Lv, P., Wang, W., Jia, Z., Wang, S., et al. (2015). Transcriptional factor DLX3 promotes the gene expression of enamel matrix proteins during amelogenesis. *PLoS One*, *10*(3), e0121288. http://dx.doi.org/10.1371/journal.pone.0121288.

Zhao, H., Feng, J., Seidel, K., Shi, S., Klein, O., Sharpe, P., et al. (2014). Secretion of shh by a neurovascular bundle niche supports mesenchymal stem cell homeostasis in the adult mouse incisor. *Cell Stem Cell*, *14*(2), 160–173. http://dx.doi.org/10.1016/j.stem.2013.12.013.

Zhao, M., Gupta, V., Raj, L., Roussel, M., & Bei, M. (2013). A network of transcription factors operates during early tooth morphogenesis. *Molecular and Cellular Biology*, *33*(16), 3099–3112. http://dx.doi.org/10.1128/MCB.00524-13.

Zhao, H., Li, S., Han, D., Kaartinen, V., & Chai, Y. (2011). Alk5-mediated transforming growth factor beta signaling acts upstream of fibroblast growth factor 10 to regulate the proliferation and maintenance of dental epithelial stem cells. *Molecular and Cellular Biology*, *31*(10), 2079–2089. http://dx.doi.org/10.1128/MCB.01439-10.

CHAPTER EIGHT

Stem Cells in Tooth Development, Growth, Repair, and Regeneration

Tian Yu[1], Ana Angelova Volponi[1], Rebecca Babb[1], Zhengwen An[1], Paul T. Sharpe[1,2]

Craniofacial Development and Stem Cell Biology, Dental Institute, Kings College London, London, United Kingdom

[2]Corresponding author: e-mail address: paul.sharpe@kcl.ac.uk

Contents

Abstract

Human teeth contain stem cells in all their mesenchymal-derived tissues, which include the pulp, periodontal ligament, and developing roots, in addition to the support tissues such as the alveolar bone. The precise roles of these cells remain poorly understood and most likely involve tissue repair mechanisms but their relative ease of harvesting makes teeth a valuable potential source of mesenchymal stem cells (MSCs) for therapeutic use. These dental MSC populations all appear to have the same developmental origins, being derived from cranial neural crest cells, a population of embryonic stem cells with multipotential properties. In rodents, the incisor teeth grow continuously throughout life, a feature that requires populations of continuously active mesenchymal and epithelial stem cells. The discrete locations of these stem cells in the incisor have rendered them amenable for study and much is being learnt about the general properties of these stem cells for the incisor as a model system. The incisor MSCs appear to be a heterogeneous population consisting of cells from different neural crest-derived tissues.

[1] All authors contributed equally.

Current Topics in Developmental Biology, Volume 115
ISSN 0070-2153
http://dx.doi.org/10.1016/bs.ctdb.2015.07.010

The epithelial stem cells can be traced directly back in development to a $Sox10^+$ population present at the time of tooth initiation. In this review, we describe the basic biology of dental stem cells, their functions, and potential clinical uses.

1. STEM CELLS IN TOOTH DEVELOPMENT

Mammalian dentition involves a complex sequence of reciprocal epithelial–mesenchymal interactions that are typically divided into initiation, morphogenesis, and differentiation phases. The mouse has a simple dentition with an incisor at the front and three molars at the back of the mouth in each quadrant. Humans have extra tooth types, such as canines and premolars, yet these teeth develop in a similar pattern as in the mouse. The main difference between humans and mice is that humans have one set of replacement teeth whereas mice only have a single set of teeth.

1.1 Developmental Origins of Dental Stem Cells

Two principal sources of cells are required for tooth development: the cranial neural crest (CNC)-derived mesenchyme and the oral ectoderm- or pharyngeal ectoderm-derived epithelium. The neural crest-derived mesenchymal cells give rise to most of the soft and hard tissues in and around the tooth, including dental pulp, dentin, alveolar bone, and periodontal ligament, as well as the facial and jaw skeletons (Bronner-Fraser, 1993; Chai et al., 2000; Lumsden, 1988). The epithelium mainly produces the enamel-secreting ameloblasts and their supporting cells (Chai et al., 2000; Rothova, Thompson, Lickert, & Tucker, 2012). Patterning of the mammalian dentition involves a three-dimensional process. The proximally expressed Fgf8 and distally expressed Bmp4 in the dental epithelium are critical for specifying the initiation sites of tooth formation as well as for the CNC cell-derived dental mesenchyme to become an incisor or molar tooth (Neubuser, Peters, Balling, & Martin, 1997; Tucker, Matthews, & Sharpe, 1998). FGF8 restricts Barx1, Dlx1, and Dlx2 expression, and BMP4 restricts Msx1 and Msx2 expression to the proximal and distal region of the ectomesenchymal domain, respectively (Thesleff & Sharpe, 1997; Tucker et al., 1998; Vainio, Karavanova, Jowett, & Thesleff, 1993).

Tooth development in the mouse begins around embryonic day (E) 10, when the first sets of signaling molecules involved in this process, such as Shh, Fgf8, and Bmp4, are expressed at the sites of tooth formation. The oral epithelium subsequently thickens into a placode, and the thickening grows as the epithelium starts to invaginate into the underlying mesenchyme. The ability to induce tooth formation resides in the epithelium at this placode stage (Lumsden, 1988; Mina & Kollar, 1987). The mesenchyme responds by condensing around the invaginating epithelium to form the dental papilla and leads to the formation of a tooth bud around E13. The epithelium and the mesenchyme continue to proliferate to form a cap around E14. At this stage, the odontogenic potential switches from the epithelium to the mesenchyme (Kollar & Baird, 1969). A signaling centre called the primary enamel knot forms in the dental epithelium during the transition from bud to cap stage and regulates the future crown shape. During the late cap stage, significant differences between the developing incisor and molar arise. In the molar, the primary enamel knots are subsequently replaced by secondary enamel knots, which determine the position of epithelial folds that correspond to the tooth cusp formation (Jernvall & Thesleff, 2000), whereas the developing incisor begins to turn to grow parallel to the jaw axis. During the bell stage beginning at E16, the final tooth crown shape becomes apparent, and tooth-specific cell types, such as enamel-producing ameloblasts and dentin-producing odontoblasts, begin to differentiate. The differentiation of these cells is regulated, as in previous stages, by interactions between the epithelium and mesenchyme.

Ameloblasts are generated from epithelial cells residing in the inner enamel epithelium (IEE), and these cells secrete enamel matrix that later mineralizes. Signals such as FGFs and BMP2 from the IEE induce the formation of odontoblasts in the dental mesenchyme (Begue-Kirn et al., 1994; Thesleff, Keranen, & Jernvall, 2001). Odontoblasts differentiate from the outermost layer of the dental papilla and secrete dentin matrix. Signals arising from the odontoblasts are able to trigger the terminal differentiation of ameloblasts in the epithelium. It has been shown that the presence of functional odontoblasts or predentin matrix is required for ameloblast differentiation (Zeichner-David et al., 1995).

1.2 Neural Crest-Derived Stem Cells During Tooth Development

Dental mesenchymal cells are derived from two different origins: the CNC and the non-CNC mesoderm. Neural crest is a transitory population of cells, emerging from the lateral ridges of the neural plate during early stages of embryogenesis and subsequently differentiating into neurons, glia, melanocytes, and mesenchymal cells. During craniofacial development, neural crest cells migrate ventrally and proliferate to form the first branchial arch. These

CNC cells can differentiate into odontoblasts, cementoblasts, fibroblasts, osteoblasts, and chondroblasts (Bronner-Fraser, 1993; Chai et al., 2000; Lumsden, 1988).

A number of signaling molecules and homeobox gene targets are involved in regulating the specification of CNC cells. FGF8 signaling from the isthmic organizer controls branchial arch patterning; transforming growth factor-β (TGF-β) regulates the differentiation of CNC cells to glial cells; and TGF-β, BMP, and Wnt together control chondrocyte differentiation (Chai, Ito, Han, 2003; Trainor, Ariza-McNaughton, & Krumlauf, 2002; Tucker, Grigoriou, Pachnis, & Sharpe, 1999).

Clonal culture experiments and *in vivo* lineage tracing have demonstrated that neural crest cells are capable of self-renewal and display unique characteristics of stem cells (Bronner, 2015; Stemple & Anderson, 1992). Neural crest cells are developmentally plastic, and thus their progenitors must be instructed by signals from other tissues to determine their fate after reaching their final destination. Baggiolini et al. traced single neural crest cells and demonstrated that the vast majority of the premigratory and migratory neural crest cells are multipotent using the Wnt1-CreER^{T} and Sox10-CreER^{T2} R26R-Confetti mouse models. Multipotent neural crest cells have also been identified in sites where postmigratory neural crest cells reside (Baggiolini et al., 2015; Lo & Anderson, 1995). Postmigratory CNC cells have been identified as MSCs. BMP and TGF-β signaling control the differentiation and fate of postmigratory CNC cells, and these cells support ectopic tooth germ survival and development via BMP signaling. Moreover, transplanted CNC cells can mediate the development of alveolar bone and teeth as a functional unit (Chung et al., 2009).

The contribution of the CNC cells to dental mesenchyme has been demonstrated by DI labeling, as well as transgenic approaches marking the neural crest cells with a LacZ reporter gene (Chai et al., 2000; Imai, Osumi-Yamashita, Ninomiya, & Eto, 1996; Serbedzija, Bronner-Fraser, & Fraser, 1992). Tissue recombination studies, in which CNC cells as well as CNC-derived ectomesenchyme but not mesoderm-derived mesenchymal tissue supported tooth formation when recombined with dental epithelium, provided additional evidence for the contribution of the CNC cells to mammalian tooth development (Lumsden, 1988). The CNC cells that migrate into the first arch between the 10- and 12-somite stage are odontogenic, either possessing instructive odontogenic potential to initiate tooth development or being able to respond to odontogenic signals such as FGF8 to support tooth formation (Zhang et al., 2003).

The shift in odontogenic potential coincides with a shift in BMP signaling from the epithelium to the mesenchyme. When Smad4 is conditionally inactivated in the neural crest cells, dental epithelium thickenings appear to be normal; however, tooth development arrests at the dental lamina stage, suggesting that Smad4-mediated BMP signaling in the CNC-derived ectomesenchyme is required for tooth development to advance into the bud stage (Ko et al., 2007).

1.3 Sox2+ Epithelial Stem Cells

Sox2 is a transcription factor expressed in the embryonic and adult stem cells and plays an important role in the maintenance of stem cell in various organs. *Sox2* is expressed in the primary dental lamina, where dental epithelial stem cells (DESCs) reside, and gives rise to all dental epithelial tissues (Juuri, Isaksson, Jussila, Heikinheimo, & Thesleff, 2013; Juuri et al., 2012; Smith, Fraser, & Mitsiadis, 2009). In the developing molar, Sox2-positive cells are detected by *in situ* hybridization from E11 in the lower jaw, and by E12, the expression is strongest along the lingual side of the dental epithelium. At E14–E16, *Sox2* expression is found in the oral epithelium of the lingual side of the tooth germ, including the dental cord and the enamel organ (Juuri, Isaksson, et al., 2013).

In the developing incisor, the labial cervical loop (CL) starts to show distinct *Sox2* expression from E14.5, and the expression in the lingual CL decreases. From E16 to E18, *Sox2* becomes increasingly restricted to the tip of the labial CL. The location of the most intense *Sox2* expression corresponds to that of the putative epithelial stem cell population, suggesting an early function for Sox2 in the specification of stem cells in the epithelium (Harada, et al., 1999; Juuri et al., 2012). Sox2 is subsequently expressed in the successional dental lamina, which is an epithelial sheet that connects the deciduous tooth germs and is embedded on their lingual side, where the replacement teeth form in all animals (Jarvinen, Tummers, & Thesleff, 2009; Smith et al., 2009). SOX2 expression has been detected in the lingual side of the dental lamina of a deciduous premolar of a 13-week human fetus (Juuri, Jussila, et al., 2013). In addition, dynamic Sox2 expression is found when mouse posterior molars develop in succession, indicating that Sox2 expression is associated with the posterior extension of dental epithelium where new teeth form. Using genetic fate mapping experiments, Sox2-positive cells of the mouse M1 were found to give rise to all epithelial cell lineages of the successionally developing M2 and to the bud of M3 (Juuri,

Jussila, et al., 2013). Furthermore, the conditional inactivation of Sox2 in the epithelium results in aberrant formation of the dental epithelium in the developing M2 and M3 of the E17.5 mouse mutant, suggesting that Sox may regulate the maintenance of the dental epithelial progenitors (Juuri, Jussila, et al., 2013).

1.3.1 Stem Cells in Tooth Pulp

Teeth have been called a "treasure chest" of MSCs since all tooth and support tissues appear to contain cells with MSC-like properties when cultured *in vitro* (Fig. 1; Balic, Aguila, Caimano, Francone, & Mina, 2010; Gronthos et al., 2002; Gronthos, Mankani, Brahim, Robey, & Shi, 2000; Huang, Gronthos, & Shi, 2009; Jo et al., 2007; Koyama, Okubo, Nakao, & Bessho, 2009; Volponi & Sharpe, 2013; Waddington, Youde, Lee, & Sloan, 2009). As with all *in vitro* studies of MSC-like cells, the differentiation potential attributed to these cells should be viewed with caution since most often formation of mineralized matrix is used to describe cells as more or less osteogenic. However, in the absence of detailed analysis of the mineral formed or *in vivo* studies, such conclusions mean very little.

Deciduous teeth are a "perfect" source of stem cells since they are plentiful and are naturally exfoliated. Culture of deciduous tooth pulp tissue results in rapid expansion of cells that have the *in vitro* characteristics of MSCs, expressing appropriate markers and being capable of multilineage

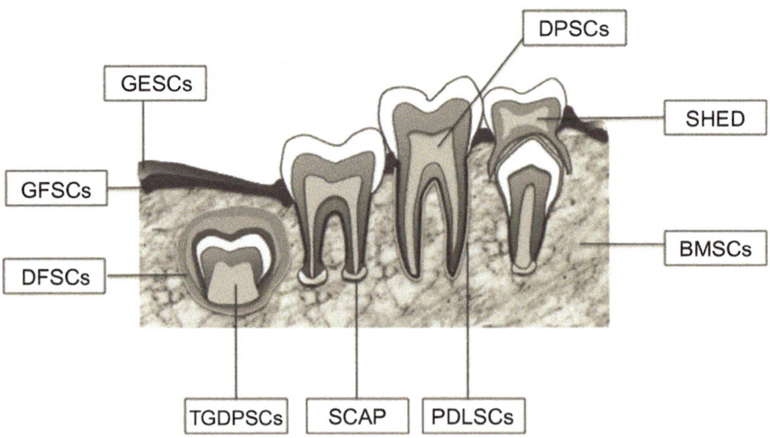

Figure 1 Teeth and supporting tissues are source for different stem cell populations. DPSCs, dental pulp stem Cells; SHED, stem cells from human exfoliated deciduous teeth; BMSCs, bone marrow stem cells; PDLSCs, periodontal ligament stem cells; SCAP, stem cells from apical papilla; TGDPSCs, tooth germ dental pulp stem cells; DFSCs, dental follicle stem cells; GFSCs, gingival fibroblast stem cells; GESCs, gingival epithelial stem cells. (A color version of this figure is located at the end of this chapter.)

mesenchymal cell differentiation into osteoblasts, chondrocytes, and adipocytes upon appropriate stimulation (Miura et al., 2003; Wang et al., 2012). These cells, called SHED (stem cells from human exfoliated deciduous teeth), can also be stimulated to form cells with properties that resemble odontoblasts and produce a dentin-like mineral (Miura et al., 2003).

Cells with virtually identical properties and of the same embryonic origin (CNC) can be isolated from the pulp of adult teeth and are generally referred to as Dental Pulp Stem cells (DPSCs). The role of these cells is to provide a source of cells to differentiate into odontoblast-like cells following dentin damage to produce reparative dentin (Feng, Mantesso, De Bari, Nishiyama, & Sharpe, 2011). It is plausible that SHED and DPSCs are the same cells. Although numerous studies have attempted to identify the origin of pulp stem cells, only one has used the accepted "gold standard" of genetic lineage tracing and shown that pericytes are a source of stem cells that can differentiate into odontoblast-like cells following dentin damage (Feng et al., 2011).

The relative ease with which pulp cells can be expanded *in vitro* suggests an obvious application in endodontics. *In vitro* experiments using tooth slice cultures suggest that pulp-like tissue can easily be generated from cultured pulp cells (Smith et al., 1995; Smith & Lesot, 2001), and *in vivo* studies similarly suggest restoration of pulp tissue and root dentin by DPSCs is possible (Huang et al., 2010; Murakami et al., 2013).

1.3.2 Periodontal Ligament Stem Cells

MSC-like cells can be isolated from periodontal ligament tissue and expanded in culture to generate populations with the properties expected of MSCs. *In vivo* studies suggest that these cells can be stimulated to differentiate into cells that are able to form cementum/PDL-like (Peiodontal Ligament) structures (Seo et al., 2004). PDL stem cells seeded as cell sheets can form PDL-like structures *in vivo* (Hasegawa, Yamato, Kikuchi, Okano, & Ishikawa, 2005), and the addition of PDL stem cells to dental implants to form an artificial PDL has been explored (Lin et al., 2011). Paradoxically, however, any PDL-like tissue around a metal implant is more likely to severely weaken rather than improve implant function. This highlights a major problem with cell-based PDL restoration: the PDL is a highly complex tissue comprising different cell types and intricate connections between the tooth and bone that are hard to imagine being fully reconstructed with cells alone. To this end, combinations of PDL stem cells and scaffolds are being tested (Morsczeck et al., 2005).

1.3.3 Root Apical Papilla and Dental Follicle Stem Cells

Not surprisingly, developing tissues such as third molar roots are an excellent source of highly proliferative MSC-like cells (often called SCAP cells) (Huang et al., 2009). These cells probably act as a source of root odontoblasts as the root grows. Cells isolated from the follicular sac, a loose connective tissue surrounding the enamel organ, of human third molars have MSC-like properties (Handa, Saito, Tsunoda, et al., 2002; Handa, Saito, Yamauchi, et al., 2002; Morsczeck et al., 2005; Park et al., 2010). These cells can differentiate into cementoblasts *in vivo* and can form cementum and PDL-like structures when cocultured with Hertwig's epithelial root sheath cells *in vitro* (Bai et al., 2011).

1.3.4 Gingival Stem Cells

The gingival tissue is composed of both epithelial and mesenchymal cells, and MSC-like cells have been cultured from gingival tissue (Fournier et al., 2010; Hsu, Huang, & Feng, 2012; Mitrano et al., 2010; Widera et al., 2009). Gingival wound repair shows similarities to scarless fetal wound repair (Schor et al., 1996; Sempowski, Borrello, Blieden, Barth, & Phipps, 1995), and the heterogeneous cell population of gingival fibroblasts plays a crucial part in the process. Gingival fibroblasts are also a convenient source of cells for iPS cell generation (Wada et al., 2011) and gingival epithelial cells are capable of forming teeth when combined with embryonic tooth mesenchymal cells (Angelova Volponi, Kawasaki, & Sharpe, 2013). Stem cells isolated from the oral epithelium have been successfully used to treat severe ocular surface disorders, and a treatment known as cultivated oral mucosal epithelial transplantation (COMET) has been introduced (Nakamura et al., 2003, 2004; Nakamura, Takeda, Inatomi, Sotozono, & Kinoshita, 2011).

2. STEM CELLS IN DENTIN REPAIR: TERTIARY DENTINOGENESIS

The dentin–pulp complex has the ability to naturally repair itself by re-initiating dentinogenesis to protect against external injuries and insults, such as caries and mechanical trauma (Smith et al., 1995). The dentin formed in response to pathological conditions is called tertiary dentin and is deposited by two distinct mechanisms, namely, reactionary and reparative dentinogenesis, depending on the severity of the injury (Fig. 2). In response to mild damage, postmitotic primary odontoblasts located at the periphery of the pulp respond by upregulating dentin secretion (reactionary

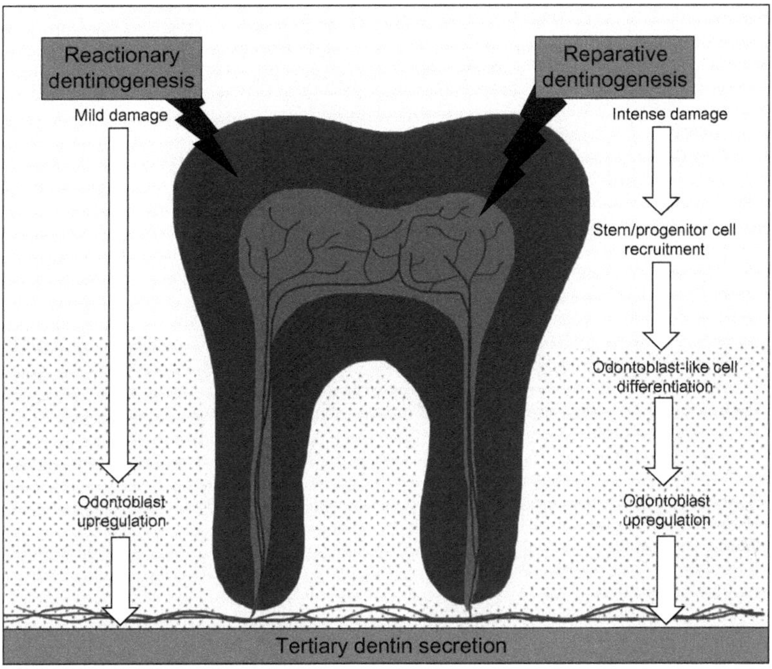

Figure 2 Diagram of a tooth showing possible sources of dental pulp stem cells in reparative dentinogenesis.

dentinogenesis). If the damage is extensive and the pulp is exposed, resulting in the death of local resident odontoblasts, stem/progenitor cells are recruited, differentiate into new odontoblast-like cells and subsequently secrete dentin.

The ontogeny, anatomical localization, and precise identity of odontogenic stem cells remain largely unknown. The cell-rich layer of Höhl, adjacent to the primary odontoblasts, contains undifferentiated MSCs that may be one stem/progenitor source of odontoblast-like cells. However, these cells are unlikely to provide a source of new odontoblasts, as their survival is closely linked to the survival of primary odontoblasts (Goldberg & Smith, 2004). Two cell niches within the dental pulp, the perivascular and neural stem cell niches, have been proposed to provide a source of stem/progenitor cells capable of generating reparative odontoblasts. Genetic-based linage tracing of pericytes using NG2, a proteoglycan that is commonly used as a marker for pericytes, revealed that differentiated odontoblasts originate from perivascular cells during incisor repair following damage (Feng et al., 2011). These $NG2^{+}$ pericyte-derived odontoblasts produce reparative dentin in response to damage. It has further been shown that $NG2^+$ pericytes represent a dental stem cell subpopulation derived from $Gli1^+$ cells, which are located on arteries and are mobilized in response to incisor damage (Zhao et al., 2014). However, not all differentiated odontoblasts are pericyte derived, indicating that the dental pulp may also contain MSC-like cells of a nonpericyte origin. Some dental stem cells have been shown to differentiate from peripheral nerve-associated glial cells (Kaukua et al., 2014). Linage tracing of Schwann cells, the principal glial cells, demonstrated that reparative odontoblasts originate from neural crest-derived Schwann cells and Schwann cell precursors. It is important to note that Schwann cells do not express pericyte markers, signifying that pericytes and Schwann cells represent two distinct dental stem cell populations.

2.1 Signaling Events Mediating Reparative Dentinogenesis

The underlying molecular and cellular events that control the recruitment, proliferation, and differentiation of dental stem/progenitor cells into reparative odontoblasts at the site of damage are poorly understood. However, from what is known, it appears that the regeneration process does not exactly recapitulate what occurs during embryogenesis to reconstitute the original tissue architecture and biological function. During embryogenesis, sequential and reciprocal interactions between the stomadial epithelial and CNC-derived mesenchyme regulate tooth morphogenesis and differentiation. The odontogenic potential that directs the differentiation of odontoblasts resides in the dental epithelium. During tooth eruption, the enamel epithelium is lost, so an alternative signaling source is required to stimulate the formation of new odontoblasts during reparative dentinogenesis.

An assortment of growth factors is fossilized within the dentin matrix during tooth development (Smith et al., 2012). These bioactive molecules include members of the TGF-β superfamily, insulin-like growth factor-1 and 2 (IGF-1 and -2), fibroblast growth factor-2 (FGF2), platelet-derived growth factor (PDGF), and vascular endothelial growth factor (VEGF). The release of these molecules into the pulp environment in response to tissue injury triggers signaling events that act on stem/progenitor cells to stimulate their recruitment, proliferation, and differentiation into reactionary odontoblasts. One particular member of the TGF-β superfamily, TGFβ-1, is a potent stimulator of odontoblast differentiation and promotes tertiary dentinogenesis in response to damage *in vivo* (Arany et al., 2014). Recently, the Wnt/β-catenin signaling pathway has been implicated in reparative dentinogenesis. Wnt signaling has been proposed to promote odontoblast survival, proliferation, and differentiation (Han et al., 2014; Hunter et al., 2015). The source and the specific Wnt ligand(s) that mediate tooth repair are currently unknown. The activation of the p38 MAP signaling pathway in primary odontoblasts is an important event that triggers the secretion of new dentin (Simon, Smith, Berdal, Lumley, & Cooper, 2010). The tissue microenvironment also plays a crucial role in determining whether reparative dentinogenesis can take place; in particular, the degree of inflammation and infection has a strong impact on the tissue environment. Inflammation is not only required to remove dead cells and debris from the site of damage, it also facilitates the repair process, although the precise nature of the interplay between repair and inflammation is unclear. If inflammation is chronic, tertiary dentinogenesis will not occur, and the tissue will not repair. Inflammatory cells recruited to an injury site release cytokines and growth factors, notably TNF-α and ROS, at low levels to promote reparative dentinogenesis (Cooper, Holder, & Smith, 2014; Cooper & Smith, 2013). MSCs have recently been shown to possess immunomodulatory properties, which they exert through direct cell–cell contact with immune cells and the secretion of soluble factors, including prostaglandin E2, indoleamine 2,3-dioxygenase, and TGF-β, to promote tissue regeneration. DPSCs have been demonstrated to elicit an anti-inflammatory effect on innate and adaptive immune cells, supporting tissue regeneration (Li et al., 2014).

3. THE CONTINUOUSLY GROWING INCISOR

3.1 A Model System for Stem Cell Biology: The Mouse Incisor

Rodent incisors, including those of mice, are capable of continuously growing throughout the lifetime of the animal. This is achieved by the stem cells resident at the proximal end of the incisors, which provide a source of cells for compensation of tissue loss and homeostasis. The existence of continuously growing rodent incisors was first documented by Fougeroux de Bondaroy (1768). In 1962, Hwang and colleagues identified histidine as being highly expressed in the enamel in growing mouse incisors using autoradiography (Hwang, Tonna, & Cronkite, 1962). A decade later, Smith identified the renewal of the cell population in rat incisors using ^{3}H-thymidine radioautography (Smith, 1980; Smith & Warshawsky, 1975). Since the landmark article on stem cell concepts by Lajtha (1979) and

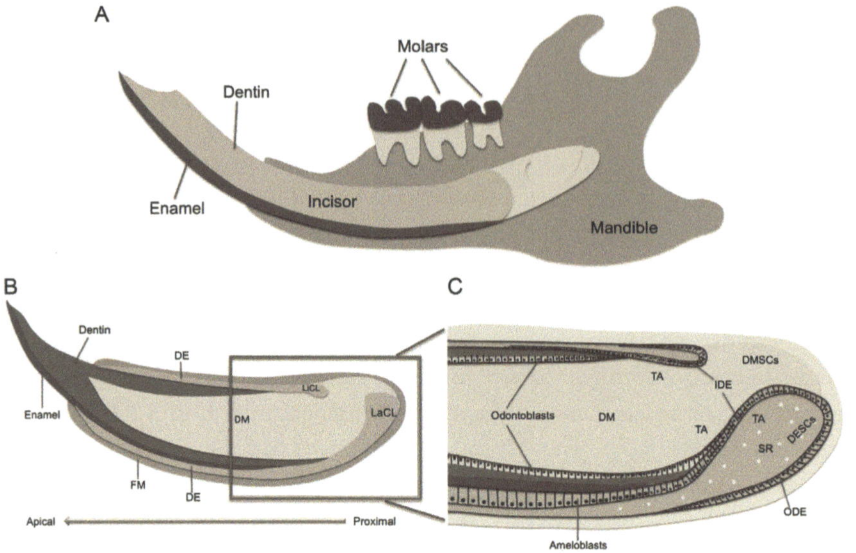

Figure 3 Schematic diagrams of the continuously growing mouse incisor. (A) Basic structure showing the mandible, incisor, and molars in adult mouse. (B) Illustrator of tissue distribution of mouse incisor on a sagittal section, and the enlarged region (C) at the proximal end of incisor showing the individual cells that constitute mouse incisor. (LaCL, labial cervical loop; LiCL, lingual cervical loop; DM, dental mesenchyme; DE, dental epithelium; FM, follicle mesenchyme; TA, transit-amplifying cells; IDE, inner dental epithelium; ODE, outer dental epithelium; SR, stellate reticulum; DESCs, dental epithelial stem cells; DMSCs, dental mesenchymal stem cells). (A color version of this figure is located at the end of this chapter.)

especially over the past 15 years, interest in rodent incisor stem cells has increased as an attractive model system for the study of adult stem cell biology (Kuang-Hsien Hu, Mushegyan, & Klein, 2014). Two basic types of stem cells are found in the mouse incisors: neural crest-derived ectomesenchymal and ectodermal epithelial (Koussoulakou, Margaritis, & Koussoulakos, 2009). Dental mesenchymal stem cells (DMSCs) give rise to pulp cells and odontoblasts which produce dentin, while DESCs only give rise to ameloblasts that produce enamel (Fig. 3).

3.2 Gene Expression in the Epithelial Stem Cell Niche of Mouse Incisors

Several signaling pathways including FGF, BMP, TGF-β, Notch, Shh, and Wnt, as well as adhesion molecules and microRNAs, have been identified as important for maintaining the continuous growth of mouse incisors (Fig. 4).

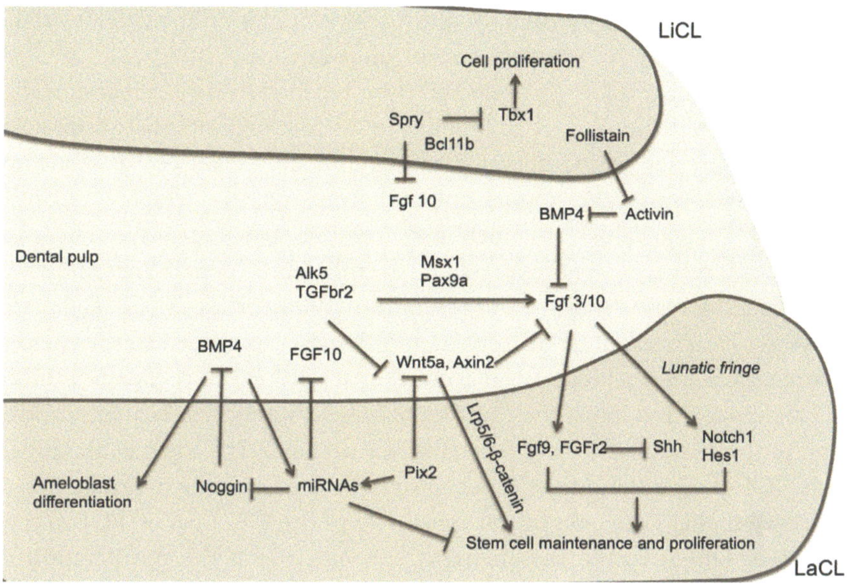

Figure 4 Gene regulatory network in a stem cell niche in the mouse incisors. Signaling molecules, transcription factors, and miRNAs expressed in the dental epithelium and mesenchyme regulate and maintain stem cells proliferation and differentiation. (A color version of this figure is located at the end of this chapter.)

Among these, FGF signaling is the most understood as crucial for epithelial stem cell proliferation and maintenance during tooth development and in adults (Kuang-Hsien Hu et al., 2014). *Fgf3* and *Fgf10* expression is restricted to the mesenchyme underlying the basal epithelial cells. FGF signaling regulates the Notch pathway in DESCs via stimulation of *lunatic fringe*, a secretory molecule that modulates Notch signaling expression (Harada, et al., 1999). Incisors in $Fgf3^{-/-}$ knockout mutants have a normal CL and $Fgf10^{-/-}$ mice die at birth, whereas $Fgf3^{-/-}$;$Fgf10^{-/-}$ compound mutants show a smaller labial CL, indicating that FGFs function cooperatively during tooth development and growth (Wang et al., 2007). *Fgf10* has been suggested to play a role in the regulation of incisor epithelial stem cells mainly through fibroblast growth factor receptor 2β (FGFR2β) (Klein et al., 2008). An inducible and reversible attenuation of FGFR2β signaling by rtTA transactivator/tetracycline promoter approaches identifies the role of FGFR2β signaling during development and adult mouse incisors. The results demonstrate that FGFR2β signaling regulates both the establishment of the incisor stem cell niche in the embryo and the regenerative capacity of incisors in the adult (Parsa et al., 2010). Other *Fgfs*, such as *Fgf9*, are expressed in the incisor epithelium during development (Porntaveetus et al., 2011) but have not yet been studied in the adult incisors.

Members of the TGF-β family and their receptors play a role in the regulation of proliferation and maintenance of incisor epithelial stem cells. BMP4 is expressed in the dental mesenchyme and represses *Fgf3* (but not *Fgf10*) expression in the lingual mesenchyme. Activin, which is strongly expressed in the labial mesenchyme, inhibits the repressive effect of BMP4 and restricts *Fgf3* expression to the labial dental mesenchyme, resulting in increased stem cell proliferation (Wang et al., 2007). *Follistatin*, an extracellular antagonist of Activin and BMP, is expressed in the mesenchyme and at higher levels in adjacent dental epithelium. *Follistain* inhibits *Fgf3* expression and limits the number of lingual stem cells, further contributing to the characteristic asymmetry of mouse incisors (Wang et al., 2007). Both TGF-β type I and type II receptors play a role in the maintenance of incisor epithelial stem cells in the apical end of the CL. Loss of TGF-β type I receptor (*Alk5*) in the mesenchyme severely affects the proliferation of transit-amplifying (TA) cells in the dental epithelium, whereas loss of *Alk5* in the dental epithelium has no effect on the CL (Zhao, Li, Han, Kaartinen, & Chai, 2011), indicating the directionality of the signaling in these mesenchymal–epithelial cell interactions. TGF-β type II receptor (*Tgfbr2*) is expressed in the dental mesenchyme and its deletion results in upregulated expression of Wnt5a and downregulated expression of *Fgf3/10* in mesenchyme, both of which synergistically enhance *Lrp5/6*-β-catenin signaling in the CL epithelium, thus causing malformed postnatal incisors (Yang et al., 2014).

The Notch pathway regulates stem cells and progenitor maintenance and proliferation in several tissues including bone marrow, intestinal epithelium, and neuronal tissues (Breunig, Silbereis, Vaccarino, Sestan, & Rakic, 2007; Fan et al., 2006; Fre et al., 2005; Sakata-Yanagimoto et al., 2008). Studies on Notch signaling in mouse incisors suggest a role of Notch receptors, ligands, and modulators in the regulation of epithelial stem cells in the continuous regeneration of the mouse incisors. *Notch1, -2,* and *-3* as well as Notch target gene *Hes1* are expressed in the stellate reticulum (SR) cells in the incisor epithelium (Felszeghy, Suomalainen, & Thesleff, 2010; Harada, et al., 1999). In addition *Notch3* is found also expressed in dental mesenchyme (Harada, et al., 1999). *Notch1* expressing SR cells in the CL of the continuously growing mouse incisor are stem cells that have the capacity to self-renew and to generate progeny that differentiate into various cell types of epithelium, especially the enamel-producing ameloblasts (Harada, et al., 1999). Using organ cultures, BrdU and *DiI* labeling experiments demonstrate that the Notch pathway in DESCs is regulated by FGF signaling in dental mesenchyme via stimulation of *lunatic fringe* expression (Harada, et al., 1999). The expression of Notch ligand, *Jagged 2* (*Jag2*), and Notch signaling modulator *lunatic fringe* is regulated by FGF and BMP signaling (Harada, et al., 1999; Mitsiadis, Graf, Luder, Gridley, & Bluteau, 2010). Notch pathway inhibitor, DAPT, inhibits Notch target gene *Hes1* expression in the CL thus causing a significant reduction in the size of the CL and massive apoptosis in the epithelial stem cell niche, suggesting Notch signaling is required for epithelial stem cell survival and enamel formation in the continuously growing mouse incisors (Felszeghy et al., 2010).

The Sonic hedgehog (*Shh*) pathway also regulates incisor epithelial stem cells in adult mice. *Shh* reporter gene *Gli1* and Hedgehog receptor Patched Homolog 1 (*Ptch1*) are expressed in both the epithelial TA cells and the dental pulp cells in the developing and growing mouse incisors (Bitgood & McMahon, 1995; Seidel et al., 2010). To identify a role of *Shh* signaling in adult stem cells, $Gli1^{lacZ}$ and $Ptch1^{lacZ}$ reporter mice are used for genetic lineage tracing of Hh-responsive cells and the fate of their progenies following continuous growth of the mouse incisors. The results reveal that high levels of *lacZ* expression in the reporter lines are found in the labial and lingual CL epithelium and in the mesenchyme between the two aspects of the CL. Inhibition of *Shh* pathway demonstrates that Hh signaling is required for the stem cells to continuously generate ameloblasts in adults but not for stem cell survival (Seidel et al., 2010). Runt-related transcription factor (*Runx*) regulates epithelial stem cell continuous proliferation and differentiation in growing incisors by maintaining of *Fgf9* and *Shh* expression, and *Fgf9* acts as antagonist of *Shh* expression (Kurosaka et al., 2011).

The Wnt signaling pathway plays a critical role in the epithelial stem cell niche in many adult tissues such as the skin (Arwert, Hoste, & Watt, 2012), hair follicle (Fujiwara et al., 2011), and the intestine (Holik et al., 2014). BAT*gal*, TOP*gal*, and $Axin2^{lacZ/lacZ}$ reporter mice are used to gain insight into the role of Wnt signaling in the regulation of the continuously growing mouse incisors. Wnt/β-catenin activity is found mainly confined to mesenchyme but completely absent from the epithelial stem cell compartment (Suomalainen & Thesleff, 2010). *Lgr5*, a Wnt target gene, is a specific epithelial stem cell marker in the hair follicle (Fuchs, 2009; Jaks et al., 2008) and intestine (Barker et al., 2007) and has been found expressed in the region of putative epithelial stem cells niche in the labial CL of mouse incisors. However, Wnt activity is not found in the *Lgr5*-positive cells in the incisor stem cell niche (Suomalainen & Thesleff, 2010). Wnt3a is expressed in the epithelial cells, and overexpression of Wnt3a inhibits amelobast formation in the postnatal mouse incisor (Millar et al., 2003). Constitutively expression of active β-catenin in the adult dental epithelium results in increased proliferation and oversized CL (Liu et al., 2010). Taken together, Wnt/β-catenin signaling may not directly regulate epithelial stem cell maintenance but might have an inhibitory role on stem cell proliferation and differentiation (Millar et al., 2003; Suomalainen & Thesleff, 2010). Recent studies suggest that mesenchymal *Tgfbr2* deletion activates dental epithelial *Lrp5*/6-β-catenin signaling by upregulation of Wnt5a and *Fgf3/10* (*Yang et al., 2014*), indicating a cross talk among TGF-β, Wnt, and *Fgf* signaling pathways in the epithelial CL.

Sprouty (*Spry*) genes are negative regulators of multiple receptor tyrosine kinases including receptors for FGF (Hacohen, Kramer, Sutherland, Hiromi, & Krasnow, 1998), EGF (Frank et al., 2009), and HGF (Lee et al., 2004). *Sprys* genes prevent the generation of lingual ameloblasts by inhibiting an FGF-mediated epithelial–mesenchymal signaling loop on the lingual side in adult mice (Boran et al., 2009; Klein et al., 2008). Transcription factors Tbx1 and *Bcl11b* have been found up- and downregulated, respectively, in $Spry2^{+/-};Spry4^{-/-}$ mutant incisors. Tbx1 maintains ameloblast progenitors in rodent incisors and loss of expression results in absence of enamel formation (Caton et al., 2009). $Bcl11b^{ko/ko}$ mice die at birth, but heterozygous $Bcl11b^{s82G/ko}$ mice suggest that attenuated *Bcl11b* activity impairs the maintenance of ameloblast progenitors in the mouse incisors (Katsuragi et al., 2013). The transcription factor Sox2 is an epithelial stem cells marker expressed in the labial aspect of the CL and contributes to ameloblast cell renewal regulated by *Fgf8* and fine tuned by specific miRNAs (Juuri et al., 2012).

Recent studies highlight the importance of adhesion molecules that are required for a given stem cell type to interact with their niche (Chen, Lewallen, & Xie, 2013). Deletion of integrin β3 resulted in a smaller CL and reduced proliferation in the TA region in adult mice (Yoshida et al., 2013). Conditional inactivation of E-cadherin causes decreased stem cell number and cell migration in the CL of the mouse incisors *via* FGF signaling regulation (Li et al., 2012). Gene microarrays identified the expression of miRNAs in the LaCL, LiCL, and ameloblasts in the adult incisors, suggesting a role of miRNAs in adult stem cell renew and differentiation during incisor growth (Jheon, Li, Wen, Michon, & Klein, 2011). Activated *miR200a-3p* is expressed in the preameloblasts and represses *Pitx2* and β-catenin expression in the labial epithelial stem cell niche, whereas *Pitx2* reciprocally activates *miR200a-3p* expression to enhance epithelial cell differentiation (Sharp et al., 2014). *Pitx2* directly binds to miR-200c/141 upregulating miR-200c and miR-203 expression and inhibiting noggin and Bmper, two antagonists of BMP, thus activating BMP signaling and regulating epithelial stem cell differentiation in the mouse incisors (Cao et al., 2013).

3.3 Identification of a MSC Niche

The continuously growing mouse incisors need a constant supply of cells to compensate tissue loss from the abrading tips, and this process requires not only epithelium stem cells that form enamel-producing ameloblasts, but also mesenchyme stem cells that form dentin-producing odontoblasts and pulp cells (Fig. 5). BrdU/EdU (5-ethynyl-2′-deoxyruidine)-based label-retaining

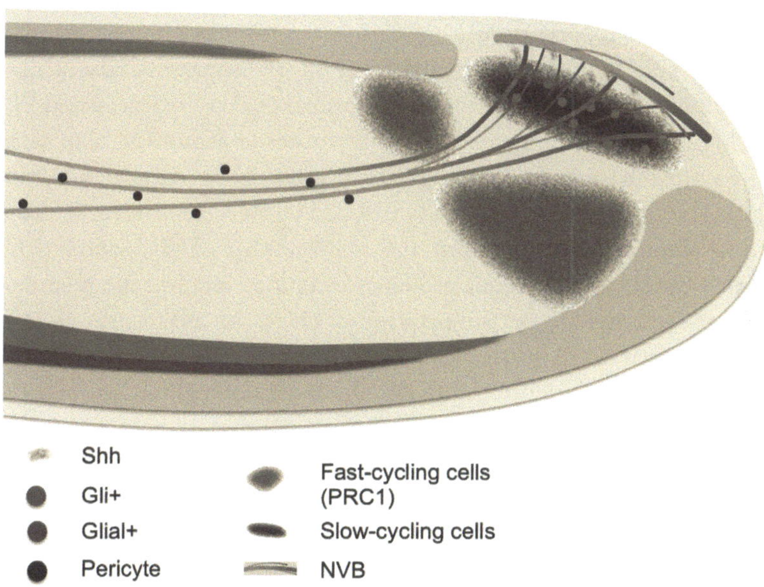

Figure 5 Identification of a mesenchymal stem cell niche in the mouse incisors. Slow-cycling stem cells locate in the mesenchyme between labial and lingual cervical loop region close to the NVB, represented by Gli+ and Glial+ expressing cells. PRC1 expressed in the mesenchyme closed to the labial and lingual cervical loop region, corresponding to the fast-cycling (TA) cells. Pericytes reside on the abluminal surface of endothelial cells in the vascularized connective tissues and contribute to MSC-derived mesenchymal cells responding to tissue damage (NVB, neurovascular bundle). (A color version of this figure is located at the end of this chapter.)

assays and H2B-GFP mice have been used to identify MSC populations in many tissues (Foudi et al., 2009; Kaukua et al., 2014; Tang et al., 2008; Tumbar et al., 2004; Zhao et al., 2014). The general concept is that stem cells divide infrequently to preserve their long-term proliferation potential and to prevent the errors occurrence during DNA replication (Fuchs, 2009), so such "slow cycling" can be identified by their ability to retain nucleosides. Progenitor cells in adult tissues divide fast and frequently, and those cells are called transit-amplifying (TA) cells or fast-cycling cells (Lapthanasupkul et al., 2012). Nucleoside labeling identifies a slow cycling cell population of mesenchymal cells that lies between the epithelial CL and a population of fast-cycling cells more distally (Kaukua et al., 2014; Lapthanasupkul et al., 2012; Zhao et al., 2014; Fig. 5).

Slow-cycling cells express Gli1 and lineage tracing using Gli1-Cre^{ERT2} mice shows that these cells gives rise to almost all odontoblasts and pulp cells during adult incisor growth. The Gli1 expressing cells are localized in a neurovascular bundle and respond to Shh ligand that appears to be neuronal derived (Zhao et al., 2014). Within this Gli1^{+} populations of slow-cycling MSCs there appears to be as yet unexplained heterogeneity. By using promoters that direct gene expression in peripheral glial cells (Sox10 and PLP), a population of slow-cycling cells that contributes to around 50% of odontoblast and pulp cell formation during incisor growth was identified (Kramann et al., 2015). Significantly, this study utilized confetti reporter mice to trace individual stem cell contributions and showed that all glial-derived MSCs are capable of forming both odontoblasts and pulp cells, but the relative proportions of each depend on the position of the stem cell in its niche. Thus, the mouse incisor MSC population consists of a glial-derived population and a nonglial-derived population both of which are Shh responsive in the neurovascular bundle. Surprisingly, an additional layer of heterogeneity also exists since a proportion of the Gli1+ cell and glial cells express Thy1 (CD90) and these cells contribute around 20% of the odontoblasts and pulp cells. The reasons for this unexpected complexity are not yet understood.

Polycomb Repressive Complex 1 (PRC1) proteins were first described as gene repressor for Hox genes in *Drosophila melanogaster* and have been identified critical roles in adult stem cells function (Luis, Morey, Di Croce, & Benitah, 2012; Morey, Aloia, Cozzuto, Benitah, & Di Croce, 2013; Morey et al., 2012). Ring1a/b are the core components of PRC1 expressed in the incisor apical mesenchyme closed to both labial and lingual CL, consistent with a location for TA (fast cycling) cells. Depletion of Ring1 in a Ring1a$^{-/-}$;Ring1b$^{cko/cko}$ incisors shows dramatically reduced cell proliferation in the apical mesenchyme and CL epithelium, and a down-regulation of FGF singling as well as downstream targets (Lapthanasupkul et al., 2012). Interestingly, *Bmi1*, one of other component of PRC1, is expressed in the CL epithelial stem cells and also expressed in the BrdU-retaining mesenchyme, indicating that those cells are also stem cells (Biehs et al., 2013).

ACKNOWLEDGMENTS

The authors acknowledge the funding support of the MRC and the NIDCR.

REFERENCES

Angelova Volponi, A., Kawasaki, M., & Sharpe, P. T. (2013). Adult human gingival epithelial cells as a source for whole-tooth bioengineering. *Journal of Dental Research, 92*, 329–334.

Arany, P. R., Cho, A., Hunt, T. D., Sidhu, G., Shin, K., Hahm, E., et al. (2014). Photoactivation of endogenous latent transforming growth factor-beta 1 directs dental stem cell differentiation for regeneration. *Science Translational Medicine, 6*, 11.

Arwert, E. N., Hoste, E., & Watt, F. M. (2012). Epithelial stem cells, wound healing and cancer. *Nature Reviews. Cancer, 12*, 170–180.

Baggiolini, A., Varum, S., Mateos, J. M., Bettosini, D., John, N., Bonalli, M., et al. (2015). Premigratory and migratory neural crest cells are multipotent in vivo. *Cell Stem Cell, 16*, 314–322.

Bai, Y., Bai, Y., Matsuzaka, K., Hashimoto, S., Fukuyama, T., Wu, L., et al. (2011). Cementum- and periodontal ligament-like tissue formation by dental follicle cell sheets co-cultured with Hertwig's epithelial root sheath cells. *Bone, 48*, 1417–1426.

Balic, A., Aguila, H. L., Caimano, M. J., Francone, V. P., & Mina, M. (2010). Characterization of stem and progenitor cells in the dental pulp of erupted and unerupted murine molars. *Bone, 46*, 1639–1651.

Barker, N., van Es, J. H., Kuipers, J., Kujala, P., van den Born, M., Cozijnsen, M., et al. (2007). Identification of stem cells in small intestine and colon by marker gene Lgr5. *Nature, 449*, 1003–1007.

Begue-Kirn, C., Smith, A. J., Loriot, M., Kupferle, C., Ruch, J. V., & Lesot, H. (1994). Comparative analysis of TGF beta s, BMPs, IGF1, msxs, fibronectin, osteonectin and bone sialoprotein gene expression during normal and in vitro-induced odontoblast differentiation. *The International Journal of Developmental Biology, 38*, 405–420.

Biehs, B., Hu, J. K., Strauli, N. B., Sangiorgi, E., Jung, H., Heber, R. P., et al. (2013). BMI1 represses Ink4a/Arf and Hox genes to regulate stem cells in the rodent incisor. *Nature Cell Biology, 15*, 846–852.

Bitgood, M. J., & McMahon, A. P. (1995). Hedgehog and Bmp genes are coexpressed at many diverse sites of cell-cell interaction in the mouse embryo. *Developmental Biology, 172*, 126–138.

Boran, T., Peterkova, R., Lesot, H., Lyons, D. B., Peterka, M., & Klein, O. D. (2009). Temporal analysis of ectopic enamel production in incisors from sprouty mutant mice. *Journal of Experimental Zoology. Part B, Molecular and Developmental Evolution, 312B*, 473–485.

Breunig, J. J., Silbereis, J., Vaccarino, F. M., Sestan, N., & Rakic, P. (2007). Notch regulates cell fate and dendrite morphology of newborn neurons in the postnatal dentate gyrus. *Proceedings of the National Academy of Sciences of the United States of America, 104*, 20558–20563.

Bronner, M. (2015). Confetti clarifies controversy: Neural crest stem cells are multipotent. *Cell Stem Cell, 16*, 217–218.

Bronner-Fraser, M. (1993). Segregation of cell lineage in the neural crest. *Current Opinion in Genetics & Development, 3*, 641–647.

Cao, H., Jheon, A., Li, X., Sun, Z., Wang, J., Florez, S., et al. (2013). The Pitx2:miR-200c/141:noggin pathway regulates Bmp signaling and ameloblast differentiation. *Development, 140*, 3348–3359.

Caton, J., Luder, H. U., Zoupa, M., Bradman, M., Bluteau, G., Tucker, A. S., et al. (2009). Enamel-free teeth: Tbx1 deletion affects amelogenesis in rodent incisors. *Developmental Biology, 328*, 493–505.

Chai, Y., Ito, Y., & Han, J. (2003). TGF beta signaling and its functional significance in regulating the fate of cranial neural crest cells. *Critical Reviews in Oral Biology and Medicine, 14*, 78–88.

Chai, Y., Jiang, X., Ito, Y., Bringas, P., Jr., Han, J., Rowitch, D. H., et al. (2000). Fate of the mammalian cranial neural crest during tooth and mandibular morphogenesis. *Development, 127*, 1671–1679.

Chen, S., Lewallen, M., & Xie, T. (2013). Adhesion in the stem cell niche: Biological roles and regulation. *Development, 140*, 255–265.

Chung, I. H., Yamaza, T., Zhao, H., Choung, P. H., Shi, S., & Chai, Y. (2009). Stem cell property of postmigratory cranial neural crest cells and their utility in alveolar bone regeneration and tooth development. *Stem Cells, 27*, 866–877.

Cooper, P. R., Holder, M. J., & Smith, A. J. (2014). Inflammation and regeneration in the dentin-pulp complex: A double-edged sword. *Journal of Endodontia, 40*, S46–S51.

Cooper, P. R., & Smith, A. J. (2013). Molecular mediators of pulp inflammation and regeneration. *Endodontic Topics, 28*, 90–105.

Fan, X., Matsui, W., Khaki, L., Stearns, D., Chun, J., Li, Y. M., et al. (2006). Notch pathway inhibition depletes stem-like cells and blocks engraftment in embryonal brain tumors. *Cancer Research, 66*, 7445–7452.

Felszeghy, S., Suomalainen, M., & Thesleff, I. (2010). Notch signalling is required for the survival of epithelial stem cells in the continuously growing mouse incisor. *Differentiation; Research in Biological Diversity, 80*, 241–248.

Feng, J., Mantesso, A., De Bari, C., Nishiyama, A., & Sharpe, P. T. (2011). Dual origin of mesenchymal stem cells contributing to organ growth and repair. *Proceedings of the National Academy of Sciences of the United States of America, 108*, 6503–6508.

Foudi, A., Hochedlinger, K., Van Buren, D., Schindler, J. W., Jaenisch, R., Carey, V., et al. (2009). Analysis of histone 2B-GFP retention reveals slowly cycling hematopoietic stem cells. *Nature Biotechnology, 27*, 84–90.

Fougeroux de Bondaroy, A. (1768). Observations anatomiques. *History of the Academy of Royal Science Paris, 1768*, 47–48.

Fournier, B. P., Ferre, F. C., Couty, L., Lataillade, J. J., Gourven, M., Naveau, A., et al. (2010). Multipotent progenitor cells in gingival connective tissue. *Tissue Engineering. Part A, 16*, 2891–2899.

Frank, M. J., Dawson, D. W., Bensinger, S. J., Hong, J. S., Knosp, W. M., Xu, L., et al. (2009). Expression of sprouty2 inhibits B-cell proliferation and is epigenetically silenced in mouse and human B-cell lymphomas. *Blood, 113*, 2478–2487.

Fre, S., Huyghe, M., Mourikis, P., Robine, S., Louvard, D., & Artavanis-Tsakonas, S. (2005). Notch signals control the fate of immature progenitor cells in the intestine. *Nature, 435*, 964–968.

Fuchs, E. (2009). The tortoise and the hair: Slow-cycling cells in the stem cell race. *Cell, 137*, 811–819.

Fujiwara, H., Ferreira, M., Donati, G., Marciano, D. K., Linton, J. M., Sato, Y., et al. (2011). The basement membrane of hair follicle stem cells is a muscle cell niche. *Cell, 144*, 577–589.

Goldberg, M., & Smith, A. J. (2004). Cells and extracellular matrices of dentin and pulp: A biological basis for repair and tissue engineering. *Critical Reviews in Oral Biology and Medicine*, *15*, 13–27.

Gronthos, S., Brahim, J., Li, W., Fisher, L. W., Cherman, N., Boyde, A., et al. (2002). Stem cell properties of human dental pulp stem cells. *Journal of Dental Research*, *81*, 531–535.

Gronthos, S., Mankani, M., Brahim, J., Robey, P. G., & Shi, S. (2000). Postnatal human dental pulp stem cells (DPSCs) in vitro and in vivo. *Proceedings of the National Academy of Sciences of the United States of America*, *97*, 13625–13630.

Hacohen, N., Kramer, S., Sutherland, D., Hiromi, Y., & Krasnow, M. A. (1998). Sprouty encodes a novel antagonist of FGF signaling that patterns apical branching of the drosophila airways. *Cell*, *92*, 253–263.

Han, N. N., Zheng, Y., Li, R., Li, X. Y., Zhou, M., Niu, Y., et al. (2014). Beta-catenin enhances odontoblastic differentiation of dental pulp cells through activation of Runx2. *PloS One*, *9*, 10.

Handa, K., Saito, M., Tsunoda, A., Yamauchi, M., Hattori, S., Sato, S., et al. (2002). Progenitor cells from dental follicle are able to form cementum matrix in vivo. *Connective Tissue Research*, *43*, 406–408.

Handa, K., Saito, M., Yamauchi, M., Kiyono, T., Sato, S., Teranaka, T., et al. (2002). Cementum matrix formation in vivo by cultured dental follicle cells. *Bone*, *31*, 606–611.

Harada, H., Kettunen, P., Jung, H. S., Mustonen, T., Wang, Y. A., & Thesleff, I. (1999). Localization of putative stem cells in dental epithelium and their association with notch and FGF signaling. *The Journal of Cell Biology*, *147*, 105–120.

Hasegawa, M., Yamato, M., Kikuchi, A., Okano, T., & Ishikawa, I. (2005). Human periodontal ligament cell sheets can regenerate periodontal ligament tissue in an athymic rat model. *Tissue Engineering*, *11*, 469–478.

Holik, A. Z., Young, M., Krzystyniak, J., Williams, G. T., Metzger, D., Shorning, B. Y., et al. (2014). Brg1 loss attenuates aberrant wnt-signalling and prevents wnt-dependent tumourigenesis in the murine small intestine. *PLoS Genetics*, *10*, e1004453.

Hsu, S. H., Huang, G. S., & Feng, F. (2012). Isolation of the multipotent MSC subpopulation from human gingival fibroblasts by culturing on chitosan membranes. *Biomaterials*, *33*, 2642–2655.

Huang, G. T., Gronthos, S., & Shi, S. (2009). Mesenchymal stem cells derived from dental tissues vs. those from other sources: Their biology and role in regenerative medicine. *Journal of Dental Research*, *88*, 792–806.

Huang, G. T., Yamaza, T., Shea, L. D., Djouad, F., Kuhn, N. Z., Tuan, R. S., et al. (2010). Stem/progenitor cell-mediated de novo regeneration of dental pulp with newly deposited continuous layer of dentin in an in vivo model. *Tissue Engineering. Part A*, *16*, 605–615.

Hunter, D. J., Bardet, C., Mouraret, S., Liu, B., Singh, G., Sadoine, J., et al. (2015). Wnt acts as a pro-survival signal to enhance dentin regeneration. *Journal of Bone and Mineral Research*, *30*(7), 1150–1159.

Hwang, W. S., Tonna, E. A., & Cronkite, E. P. (1962). Localization and distribution of tritiated histidine in growing mouse incisor. *Nature*, *193*, 896.

Imai, H., Osumi-Yamashita, N., Ninomiya, Y., & Eto, K. (1996). Contribution of early-emigrating midbrain crest cells to the dental mesenchyme of mandibular molar teeth in rat embryos. *Developmental Biology*, *176*, 151–165.

Jaks, V., Barker, N., Kasper, M., van Es, J. H., Snippert, H. J., Clevers, H., et al. (2008). Lgr5 marks cycling, yet long-lived, hair follicle stem cells. *Nature Genetics*, *40*, 1291–1299.

Jarvinen, E., Tummers, M., & Thesleff, I. (2009). The role of the dental lamina in mammalian tooth replacement. *Journal of Experimental Zoology. Part B, Molecular and Developmental Evolution*, *312B*, 281–291.

Jernvall, J., & Thesleff, I. (2000). Reiterative signaling and patterning during mammalian tooth morphogenesis. *Mechanisms of Development*, *92*, 19–29.

Jheon, A. H., Li, C. Y., Wen, T., Michon, F., & Klein, O. D. (2011). Expression of microRNAs in the stem cell niche of the adult mouse incisor. *PloS One, 6*, e24536.
Jo, Y. Y., Lee, H. J., Kook, S. Y., Choung, H. W., Park, J. Y., Chung, J. H., et al. (2007). Isolation and characterization of postnatal stem cells from human dental tissues. *Tissue Engineering, 13*, 767–773.
Juuri, E., Isaksson, S., Jussila, M., Heikinheimo, K., & Thesleff, I. (2013). Expression of the stem cell marker, SOX2, in ameloblastoma and dental epithelium. *European Journal of Oral Sciences, 121*, 509–516.
Juuri, E., Jussila, M., Seidel, K., Holmes, S., Wu, P., Richman, J., et al. (2013). Sox2 marks epithelial competence to generate teeth in mammals and reptiles. *Development, 140*, 1424–1432.
Juuri, E., Saito, K., Ahtiainen, L., Seidel, K., Tummers, M., Hochedlinger, K., et al. (2012). Sox2+ stem cells contribute to all epithelial lineages of the tooth via Sfrp5+ progenitors. *Developmental Cell, 23*, 317–328.
Katsuragi, Y., Anraku, J., Nakatomi, M., Ida-Yonemochi, H., Obata, M., Mishima, Y., et al. (2013). Bcl11b transcription factor plays a role in the maintenance of the ameloblast-progenitors in mouse adult maxillary incisors. *Mechanisms of Development, 130*, 482–492.
Kaukua, N., Shahidi, M. K., Konstantinidou, C., Dyachuk, V., Kaucka, M., Furlan, A., et al. (2014). Glial origin of mesenchymal stem cells in a tooth model system. *Nature, 513*, 551–554.
Klein, O. D., Lyons, D. B., Balooch, G., Marshall, G. W., Basson, M. A., Peterka, M., et al. (2008). An FGF signaling loop sustains the generation of differentiated progeny from stem cells in mouse incisors. *Development, 135*, 377–385.
Ko, S. O., Chung, I. H., Xu, X., Oka, S., Zhao, H., Cho, E. S., et al. (2007). Smad4 is required to regulate the fate of cranial neural crest cells. *Developmental Biology, 312*, 435–447.
Kollar, E. J., & Baird, G. R. (1969). The influence of the dental papilla on the development of tooth shape in embryonic mouse tooth germs. *Journal of Embryology and Experimental Morphology, 21*, 131–148.
Koussoulakou, D. S., Margaritis, L. H., & Koussoulakos, S. L. (2009). A curriculum vitae of teeth: Evolution, generation, regeneration. *International Journal of Biological Sciences, 5*, 226–243.
Koyama, N., Okubo, Y., Nakao, K., & Bessho, K. (2009). Evaluation of pluripotency in human dental pulp cells. *Journal of Oral and Maxillofacial Surgery: Official Journal of the American Association of Oral and Maxillofacial Surgeons, 67*, 501–506.
Kramann, R., Schneider, R. K., DiRocco, D. P., Machado, F., Fleig, S., Bondzie, P. A., et al. (2015). Perivascular Gli1+ progenitors are key contributors to injury-induced organ fibrosis. *Cell Stem Cell, 16*, 51–66.
Kuang-Hsien Hu, J., Mushegyan, V., & Klein, O. D. (2014). On the cutting edge of organ renewal: Identification, regulation, and evolution of incisor stem cells. *Genesis, 52*, 79–92.
Kurosaka, H., Islam, M. N., Kuremoto, K., Hayano, S., Nakamura, M., Kawanabe, N., et al. (2011). Core binding factor beta functions in the maintenance of stem cells and orchestrates continuous proliferation and differentiation in mouse incisors. *Stem Cells, 29*, 1792–1803.
Lajtha, L. G. (1979). Stem cell concepts. *Differentiation; Research in Biological Diversity, 14*, 23–34.
Lapthanasupkul, P., Feng, J., Mantesso, A., Takada-Horisawa, Y., Vidal, M., Koseki, H., et al. (2012). Ring1a/b polycomb proteins regulate the mesenchymal stem cell niche in continuously growing incisors. *Developmental Biology, 367*, 140–153.
Lee, C. C., Putnam, A. J., Miranti, C. K., Gustafson, M., Wang, L. M., Vande Woude, G. F., et al. (2004). Overexpression of sprouty 2 inhibits HGF/SF-mediated cell growth, invasion, migration, and cytokinesis. *Oncogene, 23*, 5193–5202.

Li, C. Y., Cha, W., Luder, H. U., Charles, R. P., McMahon, M., Mitsiadis, T. A., et al. (2012). E-cadherin regulates the behavior and fate of epithelial stem cells and their progeny in the mouse incisor. *Developmental Biology, 366*, 357–366.

Li, Z., Jiang, C. M., An, S., Cheng, Q., Huang, Y. F., Wang, Y. T., et al. (2014). Immunomodulatory properties of dental tissue-derived mesenchymal stem cells. *Oral Diseases, 20*, 25–34.

Lin, Y., Gallucci, G. O., Buser, D., Bosshardt, D., Belser, U. C., & Yelick, P. C. (2011). Bioengineered periodontal tissue formed on titanium dental implants. *Journal of Dental Research, 90*, 251–256.

Liu, F., Dangaria, S., Andl, T., Zhang, Y., Wright, A. C., Damek-Poprawa, M., et al. (2010). Beta-catenin initiates tooth neogenesis in adult rodent incisors. *Journal of Dental Research, 89*, 909–914.

Lo, L., & Anderson, D. J. (1995). Postmigratory neural crest cells expressing c-RET display restricted developmental and proliferative capacities. *Neuron, 15*, 527–539.

Luis, N. M., Morey, L., Di Croce, L., & Benitah, S. A. (2012). Polycomb in stem cells: PRC1 branches out. *Cell Stem Cell, 11*, 16–21.

Lumsden, A. G. (1988). Spatial organization of the epithelium and the role of neural crest cells in the initiation of the mammalian tooth germ. *Development, 103*(Suppl.), 155–169.

Millar, S. E., Koyama, E., Reddy, S. T., Andl, T., Gaddapara, T., Piddington, R., et al. (2003). Over- and ectopic expression of Wnt3 causes progressive loss of ameloblasts in postnatal mouse incisor teeth. *Connective Tissue Research, 44*(Suppl. 1), 124–129.

Mina, M., & Kollar, E. J. (1987). The induction of odontogenesis in non-dental mesenchyme combined with early murine mandibular arch epithelium. *Archives of Oral Biology, 32*, 123–127.

Mitrano, T. I., Grob, M. S., Carrion, F., Nova-Lamperti, E., Luz, P. A., Fierro, F. S., et al. (2010). Culture and characterization of mesenchymal stem cells from human gingival tissue. *Journal of Periodontology, 81*, 917–925.

Mitsiadis, T. A., Graf, D., Luder, H., Gridley, T., & Bluteau, G. (2010). BMPs and FGFs target notch signalling via jagged 2 to regulate tooth morphogenesis and cytodifferentiation. *Development, 137*, 3025–3035.

Miura, M., Gronthos, S., Zhao, M., Lu, B., Fisher, L. W., Robey, P. G., et al. (2003). SHED: Stem cells from human exfoliated deciduous teeth. *Proceedings of the National Academy of Sciences of the United States of America, 100*, 5807–5812.

Morey, L., Aloia, L., Cozzuto, L., Benitah, S. A., & Di Croce, L. (2013). RYBP and Cbx7 define specific biological functions of polycomb complexes in mouse embryonic stem cells. *Cell Reports, 3*, 60–69.

Morey, L., Pascual, G., Cozzuto, L., Roma, G., Wutz, A., Benitah, S. A., et al. (2012). Nonoverlapping functions of the polycomb group Cbx family of proteins in embryonic stem cells. *Cell Stem Cell, 10*, 47–62.

Morsczeck, C., Gotz, W., Schierholz, J., Zeilhofer, F., Kuhn, U., Mohl, C., et al. (2005). Isolation of precursor cells (PCs) from human dental follicle of wisdom teeth. *Matrix Biology: Journal of the International Society for Matrix Biology, 24*, 155–165.

Murakami, M., Horibe, H., Iohara, K., Hayashi, Y., Osako, Y., Takei, Y., et al. (2013). The use of granulocyte-colony stimulating factor induced mobilization for isolation of dental pulp stem cells with high regenerative potential. *Biomaterials, 34*, 9036–9047.

Nakamura, T., Endo, K., Cooper, L. J., Fullwood, N. J., Tanifuji, N., Tsuzuki, M., et al. (2003). The successful culture and autologous transplantation of rabbit oral mucosal epithelial cells on amniotic membrane. *Investigative Ophthalmology & Visual Science, 44*, 106–116.

Nakamura, T., Inatomi, T., Sotozono, C., Amemiya, T., Kanamura, N., & Kinoshita, S. (2004). Transplantation of cultivated autologous oral mucosal epithelial cells in patients with severe ocular surface disorders. *The British Journal of Ophthalmology, 88*, 1280–1284.

Nakamura, T., Takeda, K., Inatomi, T., Sotozono, C., & Kinoshita, S. (2011). Long-term results of autologous cultivated oral mucosal epithelial transplantation in the scar phase of severe ocular surface disorders. *The British Journal of Ophthalmology, 95*, 942–946.

Neubuser, A., Peters, H., Balling, R., & Martin, G. R. (1997). Antagonistic interactions between FGF and BMP signaling pathways: A mechanism for positioning the sites of tooth formation. *Cell, 90*, 247–255.

Park, C. H., Rios, H. F., Jin, Q., Bland, M. E., Flanagan, C. L., Hollister, S. J., et al. (2010). Biomimetic hybrid scaffolds for engineering human tooth-ligament interfaces. *Biomaterials, 31*, 5945–5952.

Parsa, S., Kuremoto, K., Seidel, K., Tabatabai, R., Mackenzie, B., Yamaza, T., et al. (2010). Signaling by FGFR2b controls the regenerative capacity of adult mouse incisors. *Development, 137*, 3743–3752.

Porntaveetus, T., Otsuka-Tanaka, Y., Basson, M. A., Moon, A. M., Sharpe, P. T., & Ohazama, A. (2011). Expression of fibroblast growth factors (Fgfs) in murine tooth development. *Journal of Anatomy, 218*, 534–543.

Rothova, M., Thompson, H., Lickert, H., & Tucker, A. S. (2012). Lineage tracing of the endoderm during oral development. *Developmental Dynamics: An Official Publication of the American Association of Anatomists, 241*, 1183–1191.

Sakata-Yanagimoto, M., Nakagami-Yamaguchi, E., Saito, T., Kumano, K., Yasutomo, K., Ogawa, S., et al. (2008). Coordinated regulation of transcription factors through Notch2 is an important mediator of mast cell fate. *Proceedings of the National Academy of Sciences of the United States of America, 105*, 7839–7844.

Schor, S. L., Ellis, I., Irwin, C. R., Banyard, J., Seneviratne, K., Dolman, C., et al. (1996). Subpopulations of fetal-like gingival fibroblasts: Characterisation and potential significance for wound healing and the progression of periodontal disease. *Oral Diseases, 2*, 155–166.

Seidel, K., Ahn, C. P., Lyons, D., Nee, A., Ting, K., Brownell, I., et al. (2010). Hedgehog signaling regulates the generation of ameloblast progenitors in the continuously growing mouse incisor. *Development, 137*, 3753–3761.

Sempowski, G. D., Borrello, M. A., Blieden, T. M., Barth, R. K., & Phipps, R. P. (1995). Fibroblast heterogeneity in the healing wound. *Wound Repair and Regeneration: Official Publication of the Wound Healing Society [and] the European Tissue Repair Society, 3*, 120–131.

Seo, B. M., Miura, M., Gronthos, S., Bartold, P. M., Batouli, S., Brahim, J., et al. (2004). Investigation of multipotent postnatal stem cells from human periodontal ligament. *Lancet, 364*, 149–155.

Serbedzija, G. N., Bronner-Fraser, M., & Fraser, S. E. (1992). Vital dye analysis of cranial neural crest cell migration in the mouse embryo. *Development, 116*, 297–307.

Sharp, T., Wang, J., Li, X., Cao, H., Gao, S., Moreno, M., et al. (2014). A pituitary homeobox 2 (Pitx2):microRNA-200a-3p:beta-catenin pathway converts mesenchymal cells to amelogenin-expressing dental epithelial cells. *The Journal of Biological Chemistry, 289*, 27327–27341.

Simon, S., Smith, A. J., Berdal, A., Lumley, P. J., & Cooper, P. R. (2010). The MAP kinase pathway is involved in odontoblast stimulation via p38 phosphorylation. *Journal of Endodontia, 36*, 256–259.

Smith, C. E. (1980). Cell turnover in the odontogenic organ of the rat incisor as visualized by graphic reconstructions following a single injection of 3H-thymidine. *The American Journal of Anatomy, 158*, 321–343.

Smith, A. J., Cassidy, N., Perry, H., Beguekirn, C., Ruch, J. V., & Lesot, H. (1995). Reactionary dentinogenesis. *The International Journal of Developmental Biology, 39*, 273–280.

Smith, M. M., Fraser, G. J., & Mitsiadis, T. A. (2009). Dental lamina as source of odontogenic stem cells: Evolutionary origins and developmental control of tooth generation in

gnathostomes. *Journal of Experimental Zoology. Part B, Molecular and Developmental Evolution, 312B*, 260–280.
Smith, A. J., & Lesot, H. (2001). Induction and regulation of crown dentinogenesis: Embryonic events as a template for dental tissue repair? *Critical Reviews in Oral Biology and Medicine, 12*, 425–437.
Smith, A. J., Scheuen, B. A., Takahashi, Y., Ferracane, J. L., Shelton, R. M., & Cooper, P. R. (2012). Dentine as a bioactive extracellular matrix. *Archives of Oral Biology, 57*, 109–121.
Smith, C. E., & Warshawsky, H. (1975). Cellular renewal in the enamel organ and the odontoblast layer of the rat incisor as followed by radioautography using 3H-thymidine. *The Anatomical Record, 183*, 523–561.
Stemple, D. L., & Anderson, D. J. (1992). Isolation of a stem cell for neurons and glia from the mammalian neural crest. *Cell, 71*, 973–985.
Suomalainen, M., & Thesleff, I. (2010). Patterns of Wnt pathway activity in the mouse incisor indicate absence of Wnt/beta-catenin signaling in the epithelial stem cells. *Developmental Dynamics: An Official Publication of the American Association of Anatomists, 239*, 364–372.
Tang, W., Zeve, D., Suh, J. M., Bosnakovski, D., Kyba, M., Hammer, R. E., et al. (2008). White fat progenitor cells reside in the adipose vasculature. *Science, 322*, 583–586.
Thesleff, I., Keranen, S., & Jernvall, J. (2001). Enamel knots as signaling centers linking tooth morphogenesis and odontoblast differentiation. *Advances in Dental Research, 15*, 14–18.
Thesleff, I., & Sharpe, P. (1997). Signalling networks regulating dental development. *Mechanisms of Development, 67*, 111–123.
Trainor, P. A., Ariza-McNaughton, L., & Krumlauf, R. (2002). Role of the isthmus and FGFs in resolving the paradox of neural crest plasticity and prepatterning. *Science, 295*, 1288–1291.
Tucker, A. S., Matthews, K. L., & Sharpe, P. T. (1998). Transformation of tooth type induced by inhibition of BMP signaling. *Science, 282*, 1136–1138.
Tucker, A. S., Yamada, G., Grigoriou, M., Pachnis, V., & Sharpe, P. (1999). Fgf-8 determines rostral-caudal polarity in the first branchial arch. *Development, 126*, 51–61.
Tumbar, T., Guasch, G., Greco, V., Blanpain, C., Lowry, W. E., Rendl, M., et al. (2004). Defining the epithelial stem cell niche in skin. *Science, 303*, 359–363.
Vainio, S., Karavanova, I., Jowett, A., & Thesleff, I. (1993). Identification of BMP-4 as a signal mediating secondary induction between epithelial and mesenchymal tissues during early tooth development. *Cell, 75*, 45–58.
Volponi, A. A., & Sharpe, P. T. (2013). The tooth—A treasure chest of stem cells. *British Dental Journal, 215*, 353–358.
Wada, N., Wang, B., Lin, N. H., Laslett, A. L., Gronthos, S., & Bartold, P. M. (2011). Induced pluripotent stem cell lines derived from human gingival fibroblasts and periodontal ligament fibroblasts. *Journal of Periodontal Research, 46*, 438–447.
Waddington, R. J., Youde, S. J., Lee, C. P., & Sloan, A. J. (2009). Isolation of distinct progenitor stem cell populations from dental pulp. *Cells, Tissues, Organs, 189*, 268–274.
Wang, X., Sha, X. J., Li, G. H., Yang, F. S., Ji, K., Wen, L. Y., et al. (2012). Comparative characterization of stem cells from human exfoliated deciduous teeth and dental pulp stem cells. *Archives of Oral Biology, 57*, 1231–1240.
Wang, X. P., Suomalainen, M., Felszeghy, S., Zelarayan, L. C., Alonso, M. T., Plikus, M. V., et al. (2007). An integrated gene regulatory network controls stem cell proliferation in teeth. *PLoS Biology, 5*, e159.
Widera, D., Zander, C., Heidbreder, M., Kasperek, Y., Noll, T., Seitz, O., et al. (2009). Adult palatum as a novel source of neural crest-related stem cells. *Stem Cells, 27*, 1899–1910.

Yang, G., Zhou, J., Teng, Y., Xie, J., Lin, J., Guo, X., et al. (2014). Mesenchymal TGF-beta signaling orchestrates dental epithelial stem cell homeostasis through Wnt signaling. *Stem Cells*, *32*, 2939–2948.

Yoshida, T., Iwata, T., Umemoto, T., Shiratsuchi, Y., Kawashima, N., Sugiyama, T., et al. (2013). Promotion of mouse ameloblast proliferation by Lgr5 mediated integrin signaling. *Journal of Cellular Biochemistry*, *114*, 2138–2147.

Zeichner-David, M., Diekwisch, T., Fincham, A., Lau, E., MacDougall, M., Moradian-Oldak, J., et al. (1995). Control of ameloblast differentiation. *The International Journal of Developmental Biology*, *39*, 69–92.

Zhang, Y., Wang, S., Song, Y., Han, J., Chai, Y., & Chen, Y. (2003). Timing of odontogenic neural crest cell migration and tooth-forming capability in mice. *Developmental Dynamics: An Official Publication of the American Association of Anatomists*, *226*, 713–718.

Zhao, H., Feng, J. F., Seidel, K., Shi, S. T., Klein, O., Sharpe, P., et al. (2014). Secretion of Shh by a neurovascular bundle niche supports mesenchymal stem cell homeostasis in the adult mouse incisor. *Cell Stem Cell*, *14*, 160–173.

Zhao, H., Li, S., Han, D., Kaartinen, V., & Chai, Y. (2011). Alk5-mediated transforming growth factor beta signaling acts upstream of fibroblast growth factor 10 to regulate the proliferation and maintenance of dental epithelial stem cells. *Molecular and Cellular Biology*, *31*, 2079–2089.

CHAPTER NINE

Development and Integration of the Ear

Jennifer C. Fuchs, Abigail S. Tucker[1]
Department of Craniofacial Development & Stem Cell Biology, King's College London, London, United Kingdom
[1]Corresponding author: e-mail address: abigail.tucker@kcl.ac.uk

Contents

Abstract

The perception of our environment via sensory organs plays a crucial role in survival and evolution. Hearing, one of our most developed senses, depends on the proper function of the auditory system and plays a key role in social communication, integration, and learning ability.

The ear is a composite structure, comprised of the external, middle, and inner ear. During development, the ear is formed from the integration of a number of tissues of different embryonic origin, which initiate in distinct areas of the embryo at different time points. Functional connections between the components of the hearing apparatus have to be established and maintained during development and adulthood to allow proper sound submission from the outer to the middle and inner ear. This highly organized and intimate connectivity depends on intricate spatiotemporal signaling between the various tissues that give rise to the structures of the ear. Any alterations in this chain of events can lead to the loss of integration, which can subsequently lead to conductive hearing loss, in case of outer and middle ear defects or sensorineural hearing loss, if inner ear structures are defective.

This chapter aims to review the current knowledge concerning the development of the three ear compartments as well as mechanisms and signaling pathways that have been implicated in the coordination and integration process of the ear.

Current Topics in Developmental Biology, Volume 115
ISSN 0070-2153
http://dx.doi.org/10.1016/bs.ctdb.2015.07.007

1. FUNCTION AND ANATOMY OF THE MAMMALIAN EAR

The components of the hearing apparatus are discrete but interconnected and arranged in a serial manner directed from the periphery of the head to the internal auditory regions of the central nervous system, comprising the highly specialized sound-transducing structures of the external, middle, and inner ear (Fig. 1).

The external auditory meatus (EAM) (ear canal) and its cartilaginous pinna, the visible part of the ear, work as a funnel to conduct airborne sound waves toward the middle ear via the tympanic membrane (eardrum). The tympanic membrane is a thin multilayered tissue supported by the bony tympanic ring (TR) and separates the external ear from the medially located middle ear. Air-pressure-induced vibrations of this membrane are subsequently transduced into mechanical forces by means of the ossicular chain, located in the air-filled middle ear cavity (MEC) (Novacek, 1993). The MEC is encapsuled in the auditory bulla and is commonly divided into three compartments: the attic region harboring the head of the malleus and incus, the anterior (ventral) part (hypotympanum), where the Eustachian tube (ET) inserts, and the intermediate mesotympanum, which is opposite to the eardrum (Mallo, 2001).

In contrast to nonmammalian vertebrates, which possess only one middle ear bone/ossicle (known as the stapes or columella), mammals develop three endochondrial ossicles, known as the malleus, incus, and stapes (Fig. 2). The stapes is homologous to the single ear ossicle of nonmammalian tetrapods, while the malleus and incus are homologous to the articular and palatoquadrate, which form the jaw articulation in nonmammalian jawed vertebrates (reviewed in Anthwal, Joshi, & Tucker, 2012). This is possible due to the evolution of a novel joint between the squamosal and dentary bones, which act as the jaw articulation in mammals. The presence of three ossicles in the middle ear is thus one of the defining features of mammals. The three mammalian ossicles are linked by specialized articular surfaces forming synovial joints and play a crucial role in amplifying and transmitting vibrations from the tympanic membrane across the MEC to the inner ear (Amin & Tucker, 2006).

A thin flat extension of the malleus, known as the manubrium, inserts into the large surface area of the eardrum, directly detecting air pressure changes (Fig. 2). Mechanical vibrations are then transmitted from the malleus to the head of the incus, through the v-shaped incudomalleal

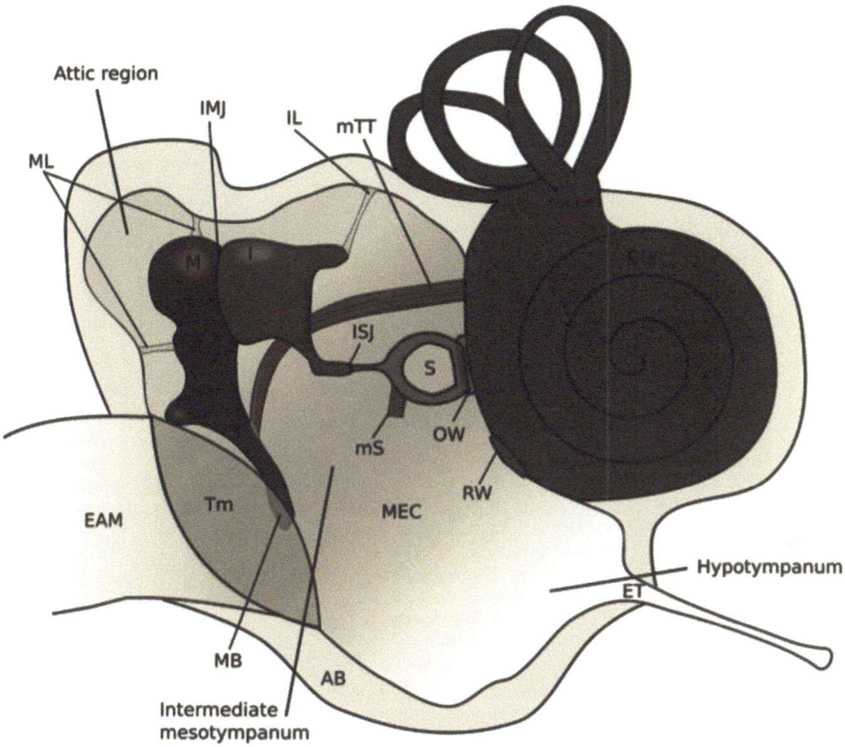

Figure 1 Schematic of the mammalian middle ear. The mammalian middle ear, surrounded by the auditory bulla (AB), is located between the external auditory meatus (EAM) and the spiral-shaped cochlea (CO) of the inner ear. The middle ear cavity (MEC) can be divided into the attic region (yellow), intermediate mesotympanum (purple), and hypotympanum (white). Three middle ear ossicles, the malleus (M), incus (I), and stapes (S) are suspended in the air-filled MEC and form a semirigid bony chain via the synovial incudomalleal joint (IMJ) and the incudostapedial joint (ISJ). The manubrium of the malleus (MB) directly inserts into the tympanic membrane (Tm) to collect sound waves, which are transmitted by the ossicular chain toward the inner ear. The stapes inserts into the membranous oval window (OW), transferring mechanical energy onto the endolymph fluid of the cochlea. The membranous round window (RW), another opening of the medial wall, allows the fluid in the cochlea to move. The ossicular chain is held in place by ligaments, such as the malleolar (ML) and incudal ligaments (IL). Stabilization of the ossicles is ensured by the stapedius muscle (mS) inserting into the crura of the stapes and the tensor tympani muscle (mTT), which is attached to the manubrium and protects the ear against loud sounds. The Eustachian tube (ET) connects the MEC to the nasopharynx and functions in pressure equalization and clearance of effusions. (A color version of this figure is located at the end of this chapter.)

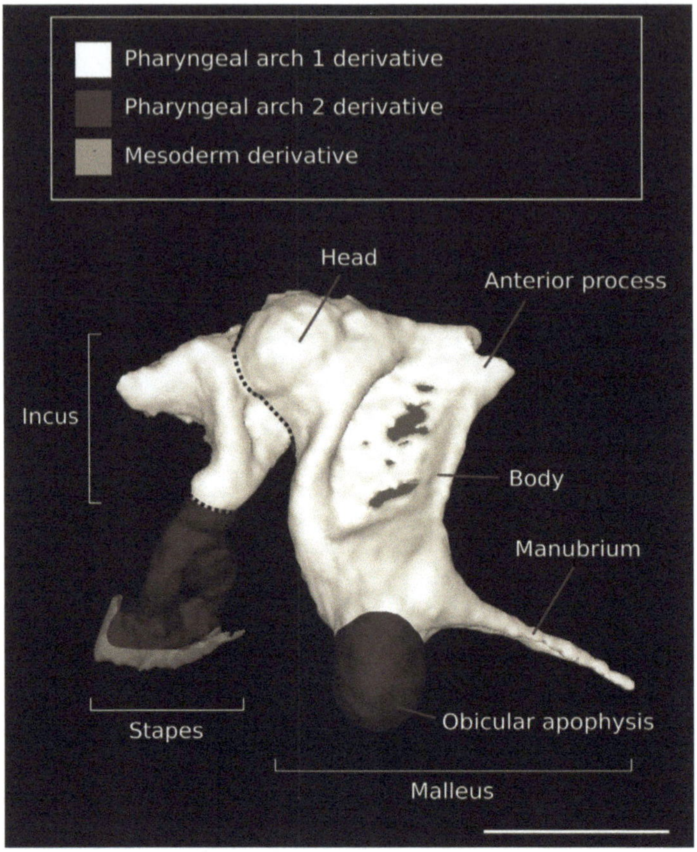

Figure 2 Micro-CT image of the ossicular chain in an adult mouse. Pseudocolored ossicular chain. Structures in green are derived from pharyngeal arch 1 mesenchyme, whereas structures in blue are pharyngeal arch 2 derivatives. The mesoderm-derived part of the stapedial footplate is highlighted in orange. Scale bar: 1 mm. (A color version of this figure is located at the end of this chapter.)

articulation. The incudostapedial joint is formed by the long process of the incus and the neck of the stirrup-shaped stapes. The footplate of the stapes, anchored by the annular ligament, inserts into the oval window of the medial wall of the spiral-shaped cochlea, which is surrounded by a protective otic capsule (Moore, 1981; Fig. 1). Compared to the tympanic membrane, the oval window only covers a small surface area, which increases the amplitude of the vibration in order to overcome the high inertia produced by the endolymph fluid of the cochlea. The organ of Corti contains mechanosensory hair cells that transduce the mechanical energy passed onto the liquid to electrochemical impulses, which are subsequently send to the brain via the auditory nerve (Lim, 1986). The tensor tympani muscle, inserting into the lateral side of the manubrium and the stapedius muscle, connecting the neck of the stapes to the roof of the MEC regulate the amplitude of sounds and hold the ossicles in position (Mallo, 1998).

As the MEC is a partially enclosed system, it faces two main issues: keeping the system clean and maintaining the air composition and pressure. These essential conditions are regulated and maintained by the middle ear epithelium and the ET, which connects the MEC to the outer world via the pharynx.

The epithelial lining of the MEC is continuous and consists of a simple unciliated epithelium covering posterior regions such as the attic and medial wall and a ciliated pseudostratisfied mucosa in the anteroinferior region close to the ET (Thompson & Tucker, 2013). The mucosa in this region contains goblet cells, which produce and secrete different types of mucins that protect the epithelium and prevent pathogen adherence (Kerschner, Li, Tsushiya, & Khampang, 2010). Both goblet cells and motile cilia have been shown to have crucial importance in protecting the middle ear from infections as they guide middle ear secretions from the MEC toward the ventrally located ET orifice (Agius, Wake, Pahor, & Smallman, 1995; Nuutinen, Kärjä, & Karjalainen, 1983).

The ET, named after Bartholomeus Eustachius (1520–1574), is an osteocartilaginous tube linking the MEC to the nasopharynx. The lateral bony part arises from the anterior wall of the MEC and the cartilaginous portion opens into the posterior (lateral) part of the nasopharynx. The ET, as a collapsible airway, is normally closed to protect the middle ear from food intake and pharyngeal pathogens. During yawning, swallowing, and mandibular movement, the tube opens in a transient sequential manner due to the contraction of the tensor veli palatini and levator veli palatini muscle (McDonald, Hoffman, Gentry, & Jiang, 2012). The mucociliary system of the ET consists of a pseudostratisfied epithelium that supports its physiological functions and maintains sterility by creating a barrier against invading pathogens (Graves & Edwards, 1944; Lim et al., 2000).

2. DEVELOPMENT OF THE PHARYNGEAL APPARATUS

The development of the mammalian middle ear and ET is closely associated with the pharyngeal apparatus, a conserved structure in all vertebrates, which first becomes apparent in the mouse at the 6–8 somite stage (embryonic day (E) 8.25). The pharyngeal apparatus is a transient embryonic structure that develops ventrally, comprising bilaterally symmetric bulges, called pharyngeal or branchial arches, that develop sequentially in a head to tail direction (Graham, 2003; Fig. 3A). The number of pharyngeal arches (PAs) differs considerably between species. Whereas lamprey and

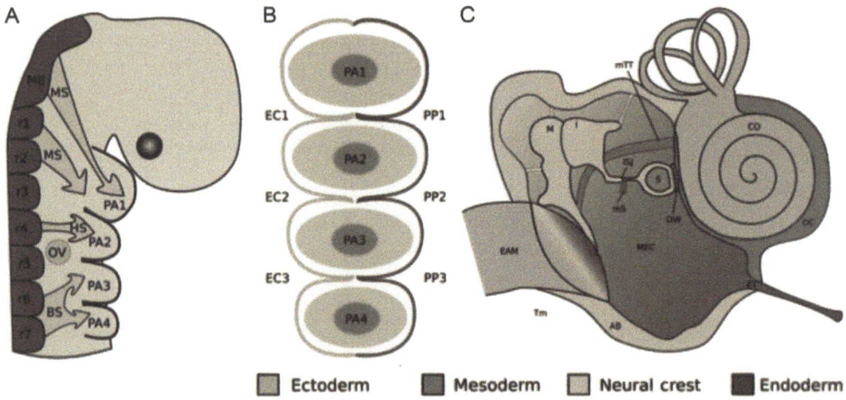

Figure 3 Schematic of mammalian middle ear development. (A) The mammalian pharyngeal apparatus is comprised of pharyngeal arches (PA). During embryonic development, streams of neural crest cells (yellow) from dorsal rhombomeres (r1–r7) and the dorsal edge of the midbrain (MB) migrate into the PA surrounding mesoderm-derived core. The mandibular stream (MS) of NCCs from MB, r1, and r2 migrate into PA1, whereas the hyoid stream (HS) from r4 populates PA2. NCCs from r6 and r7 (branchial stream; BS) migrate into PA3 and PA4. The ectoderm-derived otic vesicle (OV), between PA2 and PA3, gives rise to inner ear structures such as the cochlea (CO). (B) Each PA consists of core mesoderm, derived from paraxial mesoderm (orange), which is surrounded by NCCs (yellow). Both tissues are externally covered by ectoderm (green), which invaginates between adjacent PAs to form ectodermal clefts (ECs) and internally by endoderm (blue) forming pharyngeal pouches (PPs) by directly apposing the overlying ectoderm. (C) The matured ME is derived from all three germ layers and the neural crest. NCCs give rise to ligaments and the auditory bulla (AB) surrounding the MEC as well as the malleus (M) and incus (I) and parts of the stapes (S). The stapes, however, is of dual origin, as parts of the stapedial footplate are mesoderm derived. Other mesoderm-derived structures include both ME muscles, the mTT derived from PP1 and mS derived from PA2, most parts of the otic capsule (OC) and the oval window (OW). The Eustachian tube (ET), as an outpocketing of the PP1, as well as ventral parts of the epithelium lining the MEC are endoderm derived, whereas the mucosa covering the cochlea and attic are of NCC origin. The Tm is multilayered, consisting of the neural crest-derived lamina propia (LP), which is sandwiched in between a lateral ectodermal and a medial endodermal epithelium. The EAM, lined by PA1-derived mesenchyme is of ectoderm origin. *Panels (A and B) adapted from Lindsay (2001).* (A color version of this figure is located at the end of this chapter.)

seven-gilled sharks have nine arches, teleosts have seven arches, and amniotes have five PAs, known as PA1–PA6, with PA5 not forming (Graham, Okabe, & Quinlan, 2005).

All PAs are lined externally by ectoderm and internally by endoderm, enclosing a core of paraxial mesoderm surrounded by neural crest-derived mesenchyme (Fig. 3B). This neural crest-derived mesenchyme gives rise to skeletal elements of the jaw, connective tissue and tendons as well as pericytes, and smooth muscle cells of arch arteries, whereas endothelial cells of the PA arteries and most skeletal muscles of the head and neck, including both muscles inserting into the ossicles, derive from cranial mesoderm (Noden, 1983, 1988; Trainor, Tan, & Tam, 1994).

During PA morphogenesis, three streams of neural crest cells (NCCs), originating from the dorsal neural tube rhombomeres and the midbrain, migrate into the arches surrounding the central group of mesodermal cells (Noden, 1983). NCCs from the posterior midbrain, rhombomere 1 and 2, known as the mandibular stream, migrate into PA1 (Köntges & Lumsden, 1996). Neural crest gives rise to the incus, body of the malleus, the gonial bone, and the TR (Thompson, Ohazama, Sharpe, & Tucker, 2012). The TR is a membranous bone that forms next to the malleus and acts to support and anchor the eardrum to the skull. The hyoid stream migrates into PA2 and originates from rhombomere 4. It gives rise to the hyoid bones, the most proximal region of the lower jaw, the pinna, and the ventral process of the malleus (Minoux et al., 2013; O'Gorman, 2005; Fig. 2). In mice, this projection has been previously confused with the processus brevis, a lateral projection of the manubrium that extends into the tympanic membrane. Even though the processus brevis is found in mice, it is less pronounced. The structure commonly referred to as the murine lateral process, however, is actually the orbicular apophysis. This ventral projection is specific to microtype ears such as the murine ear and does not insert into the tympanic membrane (Mason, 2013). Interestingly, the neural crest also gives rise to the body of the stapes, but the footplate is a composite structure with the central part being NCC derived and the outer ring being of mesodermal origin (Thompson et al., 2012), making the stapes not only the smallest bone but also, together with the otic capsule, one of few compound bones in the mammalian body (Couly, Coltey, & Le Douarin, 1993). The third stream of NCCs, known as the branchial stream, migrating from rhombomere 6 and 7, populates PA3 and PA4 and gives rise to head associated cartilage, including the cricoid cartilage around the trachea and epiglottis (Hamilton & Mossman, 1972). Rhombomeres 3 and 5 contribute minimally to the hyoid stream and are mainly associated with apoptosis in the chick (Bockman, Redmond, & Kirby, 1989; Le Lièvre & Le Douarin, 1975; Lumsden & Keynes, 1989; Fig. 3A).

The pharyngeal arch NCC and mesoderm are enveloped by invaginating pharyngeal endoderm between adjacent arches, forming pharyngeal pouches (PP1–PP4), and directly opposing the overlying ectoderm, which forms ectodermal clefts (ECs). The pharyngeal endoderm gives rise to the epithelial lining of the pharynx and endocrine glands such as the thyroid, parathyroid, and thymus glands, whereas the ectoderm mainly contributes to the EAM, epidermis, otic placode, and neuronal tissues of arch-associated ganglia (Grevellec & Tucker, 2010).

3. DEVELOPMENT OF THE MIDDLE EAR

The middle ear with its associated ET and tympanic membrane is produced by all three germ layers and the neural crest of the pharyngeal apparatus. During murine development at embryonic day 12.5 (E12.5), the pharyngeal pouch 1 (PP1) elongates to form the tubotympanic recess (early ET), which is at this stage lined by an uncilitiated low columnar epithelium (Park, Ueno, & Lim, 1992). Dorsal to the PP1 and adjacent to the developing inner ear lies to the neural crest-derived mesenchyme of PP1 and PP2, in which the early ossicles start to form condensations (Kaufmann, 1992). Between E13 and E14, when the presumptive ossicles differentiate into cartilage, the PP1 extends toward the middle ear area next to the stapedial primordium (Mallo, Schrewe, Martin, Olson, & Ohnemus, 2000; Thompson & Tucker, 2013). The malleus and incus form from a single cartilaginous condensation that splits into two by the formation of the incudomalleal joint around E14, while the stapes forms from a distinct condensation (Amin & Tucker, 2006). At E16.5, the ET and MEC become distinctly segregated in a so-called bottleneck constriction process, involving the narrowing of the posterior part of the tube, which creates the ET orifice in the ventral region of the MEC. At this point, the ET is lined by a pseudostratisfied columnar epithelium, which 1 day later, shows epithelial secretory cells such as goblet cells (Park et al., 1992). At E17.5, the endodermally derived epithelium ruptures, and the future MEC becomes filled by neural crest-derived mesenchyme (Thompson & Tucker, 2013). Throughout embryonic development in the mouse, the middle ear ossicles remain surrounded by neural crest-derived mesenchyme, and therefore the ear is a nonfunctional organ until a process called cavitation occurs from postnatal day 6 (P6) (Mallo, 1998; Richter et al., 2010). During cavitation, an air-filled space develops around the ossicles allowing their free movement. Cavitation occurs in conjunction with the development of the neural crest-derived auditory bulla surrounding the MEC (Richter et al., 2010). The mesenchymal tissue within the MEC and around the ossicles retracts in a ventral-to-dorsal fashion, with the attic region of the MEC clearing last at P14, leaving the ossicles hanging in an air-filled space. The continuous epithelium of the cavitated middle ear has recently been shown to be of dual origin, as determined by lineage tracing experiments using *R26R* reporter mice crossed with either endoderm- (*Sox17-2AiCre*) or neural crest- (*Wnt1-Cre*) specific Cre recombinase mouse lines (Thompson & Tucker, 2013). The epithelial lining of the anterior wall of the MEC, where the ET enters, and the area surrounding the tympanic membrane are endoderm derived, while the dorsal attic region and medial wall overlying the cochlea are lined by a neural crest-derived epithelium that has undergone a mesenchyme-to-epithelium transition (Thompson & Tucker, 2013). While the expression of laminin, a marker of the basement membrane, and keratin14 (K14), a cytoskeletal epithelial marker, are observed in the epithelium of endoderm origin at P14, these markers are only observed in parts of the *Wnt1*-positive mucosa at P16, indicating that mesenchymal-to-epithelial transition takes place after the cavitation process of the middle ear is completed (Thompson & Tucker, 2013).

4. DEVELOPMENT OF THE EXTERNAL EAR

The external ear, consisting of the funnel-shaped auricle (pinna) and the EAM, functions in efficient sound transmission from the environment to the eardrum as well as in protecting the middle ear structures from injury. Murine external ear morphogenesis starts at E12.0 with an epithelial invagination of PA1-derived ectodermal precursor cells into neural crest-derived mesenchyme anterior to the first pharyngeal cleft (Minoux et al., 2013; Fig. 4A). This recent finding counters previous research that presumed that the EAM formed at the first PA cleft, between the first and second arches. Subsequent EAM development is associated with the forming membranous TR, which starts to condense at E13.5 ventral to the first EC1 (Mallo, 1998; Fig. 4B). The forming EAM is directed ventrally and caudally toward the presumptive TR, where it associates with the endoderm-derived tubotympanic recess, which leads to the formation of the eardrum

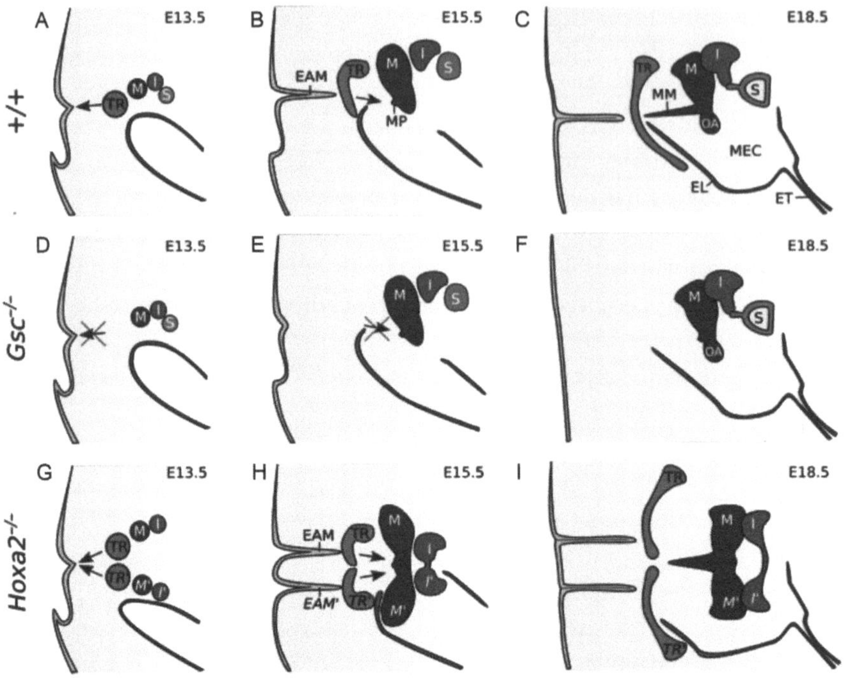

Figure 4 Developmental defects during the integration of the ear. (A–C) Wild-type (+/+). (D–F) *Goosecoid*$^{-/-}$ (*Gsc*$^{-/-}$). (G–I) *Hoxa2*$^{-/-}$. (A) At E13.5, in wild-type embryos, the mesenchymal condensation forming the early tympanic ring (TR) signals (arrow) to the overlying ectoderm (green), leading to invagination of the epithelium. The three ossicles, the malleus (M), incus (I), and stapes (S) form condensations in the surrounding neural crest-derived mesenchyme (yellow). The endoderm-derived tubotympanic recess (blue) extends from the first pharyngeal pouch toward the ossicles and the TR. (B) Two days later, the early external auditory meatus (EAM) has extended toward the TR. The endoderm ruptured, allowing mesenchyme to enter the tubotympanic recess. Signals from the TR (arrow) induce the formation of the manubrium primordium (MP). (C) Shortly before birth, the EAM has collapsed, forming a thin solid ribbon. The malleal manubrium (MM) and orbicular apophysis (OA) have formed. The ossicular chain is still surrounded by mesenchyme. The tubotympanic recess can now be divided into the Eustachian tube (ET) and the epithelial lining (EL) of the middle ear cavity (MEC). (D) The mesenchymal condensation of the early TR is missing in all *Gsc*$^{-/-}$ mice. As a result, mutants lack signals for inducing ectodermal invagination (crossed out arrow). (E) Therefore, at E15.5, the early EAM does not form in mutant mice. In addition, signals from the TR to the MP are missing (crossed out arrow). (F) The lack of inducing signals from the TR has lead to an absent EAM and malleal manubrium. (G) In *Hoxa2*$^{-/-}$ mice, second arch derivatives are transformed into first arch tissues, leading to mirror-image duplication of the tympanic ring (*TR'*), malleus (*M'*), and incus (*I'*), while the stapes is absent. Both TR condensations signal to the overlying surface ectoderm (arrows). (H) As a result of the dual signaling, mutant mice form an additional EAM (*EAM'*),

(Mallo & Gridley, 1996). This membrane develops as a trilaminar structure with apposing epithelial layers, one formed by the ectoderm of the EAM and one from the endodermal lining of the tubotympanic recess. Sandwiched in between the endoderm and ectoderm is the lamina propia, a fibrous layer derived from mesenchyme (Chin, Kurian, & Saunders, 1997; Mallo et al., 2000; Fig. 3C). During embryonic stages, the early EAM collapses to create a thin solid ribbon of tissue (Michaels & Soucek, 1989). Collapse of the EAM and fusion of the epithelia appears to involve apoptosis (Nishizaki et al., 1998). Recanalization then occurs between postnatal day 5 (P5) and P10, to reform a tube-like structure through which sound is relayed (Nishizaki et al., 1998). At this point, the EAM is positioned anterior to the cartilaginous pinna, which forms from the fusion of the so-called hillocks of His, six auricular mounds that are first visible at E11.5. The pinna, in contrast to the first arch-derived EAM, is derived from second arch neural crest-derived mesenchyme (Fig. 3C; Minoux et al., 2013).

5. DEVELOPMENT OF THE INNER EAR

The mature mammalian inner ear is comprised of an auditory organ, the spiral cochlea, and vestibular tissues, including the otolith organs utricle and saccule as well as three semicircular canals. Whereas the vestibular apparatus is important for balance and sense of movement, orientation, and gravity, auditory information is processed in the cochlea (Schoenwolf, Bleyl, Brauer, & Francis-West, 2008). The organ of Corti is the receptor organ of the cochlea and functions in the translation of vibrational energy into nervous information (Lim, 1986). During development, the inner ear including the vestibulocochlear nerve derives from an ectodermal thickening, called the otic placode, which invaginates to form the otic vesicle, as well as from neuroepithelial cells including NCCs (Anniko & Wikström, 1984; Freyer, Aggarwal, & Morrow, 2011). Induction of the otic placode is dependant on signaling from the pharyngeal mesenchyme as well as signaling from the hindbrain (Liang, Bok, & Wu, 2010). In mice, at E10.75, the otic vesicle

Figure 4—Cont'd directed to the duplicated TR. Both TRs signal to the fused malleus to induce chondrogenesis of the malleal manubrium (arrows). (I) At E18.5, *Hoxa2* mutants display two collapsed EAMs, directed to their corresponding TRs. The TRs are shortened and closely apposed. In addition, the malleus and the duplicated malleus are fused where the OA normally forms, positioning the fused and shortened manubrium between both EAMs. The incus is fused to the duplicated incus at the position of the incudostapedial joint. (A color version of this figure is located at the end of this chapter.)

forms a tube-like structure known as the endolymphatic appendage, which can be divided dorsally into the endolymphatic sac, the endolymphatic duct, and more ventrally into the developing cochlear (Morsli, Choo, Ryan, Johnson, & Wu, 1998). The mature cochlea comprises the scala vestibuli and scala tympani filled with perilymph fluid as well as the endolymph-filled scala media containing the organ of Corti. The scala tympani is separated from the scala media via the basilar membrane. An array of cell types, such as mechanosensitive hair cells, supporting cells, and nonsensory epithelial cells are located in the organ of Corti overlying the basilar membrane and project into an acellular matrix called the tectorial membrane (Lim, 1986). Fluid motion, induced by middle ear ossicles, moves the tectorial membrane resulting in the displacement of stereocilia on the hair cells (Slepecky, 1996). Inner hair cells are innervated by spiral ganglion nerve cells, transmitting the auditory information to the central nervous system. In contrast, the outer hair cells receive the majority of efferent input. Depolarization can modify the length of contractile outer hair cells, which amplifies sound waves (Schoenwolf et al., 2008; Weisz, Glowatzki, & Fuchs, 2009). The otic capsule surrounding the inner ear is of dual origin. Whereas most parts of the bony structure including the oval window are derived from mesoderm, a small region in the attic encasing the semicircular canals is of neural crest origin (Fig. 3C; Couly et al., 1993; O'Gorman, 2005; Thompson et al., 2012).

6. MECHANISM AND DEFECTS OF INTEGRATION DURING DEVELOPMENT

6.1 External Ear to Middle Ear

Proper function of the hearing apparatus is highly dependent on the integration of the ear components during development. Transmission and amplification of mechanical signals from the external ear across the MEC via the ossicular chain can only occur if the ear canal extends toward the forming TR to form the eardrum and if the manubrium of the malleus is correctly inserted into the eardrum. The formation and interaction of the tissues that establish the connection between the external and middle ear, namely the EAM ectoderm, pharyngeal endoderm, and NCCs, requires the involvement of intricate reciprocal signaling (Fig. 4A–C).

In humans, congenital defects in the development of the external ears are fairly rare, with an incidence of 1:15,000 (Farrior, 1987). EAM abnormalities, however, are commonly associated with defects in middle ears, linking the development of these two structures (Lambert & Dodson, 1996).

Since human approaches are limited, recent molecular and genetic analysis of mouse models have shed light on various genes that are involved in the patterning and formation of the distal integration sides in the mammalian ear.

From the pioneering experiments of Mallo, we know that extension of the EAM is closely linked to that of the forming TR (Mallo & Gridley, 1996). Genetically engineered mouse models, such as *Goosecoid* (*Gsc*) and *Prx* mutants, have revealed genes that are crucial for craniofacial development and coordinated ear formation (Rivera-Pérez, Mallo, Gendron-Maguire, Gridley, & Behringer, 1995; Yamada et al., 1995). *Gsc* is expressed in the mandibular component of BA1 and later in the region of the developing EAM and surrounding the malleal primordium (Gaunt, Blum, & De Robertis, 1993). In *Gsc* mutant mice, loss of the TR is correlated with loss of the EAM, indicating important signaling mechanisms between these two areas (Mallo & Gridley, 1996; Fig. 4D–F). The experimental loss of the TR by ectopic administration of retinoic acid resulted in a similar defect of the eardrum and EAC formation (Kessel, 1992; Mallo & Gridley, 1996). This intimate relationship is also emphasized by the phenotype in *Hoxa-2* null mutants, which displays a transformation of PA2-derived structures into PA1 derivates, resulting in mirror-image duplications of TRs and duplicated EAMs with absent pinnas (Mallo & Gridley, 1996; Minoux et al., 2013; Fig. 4G–I).

Other genes influencing neural crest development and therefore the TR are *Dlx5* and the transcription factor *Tbx1*. Loss of either gene results in mild abnormalities of the TR including shortening and thickening of the dermal bone (Depew, Lufkin, & Rubenstein, 2002; Moraes, Nóvoa, Jerome-Majewska, Papaioannou, & Mallo, 2005). Displacement of the TR but normal tympanic membrane development can be found in *Hoxa1* null mutants, whereas *Hoxa1/Hoxb1* double knockout results in absent TRs (Gavalas et al., 1998). *Gas1* mutant mice display a smaller but thickened TR and a concurrent hypoplastic or absent EAM (Seppala et al., 2007). The developing TR, therefore appears to signal to the surface ectoderm to control the direction of invagination of the forming ear canal. The nature of the signal/s is unknown but the phenotype of the Gas1 mutant implicates Hedgehog signaling in this process, while loss of the EAM is also observed in *Endothelin-1* knockout mice, indicating the involvement of this signaling pathway upstream of *Goosecoid* (Kurihara et al., 1994; Tucker, Yamada, Grigoriou, Pachnis, & Sharpe, 1999). At the same time, as the EAM extends toward the middle ear from the outside the tubotympanic recess extends toward the middle ear from the pharynx. The control of this movement and the later integration of this endoderm-derived structure with the ectodermal EAM to form the tympanic membrane is unknown, but may involve *Dusp6* and hence Fgf signaling. This negative feedback regulator of FGFR signaling has been shown to be expressed at the tip of the invaginating tubotympanic recess and *Dusp6* knockout mice show signs of hearing loss with variable phenotypes including a distorted opening of the tympanic membrane (Li, Scott, Hatch, Tian, & Mansour, 2007; Urness, Li, Wang, & Mansour, 2008).

In order to pick up mechanical vibrations from the external ear, the manubrium of the malleus needs to insert in and run along the tympanic membrane. The manubrium consists of a bony strut that grows out from the body of the malleus during embryonic development, extending in the direction of the forming EAM (Fig. 4B–C). The growth and proper positioning of the malleal manubrium has been linked to the development of the EAM. In human patients, where the ear canal is absent, the manubrium is also missing, while in patients where the canal is stenotic, the manubrium is misshapen and small (Ishimoto, Ito, Kondo, Yamasoba, & Kaga, 2004). In mice, if the EAM fails to form, such as in *Gsc* mutants, the result is also a failure in the development of the manubrium (Fig. 4E–F; Rivera-Pérez et al., 1995; Yamada et al., 1995), a similar loss of manubrium being shown when the middle ear is cultured without the presence of the developing EAM (Mallo et al., 2000). The expression domain of *Bapx1*, another gene implicated in skeletogenesis, partially overlaps with that of *Gsc* during murine development, and absence of *Bapx1* results in the loss of the anterior end of the TR and a reduced width of the malleus (Tucker, Watson, Lettice, Yamada, & Hill, 2004). Isolated malleal abnormalities can be observed in mice engineered to exhibit a tissue-specific ablation of the transcription factor *c-myc* in cells of neural crest origin. In these mutants, the body of the malleus appears smaller, with a shortened and altered angle of the manubrium, leading to hearing deficiencies (Wei et al., 2007). *Msx1* mutants display the absence of the orbicular apophysis of the malleus (Zhang et al., 2003). This mild anomaly, however, did not lead to obvious impairment in auditory function as the process does not insert into the tympanic membrane.

The EAM has therefore been proposed to provide signals to induce the manubrium in the underlying mesenchyme, and thereby coordinate the development of the middle and external ear so that the two structures physically interact (Mallo et al., 2000). This hypothesis is strengthened by the phenotype observed in *Gas1* mutants. In addition to the hypoplastic TR and EAM previously mentioned, these mice lack the manubrium and orbicular apophysis of the malleus (Seppala et al., 2007). *Tshz1 null* mutants show defects in dermal bones such as a shortened, thickened, and laterally displaced TR (Coré et al., 2007). *Tshz1* is expressed in neural crest-derived mesenchymal cells of BA1 and BA2 and loss of its expression results in a loss of the manubrium and orbicular apophysis, while both the incus and stapes develop normally. *Tshz1* might therefore be implicated in the formation and positioning of the condensing manubrial primordium (Coré et al., 2007). Interestingly, the human *TSHZ1* gene also maps to a 2.3 MB deletion in patients that suffer from congenital aural atresia (Feenstra et al., 2011). The etiology of the disease is characterized by conductive hearing loss and structural abnormalities of the ear including malformation of the tympanic membrane, making this gene a translational candidate gene for a human syndrome in which the integration of the ear is defective (Coré et al., 2007; Feenstra et al., 2011). Fgfs and bone morphogenetic proteins (Bmps) have been shown to be expressed in distinct regions of the early developing EAM, and may play a role in signaling to the forming manubrium, although this has yet to be tested (Mallo et al., 2000).

6.2 Middle Ear to Inner Ear (Formation of the Oval Window)

On the proximal end of the ossicular chain, mechanical vibrations are passed onto the inner ear structures via the integration of the stapes into the otic capsule, creating the oval window by formation of the annular ligament around the footplate. Defects in the insertion of the stapes into the oval window are a relatively common form of hearing loss (Bachor, Just, Wright, Pau, Hans, & Karmody, 2005). In particular, the stapes is found ankylosed to the inner ear capsule. This impairs movement of the stapes and transmission of sound into the inner ear. Such ankylosis can be congenital leading to hearing loss at birth, or progressive, leading to gradual hearing loss often associated with aging and pathological bone remodeling (Karosi et al., 2009). The oval window and footplate of the stapes share the same embryonic origin (Thompson et al., 2012). In a study of 51 ears with congenital absence of the oval window, a normal stapes was only observed in 4% of patients (Jahrsdoerfer, 1977), confirming the important relationship between these two structures.

In normal oval window formation, growth/differentiation factors (Gdf) such as *Gdf5*, *Gdf6*, and the Bmp antagonist *Noggin* are expressed in the forming of annular ligament (Hwang & Wu, 2008). Genetic ablation of *Gdf6* in mouse mutants results in defects in the articular surfaces of the ossicles and the annular ligament, which are accompanied by a decrease in proliferation (Settle et al., 2003). Mice with half the dose of *Noggin* display conductive hearing loss due to chondrocyte hyperplasia of the stapes, leading to bone formation between the stapes and the adjacent temporal bone (Hwang & Wu, 2008). Mutations in NOGGIN are also associated with human cases of congenital stapes ankylosis (Ganaha et al., 2015).

In the chick, the paraxial mesoderm has been shown to signal to both the columella and inner ear, coordinating the early development of these two structures, and Fgf signaling has been implicated (Zou et al., 2012). The murine stapes and the avian columella have two embryonic origins, neural crest (like the other ossicles) with a mesodermal footplate, which is initially continuous with the mesodermal capsule (Noden, 1986; Thompson et al., 2012; Fig. 2). How the stapes become distinct from the capsule and what positions the footplate of the stapes within the otic capsule remains elusive. Removal of the neural crest-derived tissues of the stapes leads to a loss of the footplate but a remnant of the oval window still forms, indicating that its position is not completely dependent on the presence of the stapes (Thompson et al., 2012). As the stapes insert adjacent to the vestibular epithelium of the inner ear, the presence of this structure may play a role in induction of the oval window, indicating that signals from the inner ear may pattern the middle ear. Such phenotypes highlight the integral relationships in action between the different parts of the ear; defects in one part impacting on the formation of the other (Chapman, 2011; Zou et al., 2012).

7. CONCLUSION

In conclusion, the development and integration of the mammalian ear is an intricate process, which culminates in maturation of all three parts of the ear to allow hearing. In the mouse, this occurs by postnatal day 14 with the almost simultaneous opening of the external ear canal, final clearance of the middle ear, and differentiation of the inner ear. Due to the various tissues that contribute to the ear and their different developmental timings of induction and differentiation, creating a functioning ear requires a highly coordinated system of signaling pathways and reciprocal interactions, the nature of which we are only just starting to explore. Mouse mutants have helped to identify key player in the formation of functional connections within the ear and will be an important part of future research. Conductive hearing loss due to the loss of connectivity of the ear is a major burden for society, and therefore future research is essential to shed light on more signals involved in ear development to improve treatment and pioneer preventive measures.

REFERENCES

Agius, A. M., Wake, M., Pahor, A. L., & Smallman, L. A. (1995). Smoking and middle ear ciliary beat frequency in otitis media with effusion. *Acta Oto-Laryngologica*, *115*(1), 44–49.

Amin, S., & Tucker, A. S. (2006). Joint formation in the middle ear: Lessons from the mouse and guinea pig. *Developmental Dynamics*, *235*(5), 1326–1333. http://dx.doi.org/10.1002/dvdy.20666.

Anniko, M., & Wikström, S. O. (1984). Pattern formation of the otic placode and morphogenesis of the otocyst. *American Journal of Otolaryngology*, *5*(6), 373–381.

Anthwal, N. L., Joshi, L., & Tucker, A. S. (2012). Evolution of the mammalian middle ear and jaw: Adaptations and novel structure. *Journal of Anatomy*, *222*, 147.

Bachor, E., Just, T., Wright, C., Pau, H., Hans, W., & Karmody, C. (2005). Fixation of the stapes footplate in children: A clinical and temporal bone histopathologic study. *Ototlogy & Neurotology*, *26*(5), 866–873.

Bockman, D. E., Redmond, M. E., & Kirby, M. L. (1989). Alteration of early vascular development after ablation of cranial neural crest. *The Anatomical Record*, *225*, 209–217. http://dx.doi.org/10.1002/ar.1092250306.

Chapman, S. (2011). Can you hear me now? Understanding vertebrate middle ear development. *Frontiers in Bioscience*, *16*, 1675–1692.

Chin, K., Kurian, R., & Saunders, J. C. (1997). Maturation of tympanic membrane layers and collagen in the embryonic and post-hatch chick (Gallus domesticus). *Journal of Morphology*, *233*(3), 257–266.

Coré, N., Caubit, X., Metchat, A., Boned, A., Djabali, M., & Fasano, L. (2007). Tshz1 is required for axial skeleton, soft palate and middle ear development in mice. *Developmental Biology*, *308*(2), 407–420. http://dx.doi.org/10.1016/j.ydbio.2007.05.038.

Couly, G. F., Coltey, P. M., & Le Douarin, N. M. (1993). The triple origin of skull in higher vertebrates: A study in quail-chick chimeras. *Development*, *117*(2), 409–429.

Depew, M. J., Lufkin, T., & Rubenstein, J. L. R. (2002). Specification of jaw subdivisions by Dlx genes. *Science*, *298*(5592), 381–385. http://dx.doi.org/10.1126/science.1075703.

Farrior, J. B. (1987). Surgical management of congenital conductive deafness. *Southern Medical Journal*, *80*(4), 450–453.

Feenstra, I., Vissers, L. E., Pennings, R. J., Nillessen, W., Pfundt, R., Kunst, H. P., et al. (2011). Disruption of teashirt zinc finger homeobox 1 is associated with congenital aural atresia in humans. *American Journal of Human Genetics*, *89*, 813–819.

Freyer, L., Aggarwal, V., & Morrow, B. E. (2011). Dual embryonic origin of the mammalian otic vesicle forming the inner ear. *Development (Cambridge, England)*, *138*(24), 5403–5414. http://dx.doi.org/10.1242/dev.069849.

Ganaha, A., Kaname, T., Akazawa, Y., Higa, T., Shinjou, A., Naritomi, K., et al. (2015). Identification of two novel mutations in th NOG gene associated with congenital stapes ankylosis and symphalangism. *Journal of Human Genetics*, *60*, 27–34.

Gaunt, S. J., Blum, M., & De Robertis, E. M. (1993). Expression of the mouse goosecoid gene during mid-embryogenesis may mark mesenchymal cell lineages in the developing head, limbs and body wall. *Development*, *117*(2), 769–778.

Gavalas, A., Studer, M., Lumsden, A., Rijli, F. M., Krumlauf, R., & Chambon, P. (1998). Hoxa1 and Hoxb1 synergize in patterning the hindbrain, cranial nerves and second pharyngeal arch. *Development*, *125*(6), 1123–1136.

Graham, A. (2003). Development of the pharyngeal arches. *American Journal of Medical Genetics. Part A*, *119A*(3), 251–256. http://dx.doi.org/10.1002/ajmg.a.10980.

Graham, A., Okabe, M., & Quinlan, R. (2005). The role of the endoderm in the development and evolution of the pharyngeal arches. *Journal of Anatomy, 207*(5), 479–487. http://dx.doi.org/10.1111/j.1469-7580.2005.00472.x.

Graves, G. O., & Edwards, L. F. (1944). The Eustachian tube: A review of its descriptive, microscopic and clinical anatomy. *Archives of Otolaryngology - Head and Neck Surgery, 39*(5), 359–397. http://dx.doi.org/10.1001/archotol.1944.00680010374001.

Grevellec, A., & Tucker, A. S. (2010). The pharyngeal pouches and clefts: Development, evolution, structure and derivatives. *Seminars in Cell & Developmental Biology, 21*(3), 325–332. http://dx.doi.org/10.1016/j.semcdb.2010.01.022.

Hamilton, W. J., & Mossman, H. W. (1972). *Hamilton, Boyd and Mossman's human embryology* (4th ed., p. 493). Cambridge: Heffer.

Hwang, C.-H., & Wu, D. K. (2008). Noggin heterozygous mice: An animal model for congenital conductive hearing loss in humans. *Human Molecular Genetics, 17*(6), 844–853. http://dx.doi.org/10.1093/hmg/ddm356.

Ishimoto, S.-I., Ito, K., Kondo, K., Yamasoba, T., & Kaga, K. (2004). The role of the external auditory canal in the development of the malleal manubrium in humans. *Archives of Otolaryngology—Head & Neck Surgery, 130*(8), 913–916. http://dx.doi.org/10.1001/archotol.130.8.913.

Jahrsdoerfer, R. A. (1977). Congenital absence of the oval window. *Transactions. Section on Otolaryngology. American Academy of Ophthalmology and Otolaryngology, 84*(5), ORL904–ORL914.

Karosi, T., Csomor, P., Petko, M., Liktor, B., Szabo, L. Z., Pytel, J., et al. (2009). Histopathology of non-otosclerotic stapes fixation. *Otology & Neurotology, 30*, 1058–1066.

Kaufmann, M. H. (1992). *The atlas of mouse development* (1st ed.). London: Academic Press.

Kerschner, J. E., Li, J., Tsushiya, K., & Khampang, P. (2010). Mucin gene expression and mouse middle ear epithelium. *International Journal of Pediatric Otorhinolaryngology, 74*(8), 864–868. http://dx.doi.org/10.1016/j.ijporl.2010.04.014.

Kessel, M. (1992). Respecification of vertebral identities by retinoic acid. *Development (Cambridge, England), 115*(2), 487–501.

Köntges, G., & Lumsden, A. (1996). Rhombencephalic neural crest segmentation is preserved throughout craniofacial ontogeny. *Development, 122*(10), 3229–3242.

Kurihara, Y., Kurihara, H., Suzuki, H., Kodama, T., Maemura, K., Nagia, R., et al. (1994). Elevated blood-pressure and craniofacial abnormalities in mice deficient in endothelin-1. *Nature, 368*, 703–710.

Lambert, P. R., & Dodsen, E. E. (1996). Congenital malformations of the external auditory canal. *Otolaryngologic Clinics of North America, 29*, 741–760.

Le Lièvre, C. S., & Le Douarin, N. M. (1975). Mesenchymal derivatives of the neural crest: Analysis of chimaeric quail and chick embryos. *Journal of Embryology and Experimental Morphology, 34*, 125–154.

Li, C., Scott, D. A., Hatch, E., Tian, X., & Mansour, S. L. (2007). Dusp6 (Mkp3) is a negative feedback regulator of FGF stimulated ERK signalling during mouse development. *Development, 134*, 167–176.

Liang, J. K., Bok, J., & Wu, D. K. (2010). Distinct contributions from the hindbrain and mesenchyme to inner ear morphogenesis. *Developmental Biology, 337*(2), 324–334. http://dx.doi.org/10.1016/j.ydbio.2009.11.001.

Lim, D. J. (1986). Functional structure of the organ of Corti: A review. *Hearing Research, 22*, 117–146.

Lim, D. J., Chun, Y. M., Lee, H. Y., Moon, S. K., Chang, K. H., Li, J. D., et al. (2000). Cell biology of tubotympanum in relation to pathogenesis of otitis media—A review. *Vaccine, 19*(Suppl. 1), S17–S25.

Lindsay, E. A. (2001). Chromosomal microdeletions: Dissectiog del 22q11 syndrome. *Nature Reviews Genetics*, (2), 858–868.

Lumsden, A., & Keynes, R. (1989). Segmental patterns of neuronal development in the chick hindbrain. *Nature*, *337*, 424–428.
Mallo, M. (1998). Embryological and genetic aspects of middle ear development. *The International Journal of Developmental Biology*, *42*(1), 11–22.
Mallo, M. (2001). Formation of the middle ear: Recent progress on the developmental and molecular mechanisms. *Developmental Biology*, *231*(2), 410–419. http://dx.doi.org/10.1006/dbio.2001.0154.
Mallo, M., & Gridley, T. (1996). Development of the mammalian ear: Coordinate regulation of formation of the tympanic ring and the external acoustic meatus. *Development*, *122*(1), 173–179.
Mallo, M., Schrewe, H., Martin, J. F., Olson, E. N., & Ohnemus, S. (2000). Assembling a functional tympanic membrane: Signals from the external acoustic meatus coordinate development of the malleal manubrium. *Development*, *127*, 4127–4136.
Mason, M. J. (2013). Of mice, moles and guinea pigs: Functional morphology of the middle ear in living mammals. *Hearing Research*, *301*, 4–18. http://dx.doi.org/10.1016/j.heares.2012.10.004.
McDonald, M. H., Hoffman, M. R., Gentry, L. R., & Jiang, J. J. (2012). New insights into mechanism of Eustachian tube ventilation based on cine computed tomography images. *European Archives of Oto-Rhino-Laryngology—Head & Neck Surgery*, *269*(8), 1901–1907. http://dx.doi.org/10.1007/s00405-011-1829-y.
Michaels, L., & Soucek, S. (1989). Development of the stratified squamous epithelium of the human tympanic membrane and external canal: The origin of auditory epithelial migration. *The American Journal of Anatomy*, *184*(4), 334–344. http://dx.doi.org/10.1002/aja.1001840408.
Minoux, M., Kratochwil, C. F., Ducret, S., Amin, S., Kitazawa, T., Kurihara, H., et al. (2013). Mouse Hoxa2 mutations provide a model for microtia and auricle duplication. *Development*, *140*(21), 4386–4397. http://dx.doi.org/10.1242/dev.098046.
Moore, W. J. (1981). *The mammalian skull*. Cambridge: Cambridge University Press.
Moraes, F., Nóvoa, A., Jerome-Majewska, L. A., Papaioannou, V. E., & Mallo, M. (2005). Tbx1 is required for proper neural crest migration and to stabilize spatial patterns during middle and inner ear development. *Mechanisms of Development*, *122*(2), 199–212. http://dx.doi.org/10.1016/j.mod.2004.10.004.
Morsli, H., Choo, D., Ryan, A., Johnson, R., & Wu, D. K. (1998). Development of the mouse inner ear and origin of its sensory organs. *The Journal of Neuroscience*, *18*(9), 3327–3335.
Nishizaki, K., Anniko, M., Orita, Y., Masuda, Y., Yoshino, T., Kanda, S., et al. (1998). Programmed cell death in the development of the mouse external auditory canal. *The Anatomical Record*, *252*(3), 378–382.
Noden, D. M. (1983). The embryonic origins of avian cephalic and cervical muscles and associated connective tissues. *The American Journal of Anatomy*, *168*, 257–276. http://dx.doi.org/10.1002/aja.1001680302.
Noden, D. M. (1986). Origins and patterning of craniofacial mesenchymal tissues. *Journal of Craniofacial Genetics and Developmental Biology. Supplement*, *2*, 5–31.
Noden, D. M. (1988). Interactions and fates of avian craniofacial mesenchyme. *Development (Cambridge, England)*, *103*(Suppl.), 121–140.
Novacek, M. J. (1993). Patterns of diversity in the mammalian skull. *In the skull: Vol. 2.* (pp. 438–545). Chicago: University of Chicago Press.
Nuutinen, J., Kärjä, J., & Karjalainen, P. (1983). Measurement of mucociliary function of the Eustachian tube. *Archives of Otolaryngology (Chicago, Ill: 1960)*, *109*(10), 669–672.
O'Gorman, S. (2005). Second branchial arch lineages of the middle ear of wild-type and Hoxa2 mutant mice. *Developmental Dynamics: An Official Publication of the American Association of Anatomists*, *234*(1), 124–131. http://dx.doi.org/10.1002/dvdy.20402.

Park, K., Ueno, K., & Lim, D. J. (1992). Developmental anatomy of the Eustachian tube and middle ear in mice. *American Journal of Otolaryngology*, *13*(2), 93–100.

Richter, C. A., Amin, S., Linden, J., Dixon, J., Dixon, M. J., & Tucker, A. S. (2010). Defects in middle ear cavitation cause conductive hearing loss in the Tcof1 mutant mouse. *Human Molecular Genetics*, *19*(8), 1551–1560. http://dx.doi.org/10.1093/hmg/ddq028.

Rivera-Pérez, J. A., Mallo, M., Gendron-Maguire, M., Gridley, T., & Behringer, R. R. (1995). Goosecoid is not an essential component of the mouse gastrula organizer but is required for craniofacial and rib development. *Development*, *121*(9), 3005–3012.

Schoenwolf, G., Bleyl, S., Brauer, P., & Francis-West, P. (2008). *Larsen's human embryology*. (4th ed.). Churchill Livings.

Seppala, M., Depew, M. J., Martinelli, D. C., Fan, C.-M., Sharpe, P. T., & Cobourne, M. T. (2007). Gas1 is a modifier for holoprosencephaly and genetically interacts with sonic hedgehog. *The Journal of Clinical Investigation*, *117*(6), 1575–1584. http://dx.doi.org/10.1172/JCI32032.

Settle, S. H., Rountree, R. B., Sinha, A., Thacker, A., Higgins, K., & Kingsley, D. M. (2003). Multiple joint and skeletal patterning defects caused by single and double mutations in the mouse Gdf6 and Gdf5 genes. *Developmental Biology*, *254*(1), 116–130. http://dx.doi.org/10.1016/S0012-1606(02)00022-2.

Slepecky, N. B. (1996). Structure of the mammalian cochlea. In P. Dallos, A. N. Popper, & R. R. Fay (Eds.), *The cochlea: Vol. 8.* (pp. 44–129). New York: Springer New York. http://dx.doi.org/10.1007/978-1-4612-0757-3.

Thompson, H., Ohazama, A., Sharpe, P. T., & Tucker, A. S. (2012). The origin of the stapes and relationship to the otic capsule and oval window. *Developmental Dynamics*, *241*(9), 1396–1404. http://dx.doi.org/10.1002/dvdy.23831.

Thompson, H., & Tucker, A. S. (2013). Dual origin of the epithelium of the mammalian middle ear. *Science (New York, N.Y.)*, *339*(6126), 1453–1456. http://dx.doi.org/10.1126/science.1232862.

Trainor, P. A., Tan, S. S., & Tam, P. P. (1994). Cranial paraxial mesoderm: Regionalisation of cell fate and impact on craniofacial development in mouse embryos. *Development*, *120*(9), 2397–2408.

Tucker, A. S., Watson, R. P., Lettice, L. A., Yamada, G., & Hill, R. E. (2004). Bapx1 regulates patterning in the middle ear: Altered regulatory role in the transition from the proximal jaw during vertebrate evolution. *Development*, *131*(6), 1235–1245. http://dx.doi.org/10.1242/dev.01017.

Tucker, A. S., Yamada, G., Grigoriou, M., Pachnis, V., & Sharpe, P. T. (1999). Fgf8 determines rostral-causal polarity in the first branchial arch. *Development*, *126*, 51–61.

Urness, L. D., Li, C., Wang, X., & Mansour, S. L. (2008). Expression of ERK signalling inhibitors Dusp6,7,9 during mouse ear development. *Developmental Dynamics*, *237*, 163–169.

Wei, K., Chen, J., Akrami, K., Galbraith, G. C., Lopez, I. A., & Chen, F. (2007). Neural crest cell deficiency of c-myc causes skull and hearing defects. *Genesis (New York, N.Y.: 2000)*, *45*(6), 382–390. http://dx.doi.org/10.1002/dvg.20304.

Weisz, C., Glowatzki, E., & Fuchs, P. (2009). The postsynaptic function of type II cochlear afferents. *Nature*, *461*(7267), 1126–1129.

Yamada, G., Mansouri, A., Torres, M., Stuart, E. T., Blum, M., Schultz, M., et al. (1995). Targeted mutation of the murine goosecoid gene results in craniofacial defects and neonatal death. *Development*, *121*(9), 2917–2922.

Zhang, Z., Zhang, X., Avniel, W. A., Song, Y., Jones, S. M., Jones, T. A., et al. (2003). Malleal processus brevis is dispensable for normal hearing in mice. *Developmental Dynamics*, *227*(1), 69–77. http://dx.doi.org/10.1002/dvdy.10288.

Zou, Y., Mak, S.-S., Liu, H. Z., Han, D. Y., Zhuang, H. X., Yang, S. M., et al. (2012). Induction of the chick columella and its integration with the inner ear. *Developmental Dynamics*, *241*(6), 1104–1110. http://dx.doi.org/10.1002/dvdy.23788.

PART 2

Craniofacial Patterning and Signaling Mechanisms

CHAPTER TEN

Zebrafish Craniofacial Development: A Window into Early Patterning

Lindsey Mork, Gage Crump[1]
Eli and Edythe Broad CIRM Center for Regenerative Medicine and Stem Cell Research at the University of Southern California, Los Angeles, California, USA
[1]Corresponding author: e-mail address: gcrump@usc.edu

Contents

Abstract

The formation of the face and skull involves a complex series of developmental events mediated by cells derived from the neural crest, endoderm, mesoderm, and ectoderm. Although vertebrates boast an enormous diversity of adult facial morphologies, the fundamental signaling pathways and cellular events that sculpt the nascent craniofacial skeleton in the embryo have proven to be highly conserved from fish to man. The zebrafish *Danio rerio*, a small freshwater cyprinid fish from eastern India, has served as a popular model of craniofacial development since the 1990s. Unique strengths of the zebrafish model include a simplified skeleton during larval stages, access to rapidly developing embryos for live imaging, and amenability to transgenesis and complex

Current Topics in Developmental Biology, Volume 115
ISSN 0070-2153
http://dx.doi.org/10.1016/bs.ctdb.2015.07.001

genetics. In this chapter, we describe the anatomy of the zebrafish craniofacial skeleton; its applications as models for the mammalian jaw, middle ear, palate, and cranial sutures; the superior imaging technology available in fish that has provided unprecedented insights into the dynamics of facial morphogenesis; the use of the zebrafish to decipher the genetic underpinnings of craniofacial biology; and finally a glimpse into the most promising future applications of zebrafish craniofacial research.

1. ZEBRAFISH CRANIOFACIAL ANATOMY

The skull of an adult zebrafish consists of 43 bones derived from a cartilage template and 30 directly ossifying intramembranous bones (Fig. 1D–F; Cubbage & Mabee, 1996), significantly exceeding the approximately 22 skull bones of an adult mammal. However, many of these bones, particularly those forming the cranial vault, develop and ossify long after the embryonic period, such that the skull of a zebrafish larva that is just acquiring the capacity to feed (~5 days postfertilization (dpf)) is, by contrast, quite simple (Fig. 1A) compared with that of newborn mammals. At this stage, the craniofacial skeleton consists of a neurocranium, which provides support for the brain and sensory systems, and the viscerocranium, which serves as the feeding and respiratory apparatus (Fig. 1A and B). Both are primarily composed of cartilage elements that are later replaced by bone, as well as a few intramembranous bones such as those that support the gill covering. Larval cartilages are relatively simple, often only one or a few cell layers thick, and can be easily visualized in fixed individuals by staining with Alcian blue (Fig. 1A and B) or in living larvae using fluorescent transgenic reporters expressed in chondrocytes (Fig. 3C).

The layout of the larval zebrafish craniofacial skeleton has been extensively described in anatomical studies (Cubbage & Mabee, 1996; Piotrowski et al., 1996; Schilling & Kimmel, 1997; Schilling et al., 1996). The viscerocranium is derived from cranial neural crest cells (NCCs) that populate the seven pharyngeal arches of the zebrafish embryo, with multiple skeletal elements forming along the dorsal–ventral axis of each arch (Fig. 1C). The first, or mandibular, arch gives rise to the ventrally positioned bilateral Meckel's cartilages, which join at the midline and serve as the lower jaw in the larva, as well as the more dorsal palatoquadrates, which articulate with the posterior end of Meckel's to form the jaw joint. The pterygoid processes are anterior extensions of the palatoquadrate that contact the ethmoid plate, the most anterior part of the neurocranium, to form the larval upper

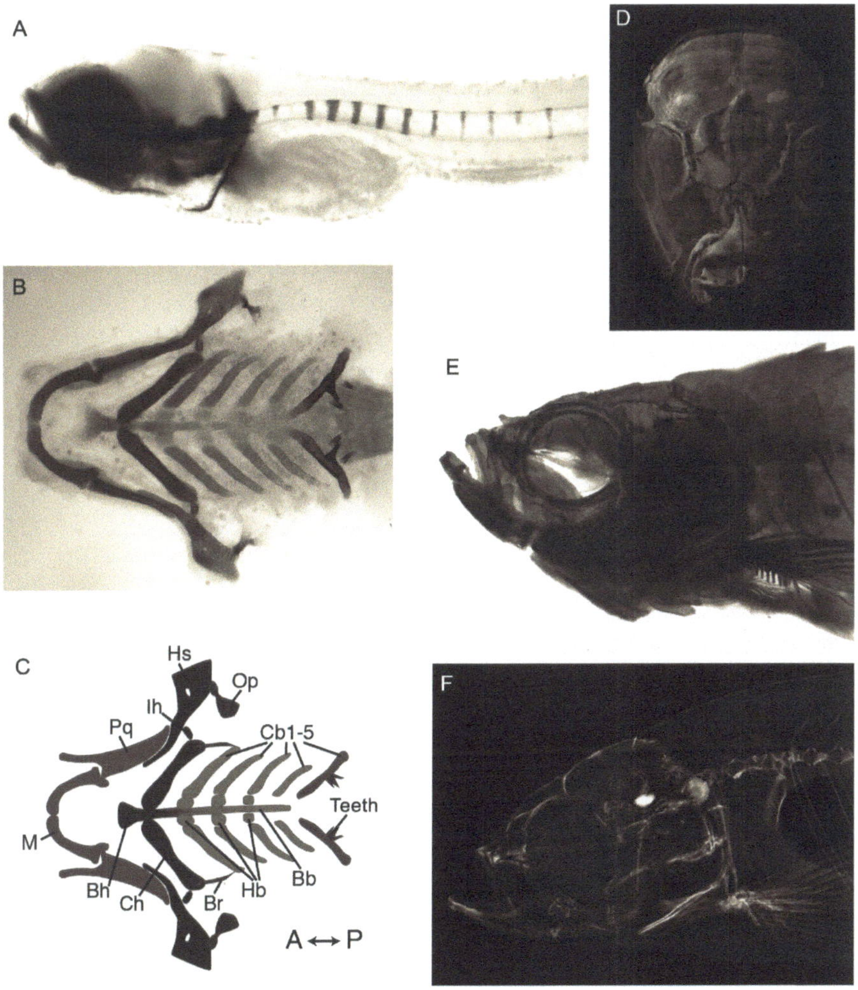

Figure 1 Zebrafish larval and adult craniofacial skeletons. (A) Larval zebrafish (6 dpf) stained with Alcian blue for cartilage and Alizarin red for mineralized bone reveals the early formation of the craniofacial skeleton. (B) Ventral view of an Alcian blue/ Alizarin red-stained larval viscerocranium dissected away from the rest of the head, showing the skeletal derivatives of the seven pharyngeal arches. (C) Schematic of the larval viscerocranium. First-arch structures (blue): Pq, palatoquadrate; M, Meckel's; second-arch structures (purple): Hs, hyosymplectic; Ih, interhyal; Ch, ceratohyal; Bh, basihyal; Op, opercle (bone); Br, branchiostegal ray (bone); arches 3–7 (teal): Cb, ceratobranchial; Hb, hypobranchial; Bb, basibranchial. (D–F) The complex adult zebrafish craniofacial skeleton can be visualized in multiple modalities. (D) Micro-CT projection of an adult zebrafish head (frontal view). (F) Adult zebrafish head stained with Alcian blue and Alizarin red (lateral view). (E) X-ray of an adult zebrafish head (lateral view). Images not to scale. Anterior is to the left in A–C, E, and F. *Panel (D) Image courtesy of S. Ruffins and Panel (E) photo courtesy of S. Paul.* (A color version of this figure is located at the end of this chapter.)

jaw. The second, or hyoid, arch also forms two large cartilages, the dorsal hyosymplectics and the ventral ceratohyals, with a smaller interhyal cartilage linking the two as part of the hyoid joint. The ceratohyals each contact the unpaired basihyal cartilage at the midline. The hyosymplectic, a fusion of the plate-like hyomandibula with the symplectic rod, is connected to the neurocranium via the anterior part of the otic cartilage and thereby secures the jaw skeleton to the rest of the head. Moving posteriorly, arches 3–7 produce paired ventrolateral rod-like ceratobranchial cartilages affixed to small hypobranchials, which in turn attach to unpaired basibranchials at the ventral midline. The ceratobranchials support the densely branched, delicate gill tissues. Unique to the larval viscerocranium, the fifth pair of ceratobranchials, from arch 7, forms several ossified pharyngeal teeth. The larval viscerocranium also contains several directly ossifying bones, the most notable of which is the fan-shaped opercle bone that projects off of the dorsal hyomandibula. This bone grows dramatically in size during the first weeks of life (Kimmel, DeLaurier, Ullmann, Dowd, & McFadden, 2010) and functions to support and protect the gills.

1.1 Zebrafish Pharyngeal Arches as Models of the Mammalian Jaw and Middle Ear

The pharyngeal arches are transient and morphologically complex embryonic structures positioned posterior to the eye and below the ear. In all vertebrates, the arches consist of cylinders of cranial NCCs surrounding a core of mesoderm, lined externally by ectoderm and internally by pharyngeal endoderm. The arches are extensively patterned by signals from the surrounding epithelia, as well as by NCC-intrinsic information carried over from their neuroepithelial origins; integration of this information determines the major anterior–posterior (Hox genes) and dorsal–ventral axes (Endothelin-1, Jagged-Notch, Bmp) as well as fine-tunes exactly what parts of each arch need to produce cartilage or bone (Fgf, Hh) to establish domain-specific skeletal bauplans. The general functions of these pathways in patterning the pharyngeal arches prior to skeletal development have proven to be highly conserved across vertebrates (reviewed by Medeiros & Crump, 2012), despite the differences in the actual shapes and numbers of arch-derived skeletal elements in different animals. For example, in mice and zebrafish, mutations in the Endothelin-1 pathway result in drastic changes specifically to the anterior lower face, manifesting as reductions and transformations of Meckel's and the ceratohyals in zebrafish *edn1* mutants (Miller, Schilling, Lee, Parker, & Kimmel, 2000), extensive abnormalities including a severely reduced mandible in *Edn1* mouse mutants (Kurihara et al., 1994) and a homeotic transformation of the lower jaw to an upper jaw, complete with whiskers, in mice mutant for *Dlx5/6*, two key Edn1 target genes (Depew, Lufkin, & Rubenstein, 2002). This deep conservation across vertebrates in early pharyngeal arch patterning has been demonstrated repeatedly and justifies the use of the zebrafish as a model to identify additional genes and better understand their functions in this process.

Skeletal elements in the fish skull do not always have recognizable homologs in mammals (or vice versa), but there are some that clearly do. In a fascinating evolutionary story, the middle ear bones of mammals have their origins in the bones that form the jaw and connect the viscerocranium to the ear in fish (Fig. 2A; reviewed by Anthwal, Joshi, & Tucker, 2012; Gould, 1990). Anatomical, developmental, and evolutionary analyses have confirmed that the posterior end of Meckel's cartilage and part of the palatoquadrate are evolutionarily homologous to the first arch-derived malleus and incus of the mammalian middle ear, with the second arch-derived hyomandibula representing the stapes homolog (Fig. 2A; Liem & Walker, 2001). The connection between the hyomandibula and the otic cartilage in fish is reminiscent of the stapes' attachment to the oval window in the otic capsule housing the mammalian inner ear. This homology between the mammalian middle ear and the fish jaw means that we can exploit the powerful genetic tools available in the zebrafish system to identify new genes that may shape these skeletal homologs in both systems. This represents an important opportunity because while a number of mouse models with middle ear abnormalities have been described, the developmental basis of the skeletal defects remains unresolved in most instances, and there are many reported clinical cases of hearing loss where one or more of the middle ear bones present with abnormal morphology or are entirely absent (e.g., Park & Choung, 2009; Park, Choung, Shin, & Hong, 2007; Rahbar, Neault, & Kenna, 2002).

1.2 The Fish Neurocranium as a Model for the Mammalian Palate

Unlike the viscerocranium, the larval neurocranium is of mixed cranial neural crest and mesoderm origins: the anterior part of the neurocranium, including the ethmoid plate and paired trabeculae, and parts of the bilateral otic capsules are formed by cranial NCCs, while the posterior and medial components that develop around the cranial end of the notochord are derived from mesoderm (Fig. 2B; Kague et al., 2012; Mongera et al., 2013).

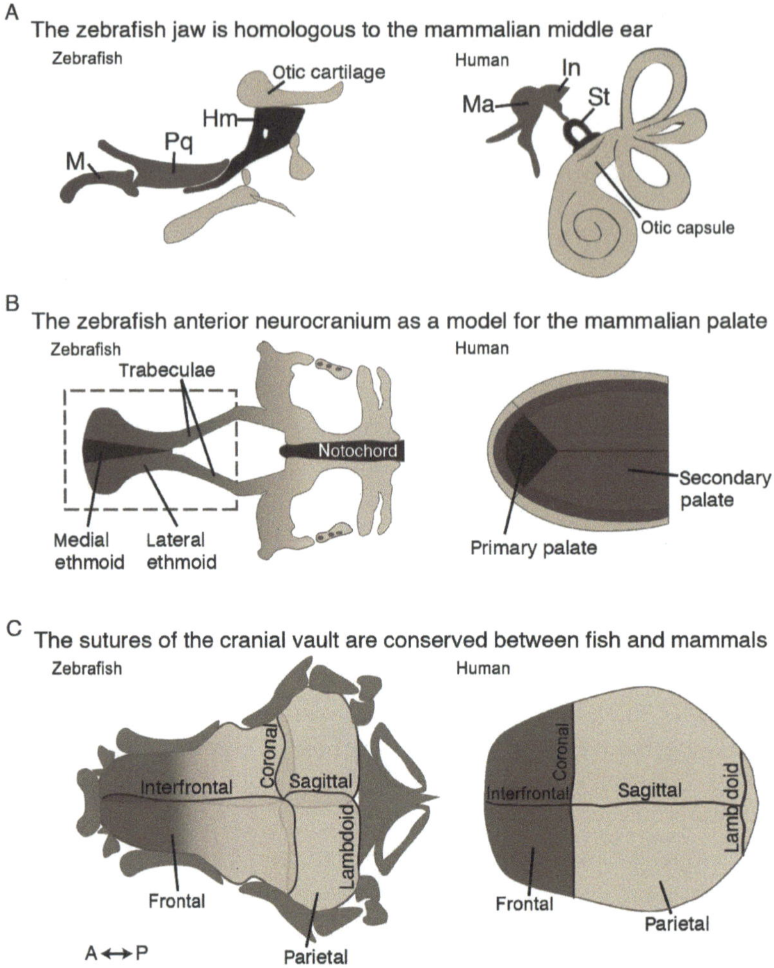

Figure 2 Zebrafish models of mammalian craniofacial development. (A) The larval zebrafish jaw is evolutionarily homologous to the mammalian middle ear ossicles. The posterior parts of the first arch-derived Meckel's (M) and palatoquadrate (Pq) cartilages (blue), which articulate the jaw in fish, correspond to the malleus (Ma) and incus (In) of mammals. The second arch-derived hyomandibula (purple) (Hm; the dorsal component of the hyosymplectic), which affixes the viscerocranium to the neurocranium via the otic cartilage, is the homolog of the mammalian stapes (St). (B) The anterior end of the larval fish neurocranium is a model for the mammalian hard palate. The medial ethmoid plate (dark blue) derives from frontonasal NCCs, with the lateral ethmoid and trabeculae (light blue) formed by medially migrating anterior maxillary NCCs (left panel), akin to the origins of the primary and secondary hard palate, respectively, in mammals (right panel). The posterior zebrafish neurocranium derives from mesoderm (green). (C) The overlapping calvarial bones of the adult zebrafish form the same suture pattern as observed in mammals. The two frontal bones are separated from each other by the interfrontal (or metopic) suture and from the parietal bones by the bilateral coronal sutures. The parietals are divided along the midline by the sagittal suture and posteriorly bounded by the lambdoid suture. The boundary between neural crest (blue) and mesoderm (green) falls within the frontal bones in fish and along the coronal suture in mammals. Anterior is to the left. (A color version of this figure is located at the end of this chapter.)

The anterior neurocranium, particularly the ethmoid, has emerged as a genetically tractable model of the mammalian hard palate (Fig. 2B). Both the ethmoid and the mammalian secondary palate derive from bilateral groups of anterior maxillary cranial NCCs (termed the "palatine shelves" in mammals) that migrate medially to fuse at the midline and form the roof of the mouth. Impairments in the migration and midline fusion processes result in cleft palate in mammals (Bush & Jiang, 2012). Comparable phenotypes have been observed in zebrafish "midline" mutants, ranging from a complete absence of the anterior neurocranium to the formation of two unfused, parallel rods or a single rod instead of a plate (Eberhart et al., 2008; Eberhart, Swartz, Crump, & Kimmel, 2006; Kimmel, Miller, & Moens, 2001; Swartz, Sheehan-Rooney, Dixon, & Eberhart, 2011; Wada et al., 2005). Importantly, zebrafish mutant for Pdgf, Shh, or Bmp pathway components, or deficient in *irf6*, all show cleft or midline phenotypes resembling those of their mammalian counterparts (Table 1; Dougherty et al., 2013; Eberhart et al., 2008, 2006; Swartz et al., 2011; Wada et al., 2005). These findings indicate that a conserved genetic network regulates the formation of this part of the face across vertebrates and support the use of the zebrafish anterior neurocranium as a model for the mammalian hard palate. However, there are significant morphogenetic differences between the two systems, particularly the fact that the mammalian palatal bone directly ossifies from mesenchymal cells while the zebrafish anterior neurocranium passes through a cartilaginous template, and so caution is advised when extrapolating zebrafish findings to mammals (Bush & Jiang, 2012). Nonetheless, zebrafish will likely be a good model for early events regulating the growth and midline fusion of the palatine shelves, whereas later processes such as ossification and development of the lip and soft palate may prove more difficult to model in fish.

1.3 Conservation of Cranial Sutures from Fish to Man

Though the bulk of research on zebrafish craniofacial biology has focused on the embryonic and early larval periods, this system can also be utilized to model other processes that are more relevant to adult stages, particularly the formation and maintenance of sutures (Grova et al., 2012; Holmes, 2012). As in mammals, the skull vault of adult zebrafish is formed by thin intramembranous bony plates separated by sutures of dense connective tissue (Fig. 2C; Quarto & Longaker, 2005). In fish, these bones progressively ossify over the brain during the second month of life (Quarto & Longaker, 2005), well after the onset of ossification in the cartilaginous templates of the

Table 1 Human Genetic Diseases with Craniofacial Abnormalities Modeled in Zebrafish

Human Disease	Zebrafish Homolog	Type of Model	References
Alagille syndrome	*jag1b*	Mutant	Zuniga et al. (2010)
Axenfeld–Rieger syndrome	*pitx2*	Morpholino	Bohnsack, Kasprick, Kish, Goldman, and Kahana (2012) and Liu and Semina (2012)
Auriculocondylar syndrome	*edn1*	Mutant	Miller et al. (2000)
	furinA	Mutant	Walker et al. (2006)
	mef2ca	Mutant	Miller et al. (2007)
	plcb3	Mutant	Walker et al. (2007)
Bamforth–Lazarus syndrome	*foxe1*	Morpholino	Nakada, Iida, Tabata, and Watanabe (2009)
Branchio-oculo-facial syndrome	*tfap2a*	Mutant, morpholino	Gestri et al. (2009)
Campomelic dysplasia	*sox9a*, *sox9b*	Mutants	Yan et al. (2002, 2005)
Cardiofaciocutaneous syndrome	*braf*	Injection of mutant mRNA	Anastasaki, Rauen, and Patton (2012)
CHARGE syndrome	*chd7*	Morpholino	Patten et al. (2012)
Cleft palate	*faf1*	Morpholino	Ghassibe-Sabbagh et al. (2011)
	mir140, *pdgfra*	miRNA injection; mutant	Eberhart et al. (2008)
	tgfb3	Morpholino	Cheah, Winkler, Jabs, and Chong (2010)
	smad5	Mutant	Swartz et al. (2011)
	tgfb2, *fgf10a*	Morpholino	Swartz et al. (2011)
	wwp2, *med25*	Morpholino	Nakamura et al. (2011)
	crispld2	Morpholino	Yuan et al. (2012)
	irf6	Dominant-negative transgene	Dougherty et al. (2013)
Cleidocranial dysplasia	*runx2b*	Morpholino	Flores, Lam, Crosier, and Crosier (2006)
Cornelia de Lange syndrome	*nipbla/nipblb*	Morpholino	Muto, Calof, Lander, and Schilling (2011)

Table 1 Human Genetic Diseases with Craniofacial Abnormalities Modeled in Zebrafish—cont'd

Human Disease	Zebrafish Homolog	Type of Model	References
Coronal craniosynostosis	*cyp26b1*	Mutant	Laue et al. (2011)
Cranio-lenticulo-sutural dysplasia	*sec23a*	Mutant	Lang, Lapierre, Frotscher, Goldenring, and Knapik (2006)
Diamond–Blackfan anemia	*rpl11*	Mutant	Danilova, Sakamoto, and Lin (2011) and Zhang et al. (2014)
	rps19	Mutant	Zhang et al. (2014)
	rps24	Morpholino	Song et al. (2014)
DiGeorge syndrome	*tbx1*	Mutant	Piotrowski et al. (2003) and Zhang, Gui, Wang, Jiang, and Song (2010)
			Choudhry and Trede (2013) and Choe and Crump (2014)
Frontonasal dysplasia	*alx1*	Morpholino	Dee, Szymoniuk, Mills, and Takahashi (2013)
Hereditary multiple exocytoses	*ext2*, *slc35b2*	Mutants	Clement et al. (2008)
Holoprosencephaly	*shha*, *smo*, *disp1*	Mutants	Brand et al. (1996), Kimmel et al. (2001), and Wada et al. (2005)
	dzip1, *gli1*, *gli2*		Eberhart et al. (2006) and Schwend and Ahlgren (2009)
	six3	Injection of mutant mRNA	Geng et al. (2008)
Jeune and Mainzer-Saldino syndromes	*ift172*	Morpholino	Halbritter et al. (2013)
Leri–Weill dyschondrosteosis	*shox*	Morpholino	Sawada, Kamei, Hakuno, Takahashi, and Shimizu (2015)
Meier–Gorlin syndrome	*orc1*	Morpholino	Bicknell et al. (2011)

Continued

Table 1 Human Genetic Diseases with Craniofacial Abnormalities Modeled in Zebrafish—cont'd

Human Disease	Zebrafish Homolog	Type of Model	References
Noonan syndrome	*raf1*	Morpholino	Razzaque et al. (2007)
	kras	Injection of mutant mRNA	Razzaque et al. (2012)
	ptpn11a, *ptpn11b*	Mutant	Bonetti et al. (2014)
Oblique facial clefts	*specc11*	Morpholino	Saadi et al. (2011) and Gfrerer et al. (2014)
Oculofaciocardiodental syndrome	*bcor*	Morpholino	Ng et al. (2004)
Osteogenesis imperfecta	*sec24d*	Mutant	Sarmah et al. (2010) and Garbes et al. (2015)
Oro-facial-digital type 1 syndrome	*ofd1*	Morpholino	Ferrante et al. (2009)
Pierre Robin Sequence	*satb2 cis*-regulatory	Transgenic enhancer lines	Rainger et al. (2014)
	sox9a cis-regulatory	Transgenic enhancer lines	Gordon et al. (2014)
Richieri–Costa–Pereira syndrome	*eif4a3*	Morpholino	Favaro et al. (2014)
Roberts syndrome	*esco2*	Morpholino	Monnich, Kuriger, Print, and Horsfield (2011)
Robinow syndrome	*wnt5a*	Morpholino	Huang et al. (2014)
Shprintzen–Goldberg syndrome	*skia*, *skib*	Morpholinos	Doyle et al. (2012)
Simpson–Golabi–Behmel syndrome	*gpc4*	Mutant	Wiweger et al. (2011)
Stickler/Marshall syndrome	*col11a1a/b*	Morpholino	Baas, Malbouyres, Haftek-Terreau, Le Guellec, and Ruggiero (2009)
Treacher–Collins syndrome	*tcof1*	Morpholino	Weiner, Scampoli, and Calcaterra (2012)
X-linked mental retardation	*phf8*	Morpholino	Qi et al. (2010)

viscerocranium and neurocranium starting at 5–6 dpf. In mammals, the coronal suture, which lies between the frontal and parietal bones, delineates the border between anterior, neural crest-derived bones and posterior, mesoderm-derived bones. In fish, as appears to be the case in avians (Evans & Noden, 2006; Noden, 1984), this border between neural crest and mesoderm actually exists as a cryptic boundary within the frontal bone, with the bones flanking the coronal suture both derived from mesoderm (Fig. 2C; Kague et al., 2012; Mongera et al., 2013). Despite this important difference in embryological contributions, the coronal sutures of zebrafish and mammals have similar anatomical structures (Quarto & Longaker, 2005).

Adult zebrafish continually grow throughout their lifetime, and their cranial sutures remain patent to allow for skull expansion (Quarto & Longaker, 2005), similar to rodents (Opperman, 2000) but unlike sutures in humans, which fuse in adulthood (Meindl & Lovejoy, 1985). Premature or ectopic suture fusion, or craniosynostosis, can occur in fish as well as mammals and leads to gross distortions in cranial shape. Though the potential of zebrafish as a model for suture formation and maintenance has only begun to be explored, recent pioneering studies have reported new unmapped adult zebrafish mutants with cranial bone fusions (Andreeva et al., 2011), genes with conserved suture expression patterns (*il11ra*; Keupp et al., 2013), and craniosynostosis phenotypes in fish mutant for a gene required for suture maintenance in mammals (*cyp26b1*; Laue et al., 2011), indicating conservation of gene function throughout evolution.

2. IMAGING CRANIOFACIAL DEVELOPMENT

The ease with which living embryos and larvae can be accessed and imaged gives an enormous advantage to the zebrafish system. Relative to mouse and chick, zebrafish embryos are transparent, develop both rapidly and externally, and can be repeatedly anesthetized and subjected to confocal imaging for extended periods while continuing to develop without major abnormalities or delays. Furthermore, both the pharyngeal arches and the later-arising skeletal elements of the zebrafish face are significantly smaller and contain many fewer cells than the corresponding structures in mouse and chick (Yelick & Schilling, 2002), making this system particularly suitable for high-resolution imaging to deconstruct skeletal shaping at the level of individual cells.

2.1 Fate-Mapping Facial Development

Fate-mapping analyses allow researchers to trace cell lineages to better understand how cells of one developmental stage (e.g., premigratory NCCs) are related to those of another (e.g., NCCs in the pharyngeal arches; Fig. 3B). Numerous tools have been used for short-term fate-mapping analyses in the zebrafish pharyngeal arches, including microinjection of fluorescent dyes into single cells (Crump, Swartz, Eberhart, & Kimmel, 2006; Crump, Swartz, & Kimmel, 2004; Kimmel et al., 1998; Wada et al., 2005) and laser-induced local activation of photoconvertible proteins (e.g., kaede; Dougherty et al., 2012; Eberhart et al., 2006) or kikGR (Balczerski et al., 2012; Swartz et al., 2011). These tools have allowed zebrafish researchers to push past the limitations of the avian and mouse models to perform extremely high-resolution fate-mapping studies of both pre- and postmigratory NCCs.

These fate maps have revealed that a cell's *premigratory* position along the anterior–posterior axis is related to both its migratory route and destination in the arches and that a *postmigratory* cell's location within its arch determines what part of the skeleton it will make. Within the first stream of neural crest, the initial, premigratory rostral–caudal position alongside the brain determines whether NCCs will migrate over or behind the eye (Dougherty et al., 2012; Wada et al., 2005), as well as whether their progeny will contribute to the anterior neurocranium or the upper or lower jaw (Dougherty et al., 2013, 2012; Eberhart et al., 2006; Wada et al., 2005). Additional analyses demonstrated that NCCs at certain anterior–posterior and mediolateral coordinates within each arch reproducibly contribute not only to particular skeletal elements but also to specific domains of said elements (Fig. 3B; Balczerski et al., 2012; Crump et al., 2006, 2004; Eberhart et al., 2006; Swartz et al., 2011; Wada et al., 2005), implying that the complex skeletal shapes of the face form as amalgamations of multiple subpopulations of arch NCCs that may be independently regulated.

These fate maps have also been particularly useful for deconstructing the developmental basis of skeletal phenotypes of mutant or transgenic lines: researchers have used these maps to hone in on the specific population of arch NCCs known to give rise to the affected element and then determine what aspect of their developmental trajectory (e.g., survival, migration, chondrocyte, or osteoblast differentiation) has been derailed in the mutant to produce the observed phenotype (Balczerski et al., 2012; Crump et al., 2006, 2004; Eberhart et al., 2006; Wada et al., 2005). For example, in

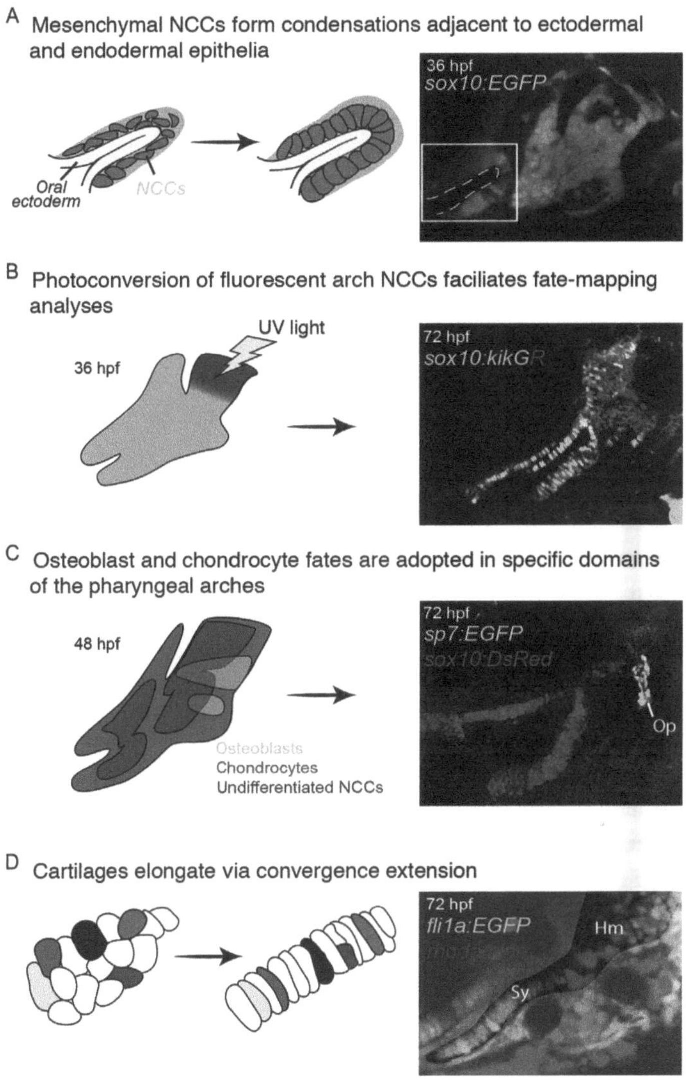

Figure 3 Morphogenetic processes that shape the facial skeleton. (A) Mesenchymal NCCs form tight condensations next to ectodermal or endodermal epithelia, including the stomodeum (i.e., oral ectoderm). The *sox10:EGFP* transgene (green) is expressed by all NCCs. Boxed area shows NCCs condensed around the oral ectoderm (dashed line). (B) Fate-mapping analyses demonstrated that NCCs at certain locations within the arches reproducibly give rise to specific parts of facial skeletal elements. For example, arch NCCs in the dorsal half of the second arch make the hyomandibular portion of the hyosymplectic. One way to track cells is to use focused applications of ultraviolet light to convert fluorescent proteins such as kikGR from green to red in specific cell populations

the *moz* mutant, which forms a set of mirror-image first-arch elements within the second arch (Miller, Maves, & Kimmel, 2004), the fate map of the second arch is completely reorganized, such that cells that normally contribute to the anterior hyomandibula no longer make any cartilage at all (Crump et al., 2006).

For longer-term lineage tracing of arch NCCs, the only Cre lines available to date are *Sox10*-based constructs that are broadly expressed in all neural crest derivatives (Kague et al., 2012; Mongera et al., 2013). While these lines have been extraordinarily useful for determining what components of the skeleton arise from neural crest versus mesoderm, the field will need to develop more specific Cre lines that label only subsets of premigratory or arch NCCs in order to characterize, for example, the origins of skeletal elements that develop after the early larval period.

2.2 Visualizing Morphogenesis and Signaling in Real-Time

Each element of the facial skeleton has a unique three-dimensional shape and position within the skull. How cells in nascent skeletal elements coordinate with each other to attain these precise shapes is a long-standing question that is still far from being fully answered, even in the simplified zebrafish larval face. However, the ability to perform extended time-lapse imaging of craniofacial development in the zebrafish embryo has brought researchers closer than ever to grasping the complexity of facial morphogenesis at a cellular level.

Several transgenic reporters, most notably *sox10:EGFP* (Wada et al., 2005) and *fli1a:EGFP* (Lawson & Weinstein, 2002), are expressed in arch NCCs as well as the forming skeleton and have been widely used to image transitions between key developmental stages, as well as for side-by-side

Figure 3—cont'd at early stages. The labeled cells are then reimaged at a later time point. (C) Following mesenchymal condensation, chondrocyte and osteoblast fates are adopted within specific domains of the pharyngeal arches around 48 hpf. At 72 hpf, the *sp7:EGFP* transgene (green) is expressed in osteoblasts, while *sox10:DsRed* (red) is specifically expressed by chondrocytes. Op, opercle. Nuclei are labeled with Hoechst (blue). (D) Convergence-extension movements of differentiating prechondrocytes, rather than cell proliferation, drive the elongation of cartilages from bulky precartilaginous condensations into stacks of disc-shaped chondrocytes. The symplectic rod (Sy) lengthens anteriorly from the plate-like hyomandibula (Hm) in this manner. The right panel shows a *fli1a:EGFP* (green) transgenic embryo with mosaic rhodamine-dextran injection (red), illustrating the intercalation of dye-labeled cells into the elongating Sy. (A color version of this figure is located at the end of this chapter.)

comparisons of mutant and control embryos (Crump et al., 2006, 2004; Eames et al., 2013; Eberhart et al., 2008, 2006; Le Pabic, Ng, & Schilling, 2014; Wada et al., 2005). These movies were captured at high spatial and temporal resolution, providing unprecedented access to the cellular processes that underlie the formation of the larval craniofacial skeleton (McGurk, Lovely, & Eberhart, 2014). One such process that was illuminated by time-lapse imaging in fish is the active condensation of mesenchymal NCCs (Fig. 3A). Arch NCCs form very tightly packed condensations at several locations throughout the arches, including around the oral epithelium and posterior to the first pharyngeal pouch. Disruption of signaling between these epithelia and the NCCs in *smo* and *itga5* mutants, respectively, causes the disintegration of these condensations, readily visualized by time-lapse imaging, resulting in losses of the specific cartilage subdomains that form from these cells (Crump et al., 2004; Eberhart et al., 2006). Another process important for shaping craniofacial cartilages, revealed at a slightly later stage of development, is convergence-extension (Fig. 3D). Both time-lapse movies and dye-based fate maps have demonstrated that cell intercalation rather than proliferation accounts for the elongation of bulky ventral precartilaginous condensations into long, rod-shaped cartilages consisting of stacks of polarized disc-shaped chondrocytes (Kimmel et al., 1998; Le Pabic et al., 2014). This process was recently shown to be regulated by the Fat/Daschous planar cell polarity pathway (Le Pabic et al., 2014).

Live imaging also offers the opportunity to correlate morphogenesis with intercellular signaling. Fluorescent transgenic lines that provide transcriptional readouts of Bmp (*BRE:EGFP*; Alexander et al., 2011), Fgf (*dusp6:EGFP*; Molina, Watkins, & Tsang, 2007), Notch (*Tp1:EGFP*; Parsons et al., 2009), Wnt (*7xTCF:EGFP*; Moro et al., 2012), and Hedgehog (*Gli-d:mCherry*; Schwend, Loucks, & Ahlgren, 2010) signaling activities can be used to identify the specific sets of NCCs that receive a given signal. For example, strong expression of *BRE:EGFP* was detected in the NCCs of the ventral first and second arches (Alexander et al., 2011), adjacent to the ventral ectoderm that expresses *bmp4* (Zuniga, Rippen, Alexander, Schilling, & Crump, 2011). The conclusion that these cells were responding to a Bmp signal was supported by the phenotype of embryos transgenic for a dominant-negative Bmp receptor that presented malformations in skeletal elements that form from ventral arch NCCs (Alexander et al., 2011). Other useful reporters for live imaging include fusions between a fluorescent protein and signaling transducers like Disheveled (Choe et al., 2013) or mediators such as the immunoglobulin-domain protein Alcama

(Choe et al., 2013), which change their cellular localization upon the receipt of a particular signal. Live imaging of these signaling reporters in conjunction with reporters that label all NCCs can reveal how signaling dynamics align with the many morphogenetic processes that transform the pharyngeal arches into the craniofacial skeleton.

2.3 Imaging Skeletal Differentiation

Time-lapse imaging of zebrafish craniofacial development also offers the opportunity to visualize the onset of differentiation of specific populations of NCCs into chondrocytes, osteoblasts, tendons/ligaments, and connective tissues (Fig. 3C). In addition to the transgenic reporters expressed in both arch NCCs and the early skeleton, the field has also developed a number of transgenic markers that are activated upon the differentiation of specific skeletal cell types, including *sox9a:EGFP* (Eames et al., 2013) and *col2a1a:EGFP* (Dale & Topczewski, 2011) for chondrocytes and *sp7:EGFP* (DeLaurier et al., 2010) and *RUNX2:EGFP* (Knopf et al., 2011) for osteoblasts. The utility of the osteoblast lines has been amply demonstrated; for example, time-lapse analyses have demonstrated that growth of the opercular bone occurs through Indian hedgehog-mediated recruitment of proliferating, *sp7:EGFP*-negative NCC precursors (Huycke, Eames, & Kimmel, 2012). In the future, a larger toolbox of differently hued fluorescent reporters should allow researchers to visualize specific stages of osteoblast and chondrocyte differentiation in the intact embryo, which when combined with mutant analyses should provide precise insights into how craniofacial genes regulate distinct stages of cartilage and bone differentiation.

3. GENETIC ANALYSES OF CRANIOFACIAL DEVELOPMENT

3.1 Forward Genetic Screens for Craniofacial Mutants

Because zebrafish can be maintained at high densities for relatively low cost and are capable of producing hundreds of offspring on a regular basis, their potential for large-scale forward genetic screens was quickly recognized in the early days of zebrafish research (Grunwald & Streisinger, 1992; Kimmel, 1989). The first large-scale screens for mutations affecting zebrafish larval craniofacial development were conducted in the Nusslein-Volhard lab at Tübingen (Piotrowski et al., 1996; Schilling et al., 1996) and the Driever lab at Harvard (Driever et al., 1996; Neuhauss et al., 1996). Here, potential mutants were identified based on the gross morphology of the head at larval stages and only later stained to illuminate cartilage morphology, so these screens favored mutants with severe morphological abnormalities over those with more subtle patterning defects. A later screen for craniofacial defects within a set of embryonic lethal insertional mutant lines in the Hopkins lab at MIT (Amsterdam et al., 1999) identified additional loci that, by definition, also had severe craniofacial and/or pleiotropic effects that precluded the detection of milder phenotypes (Nissen, Amsterdam, & Hopkins, 2006). These screens were nevertheless extraordinarily fruitful, identifying more than 80 loci (some redundantly) required for the development of cartilage elements in the viscerocranium and/or neurocranium, or more generally for chondrocyte differentiation and organization. Subsequent screens performed at the University of Oregon using direct staining of cartilage and bone with Alcian blue and Alizarin red, respectively, identified additional mutants with craniofacial patterning phenotypes, including many that are likely too subtle to have been detected in the initial screens (e.g., Cox et al., 2012; Crump et al., 2004; Eames et al., 2011; Eberhart et al., 2008; Miller et al., 2004; Zuniga, Stellabotte, & Crump, 2010).

Draft versions of the zebrafish genome first became available in the early 2000s (Jekosch, 2004), facilitating ongoing mapping efforts. To the delight of the community, many of the first mutant loci to be mapped were in genes or pathways previously identified in mice or humans as critical for early neural crest (e.g., *tfap2a*; Barrallo-Gimeno, Holzschuh, Driever, & Knapik, 2004; Knight et al., 2003) or craniofacial development (e.g., the Endothelin-1 pathway; *edn1*, Miller et al., 2000; *plcb3*, Walker, Miller, Swartz, Eberhart, & Kimmel, 2007; *furina*, Walker, Miller, Coffin Talbot, Stock, & Kimmel, 2006; *mef2ca*, Miller et al., 2007) and the Sonic Hedgehog pathway (*shh*, Schauerte et al., 1998; *smo*, Varga et al., 2001; *disp1*, Nakano et al., 2004; *gli1*, Karlstrom et al., 2003; *gli2*, Karlstrom, Talbot, & Schier, 1999). This impressive degree of conservation in itself validates the use of zebrafish as a model for investigating the genetic control of mammalian craniofacial development. Other loci were mapped to genes not previously recognized as being important for neural crest or skeletal patterning (e.g., the integrin *itga5*, Crump et al., 2004; the Notch ligand *jag1b*, Zuniga et al., 2010; and the variant histone *h3f3a*, Cox et al., 2012), opening entirely new avenues of research in the field. Another mutant, which entirely lacked endoderm (*sox32*), also failed to develop any jaw cartilages, revealing a crucial nonautonomous role for endoderm in facial cartilage formation (David, Saint-Etienne, Tsang, Schilling, & Rosa, 2002). As many of the mutants identified in these screens have still not been mapped to a specific genetic locus and are represented by only a single allele (suggesting a lack of screen saturation), the utility of the forward mutagenesis approach for identifying new genes required for early craniofacial development has yet to be fully realized.

While initial screens focused on mutations affecting early larval cartilage and bone, more recent screens have begun to identify a number of mutations affecting later development of the skull, including in adults (Andreeva et al., 2011; Harris et al., 2008; Huitema et al., 2012). As opposed to larval lethal mutants, these viable adult craniofacial mutants may better replicate the types of nonlethal birth defects seen in humans. For example, loss of the ectodysplasin A receptor (*edar*) in adult zebrafish was found to result in a loss of fins as well as scales, pharyngeal teeth, and bony gill rakers (Harris et al., 2008), mimicking the ectodermal organ defects (e.g., defective teeth, hair, sweat glands) seen in human patients with *EDA* or *EDAR* mutations (Kere et al., 1996; Monreal et al., 1999). Another common class of viable craniofacial zebrafish mutants involves misregulated ossification of cartilage-derived elements (Apschner, Huitema, Ponsioen, Peterson-Maduro, & Schulte-Merker, 2014; Eames et al., 2011; Huitema et al., 2012). In particular, a number of mutations in enzymes that add sugar groups to extracellular matrix proteins (e.g., *xylt1*, *fam20b*) have revealed important roles for cartilage-derived proteoglycans in suppressing premature osteoblast differentiation within the periosteum (Eames et al., 2011). Continued analyses of these viable craniofacial mutants will provide new insights into the development of bones and cartilage that develop after the larval period, as well as how the craniofacial complex maintains its structure during growth and aging.

An advantage of forward screens is its ability to identify partial loss-of-function alleles (hypomorphic alleles) and dominant alleles. For example, in cases where null alleles are early lethal, hypomorphic alleles can reveal dose-dependent effects of gene loss on craniofacial development, and dominant alleles can disrupt multiple members of a gene family (e.g., the $h3f3a^{db1092}$ mutation, which disrupts multiple redundant H3.3 histones during neural crest specification; Cox et al., 2012). On the other hand, forward screens will often fail to uncover gene function in cases of functional redundancy. A complication of teleost fish is that they experienced a whole-genome duplication following their divergence from the lineage that gave rise to tetrapods (reviewed by Braasch & Postlethwait, 2012), and thus zebrafish often possess two copies of what exists as a single-copy gene in mammals. However, this issue of "genome duplication" is often exaggerated, with current estimates suggesting that only 20–30% of mammalian genes have two paralogs in zebrafish (Postlethwait, Amores, Cresko, Singer, & Yan, 2004; Postlethwait et al., 2000; Woods et al., 2005). In addition, many of these paralogs are expected to have diverged in expression pattern and/or function throughout evolution, a process referred to as "subfunctionalization" (Braasch & Postlethwait, 2012; Glasauer & Neuhauss, 2014). For example, although mammalian Jag1 has both Jag1a and Jag1b paralogs in zebrafish (Zecchin, Conigliaro, Tiso, Argenton, & Bortolussi, 2005), loss of *jag1b* is sufficient to cause patterning defects of the upper face in zebrafish, with *jag1a* expression being lost from the pharyngeal arches during teleost evolution (Zecchin et al., 2005; Zuniga et al., 2010). Nonetheless, some paralog pairs or related members of a gene family may still function redundantly, such that a phenotype only manifests when both genes are mutated. These loci would therefore be difficult to uncover in forward screens.

3.2 Reverse Genetics to Test Candidate Craniofacial Genes

In the early 2000s, two new approaches were developed that allowed for targeted mutagenesis (reverse genetics) or gene knockdown in zebrafish: Targeting Induced Local Lesions in Genomes (TILLING) and morpholinos. With TILLING, researchers screen chemically mutagenized zebrafish genomes with sensitive PCR-based methods to detect point mutations in genes of interest (Wienholds et al., 2003). Enhanced TILLING methods, combined with next-generation sequencing, have been employed by the Zebrafish Mutation Project at the Sanger Institute, with the goal of identifying nonsense or essential splice-site mutations in the vast majority of predicted genes (Kettleborough et al., 2013).

Stable morpholino oligonucleotides (MOs) that interfere with splicing or mRNA translation were also developed as a novel knockdown approach (Nasevicius & Ekker, 2000). The great appeal of MOs rested in the speed of their applications, allowing researchers to rapidly test the function of candidate genes for craniofacial development. While many MOs were predicted to yield only hypomorphic phenotypes based on their knockdown efficiency, several were shown to effectively phenocopy their corresponding null mutants, including those targeting *edn1* (Miller & Kimmel, 2001), *sox9a* (Yan et al., 2002), *tfap2a* (Arduini, Bosse, & Henion, 2009), and *dlx5a* (Talbot, Johnson, & Kimmel, 2010). However, it was also recognized that MOs can cause nonspecific artifacts due to toxicity or off-target effects related to the great excess of MO injected relative to the amount of endogenous RNA in the embryo (reviewed by Schulte-Merker & Stainier, 2014); researchers thus employed a number of controls (e.g., coinjecting a MO against p53 to block cell death, Robu et al., 2007, or rescuing the phenotype by coinjecting wild-type RNA) to improve the validity of their findings. Dozens of studies have since identified craniofacial phenotypes in different morphants, ranging from general reductions in the size or elongation of the viscerocranium to defects in cartilage differentiation or malformations of specific cartilage elements. Unfortunately, subsequent identification of null alleles in the same genes has revealed that the vast majority of morphant phenotypes (>80% or more; Kok et al., 2015) may be due to nonspecific toxicity effects and/or reductions of more than just the target gene. In some cases, MOs cause the same phenotype even when the endogenous target gene is deleted (Kok et al., 2015), a strong argument for nonspecific toxicity. In more worrying cases, even when morphant phenotypes are properly controlled for by rescue with nontargeted mRNA, they still fail to be replicated in true null alleles (Law & Sargent, 2014; van Impel et al., 2014). While there is still much to be learned about the nature of these discrepancies, great caution must be taken when interpreting morphant phenotypes in the absence of a genetic mutant. Indeed, given the recent ability to rapidly generate null alleles in virtually any gene (see below), there appears to be little justification to continue to use MOs in zebrafish research.

In the last few years, newer tools for targeted mutagenesis of candidate genes, including zinc-finger nucleases (ZFNs; reviewed by Porteus & Carroll, 2005), TALENs (Christian et al., 2010), and CRISPR/Cas9 (Cong et al., 2013), have begun to supersede the use of morpholinos. Both ZFNs and TALENs are chimeric proteins consisting of sequence-specific DNA-binding domains fused to DNA endonucleases, while the CRISPR/Cas9 system targets the Cas9 endonuclease to specific sites in the genome using specially designed guide RNAs. All of these techniques produce short indels at their target sites, which, if located in an exon, can disrupt the reading frame and lead to a nonfunctional protein. Stable lines carrying these purported null mutant alleles can be constructed within two fish generations, though phenotypes can sometimes also be observed in larvae injected with highly efficient TALENs or CRISPRs (e.g., Bedell et al., 2012; Dahlem et al., 2012). CRISPR/Cas9, the most recently developed technology, is far more cost-effective and efficient than the proprietary ZFN technology and even easier and more amenable to high-throughput analyses than TALENs, making it the current system of choice for targeted mutagenesis in zebrafish.

As described above, zebrafish researchers are increasingly finding that many of these new targeted mutant lines do not exhibit the same phenotypes as observed in the corresponding morphants (Kok et al., 2015; Schulte-Merker & Stainier, 2014). One concern is that the small lesions generated by TALENs and CRISPRs could lead to residual translated protein, for example, by alternative mRNA splicing or translational frameshifting. However, in the few cases in which this has been directly examined using antibodies against the gene product, residual protein translation has not been found to account for the lack of a phenotype previously described for morphants (Kok et al., 2015; Law & Sargent, 2014). Nonetheless, due to concerns about residual gene function, some researchers are choosing to use pairs of CRISPRs to delete all or most of the open reading frame, thus ensuring that no residual functional protein is made.

With all of the new mutant lines now available, zebrafish are increasingly being used for complex genetic experiments not technically feasible in any other vertebrate model. A pair of zebrafish can produce a hundred embryos at a time on a near-weekly basis, so sufficient numbers of those rare individuals homozygous for mutations in two (1/16 offspring), three (1/64), or even four genes (1/256) can be identified in a relatively short time. The ability to perform such complex crosses will allow researchers to test for redundancy among multiple members of a gene family, as well as to determine epistatic relationships between groups of key patterning factors and build gene interaction networks.

3.3 Transgenic Technology to Manipulate Developmental Patterning

Overexpression studies are often used to complement knockdown or mutant analyses of gene function. Embryo-wide overexpression can be achieved by injecting mRNAs into single-celled embryos, though in these cases early embryonic defects often cloud analyses of later craniofacial changes. Alternatively, DNA constructs can be used that allow transgenic expression of gene products under control of a heat shock-inducible promoter (e.g., *hsp70I:dnfgfR1*; Lee, Grill, Sanchez, Murphy-Ryan, & Poss, 2005), thus temporally restricting gene expression to craniofacial stages. While injected mRNAs are usually fairly even distributed among cells of the later embryo, DNA constructs are sparsely and mosaically distributed, which most often necessitates the creation of stable transgenic lines. However, use of the I-SceI (Grabher, Joly, & Wittbrodt, 2004) or Tol2 transposon systems (reviewed by Kawakami, 2007) makes the generation of stable transgenic lines in zebrafish a highly efficient process, with the major drawback being the 3-month generation time to generate stable lines.

The Gal4/UAS system has also been used to achieve tissue-specific or temporally regulated overexpression in fish. In these experiments, the Gal4 transcriptional activator is expressed under the control of a tissue-specific or heat shock-inducible promoter, where it binds to the UAS recognition sequence and activates the expression of the target gene in only the cell type or stage of interest. Gal4 drivers expressed in different subsets of arch tissues have been developed, including *sox10:GAL4VP16* (all NCCs; Das & Crump, 2012), *hand2:GAL4VP16* (ventral arch NCCs; Nichols, Pan, Moens, & Kimmel, 2013), and *nkx2.3:GAL4VP16* (pharyngeal endoderm) and *nkx2.5:GAL4VP16* (arch mesoderm; Choe et al., 2013). The primary problem with the Gal4/UAS system is that the UAS sequence is CpG-rich and thus becomes methylated and silenced over successive generations, interfering with transcriptional activation (Goll, Anderson, Stainier, Spradling, & Halpern, 2009). The Q transcriptional regulatory system, which relies on a non-CpG-rich activating sequence, has been proposed as an alternative (Subedi et al., 2014). Nonetheless, the Gal4/UAS technique has been particularly useful for manipulating signaling pathways in the arches, e.g., researchers can activate a pathway by overexpressing a ligand or a constitutively active version of a receptor or inhibit the same pathway by overexpression of a dominant-negative receptor or competing ligand. For example, a key role for Bmp signaling in patterning the ventral-most elements of the viscerocranium was demonstrated by overexpressing Bmp4 with a heat shock-inducible Gal4 driving *UAS:Bmp4* and then blocking the pathway using Gal4-mediated activation of Grem2, a secreted BMP inhibitor (Alexander et al., 2011; Zuniga et al., 2011).

Both standard and tamoxifen-inducible Cre lines have also been successfully established in zebrafish (Mosimann & Zon, 2011), including a Cre expressed under a human *SOX10* enhancer specific to the neural crest (Kague et al., 2012) and an ERT2-Cre regulated by the zebrafish *sox10* promoter, which labels both the neural crest lineage and all chondrocytes (Mongera et al., 2013). To date, these lines have only been used to recombine transgenic constructs for the purposes of lineage tracing or conditional expression of a transgene. Conditional knockouts have been hampered by the field's inability to efficiently insert loxP sites into endogenous loci, though many new methods for targeted insertions are currently under development (reviewed by Auer, Duroure, Concordet, & Del Bene, 2014), and floxed endogenous genes are certainly on the horizon.

4. FUTURE DIRECTIONS

4.1 Gene Editing to Model Human Craniofacial Birth Defects in Zebrafish

In the past two decades, many groups have used morpholinos or mutants identified in forward screens to develop new zebrafish models of human craniofacial birth defects. The speed of zebrafish research has made this model attractive for testing new loci identified by GWAS or patient genome sequencing. Demonstrating that a gene or noncoding sequence suspected to underlie a specific human condition has a conserved function in zebrafish can help to validate its status as a human disease gene. Furthermore, the myriad imaging and genetic tools available for zebrafish can facilitate efforts to determine the root molecular or developmental basis of the human craniofacial phenotype. There are now reports of zebrafish models of at least 35 human genetic diseases with craniofacial abnormalities (Table 1). In many of these cases, the zebrafish models phenocopied, at least in some respects, the human phenotype, exhibiting symptoms recognizable as cyclopia, cleft palate, craniosynostosis, etc. In other cases, particularly in the morpholino models, the affected fish showed minor or nonspecific skeletal abnormalities or losses that were more difficult to interpret as resembling the mammalian defect.

In the next few years, as tools for targeted gene editing become even more sophisticated, we will be able to very precisely alter the zebrafish genome to not only knock out genes of interest but also edit endogenous genes to match a patient-specific allele or single-nucleotide polymorphism. This will not only improve the accuracy of the disease models but will also reveal whether a particular amino acid substitution results in gain or loss of function of the affected protein, thereby providing *in vivo* validation of computational predictions of protein function. A current limitation is that precise homologous recombination to introduce polymorphisms, even when stimulated by TALEN/CRISPR-mediated double-strand breaks, remains fairly inefficient in zebrafish (e.g., Auer, Duroure, De Cian, Concordet, & Del Bene, 2014; Zu et al., 2013), although there is hope that rapid developments in the gene-editing field will lead to improvements in the near future. In addition, not all polymorphisms found in human patients will affect regions conserved in fish, though, encouragingly, mutagenic lesions do tend to occur in protein residues and enhancers that are highly conserved throughout vertebrates. It therefore seems likely that zebrafish will become the first-pass organism of choice for testing the exponentially increasing number of putative disease alleles being identified by next-generation sequencing projects.

4.2 Adult Skeletal Homeostasis and Regeneration

Bone is a highly dynamic organ, constantly remodeling in response to mechanical forces and undergoing limited repair after fractures or minor injuries. In humans, defects in bone homeostasis can lead to osteoporosis, particularly in postmenopausal women, and birth defects such as Paget's disease have their underlying etiology in defective bone remodeling. As in mammals, larval cartilages in zebrafish are organized into growth plates, with zones of proliferative and hypertrophic chondrocytes surrounded by a distinct perichondrium and periosteum. However, unlike in mammals, zebrafish cartilage-replacement bone forms largely around the developing cartilage and not within it, and hence is best referred to as perichondrial rather than endochondral bone. Nonetheless, many of the pathways involved in mammalian endochondral ossification, such as Ihh and parathyroid hormone receptor-like signaling, appear to play analogous roles in fish perichondrial ossification (Bhattacharya, Yan, Postlethwait, & Rubin, 2011; Hammond & Schulte-Merker, 2009; Yan et al., 2012). Zebrafish also have abundant osteoclasts that promote bone remodeling through the degradation of existing bony matrix (Witten, Hansen, & Hall, 2001). While fish are not subjected to the same gravitational forces that tetrapod bones experience, load experienced by water current and respiratory and biting forces would be expected to similarly promote bone remodeling in adults, particularly in the jaw. Hence, as with developmental patterning, the relatively simple architecture of fish growth plates, combined with the likely remodeling of craniofacial bones through adulthood, makes zebrafish an interesting new model to visualize how interactions between chondrocytes, osteoblasts, and osteoclasts regulate bone remodeling in both normal and pathological contexts.

Another attractive feature of zebrafish is their remarkable regenerative ability, and the skeleton is no exception. Most work on skeletal regeneration has been conducted in the context of the fin, which in adults is able to fully regenerate all its tissues in about 1 week (reviewed by Poss, Keating, & Nechiporuk, 2003). Fin bone regeneration appears to proceed through at least two routes: dedifferentiation, proliferation, and redifferentiation of existing osteocytes and *de novo* osteoblast differentiation (Knopf et al., 2011; Sousa et al., 2011). Time-lapse imaging in adults has beautifully demonstrated this unusual dedifferentiation process in real time (Knopf et al., 2011), with more recent work showing a similar type of regeneration in the calvarial bones of fish (Geurtzen et al., 2014). In contrast, regeneration of the lower jaw proceeds through a cartilage intermediate (S. Paul & G. D. Crump, unpublished; Wang et al., 2012; Zhang et al., 2015), much like fracture repair in humans. In the future, it will be interesting to examine the extent to which zebrafish skeletal tissues regenerate better than their mammalian counterparts, as well as the genetic and cellular basis of such regeneration. Insights gained could then be used to develop novel therapies for improving cartilage and bone healing in patients.

5. CONCLUSIONS

The zebrafish has become an immensely useful and popular genetic and developmental model for studying neural crest biology and the development of the skull. With its rapidly expanding repertoire of cutting-edge genetic and live-imaging tools, this model has opened an extraordinary window into the complexities of craniofacial morphogenesis that is unparalleled by any other model system. In addition, the more we learn about craniofacial and skeletal biology in the zebrafish, the more we can appreciate that the underlying development of many human craniofacial features (sutures, palate, jaw, middle ear, cartilage, and bone differentiation) is highly conserved in fish. Nevertheless, the amount of effort put into demonstrating the high degree of conservation in craniofacial development between zebrafish and mammals tends to downplay the important areas in which they significantly differ. The face of a fish, after all, looks quite different from that of a mammal, or even an amphibian. Far from invalidating the zebrafish as a model for studying craniofacial development, uncovering the biology that underlies the differences in facial structure between fish and tetrapods will only enrich our understanding of the developmental basis for the breathtaking diversity of vertebrate skulls.

REFERENCES

Alexander, C., Zuniga, E., Blitz, I. L., Wada, N., Le Pabic, P., Javidan, Y., et al. (2011). Combinatorial roles for bmps and endothelin 1 in patterning the dorsal-ventral axis of the craniofacial skeleton. *Development*, *138*, 5135–5146.

Amsterdam, A., Burgess, S., Golling, G., Chen, W., Sun, Z., Townsend, K., et al. (1999). A large-scale insertional mutagenesis screen in zebrafish. *Genes and Development*, *13*, 2713–2724.

Anastasaki, C., Rauen, K. A., & Patton, E. E. (2012). Continual low-level MEK inhibition ameliorates cardio-facio-cutaneous phenotypes in zebrafish. *Disease Models & Mechanisms, 5*, 546–552.

Andreeva, V., Connolly, M. H., Stewart-Swift, C., Fraher, D., Burt, J., Cardarelli, J., et al. (2011). Identification of adult mineralized tissue zebrafish mutants. *Genesis, 49*, 360–366.

Anthwal, N., Joshi, L., & Tucker, A. S. (2012). Evolution of the mammalian middle ear and jaw: Adaptations and novel structures. *Journal of Anatomy, 222*, 147–160.

Apschner, A., Huitema, L. F., Ponsioen, B., Peterson-Maduro, J., & Schulte-Merker, S. (2014). Zebrafish enpp1 mutants exhibit pathological mineralization, mimicking features of generalized arterial calcification of infancy (gaci) and pseudoxanthoma elasticum (pxe). *Disease Models & Mechanisms*, 7, 811–822.

Arduini, B. L., Bosse, K. M., & Henion, P. D. (2009). Genetic ablation of neural crest cell diversification. *Development, 136*, 1987–1994.

Auer, T. O., Duroure, K., Concordet, J. P., & Del Bene, F. (2014). Crispr/cas9-mediated conversion of egfp- into gal4-transgenic lines in zebrafish. *Nature Protocols, 9*, 2823–2840.

Auer, T. O., Duroure, K., De Cian, A., Concordet, J. P., & Del Bene, F. (2014). Highly efficient crispr/cas9-mediated knock-in in zebrafish by homology-independent DNA repair. *Genome Research, 24*, 142–153.

Baas, D., Malbouyres, M., Haftek-Terreau, Z., Le Guellec, D., & Ruggiero, F. (2009). Craniofacial cartilage morphogenesis requires zebrafish col11a1 activity. *Matrix Biology, 28*, 490–502.

Balczerski, B., Matsutani, M., Castillo, P., Osborne, N., Stainier, D. Y., & Crump, J. G. (2012). Analysis of sphingosine-1-phosphate signaling mutants reveals endodermal requirements for the growth but not dorsoventral patterning of jaw skeletal precursors. *Developmental Biology, 362*, 230–241.

Barrallo-Gimeno, A., Holzschuh, J., Driever, W., & Knapik, E. W. (2004). Neural crest survival and differentiation in zebrafish depends on mont blanc/tfap2a gene function. *Development, 131*, 1463–1477.

Bedell, V. M., Wang, Y., Campbell, J. M., Poshusta, T. L., Starker, C. G., Krug Ii, R. G., et al. (2012). In vivo genome editing using a high-efficiency talen system. *Nature, 491*, 114–118.

Bhattacharya, P., Yan, Y. L., Postlethwait, J., & Rubin, D. A. (2011). Evolution of the vertebrate pth2 (tip39) gene family and the regulation of pth type 2 receptor (pth2r) and its endogenous ligand pth2 by hedgehog signaling in zebrafish development. *Journal of Endocrinology, 211*, 187–200.

Bicknell, L. S., Walker, S., Klingseisen, A., Stiff, T., Leitch, A., Kerzendorfer, C., et al. (2011). Mutations in ORC1, encoding the largest subunit of the origin recognition complex, cause microcephalic primordial dwarfism resembling Meier-Gorlin syndrome. *Nature Genetics, 43*, 350–355.

Bohnsack, B. L., Kasprick, D. S., Kish, P. E., Goldman, D., & Kahana, A. (2012). A zebrafish model of axenfeld-rieger syndrome reveals that pitx2 regulation by retinoic acid is essential for ocular and craniofacial development. *Investigative Ophthalmology & Visual Science, 53*, 7–22.

Bonetti, M., Rodriguez-Martinez, V., Paardekooper Overman, J., Overvoorde, J., van Eekelen, M., Jopling, C., et al. (2014). Distinct and overlapping functions of ptpn11 genes in Zebrafish development. *PloS One, 9*, e94884.

Braasch, I., & Postlethwait, J. (2012). Polyploidy in fish and the teleost genome duplication. In P. S. Soltis & D. E. Soltis (Eds.), *Polyploidy and genome evolution* (pp. 341–383). Berlin, Heidelberg: Springer.

Brand, M., Heisenberg, C. P., Warga, R. M., Pelegri, F., Karlstrom, R. O., Beuchle, D., et al. (1996). Mutations affecting development of the midline and general body shape during zebrafish embryogenesis. *Development, 123*, 129–142.

Bush, J. O., & Jiang, R. (2012). Palatogenesis: Morphogenetic and molecular mechanisms of secondary palate development. *Development*, *139*, 231–243.

Cheah, F. S., Winkler, C., Jabs, E. W., & Chong, S. S. (2010). Tgfbeta3 regulation of chondrogenesis and osteogenesis in zebrafish is mediated through formation and survival of a subpopulation of the cranial neural crest. *Mechanisms of Development*, *127*, 329–344.

Choe, C. P., Collazo, A., Trinh le, A., Pan, L., Moens, C. B., & Crump, J. G. (2013). Wnt-dependent epithelial transitions drive pharyngeal pouch formation. *Developmental Cell*, *24*, 296–309.

Choe, C. P., & Crump, J. G. (2014). Tbx1 controls the morphogenesis of pharyngeal pouch epithelia through mesodermal Wnt11r and Fgf8a. *Development*, *141*, 3583–3593.

Choudhry, P., & Trede, N. S. (2013). DiGeorge syndrome gene tbx1 functions through wnt11r to regulate heart looping and differentiation. *PloS One*, *8*, e58145.

Christian, M., Cermak, T., Doyle, E. L., Schmidt, C., Zhang, F., Hummel, A., et al. (2010). Targeting DNA double-strand breaks with TAL effector nucleases. *Genetics*, *186*, 757–761.

Clement, A., Wiweger, M., von der Hardt, S., Rusch, M. A., Selleck, S. B., Chien, C. B., et al. (2008). Regulation of zebrafish skeletogenesis by ext2/dackel and papst1/pinscher. *PLoS Genetics*, *4*, e1000136.

Cong, L., Ran, F. A., Cox, D., Lin, S., Barretto, R., Habib, N., et al. (2013). Multiplex genome engineering using crispr/cas systems. *Science*, *339*, 819–823.

Cox, S. G., Kim, H., Garnett, A. T., Medeiros, D. M., An, W., & Crump, J. G. (2012). An essential role of variant histone h3.3 for ectomesenchyme potential of the cranial neural crest. *PLoS Genetics*, *8*, e1002938.

Crump, J. G., Swartz, M. E., Eberhart, J. K., & Kimmel, C. B. (2006). Moz-dependent hox expression controls segment-specific fate maps of skeletal precursors in the face. *Development*, *133*, 2661–2669.

Crump, J. G., Swartz, M. E., & Kimmel, C. B. (2004). An integrin-dependent role of pouch endoderm in hyoid cartilage development. *PLoS Biology*, *2*, E244.

Cubbage, C. C., & Mabee, P. M. (1996). Development of the cranium and paired fins in the zebrafish Danio rerio (Ostariophysi, Cyprinidae). *Journal of Morphology*, *229*, 121–160.

Dahlem, T. J., Hoshijima, K., Jurynec, M. J., Gunther, D., Starker, C. G., Locke, A. S., et al. (2012). Simple methods for generating and detecting locus-specific mutations induced with TALENs in the zebrafish genome. *PLoS Genetics*, *8*, e1002861.

Dale, R. M., & Topczewski, J. (2011). Identification of an evolutionarily conserved regulatory element of the zebrafish col2a1a gene. *Developmental Biology*, *357*, 518–531.

Danilova, N., Sakamoto, K. M., & Lin, S. (2011). Ribosomal protein L11 mutation in zebrafish leads to haematopoietic and metabolic defects. *British Journal of Haematology*, *152*, 217–228.

Das, A., & Crump, J. G. (2012). Bmps and id2a act upstream of twist1 to restrict ectomesenchyme potential of the cranial neural crest. *PLoS Genetics*, *8*, e1002710.

David, N. B., Saint-Etienne, L., Tsang, M., Schilling, T. F., & Rosa, F. M. (2002). Requirement for endoderm and fgf3 in ventral head skeleton formation. *Development*, *129*, 4457–4468.

Dee, C. T., Szymoniuk, C. R., Mills, P. E., & Takahashi, T. (2013). Defective neural crest migration revealed by a Zebrafish model of Alx1-related frontonasal dysplasia. *Human Molecular Genetics*, *22*, 239–251.

DeLaurier, A., Eames, B. F., Blanco-Sanchez, B., Peng, G., He, X., Swartz, M. E., et al. (2010). Zebrafish sp7:Egfp: A transgenic for studying otic vesicle formation, skeletogenesis, and bone regeneration. *Genesis*, *48*, 505–511.

Depew, M. J., Lufkin, T., & Rubenstein, J. L. (2002). Specification of jaw subdivisions by dlx genes. *Science*, *298*, 381–385.

Dougherty, M., Kamel, G., Grimaldi, M., Gfrerer, L., Shubinets, V., Ethier, R., et al. (2013). Distinct requirements for wnt9a and irf6 in extension and integration mechanisms during zebrafish palate morphogenesis. *Development*, *140*, 76–81.

Dougherty, M., Kamel, G., Shubinets, V., Hickey, G., Grimaldi, M., & Liao, E. C. (2012). Embryonic fate map of first pharyngeal arch structures in the sox10: Kaede zebrafish transgenic model. *The Journal of Craniofacial Surgery*, *23*, 1333–1337.

Doyle, A. J., Doyle, J. J., Bessling, S. L., Maragh, S., Lindsay, M. E., Schepers, D., et al. (2012). Mutations in the TGF-beta repressor SKI cause Shprintzen-Goldberg syndrome with aortic aneurysm. *Nature Genetics*, *44*, 1249–1254.

Driever, W., Solnica-Krezel, L., Schier, A. F., Neuhauss, S. C., Malicki, J., Stemple, D. L., et al. (1996). A genetic screen for mutations affecting embryogenesis in zebrafish. *Development*, *123*, 37–46.

Eames, B. F., Delaurier, A., Ullmann, B., Huycke, T. R., Nichols, J. T., Dowd, J., et al. (2013). Fishface: Interactive atlas of zebrafish craniofacial development at cellular resolution. *BMC Developmental Biology*, *13*, 23.

Eames, B. F., Yan, Y. L., Swartz, M. E., Levic, D. S., Knapik, E. W., Postlethwait, J. H., et al. (2011). Mutations in fam20b and xylt1 reveal that cartilage matrix controls timing of endochondral ossification by inhibiting chondrocyte maturation. *PLoS Genetics*, *7*, e1002246.

Eberhart, J. K., He, X., Swartz, M. E., Yan, Y. L., Song, H., Boling, T. C., et al. (2008). MicroRNA Mirn140 modulates Pdgf signaling during palatogenesis. *Nature Genetics*, *40*, 290–298.

Eberhart, J. K., Swartz, M. E., Crump, J. G., & Kimmel, C. B. (2006). Early hedgehog signaling from neural to oral epithelium organizes anterior craniofacial development. *Development*, *133*, 1069–1077.

Evans, D. J., & Noden, D. M. (2006). Spatial relations between avian craniofacial neural crest and paraxial mesoderm cells. *Developmental Dynamics*, *235*, 1310–1325.

Favaro, F. P., Alvizi, L., Zechi-Ceide, R. M., Bertola, D., Felix, T. M., de Souza, J., et al. (2014). A noncoding expansion in EIF4A3 causes Richieri-Costa-Pereira syndrome, a craniofacial disorder associated with limb defects. *American Journal of Human Genetics*, *94*, 120–128.

Ferrante, M. I., Romio, L., Castro, S., Collins, J. E., Goulding, D. A., Stemple, D. L., et al. (2009). Convergent extension movements and ciliary function are mediated by ofd1, a zebrafish orthologue of the human oral-facial-digital type 1 syndrome gene. *Human Molecular Genetics*, *18*, 289–303.

Flores, M. V., Lam, E. Y., Crosier, P., & Crosier, K. (2006). A hierarchy of Runx transcription factors modulate the onset of chondrogenesis in craniofacial endochondral bones in zebrafish. *Developmental Dynamics*, *235*, 3166–3176.

Garbes, L., Kim, K., Riess, A., Hoyer-Kuhn, H., Beleggia, F., Bevot, A., et al. (2015). Mutations in SEC24D, encoding a component of the COPII machinery, cause a syndromic form of osteogenesis imperfecta. *American Journal of Human Genetics*, *96*, 432–439.

Geng, X., Speirs, C., Lagutin, O., Inbal, A., Liu, W., Solnica-Krezel, L., et al. (2008). Haploinsufficiency of Six3 fails to activate Sonic hedgehog expression in the ventral forebrain and causes holoprosencephaly. *Developmental Cell*, *15*, 236–247.

Gestri, G., Osborne, R. J., Wyatt, A. W., Gerrelli, D., Gribble, S., Stewart, H., et al. (2009). Reduced TFAP2A function causes variable optic fissure closure and retinal defects and sensitizes eye development to mutations in other morphogenetic regulators. *Human Genetics*, *126*, 791–803.

Geurtzen, K., Knopf, F., Wehner, D., Huitema, L. F., Schulte-Merker, S., & Weidinger, G. (2014). Mature osteoblasts dedifferentiate in response to traumatic bone injury in the zebrafish fin and skull. *Development*, *141*, 2225–2234.

Gfrerer, L., Shubinets, V., Hoyos, T., Kong, Y., Nguyen, C., Pietschmann, P., et al. (2014). Functional analysis of SPECC1L in craniofacial development and oblique facial cleft pathogenesis. *Plastic and Reconstructive Surgery*, *134*, 748–759.

Ghassibe-Sabbagh, M., Desmyter, L., Langenberg, T., Claes, F., Boute, O., Bayet, B., et al. (2011). FAF1, a gene that is disrupted in cleft palate and has conserved function in zebrafish. *American Journal of Human Genetics*, *88*, 150–161.

Glasauer, S. M., & Neuhauss, S. C. (2014). Whole-genome duplication in teleost fishes and its evolutionary consequences. *Molecular Genetics & Genomics*, *289*, 1045–1060.

Goll, M. G., Anderson, R., Stainier, D. Y., Spradling, A. C., & Halpern, M. E. (2009). Transcriptional silencing and reactivation in transgenic zebrafish. *Genetics*, *182*, 747–755.

Gordon, C. T., Attanasio, C., Bhatia, S., Benko, S., Ansari, M., Tan, T. Y., et al. (2014). Identification of novel craniofacial regulatory domains located far upstream of SOX9 and disrupted in Pierre Robin sequence. *Human Mutation*, *35*, 1011–1020.

Gould, S. J. (1990). An earful of jaw. *Natural History*, *3*, 12–23.

Grabher, C., Joly, J. S., & Wittbrodt, J. (2004). Highly efficient zebrafish transgenesis mediated by the meganuclease I-SceI. *Methods in Cell Biology*, 77, 381–401.

Grova, M., Lo, D. D., Montoro, D., Hyun, J. S., Chung, M. T., Wan, D. C., et al. (2012). Models of cranial suture biology. *The Journal of Craniofacial Surgery*, *23*, 1954–1958.

Grunwald, D. J., & Streisinger, G. (1992). Induction of recessive lethal and specific locus mutations in the zebrafish with ethyl nitrosourea. *Genetical Research*, *59*, 103–116.

Halbritter, J., Bizet, A. A., Schmidts, M., Porath, J. D., Braun, D. A., Gee, H. Y., et al. (2013). Defects in the IFT-B component IFT172 cause Jeune and Mainzer-Saldino syndromes in humans. *American Journal of Human Genetics*, *93*, 915–925.

Hammond, C. L., & Schulte-Merker, S. (2009). Two populations of endochondral osteoblasts with differential sensitivity to hedgehog signalling. *Development*, *136*, 3991–4000.

Harris, M. P., Rohner, N., Schwarz, H., Perathoner, S., Konstantinidis, P., & Nusslein-Volhard, C. (2008). Zebrafish eda and edar mutants reveal conserved and ancestral roles of ectodysplasin signaling in vertebrates. *PLoS Genetics*, *4*, e1000206.

Holmes, G. (2012). The role of vertebrate models in understanding craniosynostosis. *Childs Nervous System*, *28*, 1471–1481.

Huang, L., Xiao, A., Choi, S. Y., Kan, Q., Zhou, W., Chacon-Heszele, M. F., et al. (2014). Wnt5a is necessary for normal kidney development in zebrafish and mice. *Nephron. Experimental Nephrology*, *128*, 80–88.

Huitema, L. F., Apschner, A., Logister, I., Spoorendonk, K. M., Bussmann, J., Hammond, C. L., et al. (2012). Entpd5 is essential for skeletal mineralization and regulates phosphate homeostasis in zebrafish. *Proceedings of the National Academy of Sciences of the United States of America*, *109*, 21372–21377.

Huycke, T. R., Eames, B. F., & Kimmel, C. B. (2012). Hedgehog-dependent proliferation drives modular growth during morphogenesis of a dermal bone. *Development*, *139*, 2371–2380.

Jekosch, K. (2004). The zebrafish genome project: Sequence analysis and annotation. *Methods in Cell Biology*, 77, 225–239.

Kague, E., Gallagher, M., Burke, S., Parsons, M., Franz-Odendaal, T., & Fisher, S. (2012). Skeletogenic fate of zebrafish cranial and trunk neural crest. *PloS One*, 7, e47394.

Karlstrom, R. O., Talbot, W. S., & Schier, A. F. (1999). Comparative synteny cloning of zebrafish you-too: Mutations in the Hedgehog target gli2 affect ventral forebrain patterning. *Genes and Development*, *13*, 388–393.

Karlstrom, R. O., Tyurina, O. V., Kawakami, A., Nishioka, N., Talbot, W. S., Sasaki, H., et al. (2003). Genetic analysis of zebrafish gli1 and gli2 reveals divergent requirements for gli genes in vertebrate development. *Development*, *130*, 1549–1564.

Kawakami, K. (2007). Tol2: A versatile gene transfer vector in vertebrates. *Genome Biology*, *8*(Suppl. 1), S7.

Kere, J., Srivastava, A. K., Montonen, O., Zonana, J., Thomas, N., Ferguson, B., et al. (1996). X-linked anhidrotic (hypohidrotic) ectodermal dysplasia is caused by mutation in a novel transmembrane protein. *Nature Genetics*, *13*, 409–416.

Kettleborough, R. N., Busch-Nentwich, E. M., Harvey, S. A., Dooley, C. M., de Bruijn, E., van Eeden, F., et al. (2013). A systematic genome-wide analysis of zebrafish protein-coding gene function. *Nature*, *496*, 494–497.

Keupp, K., Li, Y., Vargel, I., Hoischen, A., Richardson, R., Neveling, K., et al. (2013). Mutations in the interleukin receptor IL11RA cause autosomal recessive Crouzon-like craniosynostosis. *Molecular Genetics & Genomic Medicine*, *1*, 223–237.

Kimmel, C. B. (1989). Genetics and early development of zebrafish. *Trends in Genetics*, *5*, 283–288.

Kimmel, C. B., DeLaurier, A., Ullmann, B., Dowd, J., & McFadden, M. (2010). Modes of developmental outgrowth and shaping of a craniofacial bone in zebrafish. *PloS One*, *5*, e9475.

Kimmel, C. B., Miller, C. T., Kruze, G., Ullmann, B., BreMiller, R. A., Larison, K. D., et al. (1998). The shaping of pharyngeal cartilages during early development of the zebrafish. *Developmental Biology*, *203*, 245–263.

Kimmel, C. B., Miller, C. T., & Moens, C. B. (2001). Specification and morphogenesis of the zebrafish larval head skeleton. *Developmental Biology*, *233*, 239–257.

Knight, R. D., Nair, S., Nelson, S. S., Afshar, A., Javidan, Y., Geisler, R., et al. (2003). Lockjaw encodes a zebrafish tfap2a required for early neural crest development. *Development*, *130*, 5755–5768.

Knopf, F., Hammond, C., Chekuru, A., Kurth, T., Hans, S., Weber, C. W., et al. (2011). Bone regenerates via dedifferentiation of osteoblasts in the zebrafish fin. *Developmental Cell*, *20*, 713–724.

Kok, F. O., Shin, M., Ni, C. W., Gupta, A., Grosse, A. S., van Impel, A., et al. (2015). Reverse genetic screening reveals poor correlation between morpholino-induced and mutant phenotypes in zebrafish. *Developmental Cell*, *32*, 97–108.

Kurihara, Y., Kurihara, H., Suzuki, H., Kodama, T., Maemura, K., Nagai, R., et al. (1994). Elevated blood pressure and craniofacial abnormalities in mice deficient in endothelin-1. *Nature*, *368*, 703–710.

Lang, M. R., Lapierre, L. A., Frotscher, M., Goldenring, J. R., & Knapik, E. W. (2006). Secretory COPII coat component Sec23a is essential for craniofacial chondrocyte maturation. *Nature Genetics*, *38*, 1198–1203.

Laue, K., Pogoda, H. M., Daniel, P. B., van Haeringen, A., Alanay, Y., von Ameln, S., et al. (2011). Craniosynostosis and multiple skeletal anomalies in humans and zebrafish result from a defect in the localized degradation of retinoic acid. *American Journal of Human Genetics*, *89*, 595–606.

Law, S. H., & Sargent, T. D. (2014). The serine-threonine protein kinase pak4 is dispensable in zebrafish: Identification of a morpholino-generated pseudophenotype. *PloS One*, *9*, e100268.

Lawson, N. D., & Weinstein, B. M. (2002). In vivo imaging of embryonic vascular development using transgenic zebrafish. *Developmental Biology*, *248*, 307–318.

Le Pabic, P., Ng, C., & Schilling, T. F. (2014). Fat-Dachsous signaling coordinates cartilage differentiation and polarity during craniofacial development. *PLoS Genetics*, *10*, e1004726.

Lee, Y., Grill, S., Sanchez, A., Murphy-Ryan, M., & Poss, K. D. (2005). Fgf signaling instructs position-dependent growth rate during zebrafish fin regeneration. *Development*, *132*, 5173–5183.

Liem, K. F., & Walker, W. F. (2001). *Functional anatomy of the vertebrates: An evolutionary perspective* (3rd ed.). Fort Worth: Harcourt College Publishers.

Liu, Y., & Semina, E. V. (2012). pitx2 Deficiency results in abnormal ocular and craniofacial development in zebrafish. *PloS One*, *7*, e30896.

McGurk, P. D., Lovely, C. B., & Eberhart, J. K. (2014). Analyzing craniofacial morphogenesis in zebrafish using 4d confocal microscopy. *Journal of Visualized Experiments*, (83), e51190.
Medeiros, D. M., & Crump, J. G. (2012). New perspectives on pharyngeal dorsoventral patterning in development and evolution of the vertebrate jaw. *Developmental Biology*, *371*, 121–135.
Meindl, R. S., & Lovejoy, C. O. (1985). Ectocranial suture closure: A revised method for the determination of skeletal age at death based on the lateral-anterior sutures. *American Journal of Physical Anthropology*, *68*, 57–66.
Miller, C. T., & Kimmel, C. B. (2001). Morpholino phenocopies of endothelin 1 (sucker) and other anterior arch class mutations. *Genesis*, *30*, 186–187.
Miller, C. T., Maves, L., & Kimmel, C. B. (2004). Moz regulates hox expression and pharyngeal segmental identity in zebrafish. *Development*, *131*, 2443–2461.
Miller, C. T., Schilling, T. F., Lee, K., Parker, J., & Kimmel, C. B. (2000). Sucker encodes a zebrafish endothelin-1 required for ventral pharyngeal arch development. *Development*, *127*, 3815–3828.
Miller, C. T., Swartz, M. E., Khuu, P. A., Walker, M. B., Eberhart, J. K., & Kimmel, C. B. (2007). Mef2ca is required in cranial neural crest to effect endothelin1 signaling in zebrafish. *Developmental Biology*, *308*, 144–157.
Molina, G. A., Watkins, S. C., & Tsang, M. (2007). Generation of FGF reporter transgenic zebrafish and their utility in chemical screens. *BMC Developmental Biology*, *7*, 62.
Mongera, A., Singh, A. P., Levesque, M. P., Chen, Y. Y., Konstantinidis, P., & Nusslein-Volhard, C. (2013). Genetic lineage labeling in zebrafish uncovers novel neural crest contributions to the head, including gill pillar cells. *Development*, *140*, 916–925.
Monnich, M., Kuriger, Z., Print, C. G., & Horsfield, J. A. (2011). A zebrafish model of Roberts syndrome reveals that Esco2 depletion interferes with development by disrupting the cell cycle. *PloS One*, *6*, e20051.
Monreal, A. W., Ferguson, B. M., Headon, D. J., Street, S. L., Overbeek, P. A., & Zonana, J. (1999). Mutations in the human homologue of mouse dl cause autosomal recessive and dominant hypohidrotic ectodermal dysplasia. *Nature Genetics*, *22*, 366–369.
Moro, E., Ozhan-Kizil, G., Mongera, A., Beis, D., Wierzbicki, C., Young, R. M., et al. (2012). In vivo Wnt signaling tracing through a transgenic biosensor fish reveals novel activity domains. *Developmental Biology*, *366*, 327–340.
Mosimann, C., & Zon, L. I. (2011). Advanced zebrafish transgenesis with Tol2 and application for Cre/lox recombination experiments. *Methods in Cell Biology*, *104*, 173–194.
Muto, A., Calof, A. L., Lander, A. D., & Schilling, T. F. (2011). Multifactorial origins of heart and gut defects in nipbl-deficient zebrafish, a model of Cornelia de Lange Syndrome. *PLoS Biology*, *9*, e1001181.
Nakada, C., Iida, A., Tabata, Y., & Watanabe, S. (2009). Forkhead transcription factor foxe1 regulates chondrogenesis in zebrafish. *Journal of Experimental Zoology. Part B, Molecular and Developmental Evolution*, *312*, 827–840.
Nakamura, Y., Yamamoto, K., He, X., Otsuki, B., Kim, Y., Murao, H., et al. (2011). Wwp2 is essential for palatogenesis mediated by the interaction between Sox9 and mediator subunit 25. *Nature Communications*, *2*, 251.
Nakano, Y., Kim, H. R., Kawakami, A., Roy, S., Schier, A. F., & Ingham, P. W. (2004). Inactivation of dispatched 1 by the chameleon mutation disrupts Hedgehog signalling in the zebrafish embryo. *Developmental Biology*, *269*, 381–392.
Nasevicius, A., & Ekker, S. C. (2000). Effective targeted gene 'knockdown' in zebrafish. *Nature Genetics*, *26*, 216–220.
Neuhauss, S. C., Solnica-Krezel, L., Schier, A. F., Zwartkruis, F., Stemple, D. L., Malicki, J., et al. (1996). Mutations affecting craniofacial development in zebrafish. *Development*, *123*, 357–367.

Ng, D., Thakker, N., Corcoran, C. M., Donnai, D., Perveen, R., Schneider, A., et al. (2004). Oculofaciocardiodental and Lenz microphthalmia syndromes result from distinct classes of mutations in BCOR. *Nature Genetics, 36*, 411–416.

Nichols, J. T., Pan, L., Moens, C. B., & Kimmel, C. B. (2013). Barx1 represses joints and promotes cartilage in the craniofacial skeleton. *Development, 140*, 2765–2775.

Nissen, R. M., Amsterdam, A., & Hopkins, N. (2006). A zebrafish screen for craniofacial mutants identifies wdr68 as a highly conserved gene required for endothelin-1 expression. *BMC Developmental Biology, 6*, 28.

Noden, D. M. (1984). Craniofacial development: New views on old problems. *The Anatomical Record, 208*, 1–13.

Opperman, L. A. (2000). Cranial sutures as intramembranous bone growth sites. *Developmental Dynamics, 219*, 472–485.

Park, K., & Choung, Y. H. (2009). Isolated congenital ossicular anomalies. *Acta Oto-Laryngologica, 129*, 419–422.

Park, K., Choung, Y. H., Shin, Y. R., & Hong, S. P. (2007). Conductive deafness with normal eardrum: Absence of the long process of the incus. *Acta Oto-Laryngologica, 127*, 816–820.

Parsons, M. J., Pisharath, H., Yusuff, S., Moore, J. C., Siekmann, A. F., Lawson, N., et al. (2009). Notch-responsive cells initiate the secondary transition in larval zebrafish pancreas. *Mechanisms of Development, 126*, 898–912.

Patten, S. A., Jacobs-McDaniels, N. L., Zaouter, C., Drapeau, P., Albertson, R. C., & Moldovan, F. (2012). Role of Chd7 in zebrafish: A model for CHARGE syndrome. *PloS One, 7*, e31650.

Piotrowski, T., Ahn, D. G., Schilling, T. F., Nair, S., Ruvinsky, I., Geisler, R., et al. (2003). The zebrafish van gogh mutation disrupts tbx1, which is involved in the DiGeorge deletion syndrome in humans. *Development, 130*, 5043–5052.

Piotrowski, T., Schilling, T. F., Brand, M., Jiang, Y. J., Heisenberg, C. P., Beuchle, D., et al. (1996). Jaw and branchial arch mutants in zebrafish II: Anterior arches and cartilage differentiation. *Development, 123*, 345–356.

Porteus, M. H., & Carroll, D. (2005). Gene targeting using zinc finger nucleases. *Nature Biotechnology, 23*, 967–973.

Poss, K. D., Keating, M. T., & Nechiporuk, A. (2003). Tales of regeneration in zebrafish. *Developmental Dynamics, 226*, 202–210.

Postlethwait, J., Amores, A., Cresko, W., Singer, A., & Yan, Y. L. (2004). Subfunction partitioning, the teleost radiation and the annotation of the human genome. *Trends in Genetics, 20*, 481–490.

Postlethwait, J. H., Woods, I. G., Ngo-Hazelett, P., Yan, Y. L., Kelly, P. D., Chu, F., et al. (2000). Zebrafish comparative genomics and the origins of vertebrate chromosomes. *Genome Research, 10*, 1890–1902.

Qi, H. H., Sarkissian, M., Hu, G. Q., Wang, Z., Bhattacharjee, A., Gordon, D. B., et al. (2010). Histone H4K20/H3K9 demethylase PHF8 regulates zebrafish brain and craniofacial development. *Nature, 466*, 503–507.

Quarto, N., & Longaker, M. T. (2005). The zebrafish (Danio rerio): A model system for cranial suture patterning. *Cells, Tissues, Organs, 181*, 109–118.

Rahbar, R., Neault, M. W., & Kenna, M. A. (2002). Congenital absence of the incus bilaterally without other otologic anomalies: A new case report. *Ear, Nose, and Throat Journal, 81*(274–276), 278.

Rainger, J. K., Bhatia, S., Bengani, H., Gautier, P., Rainger, J., Pearson, M., et al. (2014). Disruption of SATB2 or its long-range cis-regulation by SOX9 causes a syndromic form of Pierre Robin sequence. *Human Molecular Genetics, 23*, 2569–2579.

Razzaque, M. A., Komoike, Y., Nishizawa, T., Inai, K., Furutani, M., Higashinakagawa, T., et al. (2012). Characterization of a novel KRAS mutation identified in Noonan syndrome. *American Journal of Medical Genetics. Part A, 158A*, 524–532.

Razzaque, M. A., Nishizawa, T., Komoike, Y., Yagi, H., Furutani, M., Amo, R., et al. (2007). Germline gain-of-function mutations in RAF1 cause Noonan syndrome. *Nature Genetics*, *39*, 1013–1017.

Robu, M. E., Larson, J. D., Nasevicius, A., Beiraghi, S., Brenner, C., Farber, S. A., et al. (2007). P53 activation by knockdown technologies. *PLoS Genetics*, *3*, e78.

Saadi, I., Alkuraya, F. S., Gisselbrecht, S. S., Goessling, W., Cavallesco, R., Turbe-Doan, A., et al. (2011). Deficiency of the cytoskeletal protein SPECC1L leads to oblique facial clefting. *American Journal of Human Genetics*, *89*, 44–55.

Sarmah, S., Barrallo-Gimeno, A., Melville, D. B., Topczewski, J., Solnica-Krezel, L., & Knapik, E. W. (2010). Sec24D-dependent transport of extracellular matrix proteins is required for zebrafish skeletal morphogenesis. *PloS One*, *5*, e10367.

Sawada, R., Kamei, H., Hakuno, F., Takahashi, S., & Shimizu, T. (2015). In vivo loss of function study reveals the short stature homeobox-containing (shox) gene plays indispensable roles in early embryonic growth and bone formation in zebrafish. *Developmental Dynamics*, *244*, 146–156.

Schauerte, H. E., van Eeden, F. J., Fricke, C., Odenthal, J., Strahle, U., & Haffter, P. (1998). Sonic hedgehog is not required for the induction of medial floor plate cells in the zebrafish. *Development*, *125*, 2983–2993.

Schilling, T. F., & Kimmel, C. B. (1997). Musculoskeletal patterning in the pharyngeal segments of the zebrafish embryo. *Development*, *124*, 2945–2960.

Schilling, T. F., Piotrowski, T., Grandel, H., Brand, M., Heisenberg, C. P., Jiang, Y. J., et al. (1996). Jaw and branchial arch mutants in zebrafish I: Branchial arches. *Development*, *123*, 329–344.

Schulte-Merker, S., & Stainier, D. Y. (2014). Out with the old, in with the new: Reassessing morpholino knockdowns in light of genome editing technology. *Development*, *141*, 3103–3104.

Schwend, T., & Ahlgren, S. C. (2009). Zebrafish con/disp1 reveals multiple spatiotemporal requirements for Hedgehog-signaling in craniofacial development. *BMC Developmental Biology*, *9*, 59.

Schwend, T., Loucks, E. J., & Ahlgren, S. C. (2010). Visualization of Gli activity in craniofacial tissues of hedgehog-pathway reporter transgenic zebrafish. *PloS One*, *5*, e14396.

Song, B., Zhang, Q., Zhang, Z., Wan, Y., Jia, Q., Wang, X., et al. (2014). Systematic transcriptome analysis of the zebrafish model of diamond-blackfan anemia induced by RPS24 deficiency. *BMC Genomics*, *15*, 759.

Sousa, S., Afonso, N., Bensimon-Brito, A., Fonseca, M., Simoes, M., Leon, J., et al. (2011). Differentiated skeletal cells contribute to blastema formation during zebrafish fin regeneration. *Development*, *138*, 3897–3905.

Subedi, A., Macurak, M., Gee, S. T., Monge, E., Goll, M. G., Potter, C. J., et al. (2014). Adoption of the Q transcriptional regulatory system for zebrafish transgenesis. *Methods*, *66*, 433–440.

Swartz, M. E., Sheehan-Rooney, K., Dixon, M. J., & Eberhart, J. K. (2011). Examination of a palatogenic gene program in zebrafish. *Developmental Dynamics*, *240*, 2204–2220.

Talbot, J. C., Johnson, S. L., & Kimmel, C. B. (2010). Hand2 and dlx genes specify dorsal, intermediate and ventral domains within zebrafish pharyngeal arches. *Development*, *137*, 2507–2517.

van Impel, A., Zhao, Z., Hermkens, D. M., Roukens, M. G., Fischer, J. C., Peterson-Maduro, J., et al. (2014). Divergence of zebrafish and mouse lymphatic cell fate specification pathways. *Development*, *141*, 1228–1238.

Varga, Z. M., Amores, A., Lewis, K. E., Yan, Y. L., Postlethwait, J. H., Eisen, J. S., et al. (2001). Zebrafish smoothened functions in ventral neural tube specification and axon tract formation. *Development*, *128*, 3497–3509.

Wada, N., Javidan, Y., Nelson, S., Carney, T. J., Kelsh, R. N., & Schilling, T. F. (2005). Hedgehog signaling is required for cranial neural crest morphogenesis and chondrogenesis at the midline in the zebrafish skull. *Development, 132*, 3977–3988.

Walker, M. B., Miller, C. T., Coffin Talbot, J., Stock, D. W., & Kimmel, C. B. (2006). Zebrafish furin mutants reveal intricacies in regulating Endothelin1 signaling in craniofacial patterning. *Developmental Biology, 295*, 194–205.

Walker, M. B., Miller, C. T., Swartz, M. E., Eberhart, J. K., & Kimmel, C. B. (2007). Phospholipase c, beta 3 is required for Endothelin1 regulation of pharyngeal arch patterning in zebrafish. *Developmental Biology, 304*, 194–207.

Wang, X., He, H., Tang, W., Zhang, X. A., Hua, X., & Yan, J. (2012). Two origins of blastemal progenitors define blastemal regeneration of zebrafish lower jaw. *PloS One*, 7, e45380.

Weiner, A. M., Scampoli, N. L., & Calcaterra, N. B. (2012). Fishing the molecular bases of Treacher Collins syndrome. *PloS One*, 7, e29574.

Wienholds, E., van Eeden, F., Kosters, M., Mudde, J., Plasterk, R. H., & Cuppen, E. (2003). Efficient target-selected mutagenesis in zebrafish. *Genome Research, 13*, 2700–2707.

Witten, P. E., Hansen, A., & Hall, B. K. (2001). Features of mono- and multinucleated bone resorbing cells of the zebrafish Danio rerio and their contribution to skeletal development, remodeling, and growth. *Journal of Morphology, 250*, 197–207.

Wiweger, M. I., Avramut, C. M., de Andrea, C. E., Prins, F. A., Koster, A. J., Ravelli, R. B., et al. (2011). Cartilage ultrastructure in proteoglycan-deficient zebrafish mutants brings to light new candidate genes for human skeletal disorders. *The Journal of Pathology, 223*, 531–542.

Woods, I. G., Wilson, C., Friedlander, B., Chang, P., Reyes, D. K., Nix, R., et al. (2005). The zebrafish gene map defines ancestral vertebrate chromosomes. *Genome Research, 15*, 1307–1314.

Yan, Y. L., Bhattacharya, P., He, X. J., Ponugoti, B., Marquardt, B., Layman, J., et al. (2012). Duplicated zebrafish co-orthologs of parathyroid hormone-related peptide (PTHrP, Pthlh) play different roles in craniofacial skeletogenesis. *Journal of Endocrinology, 214*, 421–435.

Yan, Y. L., Miller, C. T., Nissen, R. M., Singer, A., Liu, D., Kirn, A., et al. (2002). A zebrafish sox9 gene required for cartilage morphogenesis. *Development, 129*, 5065–5079.

Yan, Y. L., Willoughby, J., Liu, D., Crump, J. G., Wilson, C., Miller, C. T., et al. (2005). A pair of Sox: Distinct and overlapping functions of zebrafish sox9 co-orthologs in craniofacial and pectoral fin development. *Development, 132*, 1069–1083.

Yelick, P. C., & Schilling, T. F. (2002). Molecular dissection of craniofacial development using zebrafish. *Critical Reviews in Oral Biology and Medicine, 13*, 308–322.

Yuan, Q., Chiquet, B. T., Devault, L., Warman, M. L., Nakamura, Y., Swindell, E. C., et al. (2012). Craniofacial abnormalities result from knock down of nonsyndromic clefting gene, crispld2, in zebrafish. *Genesis, 50*, 871–881.

Zecchin, E., Conigliaro, A., Tiso, N., Argenton, F., & Bortolussi, M. (2005). Expression analysis of jagged genes in zebrafish embryos. *Developmental Dynamics, 233*, 638–645.

Zhang, Y., Ear, J., Yang, Z., Morimoto, K., Zhang, B., & Lin, S. (2014). Defects of protein production in erythroid cells revealed in a zebrafish Diamond-Blackfan anemia model for mutation in RPS19. *Cell Death & Disease, 5*, e1352.

Zhang, L. F., Gui, Y. H., Wang, Y. X., Jiang, Q., & Song, H. Y. (2010). Effect of Tbx1 knock-down on cardiac performance in zebrafish. *Chinese Medical Journal, 123*, 1182–1189.

Zhang, H., Wang, X., Lyu, K., Gao, S., Wang, G., Fan, C., et al. (2015). Time point-based integrative analyses of deep-transcriptome identify four signal pathways in blastemal regeneration of zebrafish lower jaw. *Stem Cells, 33*, 806–818.

Zu, Y., Tong, X., Wang, Z., Liu, D., Pan, R., Li, Z., et al. (2013). Talen-mediated precise genome modification by homologous recombination in zebrafish. *Nature Methods*, *10*, 329–331.

Zuniga, E., Rippen, M., Alexander, C., Schilling, T. F., & Crump, J. G. (2011). Gremlin 2 regulates distinct roles of bmp and endothelin 1 signaling in dorsoventral patterning of the facial skeleton. *Development*, *138*, 5147–5156.

Zuniga, E., Stellabotte, F., & Crump, J. G. (2010). Jagged-Notch signaling ensures dorsal skeletal identity in the vertebrate face. *Development*, *137*, 1843–1852.

CHAPTER ELEVEN

Regulation of Jaw Length During Development, Disease, and Evolution

Richard A. Schneider[1]
Department of Orthopaedic Surgery, University of California at San Francisco, San Francisco, California, USA
[1]Corresponding author: e-mail address: rich.schneider@ucsf.edu

Contents

Abstract

Molecular and cellular mechanisms that control jaw length are becoming better understood. This is significant since the jaws are not only critical for species-specific adaptation and survival, but they are often affected by a variety of size-related anomalies including mandibular hypoplasia, retrognathia, asymmetry, and clefting. This chapter overviews how jaw length is established during the allocation, proliferation, differentiation, and growth of jaw precursor cells, which originate from neural crest mesenchyme (NCM). The focus is mainly on results from experiments transplanting NCM between quail and duck embryos. Quail have short jaws whereas those of duck are relatively long. Quail–duck chimeras reveal that the determinants of jaw length are NCM mediated throughout development and include species-specific differences in jaw progenitor number, differential regulation of various signaling pathways, and the autonomous activation of programs for skeletal matrix deposition and resorption. Such insights help make the goal of devising new therapies for birth defects, diseases, and injuries to the jaw skeleton seem ever more likely.

"The most obvious differences between different animals are differences of size, but for some reason the zoologists have paid singularly little attention to them....For every type of animal there is a most convenient size, and a large change in size inevitably carries with it a change of form."

Haldane (1926)

Current Topics in Developmental Biology, Volume 115
ISSN 0070-2153
http://dx.doi.org/10.1016/bs.ctdb.2015.08.002

1. INTRODUCTION: ON BEING THE RIGHT JAW SIZE DURING DEVELOPMENT AND EVOLUTION

As so eloquently expressed by Haldane in his classic essay, "On Being the Right Size," every animal achieves its own individual size, which is closely tied to form, function, and fitness (Haldane, 1926). In this context, size must be precisely controlled throughout development in order for animals and their constituent parts to attain proper structural integration and adaptation. Nowhere is this notion truer than in the craniofacial complex, where size-related malformations are some of the most common human birth defects (Gorlin, Cohen, & Levin, 1990; Smith, 1997). The jaw skeleton, in particular, displays a range of anomalies in size including hypo- and hyperplasia, pro- and retrognathia, micro- and macrognathia, asymmetry, and clefting. Such variation in jaw length during development can often produce a spectrum of debilitating to life-threatening conditions. Nonetheless, variation in jaw length during evolution has also been essential for the adaptive radiation of vertebrates. Thus, identifying molecular and cellular mechanisms that both control jaw length and generate species-specific variation is critical to understanding disease and evolution.

The jaws are among the most precisely adapted and highly modified structures of vertebrates, and they facilitate complex species-specific behaviors related to feeding, respiration, predation, vocalization, mating, and grooming. Such fundamental connections between jaw size and jaw function provided a foundation for early theories of evolution by natural selection best exemplified by the beaks of Darwin's finches (Darwin, 1859). Since then, a wide array of genetic and embryological studies have shown that the establishment of jaw size is a complex process involving numerous gene regulatory networks, reciprocal signaling interactions, and hierarchical levels of control. Yet what has remained unclear are the particular determinants of jaw size that may play a role during the induction, allocation, proliferation, differentiation, and growth of neural crest mesenchyme (NCM), which serve as the progenitors of the jaw. For example, mandibular hypoplasia and cleft palate may have as part of their etiology disruptions to the rate of proliferation or timing of differentiation in NCM (Dudas et al., 2006; Dudas, Sridurongrit, Nagy, Okazaki, & Kaartinen, 2004; Ito et al., 2003; Oka et al., 2007; Satokata & Maas, 1994; Sharpe & Ferguson, 1988). Identifying molecular and cellular mechanisms through which the jaw skeleton achieves its proper length is crucial for devising new and efficacious treatments that could ultimately prevent birth defects. This lack of knowledge is significant since a major clinical objective is to devise molecular and cell-based strategies to lengthen the jaw in cases of mandibular hypoplasia, asymmetry, or malocclusion.

Jaw length defects can arise from a wide range of genetic or environmental perturbations. For example, disruptions to sonic hedgehog (SHH) pathway members including *Shh*, *Ptch1*, *Gas1*, *Gli2*, *Gli3*, and *Hhat* contribute to micrognathia associated with conditions such as holoprosencephaly (Allen, Tenzen, & McMahon, 2007; Dennis et al., 2012; Hui & Angers, 2011; Melnick, Witcher, Bringas, Carlsson, & Jaskoll, 2005; Mo et al., 1997; Pineda-Alvarez et al., 2012; Roessler & Muenke, 2010). Mutations in fibroblast growth factor (FGF) pathway members including *Fgf8*, *Fgfr1*, *Fgfr2*, and *Fgfr3* also cause jaw length defects, especially in Crouzon and Apert syndromes (Martinez-Abadias, Holmes, et al., 2013; Martinez-Abadias, Motch, et al., 2013; Stanier & Pauws, 2012; Trumpp, Depew, Rubenstein, Bishop, & Martin, 1999). Mutations in bone morphogenetic protein (BMP) pathway members such as *Bmp4*, *Bmp7*, *Chordin*, *Noggin*, *Twist1*, and *Msx2* cause mandibular hypoplasia (Boell et al., 2013; Foerst-Potts & Sadler, 1997; Stottmann, Anderson, & Klingensmith, 2001; Zhang et al., 2012; Zouvelou, Luder, Mitsiadis, & Graf, 2009). Disruptions to the transforming growth factor-beta (TGFβ) signaling pathway also affect jaw length. Mutations in *TGFβ2* and *TGFBR1* cause microretrognathia in Loeys–Dietz Syndrome and when restricted to NCM (Loeys et al., 2005; Sanford et al., 1997; Zhao, Oka, Bringas, Kaartinen, & Chai, 2008). Jaw length defects are also associated with mutations in *TGFBR2* and *Smad2*, and *Matrix metalloproteinase 2 (Mmp2)* in Torg–Winchester Syndrome, *Mmp13* in Spondyloepimetaphyseal dysplasia, and *Osteoprotegerin* (*Opg*) in Juvenile Paget's disease (Gorlin et al., 1990; Nomura & Li, 1998; Oka et al., 2008, 2007).

Moreover, studies in birds have identified some factors that influence jaw size. For example, differential expression of *Bmp4* in jaw progenitor cells influences variation in jaw depth and width among birds including Darwin's finches, chicks, ducks, and cockatiels (Abzhanov, Protas, Grant, Grant, & Tabin, 2004; Wu, Jiang, Shen, Widelitz, & Chuong, 2006; Wu, Jiang, Suksaweang, Widelitz, & Chuong, 2004), whereas jaw length appears to be regulated separately through a calmodulin-dependent pathway (Abzhanov et al., 2006; Schneider, 2007). Likewise, factors such as SHH, FGFs, Wingless-related (WNTs), and BMPs that are secreted from adjacent epithelial tissues have also been implicated in mediating the shape and outgrowth of the jaw and facial skeletons (Abzhanov & Tabin, 2004; Ashique, Fu, & Richman, 2002; Bhullar et al., 2015; Brugmann et al., 2007, 2010; Doufexi & Mina, 2008; Foppiano, Hu, & Marcucio, 2007; Grant, Grant, & Abzhanov, 2006; Havens et al., 2008; Hu & Marcucio, 2009, 2012; Hu, Young, Li, et al., 2015; Hu, Young, Xu, et al., 2015; MacDonald, Abbott, & Richman, 2004; Mina, Wang, Ivanisevic, Upholt, & Rodgers, 2002; Richman, Herbert, Matovinovic, & Walin, 1997; Rowe, Richman, & Brickell, 1992; Schneider, Hu, & Helms, 1999; Schneider, Hu, Rubenstein, Maden, & Helms, 2001; Szabo-Rogers, Geetha-Loganathan, Nimmagadda, Fu, & Richman, 2008; Wu et al., 2006; Young et al., 2014). But precisely, how these pathways are regulated by NCM and how alterations to their regulation affect jaw length still needs to be clarified. Thus, an important and clinically relevant research goal is to address the question of jaw length on multiple hierarchical levels, and to manipulate developmental programs in ways that test the potential efficacy of molecular strategies for modulating jaw length. Also, observing how embryos respond to these changes is critical to devising new treatments for craniofacial defects. Currently, invasive surgery is the only option and is often needed on several occasions during childhood (Albanese & Harrison, 1998; Cordero, Schneider, & Helms, 2002; Joshi, Hamdan, & Fakhouri, 2014).

Experiments in tissue regeneration and transplantation demonstrate that organs have an intrinsic capacity to "know" their proper size and to regulate growth accordingly (Leevers & McNeill, 2005). But how intrinsic molecular and cellular programs operate within the local environment to modulate growth, and how the range of normal to abnormal phenotypic variation in size arises, remain poorly understood. Also, unclear are those mechanisms that serve as the targets of natural selection for evolutionary changes in organ size. To address the question of how the jaw skeleton achieves its proper size and shape during development, we have been using a unique avian chimeric transplantation system that exploits species-specific differences between Japanese quail and white Pekin duck (Ealba & Schneider, 2013; Fish & Schneider, 2014a; Jheon & Schneider, 2009; Lwigale & Schneider, 2008; Schneider, 2005, 2007; Schneider & Helms, 2003). In particular, we have been asking the question: how do quail and duck achieve their remarkably different jaw sizes? Quail have short jaws compared to those of duck, which are relatively long (Fig. 1A). We have focused on the lower jaw skeleton, which forms from the paired mandibular primordia. NCM that migrates out of the caudal midbrain and rostral hindbrain is the only source of

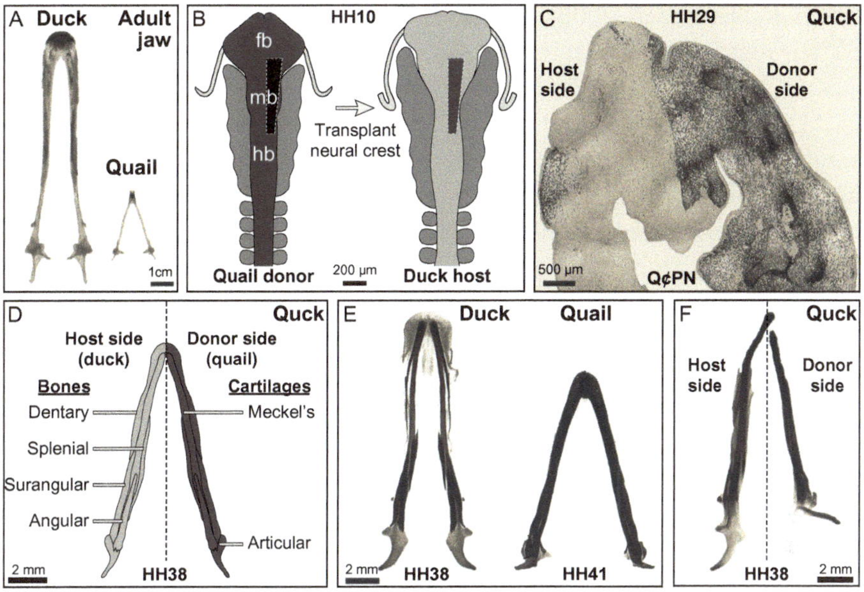

Figure 1 The quail–duck chimeric system for investigating the origins of species-specific jaw length. (A) Lower jaw skeletons of adult Japanese quail (*Coturnix coturnix japonica*) and white Pekin duck (*Anas platyrhyncos*). (B) Schematic of rostral neural tube at embryonic stage (HH) 9.5, depicting the levels of NCM destined for the jaw primordia and grafted from quail (red) to duck (blue). (C) Horizontal section through the mandibular primordium of a HH29 chimeric quck embryo (rostral at top), which will give rise to the lower jaw skeleton. Quail donor mesenchyme (black), stained with a quail-specific antibody (Q¢PN), is distributed throughout the transplanted side, while only a few quail cells are found on the contralateral duck host side. (D) Schematic of a lower jaw skeleton in a chimeric quck embryo at HH38, showing the contributions of transplanted quail donor NCM (red) to cartilage and bone. (E) The lower jaw skeletons of quail and duck display species-specific differences in size and shape with duck being longer and more curved. Meckel's cartilage is stained with Alcian blue and the bones are stained with Alizarin red. (F) In quck mandibles, the quail donor-derived jaw skeleton is shorter and straighter than that observed for the contralateral duck host-derived jaw skeleton, which is longer and curved. *Panels (A–E) modified from Eames and Schneider (2008) and Panel (F) modified from Fish and Schneider (2014a).* (A color version of this figure is located at the end of this chapter.)

skeletogenic mesenchyme within the mandibular primordia (Couly, Coltey, & Le Douarin, 1993; Köntges & Lumsden, 1996; Le Lièvre & Le Douarin, 1975; Noden, 1978; Noden & Schneider, 2006b). Our work has revealed that the orchestration of developmental programs regulating jaw length is under the regulatory control of NCM (Eames & Schneider, 2008; Jheon & Schneider, 2009; Schneider, 2005; Schneider & Helms, 2003;

Tokita & Schneider, 2009), but how NCM carries out this complicated task has remained unclear.

Our experimental approach is relatively straightforward: premigratory NCM is exchanged between quail and duck embryos at the level of the neural tube. Depending on the experimental design and desired outcome measures, we can transplant NCM unilaterally so that donor cells fill one side of the host jaw skeleton (Fig. 1B). This maintains the nonsurgical side of the host embryo as an internal control, and allows us to compare donor- and host-derived tissues directly in the same chimeric embryo (Eames & Schneider, 2005, 2008; Fish & Schneider, 2014a; Lwigale & Schneider, 2008; Solem, Eames, Tokita, & Schneider, 2011; Tokita & Schneider, 2009; Tucker & Lumsden, 2004). Quail embryos mature at a much quicker rate than do duck (17 vs. 28 days from fertilization to hatching), which causes faster-developing quail cells and relatively slower-maturing duck cells to interact with one another as they become progressively asynchronous. Such different developmental trajectories also provide a means to screen for the effects of donor cells on the host by looking for species-specific changes to the timing of gene expression, cell differentiation, and tissue formation. Additionally, we can use an anti-quail antibody (Q¢PN), which does not recognize duck cells, in order to distinguish donor from host contributions (Fig. 1C). We can also quantify the proportion of quail versus duck cells on the molecular level by applying a PCR-based strategy (Ealba & Schneider, 2013), which is particularly useful for gene expression studies (Ealba et al., 2015; Fish, Sklar, Woronowicz, & Schneider, 2014; Hall et al., 2014).

Once quail and duck cells are intertwined with one another, resulting chimeras become challenged to integrate two distinct morphogenetic programs for species-specific size and shape. This allows us to pinpoint mechanisms underlying the patterning of the jaw skeleton through an empirical strategy where we (1) characterize donor-mediated transformations to jaw size and shape; (2) look for changes to the timing and location of developmental events underlying skeletogenesis such as mesenchymal condensation and differentiation; (3) evaluate effects of NCM on host derivatives involved in skeletogenesis including epithelia, blood vessels, muscles, and osteoclasts; (4) assay for genes that become differentially expressed in chimeras; and (5) modulate the expression of these genes (i.e., perform gain- and loss-of-function experiments) to test the extent to which they regulate skeletal pattern and account for the chimeric phenotype (Eames & Schneider, 2005, 2008; Hall et al., 2014; Merrill, Eames, Weston, Heath, & Schneider, 2008; Noden & Schneider, 2006b; Solem et al., 2011; Tokita & Schneider, 2009). Overall, a major strength of this chimeric system is its ability to reveal in a relatively normal physiologic context those signals that mediate interactions between donor NCM and host tissues, and ultimately lead to the establishment of species-specific size and shape.

The use of the quail–duck chimeric system has led us to postulate that NCM employs a variety of very precise mechanisms to govern jaw length through three principal phases of development. Initially, during migration and allocation of NCM, quail and duck have distinct numbers of progenitors destined to form the jaw skeleton, with duck having significantly more cells (Fish et al., 2014). Then, as these populations expand, there is species-specific regulation of, and response to, various signaling pathways (Eames & Schneider, 2008; Hall et al., 2014; Merrill et al., 2008). Finally, when these progenitors begin to differentiate into the cartilages and bones of the jaw skeleton, they execute autonomous molecular and cellular programs for matrix deposition and resorption through patterns and processes that are intrinsic to each species (Ealba et al., 2015; Eames & Schneider, 2008; Hall et al., 2014; Merrill et al., 2008; Mitgutsch, Wimmer, Sanchez-Villagra, Hahnloser, & Schneider, 2011). A long-term goal is to understand the way these mechanisms affect jaw length, how they are regulated, and the extent to which they can be targeted. Much work points to the SHH, FGF, BMP, and TGFβ pathways as crucial players, and numerous pathway members and targets become altered in our quail–duck chimeras. We are finding that NCM differentially regulates and responds to SHH, FGF, BMP, and TGFβ signaling in a species-specific manner, which likely modulates the proliferation, differentiation, and growth of jaw progenitors and generates variation in jaw length. This provides us with insight into how these pathways empower NCM with its regulatory abilities during development, disease, and evolution. Our expectation is that using highly divergent bird species to illuminate the determinants of jaw length will provide enough resolution to detect equivalent but likely much more subtle mechanisms generating normal and abnormal variation in humans. In this framework, we have been striving to define developmental periods when cells and tissues are responsive to inductive signals, which we hope will eventually help move the standard of care toward treating craniofacial defects *in utero*.

2. PART 1: EARLY DETERMINANTS OF JAW LENGTH

The generation of NCM involves multiple and sequential developmental events, starting with induction at the boundary between neural and nonneural ectoderm, regional specification along the dorsal neural tube, maintenance of multipotency and cell cycle control, transition from epithelium to mesenchyme (EMT), and migration (Betancur, Bronner-Fraser, & Sauka-Spengler, 2010; Nikitina, Sauka-Spengler, & Bronner-Fraser, 2008). NCM that emigrates from the midbrain through the first and second rhombomeres of the hindbrain populates the mandibular primordia (Couly et al., 1993; Köntges & Lumsden, 1996; Le Lièvre & Le Douarin, 1975; Noden, 1978). Much has been written about the ways in which the gene regulatory networks and developmental programs that control these events have remained highly conserved across vertebrates, and especially function as a mechanism for the elaboration of the vertebrate head (Bronner-Fraser, 2008; Depew & Olsson, 2008; Nikitina et al., 2008; Northcutt, 2005). Yet there is very little known about how changes to these programs can occur in ways that account for the evolution of species-specific morphology.

In this context, we asked when, where, and how do duck embryos generate their long bills compared to quail embryos who make short beaks. We started with the simple analogy that building a bigger structure such as a wall might involve using more bricks, as opposed to bigger bricks (Fish & Schneider, 2014c). Therefore, we concentrated on determining the number of jaw precursor cells, which are the NCM that migrate into the mandibular primordia. We began by counting NCM at key embryonic stages (Fish et al., 2014). At an early stage, when NCM is specified at the level of the neural folds, quail and duck appear to have equivalent amounts of NCM. However, shortly thereafter, when NCM accumulates along the dorsal neural tube, duck have approximately 15% more NCM in the midbrain and rostral hindbrain, which is the population destined to migrate into the presumptive jaw region (Fig. 2B and C). Moreover, slightly thereafter, the jaw primordia of duck contain twice as many cells as do quail. To explain how an initial 15% difference could allow the population to double, we assayed for specific-specific variation in cell proliferation and cell cycle length (Fig. 2D and E). We found that while duck have a longer cell cycle, once embryonic stage is taken into account over absolute time, then duck cells actually proliferate more than those of quail, and in so doing provide duck with a cellular mechanism to increase their jaw length progressively throughout development.

To search for molecular mechanisms through which duck might possibly generate more midbrain NCM that can migrate into the jaw primordia, we assayed for species-specific differences in the expression of genes known to be involved in the regionalization of the brain. We looked at *Pax6*

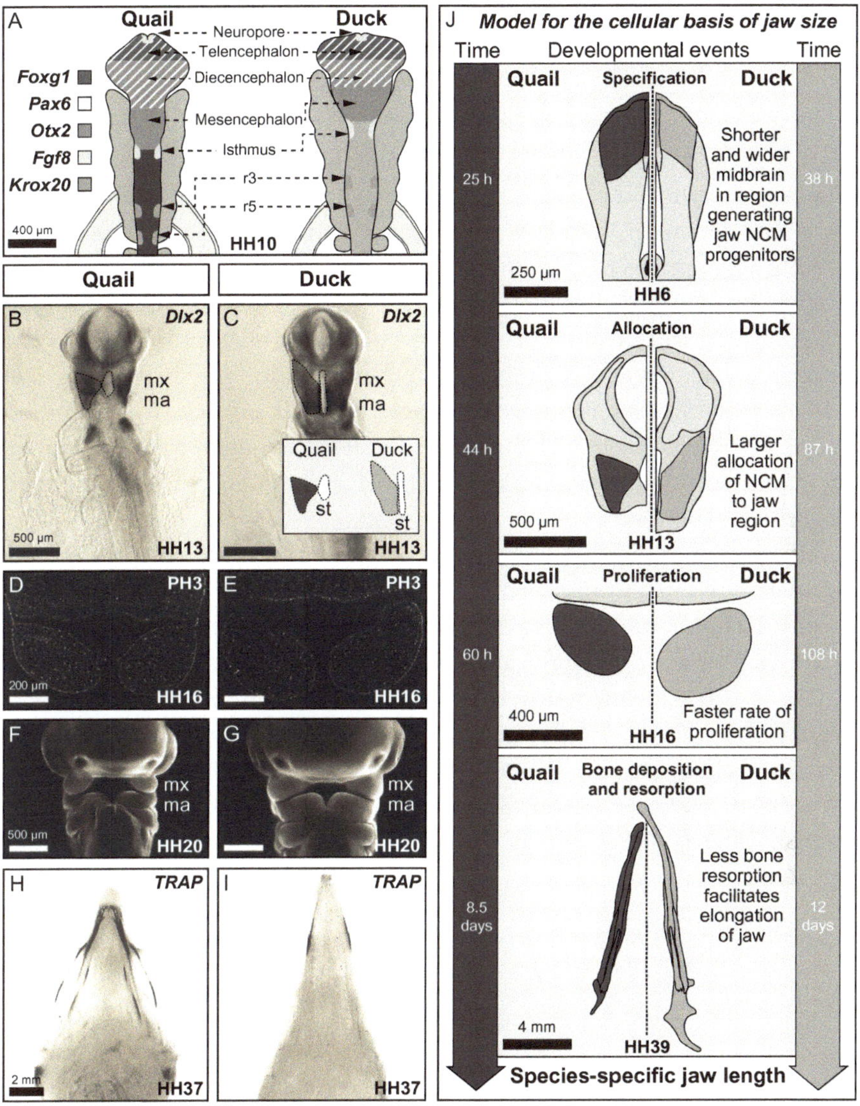

Figure 2 Molecular and cellular mechanisms regulating jaw length. (A) Quail and duck have distinct head shapes and species-specific regionalization of the neural tube at embryonic stage (HH) 10. Duck have a foreshortened and mediolaterally broader midbrain (mesencephalon). Genes including *Foxg1*, *Pax6*, *Otx2*, *Fgf8*, and *Krox20* are expressed in domains at HH10 with each being shifted more anteriorly in duck versus quail. (B, C) Differences in the allocation of NCM to the maxillary (mx) and mandibular (ma) primordia of the presumptive jaw region can be seen following *in situ* hybridization for *Dlx2* at HH13 in quail versus duck. The *Dlx2*-positive NCM domain and stomodeum (st) demonstrate that duck (blue) have a larger population of NCM relative

(Continued)

expression in the forebrain, *Otx2* in the forebrain and midbrain, *Fgf8* at the midbrain–hindbrain boundary, and *Krox20* in rhombomeres 3 and 5 of the hindbrain (Fig. 2A). We compared duck and quail embryos at the time of neurulation and identified species-specific differences in brain shape and spatial domains of gene expression. In particular, we observed that the midbrain of duck is shorter and broader, which is also evidenced by a distinct pattern of *Otx2* expression. Ostensibly, this broader midbrain of duck enables more NCM to aggregate in the region that will ultimately populate the jaw region. Surprisingly, we also detected differences in the *Otx2* expression domain between duck and quail embryos even before neurulation, indicating that essential species-specific patterning mechanisms that affect jaw size may operate at the earliest developmental stages. Overall, our results demonstrate precisely where and when changes to early developmental programs underlying the allocation and proliferation of NCM have likely played a role in the evolution of jaw size.

Although we find that early differences in NCM number appear to be important for establishing species-specific jaw length, we also discovered that if we reduce or augment the amount of jaw progenitors (up to 25%), we do not observe a significant effect on jaw length prior to hatching (Fish et al., 2014). Our results support other observations that the jaw can revert to its normal length after neural fold extirpation (Couly, Grapin-Botton, Coltey, & Le Douarin, 1996; Hunt, Ferretti, Krumlauf, & Thorogood, 1995; Scherson, Serbedzija, Fraser, & Bronner-Fraser, 1993; Sechrist, Nieto, Zamanian, & Bronner-Fraser, 1995). In these previous

Figure 2—Cont'd to quail (red). (D, E) Phosphohistone H3 (PH3) identifies mitotic cells at HH16. While duck develop at a slower rate (taking about 45 h to progress from HH13 to HH20 vs. 32 h in quail), duck NCM completes roughly 3.3 cycles during this developmental window compared to an average of 2.9 cycles for quail. Thus, the rate of proliferation relative to the rate of development is faster in duck. (F, G) As a result by HH20, the jaw of duck is approximately twice the size of that of quail. (H, I) Staining for tartrate-resistant acid phosphatase (TRAP) in the jaw skeleton reveals less bone resorption in duck versus quail at HH37. (J) Multiple developmental events regulate jaw length. Specification at the neural plate establishes a shorter and wider midbrain in duck. This difference, evidenced by distinct *Otx2* expression domains, is evident by HH6 and leads to a larger allocation of NCM to the jaw primordia by HH13. Duck NCM have a higher proliferation rate due to differences in developmental rate (time arrows on side). Finally, lower amounts of bone resorption result in differential growth and elongation of the duck jaw skeleton. *Panels (A–G) and (J) from Fish et al. (2014) and Panels (H) and (I) from Ealba et al. (2015).* (A color version of this figure is located at the end of this chapter.)

reports, however, normal jaw length was argued to result from regeneration of NCM at the neural tube, either by respecification of the residual dorsal neuroepithelium (Hunt et al., 1995; Sechrist et al., 1995) or by an expansion of NCM produced by adjacent neural folds (Couly et al., 1996; Scherson et al., 1993). In contrast, we find that NCM does not regenerate at the level of the neural tube and thus, the return to normal jaw length requires some other compensatory mechanism likely involving signaling interactions with adjacent epithelia. In other words, normal jaw length may be achieved by local regulation of proliferation within the postmigratory environment of the jaw primordia. Importantly, such regulative development in the local environment allows for compensation of deficiencies in NCM up to some prespecified species-specific population size, a capacity that could potentially be harnessed to supplement NCM number and restore normal jaw length in cases of human disease or injury.

3. PART 2: DETERMINATES OF JAW LENGTH DURING SKELETAL DIFFERENTIATION

The relationship between size and shape has long been a focus of developmental and evolutionary biology. Early size and shape studies focused principally on proportional scaling or "allometry" of anatomical structures that occurs ontogenetically during growth or phylogenetically across species (Huxley, 1932; Thompson, 1917). This type of research led to the field of geometric morphometrics, which has combined multivariate methods and computer-based algorithms to quantify and display ontogenetic and phylogenetic differences in size and shape (Benson, Chapman, & Siegel, 1982; Bookstein, 1978, 1990; Hu, Young, Xu, et al., 2015; Marcucio, Young, Hu, & Hallgrimsson, 2011; Siegel & Benson, 1982; Smith et al., 2015; Young, Chong, Hu, Hallgrimsson, & Marcucio, 2010; Young et al., 2014). Often morphometric data have been contextualized with quantitative genetics or evolutionary developmental theories like heterochrony, as a way to explain changes in size and shape during ontogeny and phylogeny (Alberch, Gould, Oster, & Wake, 1979; Atchley, 1981; Atchley & Hall, 1991; Gould, 1966; Lande, 1979; McKinney, 1988). We have combined the quail–duck chimeric system with morphometric and molecular analyses to study the development of Meckel's cartilage in the lower jaw skeleton (Fig. 1D), and in so doing have found that NCM controls both stage-specific and species-specific size and shape (Eames & Schneider, 2008).

The foundation for our work is built upon many other studies of size and shape in the vertebrate skull (de Beer, 1937; Hanken & Hall, 1993), primarily in relation to genetic specification of skeletal element identity (Balling, Mutter, Gruss, & Kessel, 1989; Creuzet, Couly, Vincent, & Le Douarin, 2002; Depew, Lufkin, & Rubenstein, 2002; Gendron-Maguire, Mallo, Zhang, & Gridley, 1993; Grammatopoulos, Bell, Toole, Lumsden, & Tucker, 2000; Hunt, Clarke, Buxton, Ferretti, & Thorogood, 1998; Kimmel et al., 2005; Lufkin et al., 1992; Pasqualetti, Ori, Nardi, & Rijli, 2000; Qiu et al., 1997; Rijli et al., 1993; Schilling, 1997; Smith & Schneider, 1998), epithelial–mesenchymal signaling interactions that are essential for the differentiation of cartilage and bone (Bee & Thorogood, 1980; Couly, Creuzet, Bennaceur, Vincent, & Le Douarin, 2002; Dunlop & Hall, 1995; Ferguson, Tucker, & Sharpe, 2000; Francis-West, Robson, & Evans, 2003; Hall, 1980, 1982, 1987; Richman & Tickle, 1989, 1992; Schowing, 1968; Shigetani, Nobusada, & Kuratani, 2000; Thorogood, 1987; Thorogood, Bee, & von der Mark, 1986; Tyler, 1978, 1983), secreted molecules that regulate skeletal polarity and dimensional growth (Abzhanov et al., 2004; Abzhanov & Tabin, 2004; Barlow & Francis-West, 1997; Crump, Maves, Lawson, Weinstein, & Kimmel, 2004; Francis-West, Ladher, Barlow, & Graveson, 1998; Hu, Marcucio, & Helms, 2003; Liu et al., 2005; Marcucio, Cordero, Hu, & Helms, 2005; Schneider et al., 2001; Wilson & Tucker, 2004; Wu et al., 2006, 2004), and mesenchymal control of species-specific skeletal morphology (Andres, 1949; Mitsiadis, Caton, & Cobourne, 2006; Noden, 1983; Schneider & Helms, 2003; Tucker & Lumsden, 2004; Wagner, 1959).

Historically, the ability of NCM to convey species-specific pattern has been revealed mostly through interspecific grafting experiments (Lwigale & Schneider, 2008; Noden & Schneider, 2006a). Employing quail–duck chimeras has been a powerful means to understand how bones and cartilages in the face and jaws acquire their species-specific pattern (Jheon & Schneider, 2009; Schneider, 2005; Schneider & Helms, 2003; Tucker & Lumsden, 2004). Chimeric "quck" embryos, which are duck hosts with quail donor cells, possess quail-like beaks and jaw joints, whereas chimeric "duail" exhibit duck-derived morphology in quail hosts (Fig. 1F). We have spent the past decade or so trying to pin down the precise molecular mechanisms through which NCM accomplishes this complex task, and we have found most strikingly that donor NCM controls its own gene expression, cell cycle, and differentiation, as well as regulates certain aspects of the developmental programs of adjacent host tissues such as epithelia and muscles (Ealba & Schneider, 2013; Eames & Schneider, 2005, 2008; Fish & Schneider, 2014b; Fish et al., 2014; Hall et al., 2014; Merrill et al., 2008; Schneider, 2005, 2007; Schneider & Helms, 2003; Solem et al., 2011; Tokita & Schneider, 2009).

To identify developmental mechanisms that generate skeletal size and shape, we focused on the differentiation and growth of Meckel's cartilage (Eames & Schneider, 2008). Meckel's cartilage of quail is substantially smaller than that of stage-matched duck and becomes distinctly shaped over time. Again, because quail embryos develop at a faster rate than do duck embryos, chimeras reveal those aspects of size and shape regulation that are NCM dependent. We have found that NCM establishes both stage-specific and species-specific size and shape, and does so by exerting spatiotemporal control over molecular and cellular programs for chondrogenesis. NCM on the quail donor side of quck mandibles differentiated into chondrocytes on the time frame of quail controls as opposed to that observed on the contralateral duck host side. Donor-mediated shifts in cartilage differentiation were observed from the earliest stage of chondrogenesis. Both *Sox9*, which is the earliest known molecular marker of chondrogenic condensations (Eames, de la Fuente, & Helms, 2003; Eames, Sharpe, & Helms, 2004; Healy, Uwanogho, & Sharpe, 1996; Zhao, Eberspaecher, Lefebvre, & De Crombrugghe, 1997), and *Col2a1*, which is regulated directly by *Sox9* (Bell et al., 1997), were expressed prematurely by quail donor NCM relative to duck host NCM on the contralateral side. Additionally, we determined that FGF signaling, which functions upstream of *Sox9* and is essential for chondrogenesis (Bobick, Thornhill, & Kulyk, 2007; de Crombrugghe et al., 2000; Eames et al., 2004; Govindarajan & Overbeek, 2006; Healy, Uwanogho, & Sharpe, 1999; Murakami, Kan, McKeehan, & de Crombrugghe, 2000; Petiot, Ferretti, Copp, & Chan, 2002), is also regulated by NCM. While the secreted ligands *Fgf4* and *Fgf8* were expressed continuously by duck host epithelium prior to and during chondrogenesis, the receptor *Fgfr2* was expressed prematurely only by quail donor NCM relative to duck host NCM on the contralateral side. When we inhibited FGF signaling during this brief window of receptor activation, we blocked the formation of Meckel's cartilage. Therefore, by controlling the timing of FGF signaling as well as the expression of *Sox9* and *Col2a1*, NCM most likely conveys information for stage-specific and species-specific size and shape to Meckel's cartilage.

In terms of evolutionary developmental biology, one exciting aspect of this work is the insight about how NCM keeps track of both stage-specific and species-specific size and shape simultaneously. Seemingly, quail NCM makes a smaller jaw skeleton by shifting the timing of developmental events in the duck to resemble that found in the quail. This is because quail NCM orchestrates its spatiotemporal programs for chondrogenesis autonomously and in so doing provides size and shape information across embryonic stages and between species in parallel. Ultimately, this reveals that the developmental programs under the regulatory control of NCM link ontogeny to phylogeny mechanistically, and likely play a generative role in morphological evolution, which is a concept central to the field of evolutionary developmental biology (Alberch, 1980, 1982; Alberch et al., 1979; Eames & Schneider, 2008; Gould, 1966, 1977; Hall & Olson, 2003; Schneider, 2005, 2007).

Similarly, for bone formation in the lower jaw, we have found that quail NCM, when transplanted into duck, maintains its faster time-table for development, and autonomously executes molecular and cellular programs for osteogenesis, including expression of essential transcription factors such as *Runx2* (Ealba & Schneider, 2013; Eames & Schneider, 2008; Hall et al., 2014; Merrill et al., 2008). Our experiments show that NCM establishes the timing of bone formation in the jaw skeleton by regulating cell cycle progression in a stage- and species-specific manner. Such work has led us to propose that NCM controls the timing of osteogenic induction, proliferation, differentiation, and matrix deposition through targets of TGFβ and BMP signaling, especially *Runx2*. We have found that quail NCM, when transplanted into duck, maintains its faster time-table for development and autonomously executes molecular and cellular programs for osteogenesis, including premature expression of matrix-producing genes such as *Col1a1*. In contrast, the duck host systemic environment appears to be relatively permissive and supports osteogenesis independently by providing circulating minerals and a vascular network. Taken together, our studies have revealed that NCM dictates when bone forms by controlling the timing of cell cycle progression and mediating the transition from cell proliferation to differentiation. Transiently altering the cell cycle during early development can mimic chimeras by accelerating expression of *Runx2* and *Col1a1* (Hall et al., 2014). We also serendipitously discovered that *Runx2* expression might relate to jaw size in quail versus duck, since we observed higher endogenous expression of *Runx2* in quail coincident with their smaller head skeletons. By the time, the jaw is becoming mineralized, *Runx2* levels in quail rise to more than double those of duck. By experimentally increasing the levels of *Runx2*, we were able to decrease the size of the beak skeleton, and in effect mirror the relationship between species-specific beak size and endogenous *Runx2* levels. Other studies have also made a connection between expected *Runx2* expression levels (based on numbers of tandem repeats) and facial length such as in adult dogs and other mammals (Fondon & Garner, 2004; Pointer et al., 2012; Sears, Goswami, Flynn, & Niswander, 2007). These observations specifically point to precise control over the levels of key transcription factors and the timing of skeletal cell differentiation as a potential developmental mechanism through which NCM can affect jaw length during development, disease, and evolution.

4. PART 3: DETERMINATES OF JAW LENGTH DURING LATE-STAGE GROWTH

While much of our work demonstrates that NCM conveys species-specific jaw size and shape by regulating the molecular and cellular programs that underlie the induction and deposition of cartilage and bone, we have also discovered that a previously unrecognized but equivalently important mechanism for regulating jaw length is the ability of NCM to mediate the process of bone resorption (Ealba et al., 2015). In adults, bone resorption is linked to bone deposition as a mechanism for maintaining homeostasis throughout the skeleton (Buckwalter, Glimcher, Cooper, & Recker, 1996; Filvaroff & Derynck, 1998; Hall, 2005; Nguyen, Tang, Nguyen, & Alliston, 2013; O'Brien et al., 2008; Teitelbaum, 2000; Teitelbaum, Tondravi, & Ross, 1997). Yet the role and regulation of bone resorption during formation of the embryonic skeleton are less well understood.

Bone resorption occurs following the actions of two cell types that are distinguished by their different embryological lineages and morphology. Osteoclasts, which are derived from the mesodermal hematopoietic lineage (Jotereau & Le Douarin, 1978; Kahn et al., 2009), have historically been considered the predominant bone-resorbing cells (Boyle, Simonet, & Lacey, 2003; Filvaroff & Derynck, 1998; Hancox, 1949; Martin & Ng, 1994; Teitelbaum, 2000; Teitelbaum et al., 1997). Osteoclasts are multinucleated cells with ruffled borders and large and irregular morphology. In our quail–duck chimeras, osteoclasts are derived solely from host mesoderm. However, osteocytes, which in the skeleton of the jaws and face arise entirely from NCM (Helms & Schneider, 2003; Le Lièvre, 1978; Noden, 1978), also resorb bone (Belanger, 1969; O'Brien et al., 2008; Qing et al., 2012; Tang, Herber, Ho, & Alliston, 2012; Xiong & O'Brien, 2012; Xiong et al., 2014). Osteocytes typically are small, star-shaped cells with long cytoplasmic extensions. When osteoclasts and osteocytes resorb bone, they both secrete tartrate-resistant acid phosphatase (TRAP; Minkin, 1982; Qing et al., 2012; Tang et al., 2012). Additionally, each expresses distinct molecular markers such as *Mmp9*, which is found in osteoclasts (Engsig et al., 2000; Reponen, Sahlberg, Munaut, Thesleff, & Tryggvason, 1994), and *Mmp13*, which is detected in osteocytes (Behonick et al., 2007; Johansson et al., 1997; Sasano et al., 2002). When cartilage is replaced by bone during endochondral ossification, *Mmp9* and *Mmp13* also become expressed by hypertrophic chondrocytes (Colnot & Helms, 2001). However, there is essentially no endochondral ossification in the lower jaw of birds since Meckel's cartilage persists as a permanent cartilage (de Beer, 1937; Eames et al., 2004; Ekanayake & Hall, 1994; Kavumpurath & Hall, 1990). The only replacement of cartilage by bone in birds occurs in the proximal-most region within the articular cartilage beginning shortly before hatching (Mitgutsch et al., 2011; Starck, 1989). The remaining bone in the lower jaw differentiates directly from NCM through intramembranous ossification (Helms & Schneider, 2003; Noden, 1978, 1982; Noden & Schneider, 2006a; Noden & Trainor, 2005). Thus, within the lower jaw of chimeric quck following transplant of NCM, *Mmp9* would be almost entirely expressed by duck host-derived osteoclasts and *Mmp13* by quail donor-derived osteocytes.

When we compare the process of bone resorption in short-beaked quail versus long-billed duck, we find that quail have dramatically higher levels of TRAP (Fig. 2H and I), *Mmp9*, and *Mmp13*. Similarly, our chimeric quck develop quail-like jaw skeletons coincident with higher quail-like levels of TRAP, *Mmp9*, and *Mmp13*. This means that in chimeric quck, quail donor NCM not only continues to act out its own intrinsic species-specific program for bone resorption via higher *Mmp13* expression and TRAP activity but also upregulates the expression of *Mmp9* in duck host osteoclasts. This reveals an unexpected NCM-mediated mechanism through which quail and chimeric quck acquire their shorter jaws. In other words, the amount of bone resorption in birds appears to be inversely proportional to jaw length. This conclusion is substantiated by the fact that either blocking or activating bone resorption with drugs (e.g., bisphosphonates), recombinant proteins (e.g., rOPG or rRANKL), or small molecule inhibitors, can significantly lengthen or shorten the jaw.

Thus, quail and duck express species-specific molecular programs underlying bone resorption, and these programs are governed by NCM. Such experiments point to a novel function for bone resorption, which is to help establish species-specific jaw length, and they build upon prior work on Darwin's finches and other species, which contend that a critical regulator of beak length is the calcium-binding protein, *calmodulin* (Abzhanov et al., 2006; Gunter, Koppermann, & Meyer, 2014; Schneider, 2007). *Calmodulin* has been shown to control osteocytes and osteoclasts locally (Choi, Ann, et al., 2013; Choi, Choi, Oh, & Lee, 2013; Seales, Micoli, & McDonald, 2006; Zayzafoon, 2006). In this regard, calcium signaling and its effects on bone resorption (Hwang & Putney, 2011; Kajiya, 2012; Xia & Ferrier, 1996; Xiong et al., 2014) may function as a developmental mechanism that facilitates the evolvability of the avian beak more generally (Kirschner & Gerhart, 1998) and dictates jaw length more specifically (Gunter et al., 2014; Parsons & Albertson, 2009). Furthermore, taken together, these studies suggest that bone resorption may function like a rheostat during jaw length evolution, and one that is particularly sensitive to the availability of dietary calcium in varying ecological niches, the endocrine effects of calcium-dependent hormones, and the temporal and spatial modulation of calcium signaling within the primordia of the developing jaw (Schneider, 2007).

Such conclusions are in agreement with previous work postulating that differential fields of deposition and resorption lead to changes in size and shape during growth of the jaw skeleton in humans (Enlow, Moyers, & Merow, 1975; Moore, 1981; Radlanski & Klarkowski, 2001; Radlanski, Renz, Lajvardi, & Schneider, 2004). These findings also help explain the basis for abnormal snouts in mice with mutations in genes known to affect resorption such as *Mmp2* (Egeblad et al., 2007), and they provide insights into the etiologies of jaw length defects in humans with conditions such as Spondyloepimetaphyseal dysplasia (i.e., Mmp13), Juvenile Paget's disease (i.e., Opg), and after treatments with high doses of bisphosphonates such as zoledronic acid, which inhibit bone resorption (Gorlin et al., 1990; Lezot et al., 2014). Based on these types of experiments, we have become increasingly optimistic that precise pharmacological strategies can be devised to target and carefully modulate bone resorption as a noninvasive, nonsurgical means for treating human defects in jaw length such as malocclusion or even mandibular hypoplasia. Overall, the extraordinary ability of NCM to exert spatiotemporal control over the induction, differentiation, deposition, mineralization, and resorption of bone (Eames & Schneider, 2008; Hall et al., 2014; Merrill et al., 2008; Schneider & Helms, 2003) is what integrates the molecular and cellular determinants of jaw length throughout embryonic development (Fig. 2J), and is what endows NCM with its unique ability to generate variation in jaw length during disease and evolution.

REFERENCES

Abzhanov, A., Kuo, W. P., Hartmann, C., Grant, B. R., Grant, P. R., & Tabin, C. J. (2006). The calmodulin pathway and evolution of elongated beak morphology in Darwin's finches. *Nature, 442*, 563–567.

Abzhanov, A., Protas, M., Grant, B. R., Grant, P. R., & Tabin, C. J. (2004). Bmp4 and morphological variation of beaks in Darwin's finches. *Science, 305*, 1462–1465.

Abzhanov, A., & Tabin, C. J. (2004). Shh and Fgf8 act synergistically to drive cartilage outgrowth during cranial development. *Developmental Biology, 273*, 134–148.

Albanese, C. T., & Harrison, M. R. (1998). Surgical treatment for fetal disease. The state of the art. *Annals of the New York Academy of Sciences, 847*, 74–85.

Alberch, P. (1980). Ontogenesis and morphological diversification. *American Zoologist, 20*, 653–667.

Alberch, P. (1982). The generative and regulatory roles of development in evolution. In D. Mossakowski & G. Roth (Eds.), *Environmental adaptation and evolution: A theoretical and empirical approach* (pp. 19–36). Stuttgart: G. Fischer-Verlag.

Alberch, P., Gould, S. J., Oster, G. F., & Wake, D. B. (1979). Size and shape in ontogeny and phylogeny. *Paleobiology, 5*, 296–317.

Allen, B. L., Tenzen, T., & McMahon, A. P. (2007). The Hedgehog-binding proteins Gas1 and Cdo cooperate to positively regulate Shh signaling during mouse development. *Genes & Development, 21*, 1244–1257.

Andres, G. (1949). Untersuchungen an Chimären von Triton und Bombinator. *Genetica, 24*, 387–534.

Ashique, A. M., Fu, K., & Richman, J. M. (2002). Endogenous bone morphogenetic proteins regulate outgrowth and epithelial survival during avian lip fusion. *Development, 129*, 4647–4660.

Atchley, W. R. (1981). Genetic components of size and shape. II. Multivariate covariance patterns in the rat and mouse skull. *Evolution, 35*, 1037–1055.

Atchley, W. R., & Hall, B. K. (1991). A model for development and evolution of complex morphological structures. *Biological Reviews of the Cambridge Philosophical Society, 66*, 101–157.

Balling, R., Mutter, G., Gruss, P., & Kessel, M. (1989). Craniofacial abnormalities induced by ectopic expression of the homeobox gene Hox-1.1 in transgenic mice. *Cell, 58*, 337–347.

Barlow, A. J., & Francis-West, P. H. (1997). Ectopic application of recombinant BMP-2 and BMP-4 can change patterning of developing chick facial primordia. *Development, 124*, 391–398.

Bee, J., & Thorogood, P. (1980). The role of tissue interactions in the skeletogenic differentiation of avian neural crest cells. *Developmental Biology, 78*, 47–66.

Behonick, D. J., Xing, Z., Lieu, S., Buckley, J. M., Lotz, J. C., Marcucio, R. S., et al. (2007). Role of matrix metalloproteinase 13 in both endochondral and intramembranous ossification during skeletal regeneration. *PLoS One, 2*, e1150.

Belanger, L. F. (1969). Osteocytic osteolysis. *Calcified Tissue Research, 4*, 1–12.

Bell, D. M., Leung, K. K., Wheatley, S. C., Ng, L. J., Zhou, S., Ling, K. W., et al. (1997). SOX9 directly regulates the type-II collagen gene. *Nature Genetics, 16*, 174–178.

Benson, R. H., Chapman, R. E., & Siegel, A. F. (1982). On the measurement of morphology and its change. *Paleobiology, 8*, 328–339.

Betancur, P., Bronner-Fraser, M., & Sauka-Spengler, T. (2010). Assembling neural crest regulatory circuits into a gene regulatory network. *Annual Review of Cell and Developmental Biology, 26*, 581–603.

Bhullar, B. A., Morris, Z. S., Sefton, E. M., Tok, A., Tokita, M., Namkoong, B., et al. (2015). A molecular mechanism for the origin of a key evolutionary innovation, the bird beak and palate, revealed by an integrative approach to major transitions in vertebrate history. *Evolution*, *69*, 1665–1677.

Bobick, B. E., Thornhill, T. M., & Kulyk, W. M. (2007). Fibroblast growth factors 2, 4, and 8 exert both negative and positive effects on limb, frontonasal, and mandibular chondrogenesis via MEK-ERK activation. *Journal of Cellular Physiology*, *211*, 233–243.

Boell, L., Pallares, L. F., Brodski, C., Chen, Y., Christian, J. L., Kousa, Y. A., et al. (2013). Exploring the effects of gene dosage on mandible shape in mice as a model for studying the genetic basis of natural variation. *Development Genes and Evolution*, *223*, 279–287.

Bookstein, F. L. (1978). *The measurement of biological shape and shape change*. New York: Springer-Verlag.

Bookstein, F. L. (1990). Multivariate methods. In F. J. Rohlf & F. L. Bookstein (Eds.), *Proceedings of the Michigan morphometrics workshop* (pp. 75–76). Ann Arbor: University of Michigan Museum of Zoology.

Boyle, W. J., Simonet, W. S., & Lacey, D. L. (2003). Osteoclast differentiation and activation. *Nature*, *423*, 337–342.

Bronner-Fraser, M. (2008). On the trail of the 'new head' in Les Treilles. *Development*, *135*, 2995–2999.

Brugmann, S. A., Goodnough, L. H., Gregorieff, A., Leucht, P., ten Berge, D., Fuerer, C., et al. (2007). Wnt signaling mediates regional specification in the vertebrate face. *Development*, *134*, 3283–3295.

Brugmann, S. A., Powder, K. E., Young, N. M., Goodnough, L. H., Hahn, S. M., James, A. W., et al. (2010). Comparative gene expression analysis of avian embryonic facial structures reveals new candidates for human craniofacial disorders. *Human Molecular Genetics*, *19*, 920–930.

Buckwalter, J. A., Glimcher, M. J., Cooper, R. R., & Recker, R. (1996). Bone biology. II: Formation, form, modeling, remodeling, and regulation of cell function. *Instructional Course Lectures*, *45*, 387–399.

Choi, Y. H., Ann, E. J., Yoon, J. H., Mo, J. S., Kim, M. Y., & Park, H. S. (2013). Calcium/calmodulin-dependent protein kinase IV (CaMKIV) enhances osteoclast differentiation via the up-regulation of Notch1 protein stability. *Biochimica et Biophysica Acta*, *1833*, 69–79.

Choi, Y. H., Choi, J. H., Oh, J. W., & Lee, K. Y. (2013). Calmodulin-dependent kinase II regulates osteoblast differentiation through regulation of Osterix. *Biochemical and Biophysical Research Communications*, *432*, 248–255.

Colnot, C. I., & Helms, J. A. (2001). A molecular analysis of matrix remodeling and angiogenesis during long bone development. *Mechanisms of Development*, *100*, 245–250.

Cordero, D. R., Schneider, R. A., & Helms, J. A. (2002). Morphogenesis of the face. In K. Y. Lin, R. C. Ogle, & J. A. Jane (Eds.), *Craniofacial surgery: Science & surgical technique* (pp. 75–83). Philadelphia: W.B. Saunders Company.

Couly, G. F., Coltey, P. M., & Le Douarin, N. M. (1993). The triple origin of skull in higher vertebrates: A study in quail-chick chimeras. *Development*, *117*, 409–429.

Couly, G., Creuzet, S., Bennaceur, S., Vincent, C., & Le Douarin, N. M. (2002). Interactions between Hox-negative cephalic neural crest cells and the foregut endoderm in patterning the facial skeleton in the vertebrate head. *Development*, *129*, 1061–1073.

Couly, G., Grapin-Botton, A., Coltey, P., & Le Douarin, N. M. (1996). The regeneration of the cephalic neural crest, a problem revisited: The regenerating cells originate from the contralateral or from the anterior and posterior neural fold. *Development*, *122*, 3393–3407.

Creuzet, S., Couly, G., Vincent, C., & Le Douarin, N. M. (2002). Negative effect of Hox gene expression on the development of the neural crest-derived facial skeleton. *Development*, *129*, 4301–4313.

Crump, J. G., Maves, L., Lawson, N. D., Weinstein, B. M., & Kimmel, C. B. (2004). An essential role for Fgfs in endodermal pouch formation influences later craniofacial skeletal patterning. *Development*, *131*, 5703–5716.

Darwin, C. (1859). *The origin of species* (1962 ed.). New York: The Crowell-Collier Publishing Co.

de Beer, G. R. (1937). *The development of the vertebrate skull*. Chicago: University of Chicago Press.

de Crombrugghe, B., Lefebvre, V., Behringer, R. R., Bi, W., Murakami, S., & Huang, W. (2000). Transcriptional mechanisms of chondrocyte differentiation. *Matrix Biology*, *19*, 389–394.

Dennis, J. F., Kurosaka, H., Iulianella, A., Pace, J., Thomas, N., Beckham, S., et al. (2012). Mutations in Hedgehog acyltransferase (Hhat) perturb Hedgehog signaling, resulting in severe acrania-holoprosencephaly-agnathia craniofacial defects. *PLoS Genetics*, *8*, e1002927.

Depew, M. J., Lufkin, T., & Rubenstein, J. L. (2002). Specification of jaw subdivisions by Dlx genes. *Science*, *298*, 381–385.

Depew, M. J., & Olsson, L. (2008). Symposium on the evolution and development of the vertebrate head. *Journal of Experimental Zoology. Part B, Molecular and Developmental Evolution*, *310*, 287–293.

Doufexi, A. E., & Mina, M. (2008). Signaling pathways regulating the expression of Prx1 and Prx2 in the chick mandibular mesenchyme. *Developmental Dynamics*, *237*, 3115–3127.

Dudas, M., Kim, J., Li, W. Y., Nagy, A., Larsson, J., Karlsson, S., et al. (2006). Epithelial and ectomesenchymal role of the type I TGF-beta receptor ALK5 during facial morphogenesis and palatal fusion. *Developmental Biology*, *296*, 298–314.

Dudas, M., Sridurongrit, S., Nagy, A., Okazaki, K., & Kaartinen, V. (2004). Craniofacial defects in mice lacking BMP type I receptor Alk2 in neural crest cells. *Mechanisms of Development*, *121*, 173–182.

Dunlop, L. L., & Hall, B. K. (1995). Relationships between cellular condensation, preosteoblast formation and epithelial-mesenchymal interactions in initiation of osteogenesis. *The International Journal of Developmental Biology*, *39*, 357–371.

Ealba, E. L., & Schneider, R. A. (2013). A simple PCR-based strategy for estimating species-specific contributions in chimeras and xenografts. *Development*, *140*, 3062–3068.

Ealba, E. L., Jheon, A. H., Hall, J., Curantz, C., Butcher, K. D., & Schneider, R. A. (2015). Neural crest-mediated bone resorption is a determinant of species-specific jaw length. *Developmental Biology*. in press. http://dx.doi.org/10.1016/j.ydbio.2015.10.001.

Eames, B. F., de la Fuente, L., & Helms, J. A. (2003). Molecular ontogeny of the skeleton. *Birth Defects Research. Part C, Embryo Today: Reviews*, *69*, 93–101.

Eames, B. F., & Schneider, R. A. (2005). Quail-duck chimeras reveal spatiotemporal plasticity in molecular and histogenic programs of cranial feather development. *Development*, *132*, 1499–1509.

Eames, B. F., & Schneider, R. A. (2008). The genesis of cartilage size and shape during development and evolution. *Development*, *135*, 3947–3958.

Eames, B. F., Sharpe, P. T., & Helms, J. A. (2004). Hierarchy revealed in the specification of three skeletal fates by Sox9 and Runx2. *Developmental Biology*, *274*, 188–200.

Egeblad, M., Shen, H. C., Behonick, D. J., Wilmes, L., Eichten, A., Korets, L. V., et al. (2007). Type I collagen is a genetic modifier of matrix metalloproteinase 2 in murine skeletal development. *Developmental Dynamics*, *236*, 1683–1693.

Ekanayake, S., & Hall, B. K. (1994). Hypertrophy is not a prerequisite for type X collagen expression or mineralization of chondrocytes derived from cultured chick mandibular ectomesenchyme. *The International Journal of Developmental Biology*, *38*, 683–694.

Engsig, M. T., Chen, Q. J., Vu, T. H., Pedersen, A. C., Therkidsen, B., Lund, L. R., et al. (2000). Matrix metalloproteinase 9 and vascular endothelial growth factor are essential for osteoclast recruitment into developing long bones. *The Journal of Cell Biology*, *151*, 879–889.

Enlow, D. H., Moyers, R. E., & Merow, W. W. (1975). *Handbook of facial growth*. Philadelphia: Saunders.

Ferguson, C. A., Tucker, A. S., & Sharpe, P. T. (2000). Temporospatial cell interactions regulating mandibular and maxillary arch patterning. *Development*, *127*, 403–412.

Filvaroff, E., & Derynck, R. (1998). Bone remodelling: A signalling system for osteoclast regulation. *Current Biology*, *8*, R679–R682.

Fish, J. L., & Schneider, R. A. (2014a). Assessing species-specific contributions to craniofacial development using quail-duck chimeras. *Journal of Visualized Experiments*, *87*, 1–6.

Fish, J. L., & Schneider, R. A. (2014b). Chapter 6—Neural crest-mediated tissue interactions during craniofacial development: The origins of species-specific pattern. In P. A. Trainor (Ed.), *Neural crest cells* (pp. 101–124). Boston: Academic Press.

Fish, J. L., & Schneider, R. A. (2014c). *On the origins of species-specific size*. The Node, February 25th ed Cambridge: The Company of Biologists.

Fish, J. L., Sklar, R. S., Woronowicz, K. C., & Schneider, R. A. (2014). Multiple developmental mechanisms regulate species-specific jaw size. *Development*, *141*, 674–684.

Foerst-Potts, L., & Sadler, T. W. (1997). Disruption of *Msx-1* and *Msx-2* reveals roles for these genes in craniofacial, eye, and axial development. *Developmental Dynamics*, *209*, 70–84.

Fondon, J. W., 3rd, & Garner, H. R. (2004). Molecular origins of rapid and continuous morphological evolution. *Proceedings of the National Academy of Sciences of the United States of America*, *101*, 18058–18063.

Foppiano, S., Hu, D., & Marcucio, R. S. (2007). Signaling by bone morphogenetic proteins directs formation of an ectodermal signaling center that regulates craniofacial development. *Developmental Biology*, *312*, 103–114.

Francis-West, P., Ladher, R., Barlow, A., & Graveson, A. (1998). Signalling interactions during facial development. *Mechanisms of Development*, *75*, 3–28.

Francis-West, P. H., Robson, L., & Evans, D. J. (2003). Craniofacial development: The tissue and molecular interactions that control development of the head. *Advances in Anatomy, Embryology, and Cell Biology*, *169*, III–VI. 1–138.

Gendron-Maguire, M., Mallo, M., Zhang, M., & Gridley, T. (1993). Hoxa-2 mutant mice exhibit homeotic transformation of skeletal elements derived from cranial neural crest. *Cell*, *75*, 1317–1331.

Gorlin, R. J., Cohen, M. M., & Levin, L. S. (1990). *Syndromes of the head and neck* (3rd ed.). New York: Oxford University Press.

Gould, S. J. (1966). Allometry and size in ontogeny and phylogeny. *Biological Reviews*, *41*, 587–640.

Gould, S. J. (1977). *Ontogeny and phylogeny*. Cambridge: Harvard University Press.

Govindarajan, V., & Overbeek, P. A. (2006). FGF9 can induce endochondral ossification in cranial mesenchyme. *BMC Developmental Biology*, *6*, 7.

Grammatopoulos, G. A., Bell, E., Toole, L., Lumsden, A., & Tucker, A. S. (2000). Homeotic transformation of branchial arch identity after Hoxa2 overexpression. *Development*, *127*, 5355–5365.

Grant, P. R., Grant, B. R., & Abzhanov, A. (2006). A developing paradigm for the development of bird beaks. *Biological Journal of the Linnean Society*, *88*, 17–22.

Gunter, H. M., Koppermann, C., & Meyer, A. (2014). Revisiting de Beer's textbook example of heterochrony and jaw elongation in fish: Calmodulin expression reflects heterochronic growth, and underlies morphological innovation in the jaws of belonoid fishes. *EvoDevo*, *5*, 8.

Haldane, J. B. S. (1926). On being the right size. *Harper's Magazine*.
Hall, B. K. (1980). Tissue interactions and the initiation of osteogenesis and chondrogenesis in the neural crest-derived mandibular skeleton of the embryonic mouse as seen in isolated murine tissues and in recombinations of murine and avian tissues. *Journal of Embryology and Experimental Morphology*, *58*, 251–264.
Hall, B. K. (1982). The role of tissue interactions in the growth of bone. In A. D. Dixon & B. G. Sarnat (Eds.), *Factors and mechanisms influencing bone growth* (pp. 205–215). New York: Alan R. Liss, Inc.
Hall, B. K. (1987). Tissue interactions in the development and evolution of the vertebrate head. In P. F. A. Maderson (Ed.), *Developmental and evolutionary aspects of the neural crest* (pp. 215–260). New York: John Wiley & Sons.
Hall, B. K. (2005). *Bones and cartilage: Developmental and evolutionary skeletal biology*. San Diego, CA: Elsevier Academic Press.
Hall, J., Jheon, A. H., Ealba, E. L., Eames, B. F., Butcher, K. D., Mak, S. S., et al. (2014). Evolution of a developmental mechanism: Species-specific regulation of the cell cycle and the timing of events during craniofacial osteogenesis. *Developmental Biology*, *385*, 380–395.
Hall, B. K., & Olson, W. (2003). *Keywords and concepts in evolutionary developmental biology*. Cambridge, MA: Harvard University Press.
Hancox, N. M. (1949). The osteoclast. *Biological Reviews*, *24*, 448–471.
Hanken, J., & Hall, B. K. (1993). *The skull*. Chicago: University of Chicago Press.
Havens, B. A., Velonis, D., Kronenberg, M. S., Lichtler, A. C., Oliver, B., & Mina, M. (2008). Roles of FGFR3 during morphogenesis of Meckel's cartilage and mandibular bones. *Developmental Biology*, *316*, 336–349.
Healy, C., Uwanogho, D., & Sharpe, P. T. (1996). Expression of the chicken Sox9 gene marks the onset of cartilage differentiation. *Annals of the New York Academy of Sciences*, *785*, 261–262.
Healy, C., Uwanogho, D., & Sharpe, P. T. (1999). Regulation and role of Sox9 in cartilage formation. *Developmental Dynamics*, *215*, 69–78.
Helms, J. A., & Schneider, R. A. (2003). Cranial skeletal biology. *Nature*, *423*, 326–331.
Hu, D., & Marcucio, R. S. (2009). Unique organization of the frontonasal ectodermal zone in birds and mammals. *Developmental Biology*, *325*, 200–210.
Hu, D., & Marcucio, R. S. (2012). Neural crest cells pattern the surface cephalic ectoderm during FEZ formation. *Developmental Dynamics*, *241*, 732–740.
Hu, D., Marcucio, R. S., & Helms, J. A. (2003). A zone of frontonasal ectoderm regulates patterning and growth in the face. *Development*, *130*, 1749–1758.
Hu, D., Young, N. M., Li, X., Xu, Y., Hallgrimsson, B., & Marcucio, R. S. (2015a). A dynamic Shh expression pattern, regulated by SHH and BMP signaling, coordinates fusion of primordia in the amniote face. *Development*, *142*, 567–574.
Hu, D., Young, N. M., Xu, Q., Jamniczky, H., Green, R. M., Mio, W., et al. (2015b). Signals from the brain induce variation in avian facial shape. *Developmental Dynamics*, *244*, 1133–1143.
Hui, C. C., & Angers, S. (2011). Gli proteins in development and disease. *Annual Review of Cell and Developmental Biology*, *27*, 513–537.
Hunt, P., Clarke, J. D., Buxton, P., Ferretti, P., & Thorogood, P. (1998). Stability and plasticity of neural crest patterning and branchial arch Hox code after extensive cephalic crest rotation. *Developmental Biology*, *198*, 82–104.
Hunt, P., Ferretti, P., Krumlauf, R., & Thorogood, P. (1995). Restoration of normal Hox code and branchial arch morphogenesis after extensive deletion of hindbrain neural crest. *Developmental Biology*, *168*, 584–597.
Huxley, J. S. (1932). *Problems of relative growth*. London: Methuen.

Hwang, S. Y., & Putney, J. W., Jr. (2011). Calcium signaling in osteoclasts. *Biochimica et Biophysica Acta, 1813*, 979–983.
Ito, Y., Yeo, J. Y., Chytil, A., Han, J., Bringas, P., Jr., Nakajima, A., et al. (2003). Conditional inactivation of Tgfbr2 in cranial neural crest causes cleft palate and calvaria defects. *Development, 130*, 5269–5280.
Jheon, A. H., & Schneider, R. A. (2009). The cells that fill the bill: Neural crest and the evolution of craniofacial development. *Journal of Dental Research, 88*, 12–21.
Johansson, N., Saarialho-Kere, U., Airola, K., Herva, R., Nissinen, L., Westermarck, J., et al. (1997). Collagenase-3 (MMP-13) is expressed by hypertrophic chondrocytes, periosteal cells, and osteoblasts during human fetal bone development. *Developmental Dynamics, 208*, 387–397.
Joshi, N., Hamdan, A. M., & Fakhouri, W. D. (2014). Skeletal malocclusion: A developmental disorder with a life-long morbidity. *Journal of Clinical Medicine Research, 6*, 399–408.
Jotereau, F. V., & Le Douarin, N. M. (1978). The development relationship between osteocytes and osteoclasts: A study using the quail-chick nuclear marker in endochondral ossification. *Developmental Biology, 63*, 253–265.
Kahn, J., Shwartz, Y., Blitz, E., Krief, S., Sharir, A., Breitel, D. A., et al. (2009). Muscle contraction is necessary to maintain joint progenitor cell fate. *Developmental Cell, 16*, 734–743.
Kajiya, H. (2012). Calcium signaling in osteoclast differentiation and bone resorption. *Advances in Experimental Medicine and Biology, 740*, 917–932.
Kavumpurath, S., & Hall, B. K. (1990). Lack of either chondrocyte hypertrophy or osteogenesis in Meckel's cartilage of the embryonic chick exposed to epithelia and to thyroxine in vitro. *Journal of Craniofacial Genetics and Developmental Biology, 10*, 263–275.
Kimmel, C. B., Ullmann, B., Walker, C., Wilson, C., Currey, M., Phillips, P. C., et al. (2005). Evolution and development of facial bone morphology in threespine sticklebacks. *Proceedings of the National Academy of Sciences of the United States of America, 102*, 5791–5796.
Kirschner, M., & Gerhart, J. (1998). Evolvability. *Proceedings of the National Academy of Sciences of the United States of America, 95*, 8420–8427.
Köntges, G., & Lumsden, A. (1996). Rhombencephalic neural crest segmentation is preserved throughout craniofacial ontogeny. *Development, 122*, 3229–3242.
Lande, R. (1979). Quantitative genetic analysis of multivariate evolution, applied to brain: Body size allometry. *Evolution, 33*, 402–416.
Le Lièvre, C. S. (1978). Participation of neural crest-derived cells in the genesis of the skull in birds. *Journal of Embryology and Experimental Morphology, 47*, 17–37.
Le Lièvre, C. S., & Le Douarin, N. M. (1975). Mesenchymal derivatives of the neural crest: Analysis of chimeric quail and chick embryos. *Journal of Embryology and Experimental Morphology, 34*, 125–154.
Leevers, S. J., & McNeill, H. (2005). Controlling the size of organs and organisms. *Current Opinion in Cell Biology, 17*, 604–609.
Lezot, F., Chesneau, J., Battaglia, S., Brion, R., Castaneda, B., Farges, J. C., et al. (2014). Preclinical evidence of potential craniofacial adverse effect of zoledronic acid in pediatric patients with bone malignancies. *Bone, 68*, 146–152.
Liu, W., Selever, J., Murali, D., Sun, X., Brugger, S. M., Ma, L., et al. (2005). Threshold-specific requirements for Bmp4 in mandibular development. *Developmental Biology, 283*, 282–293.
Loeys, B. L., Chen, J., Neptune, E. R., Judge, D. P., Podowski, M., Holm, T., et al. (2005). A syndrome of altered cardiovascular, craniofacial, neurocognitive and skeletal development caused by mutations in TGFBR1 or TGFBR2. *Nature Genetics, 37*, 275–281.

Lufkin, T., Mark, M., Hart, C., Dollé, P., Lemeur, M., & Chambon, P. (1992). Homeotic transformation of the occipital bones of the skull by ectopic expression of a homeobox gene. *Nature, 359*, 835–841.

Lwigale, P. Y., & Schneider, R. A. (2008). Other chimeras: Quail-duck and mouse-chick. *Methods in Cell Biology, 87*, 59–74.

MacDonald, M. E., Abbott, U. K., & Richman, J. M. (2004). Upper beak truncation in chicken embryos with the cleft primary palate mutation is due to an epithelial defect in the frontonasal mass. *Developmental Dynamics, 230*, 335–349.

Marcucio, R. S., Cordero, D. R., Hu, D., & Helms, J. A. (2005). Molecular interactions coordinating the development of the forebrain and face. *Developmental Biology, 284*, 48–61.

Marcucio, R. S., Young, N. M., Hu, D., & Hallgrimsson, B. (2011). Mechanisms that underlie co-variation of the brain and face. *Genesis, 49*, 177–189.

Martin, T. J., & Ng, K. W. (1994). Mechanisms by which cells of the osteoblast lineage control osteoclast formation and activity. *Journal of Cellular Biochemistry, 56*, 357–366.

Martinez-Abadias, N., Holmes, G., Pankratz, T., Wang, Y., Zhou, X., Jabs, E. W., et al. (2013). From shape to cells: Mouse models reveal mechanisms altering palate development in Apert syndrome. *Disease Models & Mechanisms, 6*, 768–779.

Martinez-Abadias, N., Motch, S. M., Pankratz, T. L., Wang, Y., Aldridge, K., Jabs, E. W., et al. (2013). Tissue-specific responses to aberrant FGF signaling in complex head phenotypes. *Developmental Dynamics, 242*, 80–94.

McKinney, M. L. (1988). Classifying heterochrony: Allometry, size, and time. In M. L. McKinney (Ed.), *Heterochrony in evolution: A multidisciplinary approach* (pp. 17–34). New York: Plenum Press.

Melnick, M., Witcher, D., Bringas, P., Jr., Carlsson, P., & Jaskoll, T. (2005). Meckel's cartilage differentiation is dependent on hedgehog signaling. *Cells, Tissues, Organs, 179*, 146–157.

Merrill, A. E., Eames, B. F., Weston, S. J., Heath, T., & Schneider, R. A. (2008). Mesenchyme-dependent BMP signaling directs the timing of mandibular osteogenesis. *Development, 135*, 1223–1234.

Mina, M., Wang, Y. H., Ivanisevic, A. M., Upholt, W. B., & Rodgers, B. (2002). Region- and stage-specific effects of FGFs and BMPs in chick mandibular morphogenesis. *Developmental Dynamics, 223*, 333–352.

Minkin, C. (1982). Bone acid phosphatase: Tartrate-resistant acid phosphatase as a marker of osteoclast function. *Calcified Tissue International, 34*, 285–290.

Mitgutsch, C., Wimmer, C., Sanchez-Villagra, M. R., Hahnloser, R., & Schneider, R. A. (2011). Timing of ossification in duck, quail, and zebra finch: Intraspecific variation, heterochronies, and life history evolution. *Zoological Science, 28*, 491–500.

Mitsiadis, T. A., Caton, J., & Cobourne, M. (2006). Waking-up the sleeping beauty: Recovery of the ancestral bird odontogenic program. *Journal of Experimental Zoology. Part B, Molecular and Developmental Evolution, 306*, 227–233.

Mo, R., Freer, A. M., Zinyk, D. L., Crackower, M. A., Michaud, J., Heng, H. H., et al. (1997). Specific and redundant functions of Gli2 and Gli3 zinc finger genes in skeletal patterning and development. *Development, 124*, 113–123.

Moore, W. J. (1981). *The mammalian skull*. Cambridge: Cambridge University Press.

Murakami, S., Kan, M., McKeehan, W. L., & de Crombrugghe, B. (2000). Up-regulation of the chondrogenic Sox9 gene by fibroblast growth factors is mediated by the mitogen-activated protein kinase pathway. *Proceedings of the National Academy of Sciences of the United States of America, 97*, 1113–1118.

Nguyen, J., Tang, S. Y., Nguyen, D., & Alliston, T. (2013). Load regulates bone formation and Sclerostin expression through a TGFbeta-dependent mechanism. *PLoS One, 8*, e53813.

Nikitina, N., Sauka-Spengler, T., & Bronner-Fraser, M. (2008). Dissecting early regulatory relationships in the lamprey neural crest gene network. *Proceedings of the National Academy of Sciences of the United States of America, 105*, 20083–20088.
Noden, D. M. (1978). The control of avian cephalic neural crest cytodifferentiation. I. Skeletal and connective tissues. *Developmental Biology, 67*, 296–312.
Noden, D. M. (1982). Patterns and organization of craniofacial skeletogenic mesenchyme: A perspective. In A. D. Dixon & B. G. Sarnat (Eds.), *Factors and mechanisms influencing bone growth* (pp. 168–203). New York: Alan R. Liss.
Noden, D. M. (1983). The role of the neural crest in patterning of avian cranial skeletal, connective, and muscle tissues. *Developmental Biology, 96*, 144–165.
Noden, D., & Schneider, R. A. (2006a). Neural crest cells and the community of plan for craniofacial development: Historical debates and current perspectives. In J.-P. Saint-Jeannet (Ed.), *Neural crest induction and differentiation* (pp. 1–23). Georgetown, TX: Landes Bioscience.
Noden, D. M., & Schneider, R. A. (2006b). Neural crest cells and the community of plan for craniofacial development: Historical debates and current perspectives. *Advances in Experimental Medicine and Biology, 589*, 1–23.
Noden, D. M., & Trainor, P. A. (2005). Relations and interactions between cranial mesoderm and neural crest populations. *Journal of Anatomy, 207*, 575–601.
Nomura, M., & Li, E. (1998). Smad2 role in mesoderm formation, left-right patterning and craniofacial development. *Nature, 393*, 786–790.
Northcutt, R. G. (2005). The new head hypothesis revisited. *Journal of Experimental Zoology. Part B, Molecular and Developmental Evolution, 304B*, 274–297.
O'Brien, C. A., Plotkin, L. I., Galli, C., Goellner, J. J., Gortazar, A. R., Allen, M. R., et al. (2008). Control of bone mass and remodeling by PTH receptor signaling in osteocytes. *PLoS One, 3*, e2942.
Oka, K., Oka, S., Hosokawa, R., Bringas, P., Jr., Brockhoff, H. C., 2nd, Nonaka, K., et al. (2008). TGF-beta mediated Dlx5 signaling plays a crucial role in osteo-chondroprogenitor cell lineage determination during mandible development. *Developmental Biology, 321*, 303–309.
Oka, K., Oka, S., Sasaki, T., Ito, Y., Bringas, P., Jr., Nonaka, K., et al. (2007). The role of TGF-beta signaling in regulating chondrogenesis and osteogenesis during mandibular development. *Developmental Biology, 303*, 391–404.
Parsons, K. J., & Albertson, R. C. (2009). Roles for Bmp4 and CaM1 in shaping the jaw: Evo-devo and beyond. *Annual Review of Genetics, 43*, 369–388.
Pasqualetti, M., Ori, M., Nardi, I., & Rijli, F. M. (2000). Ectopic Hoxa2 induction after neural crest migration results in homeosis of jaw elements in Xenopus. *Development, 127*, 5367–5378.
Petiot, A., Ferretti, P., Copp, A. J., & Chan, C. T. (2002). Induction of chondrogenesis in neural crest cells by mutant fibroblast growth factor receptors. *Developmental Dynamics, 224*, 210–221.
Pineda-Alvarez, D. E., Roessler, E., Hu, P., Srivastava, K., Solomon, B. D., Siple, C. E., et al. (2012). Missense substitutions in the GAS1 protein present in holoprosencephaly patients reduce the affinity for its ligand, SHH. *Human Genetics, 131*, 301–310.
Pointer, M. A., Kamilar, J. M., Warmuth, V., Chester, S. G., Delsuc, F., Mundy, N. I., et al. (2012). RUNX2 tandem repeats and the evolution of facial length in placental mammals. *BMC Evolutionary Biology, 12*, 103.
Qing, H., Ardeshirpour, L., Pajevic, P. D., Dusevich, V., Jahn, K., Kato, S., et al. (2012). Demonstration of osteocytic perilacunar/canalicular remodeling in mice during lactation. *Journal of Bone and Mineral Research, 27*, 1018–1029.
Qiu, M., Bulfone, A., Ghattas, I., Meneses, J. J., Christensen, L., Sharpe, P. T., et al. (1997). Role of the Dlx homeobox genes in proximodistal patterning of the branchial arches:

Mutations of Dlx-1, Dlx-2, and Dlx-1 and -2 alter morphogenesis of proximal skeletal and soft tissue structures derived from the first and second arches. *Developmental Biology, 185*, 165–184.

Radlanski, R. J., & Klarkowski, M. C. (2001). Bone remodeling of the human mandible during prenatal development. *Journal of Orofacial Orthopedics, 62*, 191–201.

Radlanski, R. J., Renz, H., Lajvardi, S., & Schneider, R. A. (2004). Bone remodeling during prenatal morphogenesis of the human mental foramen. *European Journal of Oral Sciences, 112*, 301–310.

Reponen, P., Sahlberg, C., Munaut, C., Thesleff, I., & Tryggvason, K. (1994). High expression of 92-kD type IV collagenase (gelatinase B) in the osteoclast lineage during mouse development. *Journal of Cell Biology, 124*, 1091–1102.

Richman, J. M., Herbert, M., Matovinovic, E., & Walin, J. (1997). Effect of fibroblast growth factors on outgrowth of facial mesenchyme. *Developmental Biology, 189*, 135–147.

Richman, J. M., & Tickle, C. (1989). Epithelia are interchangeable between facial primordia of chick embryos and morphogenesis is controlled by the mesenchyme. *Developmental Biology, 136*, 201–210.

Richman, J. M., & Tickle, C. (1992). Epithelial-mesenchymal interactions in the outgrowth of limb buds and facial primordia in chick embryos. *Developmental Biology, 154*, 299–308.

Rijli, F. M., Mark, M., Lakkaraju, S., Dierich, A., Dolle, P., & Chambon, P. (1993). A homeotic transformation is generated in the rostral branchial region of the head by disruption of *Hoxa-2*, which acts as a selector gene. *Cell, 75*, 1333–1349.

Roessler, E., & Muenke, M. (2010). The molecular genetics of holoprosencephaly. *American Journal of Medical Genetics. Part C, Seminars in Medical Genetics, 154C*, 52–61.

Rowe, A., Richman, J. M., & Brickell, P. M. (1992). Development of the spatial pattern of retinoic acid receptor-beta transcripts in embryonic chick facial primordia. *Development, 114*, 805–813.

Sanford, L. P., Ormsby, I., Gittenberger-de Groot, A. C., Sariola, H., Friedman, R., Boivin, G. P., et al. (1997). TGFbeta2 knockout mice have multiple developmental defects that are non-overlapping with other TGFbeta knockout phenotypes. *Development, 124*, 2659–2670.

Sasano, Y., Zhu, J. X., Tsubota, M., Takahashi, I., Onodera, K., Mizoguchi, I., et al. (2002). Gene expression of MMP8 and MMP13 during embryonic development of bone and cartilage in the rat mandible and hind limb. *The Journal of Histochemistry and Cytochemistry, 50*, 325–332.

Satokata, I., & Maas, R. (1994). *Msx1* deficient mice exhibit cleft palate and abnormalities of craniofacial and tooth development. *Nature Genetics, 6*, 348–356.

Scherson, T., Serbedzija, G., Fraser, S., & Bronner-Fraser, M. (1993). Regulative capacity of the cranial neural tube to form neural crest. *Development, 118*, 1049–1062.

Schilling, T. F. (1997). Genetic analysis of craniofacial development in the vertebrate embryo. *Bioessays, 19*, 459–468.

Schneider, R. A. (2005). Developmental mechanisms facilitating the evolution of bills and quills. *Journal of Anatomy, 207*, 563–573.

Schneider, R. A. (2007). How to tweak a beak: Molecular techniques for studying the evolution of size and shape in Darwin's finches and other birds. *Bioessays, 29*, 1–6.

Schneider, R. A., & Helms, J. A. (2003). The cellular and molecular origins of beak morphology. *Science, 299*, 565–568.

Schneider, R. A., Hu, D., & Helms, J. A. (1999). From head to toe: Conservation of molecular signals regulating limb and craniofacial morphogenesis. *Cell and Tissue Research, 296*, 103–109.

Schneider, R. A., Hu, D., Rubenstein, J. L., Maden, M., & Helms, J. A. (2001). Local retinoid signaling coordinates forebrain and facial morphogenesis by maintaining FGF8 and SHH. *Development, 128*, 2755–2767.

Schowing, J. (1968). Influence inductrice de l'encéphale embryonnaire sur le développement du crâne chez le Poulet. *Journal of Embryology and Experimental Morphology*, *19*, 9–32.

Seales, E. C., Micoli, K. J., & McDonald, J. M. (2006). Calmodulin is a critical regulator of osteoclastic differentiation, function, and survival. *Journal of Cellular Biochemistry*, *97*, 45–55.

Sears, K. E., Goswami, A., Flynn, J. J., & Niswander, L. A. (2007). The correlated evolution of Runx2 tandem repeats, transcriptional activity, and facial length in carnivora. *Evolution & Development*, *9*, 555–565.

Sechrist, J., Nieto, M. A., Zamanian, R. T., & Bronner-Fraser, M. (1995). Regulative response of the cranial neural tube after neural fold ablation: Spatiotemporal nature of neural crest regeneration and up-regulation of slug. *Development*, *121*, 4103–4115.

Sharpe, P. M., & Ferguson, M. W. (1988). Mesenchymal influences on epithelial differentiation in developing systems. *Journal of Cell Science. Supplement*, *10*, 195–230.

Shigetani, Y., Nobusada, Y., & Kuratani, S. (2000). Ectodermally derived FGF8 defines the maxillomandibular region in the early chick embryo: Epithelial-mesenchymal interactions in the specification of the craniofacial ectomesenchyme. *Developmental Biology*, *228*, 73–85.

Siegel, A. F., & Benson, R. H. (1982). A robust comparison of biological shapes. *Biometrics*, *38*, 341–350.

Smith, D. W. (1997). *Smith's recognizable patterns of human malformation* (5th ed.). Philadelphia: W.B. Saunders.

Smith, F. J., Percival, C. J., Young, N. M., Hu, D., Schneider, R. A., Marcucio, R. S., et al. (2015). Divergence of craniofacial developmental trajectories among avian embryos. *Developmental Dynamics*, *244*, 1158–1167.

Smith, K. K., & Schneider, R. A. (1998). Have gene knockouts caused evolutionary reversals in the mammalian first arch? *BioEssays*, *20*, 245–255.

Solem, R. C., Eames, B. F., Tokita, M., & Schneider, R. A. (2011). Mesenchymal and mechanical mechanisms of secondary cartilage induction. *Developmental Biology*, *356*, 28–39.

Stanier, P., & Pauws, E. (2012). Development of the lip and palate: FGF signalling. *Frontiers of Oral Biology*, *16*, 71–80.

Starck, J. M. (1989). Zeitmuster der Ontogenesen bei nestflüchtenden und nesthockenden Vögeln. *Courier Forschungsinstitut Senckenberg*, *114*, 1–319.

Stottmann, R. W., Anderson, R. M., & Klingensmith, J. (2001). The BMP antagonists Chordin and Noggin have essential but redundant roles in mouse mandibular outgrowth. *Developmental Biology*, *240*, 457–473.

Szabo-Rogers, H. L., Geetha-Loganathan, P., Nimmagadda, S., Fu, K. K., & Richman, J. M. (2008). FGF signals from the nasal pit are necessary for normal facial morphogenesis. *Developmental Biology*, *318*, 289–302.

Tang, S. Y., Herber, R. P., Ho, S. P., & Alliston, T. (2012). Matrix metalloproteinase-13 is required for osteocytic perilacunar remodeling and maintains bone fracture resistance. *Journal of Bone and Mineral Research*, *27*, 1936–1950.

Teitelbaum, S. L. (2000). Bone resorption by osteoclasts. *Science*, *289*, 1504–1508.

Teitelbaum, S. L., Tondravi, M. M., & Ross, F. P. (1997). Osteoclasts, macrophages, and the molecular mechanisms of bone resorption. *Journal of Leukocyte Biology*, *61*, 381–388.

Thompson, D. A. W. (1917). *On growth and form*. Cambridge, England: University Press.

Thorogood, P. (1987). Mechanisms of morphogenetic specification in skull development. In J. R. Wolff, J. Sievers, & M. Berry (Eds.), *Mesenchymal-epithelial interactions in neural development* (pp. 141–152). Berlin: Springer-Verlag.

Thorogood, P., Bee, J., & von der Mark, K. (1986). Transient expression of collagen type II at epitheliomesenchymal interfaces during morphogenesis of the cartilaginous neurocranium. *Developmental Biology*, *116*, 497–509.

Tokita, M., & Schneider, R. A. (2009). Developmental origins of species-specific muscle pattern. *Developmental Biology, 331*, 311–325.

Trumpp, A., Depew, M. J., Rubenstein, J. L., Bishop, J. M., & Martin, G. R. (1999). Cre-mediated gene inactivation demonstrates that FGF8 is required for cell survival and patterning of the first branchial arch. *Genes and Development, 13*, 3136–3148.

Tucker, A. S., & Lumsden, A. (2004). Neural crest cells provide species-specific patterning information in the developing branchial skeleton. *Evolution & Development, 6*, 32–40.

Tyler, M. S. (1978). Epithelial influences on membrane bone formation in the maxilla of the embryonic chick. *Anatomical Record, 192*, 225–233.

Tyler, M. S. (1983). Development of the frontal bone and cranial meninges in the embryonic chick: An experimental study of tissue interactions. *Anatomical Record, 206*, 61–70.

Wagner, G. (1959). Untersuchungen an Bombinator-Triton-Chimaeren. *Roux' Archiv für Entwicklungsmechanik der Organismen, 151*, 136–158.

Wilson, J., & Tucker, A. S. (2004). Fgf and Bmp signals repress the expression of Bapx1 in the mandibular mesenchyme and control the position of the developing jaw joint. *Developmental Biology, 266*, 138–150.

Wu, P., Jiang, T. X., Shen, J. Y., Widelitz, R. B., & Chuong, C. M. (2006). Morphoregulation of avian beaks: Comparative mapping of growth zone activities and morphological evolution. *Developmental Dynamics, 235*, 1400–1412.

Wu, P., Jiang, T. X., Suksaweang, S., Widelitz, R. B., & Chuong, C. M. (2004). Molecular shaping of the beak. *Science, 305*, 1465–1466.

Xia, S. L., & Ferrier, J. (1996). Localized calcium signaling in multinucleated osteoclasts. *Journal of Cellular Physiology, 167*, 148–155.

Xiong, J., & O'Brien, C. A. (2012). Osteocyte RANKL: New insights into the control of bone remodeling. *Journal of Bone and Mineral Research, 27*, 499–505.

Xiong, J., Piemontese, M., Thostenson, J. D., Weinstein, R. S., Manolagas, S. C., & O'Brien, C. A. (2014). Osteocyte-derived RANKL is a critical mediator of the increased bone resorption caused by dietary calcium deficiency. *Bone, 66*, 146–154.

Young, N. M., Chong, H. J., Hu, D., Hallgrimsson, B., & Marcucio, R. S. (2010). Quantitative analyses link modulation of sonic hedgehog signaling to continuous variation in facial growth and shape. *Development, 137*, 3405–3409.

Young, N. M., Hu, D., Lainoff, A. J., Smith, F. J., Diaz, R., Tucker, A. S., et al. (2014). Embryonic bauplans and the developmental origins of facial diversity and constraint. *Development, 141*, 1059–1063.

Zayzafoon, M. (2006). Calcium/calmodulin signaling controls osteoblast growth and differentiation. *Journal of Cellular Biochemistry, 97*, 56–70.

Zhang, Y., Blackwell, E. L., McKnight, M. T., Knutsen, G. R., Vu, W. T., & Ruest, L. B. (2012). Specific inactivation of Twist1 in the mandibular arch neural crest cells affects the development of the ramus and reveals interactions with hand2. *Developmental Dynamics, 241*, 924–940.

Zhao, Q., Eberspaecher, H., Lefebvre, V., & De Crombrugghe, B. (1997). Parallel expression of Sox9 and Col2a1 in cells undergoing chondrogenesis. *Developmental Dynamics, 209*, 377–386.

Zhao, H., Oka, K., Bringas, P., Kaartinen, V., & Chai, Y. (2008). TGF-beta type I receptor Alk5 regulates tooth initiation and mandible patterning in a type II receptor-independent manner. *Developmental Biology, 320*, 19–29.

Zouvelou, V., Luder, H. U., Mitsiadis, T. A., & Graf, D. (2009). Deletion of BMP7 affects the development of bones, teeth, and other ectodermal appendages of the orofacial complex. *Journal of Experimental Zoology. Part B, Molecular and Developmental Evolution, 312B*, 361–374.

CHAPTER TWELVE

Facial Morphogenesis: Physical and Molecular Interactions Between the Brain and the Face

Ralph Marcucio*,[1], Benedikt Hallgrimsson[†,‡], Nathan M. Young*
*Department of Orthopaedic Surgery, San Francisco General Hospital, Orthopaedic Trauma Institute, University of California San Francisco, San Francisco, California, USA
[†]Department of Cell Biology and Anatomy, McCaig Bone and Joint Institute, University of Calgary, Calgary, Alberta, Canada
[‡]Alberta Children's Hospital, Research Institute, University of Calgary, Calgary, Alberta, Canada
[1]Corresponding author: e-mail address: ralph.marcucio@ucsf.edu

Contents

Abstract

Morphogenesis of the brain and face is intrinsically linked by a number of factors. These include: origins of tissues, adjacency allowing their physical interactions, and molecular cross talk controlling growth. Neural crest cells that form the facial primordia originate on the dorsal neural tube. In the caudal pharyngeal arches, a Homeobox code regulates arch identity. In anterior regions, positional information is acquired locally. Second, the brain is a structural platform that influences positioning of the facial primordia, and brain growth influences the timing of primordia fusion. Third, the brain helps induce a signaling center, the frontonasal ectodermal zone, in the ectoderm, which participates in patterned growth of the upper jaw. Similarly, signals from neural crest cells regulate expression of fibroblast growth factor 8 in the anterior neural ridge, which controls

Current Topics in Developmental Biology, Volume 115
ISSN 0070-2153
http://dx.doi.org/10.1016/bs.ctdb.2015.09.001

growth of the anterior forebrain. Disruptions to these interactions have significant consequences for normal development of the craniofacial complex, leading to structural malformations and birth defects.

1. INTRODUCTION

Clinicians have long recognized that facial malformations are often associated with defects in the underlying brain. In 1964, DeMeyer coined the phrase "the face predicts the brain" after studying fetuses with holoprosencephaly (HPE; DeMyer, 1964), a common and severe malformation that affects both the brain and the face. Notably, DeMeyer observed that HPE has a phenotypic spectrum ranging from cyclopia to mid-facial hypoplasia, a missing central incisor, or even normal phenotype, and these malformations are accompanied by underlying brain anomalies that likewise range from failure of the forebrain to separate into right and left lobes to normal septation (Heussler, Suri, Young, & Muenke, 2002; Ming & Muenke, 1998; Nanni, Schelper, & Muenke, 2000). These observations highlight a critical feature of brain and facial morphogenesis: the developmental processes that control their formation are highly integrated, meaning that changes in one often have downstream effects on the other. These effects can be divided into three general categories. First, the brain and face have a common origin. Notably, the neural crest cells that give rise to the facial skeleton originate from the dorsal neural tube, and information about their spatial patterning is based on their anterior–posterior level of origin. Second, the brain serves as a structural platform, and as such imparts physical forces on adjacent tissues that help shape the location and directional growth of facial primordia. Third, the brain and facial tissues "talk to each other" via molecular interactions that control the cellular mechanisms responsible for facial morphogenesis. These three types of interactions suggest that the brain exerts multiple effects on facial morphogenesis, and these are both intrinsic to brain development itself and the paracrine interactions between the brain and the adjacent tissues. Thus, the development of the face is highly contingent on normal brain development, and consequently, events and disruptions to the processes that regulate the brain often produce facial malformations. These interactions, which are both physical and molecular, form the basis of this chapter. We describe each of these in more detail below.

2. RELATIONSHIP BETWEEN THE BRAIN AND THE FACE

2.1 Origins of the Brain and Face

The brain and face begin to take shape during the earliest stages of development. After gastrulation, the anterior neural plate undergoes neurulation, morphogenetic movements that turn the sheet of neuroepithelial cells into a tube and physically separate it from the surface ectoderm. In the anterior-most region of the neural tube, the anterior neural pore forms (Fig. 1). At this region, the neural and surface ectoderm are continuous, and this tissue expresses *fibroblast growth factor 8* (*Fgf8*). After the anterior neural pore closes, the neural and surface ectoderm are segregated, but *Fgf8* expression is

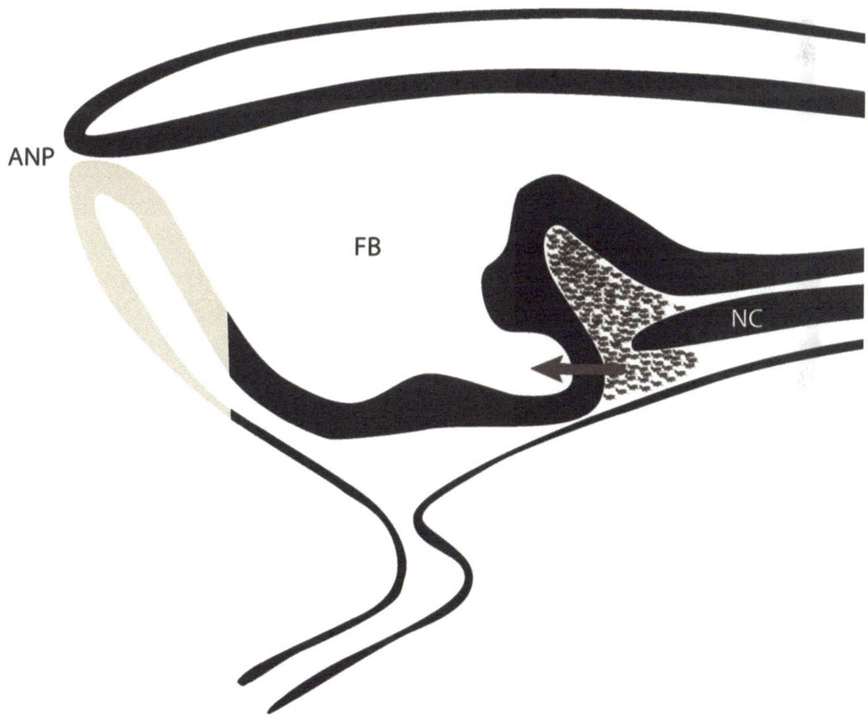

Figure 1 Shared expression domains between the brain and the face. Closure of the anterior neural pore segregates anterior neural and surface ectoderm. At this time, the *Fgf8* domain (yellow) that spans the anterior neural plate and the anterior surface ectoderm becomes two distinct domains. ANP, anterior neuropore; FB, forebrain; red arrow, signaling from mesendoderm to forebrain. (A color version of this figure is located at the end of this chapter.)

maintained in both ectodermal compartments (Ohkubo, Chiang, & Rubenstein, 2002; see Fig. 1). In this way, a domain of gene expression that was initially continuous through the neural and surface ectoderm is segregated into two discrete expression domains by closure of the anterior neural pore. The *Fgf8*-expressing cells in the forebrain will form the anterior neural ridge, and the *Fgf8*-expressing cells in the surface ectoderm will form the dorsal aspect of the frontonasal ectodermal zone (FEZ), a signaling center that regulates patterned growth of the upper jaw anlagen (Hu, Marcucio, & Helms, 2003).

Similarly, the neural crest cells, which are the connective tissue progenitor cells that form the facial skeleton, originate at the border of the neural and surface ectoderm, and their location of origin along the anterior–posterior axis of the brain plays a large role in the fate of the neural crest cells. Homeobox genes (Hox) along the dorsal axis of the neural tube exhibit unique anterior–posterior patterns of expression (Fienberg et al., 1987; Holland & Hogan, 1988; McGinnis & Krumlauf, 1992). The pattern of Hox gene expression, or Hox "code," is imparted on the neural crest cells, while they are part of the dorsal neuroepithelium, and this code is carried into the developing facial primordia, with significant implications for facial development (Minoux & Rijli, 2010; Rijli et al., 1993). The anterior-most region of the neural tube, including the prosencephalon and mesencephalon, does not express members of classic Hox gene family. Consequently, the neural crest cells derived from these regions do not have a Hox code, so when they arrive in the frontonasal process (FNP), maxillary process and mandibular process, these cell populations are interchangeable with each other (Noden, 1983, 1984, 1986). In contrast, the hindbrain expresses Hox genes and neural crest cells contain a Hox code, which they carry with them to the facial primordia (Minoux & Rijli, 2010). When neural crest progenitors derived from HOX-free regions of the neural tube are transplanted into posterior regions, where there normally is a Hox code, development is disrupted. In a classic series of experiments, Noden transplanted crest destined for the first pharyngeal arch in place of neural crest cells destined for the second arch, and the resulting skeletal structure resembled the first arch (Noden, 1983). Similarly, ablation of *Hoxa2* in mice transformed the second arch skeleton into a first arch skeleton (Rijli et al., 1993), and work in avian embryos demonstrated that this patterning mechanism is conserved in birds (Grammatopoulos, Bell, Toole, Lumsden, & Tucker, 2000). Thus, in parts of the head skeleton, the brain imparts patterning information on the connective tissue progenitors as a molecular program within the neural crest cells, but this is significantly influenced by environmental signals that neural crest cells encounter (Trainor, Ariza-McNaughton, & Krumlauf, 2002; Trainor & Krumlauf, 2000).

As described above, not all neural crest cells in the developing head contain an intrinsic Hox code. In the developing lower and upper jaw anlagen, the neural crest cells do not express Hox genes, and these cells receive patterning information from their environment. This has been well described in the maxillary and mandibular processes, where a nested pattern of Dlx gene expression within the pharyngeal arch mesenchyme confers arch identity to the developing facial primordia (Depew, Simpson, Morasso, & Rubenstein, 2005). Results of this work led to the proposal of the "hinge and caps" model of jaw development in which signaling centers located at the proximal (hinge) and distal (caps) regions of the upper and lower jaw act to control development of these two structures in an integrated fashion (Depew & Compagnucci, 2008; Depew & Simpson, 2006).

After the neural crest cells have completed their migration into the distinct facial primordia, facial morphogenesis begins through a process of growth and patterning. The FNP and paired maxillary processes grow out and fuse to form the upper jaw and midface, the paired mandibular processes form the lower jaw, and the more posterior pharyngeal arches give rise to the more posterior skeletal structures. Neural crest cells are a transient mesenchymal population of progenitor cells that are generated in the dorsal neural tube at the junction between the neuroepithelium and the surface ectoderm (Bronner & LeDouarin, 2012), and migrate laterally and ventrally to occupy the rostral region of the developing head to form these primordia *de novo* (Chambers & McGonnell, 2002; Minoux & Rijli, 2010; Noden & Trainor, 2005). In the upper jaw, the neural crest mesenchyme is encased by epithelia derived from neural and surface ectoderm, while in the lower jaw, pharyngeal endoderm and surface ectoderm encase the mandibular processes. The interactions among the distinct tissue types that comprise each of these primordia direct the growth and patterning of the distinct regions of the developing face (reviewed in Chai & Maxson, 2006; Graham, Okabe, & Quinlan, 2005; Richman & Lee, 2003; Szabo-Rogers, Smithers, Yakob, & Liu, 2010). In the lower jaw, signals among the pharyngeal endoderm, surface ectoderm, and neural crest mesenchyme control patterning and growth (Brito, Teillet, & Le Douarin, 2006; Graham et al., 2005; Graham & Smith, 2001; Swartz, Nguyen, McCarthy, & Eberhart, 2012), while in the upper jaw, physical and molecular interactions among the forebrain, surface cephalic ectoderm, and neural crest mesenchyme control patterned growth (Marcucio, Young, Hu, & Hallgrimsson, 2011).

2.2 Physical Interactions Between the Brain and the Face

The brain is the foundation upon which the face is built. In addition to sharing embryonic origins of tissues, the brain and face share a common space in the developing embryo. As the brain undergoes its own morphogenesis, the developing face is shaped in response. Both the pattern and the rate of growth of the brain provide unique influences on facial morphogenesis.

During early stages of development, the physical influences of the brain on the face are easily appreciated. The brain is essentially a large tube surrounded by mesenchymal cells of mesodermal and neural crest origin. As described above, during development of the upper jaw and midface, neural crest cells delaminate from the posterior forebrain and anterior midbrain and migrate along the surface of the brain between it and the surface ectoderm. This population of cells is initially small compared to the expanding neural tube, but once the neural crest cells arrive at their final destination, the population expands rapidly to form the facial primordia. At the same time, the brain continues to grow, changing both its size and shape. A series of studies by Diewert and colleagues showed that this process directly impacts the positioning of the facial primordia relative to the eyes and the brain (Diewert & Lozanoff, 1993a, 1993b; Diewert, Lozanoff, & Choy, 1993).

The rate of brain growth also influences these early stages of facial development. As the brain expands, it displaces the adjacent facial tissues, which must themselves compensate with their own growth to maintain their relative position to each other (Fig. 2). Variation in the coordination of the growth rates of the brain and face can have significant effects. For example, the brain in different strains of mice can exhibit different growth rates, and the growth rate of the facial tissues appears to be largely independent of the rate of brain growth. This independent growth likely results from distinct signaling centers located in the brain and face that contribute to regulating growth of each tissue. As the rate of brain growth slows among strains of mice, the face appears older at any given time point, such that slower brain growth produces a longer, more mature-looking face (Boughner et al., 2008). This occurs because when the brain is smaller relative to the facial prominences, then they will be more prognathic and more closely apposed. This effect results from the spatial constraints imposed by the brain on the developing face, and it illustrates that the brain acts as a foundation to shape

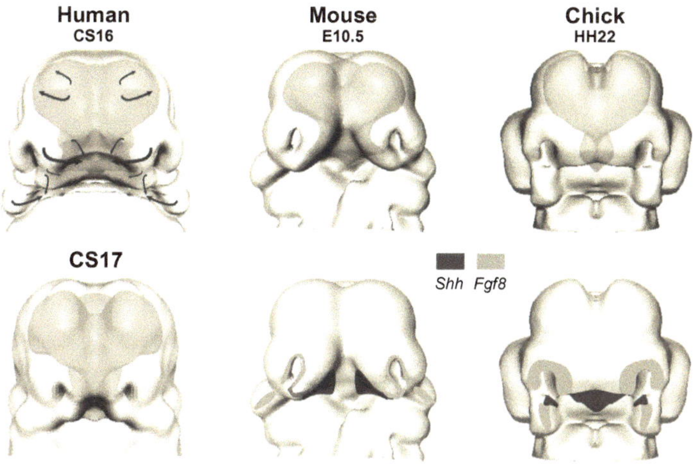

Figure 2 Growth of the brain and face. Relationship of the brain (shaded) to the face during facial morphogenesis shown at comparable stages in human (CS16), mouse (E10.5), and chicken (HH22). Facial prominences grow toward each other (arrows), converging at the midline, while the brain becomes relatively wider and divides into two distinct lobes. (A color version of this figure is located at the end of this chapter.)

the face. Thus, a small brain may lead to a relatively larger and more prognathic face. Conversely, a larger brain will lead to a flattened face. We have conjectured based on this relationship that a larger brain relative to the facial prominences may narrow the timing window of opportunity for facial prominence fusion and thus increase the likelihood of clefting (Boughner et al., 2008; Marcucio et al., 2011). This builds on similar arguments that were made earlier by Diewert and collaborators (Diewert & Lozanoff, 1993a, 1993b; Wang & Diewert, 1992; Wang, Juriloff, & Diewert, 1995).

Recent work has shown that the brain controls some of the temporal aspects of facial development. The effect of the brain on the timing of facial development can be observed by unilaterally transplanting the brain from the faster developing duck into the slower developing chick embryo. In this case, neural crest migration into the anterior regions of the face appears to be slowed, fusion of the primary palate is delayed, and development of the scleral papillae on the eye is delayed on the transplanted side compared to the control side (Hu et al., 2015). At this point, these data are phenomenological, and the molecular mechanism(s) underlying these delays are not known. Presumably, the temporal pattern of gene expression within the brain transmits some type of timing information to the adjacent tissues, but the nature of these signals, if they even exist, remains completely hypothetical.

These early influences of the brain on the face may be partially erased during later stages of development (Hallgrímsson et al., 2009). Some processes may exert different influences at different times and may have greater or lesser overall roles in development of these tissues. For example, while the brain has a tremendous influence during early stages of normal facial development, the contribution of the brain to the final form may not be as important as other processes intrinsic to the face. As the cartilages and bones form within the facial skeleton, the morphogenetic processes that shape the skeletal elements have a profound influence on facial shape and are largely responsible for sculpting the final form of the face (Aldridge et al., 2005; Diewert, 1983, 1985; Richtsmeier et al., 2006; Richtsmeier & Deleon, 2009). Nonetheless, it is easy to see how facial dysmorphology may arise from aberrant physical interactions between the brain and the face, for example, due to a growth rate that is too rapid, leading to a larger platform for the face to grow on and leaving the face relatively small. This configuration may lead to structural defects for which later developmental processes cannot fully compensate (discussed further below).

2.3 Molecular Interactions Between the Brain and the Face

In addition to the direct physical forces that the brain imparts on the developing face, there are also molecular interactions between the brain and the face that regulate morphogenesis of the upper jaw (Marcucio et al., 2011). These interactions begin during the earliest stages of embryonic development and continue throughout the period of patterned growth of the facial primordia that comprise the midface and upper jaw.

The prechordal plate (mesendoderm) is located beneath the prospective forebrain epithelium, and signals from the mesendoderm establish the basal plate of the forebrain (Fig. 1; Aoto et al., 2009; Muller et al., 2000). This signaling interaction is essential for proper development of the midface. SHH signaling from the mesendoderm to the neuroepithelium induces expression of Nk2.1 and Shh in the basal forebrain (Aoto et al., 2009). This domain eventually becomes the basal diencephalon. Then a signaling relay mechanism induces *Nkx2.1* and *Shh* expression in the more anteriorly located basal telencephalon (Marcucio, Cordero, Hu, & Helms, 2005). Disrupting this signal relay by blocking SHH signaling in the brain disrupts the induction of *Shh* in the telencephalon and produces severe malformations of the face (Marcucio et al., 2005). These early interactions are essential for establishing the basis for the molecular interactions that will occur during later periods of development.

Within the developing midface, a signaling center known as the FEZ regulates growth and patterning of the upper jaw (Hu et al., 2003). The FEZ, first identified in the chicken and later in mammals (Hu & Marcucio, 2009a, 2009b; Young et al., 2014), forms as a boundary in the surface cephalic ectoderm between cells expressing *Fgf8* and *Shh* (Fig. 2). The boundary between these two gene expression domains marks the distal tip of the upper jaw, and the ectoderm flanking the boundary corresponds to the dorsal and ventral aspect of the upper jaw. Subsequent research has shown that a series of signaling molecules including Wnts and bone morphogenetic protein (BMP) family members are also expressed in the FEZ (Ashique, Fu, & Richman, 2002; Barlow & Francis-West, 1997; Francis-West, Ladher, Barlow, & Graveson, 1998; Geetha-Loganathan et al., 2009; Lan et al., 2006; Reid, Yang, Melvin, Taketo, & Williams, 2011), and these signaling pathways are very likely to mediate aspects of FEZ function. The FEZ signals to the adjacent mesenchymal tissues to regulate proliferation to control directed growth of the upper jaw anlage. Shh is a very important component of FEZ function. Conditionally, removing Smoothened from the neural crest cells leads to severe defects of the upper jaw (Jeong, Mao, Tenzen, Kottmann, & McMahon, 2004), illustrating the importance of SHH signaling for development of this part of the face.

Formation and patterning of the FEZ appears to require signals from the brain. Research has shown that SHH signaling within the brain is responsible for inducing the pattern of *Shh* expression observed in the ventral forebrain. These experiments showed that blocking SHH signaling in the brain prevented *Shh* expression in the FEZ (Marcucio et al., 2005). The absence of *Shh* expression in the FEZ led to a truncation of the upper jaw due to a failure of outgrowth. This appears to be a direct effect of SHH on the facial ectoderm itself as removing Smoothened from the neural crest cells does not appear to affect *Shh* expression in the FEZ (Chong et al., 2012).

While SHH signaling to the cephalic ectoderm is required for *Shh* expression in the FEZ, the actual role that SHH signaling plays is not entirely clear. Neural crest cells are also required for induction of *Shh* in the FEZ. In experiments that led to large amounts of apoptosis of neural crest cells, *Shh* expression in the FEZ was never initiated, but *Shh* expression could be restored by transplantation of neural crest cells into the FNP (Hu & Marcucio, 2012). Furthermore, blocking signaling by BMPs within the neural crest cells also prevented *Shh* expression in the FEZ (Foppiano, Hu, & Marcucio, 2007), and this could not be restored by addition of SHH back to the tissue (Chong et al., 2012). Together, these data suggest that a two-step process may function whereby SHH induces a "competence" in the presumptive FEZ ectoderm, and the neural crest cells induce the expression of *Shh* in a BMP-dependent manner. However, the exact molecular mechanisms that underlie this induction process are unknown.

2.4 Signaling from the Face to the Brain

Up to this point, we have described signals as emanating from the brain to control the face. However, evidence has emerged that signaling is not unidirectional, but rather that the neural crest cells also play a role in shaping the brain (Etchevers, Couly, Vincent, & Le Douarin, 1999; Tucker et al., 2008). Just as the brain regulates facial development, it is not surprising that there is reciprocal signaling from the facial tissues that contribute to brain development since the brain and facial skeleton coevolved (Gans & Northcutt, 1983). Nonetheless, the nature of these molecular interactions is just being discovered (Le Douarin, Couly, & Creuzet, 2012).

As mentioned above, the neural crest cells that form the skeleton of the upper jaw do not express Hox genes, and misexpression of Hox genes causes severe craniofacial malformations and brain defects (Creuzet, Couly, Vincent, & Le Douarin, 2002). This work has shed light on the molecular mechanisms by which the face contributes to brain development by uncovering a signaling network involving six family transcription factors, Bmp signaling, and *Fgf8* expression in the anterior neural ridge. It appears that in the presence of *HoxA2*, the transcription factors *Six1*, *Six2*, and *Six4*, which are normally expressed in the neural crest mesenchyme in this part of the face, are downregulated. This downregulation is accompanied by a decrease in expression of Bmp antagonists in the neural crest mesenchyme and a decrease in *Fgf8* expression in the anterior neural ridge. These results, along with other work by these investigators, have led to the idea that expression of Bmp antagonists, such as Gremlin and Noggin, are regulated by the six family of transcription factors, which in turn contribute to regulation of *Fgf8* expression levels in the anterior neural ridge (Aguiar, Sghari, & Creuzet, 2014; Creuzet, Martinez, & Le Douarin, 2006). This work is likely to lead to further discoveries that show the importance of signaling by neural crest cells to the developing brain, such as the importance of Wnt antagonism for expression of *Foxg1* in the developing telencephalon (Garcez, Le Douarin, & Creuzet, 2014).

3. IMPLICATIONS FOR STRUCTURAL DISEASES OF THE FACE

Altering the interactions between the brain and the face can produce variable morphologies that natural selection may have acted on to produce the diversity of facial form that we see among vertebrates. Hence, it is very likely that altering these interactions may also produce disease phenotypes in the craniofacial complex. Indeed, there are many examples that illustrate this point.

3.1 Holopresencephaly

One of the most commonly studied diseases in this regard is HPE. HPE is a structural disease of the brain that affects facial development. In HPE, the telencephalon undergoes varying degrees of septation, and in its most severe presentation, patients with HPE have cyclopia, a single median eye located beneath the proboscis. However, phenotypes range from this most severe form through mid-facial hypoplasia, clefting, and hypotelorism with a single central upper incisor, and some individuals with causative mutations may appear unaffected (Heussler et al., 2002; Ming & Muenke, 1998; Nanni et al., 2000).

The molecular etiology of HPE has been largely attributed to alterations in signaling by the Hedgehog pathway during brain development (Belloni et al., 1996; Brown et al., 1998; Gripp et al., 2000; Jeong, El-Jaick, Roessler, Muenke, & Epstein, 2006; Jeong et al., 2008; Ming et al., 2002; Ming & Muenke, 1998; Muenke & Beachy, 2000; Nanni et al., 2001; Roessler et al., 1996, 1997; Roessler, El-Jaick, et al., 2009; Roessler, Lacbawan, et al., 2009; Roessler, Ma, et al., 2009; Roessler & Muenke, 1998, 1999, 2010; Schell-Apacik et al., 2003; Solomon et al., 2009; Wallis & Muenke, 1999; Wallis et al., 1999). *In utero* exposure to alcohol or other teratogens also causes malformations in the face, and these resemble HPE (Hong & Krauss, 2012; Sulik, 2005; Sulik, Cook, & Webster, 1988; Welch et al., 2009). However, the cause of the variable phenotypic presentation is not known. Mice homozygous for a null mutation in *Shh* have a very severe phenotype that includes cyclopia or anencephaly of the anterior neural tube (Chiang et al., 1996), but these animals do not display the less severe forms observed in patients. How this spectrum of outcomes is observed in patients has been under intense investigation for a long time, and the work in mice suggests that thresholds of SHH signaling may be an important factor in phenotypic outcome. In the Shh heterozygotes, the level of SHH signaling is above a threshold that causes disease, but why does haploinsufficiency in people cause the range of phenotypes that are observed? One suggested possibility is that individuals with more severe forms of HPE have a second genetic alteration that modifies the primary mutation to produce a more severe phenotype (Ming & Muenke, 2002). This second-hit model has received wide attention and has been observed in a set of HPE patients (Mercier et al., 2011). Further, experimental evidence supports this idea in mice. Mutations in the SHH coreceptor *Cdo* (Tenzen et al., 2006) lead to HPE in some strains of mice, but not in others. However, mouse strains that are normally resistant to HPE have a phenotype if they also have a mutation in the related gene *Boc* (Tenzen et al., 2006; Zhang, Hong, Bae, Kang, & Krauss, 2011). Further, mutations can sensitize animals to teratogenic insults. Homozygous null mutations in the putative SHH coreceptor Cdo do not produce HPE in 129S6 mice but does sensitize mice to other perturbations that by themselves would not produce disease. For example, in elegant work, when subteratogenic doses of alcohol were administered to these mice, a range of phenotypes was produced that reflected the spectrum of HPE phenotypes seen in patients, suggesting that a sensitizing mutation synergizes with, in this case, a subteratogenic environmental insult to produce HPE (Hong & Krauss, 2012), and this can be rescued by reducing levels of the negative regulator of SHH signaling PTCH1 (Hong & Krauss, 2013). The extent to which the two-hit method of modifying phenotypic severity occurs in a large number of HPE patients is unknown, but this is a very intriguing possibility.

An alternative model may also help explain the variation in phenotype observed in patients; variation can be produced as a result of the function of signaling pathways in general. Modeling experiments have shown that in individual cells, the SHH pathway appears to be a binary switch (Lai, Robertson, & Schaffer, 2004). A cell has either responded to SHH or it has not, and the probability that a cell has responded is related to the concentration of SHH ligand. At low concentrations, cells are unlikely to respond, while at high concentrations, cells are very likely to respond. However, at intermediate concentrations of SHH ligand, large amounts of variation in responsiveness can be observed. This leads to nonlinear response of cells to various ligand concentrations (Fig. 3). Using a dose–response

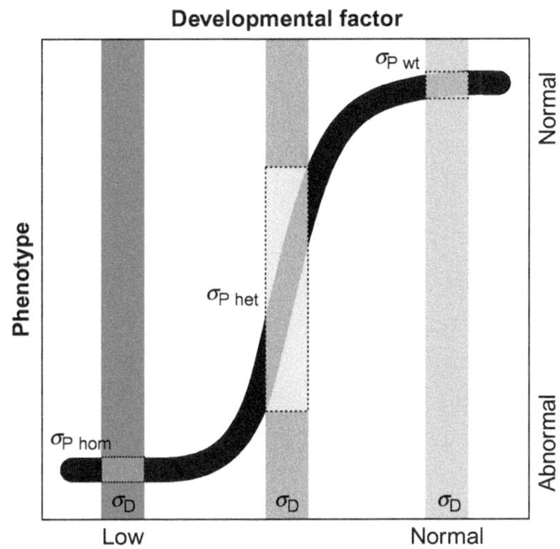

Figure 3 Nonlinearity in the relationship between development and phenotypic outcomes. In a nonlinear relationship, in this case represented as an S-shaped curve, there is a continuous relationship between a genetic or developmental factor and the phenotypic outcome; however, the variance in phenotypic outcomes differs for the same developmental variance. In this example, at normal levels of a developmental factor, phenotypic outcomes tend to be similar. Likewise, when the developmental factor is completely removed, phenotypic outcomes are also similar, such as the case in a highly penetrant genetic knockout. In contrast, when the developmental factor is reduced, such as in a heterozygote animal, the variance in phenotypic outcomes is much greater for the same developmental variance. (A color version of this figure is located at the end of this chapter.)

experiment, avian embryos were subjected to varying degrees of SHH signal blockade in the developing brain. A similar nonlinear relationship was observed between the level of blockade and facial shape (Young, Chong, Hu, Hallgrimsson, & Marcucio, 2010). This indicates that for small changes in ligand concentration, large changes in facial shape can result. Hence, nonlinear relationships arising as a result of the function of signaling pathways may also contribute to phenotypic variation in patients.

3.2 Cleft Lip/Palate

In addition to HPE, which produces a combined brain and face malformation, other structural malformations of the face may result from disruptions to the interactions of the brain and face during development. One particularly important area of exploration is related to nonsyndromic cleft lip and palate. These forms of clefting occur in isolation from other malformations and often have no known underlying genetic cause. However, it seems apparent that alterations in the rate of brain growth could potentially lead to clefting malformations due to the influence that the brain has on positioning of the facial primordia during development in some individuals (Diewert & Lozanoff, 1993a, 1993b; Diewert et al., 1993; Jiang, Bush, & Lidral, 2006).

The developing facial primordia must exhibit a concerted pattern of growth that brings them in apposition with one another in order to fuse and form the primary and secondary palates. Failure to meet at the appropriate times in development leads to fusion failure and orofacial clefting. If the growth rate of the brain is more rapid, the facial primordia may be positioned farther apart. Hence, the primordia may have reduced likelihood of achieving apposition at the times required for proper fusion.

A significant relationship between brain shape and incidence of orofacial clefting has been observed in some studies, suggesting that alterations in the brain may contribute to the incidence and/or severity of nonsyndromic clefting in the face. Nonpathological changes in the volume of various brain regions have been observed in some patients with nonsyndromic cleft lip and palate (Nopoulos et al., 2002). These individuals exhibited an enlarged cerebrum and changes in the shape of the corpus callosum, among other changes, suggesting that a change in the brain may play a role in clefting (Weinberg, Andreasen, & Nopoulos, 2009). Alterations to midline development of the brain may be a common event associated with facial malformations. However, the causal relationships between these brain alterations and cleft lip and palate remain unexplored, and the extent to which these changes are observed in larger cohorts of affected individuals is not known.

3.3 Shape Correlations with Brain Pathology

While most work focuses on the structural relationships between the developing brain and the face, there also appears to be a relationship between brain pathologies and facial shape. For example, recent investigations have revealed significant differences in the shape of the faces of people with psychological disorders. In patients with bipolar disorder, significant differences in facial shape were observed. The changes were primarily found in the frontonasal region and consisted of facial widening, vertical shortening, and alterations to the nose and mouth (Hennessy, Baldwin, Browne, Kinsella, & Waddington, 2007, 2010). Interestingly, these same general changes were observed with individuals diagnosed with schizophrenia.

While the exact mechanisms underlying these facial changes are unknown, it is worthy to note that the changes in facial shape occur in regions that are highly influenced by molecular signaling from the brain to the face. Whether these signaling interactions are altered in these patients is not known but deserves consideration.

4. THE ROLE OF BRAIN–FACE INTERACTIONS IN GENERATING EVOLUTIONARY VARIATION

Given the impact of the brain on the face, variation in these interactions is likely to contribute to evolutionary differences in morphology. Indeed, the FEZ is present in all assayed amniotes (Hu & Marcucio, 2009a, 2009b; Odent et al., 1999; Young et al., 2014), and its spatial organization appears to vary with differences in facial morphology. For example, in mice, *Shh* expression is divided into medial and lateral domains, while in birds, *Shh* expression is observed as a single domain that spans the primordial upper jaw (Hu & Marcucio, 2009a, 2009b; Reid et al., 2011). Altering SHH signaling within the forebrain leads to changes in the spatial organization of the *Shh* expression domain in the FEZ. Activating SHH signaling in the forebrain of early stage chick embryos superimposes a mammalian pattern of *Shh* expression in the FEZ and similarly alters the morphology of the developing upper jaw (Hu & Marcucio, 2009a, 2009b). This result suggests that SHH signaling within the brain controls the spatial organization of *Shh* expression in the FEZ, and altering this signaling system may generate variation in natural systems.

The differences in the patterns of *Shh* expression in the FEZ of birds and mice are dramatic and obvious. However, more subtle differences in the shape of the *Shh* expression domains in the FEZ among closely related animals are closely related to the morphology of the animal's face. Novel methods to quantify the shapes of *Shh* expression domains have revealed that differences in the shapes of the *Shh* expression domains in the FEZ of chicken, quail, and duck embryos are detectable and that the shapes of these domains are highly correlated with the facial shapes of the developing embryos (Hu et al., 2015).

The exact mechanisms by which these different patterns are generated are not known. Experimental evidence indicates that the brain plays some role in this process, possibly due to subtle differences in the *Shh* expression patterns within the brain of each embryo. In the basal forebrain of duck embryos, the *Shh* expression domain is wider compared to that in chick embryos, and transplanting the basal forebrain from the duck to the chick embryo shifted the morphology of the chimera in the direction of the duck. This was accompanied by a shift in the shape of the *Shh* expression domain in the chimera (Hu et al., 2015). However, it must be noted that the changes in the shape of the *Shh* expression domain in the FEZ were not statistically significant, but there was likely experimental variability that precluded detecting these differences. Nonetheless, the shapes were shifted in the right direction in the experimental embryos, and importantly, the shape of the *Shh* expression domain in the FEZ was highly correlated with the shape of the developing midface. Together, these data indicate that the brain plays some, as yet undefined, role in patterning the shape of the *Shh* expression domain, and the shape of the *Shh* expression domain contributes to the pattern of morphogenesis observed in the developing midface.

5. CONCLUSIONS

Studies performed in a variety of animal models and on human patients have uncovered a series of interactions between the brain and the face that regulate the coordinated development of these structures. The connective tissue precursors of the facial skeleton are derived from the brain during early stages of development. The brain exerts an epigenetic influence on the developing face through the effect of its expansion on position and subsequent growth of the facial anlagen, and these epigenetic influences may have profound consequences on the incidence and severity of structural malformations, such as cleft lip and palate, of the developing face. Finally, a series of molecular signaling interactions between the brain and the facial tissues appear to coordinate the growth of the brain and the face. Understanding these interactions will lead to a greater understanding of developmental mechanisms, and the etiologies of birth defects that affect the brain and the face. Ultimately, this understanding may lead to improved patient care through a better appreciation of the human condition, improved diagnostic tools, and/or development of new therapeutic regimens to directly treat patients.

REFERENCES

Aguiar, D. P., Sghari, S., & Creuzet, S. (2014). The facial neural crest controls fore- and midbrain patterning by regulating Foxg1 expression through Smad1 activity. *Development*, *141*(12), 2494–2505.

Aldridge, K., Kane, A. A., Marsh, J. L., Yan, P., Govier, D., & Richtsmeier, J. T. (2005). Relationship of brain and skull in pre- and postoperative sagittal synostosis. *Journal of Anatomy*, *206*(4), 373–385.

Aoto, K., Shikata, Y., Imai, H., Matsumaru, D., Tokunaga, T., Shioda, S., et al. (2009). Mouse Shh is required for prechordal plate maintenance during brain and craniofacial morphogenesis. *Developmental Biology, 327*(1), 106–120.

Ashique, A. M., Fu, K., & Richman, J. M. (2002). Endogenous bone morphogenetic proteins regulate outgrowth and epithelial survival during avian lip fusion. *Development, 129*(19), 4647–4660.

Barlow, A. J., & Francis-West, P. H. (1997). Ectopic application of recombinant BMP-2 and BMP-4 can change patterning of developing chick facial primordia. *Development, 124*(2), 391–398.

Belloni, E., Muenke, M., Roessler, E., Traverso, G., Siegel-Bartelt, J., Frumkin, A., et al. (1996). Identification of Sonic hedgehog as a candidate gene responsible for holoprosencephaly. *Nature Genetics, 14*(3), 353–356.

Boughner, J. C., Wat, S., Diewert, V. M., Young, N. M., Browder, L. W., & Hallgrimsson, B. (2008). Short-faced mice and developmental interactions between the brain and the face. *Journal of Anatomy, 213*(6), 646–662.

Brito, J. M., Teillet, M. A., & Le Douarin, N. M. (2006). An early role for sonic hedgehog from foregut endoderm in jaw development: Ensuring neural crest cell survival. *Proceedings of the National Academy of Sciences of the United States of America, 103*(31), 11607–11612.

Bronner, M. E., & LeDouarin, N. M. (2012). Development and evolution of the neural crest: An overview. *Developmental Biology, 366*(1), 2–9.

Brown, S. A., Warburton, D., Brown, L. Y., Yu, C. Y., Roeder, E. R., Stengel-Rutkowski, S., et al. (1998). Holoprosencephaly due to mutations in ZIC2, a homologue of Drosophila odd-paired. *Nature Genetics, 20*(2), 180–183.

Chai, Y., & Maxson, R. E., Jr. (2006). Recent advances in craniofacial morphogenesis. *Developmental Dynamics, 235*(9), 2353–2375.

Chambers, D., & McGonnell, I. M. (2002). Neural crest: Facing the facts of head development. *Trends in Genetics, 18*(8), 381–384.

Chiang, C., Litingtung, Y., Lee, E., Young, K. E., Corden, J. L., Westphal, H., et al. (1996). Cyclopia and defective axial patterning in mice lacking Sonic hedgehog gene function. *Nature, 383*(6599), 407–413.

Chong, H. J., Young, N. M., Hu, D., Jeong, J., McMahon, A. P., Hallgrimsson, B., et al. (2012). Signaling by SHH rescues facial defects following blockade in the brain. *Developmental Dynamics: An Official Publication of the American Association of Anatomists, 241*(2), 247–256.

Creuzet, S., Couly, G., Vincent, C., & Le Douarin, N. M. (2002). Negative effect of Hox gene expression on the development of the neural crest-derived facial skeleton. *Development, 129*(18), 4301–4313.

Creuzet, S. E., Martinez, S., & Le Douarin, N. M. (2006). The cephalic neural crest exerts a critical effect on forebrain and midbrain development. *Proceedings of the National Academy of Sciences of the United States of America, 103*(38), 14033–14038.

DeMyer, W. (1964). The face predicts the brain: Diagnostic significance of median facial anomalies for holoprosencephaly (arhinencephay). *Pediatrics, 34*, 256–263.

Depew, M. J., & Compagnucci, C. (2008). Tweaking the hinge and caps: Testing a model of the organization of jaws. *Journal of Experimental Zoology. Part B, Molecular and Developmental Evolution, 310*(4), 315–335.

Depew, M. J., & Simpson, C. A. (2006). 21st century neonatology and the comparative development of the vertebrate skull. *Developmental Dynamics, 235*(5), 1256–1291.

Depew, M. J., Simpson, C. A., Morasso, M., & Rubenstein, J. L. (2005). Reassessing the Dlx code: The genetic regulation of branchial arch skeletal pattern and development. *Journal of Anatomy, 207*(5), 501–561.

Diewert, V. M. (1983). A morphometric analysis of craniofacial growth and changes in spatial relations during secondary palatal development in human embryos and fetuses. *The American Journal of Anatomy, 167*(4), 495–522.

Diewert, V. M. (1985). Growth movements during prenatal development of human facial morphology. *Progress in Clinical and Biological Research*, *187*, 57–66.

Diewert, V. M., & Lozanoff, S. (1993a). Growth and morphogenesis of the human embryonic midface during primary palate formation analyzed in frontal sections. *Journal of Craniofacial Genetics and Developmental Biology*, *13*(3), 162–183.

Diewert, V. M., & Lozanoff, S. (1993b). A morphometric analysis of human embryonic craniofacial growth in the median plane during primary palate formation. *Journal of Craniofacial Genetics and Developmental Biology*, *13*(3), 147–161.

Diewert, V. M., Lozanoff, S., & Choy, V. (1993). Computer reconstructions of human embryonic craniofacial morphology showing changes in relations between the face and brain during primary palate formation. *Journal of Craniofacial Genetics and Developmental Biology*, *13*(3), 193–201.

Etchevers, H. C., Couly, G., Vincent, C., & Le Douarin, N. M. (1999). Anterior cephalic neural crest is required for forebrain viability. *Development*, *126*(16), 3533–3543.

Fienberg, A. A., Utset, M. F., Bogarad, L. D., Hart, C. P., Awgulewitsch, A., Ferguson-Smith, A., et al. (1987). Homeo box genes in murine development. *Current Topics in Developmental Biology*, *23*, 233–256.

Foppiano, S., Hu, D., & Marcucio, R. S. (2007). Signaling by bone morphogenetic proteins directs formation of an ectodermal signaling center that regulates craniofacial development. *Developmental Biology*, *312*(1), 103–114.

Francis-West, P., Ladher, R., Barlow, A., & Graveson, A. (1998). Signalling interactions during facial development. *Mechanisms of Development*, *75*(1–2), 3–28.

Gans, C., & Northcutt, R. (1983). Neural crest and the origin of vertebrates: A new head. *Science*, *220*, 268–274.

Garcez, R. C., Le Douarin, N. M., & Creuzet, S. E. (2014). Combinatorial activity of Six1-2-4 genes in cephalic neural crest cells controls craniofacial and brain development. *Cellular and Molecular Life Sciences*, *71*(11), 2149–2164.

Geetha-Loganathan, P., Nimmagadda, S., Antoni, L., Fu, K., Whiting, C. J., Francis-West, P., et al. (2009). Expression of WNT signalling pathway genes during chicken craniofacial development. *Developmental Dynamics*, *238*(5), 1150–1165.

Graham, A., Okabe, M., & Quinlan, R. (2005). The role of the endoderm in the development and evolution of the pharyngeal arches. *Journal of Anatomy*, *207*(5), 479–487.

Graham, A., & Smith, A. (2001). Patterning the pharyngeal arches. *Bioessays*, *23*(1), 54–61.

Grammatopoulos, G. A., Bell, E., Toole, L., Lumsden, A., & Tucker, A. S. (2000). Homeotic transformation of branchial arch identity after Hoxa2 overexpression. *Development*, *127*(24), 5355–5365.

Gripp, K. W., Wotton, D., Edwards, M. C., Roessler, E., Ades, L., Meinecke, P., et al. (2000). Mutations in TGIF cause holoprosencephaly and link NODAL signalling to human neural axis determination. *Nature Genetics*, *25*(2), 205–208.

Hallgrímsson, B., Jamniczky, H., Young, N. M., Rolian, C., Parsons, T. E., Boughner, J. C., et al. (2009). Deciphering the palimpsest: Studying the relationship between morphological integration and phenotypic covariation. *Evolutionary Biology*, *36*(4), 355–376.

Hennessy, R. J., Baldwin, P. A., Browne, D. J., Kinsella, A., & Waddington, J. L. (2007). Three-dimensional laser surface imaging and geometric morphometrics resolve frontonasal dysmorphology in schizophrenia. *Biological Psychiatry*, *61*(10), 1187–1194.

Hennessy, R. J., Baldwin, P. A., Browne, D. J., Kinsella, A., & Waddington, J. L. (2010). Frontonasal dysmorphology in bipolar disorder by 3D laser surface imaging and geometric morphometrics: Comparisons with schizophrenia. *Schizophrenia Research*, *122*(1–3), 63–71.

Heussler, H. S., Suri, M., Young, I. D., & Muenke, M. (2002). Extreme variability of expression of a Sonic Hedgehog mutation: Attention difficulties and holoprosencephaly. *Archives of Disease in Childhood*, *86*(4), 293–296.

Holland, P. W., & Hogan, B. L. (1988). Spatially restricted patterns of expression of the homeobox-containing gene Hox 2.1. during mouse embryogenesis. *Development, 102*(1), 159–174.

Hong, M., & Krauss, R. S. (2012). Cdon mutation and fetal ethanol exposure synergize to produce midline signaling defects and holoprosencephaly spectrum disorders in mice. *PLoS Genetics, 8*(10), e1002999.

Hong, M., & Krauss, R. S. (2013). Rescue of holoprosencephaly in fetal alcohol-exposed Cdon mutant mice by reduced gene dosage of Ptch1. *PLoS One, 8*(11), e79269.

Hu, D., & Marcucio, R. S. (2009a). A SHH-responsive signaling center in the forebrain regulates craniofacial morphogenesis via the facial ectoderm. *Development, 136*(1), 107–116.

Hu, D., & Marcucio, R. S. (2009b). Unique organization of the frontonasal ectodermal zone in birds and mammals. *Developmental Biology, 325*(1), 200–210.

Hu, D., & Marcucio, R. S. (2012). Neural crest cells pattern the surface cephalic ectoderm during FEZ formation. *Developmental Dynamics, 241*(4), 732–740.

Hu, D., Marcucio, R., & Helms, J. A. (2003). A zone of frontonasal ectoderm regulates patterning and growth in the face. *Development, 130*(9), 1749–1758.

Hu, D., Young, N. M., Xu, Q., Jamniczky, H., Green, R. M., Mio, W., et al. (2015). Signals from the brain induce variation in avian facial shape. *Developmental Dynamics, 244*(9), 1133–1143.

Jeong, Y., El-Jaick, K., Roessler, E., Muenke, M., & Epstein, D. J. (2006). A functional screen for sonic hedgehog regulatory elements across a 1 Mb interval identifies long-range ventral forebrain enhancers. *Development, 133*(4), 761–772.

Jeong, Y., Leskow, F. C., El-Jaick, K., Roessler, E., Muenke, M., Yocum, A., et al. (2008). Regulation of a remote Shh forebrain enhancer by the Six3 homeoprotein. *Nature Genetics, 40*(11), 1348–1353.

Jeong, J., Mao, J., Tenzen, T., Kottmann, A. H., & McMahon, A. P. (2004). Hedgehog signaling in the neural crest cells regulates the patterning and growth of facial primordia. *Genes & Development, 18*(8), 937–951.

Jiang, R., Bush, J. O., & Lidral, A. C. (2006). Development of the upper lip: Morphogenetic and molecular mechanisms. *Developmental Dynamics: An Official Publication of the American Association of Anatomists, 235*(5), 1152–1166.

Lai, K., Robertson, M. J., & Schaffer, D. V. (2004). The sonic hedgehog signaling system as a bistable genetic switch. *Biophysical Journal, 86*(5), 2748–2757.

Lan, Y., Ryan, R. C., Zhang, Z., Bullard, S. A., Bush, J. O., Maltby, K. M., et al. (2006). Expression of *Wnt9b* and activation of canonical Wnt signaling during midfacial morphogenesis in mice. *Developmental Dynamics, 235*, 1448–1454.

Le Douarin, N. M., Couly, G., & Creuzet, S. E. (2012). The neural crest is a powerful regulator of pre-otic brain development. *Developmental Biology, 366*(1), 74–82.

Marcucio, R. S., Cordero, D. R., Hu, D., & Helms, J. A. (2005). Molecular interactions coordinating the development of the forebrain and face. *Developmental Biology, 284*(1), 48–61.

Marcucio, R. S., Young, N. M., Hu, D., & Hallgrimsson, B. (2011). Mechanisms that underlie co-variation of the brain and face. *Genesis, 49*(4), 177–189.

McGinnis, W., & Krumlauf, R. (1992). Homeobox genes and axial patterning. *Cell, 68*, 283–302.

Mercier, S., Dubourg, C., Garcelon, N., Campillo-Gimenez, B., Gicquel, I., Belleguic, M., et al. (2011). New findings for phenotype-genotype correlations in a large European series of holoprosencephaly cases. *Journal of Medical Genetics, 48*(11), 752–760.

Ming, J. E., Kaupas, M. E., Roessler, E., Brunner, H. G., Golabi, M., Tekin, M., et al. (2002). Mutations in PATCHED-1, the receptor for SONIC HEDGEHOG, are associated with holoprosencephaly. *Human Genetics, 110*(4), 297–301.

Ming, J. E., & Muenke, M. (1998). Holoprosencephaly: From Homer to hedgehog. *Clinical Genetics*, *53*(3), 155–163.
Ming, J. E., & Muenke, M. (2002). Multiple hits during early embryonic development: Digenic diseases and holoprosencephaly. *American Journal of Human Genetics*, *71*(5), 1017–1032.
Minoux, M., & Rijli, F. M. (2010). Molecular mechanisms of cranial neural crest cell migration and patterning in craniofacial development. *Development*, *137*(16), 2605–2621.
Muenke, M., & Beachy, P. A. (2000). Genetics of ventral forebrain development and holoprosencephaly. *Current Opinion in Genetics and Development*, *10*(3), 262–269.
Muller, F., Albert, S., Blader, P., Fischer, N., Hallonet, M., & Strahle, U. (2000). Direct action of the nodal-related signal cyclops in induction of sonic hedgehog in the ventral midline of the CNS. *Development*, *127*(18), 3889–3897.
Nanni, L., Ming, J. E., Du, Y., Hall, R. K., Aldred, M., Bankier, A., et al. (2001). SHH mutation is associated with solitary median maxillary central incisor: A study of 13 patients and review of the literature. *American Journal of Medical Genetics*, *102*(1), 1–10.
Nanni, L., Schelper, R. L., & Muenke, M. T. (2000). Molecular genetics of holoprosencephaly. *Frontiers in Bioscience*, *5*(8), D334–D342.
Noden, D. M. (1983). The role of the neural crest in patterning of avian cranial skeletal, connective, and muscle tissues. *Developmental Biology*, *96*(1), 144–165.
Noden, D. M. (1984). Craniofacial development: New views on old problems. *The Anatomical Records*, *208*, 1–13.
Noden, D. M. (1986). Origins and patterning of craniofacial mesenchymal tissues. *Journal of Craniofacial Genetics and Developmental Biology*, *2*, 15–31.
Noden, D. M., & Trainor, P. A. (2005). Relations and interactions between cranial mesoderm and neural crest populations. *Journal of Anatomy*, *207*(5), 575–601.
Nopoulos, P., Berg, S., Canady, J., Richman, L., Van Demark, D., & Andreasen, N. C. (2002). Structural brain abnormalities in adult males with clefts of the lip and/or palate. *Genetics in Medicine*, *4*(1), 1–9.
Odent, S., Atti-Bitach, T., Blayau, M., Mathieu, M., Aug, J., Delezo de, A. L., et al. (1999). Expression of the Sonic hedgehog (SHH) gene during early human development and phenotypic expression of new mutations causing holoprosencephaly. *Human Molecular Genetics*, *8*(9), 1683–1689.
Ohkubo, Y., Chiang, C., & Rubenstein, J. L. (2002). Coordinate regulation and synergistic actions of BMP4, SHH and FGF8 in the rostral prosencephalon regulate morphogenesis of the telencephalic and optic vesicles. *Neuroscience*, *111*(1), 1–17.
Reid, B. S., Yang, H., Melvin, V. S., Taketo, M. M., & Williams, T. (2011). Ectodermal Wnt/beta-catenin signaling shapes the mouse face. *Developmental Biology*, *349*(2), 261–269.
Richman, J. M., & Lee, S. H. (2003). About face: Signals and genes controlling jaw patterning and identity in vertebrates. *Bioessays*, *25*(6), 554–568.
Richtsmeier, J. T., Aldridge, K., DeLeon, V. B., Panchal, J., Kane, A. A., Marsh, J. L., et al. (2006). Phenotypic integration of neurocranium and brain. *Journal of Experimental Zoology. Part B, Molecular and Developmental Evolution*, *306*(4), 360–378.
Richtsmeier, J. T., & Deleon, V. B. (2009). Morphological integration of the skull in craniofacial anomalies. *Orthodontics & Craniofacial Research*, *12*(3), 149–158.
Rijli, F. M., Mark, M., Lakkaraju, S., Dierich, A., Dolle, P., & Chambon, P. (1993). A homeotic transformation is generated in the rostral branchial region of the head by disruption of *Hoxa-2*, which acts as a selector gene. *Cell*, *75*(7), 1333–1349.
Roessler, E., Belloni, E., Gaudenz, K., Jay, P., Berta, P., Scherer, S. W., et al. (1996). Mutations in the human Sonic Hedgehog gene cause holoprosencephaly. *Nature Genetics*, *14*(3), 357–360.

Roessler, E., Belloni, E., Gaudenz, K., Vargas, F., Scherer, S. W., Tsui, L. C., et al. (1997). Mutations in the C-terminal domain of Sonic Hedgehog cause holoprosencephaly. *Human Molecular Genetics*, *6*(11), 1847–1853.
Roessler, E., El-Jaick, K. B., Dubourg, C., Velez, J. I., Solomon, B. D., Pineda-Alvarez, D. E., et al. (2009). The mutational spectrum of holoprosencephaly-associated changes within the SHH gene in humans predicts loss-of-function through either key structural alterations of the ligand or its altered synthesis. *Human Mutation*, *30*(10), E921–E935.
Roessler, E., Lacbawan, F., Dubourg, C., Paulussen, A., Herbergs, J., Hehr, U., et al. (2009). The full spectrum of holoprosencephaly-associated mutations within the ZIC2 gene in humans predicts loss-of-function as the predominant disease mechanism. *Human Mutation*, *30*(4), E541–E554.
Roessler, E., Ma, Y., Ouspenskaia, M. V., Lacbawan, F., Bendavid, C., Dubourg, C., et al. (2009). Truncating loss-of-function mutations of DISP1 contribute to holoprosencephaly-like microform features in humans. *Human Genetics*, *125*(4), 393–400.
Roessler, E., & Muenke, M. (1998). Holoprosencephaly: A paradigm for the complex genetics of brain development. *Journal of Inherited Metabolic Disease*, *21*(5), 481–497.
Roessler, E., & Muenke, M. (1999). The molecular genetics of holoprosencephaly: A model of brain development for the next century. *Childs Nervous System*, *15*(11–12), 646–651.
Roessler, E., & Muenke, M. (2010). The molecular genetics of holoprosencephaly. *American Journal of Medical Genetics. Part C, Seminars in Medical Genetics*, *154C*(1), 52–61.
Schell-Apacik, C., Rivero, M., Knepper, J. L., Roessler, E., Muenke, M., & Ming, J. E. (2003). SONIC HEDGEHOG mutations causing human holoprosencephaly impair neural patterning activity. *Human Genetics*, *113*, 170–177.
Solomon, B. D., Lacbawan, F., Jain, M., Domene, S., Roessler, E., Moore, C., et al. (2009). A novel SIX3 mutation segregates with holoprosencephaly in a large family. *American Journal of Medical Genetics Part A*, *149A*(5), 919–925.
Sulik, K. K. (2005). Genesis of alcohol-induced craniofacial dysmorphism. *Experimental Biology and Medicine*, *230*(6), 366–375.
Sulik, K. K., Cook, C. S., & Webster, W. S. (1988). Teratogens and craniofacial malformations: Relationships to cell death. *Development*, *103*(Suppl.), 213–232.
Swartz, M. E., Nguyen, V., McCarthy, N. Q., & Eberhart, J. K. (2012). Hh signaling regulates patterning and morphogenesis of the pharyngeal arch-derived skeleton. *Developmental Biology*, *369*(1), 65–75.
Szabo-Rogers, H. L., Smithers, L. E., Yakob, W., & Liu, K. J. (2010). New directions in craniofacial morphogenesis. *Developmental Biology*, *341*(1), 84–94.
Tenzen, T., Allen, B. L., Cole, F., Kang, J. S., Krauss, R. S., & McMahon, A. P. (2006). The cell surface membrane proteins Cdo and Boc are components and targets of the Hedgehog signaling pathway and feedback network in mice. *Developmental Cell*, *10*(5), 647–656.
Trainor, P. A., Ariza-McNaughton, L., & Krumlauf, R. (2002). Role of the isthmus and FGFs in resolving the paradox of neural crest plasticity and prepatterning. *Science*, *295*(5558), 1288–1291.
Trainor, P. A., & Krumlauf, R. (2000). Patterning the cranial neural crest: Hindbrain segmentation and Hox gene plasticity. *Nature Reviews. Neuroscience*, *1*(2), 116–124.
Tucker, E. S., Segall, S., Gopalakrishna, D., Wu, Y., Vernon, M., Polleux, F., et al. (2008). Molecular specification and patterning of progenitor cells in the lateral and medial ganglionic eminences. *The Journal of Neuroscience*, *28*(38), 9504–9518.
Wallis, D. E., & Muenke, M. (1999). Molecular mechanisms of holoprosencephaly. *Molecular Genetics and Metabolism*, *68*(2), 126–138.
Wallis, D. E., Roessler, E., Hehr, U., Nanni, L., Wiltshire, T., Richieri-Costa, A., et al. (1999). Mutations in the homeodomain of the human SIX3 gene cause holoprosencephaly. *Nature Genetics*, *22*(2), 196–198.

Wang, K. Y., & Diewert, V. M. (1992). A morphometric analysis of craniofacial growth in cleft lip and noncleft mice. *Journal of Craniofacial Genetics and Developmental Biology*, *12*(3), 141–154.

Wang, K. Y., Juriloff, D. M., & Diewert, V. M. (1995). Deficient and delayed primary palatal fusion and mesenchymal bridge formation in cleft lip-liable strains of mice. *Journal of Craniofacial Genetics and Developmental Biology*, *15*(3), 99–116.

Weinberg, S. M., Andreasen, N. C., & Nopoulos, P. (2009). Three-dimensional morphometric analysis of brain shape in nonsyndromic orofacial clefting. *Journal of Anatomy*, *214*(6), 926–936.

Welch, K. D., Panter, K. E., Lee, S. T., Gardner, D. R., Stegelmeier, B. L., & Cook, D. (2009). Cyclopamine-induced synophthalmia in sheep: Defining a critical window and toxicokinetic evaluation. *Journal of Applied Toxicology*, *29*(5), 414–421.

Young, N. M., Chong, H. J., Hu, D., Hallgrimsson, B., & Marcucio, R. S. (2010). Quantitative analyses link modulation of sonic hedgehog signaling to continuous variation in facial growth and shape. *Development*, *137*(20), 3405–3409.

Young, N. M., Hu, D., Lainoff, A. J., Smith, F. J., Diaz, R., Tucker, A. S., et al. (2014). Embryonic bauplans and the developmental origins of facial diversity and constraint. *Development*, *141*(5), 1059–1063.

Zhang, W., Hong, M., Bae, G. U., Kang, J. S., & Krauss, R. S. (2011). Boc modifies the holoprosencephaly spectrum of Cdo mutant mice. *Disease Models & Mechanisms*, *4*(3), 368–380.

CHAPTER THIRTEEN

Developmental Plasticity of Patterned and Regenerating Oral Organs

J. Todd Streelman[1], Ryan F. Bloomquist, Teresa E. Fowler
School of Biology, Petit Institute for Bioengineering and Bioscience, Georgia Institute of Technology, Atlanta, Georgia, USA
[1]Corresponding author: e-mail address: todd.streelman@biology.gatech.edu

Contents

Abstract

In many aquatic vertebrates, including bony and cartilaginous fishes, teeth and taste buds colocalize on jaw elements. In these animals, taste buds are renewed continuously throughout life, whereas teeth undergo cycled whole-organ replacement by various means. Recently, studies of cichlid fishes have yielded new insights into the development and regeneration of these dental and sensory oral organs. Tooth and taste bud densities covary positively across species with different feeding strategies, controlled by common regions of the genome and integrated molecular signals. Developing teeth and taste buds share a bipotent epithelium during early patterning stages, from which dental and taste fields are specified. Moreover, these organs share a common epithelial ribbon that supports label-retaining cells during later stages of regeneration. During both patterning and regeneration stages, dental organs can be converted to taste bud fate by manipulation of BMP signaling. These observations highlight a surprising long-term plasticity between dental and sensory organ types. Here, we review these findings and discuss the implications of developmental plasticity that spans the continuum of craniofacial organ patterning and regeneration.

Current Topics in Developmental Biology, Volume 115
ISSN 0070-2153
http://dx.doi.org/10.1016/bs.ctdb.2015.07.005

1. INTRODUCTION AND BACKGROUND

Vertebrate dentitions differ in their location, pattern, and regenerative capacity. Birds and turtles lack true teeth entirely; some bony fishes such as zebrafish possess teeth only in the pharynx, whereas others (e.g., cichlids) have numerous rows of teeth on both oral and pharyngeal jaws (Fraser et al., 2009). Sharks and rays exhibit teeth (often in multiple rows) or tooth plates on their cartilaginous oral jaws, and mammalian teeth are confined to a single row at the oral jaw margin. Mammalian teeth may vary in shape and function along the tooth row (e.g., incisor, molar) but nonmammalian dentitions tend to be homodont (i.e., each tooth is the same shape in a row; tooth shape may differ across rows). Nonmammalian teeth are replaced *de novo* throughout life via various mechanisms (Tucker & Fraser, 2014), but most mammals have lost this regenerative ability. For example, humans replace each tooth only once and mice never replace their teeth.

In sharks, snakes, lizards, crocodiles, and mammals that undergo limited replacement, the earliest embryonic teeth are typically smaller versions of their adult counterparts. However, in ray-finned fishes, lungfishes, and salamanders, first-generation teeth are small and simple unicuspids (i.e., one-cusp teeth), lacking neurovascular bundles [NVBs] (Sire, Davit-Beal, Delgado, Van Der Heyden, & Huysseune, 2002). In these latter dentitions, shape and increased complexity (e.g., size, curvature, multiple cusps) arise through the process of replacement. In cichlid fishes, the particular oral jaw tooth pattern—that is, the size and spacing of teeth—is set with the initiation of the first generation, prior to the development of complex tooth shape (Fraser, Bloomquist, & Streelman, 2008). This occurs differently in the pharyngeal dentition of sticklebacks (Cleves et al., 2014). Cichlids exhibit one-for-one tooth replacement (like humans, unlike sharks), with each oral tooth replaced in 30–100-day cycles (depending on age and size), never in adjoining tooth positions (Tuisku & Hildebrand, 1994). This establishes a continuum of lifelong dental patterning and regeneration that integrates key molecular pathways of odontogenesis and organ renewal (Fraser, Bloomquist, & Streelman, 2013).

Mammalian and nonmammalian dentitions differ in at least one other important aspect. In mammals, teeth are confined to a single row on the oral jaw, and taste buds are located separately on the tongue and palate. In other vertebrate groups, including bony and cartilaginous fishes, teeth and taste buds are often colocalized on jaw elements (Fig. 1).

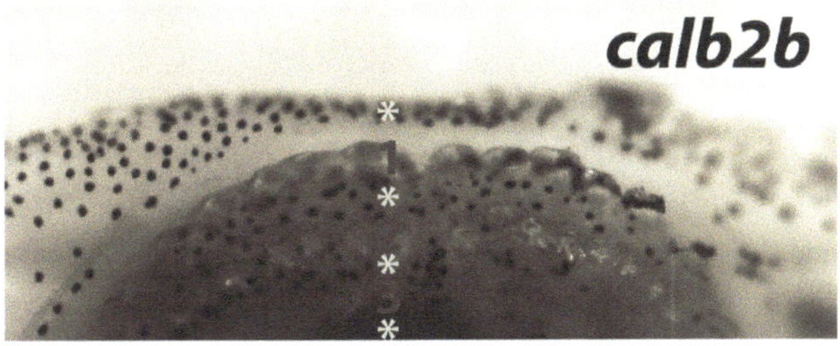

Figure 1 Comingling of teeth and taste buds on the oral jaws of Lake Malawi cichlid fishes. ISH of *calb2* in the cichlid lower jaw, at 4 months of age. Yellow asterisk marks rows of expression in taste buds, while rows of teeth are marked by red numerals. Note that tooth row three has not yet erupted. Dorsal view, labial to bottom of page, scale = 100 μm. (A color version of this figure is located at the end of this chapter.)

We highlight recent advances in our understanding of the development and regeneration of these colocalized oral organs, focusing on cichlid fishes from Lake Malawi. These studies reveal a surprising and long-lasting plasticity between dental and sensory organs that spans the continuum of craniofacial organ patterning and regeneration. In what follows, we review recent findings and discuss the implications of long-term developmental plasticity for the evolution and bioinspired design of oral organs.

2. THE CONTINUUM OF ORAL ORGAN PATTERNING AND REGENERATION

Whole-organ replacement of dentitions in a one-for-one fashion, while maintaining fidelity of pattern, requires (i) signals of dental competence to specify tooth versus nontooth, (ii) an odontogenesis program that can be repeatedly recruited within a tooth family, (iii) signals/sources of renewal that can be cyclically and spatially restricted, and (iv) a clock mechanism to coordinate timing in (alternate) tooth positions. Cichlid teeth execute largely conserved developmental programs of competence and morphogenesis (Fraser et al., 2013) in an environment filled with taste buds (Fig. 1).

Recently, the evolutionary developmental interplay between these placode-derived craniofacial organs was examined. There is an adaptive positive correlation between oral jaw densities of teeth and taste buds in adult Lake Malawi cichlids. Piscivorous and planktivorous species, which hunt with acute vision and swallow prey without processing it in the mouth, exhibit reduced tooth and taste bud densities. Algivores and insectivores, on the other hand, which actively search for appropriate food and acquire it with the mouth, possess many teeth and taste buds. Using an intercross design (i.e., a cross between a planktivore and an algivore species) and genetic linkage mapping, Bloomquist and colleagues identified common regions of the genome that contribute to this phenotypic correlation (Bloomquist et al., 2015a).

To explore this association during development, they next used *in situ* hybridization (ISH) at 5 days postfertilization (5 dpf), when the oral jaws first become apparent, to chart the spatial activity of tooth and taste bud markers. The earliest oral lamina expresses *pitx2* (a marker of dental epithelium), *shh* (dental and taste epithelium), and *sox2* (taste epithelium) in overlapping patterns (Fig. 2A). A day later, the first *pitx2+*, *shh+* dental placodes are detectable and *calb2+* (a marker of mature taste buds with more distinct expression than *sox2*) taste buds lingual to first-generation teeth are initiated. Subsequently, a labial band of taste buds appears flanking the first tooth row. As seen in dorsal views of whole-mount ISH jaws, *pitx2*, *shh*, and *sox2* continue to be expressed in these increasingly distinct fields (Fig. 2B).

Small-molecule manipulation of the Wnt, Hh, and BMP pathways, during the critical developmental window (5–6 dpf) when organ fields differentiate one from the other, provides clues to the regulatory logic of cichlid tooth and taste bud copatterning. Wnt signaling couples tooth and taste bud

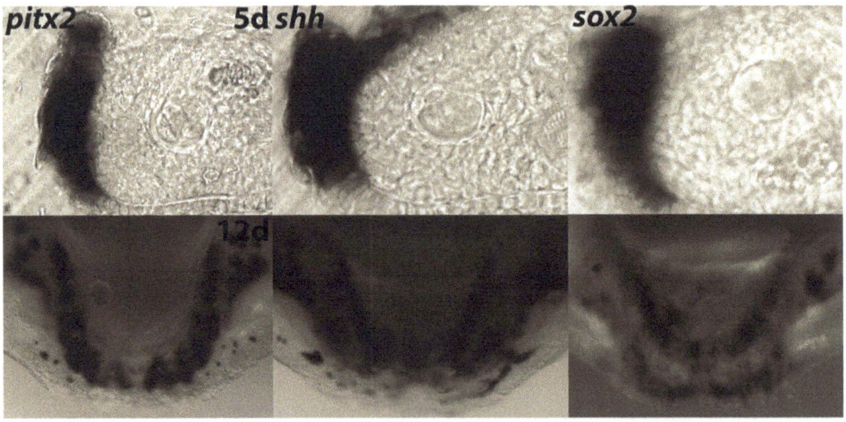

Figure 2 Teeth and taste buds share a bipotent oral lamina. ISH of dental marker *pitx2*, taste/sensory marker *sox2*, and an early marker of both, *shh*, at two representative stages of development. Top panel: sagittal section (18 μm thickness, 40× magnification) at initial stages of jaw formation, 5 dpf. Bottom panel: by 12 dpf, distinct fields of teeth (marked by *pitx2*) and taste buds (*sox2*) have emerged, while *shh* is expressed in both domains. Whole mount dorsal views, labial to bottom of page. (A color version of this figure is located at the end of this chapter.)

density and acts upstream of Hh and BMP, which together mediate organ plasticity. LDN treatment, which inhibits BMP signaling, results in a striking phenotype wherein tooth density is reduced and taste buds infiltrate the tooth field (Fig. 6A–B′). Thus, these studies and others suggest that cichlid teeth and taste buds are tightly copatterned, under common genetic and developmental control (Bloomquist et al., 2015a). Developmental plasticity or bipotency may be a general feature of organ systems like teeth and taste buds that function together—as previously noted for liver/pancreas (Xu et al., 2011) and heart/lung (Peng et al., 2013).

Although teeth and taste buds are typically studied in isolation because of their separate location in mammals, vertebrate teeth and taste buds exhibit significant gene synexpression and may share an ancestral origin (Fraser, Cerny, Soukup, Bronner-Fraser, & Streelman, 2010). Moreover, a common epithelial ribbon links cichlid teeth and taste buds as they regenerate (Fig. 3).

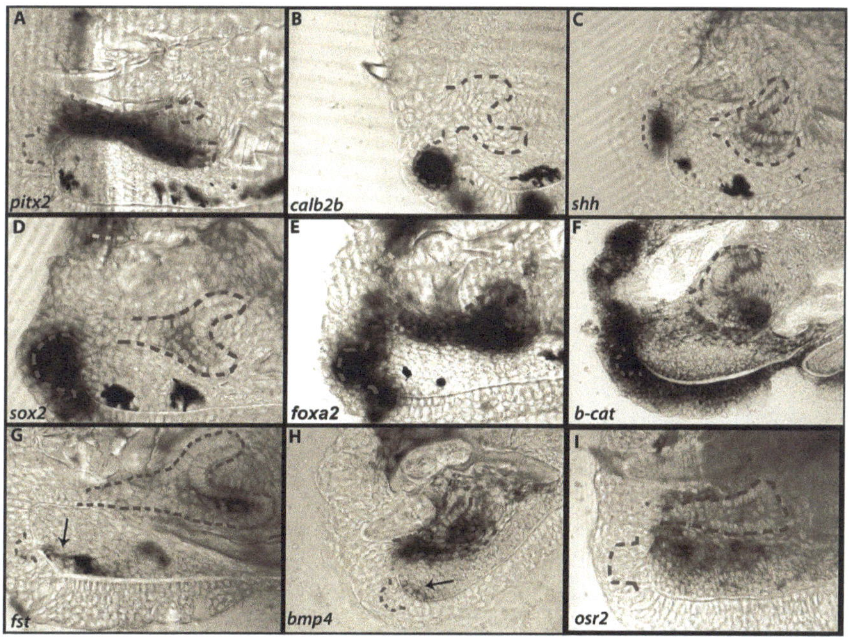

Figure 3 Teeth and taste buds share an epithelial ribbon and molecular marker expression during regeneration stages (~20 dpf). Orange dashes outline taste buds, green outlines the functional tooth, and red outlines replacement teeth. Note expression of *fst* and *bmp4* outside of and juxtaposed to the taste bud unit. These are sagittal sections, 40× magnification; labial is to the bottom of the page, rostral to the left, scale = 10 μm. (A color version of this figure is located at the end of this chapter.)

As seen in sagittal section of the lower oral jaw, the successional lamina for each replacement tooth (outlined in red) is contiguous with oral epithelium that houses taste buds (outlined in orange) and their support cells. Both organs express members of FGF, Hh, and Wnt pathways (see also Fraser et al., 2013). BMP pathway genes are biased toward expression in dental epithelium and mesenchyme, whereas BMP antagonists (e.g., *fst*, *osr2*) are active in nondental mesenchyme, subjacent to taste buds (Bloomquist et al., 2015a). Notably, at these later stages, *pitx2* remains confined to teeth and *calb2* marks taste buds, but other genes thought to be organ specific are not. *sox2* is expressed in both organs, as is *foxa2*, previously described as an endodermal taste bud marker (Beites et al., 2009; Luo, Okubo, Randell, & Hogan, 2009). *foxa2* activity in these most rostral teeth and taste buds thought to be derived from ectoderm is reminiscent of previous observations that putatively endoderm-derived pharyngeal dentitions express the ectodermal factors *eda* and *edar* (Fraser et al., 2009).

The integration of teeth and taste buds over ontogeny prompted examination of the expression of adult stem cell markers in these adjoining structures undergoing renewal (taste buds) and replacement (teeth). Chosen markers have known roles in (i) regenerating reptile teeth, (ii) the labial stem cell niche of mouse incisors, (iii) taste bud and tongue epithelium renewal, (iv) dental mesenchymal homeostasis/repair, as well as (v) intestinal villi and (vi) hair follicle regeneration (Biehs et al., 2013; Handrigan, Leung, & Richman, 2010; Juuri et al., 2013; Kaukua et al., 2014; Okubo, Clark, & Hogan, 2009; Tanaka et al., 2013; Yee et al., 2013; intestine and hair follicle data reviewed in Blanpain & Fuchs, 2014; Krieger & Simons, 2015).

Visualized in sagittal section, adult stem cell markers are expressed in the successional lamina, at the base of taste buds, in the cervical loop domains and at the tip of replacement teeth, in the dental follicle, in the dental mesenchyme where teeth will be penetrated by NVBs, and in the mesenchyme subjacent to taste buds (Fig. 4). Stem cell marker expression mirrors the location of label-retaining cells (LRCs) after long-term BrdU pulse-chase experiments. Immunohistochemical (IHC) costaining of LRCs with select adult stem factors (e.g., Sox2, Bmi1) further supports the location of multiple stem niches in adjacent taste bud and dental territories (Fig. 5; Bloomquist et al., 2015b). Overall, cichlid teeth and taste buds exhibit sustained cellular and molecular connectivity—from early patterning into regeneration stages.

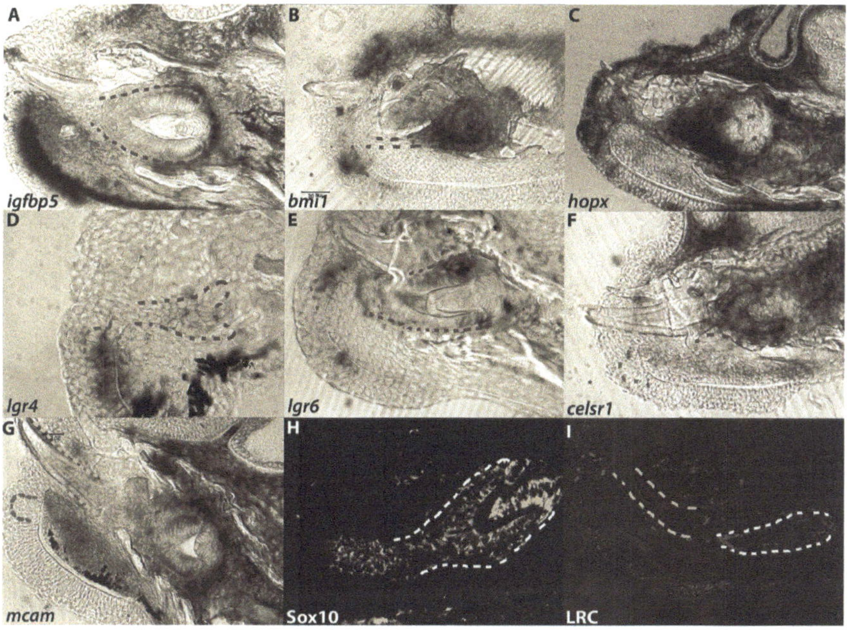

Figure 4 Expression of adult stem cell markers in cichlid tooth and taste bud renewal. In A–G, orange dashes outline taste buds, green outlines the functional tooth, and red outlines replacement teeth. These are sagittal sections, 40 ×; labial is to the bottom of the page, rostral to the left, scale in B = 20 μm. Sox10 IHC (H) in green with replacement tooth epithelium outlined in white. LRC after 100 days chase (I) marked with IHC against BrdU, where LRC clusters are seen at cervical loops, tip of tooth, and in taste bud and surrounding epithelium; replacement tooth outlined in white, 20 × objective. (A color version of this figure is located at the end of this chapter.)

3. THE PLASTICITY OF STEM-BASED ORGAN REGENERATION

Recent work in models of stem cell-driven organ renewal (e.g., tooth, intestine, hair follicle) has revealed surprising plasticity and noteworthy context dependence of cellular behavior. In mouse incisor mesenchyme for instance, pericytes can be reprogrammed to form odontoblasts upon injury (Feng, Mantesso, De Bari, Nishiyama, & Sharpe, 2011), whereas neural crest-derived glia (Kaukua et al., 2014) and Gli1-expressing periarterial cells associated with the NVB (Zhao et al., 2014) contribute to dental tissue homeostasis. Similarly in the intestinal crypt and the hair follicle, cells from differentiated organ zones can regain stem cell competence and ultimately

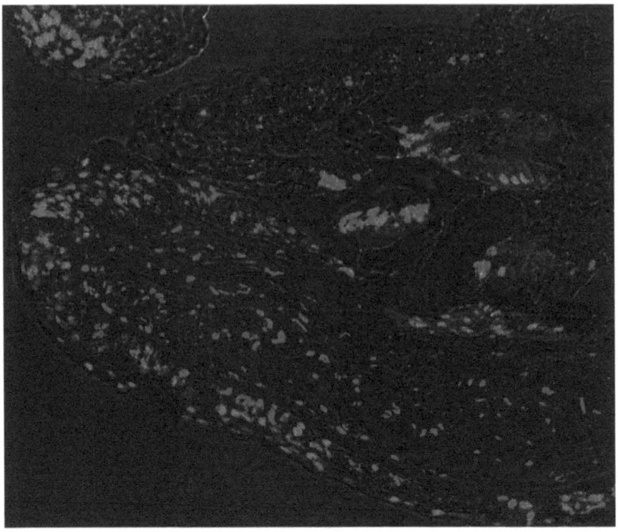

Figure 5 Identification of LRCs/stem cells in cichlid tooth and taste bud regeneration. Double label for LRC after 90 days chase (red) and IHC of adult stem marker Sox2 (blue). This is a sagittal section, 20 × objective; labial is to the bottom of the page, rostral to the left. (A color version of this figure is located at the end of this chapter.)

repopulate the organ upon targeted ablation (Blanpain & Fuchs, 2014). Notably, epithelial cells from outside the hair follicle can migrate to the follicular stem cell niche. Once in position, these cells behave like endogenous stem cells and promote regeneration (Rompolas, Mesa, & Greco, 2013). Taken together, these studies highlight the importance of the niche environment, which can impinge upon internal molecular programs to mediate reversible behavior of cells (Nieto, 2013).

Given the developmental bipotency observed for cichlid teeth and taste buds (Bloomquist et al., 2015a; Figure 2 and discussion above), recent studies addressed whether this plasticity persists into later regeneration stages. Intriguingly, after brief LDN (BMP inhibitor) treatment of juvenile animals at organ regeneration stages, *shh* expression is abrogated from the cervical loop regions of cichlid teeth and *calb2* is activated in these domains (Bloomquist et al., 2015b; Fig. 6A″–B″).

These data suggest that a change in the local environment is sufficient to switch the competence of progenitor cells from a dental to sensory fate. It is generally unknown whether the reverse is true: whether taste buds cells can adopt dental competence. Interestingly, Beites et al. (2009) showed that tongue taste buds in Follistatin (a BMP antagonist) null mice invaginated

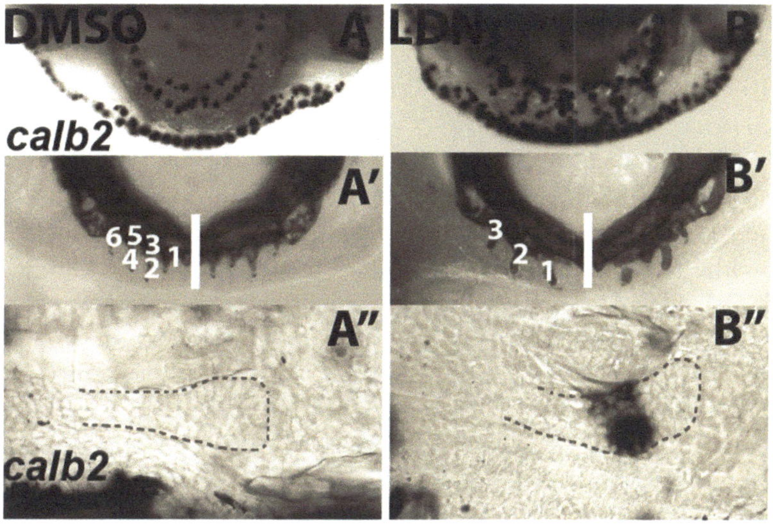

Figure 6 Interorgan plasticity between cichlid teeth and taste buds at patterning and regeneration stages. *calb2* ISH was used to mark taste buds. A–B′ show infiltration of taste buds into the tooth field and reduced tooth density in early (5–6 dpf) LDN-treated animals. In the middle panel, midline of cleared and stained fishes is marked by a white line, and teeth in each half of the dentary are marked by white numerals. A–B′, these are dorsal views of dentaries, labial to bottom of page. At later regeneration stages (20 dpf), LDN treatment results in *calb2* expression in the cervical loops of replacement tooth germs (B″), which is never present in controls (A″). A″–B″ are sagittal sections, 40× objective. For experiments at patterning stages, animals were treated for 24 h at 5–6 dpf, allowed to recover for a week, and then sacrificed and assayed. For experiments at regeneration stages, animals were treated for 48 h at 20 dpf and then immediately sacrificed and assayed. (A color version of this figure is located at the end of this chapter.)

rather than evaginated, recapitulating at least a morphogenetic hallmark of tooth development. Overall, these findings reveal a striking and potentially long-lasting plasticity between dental and taste bud tissues, raising the possibility that these oral organs can be interconverted throughout the time window from developmental patterning to adult regeneration. Such interorgan reprogramming is a unique extension of the intraorgan conversion discussed above for intestine and hair follicle models (Blanpain & Fuchs, 2014).

4. CICHLIDS COMBINE THE FEATURES OF RENEWING AND REPLACING ORAL ORGAN SYSTEMS

Teeth and taste buds of cichlid fishes are copatterned across ontogeny, from an early bipotent epithelium to cellular connectivity during regeneration stages. These organs may share developmental as well as adult progenitors, and remarkable interorgan plasticity is observed across the life span (Bloomquist et al., 2015a, 2015b; Fig. 6). Cichlid taste buds persist in the mouth and are continuously renewed, probably fueled by label-retaining stem cells located at their base and throughout the oral epithelium (see Fig. 5), similar in large part to taste bud regeneration on the mammalian tongue (Okubo et al., 2009). Cichlid taste buds are innervated throughout the lifetime of the organ (Fig. 10.3 in Bloomquist et al., 2015a; Streelman, 2013). Cichlid teeth, on the other hand, exhibit one-for-one cycled replacement, roughly every 5 weeks per tooth. Label-retaining stem cells are located in the cervical loop regions, at the tooth tip and in dental mesenchyme (Bloomquist et al., 2015b; Figs. 5 and 6). Teeth and taste buds share expression of adult stem cell markers *igfbp5*, *sox2*, *lgr4*, *lgr6*, *bmi1*, and *hopx*, also observed in other regenerating systems like intestinal villi and hair follicles.

Cichlid teeth exhibit intramedullary replacement—within a bony crypt of the jaw (Fig. 7). Whole-tooth replacement thus entails significant jaw remodeling. Given that some individuals possess hundreds of oral (and pharyngeal) teeth (Fraser et al., 2008, 2009), all undergoing patterned

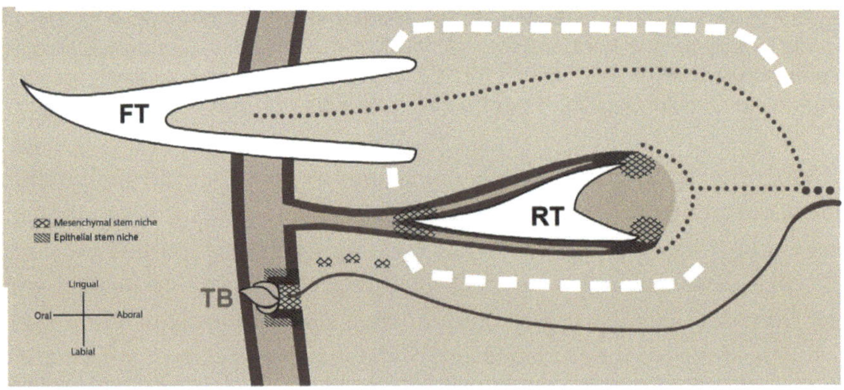

Figure 7 Cichlid teeth and taste buds combine the features of regenerating and replacing organ systems. Here, layers of epithelium are indicated in blue and green. Light pink at the base of the replacement tooth shows dental papilla. Areas of bone are depicted in white, with broken white lines representing remodeling of the bony crypt. Dark pink NVBs are shown penetrating the functional tooth and taste bud, with broken lines in areas of reinnervation. Putative stem cell niches are indicated by crosshatching (mesenchyme) and parallel lines (epithelium). Oral to left of page; labial to bottom. (A color version of this figure is located at the end of this chapter.)

replacement, cichlid jaws face a constant biomechanical demand as a result of this turnover. Whole-tooth replacement also requires NVB pathfinding and reinnervation, as functional teeth are shed and replacement organs take their place (Fig. 7).

The trigger of cycled tooth replacement and reinnervation has yet to be identified. The recent finding that hair follicle "plucking" initiates a sequential immune cascade, which mobilizes stem-based regional regeneration (Chen et al., 2015), may provide a model. Whole-tooth replacement would then be viewed as a natural case of experimental wounding or ablation (also Rompolas et al., 2013).

Generally, the proximity of cichlid teeth and taste buds puts two distinct types of regenerating organ systems together. Taste buds undergo continuous renewal but not organ replacement; teeth are cyclically replaced *in toto*. This effectively locates a system of homeostatic renewal and continuous turnover just superficial to one characterized by cycled whole-organ replacement (Fig. 7). This is analogous to placing many of the features and molecular markers of intestinal villi next to those of a hair follicle. It has been speculated that the difference between organisms with lifelong regeneration of complex dentitions (e.g., cichlids, reptiles, sharks) and those without (mammals) lies in the continued maintenance and ultimate recruitment of stem cells superficial to the successional lamina (Handrigan et al., 2010; Fraser et al., 2013). Cichlids may solve this problem by integrating the development and regeneration of copatterned sensory and dental organs.

ACKNOWLEDGMENTS

The U.S. National Institute of Dental and Craniofacial Research supports our work on tooth and taste bud evolution, development, and regeneration (2R01DE019637-06 to J.T.S., F30DE023013 to R.F.B.).

REFERENCES

Beites, C. L., Hollenbeck, P. L., Kim, J., Lovell-Badge, R., Lander, A. D., & Calof, A. L. (2009). Follistatin modulates a BMP autoregulatory loop to control the size and patterning of sensory domains in the developing tongue. *Development*, *136*, 2187–2197.

Biehs, B., Hu, J. K., Strauli, N. B., Sangiorgi, E., Jung, H., Heber, R. P., et al. (2013). Bmi1 represses Ink4a/Arf and Hox genes to regulate stem cells in the rodent incisor. *Nature Cell Biology*, *15*, 846–852.

Blanpain, C., & Fuchs, E. (2014). Plasticity of epithelial stem cells in tissue regeneration. *Science*, *344*, 1242281.

Bloomquist, R. F., Parnell, N. F., Phillips, K. A., Fowler, T. E., Yu, T. Y., Sharpe, P. T., et al. (2015a). Co-evolutionary patterning of teeth and taste buds. *PNAS*, in press.

Bloomquist, R. F., Fowler, T. E., Abdilleh, K., & Streelman, J. T. (2015b). Developmental plasticity of stem cells in tooth and taste bud renewal. In preparation.

Chen, C.-C., Wang, L., Plikus, M. V., Jiang, T. X., Murray, P. J., Ramos, R., et al. (2015). Organ-level quorum sensing directs regeneration in hair stem cell populations. *Cell*, *161*, 277–290.

Cleves, P. A., Ellis, N. A., Jimenez, M. T., Nunez, S. M., Schluter, D., Kingsley, D. M., et al. (2014). Evolved tooth gain in sticklebacks is associated with a cis-regulatory allele of Bmp6. *Proceedings of the National Academy of Sciences of the United States of America*, *111*, 13912–13917.

Feng, J., Mantesso, A., De Bari, C., Nishiyama, A., & Sharpe, P. T. (2011). Dual origin of mesenchymal stem cells contributing to organ growth and repair. *Proceedings of the National Academy of Sciences of the United States of America*, *106*, 6503–6508.

Fraser, G. J., Bloomquist, R. F., & Streelman, J. T. (2008). A periodic pattern generator for dental diversity. *BMC Biology*, *6*, 32.

Fraser, G. J., Bloomquist, R. F., & Streelman, J. T. (2013). Common developmental pathways link tooth shape to regeneration. *Developmental Biology*, *377*, 399–414.

Fraser, G. J., Cerny, R., Soukup, V., Bronner-Fraser, M., & Streelman, J. T. (2010). The odontode explosion: The origin of tooth-like structures in vertebrates. *Bioessays*, *32*, 808–817.

Fraser, G. J., Hulsey, C. D., Bloomquist, R. F., Uyesugi, K., Manley, N. R., & Streelman, J. T. (2009). An ancient gene network is co-opted for teeth on old and new jaws. *PLoS Biology*, 7(2), e1000031.

Handrigan, G. R., Leung, K. J., & Richman, J. M. (2010). Identification of putative dental epithelial stem cells in a lizard with life-long tooth replacement. *Development*, *137*, 3545–3549.

Juuri, E., Jussila, M., Seidel, K., Holmes, S., Wu, P., Richman, J., et al. (2013). Sox2 marks epithelial competence to generate teeth in mammals and reptiles. *Development*, *140*, 1424–1432.

Kaukua, N., Shahidi, M. K., Konstantinidou, C., Dyachuk, V., Kaucka, M., Furlan, A., et al. (2014). Glial origin of mesenchymal stem cells in a tooth model system. *Nature*, *513*, 551–554.

Krieger, T., & Simons, B. D. (2015). Dynamic stem cell heterogeneity. *Development*, *142*, 1396–1406.

Luo, X., Okubo, T., Randell, S., & Hogan, B. L. (2009). Culture of endodermal stem/progenitor cells of the mouse tongue. *In Vitro Cellular & Developmental Biology—Animal*, *45*, 44–54.

Nieto, M. A. (2013). Epithelial plasticity: A common theme in embryonic and cancer cells. *Science*, *342*, 1234580.

Okubo, T., Clark, C., & Hogan, B. L. (2009). Cell lineage mapping of taste bud cells and keratinocytes in the mouse tongue and soft palate. *Stem Cells*, *27*, 442–450.

Peng, T., Tian, Y., Boogerd, C. J., Lu, M. M., Kadzik, R. S., Stewart, K. M., et al. (2013). Coordination of heart and lung co-development by a multipotent cardiopulmonary progenitor. *Nature*, *500*, 589–592.

Rompolas, P., Mesa, K. R., & Greco, V. (2013). Spatial organization within a niche as a determinant of stem cell fate. *Nature*, *502*, 513–518.

Sire, J.-Y., Davit-Beal, T., Delgado, S., Van Der Heyden, C., & Huysseune, A. (2002). First-generation teeth in nonmammalian lineages: Evidence for a conserved ancestral character? *Microscopy Research and Technique*, *59*, 408–434.

Streelman, J. T. (2013). Advancing evolutionary developmental biology. In J. T. Streelman (Ed.), *Advances in evolution and development* (pp. 203–217). New York: Wiley-Blackwell.

Tanaka, T., Komai, Y., Tokuyama, Y., Yanai, H., Ohe, S., Okazaki, K., et al. (2013). Identification of stem cells that maintain and regenerate lingual keratinized epithelial cells. *Nature Cell Biology*, *15*, 511–518.

Tucker, A. S., & Fraser, G. J. (2014). Evolution and developmental diversity of tooth regeneration. *Seminars in Cell & Developmental Biology*, *25–26*, 71–80.

Tuisku, F., & Hildebrand, C. (1994). Evidence for a neural influence on tooth germ generation in a polyphyodont species. *Developmental Biology*, *165*, 1–9.

Xu, C.-R., Cole, P. A., Meyers, D. J., Kormish, J., Dent, S., & Zaret, K. S. (2011). Chromatin "prepattern" and histone modifiers in a fat choice for liver and pancreas. *Science*, *332*, 963–966.

Yee, K. K., Li, Y., Redding, K. M., Iwatsuki, K., Margolskee, R. F., & Jiang, P. (2013). Lgr5-EGFP marks taste bud stem/progenitor cells in posterior tongue. *Stem Cells*, *31*, 992–1000.

Zhao, H., Feng, J., Seidel, K., Shi, S., Klein, O., Sharpe, P. T., et al. (2014). Secretion of shh by a neurovascular bundle niche supports mesenchymal stem cell homeostasis in the adult mouse incisor. *Cell Stem Cell*, *14*, 160–173.

CHAPTER FOURTEEN

Regulating Craniofacial Development at the 3′ End: MicroRNAs and Their Function in Facial Morphogenesis

Andre L.P. Tavares, Kristin B. Artinger, David E. Clouthier[1]
Department of Craniofacial Biology, University of Colorado Anschutz Medical Campus, Aurora, Colorado, USA
[1]Corresponding author: e-mail address: david.clouthier@ucdenver.edu

Contents

Abstract

Defects in craniofacial development represent a majority of observed human birth defects, occurring at a rate as high as 1:800 live births. These defects often occur due to changes in neural crest cell (NCC) patterning and development and can affect non-NCC-derived structures due to interactions between NCCs and the surrounding cell types. Proper craniofacial development requires an intricate array of gene expression networks that are tightly controlled spatiotemporally by a number of regulatory mechanisms. One of these mechanisms involves the action of microRNAs (miRNAs), a class of noncoding RNAs that repress gene expression by binding to miRNA recognition sequences typically located in the 3′ UTR of target mRNAs. Recent evidence illustrates that miRNAs are crucial for vertebrate facial morphogenesis, with changes in miRNA

Current Topics in Developmental Biology, Volume 115
ISSN 0070-2153
http://dx.doi.org/10.1016/bs.ctdb.2015.08.001

expression leading to facial birth defects, including some in complex human syndromes such as 22q11 (DiGeorge Syndrome). In this review, we highlight the current understanding of miRNA biogenesis, the roles of miRNAs in overall craniofacial development, the impact that loss of miRNAs has on normal development and the requirement for miRNAs in the development of specific craniofacial structures, including teeth. From these studies, it is clear that miRNAs are essential for normal facial development and morphogenesis, and a potential key in establishing new paradigms for repair and regeneration of facial defects.

1. INTRODUCTION

Craniofacial morphogenesis is a complex process that is regulated by a diverse set of signaling networks. Members of these networks are expressed at precise times and locations, a fact that is requisite for the establishment and maintenance of specific developmental domains. One mechanism now recognized as being involved in regulating gene expression during development involves gene silencing by microRNAs (miRNAs or miRs) (Bartel, 2004, 2009; Ha & Kim, 2014). One of several classes of noncoding RNAs that are currently known (Mattick & Makunin, 2006), these small noncoding regulatory RNAs are approximately 22 nucleotides in length (Lagos-Quintana, Rauhut, Lendeckel, & Tuschi, 2001; Lee, Feinbaum, & Ambros, 1993) and inhibit translation of target mRNAs through several distinct mechanisms (Lee et al., 1993; Wightman, Ha, & Ruvkun, 1993). Since the discovery of the first miRNA in *Caenorhabditis elegans* that controls cell fate decision during larval development (Lee et al., 1993; Wightman et al., 1993), miRNAs have been identified in many species (Bartel, 2004, 2009; Lagos-Quintana et al., 2001; Song & Tuan, 2006). In addition, miRNA function has been implicated in many biological processes and disease mechanisms; to this point, over 37,000 references exist in PubMed that include the term "miRNA" (http://www.ncbi.nlm.nih.gov/pubmed/?term=mirna).

From some of these studies, it is now clear that miRNAs play crucial roles in facial morphogenesis, in part through exquisite fine-tuning of gene expression, thus establishing the sharp gene expression boundaries required for initiation and maintenance of development. This review focuses on miRNA function during multiple aspects of facial development, focusing on the development of teeth in addition to bone, cartilage, glandular, and connective tissue of the face.

2. CRANIOFACIAL DEVELOPMENT: CELL TYPES AND TISSUE INTERACTIONS

Craniofacial skeletal structures are derived from cranial neural crest cells (NCCs) that form at the neural plate border, undergo an epithelial-to-mesenchymal transition, and migrate into the pharyngeal arches (PAs), which are transient structures on the ventral embryo surface (Le Douarin, 1982). There the NCCs accumulate, condense, and proliferate in precise growth centers within the arches and give rise to specific craniofacial cartilages and bones. Each arch gives rise to specific skeletal derivatives: NCCs migrating into the maxillary and mandibular portions of the first arch form structures that include the maxilla, mandible, and Meckel's cartilage (Couly, Coltey, & Le Douarin, 1993; Kontges & Lumsden, 1996), in addition to the palatoquadrate in zebrafish and *Xenopus* (Kimmel, Miller, & Moens, 2001). The mouse and chick second PAs give rise to the stapes (a middle ear bone) and part of the hyoid bone, while PA3 gives rise to another part of the hyoid; PA4–6 give rise to the laryngeal or gill cartilages. The zebrafish and *Xenopus* second arch cartilages, the ceratohyal and hyosymplectic, are recruited to reinforce the jaw opening mechanism by creating a second attachment to the neurocranium (Hyman, 1979; Takechi & Kuratani, 2010). NCCs maintain intrinsic information from their origin in the dorsal mid/hindbrain (Couly, Grapin-Botton, Coltey, Ruhin, & Le Douarin, 1998; Le Douarin, 1982; Noden, 1988) while also receiving substantial information from the environment during and after migration (Baker, Bronner-Fraser, Le Douarin, & Teillet, 1997; Trainor & Krumlauf, 2000). The combinatorial readout of these extrinsic signaling events leads to the development of the craniofacial skeleton.

NCCs do not form these derivatives cell autonomously, but rather require environmental signals via tissue interactions between the overlying ectoderm, underlying endoderm, and cranial mesoderm to drive differentiation of specific skeletal derivatives. Many of the signals are known, including Endothelin-1, SHH, BMPs, WNTs, FGFs, Notch, and retinoic acid (Clouthier, Garcia, & Schilling, 2010; Cobourne & Green, 2012; Kimmel, Miller, & Keynes, 2001; Stanier & Pauws, 2012). Each arch can be divided into a proximal (dorsal), intermediate and distal (ventral) domain, with different derivatives arising from each component (Clouthier et al., 2010) (Fig. 1). During patterning of the proximal portion of the mandibular arch, PA endoderm-derived SHH induces expression of FGF8 in the arch

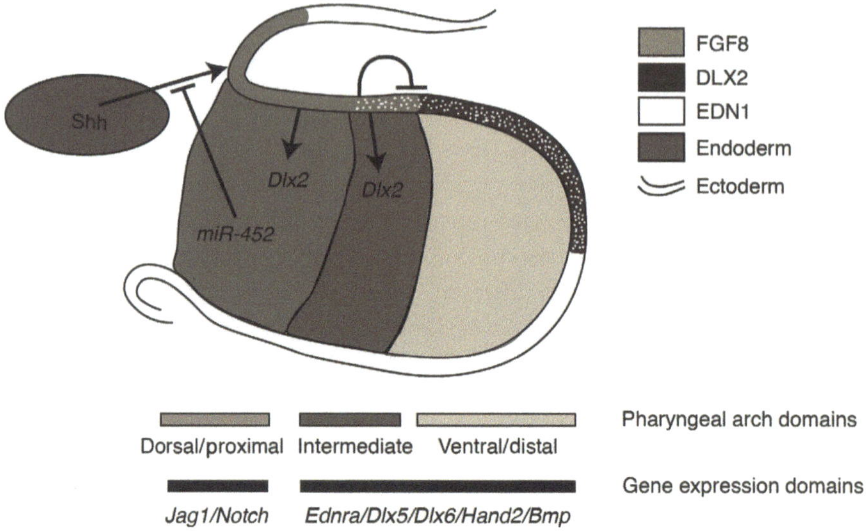

Figure 1 Pharyngeal arch organization and miRNA function. A schematized first mandibular arch with dorsal/proximal, intermediate, and ventral/distal domains demarcated. Gene regulatory networks active in these domains are listed below the domain description. The *Ednra/Dlx5/Dlx6/Hand2/Bmp* pathway is induced by ectodermal EDN1. One miRNA-influenced pathway is shown, in which SHH from the endoderm induces ectodermal *Fgf8*, which in turn induces *Dlx2* expression in the NCC-derived mesenchyme. WNT5A negatively regulates this activation of Fgf8 expression and is the target of *miR-452*. Also shown is that FGF8 negatively represses ectodermal *Dlx2* expression. (A color version of this figure is located at the end of this chapter.)

ectoderm (Sheehy, Cordes, White, Ivey, & Srivastava, 2010) (Fig. 1). FGF8 subsequently induces *Dlx2* expression in the underlying mesenchyme (Thomas, Liu, Rubenstein, & Sharpe, 2000). DLX2, in turn, is crucial for the proper morphogenesis of maxillary and rostral skull structures, as loss of *Dlx2* leads to atavistic changes in the mammalian skull (Depew, Simpson, Morasso, & Rubenstein, 2005; Qiu et al., 1995, 1997). A wingless ortholog, WNT5A, inhibits SHH-induced Fgf8 expression. In zebrafish, the dorsal domain is further delineated by Notch/Jagged signaling, where Jagged1 interacts with Notch2 to both drive dorsal development and prevent expression of genes that drive ventral and intermediate development (Zuniga, Stellabotte, & Crump, 2010). In contrast, the distal and intermediate domains are established by signaling from the Endothelin-A receptor (EDNRA), which functions with BMP4 (Alexander et al., 2011) to pattern and maintain these two domains (Clouthier et al., 2013). Loss of EDNRA signaling results in defects that include transformation of the mandible into a maxilla-like structure in mice (Ozeki, Kurihara, Tonami, Watatani, & Kurihara, 2004; Ruest, Xiang, Lim, Levi, & Clouthier, 2004) and loss of joint structures in zebrafish (Kimmel, Ullmann, Walker, Miller, & Crump, 2003; Miller & Kimmel, 2001; Miller, Schilling, Lee, Parker, & Kimmel, 2000; Nair, Li, Cornell, & Schilling, 2007). Similar defects are observed following loss of genes downstream of EDNRA signaling, including members of the Dlx (Beverdam et al., 2002; Depew, Lufkin, & Rubenstein, 2002; Talbot, Johnson, & Kimmel, 2010) and Hand (Barron et al., 2011; Talbot et al., 2010) families. In all species, the intermediate domain, demarcated by *Bapx1* in both mouse and zebrafish (Miller, Yelon, Stainier, & Kimmel, 2003; Tavares et al., 2012), gives rise to joint regions between first arch-derived bones and cartilages.

3. miRNA BIOGENESIS, STRUCTURE, AND FUNCTION

According to the latest release of miRBase (miRBase 21), 2585 human and 1899 mouse miRNAs have been identified. The actual number of functional miRNAs is unclear, since many miRNAs identified by high-throughput sequencing or bioinformatic analysis have not been experimentally validated (as opposed to decay intermediates of other mRNAs; Ha & Kim, 2014). Genes encoding miRNAs can be located in multiple genomic contexts, including intergenic regions, exons of protein coding genes, and intronic locations. In addition, miRNAs are localized in the genome as single miRNAs or as clusters (Bartel, 2004; Ha & Kim, 2014). The generation of intergenic and most intronic miRNAs occurs through what is referred to as a canonical processing pathway (Fig. 2). A miRNA gene is first transcribed by RNA Pol II, generating the primary miRNA transcript (pri-miRNA) containing a stem loop structure. This somewhat complex structure is then cleaved by the Microprocessor complex, composed of the nuclear RNaseIII endonuclease Drosha and DGCR8 (DiGeorge Syndrome Region 8), generating a ~65 bp structure referred to as a precursor-miRNA (pre-miRNA) (Lee et al., 2003). The enzymatic action of Drosha is essential to generate the pre-miRNA (Lee et al., 2003), while DGCR8 is required for Drosha to interact with the pri-miRNA (Han et al., 2006). miRNA genes located in short intronic sequences can be generated independently of Drosha/DGCR8. These miRNAs, referred to as mirtrons, are transcribed with the host gene and then spliced out (Curtis, Sibley, & Wood, 2012; Ruby, Jan, & Bartel, 2007; Westholm & Lai, 2011). Like all spliced introns, the resulting structure forms a lariat that must be debranched by lariat

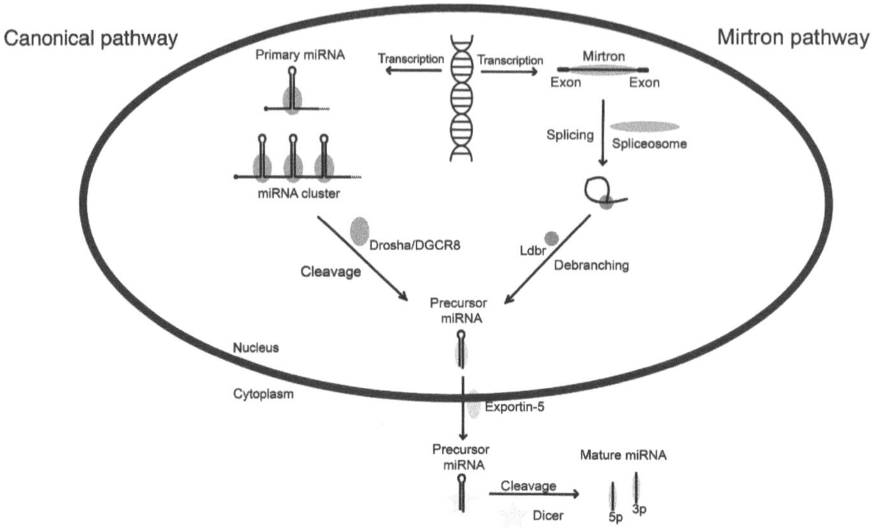

Figure 2 miRNA biogenesis pathways. The left side represents the canonical pathway. miRNA genes are transcribed into primary miRNAs (pri-miRNAs) or a miRNA cluster by RNA polymerase II. Drosha/DGCR8 complex cleaves these transcripts into precursor-miRNA (pre-miRNA) hairpins. Export by Exportin-5 transports pre-miRNAs into the cytoplasm, where a second cleavage by DICER1 produces mature miRNAs. Either strand (5p or 3p) can complex with Argonaute (AGO) and produce a functional RISC complex. The right side represents the mirtron pathway. After transcription by RNA polymerase II, splicing of a shorter intronic sequence may generate a lariat that is debranched into a pre-miRNA hairpin by lariat debranching enzyme (LDBR). This process depends on the splicing machinery and is independent of Drosha/DGCR8 function. Export and processing into mature miRNAs are similar to the canonical pathway. (A color version of this figure is located at the end of this chapter.)

debranching enzyme (LDBR) prior to the formation of the pre-miRNA (Chong et al., 2010; Okamura, Hagen, Duan, Tyler, & Lai, 2007; Ruby et al., 2007) (Fig. 2). Pre-miRNAs generated by the canonical and mirtron pathways are then translocated to the cytoplasm by exportin 5 (Bohnsack, Czaplinski, & Gorlich, 2004; Lund, Guttinger, Calado, Dahlberg, & Kutay, 2004; Yi, Qin, Macara, & Cullen, 2003) and cleaved by the RNase III endonuclease Dicer (Bernstein, Caudy, Hammond, & Hannon, 2001; Hutvagner et al., 2001; Ketting et al., 2001), which removes the stem loop and generates a duplex RNA, each strand of which is ~22 nucleotides in length (Bartel, 2009). These two miRNAs, one from the 5′ strand and one from the 3′ strand of the precursor, are denoted as miR-x-5p and miR-x-3p. This duplex forms a complex with an argonaute (AGO) protein to form the pre-RNA-induced silencing complex (pre-RISC) (Hammond,

Boettcher, Caudy, Kobayashi, & Hannon, 2001). Typically, one of the strands is more biologically active than the other and is referred to as the guide strand. The other strand is referred to as the passenger strand and is often denoted with an asterisk (miR-x*). The passenger strand is then unwound from the guide strand, generating the mature RISC. The guide strand is selected based on which strand has the highest thermodynamic stability between the two strands of duplex RNA (Khvorova, Reynolds, & Jayasena, 2003; Schwarz et al., 2003). While the passenger strand has been viewed as a relatively inactive form, it is becoming clear that they also function to silence target genes (Bhayani, Calin, & Lai, 2012; Chang et al., 2013). In addition, the selection of the guide and passenger strands for a particular miRNA can change in different tissues, suggesting another level of regulation of miRNA function (Ha & Kim, 2014).

miRNAs recognize and bind to their recognition sequence (often though not always) in the 3′ UTR of target mRNAs via the miRNA's "seed" sequence (Lai, 2002), generally considered to be the six to eight base pairs centered around bases seven to nine of the miRNA (Lewis, Shih, Jones-Rhoades, Bartel, & Burge, 2003). miRNAs with identical seed sequences are classified as being in the same miRNA family and often share the same targets. However, while miRNA family members can act redundantly on their mRNA targets, some evidence suggests that other sequences in the miRNA or in the architecture of the target 3′ UTR may provide subtle differences in target selection specificity (Hausser & Zavolan, 2014). Within a target mRNA, canonical miRNA-binding sites match the seed sequence and also pair with position 9 or have an adenosine opposite position 1 of the miRNA (Bartel, 2009; Hausser & Zavolan, 2014). These sites are typically more highly conserved and have a greater effect on the regulation of mRNA expression than sites in mRNA that pair with only six of the nucleotides in the miRNA seed sequence; these sites are referred to as atypical or marginal sites (Bartel, 2009).

The best recognized outcomes of miRNA:mRNA interaction are destabilization of the mRNA or translational repression, thus preventing or reducing target gene expression (Hausser & Zavolan, 2014). This action may act as a buffer against transcriptional "noise" by binding to mRNAs in small abundance, thus establishing sharp domains of gene expression or gene function (Hornstein & Shomron, 2006). This, in effect, would address leaky gene expression; only when mRNA levels reach a certain threshold would protein be produced. Similarly, miRNAs can function to refine gene expression, acting to keep gene expression at a specific level

(Hausser & Zavolan, 2014). It is also now clear that miRNA function can be buffered by removing miRNAs through a process referred to as miRNA sponging, in which RNAs are produced that have specific miRNA recognition sites in their 3′ UTR (Jens & Rajewsky, 2015). The function of these RNAs is to simply allow binding of specific miRNAs, thus acting as a sponge to reduce the levels of those miRNAs. This allows increased expression of other mRNAs that would otherwise be targeted by the specific miRNA.

4. *miR-140*: THE FIRST INDIVIDUAL miRNA ASSOCIATED WITH FACIAL MORPHOGENESIS

Not until 2007, was an individual miRNA demonstrated to be required for normal facial morphogenesis. The function of that miRNA, *miR-140*, is also arguably still one of the best understood in both NCC and craniofacial development. In the ground-breaking 2008 study, miR-140 was found to directly regulate *pdfgra*, a gene required for NCC migration and differentiation (Eberhart et al., 2008). In a hypomorphic mutation or Morpholino knockdown of *pdfgra*, NCCs failed to disperse both rostrally and ventrally around the eye during migration at 24 hours postfertilization (hpf), while overexpression of *pdgfra* in zebrafish resulted in bunching of NCCs near the optic stalk, a critical source of the ligand for Pdgfra, Pdgfaa (Fig. 3). In both approaches, these NCC-derived palatal precursor cells failed to reach the oral ectoderm, disrupting epithelial–mesenchymal interactions, thus resulting in palatal clefting. The ability to properly regulate Pdfgra signaling and thus balance NCC migration with the signal to stop migrating is accomplished though the action of *miR-140*, which binds to the 3′ UTR of the *pdgfra* mRNA and interferes with its translation. After reaching the optic stalk, the dependence on Pdfgra for normal migration of the palatal precursors is severed through the action of *miR-140*, thus allowing the NCCs to continue migrating to the oral ectoderm. Like many aspects of miRNA action, *miR-140* activity must be balanced to ensure that proper migration initially occurs; this could be achieved in part by spatiotemporal regulation of *miR-140* expression.

Similarly, mouse embryos lacking *Pdgfra* specifically in NCCs develop cleft palate (Tallquist & Soriano, 2003). Likewise, several coding variants have been identified with the *PDGFRA* gene (in both exons and the 3′ UTR) of human individuals with cleft palate (Rattanasopha et al., 2012). However, it is not yet known whether *miR-140* plays a similar role in NCC palate development in mammals as it does in zebrafish. *miR-140* is

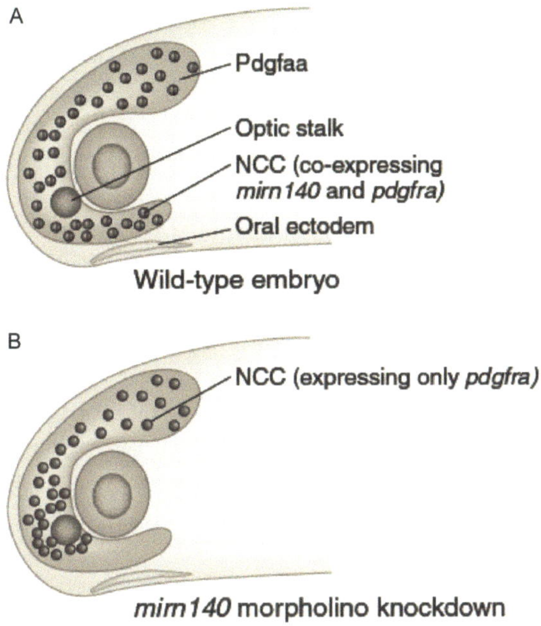

Figure 3 *miR-140* attenuates Pdgfra signaling in palatal precursors. (A) *pdgfaa* expression (blue shading) is observed along the migratory route of NCC-derived palatal precursor cells that express both *Pdgfra* and *miR-140* (red/green circles). (B) *miR-140* knockdown results in palatal precursor cells clumping around the optic stalk (yellow circle), rather than continuing to migrate to the oral ectoderm (orange stripe). This subsequently leads to palatal clefting. *From Clouthier (2008).* (A color version of this figure is located at the end of this chapter.)

expressed in mouse embryos starting at E11.5 (Li et al., 2011), a time point after NCCs have populated the PAs. Expression initiates in the developing palatal shelves by E13.5 and continues through E15.5 after fusion of the palatal shelves. By E15.5, most of the expression appears to be in the palatine bone itself, though earlier expression appears to be in the proximal palatal shelf mesenchyme. Interestingly, *miR-140* can target the mouse *Pdgfra* 3′ region in cultured murine palatal mesenchyme cells (MPMCs) (Li et al., 2011), suggesting a role in mouse palate development. A SNP in the human gene encoding *miR-140* results in reduced *miR-140* expression in MPMCs (Li et al., 2011), and this SNP is associated with increased risk of nonsyndromic cleft palate in mothers who smoke during pregnancy (Li, Meng, Jia, Zhu, & Shi, 2010). It is therefore plausible that the regulation of palate development by *miR-140* represents a conserved function of this miRNA, albeit through potentially different mechanisms.

Because the ethmoid plate (essentially the palate in zebrafish, as fish do not have a nasal cavity) forms through endochondral ossification of a cartilage anlage (Kimmel, Miller, & Moens, 2001), it is alternatively possible that the clefting observed in zebrafish is due to a later requirement of *miR-140* in chondrogenesis. Indeed, targeted deletion of *miR-140* in mouse embryos results in changes in skull structure, including defects in the cartilage composing the nasal septum (Miyaki et al., 2010). While detailed analysis of craniofacial development has not been reported, these results suggest that *miR-140* function is required at least for cranial cartilage ossification into bone. This fits well with the fact that *miR-140* expression is induced by Sox9 (Nakamura et al., 2012), a master regulator of chondrogenic fate in cranial NCCs (Mori-Akiyama, Akiyama, Rowitch, & de Crombrugghe, 2003), and is required for hypertrophic chondrocyte differentiation (Papaioannou et al., 2014). Further analysis of cartilage differentiation in the mouse skull is required to answer these questions.

5. miRNA FUNCTION IN CRANIOFACIAL DEVELOPMENT—LESSONS FROM DELETION OF *DICER1*

Because miRNA biogenesis can proceed along several pathways, perhaps the clearest evidence for the function of miRNAs in normal craniofacial morphogenesis has come from the analysis of mutants in which key miRNA processing enzymes have been deleted. One of first examples of this approach came from inactivating a conditional allele of *Dicer1* (*Dicer1*$^{flox/flox}$ or *Dicer1*$^{fl/fl}$) in NCCs using the *Wnt1-Cre* strain, which targets Cre activity to premigratory NCCs (Danielian, Muccino, Rowitch, Michael, & McMahon, 1998; Soriano, 1999). *Dicer1*$^{fl/fl}$;*Wnt1-Cre* embryos develop severe facial defects, characterized by hypoplasia (frontal, nasal, nasal capsule, basisphenoid, alisphenoid, and jugal bones) or absence (maxilla, premaxilla, mandible, presphenoid, and palatine bones) of almost all NCC-derived facial bones (Huang, Liu, Huang, Zhao, & Cheng, 2010; Zehir, Hua, Maska, Morikawa, & Cserjesi, 2010). In these mutants, NCC migration to and population of the PAs appears normal. However, NCC-derived cells later undergo cell death around the time that they begin to differentiate into bone and cartilage (Zehir et al., 2010). In addition, deletion of *Dicer1* in NCCs perturbs development of the tongue and cranial ganglion and associated nerves, likely reflecting the requirement of NCCs for the development of these structures. Increased apoptosis is also observed in the mesoderm-derived mesenchyme (Nie, Wang, & Jiao, 2011). The severity of the phenotype in *Dicer1*$^{fl/fl}$*;Wnt1-Cre* embryos appears to be background dependent, as *Dicer1;Wnt1-Cre* embryos in a CD1 background have a less severe craniofacial phenotype, in which the maxilla, mandible, and tongue are present (Oommen et al., 2012). Conditional deletion of *Dicer1* with a *Pax2-Cre* strain (in which Cre expression is also driven in NCCs; Ohyama & Groves, 2004) results in a milder phenotype, in which the mandible and maxilla are present but reduced in size (Barritt et al., 2012). The milder phenotype is likely due to the mosaic expression of the *Pax2-Cre* transgene in the PAs (Ohyama & Groves, 2004). Cleft palate is also present and appears to be due to reduced proliferation and increased apoptosis in the palatal shelf mesenchyme (Barritt et al., 2012). Inactivation of zebrafish and *Xenopus dicer* through either mutation or Morpholino knockdown results in severe defects in overall head development, suggesting miRNAs control head development in general, although the craniofacial structures were not directly assayed (Decembrini, Andreazzoli, Barsacchi, & Cremisi, 2008; Giraldez et al., 2005). These findings illustrate the complexity of miRNA function required for facial morphogenesis and suggest that miRNAs are crucial for proper patterning and differentiation. It is important to note that earlier functions of miRNAs in mammalian NCC specification are not well understood since these Cre strains delete *Dicer1* at a later time point during NCC development.

A subsequent and more detailed analysis of gene expression differences between control and *Dicer1*$^{fl/fl}$*;Wnt1-Cre* embryos using fluorescence-activated cell sorting of NCCs revealed that *miR-452* is highly expressed in NCCs that populate the mouse PAs and frontonasal prominence at E10.5 and E11.5 (Sheehy et al., 2010). The expression of *miR-452* is more profoundly affected in *Dicer1*$^{fl/fl}$*;Wnt1-Cre* embryos compared to other miRNAs also expressed in this region, including *miR-715*, *miR-425*, and *miR-675*. *In utero* inhibition of *miR-452* by injection of an miRNA inhibitor into the yolk sac results in hypoplasia of NCC-derived structures in the skull base (ala orbitalis, palatine bones, and ala temporalis), an event that requires multiple signaling pathways. As described above, SHH induces ectodermal *Fgf8* expression, which subsequently induces *Dlx2* expression in the underlying mesenchyme. The ability of WNT5A to inhibit this SHH-induced *Fgf8* expression is controlled by *miR-452*, which targets *Wnt5a* (Fig. 1) (Sheehy et al., 2010). Because FGF8 function is crucial for facial development (Abe, Maeda, & Wakisaka, 2008; Abu-Issa, Smyth, Smoak,

Yamamura, & Meyers, 2002; Abzhanov & Tabin, 2004; Albertson & Yelick, 2007; Green et al., 2015; Guo et al., 2011; Inman, Purcell, Kume, & Trainor, 2013; Jin, Turcotte, Crocker, Han, & Yoon, 2011; Shigetani, Nobusada, & Kuratani, 2000; Trummp, Depew, Rubenstein, Bishop, & Martin, 1999), these findings indicate that *miR-452* likely acts in a narrow spatiotemporal manner to inhibit *Wnt5a*. This narrow window of *Wnt5a* repression would also agree with the facts that targeted deletion of *Wnt5a* leads to widespread craniofacial defects (Person et al., 2010) and that mutations in WNT5A have been found in at least some individuals with Robinow syndrome (OMIM 180700) (Person et al., 2010).

Similar changes are observed (though less well described) following disruption of a conditional *Dgcr8* allele (*Dgcr8*$^{fl/fl}$) in mouse NCCs, also using the *Wnt1-Cre* strain. In these *Dgcr8*$^{fl/fl}$*;Wnt1-Cre* mutants, most of the facial region is absent or extremely hypoplastic at E16.5, though this is only based on gross analysis of whole embryos (Chapnik, Sasson, Blelloch, & Hornstein, 2012). Still, this finding suggests that at some point, NCCs were lost and/or failed to proliferate. Cardiovascular defects in *Dgcr8*$^{fl/fl}$*;Wnt1-Cre* mutants have been examined in more detail, with observed defects including persistent truncus arteriosus (PTA), ventricular septal defects (VSD), and interruption of the aorta type B (Chapnik et al., 2012). These changes are characteristic of disrupted cardiac NCC development (Creazzo, Godt, Leatherbury, Conway, & Kirby, 1998) and appear to occur due to elevated apoptosis of migrating NCCs, leading to a reduced number of progenitors prior to their arrival in the cardiac region (Chapnik et al., 2012). These findings further strengthen the hypothesis that the craniofacial defects result from a mechanism of cell death similar to what has been observed in cranial NCCs. Potential reasons for this are discussed below. It is not clear why loss of *Dgcr8* leads to migration defects while loss of *Dicer1* does not. It is worth noting that the phenotypic differences between *Dicer1* and *Dgcr8* mutants likely reflect differences in miRNA processing in addition to other roles these two proteins have within the cell, including chromatin modification (Chong et al., 2010; Kurzynska-Kokorniak et al., 2015).

The functions of DGCR8 have drawn additional attention due to the fact that *DGCR8* resides within the 3 Mb region of human chromosome 22q11 that is associated with 22q11 syndrome (DiGeorge Syndrome (DGS); OMIM 188400), a congenital condition that results in defects in the heart, face, and thymus, in addition to neurocognitive defects (Lindsay et al., 1995; Cohen, Chow, Weksberg, & Bassett, 1999). Facial defects associated with 22q11 syndrome are unique and include micrognathia, low-set ears, a short philtrum above a small mouth and a bulbous nose (Kobrynski & Sullivan, 2007). 22q11 syndrome occurs due to deletion of a 1.5–3.0 Mb region of human chromosome 22q11, a region containing 30–60 genes (including *DGCR8*) and 4 miRNAs. It is not currently known whether there is a correlation between disruption of miRNAs and facial defects observed in 22q11 syndrome individuals, though loss of DGCR8 would almost certainly disrupt miRNA processing (Ha & Kim, 2014). Indeed, in a cohort of 31 22q11 syndrome individuals, analysis of peripheral leukocytes found that 18 miRNAs were differentially expressed (de la Morena et al., 2013). These changes were confirmed in a later study of 30 additional 22q11 individuals (Sellier et al., 2014). All 22q11 individuals with both T-cell and heart defects showed upregulation of four miRNAs in particular (miR-15b-3p, miR-194, miR-324-5p, and miR-361-5p) and downregulation of another four (miR-185, miR-150, let-7, and miR-720). These affected miRNAs are normally involved in cell cycle control, cell survival, or heart/cardiomyocte development. In addition, similar craniofacial and cardiovascular defects are observed in *Df1/+* mice, a mouse model of 22q11 syndrome, in which a chromosomal deletion was introduced in part of mouse chromosome 16, a region that shows conserved synteny with the human 22q11 region and includes the genes encoding four miRNAs and *Dgcr8* (Botta, Lindsay, Jurecic, & Baldini, 1997; Lindsay et al., 1999; Puech et al., 1997). While numerous genes reside in the deleted segment, including *Tbx1* (previously associated with 22q11 syndrome; see Garg et al., 2001; Lindsay et al., 2001; Merscher et al., 2001; Piotrowski et al., 2003; Vitelli, Morishima, Taddei, Lindsay, & Baldini, 2002), these findings further strengthen the assertion that miRNA activity is required for facial morphogenesis.

So how might loss of miRNA function lead to the observed craniofacial defects? Interneuron migration defects contributing to neurodevelopmental deficits observed in *Df1/+* mice are accompanied by reduced levels of CXCR4/CXCL12 (SDF1), a signaling axis associated with cell migration during development and metastasis in cancer (Ghanem et al., 2014; Liekens, Schols, & Hatse, 2010; Masyuk & Brand-Saberi, 2015; Miller, Banisadr, & Bhattacharyya, 2008; Raz & Mahabaleshwar, 2009). Similar neurodevelopmental deficits are present in *Dgcr8*$^{+/-}$ mice, suggesting that failure of miRNA biogenesis may underlie these deficits. Overexpression of *Dgcr8* partially rescues the interneuron migration deficits, strengthening this assertion. Interestingly, morpholino knockdown of *Cxcr4* in zebrafish leads to defects in both cranial NCC migration and subsequent pharyngeal cartilage formation (Olesnicky-Killian, Birkholz, & Artinger, 2009). While it is not yet known if the miRNAs in the 22q11 region have individual roles in facial development, these findings support the hypothesis that disrupted miRNA production could contribute to the unique facial features of 22q11 individuals due to disruption of NCC development and survival. Determining whether this is true is somewhat difficult, as the *Df1/+* mouse does not have obvious skeletal defects (Lindsay et al., 1999). Perhaps, more sensitive analysis using morphometric analysis may be able to uncover previously unrecognized shape distortions (Percival, Green, Marcucio, & Hallgrimsson, 2014).

6. THE *miR-17∼92* FAMILY: COMPLEX AND COMPENSATING ACTIONS DURING VERTEBRATE DEVELOPMENT

One of most intriguing miRNAs involved in human facial development belongs to the miR-17∼92 family. This family encodes a total of 15 different miRNAs, including three families with the same seed sequence, many of which have been associated with adult disorders (Concepcion, Bonetti, & Ventura, 2012; Kan & Meltzer, 2009; Li, Guan, et al., 2014; Mogilyansky & Rigoutsos, 2013; Olive, Jiang, & He, 2010; Tan, Li, Lim, & Tan, 2014). This family is highly conserved in vertebrates and consists of three loci on three different chromosomes: the *miR-17∼92* locus contains *miR-17*, *18a*, *19a*, *20a*, *19b*, and *92-1* on human chromosome 13; the *miR-106a∼363* locus contains *miR106a*, *18b*, *20b*, *19b2*, *92-2*, and *363* on human chromosome X; the *miR-106b∼25* locus contains *miR-106b*, *93*, and *25* on human chromosome 7. A role for this miR family in craniofacial morphogenesis in humans was identified in a group of individuals with Feingold syndrome-2 MIM614326 (de Pontual et al., 2011), the primary features of which are microcephaly along with short stature and digit defects (Celli, van Bokhoven, & Brunner, 2003; Marcelis & de Brouwer, 1993). Comparative genomic *in situ* hybridization analysis revealed microdeletions on chromosome 13 covering the *MIR17HG* locus, which encodes the *miR-17∼92* cluster of miRNAs (de Pontual et al., 2011). A *de novo* deletion in 13q31.3 (including the *MIR17HG* locus) has also been reported in a Feingold syndrome-2 individual who presented with defects that include facial dysmorphism (Tassano, Di Rocco, Signa, & Gimelli, 2013). In addition, there are at least 10 additional published reports of individuals with deletions in 13q31, all of which result in some form of facial dysmorphism (reviewed by Kannu et al., 2013). Thus, it is reasonable to assume that the facial phenotypes observed in at least some of these individuals may be a result of disruption of the *MIR17HG* locus.

Targeted mutation of the *miR17~92* cluster in mice results in defects similar to those observed in Feingold patients (Ventura et al., 2008). Notably, mutants have craniofacial defects including hypoplasia of most skull bones, failure of palatine bone fusion (resulting in cleft palate) and reduced ossification of cranial vault bones. While the skull defects were not specifically analyzed, a separate targeted mutation of the *miR-17~92* locus also resulted in grossly smaller embryos at E18.5 (de Pontual et al., 2011). Interestingly, when the *miR-17~92* and *miR-106b~25* loci were both deleted, embryos were smaller than control littermates and had VSD (Ventura et al., 2008), an indication of disruption of cardiac NCC development (Creazzo et al., 1998; Kirby & Waldo, 1995; Yutzey & Kirby, 2002). Embryos containing targeted mutations of all three miR-17~92 loci (*miR-17~92*$^{\Delta/\Delta}$; *miR-106b-25*$^{\Delta/\Delta}$; *miR106a~363*$^{\Delta/\Delta}$) were not viable past E15 due to severe vascular developmental defects but did present with decreased overall embryonic size (Ventura et al., 2008).

The *miR-17~92* cluster appears to function, at least in part, by negatively regulating *Tbx* gene family members *Tbx1* and *Tbx3*, which are important for craniofacial development. In an independently generated *miR-17~92* mutant mouse strain, incompletely penetrant cleft lip/palate (CL/P) and mandibular hypoplasia were observed, with CL/P becoming completely penetrant in double *miR-17~92*$^{\Delta/\Delta}$; *miR-106b-25*$^{\Delta/\Delta}$ mutant mouse embryos (Wang et al., 2013). This is an example of the redundancy that can occur among miRNA family members due to conserved seed sequences in target mRNAs. These studies also revealed that the *miR-17~92* locus is regulated by Tfap2 and Bmp signaling. Both Tfap2 and BMP are involved in numerous aspects of early facial morphogenesis, including patterning of NCCs in the PAs and primary and secondary palate development, suggesting that the *miR-17~92* family may play early roles in establishing developmental expression domains of these important regulators. This possibility is supported by the fact that the CL/P observed in *miR-17~92*$^{\Delta/\Delta}$ and *miR-17~92*$^{\Delta/\Delta}$; *miR-106b-25*$^{\Delta/\Delta}$ embryos occurs in part due to earlier defects in facial prominence growth and fusion and the proliferation of the underlying NCC-derived mesenchyme (Wang et al., 2013). This may reflect altered Tgfbr2 signaling, as both *miR-17* and *miR-20a* target the 3′ UTR of TGFBR2, while *miR-18a* targets both SMAD2 and SMAD4A (Li, Shi, Zhu, & Shi, 2012). Tgfbr2 signaling is crucial for normal secondary palate and facial prominence development (Iwata et al., 2012, 2013, 2014), so disruption of the *miR-17~92* family would likely alter Tgfbr2 signaling and thus disrupt palatal development.

The BMP-*miR-17-92* cluster pathway also plays additional roles in cartilage differentiation through a reinforcing positive feedback loop that drives proliferation and survival of chondrocytes. Morpholino knockdown of *miR-92a* in zebrafish leads to disruption of cartilage morphogenesis due to reduced BMP signaling (Ning, Liu, Dai, Meng, & Wang, 2013). This occurs because *miR-92a* normally binds to the mRNA encoding the Bmp inhibitor Noggin3 (Nog3). In the absence of *miR-92a*, Nog3 antagonizes Bmp function, leading to reduced proliferation of chondrocyte progenitors and enhanced cell death in chondrocytes. It is important to note that there is one known case of a human individual with a very small (<1 Mb) microduplication in 13q31.3, which encompasses the *MIR17HG* locus (Kannu et al., 2013). Similar to individuals with a deletion of this region, the affected individual had digit defects and autistic traits, in addition to low-set ears and a prominent jaw/teeth. These findings illustrate the necessity of proper miRNA regulation, with both overexpression or loss potentially damaging in a developmental context.

7. miRNA REGULATION OF BONE AND CARTILAGE DEVELOPMENT

Other miRNAs are also involved in craniofacial and bone development. Targeted deletion of the *miR-199/214* cluster in mice leads to severe skeletal problems in axial and craniofacial structures, including microcephaly with malocclusion, defects in the spinous processes of the vertebrae and osteopenia (Watanabe et al., 2008). These miRNAs are expressed in the PAs and frontonasal prominence at E9.5 and E10.5 in mice (Watanabe et al., 2008) and *pri-miR-199-3a* and *pri-miR-214a* are enriched in the mesenchyme surrounding the developing craniofacial structures in zebrafish at 24, 48, and 72 hpf (Desvignes, Contreras, & Postlethwait, 2014). The location and timing of expression suggest that these miRNAs are important for proper NCC patterning in the developing face region and later in bone and cartilage development (Desvignes et al., 2014).

8. miRNAs AND ODONTOGENESIS

Odontogenesis is an intricate process that involves complex interactions between the PA ectoderm that will give rise to ameloblasts and the underlying NCC-derived mesenchyme that will give rise to odontoblasts and cementoblasts (Chai & Maxson, 2006; Sharpe, 2001). Each stage of odontogenesis is characterized by specific gene expression profiles in both the ectoderm and mesenchyme, with the cross talk between these two tissue layers being crucial for continuation of tooth development (Jernvall & Thesleff, 2000). Early analysis of miRNA expression in tooth germs from first mandibular molars of E15.5 embryos and P0 and P5 pups was the first to illustrate that miRNA expression is dynamic between odontogenic stages, with 31 miRNAs expressed only during developmental stages prior to birth (Jevnaker & Osmundsen, 2008). These 31 almost exclusively regulate the processes of cell division and cell growth. Comparisons of miRNA profiles between the bud, cap, early, and late bell stages of the miniature pig lower molar have revealed that 166 miRNAs have distinct expression patterns during the four stages, including *let-7f*, *miR-128*, *miR-200b*, and *miR-200c* (Li, Song, et al., 2012). Many of the identified miRNAs target mRNAs in specific signaling pathways previously linked to tooth development, including members of the TGFβ (Huang & Chai, 2010) and Wnt (Liu & Millar, 2010) signaling pathways. This is also true for miRNAs identified by stimulating *in vitro* differentiation of human dental pulp cells into odontoblasts. In this differentiation scheme, 12 miRNAs are upregulated (*miR-20b*, *miR-34a*, *miR-937*, *miR-130b*, *miR-100*, *miR-335*, *miR-944*, *miR-17*, *miR-562*, *miR-338-5p*, *miR-122*, and *miR-521*) and 10 miRNAs are downregulated (*miR-542-5p*, *miR-1224-5p*, *miR-382*, *miR431*, *miR-203*, *miR-1225-5p*, *miR-486-3p*, *miR-517c*, *miR-135b*, and *miR-371-3p*) (Gong, Wang, Jiang, Lin, & Ling, 2012). Together, these findings suggest that multiple miRNAs act during odontogenic differentiation by either maintaining cells in an undifferentiated state or driving progenitor cells toward an odontoblastic fate.

Similarly to the study of craniofacial development, conditional deletion of *Dicer1* in the tooth germ has been used to identify miRNAs involved in odontogenesis. In the first approach, *Dicer1* was conditionally deleted in ectodermal cells of the embryo using a *K14-Cre* strain (Dassule, Lewis, Bei, Maas, & McMahon, 2000). This deletion resulted in changes in tooth shape, cusp patterning of molars, ameloblast differentiation, and enamel formation, suggesting that miRNAs are crucial for odontogenesis and that their appearance during vertebrate evolution likely contributed to differences in tooth shape, number, and size (Michon, Tummers, Kyyronen, Frilander, & Thesleff, 2010). One caveat regarding this approach is that Cre activity is not observed in the epithelium of *K14-Cre* mice until E12.5, which is after the initiation of budding and prior to crown morphogenesis (Jarvinen et al., 2006). To examine earlier functions of miRNAs in odontogenesis, a second approach used the *Pitx2-Cre* strain to delete *Dicer1*, as *Pitx2* is expressed at E10.5 in the PA ectoderm (Hjalt, Semina, Amendt, & Murray, 2000; Liu, Selever, Lu, & Martin, 2003; Mucchielli et al., 1997). *Dicer1*$^{fl/fl}$*;Pitx2-Cre* embryos presented with tooth defects that included multiple thin incisors that were branched and lacked enamel, in addition to cusp defects in molars (Cao, Wang, et al., 2010). This is consistent with the earlier study and supports the idea that miRNAs expressed in the ectoderm of the developing dental organ regulate genes important for tooth shape and size. In contrast, when *Dicer1* was removed from the epithelia of the developing tooth using a *Shh-Cre* strain (Harfe et al., 2004), *Dicer1*$^{fl/fl}$*;Shh-Cre* embryos had a less severe phenotype than deletion with *Pitx2-Cre*, with extranumerary upper jaw incisors, even though *Shh* expression is first observed at sites of future dentition as early as E10.5 (Oommen et al., 2012). In these mutants, gene expression associated with incisor development was unchanged and neither cell death nor proliferation was changed from that observed in control embryos. In contrast, when *Dicer1* was conditionally deleted from the NCC-derived mesenchyme of the developing tooth germ using the *Wnt1-Cre* strain, resulting *Dicer-1*$^{fl/fl}$*;Wnt1-Cre* embryos developed maxillary teeth with abnormal morphology (incisors and molars) and mandibular tooth development arrested around the cap stage. This arrest was caused in part by down-regulation of Wnt and Bmp signaling (Oommen et al., 2012). In combination, these studies suggest that miRNAs are required for proper tooth morphogenesis, with different miRNAs regulating incisor and molar development from both the epithelium and underlying mesenchyme. Further, miRNAs differentially regulate upper and lower teeth, as demonstrated by the differential affects upon *Dicer* deletion, suggesting additional mechanisms modulate miRNA function.

A variety of approaches have subsequently been used to more closely examine the role of individual miRNAs in odontogenesis. *In vivo* transfection of a *miR-214* inhibitor into mouse molar tooth germs results in enamel hypomineralization and decreased surface roughness, due in part to decreased expression of genes encoding ameloblastin (AMBN), amelogenin X chromosome (AMELX), and enamelin (ENAM) (required for proper enamel biosynthesis) (Sehic, Risnes, Khan, Khuu, & Osmundsen, 2010). This suggests indicates that this miRNA negatively regulates an inhibitor of enamel biosynthesis, thereby promoting enamel formation. In addition, expression of *Clusterin* (*CLU*; expressed in postsecretory ameloblasts) and

TGFB1 (required for proper odontogenesis and tooth eruption) were downregulated following *miR-214* inhibition while genes involved in organ growth were upregulated, including *Dlk1*, *Grb10*, *Gtl2*, *H19*, *Igf2*, *Ndn*, and *Plagl1*, suggesting a change of developmental fate (Khan, Sehic, Khuu, Risnes, & Osmundsen, 2013). Because a *miR-214* inhibitor results in downregulation of mRNA gene expression, the normal function of this miRNA in driving epithelial cells toward an ameloblastic fate is most likely activating and therefore indirect. Further, *miR-214* does not target the 3′ UTR of any of the seven genes listed above (www.targetscan.org). This is similar to the function of *miR-214* in keratinocytes and hair follicle cell development, where it blocks differentiation by targeting *CTNNB* (*b-catenin*) and thus reducing WNT signaling (Ahmed et al., 2014).

8.1 miRNAs and Odontoblastogenesis

Transition of dental papilla cells into odontoblasts capable of secreting dentin is controlled by several signaling pathways, including the WNT, Notch, TGFB, FGF, and SHH pathways. These pathways function at specific time points during odontogenesis (Michon, 2011). During odontoblast differentiation, multiple miRNAs are upregulated, including *miR-34a*, *miR-338-3p*, *miR-27*, and *miR-663*, and function to promote odontoblast development. Specifically, *miR-34a* controls the differentiation of human dental papilla cells into functional odontoblasts by targeting *NOTCH1* and *BMP7*, two factors involved in maintaining odontoblasts in an undifferentiated state (Wan et al., 2012).

miR-338-3p acts to promote odontoblast formation by targeting *Runx2* (Sun et al., 2013). Initially expressed in osteoblast and odontoblast precursors, *Runx2* expression is reduced during odontoblast differentiation while being maintained in osteoblasts (Chen et al., 2009). A miR-338-3p inhibitor blocks differentiation of mouse dental papilla cells (mDPC6T), whereas introduction of a miR-338-3p mimic enhances differentiation (Sun et al., 2013). Similarly, *miR-338-3p* needs to be downregulated in mouse bone marrow stromal stem cells to allow osteogenic differentiation (Liu, Sun, et al., 2014), suggesting that *miR-338-3p* is crucial for commitment of mesenchymal cells toward either an osteoblastic or odontoblastic fate. Finally, mouse dental papilla odontoblastic cells (mDPC-23) transfected with a *miR-27* mimic (Park et al., 2014) or *miR-663* mimic (Kim et al., 2014) have an accelerated mineralization rate that is preceded by a decrease in both APC protein levels and translocation of β-catenin into the nucleus. Previously, *miR-27* had already been shown to target APC and promote Wnt signaling during osteoblast differentiation (Wang & Xu, 2010). These findings are in accordance with the importance of an active WNT signaling pathway during odontoblast and osteoblast differentiation (Regard, Zhong, Williams, & Yang, 2012; Zhang et al., 2013).

Conversely, downregulation of *miR-143* and *miR-145* is required for odontoblast differentiation, both in mouse primary dental papilla cells and in mouse oral tissue *in vivo*. Overexpression of *miR-143* and *miR-145* *in vitro* delays odontoblast differentiation and dentin formation by targeting the 3′ UTR of *Kruppel-like factor 4 (Klf4)* and *Osterix* (*Osx*) genes (Liu et al., 2013), two factors important for odontoblast and osteoblast differentiation (Chen et al., 2009; Lin et al., 2011; Long, 2012; Michikami et al., 2012). Interestingly, *miR-143* and *miR-145* levels decrease during differentiation of the pre-osteoblastic cell line MC3T3-E1, and following transfection of MC3T3-E1 cells with either a *miR-143* and *miR-145* mimic, osteogenic formation is inhibited by its targeting of *Osx* (Jia et al., 2013; Li, Zhang, Yuan, & Ma, 2014), suggesting that downregulation of both *miR-143* and *miR-145* is also required to promote the expression of factors required for osteoblast differentiation. Since dental pulp stem cells possess the potential to become odontoblasts or osteoblasts (Wang et al., 2014), these studies illustrate the importance of miRNAs in regulating cell fate switch during orofacial development.

8.2 miRNAs and Amelogenesis

miRNAs are also crucially involved in ameloblast differentiation, a complex morphogenetic event that, in rodent incisors, is a continual process (Caton & Tucker, 2009). This process is possible because of the presence of a stem cell niche in the apical end of the mouse incisor that continually produces ameloblasts and odontoblasts. Moreover, only the labial cervical loop in the apex of the incisor possesses stem cells that will differentiate sequentially into presecretory, secretory, and mature ameloblasts (Harada et al., 1999), all of which have differential gene expression profiles (Kuang-Hsien Hu, Mushegyan, & Klein, 2014). Comparison of miRNA profiles between the labial cervical loop and lingual cervical loop and between the labial cervical loop and ameloblasts demonstrated that *miR-31*, *miR-96*, *miR-182*, *miR-200c*, and *miR-429* are highly expressed in the labial cervical loop when compared to the lingual loop. Similarly, *miR-138*, *miR-141*, *miR-200c*, and *miR-429* are highly expressed in the labial loop when compared to ameloblasts. Bioinformatic analysis suggests that miRNAs highly expressed in ameloblasts drive these cells toward differentiation, as they target members of the FGF pathway normally involved in maintaining progenitors in an undifferentiated state in the stem cell niche (Harada et al., 2002). Conversely, miRNAs highly expressed in the labial and lingual cervical loop (where the progenitor cells reside) target mRNAs encoding proteins that normally drive progenitors toward an ameloblastic fate (Hu & Simmer, 2007). These findings illustrate that, as in other systems (Ribeiro, Schoof, Izzotti, Pereira, & Vasques, 2014), miRNAs function in amelogenesis by balancing stemness and differentiation of ameloblasts.

More recently, a role for miRNA activity in the production of ameloblasts has also been demonstrated. Progenitor cells in the labial cervical loop express *Pitx2*, which is responsible for starting the transition of progenitor cells into *Pitx2*-negative pre-ameloblasts. PITX2 initiates this transition by inducing expression of the *mir-200c/141* cluster in progenitor cells, with at least *miR-200c* repressing *Noggin* expression. This forms a positive feedback loop in which, in the absence of Noggin, BMP signaling maintains *miR-200c/141* expression in pre-ameloblasts. This subsequently leads to intercalation of the pre-ameloblasts into the enamel epithelium after *Pitx2* expression is decreased in these cells (Cao et al., 2013). BMP signaling also induces the expression of genes required for enamel formation, including *Amelx* in the dental epithelium (Gluhak-Heinrich et al., 2010). *miR-200c/141* knockout mice show defects in enamel mineralization in the lower incisors and an eruption defect of the third molar, both of which support the role of the *miR-200c/141* cluster in controlling the fate of epithelial dental stem cells *in vitro* and *in vivo* (Cao et al., 2013). Interestingly, mesenchymal cells can be converted into *Amelx*-expressing dental epithelial cells by overexpression of both *Pitx2* and *mir-200a-3p* (Sharp et al., 2014). In this context, PITX2 induces *Lef-1* and other factors that start the transition from a mesenchymal fate to an epithelial fate. *mir-200a-3p* contributes to this process by inhibiting *Zeb1* and *Zeb2*, both of which normally repress *E-cadherin* (*Cdh1*) expression, a marker of epithelial cell fate. *miR-200a-3p* also represses *Pitx2* and *Ctnnb1* (the gene encoding β-catenin) expression, which results in a decrease in their expression, a prerequisite for full epithelial differentiation (Sharp et al., 2014). Promoting differentiation seems to be a conserved role for these two miRNAs, as both *miR-141* and *miR-200b* have been previously shown to control early differentiation of osteogenic precursors into osteoblasts by directly targeting *Dlx5* and modulating BMP signaling (Itoh, Nozawa, & Akao, 2009).

Another level of control for this transition is provided by TBX1 and *miR-96-5p*. TBX1 is also expressed in the labial cervical loop, interacts with PITX2, and represses *miR-96-5p* (Gao et al., 2015). TBX1 is responsible for maintaining these cells in a high proliferative state. When pre-ameloblasts enter the enamel epithelium, they increase their expression levels of *miR-96-5p*, which in turn acts to repress *Tbx1* levels (Gao et al., 2015). This subsequently results in the upregulation of *Amelx* expression, which decreases the rate of proliferation. It is interesting to note that in 22q11 syndrome (discussed above), numerous dental defects are present, including hypodontia and abnormal tooth shape (Klingberg, Oskarsdottir, Johannesson, & Noren, 2002). Similar hypodontia and changes in cusp morphology are observed in $Tbx1^{-/-}$ mice, apparently due to the fact that TBX1 normally binds to PITX2, attenuating the transcriptional activity of PITX2 described above (Cao, Florez, et al., 2010). One can thus envision a pathway involving TBX1, PITX2, *miR-96-5p*, and *miR-200-c*, in which PITX2 primarily drives differentiation of ameloblasts while TBX1 acts to limit the rate of differentiation by promoting proliferation of the precursor cells. In the absence of TBX1, insufficient numbers of precursor cells would be produced prior to the initiation of differentiation, leading to tooth formation defects (Fig. 4). Similarly, *miR-96-5p* appears to mediate a transition from a high proliferative state to a low proliferative state in multiple other cell types, with the low expression of *miR-96-5p* associated with a poor prognosis in colon cancer patients (Ress et al., 2014).

In later ameloblast differentiation, pre-ameloblasts transition from the secretory stage to the maturation stage (Kuang-Hsien Hu et al., 2014). Comparison of miRNA profiles from secretory and maturation stage ameloblasts from rat incisors identified 59 differentially expressed miRNAs (Yin, Hacia, Zhong, & Paine, 2014). Upregulated miRNAs at the maturation stage that were confirmed by *in situ* hybridization included *miR-21*, *miR-31*, and *miR-488*, while confirmed downregulated miRNAs included *miR-153*, *miR-135a*, *miR-135b*, and *miR-298*. LAMP1, the small amelogenin receptor (Xu, Harada, & Taniguchi, 2008), was identified as a target for *miR-153*, with transfection of a miR-153 mimic into the mouse ameloblastic-like LS8 cell line causing decreased activity of a luciferase reporter containing the 3′ UTR of *Lamp1* (Yin et al., 2014). Transferrin receptor C (TFRC), expressed in odontoblasts and ameloblasts (Cam, Boukari, & Ruch, 1989), is a target for *miR-31*; introduction of an inhibitor of miR-31 caused an increase in activity of a luciferase reporter containing the 3′ UTR of

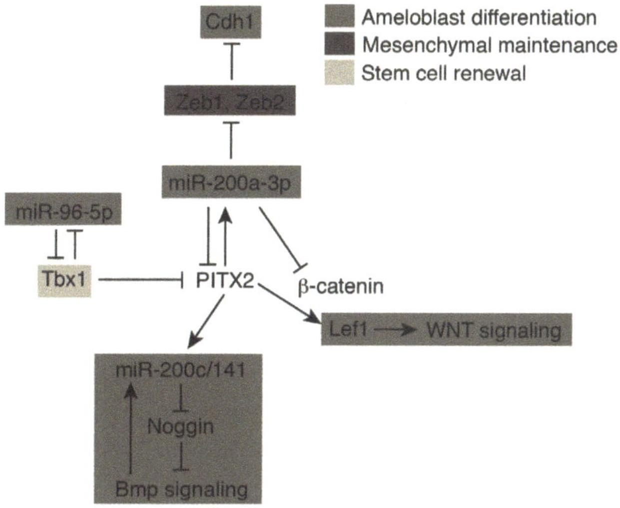

Figure 4 Regulation of ameloblast development by PITX2 and miRNAs. *Pitx2* is an essential gene expressed in ameloblastic progenitor cells before differentiation. These cells are in a high proliferative state that is maintained by Tbx1-mediated repression of *Pitx2*. Transition toward differentiation and a nonproliferative state require repression of *Tbx1* by *miR-96-5p*. Pitx2 physically interacts with β-catenin to induce the WNT and BMP pathways, both required for ameloblast differentiation. WNT signaling is activated by PITX2/β-catenin-mediated induction of *Lef1*. Bmp signaling is activated by *miR-200c*. Expression of this miRNA is driven by PITX2/β-catenin. miR-200c directly targets Noggin, allowing Bmp activity. At the same time, Pitx2 drives expression of *miR-200a-3p*, which represses Zeb1 and Zeb2, causing a sequential expression of *Cdh1* (E-cadherin) and transformation of progenitor cells into epithelial-like ameloblastic cells. *miR-200a-3p* also fine-tunes the levels of *Pitx2* and β-catenin, modulating *Lef1* expression, and WNT pathway activity in differentiated cells. (A color version of this figure is located at the end of this chapter.)

Tfrc (Yin et al., 2014). While the functions of LAMP1 and TFRC in amelogenesis are unknown, they are likely involved in protein trafficking in the cell. Similar to the role of LAMP1 in bone remodeling, LAMP1 and TFRC could act to sequester maturation factors in ameloblasts. In osteoblasts and osteocytes, LAMP1-positive vacuoles also contain RANKL, a key regulator of osteoclastogenesis (Ress et al., 2014). It is thought that these LAMP1+;RANKL+ vacuoles exists to sequester RANKL+, thus controlling the rate of osteoclast production.

9. miRNAs AND SALIVARY GLAND DEVELOPMENT

Epithelial branching is an important morphogenetic event that builds the arborized networks found in different organs, such as the kidneys, lungs, pancreas, mammary, and salivary glands (Ochoa-Espinosa & Affolter, 2012;

Varner & Nelson, 2014). These complex structures are formed by two closely related mechanisms: (1) differential growth and proliferation of the epithelium and (2) expression of different genes and/or extracellular matrix (ECM) components in the mesenchyme (Varner & Nelson, 2014). Early analysis of miRNA expression in salivary gland tissue from E15.5 mouse embryos and P0, P5, and P25 pups illustrates dynamic miRNA expression profiles between these stages, suggesting that, as has been discussed previously, miRNAs appear to fine-tune mRNA levels during morphogenesis (Jevnaker & Osmundsen, 2008). Since branching of the submandibular gland happens before E15.5 (Tucker, 2007), another miRNA profiling experiment was performed using tissue from the epithelium and mesenchyme of submandibular glands at E13.0. *miR-199b/199a-3p*, *miR-125-5p*, and *miR-214* were highly expressed in the mesenchyme of the gland, suggesting a role in branching. *miR-21* was also present in the mesenchyme of the submandibular gland and its inhibition in submandibular glands cultured *in vitro* results in a decrease in the numbers of epithelial buds (Hayashi, Koyama, Azuma, & Kashimata, 2011). This occurs because *miR-21* targets the matrix metalloproteinase (MMP) inhibitors *Reck* and *Pdcd4* (Chang, Hung, & Chang, 2008; Yang, Yue, Pfeffer, Handorf, & Pfeffer, 2011). MMP activity is crucial for branching morphogenesis (in part through altering ECM composition) and absence or antagonism of MMPs in branching organs, including the salivary gland, results in decreased branching (Patel, Rebustini, & Hoffman, 2006; Radisky & Radisky, 2010). *miR-21* activity is thus crucial to ensure normal glandular morphogenesis. This seems to be a conserved role of *miR-21*, as expression of this miRNA correlates with invasiveness of several types of cancers, also through its effect on MMP activity (Martin del Campo et al., 2015; Reis et al., 2012; Soria-Valles et al., 2014).

More recently, 264 miRNAs were found and expressed in the submandibular gland main duct, epithelium, mesenchyme, and end buds from E13 mouse embryos. *In vitro* inhibition of *miR-200c* or *miR-34a*, which are highly expressed in the end buds, caused an increase in the number of end buds, whereas inhibition of one of the two miRNAs most highly expressed in the main duct, *miR-204* and *miR-135a*, caused a decrease in branching morphogenesis. *miR-200c* appears to control branching by regulating cell proliferation in the end buds by targeting very low density lipoprotein receptor (Vldlr) signaling, which induces expression of FGF receptors required for differentiation of end buds (Rebustini et al.,

2012). This is similar to *miR-200c* function in prostate, bladder, and breast cancers, where its expression correlates with a decreased rate of proliferation and invasiveness (Knezevic et al., 2015; Liu, Qiu, et al., 2014; Shi et al., 2014). In invertebrates, miRNAs also play a role in the regulation of salivary gland development and function, where repression of I3PK2 by *miR-14* is required for developmentally regulated cell death and autophagy in the salivary gland, a crucial aspect of gland formation (Dutta & Baehrecke, 2008; Nelson, Ambros, & Baehrecke, 2014). Together, these findings illustrate that miRNAs participate in multiple aspects of gland growth.

10. miRNAs: A POTENTIAL TOOL FOR CORRECTING CRANIOFACIAL DEFECTS

One goal shared by developmental biologists and geneticists is to use knowledge of gene regulation and function in treating human birth defects. miRNA deep sequencing has been completed as part of the Facebase Consortium between E10.5 and 12.5 to determine the miRNA network required for development and morphogenesis of the face (D.E. Clouthier, K.B. Artinger, J.H. Postlethwait, unpublished; https://www.facebase.org/). These data will be important in defining the key miRNAs involved in specific processes in craniofacial development. As these data are used, novel therapeutic approaches could emerge to treat congenital craniofacial defects and injuries to facial and skull structures (Fang, Deng, Gu, & Fan, 2015). Because, insults that give rise to facial birth defects occur at or before the point when they can be easily detected using current methods, the most likely initial approach will be to treat defects after the affected structures are formed. One example currently being developed in animal models is the use of *miR-31*-expressing bone marrow-derived stem cells with or without scaffolds to treat bone defects introduced into rat calvaria (Deng, Bi, et al., 2014; Deng, Zhou, Gu, & Fan, 2014; Deng et al., 2013). Other approaches have investigated using miRNAs to enhance osteogenesis (Nguyen, Jeon, Krebs, Schapira, & Alsberg, 2014; Ochi, Nakasa, Kamei, Usman, & El Mahmoud, 2014). These approaches, and those not yet envisioned and/or utilizing miRNAs not yet investigated, will likely lead to a new wave of reparative strategies that will lower the burden of human facial birth defect syndromes for patients and their families.

11. SUMMARY

Since their discovery, miRNAs have been widely studied in embryonic and adult processes in multiple organisms. In this review, we have highlighted different aspects of miRNA function during development of bone, cartilage, teeth, glandular, and connective tissue of the face. In these processes, miRNAs appear to fine-tune expression of target genes and consequently modulate signaling pathways required for structure development. That miRNAs are required for proper craniofacial morphogenesis is not surprising, as patterning of facial structures requires spatio-temporal-specific gene expression within the developing pharyngeal arches and frontonasal prominence. miRNAs are an efficient and rapid mechanism for regulating this gene expression. The challenge now is identify all miRNAs (and their targets) involved in the development of different craniofacial structures and to then use this information to develop new strategies for repair of craniofacial defects resulting from injury or developmental malformation (Table 1).

Table 1 Description of miRNAs Involved in Craniofacial Development in Different Organisms

miRNA	Target Genes/ Pathways	Role During Craniofacial Development	Species/Cell Type	References
miR-140	*pdgfra*/PDGF signaling	Required for proper palatogenesis. SNP in human pre-miR-140 is associated with nonsyndromic cleft palate	Zebrafish/ human	Dougherty et al. (2012), Eberhart et al. (2008), He, Yan, DeLaurier, and Postlethwait (2011), and Li, Felber, Elks, Croucher, and Roehl (2009)
miR-17/92	*Tbx1*, *Tbx3*, *Fgf10*, *Shox2*, and *Osr1*	Required for proper orofacial development by promoting proliferation	Mouse	Wang et al. (2013)
miR-452	*Wnt5a*/Wnt, Shh, and Fgf signaling	Regulates signaling between the ectoderm and	Mouse	Sheehy et al. (2010)

Table 1 Description of miRNAs Involved in Craniofacial Development in Different Organisms—cont'd

miRNA	Target Genes/ Pathways	Role During Craniofacial Development	Species/Cell Type	References
		ectomesenchyme in the first pharyngeal arch		
miR-92a	*noggin3*/BMP signaling	Controls proliferation, differentiation, and survival of chondrogenic progenitors in the pharyngeal arches	Zebrafish	Ning et al. (2013)
miR-200b	*Rx1*, *Pax6*	Controls cranial cartilage size and eye formation	*Xenopus*	Gessert, Bugner, Tecza, Pinker, and Kuhl (2010)
miR-96	*Rx1*, *Pax6*	Controls cranial cartilage size and eye formation	*Xenopus*	Gessert et al. (2010)
miR-196a	*Rx1*, *Pax6*, *Emx1*, *En2*, *Krox20*	Controls cranial cartilage size and eye formation	*Xenopus*	Gessert et al. (2010)
miR-199/214	*Gli2*, *Fgfr1*/ SHH and FGF signaling	Required for proper craniofacial bone and cartilage formation	Mice and zebrafish	Desvignes et al. (2014) and Watanabe et al. (2008)
miR-873/876	*Zic2*/SHH signaling	Deletion associated with an undefined congenital craniofacial malformation	Human	Koufaris et al. (2015)
miR-214	*Ambn*, *Amelx*, *Enam*, *Clusterin*, *TGFβ1*, *Dlk1*, *Grb10*, *Gtl2*, *H19*, *Igf2*, *Ndn*, and *Plagl1*	Required for proper enamel mineralization and ameloblast differentiation	Mouse	Khan et al. (2013) and Sehic et al. (2010)
miR-34a	*NOTCH1*, *BMP7*	Controls the proper		Wan et al. (2012)

Continued

Table 1 Description of miRNAs Involved in Craniofacial Development in Different Organisms—cont'd

miRNA	Target Genes/ Pathways	Role During Craniofacial Development	Species/Cell Type	References
		differentiation of human papilla cells into odontotoblasts	Human fetal dental papilla cells	
miR-338-3p	*Runx2*	Required for proper differentiation of mouse papilla cells into odontotoblasts	Mouse dental papilla cells (mDPC6T)	Sun et al. (2013)
miR-27	*APC*/Wnt signaling	Required for proper differentiation of mouse papilla cells into odontotoblasts	Mouse dental papilla odontoblastic cells (mDPC-23)	Park et al. (2014)
miR-663	*APC*/Wnt signaling	Required for proper differentiation of mouse papilla cells into odontotoblasts	Mouse dental papilla odontoblastic cells (mDPC-23)	Kim et al. (2014)
miR-143/145	*Klf4*, *Osx*	Inhibits odontoblast differentiation and dentin formation	Mouse/mouse dental papilla cells (mDPC6T)	Liu et al. (2013)
miR-200c/141	*Noggin*/BMP signaling	Required for proper differentiation of ameloblasts	Mouse/LS-8 mouse oral epithelial-like cells	Cao et al. (2013)
miR-96-5p	*Tbx1*	Required for proper differentiation of ameloblasts	Mouse/LS-8 mouse oral epithelial-like cells	Gao et al. (2015)
miR-200a-3p	*Pitx2*, *b-catenin*, *Zeb1*, *Zeb2*	Conversion of mesenchymal cells into amelogenin-expressing dental epithelial cells	Mouse odontoblast-like cells (MDPC-23)/ LS-8 mouse	Sharp et al. (2014)

Table 1 Description of miRNAs Involved in Craniofacial Development in Different Organisms—cont'd

miRNA	Target Genes/ Pathways	Role During Craniofacial Development	Species/Cell Type	References
			oral epithelial-like cells	
miR-153	*Lamp1*	Downregulated in mature ameloblasts	Rat/LS-8 mouse oral epithelial-like cells	Yin et al. (2014)
miR-31	*Tfrc*	Upregulated in mature ameloblasts	Rat/LS-8 mouse oral epithelial-like cells	Yin et al. (2014)
miR-21	*Reck, Pdcd4*	Required for proper branching of end buds	Mouse submandibular gland culture	Hayashi et al. (2011)
miR-200c	*Vldlr*/FGF signaling	Control of cell proliferation in the end buds	Mouse submandibular gland culture	Rebustini et al. (2012)

REFERENCES

Abe, M., Maeda, T., & Wakisaka, S. (2008). Retinoic acid affects craniofacial patterning by changing Fgf8 expression in the pharyngeal ectoderm. *Development, Growth & Differentiation, 50*, 717–729.

Abu-Issa, R., Smyth, G., Smoak, I., Yamamura, K., & Meyers, E. N. (2002). Fgf8 is required for pharyngeal arch and cardiovascular development in the mouse. *Development, 129*, 4613–4625.

Abzhanov, A., & Tabin, C. J. (2004). Shh and Fgf8 act synergistically to drive cartilage outgrowth during cranial development. *Developmental Biology, 273*, 134–148.

Ahmed, M. I., Alam, M., Emelianov, V. U., Poterlowicz, K., Patel, A., Sharov, A. A., et al. (2014). MicroRNA-214 controls skin and hair follicle development by modulating the activity of the Wnt pathway. *The Journal of Cell Biology, 207*, 549–567.

Albertson, R. C., & Yelick, P. C. (2007). Fgf8 haploinsufficiency results in distinct craniofacial defects in adult zebrafish. *Developmental Biology, 306*, 505–515.

Alexander, C., Zuniga, E., Blitz, I. L., Wada, N., LePabic, P., Javidan, Y., et al. (2011). Combinatorial roles for Bmps and Endothelin 1 in patterning the dorsal-ventral axis of the craniofacial skeleton. *Development, 138*, 5135–5146.

Baker, C. V. H., Bronner-Fraser, M., Le Douarin, N. M., & Teillet, M.-A. (1997). Early- and late-migrating neural crest cell populations have equivalent developmental potential in vivo. *Development, 124*, 3077–3087.

Barritt, L. C., Miller, J. M., Scheetz, L. R., Gardner, K., Pierce, M. L., Soukup, G. A., et al. (2012). Conditional deletion of the human ortholog gene Dicer1 in Pax2-Cre expression domain impairs orofacial development. *Indian Journal of Human Genetics*, *18*, 310–319.

Barron, F., Woods, C., Kuhn, K., Bishop, J., Howard, M. J., & Clouthier, D. E. (2011). Downregulation of *Dlx5* and *Dlx6* expression by Hand2 is essential for initiation of tongue morphogenesis. *Development*, *138*, 2249–2259.

Bartel, D. P. (2004). MicroRNAs: Genomics, biogenesis, mechanism, and function. *Cell*, *116*, 281–297.

Bartel, D. P. (2009). MicroRNAs: Target recognition and regulatory functions. *Cell*, *136*, 215–233.

Bernstein, E., Caudy, A. A., Hammond, S. M., & Hannon, G. J. (2001). Role for a bidentate ribonuclease in the initiation step of RNA interference. *Nature*, *409*, 363–366.

Beverdam, A., Merlo, G. R., Paleari, L., Mantero, S., Genova, F., Barbieri, O., et al. (2002). Jaw transformation with gain of symmetry after *Dlx5/Dlx 6* inactivation: Mirror of the past. *Genesis*, *34*, 221–227.

Bhayani, M. K., Calin, G. A., & Lai, S. Y. (2012). Functional relevance of miRNA sequences in human disease. *Mutation Research*, *731*, 14–19.

Bohnsack, M. T., Czaplinski, K., & Gorlich, D. (2004). Exportin 5 is a RanGTP-dependent dsRNA-binding protein that mediates nuclear export of pre-miRNAs. *RNA*, *10*, 185–191.

Botta, A., Lindsay, E. A., Jurecic, V., & Baldini, A. (1997). Comparative mapping of the DiGeorge syndrome region in mouse shows inconsistent gene order and differential degree of gene conservation. *Genome*, *8*, 890–895.

Cam, Y., Boukari, A., & Ruch, J. V. (1989). Stimulatory effect of transferrin on the proliferation of embryonic mouse molar pre-odontoblasts and pre-ameloblasts in organ culture. *Archives of Oral Biology*, *34*, 153–159.

Cao, H., Florez, S., Amen, M., Huynh, T., Skobe, Z., Baldini, A., et al. (2010). Tbx1 regulates progenitor cell proliferation in the dental epithelium by modulating Pitx2 activation of p21. *Developmental Biology*, *347*, 289–300.

Cao, H., Jheon, A., Li, X., Sun, Z., Wang, J., Florez, S., et al. (2013). The Pitx2:miR-200c/141:noggin pathway regulates Bmp signaling and ameloblast differentiation. *Development*, *140*, 3348–3359.

Cao, H., Wang, J., Li, X., Florez, S., Huang, Z., Venugopalan, S. R., et al. (2010). MicroRNAs play a critical role in tooth development. *Journal of Dental Research*, *89*, 779–784.

Caton, J., & Tucker, A. S. (2009). Current knowledge of tooth development: Patterning and mineralization of the murine dentition. *Journal of Anatomy*, *214*, 502–515.

Celli, J., van Bokhoven, H., & Brunner, H. G. (2003). Feingold syndrome: Clinical review and genetic mapping. *American Journal of Medical Genetics Part A*, *122a*, 294–300.

Chai, Y., & Maxson, R. E., Jr. (2006). Recent advances in craniofacial morphogenesis. *Developmental Dynamics*, *235*, 2353–2375.

Chang, C. K., Hung, W. C., & Chang, H. C. (2008). The Kazal motifs of RECK protein inhibit MMP-9 secretion and activity and reduce metastasis of lung cancer cells in vitro and in vivo. *Journal of Cellular and Molecular Medicine*, *12*, 2781–2789.

Chang, K. W., Kao, S. Y., Wu, Y. H., Tsai, M. M., Tu, H. F., Liu, C. J., et al. (2013). Passenger strand miRNA miR-31* regulates the phenotypes of oral cancer cells by targeting RhoA. *Oral Oncology*, *49*, 27–33.

Chapnik, E., Sasson, V., Blelloch, R., & Hornstein, E. (2012). Dgcr8 controls neural crest cells survival in cardiovascular development. *Developmental Biology*, *362*, 50–56.

Chen, S., Gluhak-Heinrich, J., Wang, Y. H., Wu, Y. M., Chuang, H. H., Chen, L., et al. (2009). Runx2, osx, and dspp in tooth development. *Journal of Dental Research*, *88*, 904–909.

Chong, M. M., Zhang, G., Cheloufi, S., Neubert, T. A., Hannon, G. J., & Littman, D. R. (2010). Canonical and alternate functions of the microRNA biogenesis machinery. *Genes & Development*, *24*, 1951–1960.
Clouthier, D. E. (2008). MicroRNAs in facial development. *Nature Genetics*, *40*, 268–269.
Clouthier, D. E., Garcia, E., & Schilling, T. F. (2010). Regulation of facial morphogenesis by endothelin signaling: Insights from mouse and fish. *American Journal of Medical Genetics. Part A*, *152A*, 2962–2973.
Clouthier, D. E., Passos-Bueno, M. R., Tavares, A. L. P., Lyonnet, S., Amiel, J., & Gordon, C. T. (2013). Understanding the basis of Auriculocondylar syndrome: Insights from human, mouse and zebrafish studies. *American Journal of Medical Genetics Part C*, *163*, 306–317.
Cobourne, M. T., & Green, J. B. (2012). Hedgehog signalling in development of the secondary palate. *Frontiers of Oral Biology*, *16*, 52–59.
Cohen, E., Chow, E. W., Weksberg, R., & Bassett, A. S. (1999). Phenotype of adults with the 22q11 deletion syndrome: A review. *American Journal of Medical Genetics*, *86*, 359–365.
Concepcion, C. P., Bonetti, C., & Ventura, A. (2012). The microRNA-17-92 family of microRNA clusters in development and disease. *Cancer Journal*, *18*, 262–267.
Couly, G. F., Coltey, P. M., & Le Douarin, N. M. (1993). The triple origin of skull in higher vertebrates: A study in chick-quail chimeras. *Development*, *117*, 409–429.
Couly, G. F., Grapin-Botton, A., Coltey, P., Ruhin, B., & Le Douarin, N. M. (1998). Determination of the identity of the derivatives of the cephalic neural crest: Incompatibility between Hox gene expression and lower jaw development. *Development*, *125*, 3445–3459.
Creazzo, T. L., Godt, R. E., Leatherbury, L., Conway, S. J., & Kirby, M. L. (1998). Role of cardiac neural crest cells in cardiovascular development. *Annual Review of Physiology*, *60*, 267–286.
Curtis, H. J., Sibley, C. R., & Wood, M. J. (2012). Mirtrons, an emerging class of atypical miRNA. *Wiley Interdisciplinary Reviews—RNA*, *3*, 617–632.
Danielian, P. S., Muccino, D., Rowitch, D. H., Michael, S. K., & McMahon, A. P. (1998). Modification of gene activity in mouse embryos in utero by a tamoxifen-inducible form of Cre recombinase. *Current Biology*, *8*, 1323–1326.
Dassule, H. R., Lewis, P., Bei, M., Maas, R., & McMahon, A. P. (2000). Sonic hedgehog regulates growth and morphogenesis of the tooth. *Development*, *127*, 4775–4785.
de la Morena, M. T., Eitson, J. L., Dozmorov, I. M., Belkaya, S., Hoover, A. R., Anguiano, E., et al. (2013). Signature microRNA expression patterns identified in humans with 22q11.2 deletion/DiGeroge syndrome. *Clinical Immunology*, *147*, 11–22.
de Pontual, L., Yao, E., Callier, P., Faivre, L., Drouin, V., Cariou, S., et al. (2011). Germline deletion of the miR-17 approximately 92 cluster causes skeletal and growth defects in humans. *Nature Genetics*, *43*, 1026–1030.
Decembrini, S., Andreazzoli, M., Barsacchi, G., & Cremisi, F. (2008). Dicer inactivation causes heterochronic retinogenesis in *Xenopus* laevis. *The International Journal of Developmental Biology*, *52*, 1099–1103.
Deng, Y., Bi, X., Zhou, H., You, Z., Wang, Y., Gu, P., et al. (2014). Repair of critical-sized bone defects with anti-miR-31-expressing bone marrow stromal stem cells and poly(glycerol sebacate) scaffolds. *European Cells & Materials*, *27*, 13–24 (discussion 24–15).
Deng, Y., Zhou, H., Gu, P., & Fan, X. (2014). Repair of canine medial orbital bone defects with miR-31-modified bone marrow mesenchymal stem cells. *Investigative Ophthalmology & Visual Science*, *55*, 6016–6023.

Deng, Y., Zhou, H., Zou, D., Xie, Q., Bi, X., Gu, P., et al. (2013). The role of miR-31-modified adipose tissue-derived stem cells in repairing rat critical-sized calvarial defects. *Biomaterials*, *34*, 6717–6728.

Depew, M. J., Lufkin, T., & Rubenstein, J. L. (2002). Specification of jaw subdivisions by *Dlx* genes. *Science*, *298*, 381–385.

Depew, M. J., Simpson, C. A., Morasso, M., & Rubenstein, J. L. R. (2005). Reassessing the *Dlx* code: The genetic regulation of branchial arch skeletal pattern and development. *Journal of Anatomy*, *207*, 501–561.

Desvignes, T., Contreras, A., & Postlethwait, J. H. (2014). Evolution of the miR199-214 cluster and vertebrate skeletal development. *RNA Biology*, *11*, 281–294.

Dougherty, M., Kamel, G., Shubinets, V., Hickey, G., Grimaldi, M., & Liao, E. C. (2012). Embryonic fate map of first pharyngeal arch structures in the sox10: Kaede zebrafish transgenic model. *The Journal of Craniofacial Surgery*, *23*, 1333–1337.

Dutta, S., & Baehrecke, E. H. (2008). Warts is required for PI3K-regulated growth arrest, autophagy, and autophagic cell death in Drosophila. *Current Biology*, *18*, 1466–1475.

Eberhart, J. K., He, X., Swartz, M. E., Yan, Y. L., Song, H., Boling, T. C., et al. (2008). MicroRNA Mirn140 modulates Pdgf signaling during palatogenesis. *Nature Genetics*, *40*, 290–298.

Fang, S., Deng, Y., Gu, P., & Fan, X. (2015). MicroRNAs regulate bone development and regeneration. *International Journal of Molecular Sciences*, *16*, 8227–8253.

Gao, S., Moreno, M., Eliason, S., Cao, H., Li, X., Yu, W., et al. (2015). TBX1 protein interactions and microRNA-96-5p regulation controls cell proliferation during craniofacial and dental development: Implications for 22q11.2 deletion syndrome. *Human Molecular Genetics*, *24*, 2330–2348.

Garg, V., Yamagishi, C., Hu, T., Kathiriya, I. S., Yamagishi, H., & Srivastava, D. (2001). Tbx1, a DiGeorge syndrome candidate gene, is regulated by sonic hedgehog during pharyngeal arch development. *Developmental Biology*, *235*, 62–73.

Gessert, S., Bugner, V., Tecza, A., Pinker, M., & Kuhl, M. (2010). FMR1/FXR1 and the miRNA pathway are required for eye and neural crest development. *Developmental Biology*, *341*, 222–235.

Ghanem, I., Riveiro, M. E., Paradis, V., Faivre, S., de Parga, P. M., & Raymond, E. (2014). Insights on the CXCL12-CXCR4 axis in hepatocellular carcinoma carcinogenesis. *American Journal of Translational Research*, *6*, 340–352.

Giraldez, A. J., Cinalli, R. M., Glasner, M. E., Enright, A. J., Thomson, J. M., Baskerville, S., et al. (2005). MicroRNAs regulate brain morphogenesis in zebrafish. *Science*, *308*, 833–838.

Gluhak-Heinrich, J., Guo, D., Yang, W., Harris, M. A., Lichtler, A., Kream, B., et al. (2010). New roles and mechanism of action of BMP4 in postnatal tooth cytodifferentiation. *Bone*, *46*, 1533–1545.

Gong, Q., Wang, R., Jiang, H., Lin, Z., & Ling, J. (2012). Alteration of microRNA expression of human dental pulp cells during odontogenic differentiation. *Journal of Endodontia*, *38*, 1348–1354.

Green, R. M., Feng, W., Phang, T., Fish, J. L., Li, H., Spritz, R. A., et al. (2015). Tfap2a-dependent changes in mouse facial morphology result in clefting that can be ameliorated by a reduction in Fgf8 gene dosage. *Disease Models & Mechanisms*, *8*, 31–43.

Guo, C., Sun, Y., Zhou, B., Adam, R. M., Li, X., Pu, W. T., et al. (2011). A Tbx1-Six1/Eya1-Fgf8 genetic pathway controls mammalian cardiovascular and craniofacial morphogenesis. *The Journal of Clinical Investigation*, *121*, 1585–1595.

Ha, M., & Kim, V. N. (2014). Regulation of microRNA biogenesis. *Nature Reviews Molecular Cell Biology*, *15*, 509–524.

Hammond, S. M., Boettcher, S., Caudy, A. A., Kobayashi, R., & Hannon, G. J. (2001). Argonaute2, a link between genetic and biochemical analyses of RNAi. *Science*, *293*, 1146–1150.

Han, J., Lee, Y., Yeom, K. H., Nam, J. W., Heo, I., Rhee, J. K., et al. (2006). Molecular basis for the recognition of primary microRNAs by the Drosha-DGCR8 complex. *Cell*, *125*, 887–901.

Harada, H., Kettunen, P., Jung, H. S., Mustonen, T., Wang, Y. A., & Thesleff, I. (1999). Localization of putative stem cells in dental epithelium and their association with Notch and FGF signaling. *The Journal of Cell Biology*, *147*, 105–120.

Harada, H., Toyono, T., Toyoshima, K., Yamasaki, M., Itoh, N., Kato, S., et al. (2002). FGF10 maintains stem cell compartment in developing mouse incisors. *Development*, *129*, 1533–1541.

Harfe, B. D., Scherz, P. J., Nissim, S., Tian, H., McMahon, A. P., & Tabin, C. J. (2004). Evidence for an expansion-based temporal Shh gradient in specifying vertebrate digit identities. *Cell*, *118*, 517–528.

Hausser, J., & Zavolan, M. (2014). Identification and consequences of miRNA-target interactions—Beyond repression of gene expression. *Nature Reviews Genetics*, *15*, 599–612.

Hayashi, T., Koyama, N., Azuma, Y., & Kashimata, M. (2011). Mesenchymal miR-21 regulates branching morphogenesis in murine submandibular gland in vitro. *Developmental Biology*, *352*, 299–307.

He, X., Yan, Y. L., DeLaurier, A., & Postlethwait, J. H. (2011). Observation of miRNA gene expression in zebrafish embryos by in situ hybridization to microRNA primary transcripts. *Zebrafish*, *8*, 1–8.

Hjalt, T. A., Semina, E. V., Amendt, B. A., & Murray, J. C. (2000). The Pitx2 protein in mouse development. *Developmental Dynamics*, *218*, 195–200.

Hornstein, E., & Shomron, N. (2006). Canalization of development by microRNAs. *Nature Genetics*, *38*(Suppl.), S20–S24.

Hu, J. C., & Simmer, J. P. (2007). Developmental biology and genetics of dental malformations. *Orthodontics & Craniofacial Research*, *10*, 45–52.

Huang, X. F., & Chai, Y. (2010). TGF-ss signalling and tooth development. *The Chinese Journal of Dental Research*, *13*, 7–15.

Huang, T., Liu, Y., Huang, M., Zhao, X., & Cheng, L. (2010). Wnt1-cre-mediated conditional loss of Dicer results in malformation of the midbrain and cerebellum and failure of neural crest and dopaminergic differentiation in mice. *Journal of Molecular Cell Biology*, *2*, 152–163.

Hutvagner, G., McLachlan, J., Pasquinelli, A. E., Balint, E., Tuschl, T., & Zamore, P. D. (2001). A cellular function for the RNA-interference enzyme Dicer in the maturation of the let-7 small temporal RNA. *Science*, *293*, 834–838.

Hyman. (1979). *Hyman's comparative vertebrate anatomy*. Chicago: University of Chicago Press.

Inman, K. E., Purcell, P., Kume, T., & Trainor, P. A. (2013). Interaction between Foxc1 and Fgf8 during mammalian jaw patterning and in the pathogenesis of syngnathia. *PLoS Genetics*, *9*, e1003949.

Itoh, T., Nozawa, Y., & Akao, Y. (2009). MicroRNA-141 and -200a are involved in bone morphogenetic protein-2-induced mouse pre-osteoblast differentiation by targeting distal-less homeobox 5. *The Journal of Biological Chemistry*, *284*, 19272–19279.

Iwata, J., Hacia, J. G., Suzuki, A., Sanchez-Lara, P. A., Urata, M., & Chai, Y. (2012). Modulation of noncanonical TGF-beta signaling prevents cleft palate in Tgfbr2 mutant mice. *The Journal of Clinical Investigation*, *122*, 873–885.

Iwata, J., Suzuki, A., Pelikan, R. C., Ho, T. V., Sanchez-Lara, P. A., Urata, M., et al. (2013). Smad4-Irf6 genetic interaction and TGFbeta-mediated IRF6 signaling cascade are crucial for palatal fusion in mice. *Development*, *140*, 1220–1230.

Iwata, J., Suzuki, A., Yokota, T., Ho, T. V., Pelikan, R., Urata, M., et al. (2014). TGFbeta regulates epithelial-mesenchymal interactions through WNT signaling activity to control muscle development in the soft palate. *Development*, *141*, 909–917.

Jarvinen, E., Salazar-Ciudad, I., Birchmeier, W., Taketo, M. M., Jernvall, J., & Thesleff, I. (2006). Continuous tooth generation in mouse is induced by activated epithelial Wnt/

beta-catenin signaling. *Proceedings of the National Academy of Sciences of the United States of America, 103*, 18627–18632.
Jens, M., & Rajewsky, N. (2015). Competition between target sites of regulators shapes post-transcriptional gene regulation. *Nature Reviews Genetics, 16*, 113–126.
Jernvall, J., & Thesleff, I. (2000). Reiterative signaling and patterning during mammalian tooth morphogenesis. *Mechanisms of Development, 92*, 19–29.
Jevnaker, A. M., & Osmundsen, H. (2008). MicroRNA expression profiling of the developing murine molar tooth germ and the developing murine submandibular salivary gland. *Archives of Oral Biology, 53*, 629–645.
Jia, J., Tian, Q., Ling, S., Liu, Y., Yang, S., & Shao, Z. (2013). miR-145 suppresses osteogenic differentiation by targeting Sp7. *FEBS Letters, 587*, 3027–3031.
Jin, Y. R., Turcotte, T. J., Crocker, A. L., Han, X. H., & Yoon, J. K. (2011). The canonical Wnt signaling activator, R-spondin2, regulates craniofacial patterning and morphogenesis within the branchial arch through ectodermal-mesenchymal interaction. *Developmental Biology, 352*, 1–13.
Kan, T., & Meltzer, S. J. (2009). MicroRNAs in Barrett's esophagus and esophageal adenocarcinoma. *Current Opinion in Pharmacology, 9*, 727–732.
Kannu, P., Campos-Xavier, A. B., Hull, D., Martinet, D., Ballhausen, D., & Bonafe, L. (2013). Post-axial polydactyly type A2, overgrowth and autistic traits associated with a chromosome 13q31.3 microduplication encompassing miR-17-92 and GPC5. *European Journal of Medical Genetics, 56*, 452–457.
Ketting, R. F., Fischer, S. E., Bernstein, E., Sijen, T., Hannon, G. J., & Plasterk, R. H. (2001). Dicer functions in RNA interference and in synthesis of small RNA involved in developmental timing in C. elegans. *Genes & Development, 15*, 2654–2659.
Khan, Q. E., Sehic, A., Khuu, C., Risnes, S., & Osmundsen, H. (2013). Expression of Clu and Tgfb1 during murine tooth development: Effects of in-vivo transfection with anti-miR-214. *European Journal of Oral Sciences, 121*, 303–312.
Khvorova, A., Reynolds, A., & Jayasena, S. D. (2003). Functional siRNAs and miRNAs exhibit strand bias. *Cell, 115*, 209–216.
Kim, J. S., Park, M. G., Lee, S. A., Park, S. Y., Kim, H. J., Yu, S. K., et al. (2014). Downregulation of adenomatous polyposis coli by microRNA-663 promotes odontogenic differentiation through activation of Wnt/beta-catenin signaling. *Biochemical and Biophysical Research Communications, 446*, 894–900.
Kimmel, C. B., Miller, C. T., & Keynes, R. J. (2001). Neural crest patterning and the evolution of the jaw. *Journal of Anatomy, 199*, 105–119.
Kimmel, C. B., Miller, C. T., & Moens, C. B. (2001). Specification and morphogenesis of the zebrafish larval head skeleton. *Developmental Biology, 233*, 239–257.
Kimmel, C. B., Ullmann, B., Walker, M., Miller, C. T., & Crump, J. G. (2003). Endothelin 1-mediated regulation of pharyngeal bone development in zebrafish. *Development, 130*, 1339–1351.
Kirby, M. L., & Waldo, K. L. (1995). Neural crest and cardiovascular patterning. *Circulation Research*, 77, 211–215.
Klingberg, G., Oskarsdottir, S., Johannesson, E. L., & Noren, J. G. (2002). Oral manifestations in 22q11 deletion syndrome. *International Journal of Paediatric Dentistry, 12*, 14–23.
Knezevic, J., Pfefferle, A. D., Petrovic, I., Greene, S. B., Perou, C. M., & Rosen, J. M. (2015). Expression of miR-200c in claudin-low breast cancer alters stem cell functionality, enhances chemosensitivity and reduces metastatic potential. *Oncogene*, in press.
Kobrynski, L. J., & Sullivan, K. E. (2007). Velocardiofacial syndrome, DiGeorge syndrome: The chromosome 22q11.2 deletion syndromes. *Lancet, 370*, 1443–1452.
Kontges, G., & Lumsden, A. (1996). Rhombencephalic neural crest segmentation is preserved throughout craniofacial ontogeny. *Development, 122*, 3229–3242.

Koufaris, C., Papagregoriou, G., Kousoulidou, L., Moutafi, M., Tauber, M., Jouret, B., et al. (2015). Haploinsufficiency of the miR-873/miR-876 microRNA cluster is associated with craniofacial abnormalities. *Gene, 561*, 95–100.

Kuang-Hsien Hu, J., Mushegyan, V., & Klein, O. D. (2014). On the cutting edge of organ renewal: Identification, regulation, and evolution of incisor stem cells. *Genesis, 52*, 79–92.

Kurzynska-Kokorniak, A., Koralewska, N., Pokornowska, M., Urbanowicz, A., Tworak, A., Mickiewicz, A., et al. (2015). The many faces of Dicer: The complexity of the mechanisms regulating Dicer gene expression and enzyme activities. *Nucleic Acids Research, 43*, 4365–4380.

Lagos-Quintana, M., Rauhut, R., Lendeckel, W., & Tuschi, W. (2001). Identification of novel genes encoding for small expressed RNAs. *Science, 294*, 853–858.

Lai, E. C. (2002). Micro RNAs are complementary to 3′ UTR sequence motifs that mediate negative post-transcriptional regulation. *Nature Genetics, 30*, 363–364.

Le Douarin, N. M. (1982). *The neural crest.* Cambridge: Cambridge University Press.

Lee, Y., Ahn, C., Han, J., Choi, H., Kim, J., Yim, J., et al. (2003). The nuclear RNase III Drosha initiates microRNA processing. *Nature, 425*, 415–419.

Lee, R. C., Feinbaum, R. L., & Ambros, V. (1993). The *C. elegans* heterochronic gene lin-4 encodes small RNAs with antisense complementarity to lin-14. *Cell, 75*, 843–854.

Lewis, B. P., Shih, I. H., Jones-Rhoades, M. W., Bartel, D. P., & Burge, C. B. (2003). Prediction of mammalian microRNA targets. *Cell, 115*, 787–798.

Li, N., Felber, K., Elks, P., Croucher, P., & Roehl, H. H. (2009). Tracking gene expression during zebrafish osteoblast differentiation. *Developmental Dynamics, 238*, 459–466.

Li, M., Guan, X., Sun, Y., Mi, J., Shu, X., Liu, F., et al. (2014). miR-92a family and their target genes in tumorigenesis and metastasis. *Experimental Cell Research, 323*, 1–6.

Li, L., Meng, T., Jia, Z., Zhu, G., & Shi, B. (2010). Single nucleotide polymorphism associated with nonsyndromic cleft palate influences the processing of miR-140. *American Journal of Medical Genetics Part A, 152a*, 856–862.

Li, L., Shi, J. Y., Zhu, G. Q., & Shi, B. (2012). MiR-17-92 cluster regulates cell proliferation and collagen synthesis by targeting TGFB pathway in mouse palatal mesenchymal cells. *Journal of Cellular Biochemistry, 113*, 1235–1244.

Li, A., Song, T., Wang, F., Liu, D., Fan, Z., Zhang, C., et al. (2012). MicroRNAome and expression profile of developing tooth germ in miniature pigs. *PLoS One, 7*, e52256.

Li, E., Zhang, J., Yuan, T., & Ma, B. (2014). MiR-143 suppresses osteogenic differentiation by targeting Osterix. *Molecular and Cellular Biochemistry, 390*, 69–74.

Li, L., Zhu, G. Q., Meng, T., Shi, J. Y., Wu, J., Xu, X., et al. (2011). Biological and epidemiological evidence of interaction of infant genotypes at Rs7205289 and maternal passive smoking in cleft palate. *American Journal of Medical Genetics Part A, 155a*, 2940–2948.

Liekens, S., Schols, D., & Hatse, S. (2010). CXCL12-CXCR4 axis in angiogenesis, metastasis and stem cell mobilization. *Current Pharmaceutical Design, 16*, 3903–3920.

Lin, H., Xu, L., Liu, H., Sun, Q., Chen, Z., Yuan, G., et al. (2011). KLF4 promotes the odontoblastic differentiation of human dental pulp cells. *Journal of Endodontia, 37*, 948–954.

Lindsay, E. A., Bota, A., Jurecic, V., Carattini-Rivera, S., Cheah, Y.-C., Rosenblatt, H. M., et al. (1999). Congenital heart disease in mice deficient for the DiGeorge syndrome region. *Nature, 401*, 379–383.

Lindsay, E. A., Greenberg, F., Shaffer, L. G., Shapira, S. K., Scambler, P. J., & Baldini, A. (1995). Submicroscopic deletions at 22q11.2: Variablitiy of the clinical picture and delineation of a commonly deleted region. *American Journal of Medical Genetics, 56*, 191–197.

Lindsay, E. A., Vitelli, F., Su, H., Morishima, M., Huynh, T., Pramparo, T., et al. (2001). Tbx1 haploinsufficieny in the DiGeorge syndrome region causes aortic arch defects in mice. *Nature, 410*, 97–101.

Liu, H., Lin, H., Zhang, L., Sun, Q., Yuan, G., Zhang, L., et al. (2013). miR-145 and miR-143 regulate odontoblast differentiation through targeting Klf4 and Osx genes in a feedback loop. *The Journal of Biological Chemistry, 288*, 9261–9271.
Liu, F., & Millar, S. E. (2010). Wnt/beta-catenin signaling in oral tissue development and disease. *Journal of Dental Research, 89*, 318–330.
Liu, L., Qiu, M., Tan, G., Liang, Z., Qin, Y., Chen, L., et al. (2014). miR-200c inhibits invasion, migration and proliferation of bladder cancer cells through down-regulation of BMI-1 and E2F3. *Journal of Translational Medicine, 12*, 305.
Liu, W., Selever, J., Lu, M.-F., & Martin, J. F. (2003). Genetic dissection of *Pitx2* in craniofacial development uncovers new functions in branchial arch morphogenesis, late aspects of tooth morphogenesis and cell migration. *Development, 130*, 6375–6385.
Liu, H., Sun, Q., Wan, C., Li, L., Zhang, L., & Chen, Z. (2014). MicroRNA-338-3p regulates osteogenic differentiation of mouse bone marrow stromal stem cells by targeting Runx2 and Fgfr2. *Journal of Cellular Physiology, 229*, 1494–1502.
Long, F. (2012). Building strong bones: Molecular regulation of the osteoblast lineage. *Nature Reviews Molecular Cell Biology, 13*, 27–38.
Lund, E., Guttinger, S., Calado, A., Dahlberg, J. E., & Kutay, U. (2004). Nuclear export of microRNA precursors. *Science, 303*, 95–98.
Marcelis, C. L. M., & de Brouwer, A. P. M. (1993). Feingold syndrome 1. In R. A. Pagon, M. P. Adam, H. H. Ardinger, S. E. Wallace, A. Amemiya, L. J. H. Bean, T. D. Bird, C. R. Dolan, C. T. Fong, R. J. H. Smith, & K. Stephens (Eds.), *GeneReviews(R)*. Seattle: University of Washington. All rights reserved, Seattle (WA).
Martin del Campo, S. E., Latchana, N., Levine, K. M., Grignol, V. P., Fairchild, E. T., Jaime-Ramirez, A. C., et al. (2015). MiR-21 enhances melanoma invasiveness via inhibition of tissue inhibitor of metalloproteinases 3 expression: In vivo effects of MiR-21 inhibitor. *PLoS One, 10*, e0115919.
Masyuk, M., & Brand-Saberi, B. (2015). Recruitment of skeletal muscle progenitors to secondary sites: A role for CXCR4/SDF-1 signalling in skeletal muscle development. *Results and Problems in Cell Differentiation, 56*, 1–23.
Mattick, J. S., & Makunin, I. V. (2006). Non-coding RNA. *Human Molecular Genetics, 15*(Spec. No. 1), R17–R29.
Merscher, S., Funke, B., Epstein, J. A., Heyer, J., Puech, A., Lu, M. M., et al. (2001). TBX1 is responsible for cardiovascular defects in velo-cardio-facial/DiGeorge syndrome. *Cell, 104*, 619–629.
Michikami, I., Fukushi, T., Tanaka, M., Egusa, H., Maeda, Y., Ooshima, T., et al. (2012). Kruppel-like factor 4 regulates membranous and endochondral ossification. *Experimental Cell Research, 318*, 311–325.
Michon, F. (2011). Tooth evolution and dental defects: From genetic regulation network to micro-RNA fine-tuning. *Birth Defects Research, Part A: Clinical and Molecular Teratology, 91*, 763–769.
Michon, F., Tummers, M., Kyyronen, M., Frilander, M. J., & Thesleff, I. (2010). Tooth morphogenesis and ameloblast differentiation are regulated by micro-RNAs. *Developmental Biology, 340*, 355–368.
Miller, R. J., Banisadr, G., & Bhattacharyya, B. J. (2008). CXCR4 signaling in the regulation of stem cell migration and development. *Journal of Neuroimmunology, 198*, 31–38.
Miller, C. T., & Kimmel, C. B. (2001). Morpholino phenocopies of *endothelin 1* (*sucker*) and other anterior arch class mutations. *Genesis, 30*, 186–187.
Miller, C. T., Schilling, T. F., Lee, K., Parker, J., & Kimmel, C. B. (2000). sucker encodes a zebrafish Endothelin-1 required for ventral pharyngeal arch development. *Development, 127*, 3815–3828.
Miller, C. T., Yelon, D., Stainier, D. Y., & Kimmel, C. B. (2003). Two *endothelin 1* effectors, *hand2* and *bapx1*, pattern ventral pharyngeal cartilage and the jaw joint. *Development, 130*, 1353–1365.

Miyaki, S., Sato, T., Inoue, A., Otsuki, S., Ito, Y., Yokoyama, S., et al. (2010). MicroRNA-140 plays dual roles in both cartilage development and homeostasis. *Genes & Development*, *24*, 1173–1185.

Mogilyansky, E., & Rigoutsos, I. (2013). The miR-17/92 cluster: A comprehensive update on its genomics, genetics, functions and increasingly important and numerous roles in health and disease. *Cell Death and Differentiation*, *20*, 1603–1614.

Mori-Akiyama, Y., Akiyama, H., Rowitch, D. H., & de Crombrugghe, B. (2003). Sox9 is required for determination of the chondrogenic cell lineage in the cranial neural crest. *Proceedings of the National Academy of Sciences of the United States of America*, *100*, 9360–9365.

Mucchielli, M. L., Mitsiadis, T. A., Raffo, S., Brunet, J. F., Proust, J. P., & Goridis, C. (1997). Mouse Otlx2/RIEG expression in the odontogenic epithelium precedes tooth initiation and requires mesenchyme-derived signals for its maintenance. *Developmental Biology*, *189*, 275–284.

Nair, S., Li, W., Cornell, R., & Schilling, T. F. (2007). Requirements for endothelin type-A receptors and endothelin-1 signaling in the facial ectoderm for the patterning of skeletogenic neural crest cells in zebrafish. *Development*, *134*, 335–345.

Nakamura, Y., He, X., Kato, H., Wakitani, S., Kobayashi, T., Watanabe, S., et al. (2012). Sox9 is upstream of microRNA-140 in cartilage. *Applied Biochemistry and Biotechnology*, *166*, 64–71.

Nelson, C., Ambros, V., & Baehrecke, E. H. (2014). miR-14 regulates autophagy during developmental cell death by targeting ip3-kinase 2. *Molecular Cell*, *56*, 376–388.

Nguyen, M. K., Jeon, O., Krebs, M. D., Schapira, D., & Alsberg, E. (2014). Sustained localized presentation of RNA interfering molecules from in situ forming hydrogels to guide stem cell osteogenic differentiation. *Biomaterials*, *35*, 6278–6286.

Nie, X., Wang, Q., & Jiao, K. (2011). Dicer activity in neural crest cells is essential for craniofacial organogenesis and pharyngeal arch artery morphogenesis. *Mechanisms of Development*, *128*, 200–207.

Ning, G., Liu, X., Dai, M., Meng, A., & Wang, Q. (2013). MicroRNA-92a upholds Bmp signaling by targeting noggin3 during pharyngeal cartilage formation. *Developmental Cell*, *24*, 283–295.

Noden, D. M. (1988). Interactions and fates of avian craniofacial mesenchyme. *Development*, *103*, 121–140.

Ochi, M., Nakasa, T., Kamei, G., Usman, M. A., & El Mahmoud, H. (2014). Regenerative medicine in orthopedics using cells, scaffold, and microRNA. *Journal of Orthopaedic Science*, *19*, 521–528.

Ochoa-Espinosa, A., & Affolter, M. (2012). Branching morphogenesis: From cells to organs and back. *Cold Spring Harbor Perspectives in Biology*, *4*, a008243.

Ohyama, T., & Groves, A. K. (2004). Generation of Pax2-Cre mice by modification of a Pax2 bacterial artificial chromosome. *Genesis*, *38*, 195–199.

Okamura, K., Hagen, J. W., Duan, H., Tyler, D. M., & Lai, E. C. (2007). The mirtron pathway generates microRNA-class regulatory RNAs in Drosophila. *Cell*, *130*, 89–100.

Olesnicky-Killian, E. C., Birkholz, D. A., & Artinger, K. B. (2009). A role for chemokine signaling in neural crest cell migration and craniofacial development. *Developmental Biology*, *333*, 161–172.

Olive, V., Jiang, I., & He, L. (2010). mir-17-92, a cluster of miRNAs in the midst of the cancer network. *The International Journal of Biochemistry & Cell Biology*, *42*, 1348–1354.

Oommen, S., Otsuka-Tanaka, Y., Imam, N., Kawasaki, M., Kawasaki, K., Jalani-Ghazani, F., et al. (2012). Distinct roles of microRNAs in epithelium and mesenchyme during tooth development. *Developmental Dynamics*, *241*, 1465–1472.

Ozeki, H., Kurihara, Y., Tonami, K., Watatani, S., & Kurihara, H. (2004). Endothelin-1 regulates the dorsoventral branchial arch patterning in mice. *Mechanisms of Development*, *121*, 387–395.

Papaioannou, G., Mirzamohammadi, F., Lisse, T. S., Nishimori, S., Wein, M. N., & Kobayashi, T. (2014). MicroRNA-140 provides robustness to the regulation of hypertrophic chondrocyte differentiation by the PTHrP-HDAC4 pathway. *Journal of Bone and Mineral Research, 30,* 1044–1052.

Park, M. G., Kim, J. S., Park, S. Y., Lee, S. A., Kim, H. J., Kim, C. S., et al. (2014). MicroRNA-27 promotes the differentiation of odontoblastic cell by targeting APC and activating Wnt/beta-catenin signaling. *Gene, 538,* 266–272.

Patel, V. N., Rebustini, I. T., & Hoffman, M. P. (2006). Salivary gland branching morphogenesis. *Differentiation, 74,* 349–364.

Percival, C. J., Green, R., Marcucio, R., & Hallgrimsson, B. (2014). Surface landmark quantification of embryonic mouse craniofacial morphogenesis. *BMC Developmental Biology, 14,* 31.

Person, A. D., Beiraghi, S., Sieben, C. M., Hermanson, S., Neumann, A. N., Robu, M. E., et al. (2010). WNT5A mutations in patients with autosomal dominant Robinow syndrome. *Developmental Dynamics, 239,* 327–337.

Piotrowski, T., Ahn, D. G., Schilling, T. F., Nair, S., Ruvinsky, I., Geisler, R., et al. (2003). The zebrafish van gogh mutation disrupts tbx1, which is involved in the DiGeorge deletion syndrome in humans. *Development, 130,* 5043–5052.

Puech, A., Saint-Jore, B., Funke, B., Gilbert, D. J., Sirotkin, H., Copeland, N. G., et al. (1997). Comparative mapping of the human 22q11 chromosomal region and the orthologous region in mice reveals complex changes in gene organization. *Proceedings of the National Academy of Sciences, 94,* 14608–14613.

Qiu, M., Bulfone, A., Ghattas, I., Meneses, J. J., Christensen, L., Sharpe, P. T., et al. (1997). Role of the Dlx homeobox genes in proximodistal patterning of the branchial arches: Mutations of Dlx-1, Dlx-2, and Dlx-1 and -2 alter morphogenesis of proximal skeletal and soft tissue structures derived from the first and second arches. *Developmental Biology, 185,* 165–184.

Qiu, M., Bulfone, A., Martinez, S., Meneses, J. J., Shimamura, K., Pedersen, R. A., et al. (1995). Null mutation of *Dlx-2* results in abnormal morphogenesis of proximal first and second branchial arch derivatives and abnormal differentiation in the forebrain. *Genes & Development, 9,* 2523–2538.

Radisky, E. S., & Radisky, D. C. (2010). Matrix metalloproteinase-induced epithelial-mesenchymal transition in breast cancer. *Journal of Mammary Gland Biology and Neoplasia, 15,* 201–212.

Rattanasopha, S., Tongkobpetch, S., Srichomthong, C., Siriwan, P., Suphapeetiporn, K., & Shotelersuk, V. (2012). PDGFRa mutations in humans with isolated cleft palate. *European Journal of Human Genetics, 20,* 1058–1062.

Raz, E., & Mahabaleshwar, H. (2009). Chemokine signaling in embryonic cell migration: A fisheye view. *Development, 136,* 1223–1229.

Rebustini, I. T., Hayashi, T., Reynolds, A. D., Dillard, M. L., Carpenter, E. M., & Hoffman, M. P. (2012). miR-200c regulates FGFR-dependent epithelial proliferation via Vldlr during submandibular gland branching morphogenesis. *Development, 139,* 191–202.

Regard, J. B., Zhong, Z., Williams, B. O., & Yang, Y. (2012). Wnt signaling in bone development and disease: Making stronger bone with Wnts. *Cold Spring Harbor Perspectives in Biology, 4,* 1–17, a007997.

Reis, S. T., Pontes-Junior, J., Antunes, A. A., Dall'Oglio, M. F., Dip, N., Passerotti, C. C., et al. (2012). miR-21 may acts as an oncomir by targeting RECK, a matrix metalloproteinase regulator, in prostate cancer. *BMC Urology, 12,* 14.

Ress, A. L., Stiegelbauer, V., Winter, E., Schwarzenbacher, D., Kiesslich, T., Lax, S., et al. (2014). MiR-96-5p influences cellular growth and is associated with poor survival in colorectal cancer patients. *Molecular Carcinogenesis,* in press.

Ribeiro, A. O., Schoof, C. R., Izzotti, A., Pereira, L. V., & Vasques, L. R. (2014). MicroRNAs: Modulators of cell identity, and their applications in tissue engineering. *MicroRNA*, *3*, 45–53.

Ruby, J. G., Jan, C. H., & Bartel, D. P. (2007). Intronic microRNA precursors that bypass Drosha processing. *Nature*, *448*, 83–86.

Ruest, L. B., Xiang, X., Lim, K. C., Levi, G., & Clouthier, D. E. (2004). Endothelin-A receptor-dependent and -independent signaling pathways in establishing mandibular identity. *Development*, *131*, 4413–4423.

Schwarz, D. S., Hutvagner, G., Du, T., Xu, Z., Aronin, N., & Zamore, P. D. (2003). Asymmetry in the assembly of the RNAi enzyme complex. *Cell*, *115*, 199–208.

Sehic, A., Risnes, S., Khan, Q. E., Khuu, C., & Osmundsen, H. (2010). Gene expression and dental enamel structure in developing mouse incisor. *European Journal of Oral Sciences*, *118*, 118–130.

Sellier, C., Hwang, V. J., Dandekar, R., Durbin-Johnson, B., Charlet-Berguerand, N., Ander, B. P., et al. (2014). Decreased DGCR8 expression and miRNA dysregulation in individuals with 22q11.2 deletion syndrome. *PLoS One*, *9*, e103884.

Sharp, T., Wang, J., Li, X., Cao, H., Gao, S., Moreno, M., et al. (2014). A pituitary homeobox 2 (Pitx2):microRNA-200a-3p:beta-catenin pathway converts mesenchymal cells to amelogenin-expressing dental epithelial cells. *The Journal of Biological Chemistry*, *289*, 27327–27341.

Sharpe, P. T. (2001). Neural crest and tooth morphogenesis. *Advances in Dental Research*, *15*, 4–7.

Sheehy, N. T., Cordes, K. R., White, M. P., Ivey, K. N., & Srivastava, D. (2010). The neural crest-enriched microRNA miR-452 regulates epithelial-mesenchymal signaling in the first pharyngeal arch. *Development*, *137*, 4307–4316.

Shi, R., Xiao, H., Yang, T., Chang, L., Tian, Y., Wu, B., et al. (2014). Effects of miR-200c on the migration and invasion abilities of human prostate cancer Du145 cells and the corresponding mechanism. *Frontiers in Medicine*, *8*, 456–463.

Shigetani, Y., Nobusada, Y., & Kuratani, S. (2000). Ectodermally derived FGF8 defines the maxillomandibular region in the early chick embryo: epithelial-mesenchymal interactions in the specification of the craniofacial ectomesenchyme. *Developmental Biology*, *228*, 73–85.

Song, L., & Tuan, R. S. (2006). MicroRNAs and cell differentiation in mammalian dvelopment. *Birth Defects Research, Part C: Embryo Today-Reviews*, *78*, 140–149.

Soriano, P. (1999). Generalized *lacZ* expression with the ROSA26 Cre reporter strain. *Nature Genetics*, *21*, 70–71.

Soria-Valles, C., Gutierrez-Fernandez, A., Guiu, M., Mari, B., Fueyo, A., Gomis, R. R., et al. (2014). The anti-metastatic activity of collagenase-2 in breast cancer cells is mediated by a signaling pathway involving decorin and miR-21. *Oncogene*, *33*, 3054–3063.

Stanier, P., & Pauws, E. (2012). Development of the lip and palate: FGF signalling. *Frontiers of Oral Biology*, *16*, 71–80.

Sun, Q., Liu, H., Lin, H., Yuan, G., Zhang, L., & Chen, Z. (2013). MicroRNA-338-3p promotes differentiation of mDPC6T into odontoblast-like cells by targeting Runx2. *Molecular and Cellular Biochemistry*, *377*, 143–149.

Takechi, M., & Kuratani, S. (2010). History of studies on mammalian middle ear evolution: A comparative morphological and developmental biology perspective. *Journal of Experimental Zoology Part B: Molecular and Developmental Evolution*, *314*, 417–433.

Talbot, J. C., Johnson, S. L., & Kimmel, C. B. (2010). *hand2* and Dlx genes specify dorsal, intermediate and ventral domains within zebrafish pharyngeal arches. *Development*, *137*, 2507–2517.

Tallquist, M. D., & Soriano, P. (2003). Cell autonomous requirement for PDGFRa in populations of cranial and cardiac neural crest cells. *Development*, *130*, 507–518.

Tan, W., Li, Y., Lim, S. G., & Tan, T. M. (2014). miR-106b-25/miR-17-92 clusters: Polycistrons with oncogenic roles in hepatocellular carcinoma. *World Journal of Gastroenterology*, *20*, 5962–5972.

Tassano, E., Di Rocco, M., Signa, S., & Gimelli, G. (2013). De novo 13q31.1-q32.1 interstitial deletion encompassing the miR-17-92 cluster in a patient with Feingold syndrome-2. *American Journal of Medical Genetics Part A*, *161a*, 894–896.

Tavares, A. L. P., Garcia, E. L., Kuhn, K., Woods, C. M., Williams, T., & Clouthier, D. E. (2012). Ectodermal-derived Endothelin1 is required for patterning the distal and intermediate domains of the mouse mandibular arch. *Developmental Biology*, *371*, 47–56.

Thomas, B. L., Liu, J. K., Rubenstein, J. L. R., & Sharpe, P. T. (2000). Independent regulation of *Dlx2* expression in the epithelium and mesenchyme of the first branchial arch. *Development*, *127*, 217–224.

Trainor, P., & Krumlauf, R. (2000). Plasticity in mouse neural crest cells reveals a new patterning role for cranial mesoderm. *Nature Cell Biology*, *2*, 96–102.

Trummp, A., Depew, M. J., Rubenstein, J. L. R., Bishop, J. M., & Martin, G. R. (1999). Cre-mediated gene inactivation demonstrates that FGF8 is required for cell survival and patterning of the first branchial arch. *Genes & Development*, *13*, 3136–3148.

Tucker, A. S. (2007). Salivary gland development. *Seminars in Cell & Developmental Biology*, *18*, 237–244.

Varner, V. D., & Nelson, C. M. (2014). Cellular and physical mechanisms of branching morphogenesis. *Development*, *141*, 2750–2759.

Ventura, A., Young, A. G., Winslow, M. M., Lintault, L., Meissner, A., Erkeland, S. J., et al. (2008). Targeted deletion reveals essential and overlapping functions of the miR-17w92 family of miRNA clusters. *Cell*, *132*, 875–886.

Vitelli, F., Morishima, M., Taddei, I., Lindsay, E. A., & Baldini, A. (2002). *Tbx1* mutation causes multiple cardiovascular defects and disrupts neural crest and cranial nerve migratory pathways. *Human Molecular Genetics*, *11*, 915–922.

Wan, M., Gao, B., Sun, F., Tang, Y., Ye, L., Fan, Y., et al. (2012). microRNA miR-34a regulates cytodifferentiation and targets multi-signaling pathways in human dental papilla cells. *PLoS One*, *7*, e50090.

Wang, J., Bai, Y., Li, H., Greene, S. B., Klysik, E., Yu, W., et al. (2013). MicroRNA-17-92, a direct Ap-2alpha transcriptional target, modulates T-box factor activity in orofacial clefting. *PLoS Genetics*, *9*, e1003785.

Wang, J., Bai, Y., Li, N., Ye, W., Zhang, M., Greene, S. B., et al. (2014). Pitx2-microRNA pathway that delimits sinoatrial node development and inhibits predisposition to atrial fibrillation. *Proceedings of the National Academy of Sciences of the United States of America*, *111*, 9181–9186.

Wang, T., & Xu, Z. (2010). miR-27 promotes osteoblast differentiation by modulating Wnt signaling. *Biochemical and Biophysical Research Communications*, *402*, 186–189.

Watanabe, T., Sato, T., Amano, T., Kawamura, Y., Kawamura, N., Kawaguchi, H., et al. (2008). Dnm3os, a non-coding RNA, is required for normal growth and skeletal development in mice. *Developmental Dynamics*, *237*, 3738–3748.

Westholm, J. O., & Lai, E. C. (2011). Mirtrons: MicroRNA biogenesis via splicing. *Biochimie*, *93*, 1897–1904.

Wightman, B., Ha, I., & Ruvkun, G. (1993). Posttranscriptional regulation of the heterochronic gene *lin-14* by *lin-4* mediates temporal pattern formation in C. elegans. *Cell*, *75*, 855–862.

Xu, L., Harada, H., & Taniguchi, A. (2008). The effects of LAMP1 and LAMP3 on M180 amelogenin uptake, localization and amelogenin mRNA induction by amelogenin protein. *Journal of Biochemistry*, *144*, 531–537.

Yang, C. H., Yue, J., Pfeffer, S. R., Handorf, C. R., & Pfeffer, L. M. (2011). MicroRNA miR-21 regulates the metastatic behavior of B16 melanoma cells. *The Journal of Biological Chemistry*, *286*, 39172–39178.

Yi, R., Qin, Y., Macara, I. G., & Cullen, B. R. (2003). Exportin-5 mediates the nuclear export of pre-microRNAs and short hairpin RNAs. *Nature Genetics*, *38*, 356–362.

Yin, K., Hacia, J. G., Zhong, Z., & Paine, M. L. (2014). Genome-wide analysis of miRNA and mRNA transcriptomes during amelogenesis. *BMC Genomics*, *15*, 998.

Yutzey, K. E., & Kirby, M. L. (2002). Wherefore heart thou? Embryonic origins of cardiogenic mesoderm. *Developmental Dynamics*, *223*, 307–320.

Zehir, A., Hua, L. L., Maska, E. L., Morikawa, Y., & Cserjesi, P. (2010). Dicer is required for survival of differentiating neural crest cells. *Developmental Biology*, *340*, 459–467.

Zhang, Z., Gutierrez, D., Li, X., Wang, J., Andrade, K., & Amendt, B. A. (2013). Homeodomain transcription factor Lhx6: A transcriptional repressor that interacts with PITX2 to regulate odontogenesis. *The Journal of Biological Chemistry*, *288*, 2485–2500.

Zuniga, E., Stellabotte, F., & Crump, J. G. (2010). Jagged-Notch signaling ensures dorsal skeletal identity in the vertebrate face. *Development*, *137*, 1843–1852.

CHAPTER FIFTEEN

Transcription Factors in Craniofacial Development: From Receptor Signaling to Transcriptional and Epigenetic Regulation

Yongchao Gou[*,†], **Tingwei Zhang**[†,‡], **Jian Xu**[†,1]

[*]State Key Laboratory of Oral Diseases, Department of Orthodontics, West China Hospital of Stomatology, Sichuan University, Chengdu, China
[†]Center for Craniofacial Molecular Biology, School of Dentistry, University of Southern California, Los Angeles, USA
[‡]State Key Laboratory of Oral Diseases, Department of Prosthodontics, West China Hospital of Stomatology, Sichuan University, Chengdu, China
[1]Corresponding author: e-mail address: xujian@usc.edu

Contents

Abstract

Craniofacial morphogenesis is driven by spatial–temporal terrains of gene expression, which give rise to stereotypical pattern formation. Transcription factors are key cellular components that control these gene expressions. They are information hubs that

Current Topics in Developmental Biology, Volume 115
ISSN 0070-2153
http://dx.doi.org/10.1016/bs.ctdb.2015.07.009

integrate inputs from extracellular factors and environmental cues, direct epigenetic modifications, and define transcriptional status. These activities allow transcription factors to confer specificity and potency to transcription regulation during development.

1. INTRODUCTION

Face is the signature of an animal. The composition of the face and structure of the head are much more variable compared to other parts of the body. Formation of these common yet distinct head structures involves contributions from all three germ layers, ectoderm, mesoderm and endoderm, and the neural crest. Neural crest cells (NCCs) emerge at the neural plate border, undergo epithelial–mesenchymal transition (EMT) while they delaminate from the neural tube and migrate ventrally to contribute to the facial mesenchyme, eventually forming the majority of cranial bones and cartilage, as well as nerves and connective tissue. Mesoderm components contribute to facial muscles, endothelial cells, and posterior portion of the skull. The ectoderm contributes to the epidermis, nervous system, and the sense organs. The endoderm lines the pharynx and forms the glands. During development, ectoderm and endoderm signal to the adjacent facial mesenchyme and neural crest-derived cells guide mesoderm-derived tissue formation within the facial mesenchyme.

These complex morphogenetic processes are orchestrated by TGF-β, BMP, FGF, Wnt, and hedgehog (Hh) signaling pathways that respond to extracellular factors/morphogens and environmental cues. Signals are then dispatched from cell-surface receptors to transcription factors, which recruit cofactors, activators or repressors, and chromatin-modifying enzymes to drive gene expressions that determine cell fate. Integration of the transcriptional and epigenetic regulation by transcription factors is a central step in craniofacial development. In this chapter, we will focus on craniofacial transcription factors, from their signaling responses, cross-regulation, transcriptional activity, to their regulation by mechanisms including posttranslational modifications.

2. SIGNALING PATHWAYS REGULATING CRANIOFACIAL DEVELOPMENT

2.1 TGF-β Family Signaling

The TGF-β family proteins comprise TGF-βs, BMPs, activins, and growth and differentiation factors that all act through membrane-bound dual-specificity receptor kinases (Feng & Derynck, 2005). The binding of TGF-β family protein to tetrameric receptor complexes of two type II and two type I receptors enables type II receptors to phosphorylate and activate type I receptors, which then activate receptor-activated Smads (R-Smads) through C-terminal serine phosphorylation. Smad2 and Smad3 are activated by TGF-βs, while Smad1, Smad5, and Smad8 are activated by BMPs. Upon activation, R-Smads form a complex with the co-Smad, Smad4, translocate to the nucleus for transcriptional functions. The inhibitory Smads, Smad6 and Smad7, antagonize TGF-β and BMP signaling (Feng & Derynck, 2005). Complementing their signaling through Smad, TGF-βs also signal through TAK1, PI3K, Rho-like GTPase, and Ras–MEK–ERK pathways. TGF-βs induce these pathways through adaptor-mediated activation, in the case of TAK1 and PI3K phosphorylation, or through the dual-specificity function of receptor kinases, as in the activation of Ras–MEK–ERK. Likewise, BMPs signal through LIMK, PI3K, and MEK–ERK pathways (Feng & Derynck, 2005).

TGF-β and BMP signaling have multiple roles during craniofacial development. In cranial bone formation, TGF-β and BMP signaling both promote NCC-derived osteoprogenitor proliferation. BMP signaling further promotes osteogenic differentiation, but TGF-β inhibits it (Ito et al., 2003; Iwata et al., 2010). In palate formation, TGF-β signaling drives migration and median edge epithelium (MEE) degeneration, and together with BMP signaling, enhances palate mesenchymal cell proliferation, to promote palate fusion (Iwata, Suzuki, Pelikan, Ho, Sanchez-Lara, et al., 2013; Parada, Li, Iwata, Suzuki, & Chai, 2013).

2.2 FGF Signaling

The FGF family comprises 22 FGF ligands, most of which act through receptor tyrosine kinases FGFRs and the coreceptor heparin sulfate proteoglycans. FGF binding to FGFRs elicits receptor dimerization, autophosphorylation and recruitment of adaptor, and signaling protein complexes. Phosphorylated FGFRs directly activate PLCγ, which hydrolyzes phosphatidyl inositol to form diacylglycerol and inositol triphosphate (IP3), to activate PKC and intracellular calcium release. Recruitment of adaptor proteins enables FGFR to activate Ras/MEK/ERK and PI3K pathways (Dorey & Amaya, 2010).

FGF signaling promotes lineage-specific proliferation and differentiation. FGF ligands and receptors are expressed in the facial mesenchyme and regulate differentiation toward the osteoblasts or chondrocytes fate.

Mutations of FGF signaling in human cause craniosynostosis, a clinical condition of precocious suture fusion (Mishina & Snider, 2014).

2.3 Wnt Signaling

When Wnt ligands bind to the receptor Frizzled (Fzd), and coreceptor, LRP5/6, the destruction complex is inhibited, which stabilizes β-catenin. This destruction complex, composed of GSK3β, Axin, casein kinase 1, and adenomatous polyposis coli, phosphorylates serine residues in β-catenin leading to proteasomal degradation by E3 ligase β-transducin repeat-containing protein. Upon Wnt signal activation, stabilized β-catenin translocates to the nucleus and controls gene expression through an interaction with TCF/LEF1. The activation of canonical Wnt signaling can be blocked by extracellular proteins, such as Dickkopf (DKK), secreted Frizzled-related protein (SFRP), and Wnt inhibitory factor, through blocking Wnt ligand–receptor interactions. Wnt signaling also activates noncanonical β-catenin-independent pathways, including PKC and Rho (Lien & Fuchs, 2014).

Wnt signaling regulates lip, palate, and bone development (Wang et al., 2014). The canonical Wnt signaling regulates palate fusion through TGF-β3 transcription (He et al., 2011). Wnt/β-catenin is also required for the suppression of chondrocyte differentiation and the allowance of osteoblast formation during intramembranous and endochondral ossification (Hill, Spater, Taketo, Birchmeier, & Hartmann, 2005).

2.4 Hh Signaling

In canonical Hh signaling, Hh molecules bind to receptor Ptch1/2 and derepress Smoothened (Smo), which then activates the Gli proteins Gli1, Gli2, and Gli3 (Gradilla & Guerrero, 2013). In the absence of Hh, Gli2 is cleaved and degraded, while Gli3 is cleaved and the truncated form functions as a repressor. Hh is a morphogen in *Drosophila* patterning, whereby distinct transcriptional responses are elicited by different doses of Hh ligand. For example, Gli2 requires less Hh ligand than Gli3 to function as an activator (McMahon, Ingham, & Tabin, 2003; Resende et al., 2010).

Mammals have three Hh homologues, Desert hedgehog, Indian hedgehog (IHH), and Sonic hedgehog (SHH). SHH dose-dependently regulates the morphogenesis of midline facial structures. SHH signaling disturbance causes a wide spectrum of craniofacial anomalies (Hu & Helms, 1999). SHH also controls cranial suture formation in cooperation with IHH, whereby SHH increases mesenchymal proliferation and prevents suture fusion, and IHH promotes calvarial ossification and suture fusion

(Alappat, Zhang, & Chen, 2003; Jacob, Wu, Freeman, Koyama, & Kirschner, 2007; Lenton et al., 2011; Murakami, Nifuji, & Noda, 1997).

3. FAMILIES OF TRANSCRIPTION FACTORS

3.1 Homeobox Gene Family of Transcription Factors

Homeobox gene family is a large family of ancient origin. The common feature is the homeobox, which encodes the homeodomain, a motif with a helix–loop–helix–turn–helix structure that recognizes specific AT-rich sequences for DNA binding (Holland, 2013). In mammals, there are around 100 genes in the homeobox gene family, divided into 11 classes. Each class is further divided into gene families (Holland, 2013). For example, the largest class, ANTP, is divided into 37 gene families in human. Homeobox genes encode transcription factors that carry out diverse functions in morphogenesis, patterning, and differentiation (Holland, 2013). We will focus on gene families with key functions in craniofacial development, the *Dlx*, *Msx*, *Hox*, *Pax*, and *Pitx* families (Fig. 1).

3.1.1 Dlx *Homeobox Gene Family*

The vertebrate *Dlx* gene family, homologous to *Distal-less* in *Drosophila*, includes three bigenic clusters, *Dlx*1/2, Dlx3/4, and Dlx5/6. Dlx genes are expressed in a nested pattern in the pharyngeal arches to specify dorsal–ventral identity and guide muscle formation in the jaw (Acampora et al., 1999; Beverdam et al., 2002; Depew et al., 1999; Heude et al., 2010; Qiu et al., 1995, 1997; Robledo, Rajan, Li, & Lufkin, 2002).

Dlx proteins are transcriptional targets of BMP signaling, activated through Smad1/5 and p38 pathways (Hassan et al., 2006; Yang, Yuan, Li, et al., 2014). Dlx3 and Dlx5, through an interaction with Smad4, then relay BMP signal to induce Runx2 expression (Chiba et al., 2003; Hassan et al., 2006; Holleville, Mateos, Bontoux, Bollerot, & Monsoro-Burq, 2007; Trinh, Barengo, & Naora, 2011). The induced Runx2 cooperates with Dlx3 and Dlx5 to activate bone marker expression, including *Sp7*, *alkaline phosphatase* (*ALP*), *osteocalcin* (*Ocn*), and *osteoactivin* (Hassan et al., 2004, 2006; Kim et al., 2004). Dlx5/6 transcription is also activated by the transcription factor p63. This p63–Dlx5/6 axis explains the occurrence of both p63 and dlx5/6 mutations in split hand–foot malformation (Lo Iacono et al., 2008). Besides transcriptional regulation, Akt1 and CaM kinase II-mediated phosphorylation stabilizes Dlx and enhances osteogenic differentiation (Choi, Choi, Lee, & Oh, 2012; Jeong et al., 2011; Seo et al., 2009).

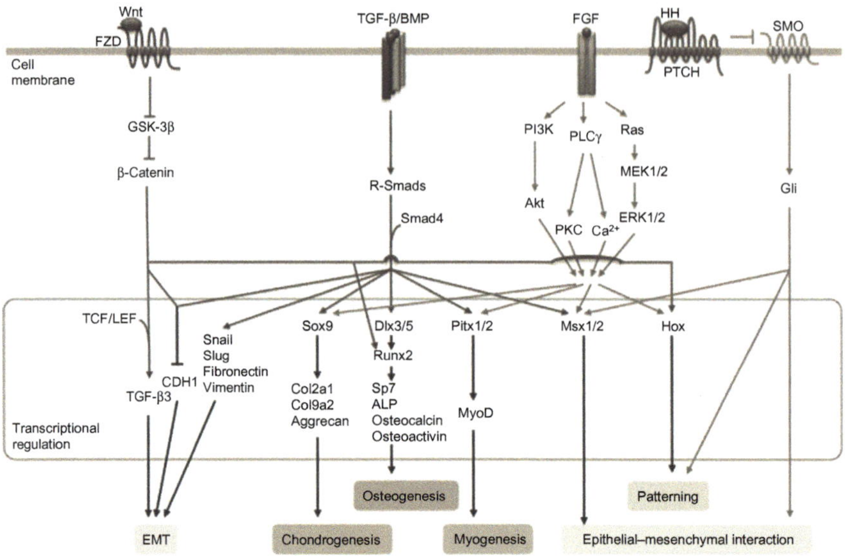

Figure 1 Key transcription factors during craniofacial development. TGF-β/BMP signaling induces palatal EMT through Smad-mediated induction of mesenchymal genes and repression of epithelial genes via cooperation with Wnt signaling. Wnt/β-catenin also induces TGF-β3 expression, further enhancing the activation loop. BMP signaling induces Dlx, which in turn induces Runx2 and interact with Runx2 to induce the osteogenic gene program. BMP and FGF signaling induce Sox9, which induce the chondrogenic gene program, and Pitx2, which induces the myogenic gene program in head muscles. BMP, FGF and Hh signaling converge on Msx1/2, which regulate epithelial-mesenchymal interaction. Wnt and FGF signaling induces Hox genes for segment specification. (A color version of this figure is located at the end of this chapter.)

3.1.2 Msx *Homeobox Gene Family*

The vertebrate *Msx* genes, homologous to *muscle segment homeobox (msh)* in *Drosophila*, includes three unrelated genes, *Msx1–3*. *Msx1* and *Msx2* are widely expressed, including prominent expression in the developing skull and teeth, while *Msx3* is restricted to the dorsal neural tube. *Msx* genes encode transcriptional repressors, which interact with the TATA-binding protein through their homeodomain to repress transcriptional machinery (Catron et al., 1995). Msx proteins also repress other homeodomain proteins through heterodimerization. For example, Msx1 forms heterodimers with Dlx2, Dlx5, Lhx2, Pax3, Pax9, and Pitx2 to modulate bone formation, tooth morphogenesis, and muscle differentiation (Bendall, Ding, Hu, Shen, & Abate-Shen, 1999; Bendall, Rincon-Limas, Botas, & Abate-Shen, 1998; Zhang et al., 1997). Further, Msx proteins recruit histone modification complexes. Msx1 recruits G9a/GLP and/or EZH2/Suz12 repressor complexes, which deposit H3K9me2 and H3K27me3 repressive marks, to the target gene regulatory region (Wang & Abate-Shen, 2012).

The functions of Msx proteins are critical in epithelium–mesenchyme interactions, whereby growth factors from oral and dental epithelium, such as BMP, FGF, and SHH, act on the subjacent mesenchyme (Vainio, Karavanova, Jowett, & Thesleff, 1993). Epithelial BMP, through Smad1/5 and Smad4, induces Msx1 and Msx2 (Yang, Yuan, Ye, Cho, & Chen, 2014), which in turn activate mesenchymal expression of *Bmp4* that is indispensible for tooth morphogenesis, alveolar bone formation, and head blood vessel maturation (Bei & Maas, 1998; Chen, Bei, Woo, Satokata, & Maas, 1996; Lopes et al., 2011). In tooth morphogenesis, Msx1 potentiates Pax9-induced *Bmp4* expression. Msx1 or Pax9 mutations that disrupt this synergy contribute to tooth agenesis (Ogawa et al., 2006; Wang et al., 2009). Msx1 also antagonizes Tbx2-mediated repression of *Bmp4* expression (Saadi et al., 2013).

The activity and stability of Msx proteins are regulated by phosphorylation and ubiquitination. Msx2 phosphorylation by PKCβ at Thr135 and Thr141 blocks Msx2 ubiquitination and facilitates Msx2-mediated *Runx2* repression (Jeong et al., 2012). While the Boston-type craniosynostosis mutation of Msx2, P148H, promotes ubiquitination by a RING finger protein Praja1 (Furumatsu, Tsuda, Taniguchi, Tajima, & Asahara, 2005; Yoon et al., 2008).

3.1.3 Hox *Homeobox Gene Family*

Hox homeobox genes are organized in clusters, *Hoxa–Hoxd*, and regulate body segment identity. Their expressions follow a similar spatial and temporal collinearity principle, which confers positional information along the body axis (Mallo & Alonso, 2013; Trainor & Krumlauf, 2001). *Hox* genes are silenced globally at early stages of development, and activated progressively according to their genomic positions, by concerted changes in chromatin structure and transcriptional regulation. *Hox* gene expression domain is further refined through cross-regulation among Hox proteins and cofactors. In the craniofacial area, *Hox* genes are silenced by the EZH2/PRC2 complex, which deposits histone H3K27 trimethylation (H3K27me3) to close the chromatin (Schwarz et al., 2014). To activate *Hox* genes, demethylases KDM4B and KDM6A/KDM6B erase H3K27me3 marks and methyltransferase, possibly MLL2, writes H3K4 marks to open up the chromatin (Agger et al., 2007; Lan et al., 2007; Lederer et al., 2012; Lindgren et al., 2013; Miyake et al., 2013; Ye et al., 2012). Subsequent signaling inputs from retinoic acid, FGFs, and Wnts bring transcriptional regulators, such as Krox20, AP-2, and Hox themselves, to the Hox promoter. The induced Hox proteins, together with a large arsenal of binding partners, elicit segment-specific gene expression (Rezsohazy, 2014).

3.1.4 Paired Box (Pax) *Homeobox Gene Family*

Pax gene family encode nine proteins, Pax1–9, all sharing a conserved DNA-binding domain called the paired domain. They are further divided into four groups based on the existence of a DNA-binding homeodomain and an octapeptide region (Blake & Ziman, 2014; Wang et al., 2008). Pax3 and Pax7 are critical for the induction, expansion, and specification of NCC (Basch, Bronner-Fraser, & Garcia-Castro, 2006; Milet, Maczkowiak, Roche, & Monsoro-Burq, 2013; Murdoch, DelConte, & Garcia-Castro, 2012; Zalc, Rattenbach, Aurade, Cadot, & Relaix, 2015). Pax3 and Zic1 induce the early NCC transcription program, including *snail1/2*, *foxd3*, *sox8*, *sox9*, and *twist* that specify NCC (Bae et al., 2014; Milet et al., 2013; Plouhinec et al., 2014; Sato, Sasai, & Sasai, 2005; Fig. 2). During

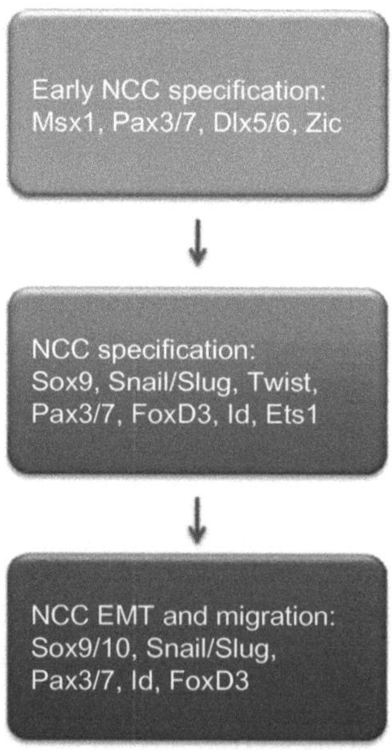

Figure 2 Regulatory factors for neural crest cell (NCC) formation. Msx1, Pax3/7, Dlx5/6 and Zic are expressed at the neural plate border prior to NCC specification. They induce Sox9, Snail/Slug, Twist, FoxD3 and Id in the nascent NCC as specifiers, resulting in a discreet cell population. These cells delaminate from the neural tube and migrate dorsolaterally. The expression of Sox9, Snail/Slug, Twist, FoxD3 and Id are maintained to initiate EMT and ensure proper cell dissemination. They further drive the expression of Sox10, which facilitates migration. (A color version of this figure is located at the end of this chapter.)

NCC expansion, Pax3 and Pax7 sustain NCC expansion by antagonizing aryl hydrocarbon receptor signaling, the pathway activated by environmental pollutants, implicating Pax3 and Pax7 as guardians from environmental stress (Zalc et al., 2015). Pax7 continues to express in migratory NCC derivatives, along with Pax9, to regulate cranial bone and teeth formation (Basch et al., 2006; Murdoch et al., 2012; Peters, Neubuser, Kratochwil, & Balling, 1998; Zalc et al., 2015).

3.1.5 LIM Homeobox *Gene Family (Lhx)*

Lhx genes encode 12 proteins featuring two LIM domains in the amino termini and a centrally located DNA-binding homeodomain. Among Lhx proteins, Lhx2, Lhx6, Lhx7, and Lhx8 are involved in craniofacial morphogenesis, through interaction with other homeodomain proteins. Lhx2 acts downstream of Pitx2 and Tbx1 for pharyngeal muscle specification (Harel et al., 2012). Lhx6 is another transcriptional target for Pitx2, which in turn represses Pitx2 activity at the odontogenic gene promoters (Zhang et al., 2013). Lhx6 and Lhx7 regulate cranial skeletal and teeth formation (Denaxa, Sharpe, & Pachnis, 2009). Lhx8 regulates secondary palate formation downstream of Wnt/β-catenin signaling (Landin Malt, Cesario, Tang, Brown, & Jeong, 2014; Zhao et al., 1999).

3.1.6 Paired-Like (Pitx) *Homeobox Gene Family*

Pitx genes, also known as *pituitary homeobox*, encode three vertebrate homologs, Pitx1, Pitx2, and Pitx3. They share a highly conserved homeodomain, responsible for DNA binding, and an OAR sequence in the carboxyl tail, which modulates homeodomain function (Amendt, Sutherland, & Russo, 1999). During facial prominence and tooth formation, BMP4 and FGF8 signaling define Pitx1 and Pitx2 expression to mandibular mesenchyme and dental epithelium. Pitx2 reciprocally represses *Fgf8* but activates *Bmp4* expression (Lin et al., 1999; Mitsiadis & Drouin, 2008; Neubuser, Peters, Balling, & Martin, 1997; St Amand et al., 2000). Dental epithelial Pitx2 interacts with β-catenin and synergistically activate odontogenic gene expression (Sharp et al., 2014). Concomitant with promotion of differentiation, Pitx2 and β-catenin activate p21 to repress proliferation. Tbx1 represses Pitx2 function through interaction with its carboxyl region, thereby sustaining epithelial proliferation (Cao et al., 2010). During palatogenesis, FGF9 activates Pitx2 and cyclin D1 and D3 expression to promote epithelial proliferation, downstream of TGF-β signaling (Iwata et al., 2012). During pharyngeal arch muscle formation, Pitx2 specifies muscle fate through cooperation with Tbx1 and Myf5 to induce MyoD (Sambasivan et al., 2009).

3.2 Runx Transcription Factors

The Runx family proteins, which consist of Runx1, Runx2, and Runx3, have a Runt domain, transactivation domain, and Ihh domain. They all share a common cofactor, Cbfβ, which protects Runx protein from proteasomal degradation and facilitate Runx binding at the DNA (Lim et al., 2015). Among the three Runx proteins, Runx2 is indispensable for the commitment of multipotent mesenchymal cells to the osteoblastic lineage (Komori, 2003). Runx2-deficient mice have no bone formation (Komori et al., 1997; Otto et al., 1997). Mesenchymal cells from these mice lack the ability to differentiate into osteoblasts, but retain the ability to differentiate into adipocytes and chondrocytes (Geoffroy, Kneissel, Fournier, Boyde, & Matthias, 2002). Therefore, human mutations in Runx2 are associated with delayed ossification (Tanaka et al., 2006; Zabeen et al., 2008). Runx2 and Runx3 also promote chondrocyte maturation (Enomoto et al., 2000; Inada et al., 1999; Kim, Otto, Zabel, & Mundlos, 1999; Komori, 2003; Ueta et al., 2001; Yoshida et al., 2004).

Cbfβ is also required for osteoblast differentiation, chondrocyte differentiation, and proliferation. Cbfβ/Runx complex regulate the expressions of *ColX* and *Osx* during chondrocyte and osteoblast development (Wu et al., 2014). Cbfβ balances chondrocyte proliferation with differentiation by upregulating *Ihh* and inhibiting *parathyroid hormone-related protein receptor* expression in postnatal cartilage and bone formation. Consistently, the expressions of Ihh target genes, including *CyclinD1*, *Ptc*, and *Pthlh*, were downregulated in Cbfβ-deficient chondrocytes (Tian et al., 2014).

3.3 SRY-Related HMG-Box Family of Transcription Factors

Sry-related HMG box (Sox) transcription factors play versatile roles in development and cell fate determination, carrying out as diverse functions as sex determination and craniofacial, neuronal and cardiovascular development. Sox proteins function as transcription factors by binding to a conserved motif (A/T)(A/T)CAA(A/T)G through their HMG domain (Badis et al., 2009; Kondoh & Kamachi, 2010).

Among the many SOX family proteins, subdivided into group A to J, SoxE group is critical for craniofacial development. SoxE group comprises Sox8, Sox9, and Sox10, all required for the specification, multipotency, and survival of NCCs in vertebrates (Haldin & LaBonne, 2010; Hong & Saint-Jeannet, 2005; Stolt & Wegner, 2010; Wegner, 2008). Sox9, detected in neural crest progenitors at the lateral edge of the neural plate and maintained in cranial NCCs as they populate the pharyngeal arches in the head region, is also the master regulator of the chondrogenic lineage (Akiyama & Lefebvre, 2011; Hong & Saint-Jeannet, 2005). Inactivation of Sox9 in CNC resulted in a complete absence of cartilages and endochondral bones derived from the CNC, without affecting the mesoderm-derived skeleton and intramembranous bone formation (Mori-Akiyama, Akiyama, Rowitch, & de Crombrugghe, 2003). Human skeletal dysmorphology syndrome, campomelic dysplasia, is caused by heterozygous mutations in and around the Sox9 gene (Foster et al., 1994; Wagner et al., 1994), which is further recapitulated in a Sox9 heterozygous mutant mouse model (Bi et al., 2001).

Sox proteins generally regulate transcription by forming complexes with partner transcription factors (Kamachi, Uchikawa, & Kondoh, 2000; Kondoh & Kamachi, 2010). All SoxE proteins possess a discrete dimerization domain, which allows homo- and heterodimerization (Bernard et al., 2003; Peirano & Wegner, 2000; Sock et al., 2003; Wissmuller, Kosian, Wolf, Finzsch, & Wegner, 2006). Sox9 often forms homodimers and requires SoxD (Sox5/Sox6) dimers bound at adjacent DNA sequences in many occasions of chondrocyte development (Han & Lefebvre, 2008; Lefebvre, Li, & de Crombrugghe, 1998). *Sox5/Sox6* double null embryo die with a severe generalized chondrodysplasia, consistent with their essential roles (Smits et al., 2001).

Sox9 activity is influenced by posttranslational modifications. In chondrocytes, Sox9 is phosphorylated on two serines (S64 and S181) by PKA, enhancing both DNA binding and transcriptional activity on the *Col2a1* enhancer (Huang, Zhou, Lefebvre, & de Crombrugghe, 2000). Mutations of these two serine sites also abolished the ability of Sox9 to initiate NC delamination (Sakai, Suzuki, Osumi, & Wakamatsu, 2006). Moreover, BMP and canonical Wnt signaling promote Sox9 phosphorylation and its cooperation with Snail2 to induce NCC delamination (Liu et al., 2013).

Another mechanism via which Sox9 could direct distinct developmental outcomes involves context-dependent interactions with key signal transduction pathways. For example, Wnt signals must be inactivated in order to promote chondrogenesis (Guo et al., 2004). Sox9, via the C-terminal region, binds to β-catenin and inhibits its transcriptional activity; via regions outside the C-terminus, promotes β-catenin phosphorylation and proteasomal degradation (Topol, Chen, Song, Day, & Yang, 2009).

Another Sox protein, Sox10 plays key roles in NCC migration. The expression of Sox10 is regulated by three transcription factors, Sox9, Ets1, and cMyb (Betancur, Simoes-Costa, Sauka-Spengler, & Bronner,

2014), via two enhancers with distinct spatiotemporal regulatory activities (Betancur, Bronner-Fraser, & Sauka-Spengler, 2010). Sox10E2 is critical for initiating Sox10 expression in newly formed NC; Sox10E1 acts in later migrating vagal and trunk crest cells (Betancur et al., 2010).

3.4 bHLH Family of Transcription Factors

Basic Helix-Loop-Helix (bHLH) family proteins are master regulator of lineage specification and differentiation. Among the many bHLH proteins, Twist1/2, E12, E47, and Ids have key roles in NCC specification and craniofacial morphogenesis. MRFs, including MyoD, Myf5, myogenin, and MRF4, have key roles in head muscle formation.

3.4.1 Twist

Twist1 and Twist2 control diverse developmental processes including NCC delamination, migration, myogenesis, and skeletogenesis (Chen & Behringer, 1995; Cripps & Olson, 1998; Hamamori, Wu, Sartorelli, & Kedes, 1997; Hebrok, Wertz, & Fuchtbauer, 1994; Miraoui & Marie, 2010; Verzi et al., 2002). *Twist1* mutations in human cause Sawthre–Chotzen syndrome (SCS), characterized by a broad spectrum of malformations including short stature and craniosynostoses (ElGhouzzi et al., 1997; Howard et al., 1997; Reardon & Winter, 1994). The craniofacial and limb abnormalities are recapitulated in hemizygous *Twist1* deletion mice (Bialek et al., 2004; Chen & Behringer, 1995; Soo et al., 2002). Twist1 initially occurs in a dorsoventral gradient pattern until the headfold stage, and then moves along the rostrocaudal axis of the embryos as mesoderm cell layer and NCC-derived tissues develop (Qin, Xu, He, Qin, & Xu, 2012). Twist1 is also detected in calvarial osteoblasts (Murray et al., 1992).

Twist1 drives NCC dispersion through induction of *Snail/Slug* and repression of *E-cadherin* (Yang et al., 2004), and guides NCC migration through interaction with EphA4 (Merrill et al., 2006; Yoshida et al., 2005, 2009). Twist1 also induces Bmi1, a polycomb-group repressor complex protein, and cooperates with Bmi1 to repress *E-cadherin* and *p16INK4a* in cancer, a mechanism that may also be employed by neural stem cells for self-renewal (Molofsky et al., 2003; Yang et al., 2010).

Twist proteins regulate organogenesis via control over FGF and BMP signaling (Connerney et al., 2006). Their activities depend on dimer composition. Twist can form homodimer or heterodimer with E12 or E47. In the osteogenic front of the cranial sutures where Twist1 and Id1 are coexpressed, genes regulated by homodimers, such as FGFR2, are expressed; the Id1 here binds to E12/47, thus favoring homodimer formation, elevating and expanding the expression of FGFR2. In the mid-suture area where Id1 is not expressed, Twist1 heterodimerizes with E12/47 to drive the expression of heterodimer-regulated genes such as thrombospondin 1 (TSP-1). In the sutures of Twist1 +/− mice, the ratio between these dimers is altered that elevated homodimer-regulated gene expression (Connerney et al., 2008). Additionally, heterodimers inhibit BMP signaling (Hayashi et al., 2007), which is predominantly active in the osteogenic front of cranial sutures. Thus, the decrease of heterodimers in the cranial sutures of Twist1 +/− mice or SCS patients also allows for an expansion of BMP signaling. BMP-induced Id expression further promotes homodimer formation (Rice et al., 2000). This positive feedback loop is likely a cause for premature closing of the sutures, leading to craniosynostosis in SCS.

Twist1 maintains mesenchymal progenitor state by negatively regulating Runx2 (Bialek et al., 2004; Rice et al., 2000). Twist1 binds to the DNA-binding domain of Runx2, and the relief of inhibition allows for osteoblast differentiation. Insufficient Runx2 inhibition in *Twist1*$^{+/-}$ mice causes premature odontoblasts differentiation, leading to extensive pulp stone formation in the tooth (Galler et al., 2007).

3.4.2 Id

ID proteins, Id1, Id2, Id3, and Id4, have intact HLH domains but lack DNA-binding domain (Ruzinova & Benezra, 2003); therefore, they heterodimerize with other bHLH proteins as dominant negative regulators. Their negative regulation also extends to the Ets and Pax family proteins (Kumar, Shadrach, Wagers, & Lassar, 2009; Ohtani et al., 2001; Roberts, Deed, Inoue, Norton, & Sharrocks, 2001; Yates, Atherton, Deed, Norton, & Sharrocks, 1999). Id proteins control the timing of cell fate determination and differentiation (Lasorella, Benezra, & Iavarone, 2014). For example, Id4 promotes osteogenic differentiation by releasing Hes1 from Hes1/Hey2 complexes. Subsequently, Hes1 stabilizes Runx2 and enhances its transcriptional activity to promote osteoblast-specific gene expression (Tokuzawa et al., 2010).

3.4.3 MRF

The myogenic regulatory factors (MRFs): myogenic factor 5 (Myf5), myogenic differentiation (MyoD), myogenic regulatory factor 4 (Mrf4), and myogenin are the core components of the myogenic pathway (Moncaut, Rigby, & Carvajal, 2013). *Myf5*, the determination factor, is expressed before the adoption of the myogenic fate. *MyoD*, induced by Myf5, drives cells to the myogenic lineage (Braun, Bober, Rudnicki, Jaenisch, & Arnold,

1994). Myogenin expression happens later and is crucial for terminal differentiation, while Mrf4 promotes both fate determination and differentiation (Hasty et al., 1993; Nabeshima et al., 1993). MyoD, Myf5, and Mrf4 orchestrate the recruitment of ATP-dependent SWI/SNF complex, including Brg1 and BAF60c, to remodel chromatin structure and initiate myogenic program (de la Serna et al., 2005; Gerber, Klesert, Bergstrom, & Tapscott, 1997; Roy, de la Serna, & Imbalzano, 2002).

Myogenic cascades in head muscle formation have distinct features compared to trunk muscles (Braun & Gautel, 2011). Head muscles, including extraocular, branchial arch, tongue, and laryngeal muscles, require Pitx2 to induce MRFs (Dong et al., 2006; Shih, Gross, & Kioussi, 2007). Branchial arch muscle formation also requires T-box factor Tbx1 to cooperate with Pitx2 in the activation of *Myf5* and *MyoD* (Dastjerdi et al., 2007; Jerome & Papaioannou, 2001; Kelly, Jerome-Majewska, & Papaioannou, 2004). The cooperation between Pitx2 and Tbx1 in head muscle replaces the central role of Pax3 in trunk muscle formation. Two other bHLH transcription factors, Msc and Tcf21 (also known as epicardin/capsulin/POD-1), also contribute to branchial arch muscle formation with overlapping functions (Lu et al., 2002; Moncaut et al., 2012; von Scheven, Bothe, Ahmed, Alvares, & Dietrich, 2006).

3.5 Snail Family of Transcription Factors

Snail family includes Snail1 and Snail2, also called Snail and Slug, and Snail3. It is one of the main families of transcription factors for EMT (Lamouille, Xu, & Derynck, 2014). Snail and Slug mark the neural crest forming region and function mainly as transcription repressors. Genetic studies suggest that Snail1 and Snail2 are dispensable for NCC specification and delamination in mice (Murray & Gridley, 2006), but required for the later development of neural crest-derived structure, through regulating EMT and migration (Murray, Oram, & Gridley, 2007). Neural crest-specific deletion of both Snail and Slug causes cleft palate and cranial skeletal defects (Murray et al., 2007). Slug represses epithelial gene expression through recruitment of PHD12/Sin3A/HDAC to the *Cad6B* promoter for H3 hyperacetylation (Strobl-Mazzulla & Bronner, 2012; Taneyhill, Coles, & Bronner-Fraser, 2007) and EZH2/PRC2 complex to E-cadherin promoter for H3K27 trimethylation (Tien et al., 2015).

The stability of Snail and Slug is regulated by ubiquitination. F-box protein Partner of paired (Ppa) and p53 promote ubiquitin-mediated proteosomal degradation of Snail and/or Slug, thereby coordinate NCC growth and differentiation (Lander, Nordin, & LaBonne, 2011; Rinon et al., 2011; Vernon & LaBonne, 2006).

3.6 Smad Transcription Factors

Smad transcription factors relay signals for the TGF-β and BMP family of growth factors to control multiple aspects of craniofacial morphogenesis. Smad1/5, together with Smad4, relay BMP signals to activate *Dlx3/5*, *Msx1/2*, *Barx1*, *Pax3*, *Sox9*, *Satb2*, *Hand1*, and *Gata3* for craniofacial patterning (Bonilla-Claudio et al., 2012; Furumatsu et al., 2005; Hassan et al., 2006; Ko et al., 2007; Nie, Deng, Wang, & Jiao, 2008; Yang, Yuan, Li, et al., 2014). During palatogenesis, Smad2, together with Smad4 and LEF1, repress *Cdh1*, the gene encoding E-cadherin, during palate fusion (Nawshad, Medici, Liu, & Hay, 2007). Mesenchymal Smad4 relays TGF-β signals to activate *Ctgf* for palatal mesenchyme proliferation, while epithelial Smad4 relays TGF-β signals to activate *Irf6* for medial edge epithelium degeneration (Iwata, Suzuki, Pelikan, Ho, & Chai, 2013; Parada et al., 2013). During tooth formation, dental epithelial Smad4 induces Shh expression in the epithelium, which subsequently induces Nfic in the dental mesenchyme through Gli1 for root development (Huang, Xu, Bringas, Hung, & Chai, 2010). Dental mesenchymal Smad4 promotes odontoblast fate and represses osteogenic gene programs partly through induction of *Dkk1* and *Sfrp1* to inhibit Wnt-mediated osteogenesis (Li et al., 2011). Finally, in tongue formation, Smad4 is required for myogenic differentiation in muscle fibers partly through induction of *Fgf4 and Fgf6* (Han et al., 2012).

3.7 β-Catenin/LEF/TCF Transcription Factors

β-catenin relays signals of the canonical Wnt pathway. Inactivation of β-catenin in NCC leads to the absence of head structures (Brault et al., 2001). Genetic manipulation of Wnt/β-catenin signal component sFRP1/2, LRP indicates roles in osteoblast proliferation and differentiation (Babij et al., 2003; Bodine et al., 2004; Gaur et al., 2005; Little et al., 2002; Morello et al., 2008; Nakanishi et al., 2008; Yao et al., 2010). β-Catenin suppresses chondrocyte fate through the repression of mesenchymal cell differentiation into Runx2- and Sox9-positive precursors, to allow osteoblast formation during both intramembranous and endochondral ossification (Hill et al., 2005). In more mature osteoblast, loss of β-catenin promotes osteoblast formation and suppresses bone resorption, indicating a complicated role for β-catenin in bone homeostasis. In palatogenesis, Wnt/β-catenin signal induces the expression of TGF-β3 expression, which is a key factor for palate development. Epithelial β-catenin/TCF4/LEF1 promotes MEE degeneration to facilitate palate fusion (Brugmann et al., 2007; Medici, Hay, & Olsen, 2008).

Wnt and BMP signaling pathways interact and influence bone development at multiple stages. In osteoblast progenitors, BMP2 antagonizes Wnt signaling by promoting an interaction between Smad1 and Dvl-1 that restricts β-catenin activity (Liu, Tang, Qiu, Cao, & Clemens, 2006). During osteoblast differentiation, BMP signaling acts synergistically with β-catenin to promote differentiation. BMP enhances Wnt ligands and receptor expression. Reciprocally, the Wnt autocrine loop mediates the induction of ALP and mineralization by BMP2 (Chen et al., 2007).

3.8 Gli Transcription Factors

The three Gli proteins are zinc-finger transcription factors of the Hh signaling pathway. Gli1 is a strong transcriptional activator, Gli2 has both activator and repressor functions, and Gli3 is mostly a repressor (Aza-Blanc, Lin, Ruiz i Altaba, & Kornberg, 2000; Bai, Stephen, & Joyner, 2004; Ruiz i Altaba, 1998, 1999; Tyurina et al., 2005). Among Gli transcriptional targets are Gli1 and Hh receptor Ptch1, which form positive feedback loops for Hh signaling (Ruiz i Altaba, 1997; Stecca & Ruiz, 2010). The transcriptional targets of Gli activator and repressor include CyclinD1, N-Myc, Bcl2, Bmi1, and Snail, which regulates of cell cycle, proliferation, survival, self-renewal, and EMT, respectively, and contribute to the function of Gli proteins in morphogenesis (McMahon et al., 2003; Resende et al., 2010).

SHH signaling dose-dependently regulates the morphogenesis of frontonasal and maxillary processes (FNP and MXPs), which give rise to the mid- and upper face (Hu & Helms, 1999). Hh signaling also controls cranial suture formation, whereby SHH increases mesenchymal proliferation and suture mesenchyme thickness via Msx2 activation in concerto with BMP4 signaling (Alappat et al., 2003), IHH, expressed by calvarial osteoblasts, induces PTCH and BMP2/4 expression to promote intramembranous bone formation at the osteogenic fronts and subsequent suture fusion (Jacob et al., 2007; Lenton et al., 2011; Murakami et al., 1997). Glis3 is highly expressed in osteoblasts and acts synergistically with BMP2 to promote osteoblast differentiation and inhibit adipocyte differentiation through regulating *FGF18* transcription (Beak, Kang, Kim, & Jetten, 2007).

The Gli proteins are regulated exquisitely by multiple mechanisms. Negative regulators induced by Hh ligands modulate Gli function. PKA, GSK3β, and CK1 phosphorylate and target Glis for cleavage or degradation (Di Marcotullio et al., 2007). Numb, an inhibitor of Notch signaling, targets Gli1 for proteasome degradation through the E3 ligase Itch (Di Marcotullio et al., 2006). Sufu and Ren inhibit Gli proteins function by sequestering them in the cytoplasm (Kogerman et al., 1999). Positive regulators of Gli function have also been identified, including the kinase Dyrk1 (Mao et al., 2002), the actin-binding protein Missing in metastasis (MIM, also called BEG4) (Callahan et al., 2004), and multiple receptor tyrosine kinase-induced signaling pathways (Brewster, Mullor, & Ruiz i Altaba, 2000; Palma et al., 2005).

3.9 Forkhead Family of Transcription Factors

The forkhead box, or Fox, gene family is an ancient family identified in species ranging from yeasts to humans (Hannenhalli & Kaestner, 2009). The unifying feature of Fox proteins is the highly conserved forkhead (FKH) DNA-binding domain (Kaestner, Knochel, & Martinez, 2000). The FKH domain is remarkably similar to the winged-helix structures of histones H1 and H5 (Clark, Halay, Lai, & Burley, 1993). FOX protein Foxi1 binds to the condensed chromosomes continuously during mitosis, which may serve to maintain transcriptional competency during development (Yan, Xu, Crawford, Wang, & Burgess, 2006).

Among over 40 members in mammals, Foxi and Foxc groups play vital roles in craniofacial development. Foxi is necessary for jaw development (Edlund, Ohyama, Kantarci, Riley, & Groves, 2014; Nissen, Yan, Amsterdam, Hopkins, & Burgess, 2003). Foxi3 mutant mice are not viable and display severe branchial arch-derived defects, due to reduced FGF8 signaling (Edlund et al., 2014). Foxc plays roles in calvarial, ocular, and jaw development (Inman, Purcell, Kume, & Trainor, 2013; Rice, Rice, Olsen, & Thesleff, 2003; Rice, Rice, & Thesleff, 2005; Smith et al., 2000). Calvarial bones remain rudimentary at the sites of initial osteogenic condensations in mutant mice that lack Foxc1 (Rice et al., 2003). Foxc1 null embryos also display jaw fusion similar to that observed in Fgf8 mutant embryos (Inman et al., 2013). Foxc2-deficient mice display defects in the aortic arches, craniofacial bones, and vertebral columns (Kim et al., 2009; Park, Kim, & Nam, 2011). Besides Foxi and Foxc group, FoxE1 and FoxF2 are involved in palate formation (Tuteja & Kaestner, 2007; Venza et al., 2009).

Forkhead factors are the recipients of many posttranslational modifications, including phosphorylation, acetylation, methylation, and ubiquitination, which affect the activity, cellular localization, and/or stability (Calnan & Brunet, 2008; Li et al., 2007; van der Horst et al., 2006; Yamagata et al., 2008). For example, FoxC1 is phosphorylated at Serine 272 by ERK1/2 MAPK, which promotes its transcriptional activation and stability (Berry, Mirzayans, & Walter, 2006; Berry, Saleem, & Walter, 2002).

3.10 Other Transcription Factors

3.10.1 IRF6

Interferon regulatory factor 6 (IRF6) belongs to a family of nine transcription factors that share a highly conserved helix–turn–helix DNA-binding domain and a less conserved protein-binding domain. Most IRFs regulate the expression of interferon after viral infection (Taniguchi, Ogasawara, Takaoka, & Tanaka, 2001). Only IRF6 is associated with craniofacial development (Blanton et al., 2005; Ghassibe et al., 2005; Kondo et al., 2002; Scapoli et al., 2005). Epithelial IRF6 controls the balance between proliferation and differentiation, partly through synergy with Smad4 in response to TGF-β (Iwata, Suzuki, Pelikan, Ho, Sanchez-Lara, et al., 2013; Richardson, Dixon, Jiang, & Dixon, 2009; Richardson et al., 2006). IRF6 is also a direct target of p63, through an enhancer element, variation within which increases susceptibility to cleft lip (Thomason et al., 2010).

3.10.2 TBX22

TBX22 is a member of T-box family with repressor activity (Laugier-Anfossi & Villard, 2000). Mutations in *TBX22* are associated with cleft palate and ankyloglossia (CPX) (Braybrook et al., 2001, 2002; Herr et al., 2003; Suphapeetiporn, Tongkobpetch, Siriwan, & Shotelersuk, 2007). Tbx22 is induced by FGF8 in the early face and repressed by BMP4 during face and palate development (Fuchs et al., 2010). The transcription factor Mn1, upstream of TBX22, coexpresses with TBX22 to regulate posterior palate development (Liu et al., 2008). TBX22 also requires SUMO1-mediated sumoylation for its repression activity (Andreou et al., 2007).

3.10.3 MEF2C

The myocyte enhancer factor 2 (MEF2) family has four members in vertebrates, MEF2A–D (Black & Olson, 1998; McKinsey, Zhang, & Olson, 2002). MEF2 proteins bind to a consensus DNA-binding element as homo- and heterodimers and interact with other transcription factors to function as both positive and negative regulators of gene expression partly through their association with class II histone deacetylases (HDACs) (McKinsey, Zhang, & Olson, 2001). MEF2C is widely appreciated for its role in the muscle lineage. In addition, its expression in early NC development is essential for craniofacial development through induction of *Dlx5*, *Dlx6*, and *Hand2* in the branchial arches and subsequent synergistic function with these factors (Verzi et al., 2007; Lin, Schwarz, Bucana, & Olson, 1997). Mice lacking MEF2C function in the neural crest have cleft palates and misshaped heads (Verzi et al., 2007).

4. QUESTIONS FOR THE FUTURE

Genetic studies using zebrafish, chick, and mouse, aided by large-scale arrays and sequencing techniques, have significantly enriched our knowledge on genetic underpinnings of craniofacial development. When combined with human patient information, we have gained valuable insights on the genetic determinants of congenital defects. But compared to the exponential growth of information on gene expression, we know surprisingly little about how gene expression is regulated during craniofacial development, by transcription factors. This is mostly limited on the technical side. Chromatin immunoprecipitation-based techniques such as ChIP-seq have accelerated research on transcription and epigenetic modification and enabled us to study transcription factor binding and activity with precision. However, tissue of interest during craniofacial morphogenesis often comes in limited quantity, which makes ChIP-seq-based approach very challenging. Current single-cell technologies have tacked a similar problem and opened new frontiers for researchers working with limited material. Future application of single-cell technology to transcription factor-oriented analysis will bring a new dimension to this area of research.

REFERENCES

Acampora, D., Merlo, G. R., Paleari, L., Zerega, B., Postiglione, M. P., Mantero, S., et al. (1999). Craniofacial, vestibular and bone defects in mice lacking the Distal-less-related gene Dlx5. *Development*, *126*(17), 3795–3809.

Agger, K., Cloos, P. A., Christensen, J., Pasini, D., Rose, S., Rappsilber, J., et al. (2007). UTX and JMJD3 are histone H3K27 demethylases involved in HOX gene regulation and development. *Nature*, *449*(7163), 731–734.

Akiyama, H., & Lefebvre, V. (2011). Unraveling the transcriptional regulatory machinery in chondrogenesis. *Journal of Bone and Mineral Metabolism*, *29*(4), 390–395.

Alappat, S., Zhang, Z. Y., & Chen, Y. P. (2003). Msx homeobox gene family and craniofacial development. *Cell Research*, *13*(6), 429–442.

Amendt, B. A., Sutherland, L. B., & Russo, A. F. (1999). Multifunctional role of the Pitx2 homeodomain protein C-terminal tail. *Molecular and Cellular Biology*, *19*(10), 7001–7010.

Andreou, A. M., Pauws, E., Jones, M. C., Singh, M. K., Bussen, M., Doudney, K., et al. (2007). TBX22 missense mutations found in patients with X-linked cleft palate affect DNA binding, sumoylation, and transcriptional repression. *American Journal of Human Genetics*, *81*(4), 700–712.

Aza-Blanc, P., Lin, H. Y., Ruiz i Altaba, A., & Kornberg, T. B. (2000). Expression of the vertebrate Gli proteins in Drosophila reveals a distribution of activator and repressor activities. *Development*, *127*(19), 4293–4301.

Babij, P., Zhao, W., Small, C., Kharode, Y., Yaworsky, P. J., Bouxsein, M. L., et al. (2003). High bone mass in mice expressing a mutant LRP5 gene. *Journal of Bone and Mineral Research*, *18*(6), 960–974.

Badis, G., Berger, M. F., Philippakis, A. A., Talukder, S., Gehrke, A. R., Jaeger, S. A., et al. (2009). Diversity and complexity in DNA recognition by transcription factors. *Science*, *324*(5935), 1720–1723.

Bae, C. J., Park, B. Y., Lee, Y. H., Tobias, J. W., Hong, C. S., & Saint-Jeannet, J. P. (2014). Identification of Pax3 and Zic1 targets in the developing neural crest. *Developmental Biology*, *386*(2), 473–483.

Bai, C. B., Stephen, D., & Joyner, A. L. (2004). All mouse ventral spinal cord patterning by hedgehog is Gli dependent and involves an activator function of Gli3. *Developmental Cell*, *6*(1), 103–115.

Basch, M. L., Bronner-Fraser, M., & Garcia-Castro, M. I. (2006). Specification of the neural crest occurs during gastrulation and requires Pax7. *Nature*, *441*(7090), 218–222.

Beak, J. Y., Kang, H. S., Kim, Y. S., & Jetten, A. M. (2007). Kruppel-like zinc finger protein Glis3 promotes osteoblast differentiation by regulating FGF18 expression. *Journal of Bone and Mineral Research*, *22*(8), 1234–1244.

Bei, M., & Maas, R. (1998). FGFs and BMP4 induce both Msx1-independent and Msx1-dependent signaling pathways in early tooth development. *Development*, *125*(21), 4325–4333.

Bendall, A. J., Ding, J., Hu, G., Shen, M. M., & Abate-Shen, C. (1999). Msx1 antagonizes the myogenic activity of Pax3 in migrating limb muscle precursors. *Development*, *126*(22), 4965–4976.

Bendall, A. J., Rincon-Limas, D. E., Botas, J., & Abate-Shen, C. (1998). Protein complex formation between Msx1 and Lhx2 homeoproteins is incompatible with DNA binding activity. *Differentiation*, *63*(3), 151–157.

Bernard, P., Tang, P., Liu, S., Dewing, P., Harley, V. R., & Vilain, E. (2003). Dimerization of SOX9 is required for chondrogenesis, but not for sex determination. *Human Molecular Genetics*, *12*(14), 1755–1765.

Berry, F. B., Mirzayans, F., & Walter, M. A. (2006). Regulation of FOXC1 stability and transcriptional activity by an epidermal growth factor-activated mitogen-activated protein kinase signaling cascade. *The Journal of Biological Chemistry*, *281*(15), 10098–10104.

Berry, F. B., Saleem, R. A., & Walter, M. A. (2002). FOXC1 transcriptional regulation is mediated by N- and C-terminal activation domains and contains a phosphorylated transcriptional inhibitory domain. *The Journal of Biological Chemistry*, *277*(12), 10292–10297.

Betancur, P., Bronner-Fraser, M., & Sauka-Spengler, T. (2010). Genomic code for Sox10 activation reveals a key regulatory enhancer for cranial neural crest. *Proceedings of the National Academy of Sciences of the United States of America*, *107*(8), 3570–3575.

Betancur, P., Simoes-Costa, M., Sauka-Spengler, T., & Bronner, M. E. (2014). Expression and function of transcription factor cMyb during cranial neural crest development. *Mechanisms of Development*, *132*, 38–43.

Beverdam, A., Merlo, G. R., Paleari, L., Mantero, S., Genova, F., Barbieri, O., et al. (2002). Jaw transformation with gain of symmetry after Dlx5/Dlx6 inactivation: Mirror of the past? *Genesis*, *34*(4), 221–227.

Bi, W., Huang, W., Whitworth, D. J., Deng, J. M., Zhang, Z., Behringer, R. R., et al. (2001). Haploinsufficiency of Sox9 results in defective cartilage primordia and premature skeletal mineralization. *Proceedings of the National Academy of Sciences of the United States of America, 98*(12), 6698–6703.
Bialek, P., Kern, B., Yang, X. L., Schrock, M., Sosic, D., Hong, N., et al. (2004). A twist code determines the onset of osteoblast differentiation. *Developmental Cell, 6*(3), 423–435.
Black, B. L., & Olson, E. N. (1998). Transcriptional control of muscle development by myocyte enhancer factor-2 (MEF2) proteins. *Annual Review of Cell and Developmental Biology, 14*, 167–196.
Blake, J. A., & Ziman, M. R. (2014). Pax genes: Regulators of lineage specification and progenitor cell maintenance. *Development, 141*(4), 737–751.
Blanton, S. H., Cortez, A., Stal, S., Mulliken, J. B., Finnell, R. H., & Hecht, J. T. (2005). Variation in IRF6 contributes to nonsyndromic cleft lip and palate. *American Journal of Medical Genetics. Part A, 137A*(3), 259–262.
Bodine, P. V., Zhao, W., Kharode, Y. P., Bex, F. J., Lambert, A. J., Goad, M. B., et al. (2004). The Wnt antagonist secreted frizzled-related protein-1 is a negative regulator of trabecular bone formation in adult mice. *Molecular Endocrinology, 18*(5), 1222–1237.
Bonilla-Claudio, M., Wang, J., Bai, Y., Klysik, E., Selever, J., & Martin, J. F. (2012). Bmp signaling regulates a dose-dependent transcriptional program to control facial skeletal development. *Development, 139*(4), 709–719.
Brault, V., Moore, R., Kutsch, S., Ishibashi, M., Rowitch, D. H., McMahon, A. P., et al. (2001). Inactivation of the beta-catenin gene by Wnt1-Cre-mediated deletion results in dramatic brain malformation and failure of craniofacial development. *Development, 128*(8), 1253–1264.
Braun, T., Bober, E., Rudnicki, M. A., Jaenisch, R., & Arnold, H. H. (1994). MyoD expression marks the onset of skeletal myogenesis in Myf-5 mutant mice. *Development, 120*(11), 3083–3092.
Braun, T., & Gautel, M. (2011). Transcriptional mechanisms regulating skeletal muscle differentiation, growth and homeostasis. *Nature Reviews. Molecular Cell Biology, 12*(6), 349–361.
Braybrook, C., Doudney, K., Marcano, A. C., Arnason, A., Bjornsson, A., Patton, M. A., et al. (2001). The T-box transcription factor gene TBX22 is mutated in X-linked cleft palate and ankyloglossia. *Nature Genetics, 29*(2), 179–183.
Braybrook, C., Lisgo, S., Doudney, K., Henderson, D., Marcano, A. C., Strachan, T., et al. (2002). Craniofacial expression of human and murine TBX22 correlates with the cleft palate and ankyloglossia phenotype observed in CPX patients. *Human Molecular Genetics, 11*(22), 2793–2804.
Brewster, R., Mullor, J. L., & Ruiz i Altaba, A. (2000). Gli2 functions in FGF signaling during antero-posterior patterning. *Development, 127*(20), 4395–4405.
Brugmann, S. A., Goodnough, L. H., Gregorieff, A., Leucht, P., ten Berge, D., Fuerer, C., et al. (2007). Wnt signaling mediates regional specification in the vertebrate face. *Development, 134*(18), 3283–3295.
Callahan, C. A., Ofstad, T., Horng, L., Wang, J. K., Zhen, H. H., Coulombe, P. A., et al. (2004). MIM/BEG4, a Sonic hedgehog-responsive gene that potentiates Gli-dependent transcription. *Genes & Development, 18*(22), 2724–2729.
Calnan, D. R., & Brunet, A. (2008). The FoxO code. *Oncogene, 27*(16), 2276–2288.
Cao, H., Florez, S., Amen, M., Huynh, T., Skobe, Z., Baldini, A., et al. (2010). Tbx1 regulates progenitor cell proliferation in the dental epithelium by modulating Pitx2 activation of p21. *Developmental Biology, 347*(2), 289–300.
Catron, K. M., Zhang, H., Marshall, S. C., Inostroza, J. A., Wilson, J. M., & Abate, C. (1995). Transcriptional repression by Msx-1 does not require homeodomain DNA-binding sites. *Molecular and Cellular Biology, 15*(2), 861–871.

Chen, Z. F., & Behringer, R. R. (1995). Twist is required in head mesenchyme for cranial neural-tube morphogenesis. *Genes & Development*, *9*(6), 686–699.

Chen, Y., Bei, M., Woo, I., Satokata, I., & Maas, R. (1996). Msx1 controls inductive signaling in mammalian tooth morphogenesis. *Development*, *122*(10), 3035–3044.

Chen, Y., Whetstone, H. C., Youn, A., Nadesan, P., Chow, E. C., Lin, A. C., et al. (2007). Beta-catenin signaling pathway is crucial for bone morphogenetic protein 2 to induce new bone formation. *The Journal of Biological Chemistry*, *282*(1), 526–533.

Chiba, S., Takeshita, K., Imai, Y., Kumano, K., Kurokawa, M., Masuda, S., et al. (2003). Homeoprotein DLX-1 interacts with Smad4 and blocks a signaling pathway from activin A in hematopoietic cells. *Proceedings of the National Academy of Sciences of the United States of America*, *100*(26), 15577–15582.

Choi, Y. H., Choi, H. J., Lee, K. Y., & Oh, J. W. (2012). Akt1 regulates phosphorylation and osteogenic activity of Dlx3. *Biochemical and Biophysical Research Communications*, *425*(4), 800–805.

Clark, K. L., Halay, E. D., Lai, E., & Burley, S. K. (1993). Co-crystal structure of the HNF-3/fork head DNA-recognition motif resembles histone H5. *Nature*, *364*(6436), 412–420.

Connerney, J., Andreeva, V., Leshem, Y., Mercado, M. A., Dowell, K., Yang, X. H., et al. (2008). Twist1 homodimers enhance FGF responsiveness of the cranial sutures and promote suture closure. *Developmental Biology*, *318*(2), 323–334.

Connerney, J., Andreeva, V., Leshem, Y., Muentener, C., Mercado, M. A., & Spicer, D. B. (2006). Twist1 dimer selection regulates cranial suture patterning and fusion. *Developmental Dynamics*, *235*(5), 1345–1357.

Cripps, R. M., & Olson, E. N. (1998). Twist is required for muscle template splitting during adult Drosophila myogenesis. *Developmental Biology*, *203*(1), 106–115.

Dastjerdi, A., Robson, L., Walker, R., Hadley, J., Zhang, Z., Rodriguez-Niedenfuhr, M., et al. (2007). Tbx1 regulation of myogenic differentiation in the limb and cranial mesoderm. *Developmental Dynamics*, *236*(2), 353–363.

de la Serna, I. L., Ohkawa, Y., Berkes, C. A., Bergstrom, D. A., Dacwag, C. S., Tapscott, S. J., et al. (2005). MyoD targets chromatin remodeling complexes to the myogenin locus prior to forming a stable DNA-bound complex. *Molecular and Cellular Biology*, *25*(10), 3997–4009.

Denaxa, M., Sharpe, P. T., & Pachnis, V. (2009). The LIM homeodomain transcription factors Lhx6 and Lhx7 are key regulators of mammalian dentition. *Developmental Biology*, *333*(2), 324–336.

Depew, M. J., Liu, J. K., Long, J. E., Presley, R., Meneses, J. J., Pedersen, R. A., et al. (1999). Dlx5 regulates regional development of the branchial arches and sensory capsules. *Development*, *126*(17), 3831–3846.

Di Marcotullio, L., Ferretti, E., Greco, A., De Smaele, E., Po, A., Sico, M. A., et al. (2006). Numb is a suppressor of Hedgehog signalling and targets Gli1 for Itch-dependent ubiquitination. *Nature Cell Biology*, *8*(12), 1415–1423.

Di Marcotullio, L., Ferretti, E., Greco, A., De Smaele, E., Screpanti, I., & Gulino, A. (2007). Multiple ubiquitin-dependent processing pathways regulate hedgehog/gli signaling: Implications for cell development and tumorigenesis. *Cell Cycle*, *6*(4), 390–393.

Dong, F. Y., Sun, X. X., Liu, W., Ai, D., Klysik, E., Lu, M. F., et al. (2006). Pitx2 promotes development of splanchnic mesoderm-derived branchiomeric muscle. *Development*, *133*(24), 4891–4899.

Dorey, K., & Amaya, E. (2010). FGF signalling: Diverse roles during early vertebrate embryogenesis. *Development*, *137*(22), 3731–3742.

Edlund, R. K., Ohyama, T., Kantarci, H., Riley, B. B., & Groves, A. K. (2014). Foxi transcription factors promote pharyngeal arch development by regulating formation of FGF signaling centers. *Developmental Biology*, *390*(1), 1–13.

ElGhouzzi, V., LeMerrer, M., PerrinSchmitt, F., Lajeunie, E., Benit, P., Renier, D., et al. (1997). Mutations of the TWIST gene in the Saethre-Chotzen syndrome. *Nature Genetics*, *15*(1), 42–46.

Enomoto, H., Enomoto-Iwamoto, M., Iwamoto, M., Nomura, S., Himeno, M., Kitamura, Y., et al. (2000). Cbfa1 is a positive regulatory factor in chondrocyte maturation. *Journal of Biological Chemistry*, *275*(12), 8695–8702.

Feng, X. H., & Derynck, R. (2005). Specificity and versatility in tgf-beta signaling through Smads. *Annual Review of Cell and Developmental Biology*, *21*, 659–693.

Foster, J. W., Dominguez-Steglich, M. A., Guioli, S., Kwok, C., Weller, P. A., Stevanovic, M., et al. (1994). Campomelic dysplasia and autosomal sex reversal caused by mutations in an SRY-related gene. *Nature*, *372*(6506), 525–530.

Fuchs, A., Inthal, A., Herrmann, D., Cheng, S., Nakatomi, M., Peters, H., et al. (2010). Regulation of Tbx22 during facial and palatal development. *Developmental Dynamics*, *239*(11), 2860–2874.

Furumatsu, T., Tsuda, M., Taniguchi, N., Tajima, Y., & Asahara, H. (2005). Smad3 induces chondrogenesis through the activation of SOX9 via CREB-binding protein/p300 recruitment. *The Journal of Biological Chemistry*, *280*(9), 8343–8350.

Galler, K. M., Yasue, A., Cavender, A. C., Bialek, P., Karsenty, G., & D'Souza, R. N. (2007). A novel role for twist-1 in pulp homeostasis. *Journal of Dental Research*, *86*(10), 951–955.

Gaur, T., Lengner, C. J., Hovhannisyan, H., Bhat, R. A., Bodine, P. V., Komm, B. S., et al. (2005). Canonical WNT signaling promotes osteogenesis by directly stimulating Runx2 gene expression. *The Journal of Biological Chemistry*, *280*(39), 33132–33140.

Geoffroy, V., Kneissel, M., Fournier, B., Boyde, A., & Matthias, P. (2002). High bone resorption in adult aging Transgenic mice overexpressing Cbfa1/Runx2 in cells of the osteoblastic lineage. *Molecular and Cellular Biology*, *22*(17), 6222–6233.

Gerber, A. N., Klesert, T. R., Bergstrom, D. A., & Tapscott, S. J. (1997). Two domains of MyoD mediate transcriptional activation of genes in repressive chromatin: A mechanism for lineage determination in myogenesis. *Genes & Development*, *11*(4), 436–450.

Ghassibe, M., Bayet, B., Revencu, N., Verellen-Dumoulin, C., Gillerot, Y., Vanwijck, R., et al. (2005). Interferon regulatory factor-6: A gene predisposing to isolated cleft lip with or without cleft palate in the Belgian population. *European Journal of Human Genetics*, *13*(11), 1239–1242.

Gradilla, A. C., & Guerrero, I. (2013). Hedgehog on the move: A precise spatial control of Hedgehog dispersion shapes the gradient. *Current Opinion in Genetics & Development*, *23*(4), 363–373.

Guo, X., Day, T. F., Jiang, X., Garrett-Beal, L., Topol, L., & Yang, Y. (2004). Wnt/beta-catenin signaling is sufficient and necessary for synovial joint formation. *Genes & Development*, *18*(19), 2404–2417.

Haldin, C. E., & LaBonne, C. (2010). SoxE factors as multifunctional neural crest regulatory factors. *The International Journal of Biochemistry & Cell Biology*, *42*(3), 441–444.

Hamamori, Y., Wu, H. Y., Sartorelli, V., & Kedes, L. (1997). The basic domain of myogenic basic helix-loop-helix (bHLH) proteins is the novel target for direct inhibition by another bHLH protein, Twist. *Molecular and Cellular Biology*, *17*(11), 6563–6573.

Han, Y., & Lefebvre, V. (2008). L-Sox5 and Sox6 drive expression of the aggrecan gene in cartilage by securing binding of Sox9 to a far-upstream enhancer. *Molecular and Cellular Biology*, *28*(16), 4999–5013.

Han, D., Zhao, H., Parada, C., Hacia, J. G., Bringas, P., Jr., & Chai, Y. (2012). A TGFbeta-Smad4-Fgf6 signaling cascade controls myogenic differentiation and myoblast fusion during tongue development. *Development*, *139*(9), 1640–1650.

Hannenhalli, S., & Kaestner, K. H. (2009). The evolution of Fox genes and their role in development and disease. *Nature Reviews. Genetics*, *10*(4), 233–240.

Harel, I., Maezawa, Y., Avraham, R., Rinon, A., Ma, H. Y., Cross, J. W., et al. (2012). Pharyngeal mesoderm regulatory network controls cardiac and head muscle morphogenesis. *Proceedings of the National Academy of Sciences of the United States of America, 109*(46), 18839–18844.

Hassan, M. Q., Javed, A., Morasso, M. I., Karlin, J., Montecino, M., van Wijnen, A. J., et al. (2004). Dlx3 transcriptional regulation of osteoblast differentiation: Temporal recruitment of Msx2, Dlx3, and Dlx5 homeodomain proteins to chromatin of the osteocalcin gene. *Molecular and Cellular Biology, 24*(20), 9248–9261.

Hassan, M. Q., Tare, R. S., Lee, S. H., Mandeville, M., Morasso, M. I., Javed, A., et al. (2006). BMP2 commitment to the osteogenic lineage involves activation of Runx2 by DLX3 and a homeodomain transcriptional network. *The Journal of Biological Chemistry, 281*(52), 40515–40526.

Hasty, P., Bradley, A., Morris, J. H., Edmondson, D. G., Venuti, J. M., Olson, E. N., et al. (1993). Muscle deficiency and neonatal death in mice with a targeted mutation in the myogenin gene. *Nature, 364*(6437), 501–506.

Hayashi, M., Nimura, K., Kashiwagi, K., Harada, T., Takaoka, K., Kato, H., et al. (2007). Comparative roles of Twist-1 and Id1 in transcriptional regulation by BMP signaling. *Journal of Cell Science, 120*(8), 1350–1357.

He, F., Xiong, W., Wang, Y., Li, L., Liu, C., Yamagami, T., et al. (2011). Epithelial Wnt/beta-catenin signaling regulates palatal shelf fusion through regulation of Tgfbeta3 expression. *Developmental Biology, 350*(2), 511–519.

Hebrok, M., Wertz, K., & Fuchtbauer, E. M. (1994). M-Twist is an inhibitor of muscle differentiation. *Developmental Biology, 165*(2), 537–544.

Herr, A., Meunier, D., Muller, I., Rump, A., Fundele, R., Ropers, H. H., et al. (2003). Expression of mouse Tbx22 supports its role in palatogenesis and glossogenesis. *Developmental Dynamics, 226*(4), 579–586.

Heude, E., Bouhali, K., Kurihara, Y., Kurihara, H., Couly, G., Janvier, P., et al. (2010). Jaw muscularization requires Dlx expression by cranial neural crest cells. *Proceedings of the National Academy of Sciences of the United States of America, 107*(25), 11441–11446.

Hill, T. P., Spater, D., Taketo, M. M., Birchmeier, W., & Hartmann, C. (2005). Canonical Wnt/beta-catenin signaling prevents osteoblasts from differentiating into chondrocytes. *Developmental Cell, 8*(5), 727–738.

Holland, P. W. (2013). Evolution of homeobox genes. *Wiley Interdisciplinary Reviews. Developmental Biology, 2*(1), 31–45.

Holleville, N., Mateos, S., Bontoux, M., Bollerot, K., & Monsoro-Burq, A. H. (2007). Dlx5 drives Runx2 expression and osteogenic differentiation in developing cranial suture mesenchyme. *Developmental Biology, 304*(2), 860–874.

Hong, C. S., & Saint-Jeannet, J. P. (2005). Sox proteins and neural crest development. *Seminars in Cell & Developmental Biology, 16*(6), 694–703.

Howard, T. D., Paznekas, W. A., Green, E. D., Chiang, L. C., Ma, N., DeLuna, R. I. O., et al. (1997). Mutations in TWIST, a basic helix-loop-helix transcription factor, in Saethre-Chotzen syndrome. *Nature Genetics, 15*(1), 36–41.

Hu, D., & Helms, J. A. (1999). The role of Sonic hedgehog in normal and abnormal craniofacial morphogenesis. *Development, 126*(21), 4873–4884.

Huang, X., Xu, X., Bringas, P., Jr., Hung, Y. P., & Chai, Y. (2010). Smad4-Shh-Nfic signaling cascade-mediated epithelial-mesenchymal interaction is crucial in regulating tooth root development. *Journal of Bone and Mineral Research, 25*(5), 1167–1178.

Huang, W., Zhou, X., Lefebvre, V., & de Crombrugghe, B. (2000). Phosphorylation of SOX9 by cyclic AMP-dependent protein kinase A enhances SOX9's ability to transactivate a Col2a1 chondrocyte-specific enhancer. *Molecular and Cellular Biology, 20*(11), 4149–4158.

Inada, M., Yasui, T., Nomura, S., Miyake, S., Deguchi, K., Himeno, M., et al. (1999). Maturational disturbance of chondrocytes in Cbfa1-deficient mice. *Developmental Dynamics*, *214*(4), 279–290.

Inman, K. E., Purcell, P., Kume, T., & Trainor, P. A. (2013). Interaction between Foxc1 and Fgf8 during mammalian jaw patterning and in the pathogenesis of syngnathia. *PLoS Genetics*, *9*(12), e1003949.

Ito, Y., Yeo, J. Y., Chytil, A., Han, J., Bringas, P., Jr., Nakajima, A., et al. (2003). Conditional inactivation of Tgfbr2 in cranial neural crest causes cleft palate and calvaria defects. *Development*, *130*(21), 5269–5280.

Iwata, J., Hosokawa, R., Sanchez-Lara, P. A., Urata, M., Slavkin, H., & Chai, Y. (2010). Transforming growth factor-beta regulates basal transcriptional regulatory machinery to control cell proliferation and differentiation in cranial neural crest-derived osteoprogenitor cells. *The Journal of Biological Chemistry*, *285*(7), 4975–4982.

Iwata, J., Suzuki, A., Pelikan, R. C., Ho, T. V., & Chai, Y. (2013). Noncanonical transforming growth factor beta (TGFbeta) signaling in cranial neural crest cells causes tongue muscle developmental defects. *The Journal of Biological Chemistry*, *288*(41), 29760–29770.

Iwata, J., Suzuki, A., Pelikan, R. C., Ho, T. V., Sanchez-Lara, P. A., Urata, M., et al. (2013). Smad4-Irf6 genetic interaction and TGFbeta-mediated IRF6 signaling cascade are crucial for palatal fusion in mice. *Development*, *140*(6), 1220–1230.

Iwata, J., Tung, L., Urata, M., Hacia, J. G., Pelikan, R., Suzuki, A., et al. (2012). Fibroblast growth factor 9 (FGF9)-pituitary homeobox 2 (PITX2) pathway mediates transforming growth factor beta (TGFbeta) signaling to regulate cell proliferation in palatal mesenchyme during mouse palatogenesis. *The Journal of Biological Chemistry*, *287*(4), 2353–2363.

Jacob, S., Wu, C., Freeman, T. A., Koyama, E., & Kirschner, R. E. (2007). Expression of Indian Hedgehog, BMP-4 and Noggin in craniosynostosis induced by fetal constraint. *Annals of Plastic Surgery*, *58*(2), 215–221.

Jeong, H. M., Jin, Y. H., Choi, Y. H., Yum, J., Choi, J. K., Yeo, C. Y., et al. (2012). PKC signaling inhibits osteogenic differentiation through the regulation of Msx2 function. *Biochimica et Biophysica Acta*, *1823*(8), 1225–1232.

Jeong, H. M., Jin, Y. H., Kim, Y. J., Yum, J., Choi, Y. H., Yeo, C. Y., et al. (2011). Akt phosphorylates and regulates the function of Dlx5. *Biochemical and Biophysical Research Communications*, *409*(4), 681–686.

Jerome, L. A., & Papaioannou, V. E. (2001). DiGeorge syndrome phenotype in mice mutant for the T-box gene, Tbx1. *Nature Genetics*, *27*(3), 286–291.

Kaestner, K. H., Knochel, W., & Martinez, D. E. (2000). Unified nomenclature for the winged helix/forkhead transcription factors. *Genes & Development*, *14*(2), 142–146.

Kamachi, Y., Uchikawa, M., & Kondoh, H. (2000). Pairing SOX off: With partners in the regulation of embryonic development. *Trends in Genetics*, *16*(4), 182–187.

Kelly, R. G., Jerome-Majewska, L. A., & Papaioannou, V. E. (2004). The del22q11.2 candidate gene Tbx1 regulates branchiomeric myogenesis. *Human Molecular Genetics*, *13*(22), 2829–2840.

Kim, S. H., Cho, K. W., Choi, H. S., Park, S. J., Rhee, Y., Jung, H. S., et al. (2009). The forkhead transcription factor Foxc2 stimulates osteoblast differentiation. *Biochemical and Biophysical Research Communications*, *386*(3), 532–536.

Kim, S. J., Jeon, S., Shin, E. Y., Kim, E. G., Park, J., & Bae, C. D. (2004). AMPA, not NMDA, activates RhoA GTPases and subsequently phosphorylates moesin. *Experimental & Molecular Medicine*, *36*(1), 98–102.

Kim, I. S., Otto, F., Zabel, B., & Mundlos, S. (1999). Regulation of chondrocyte differentiation by Cbfa1. *Mechanisms of Development*, *80*(2), 159–170.

Ko, S. O., Chung, I. H., Xu, X., Oka, S., Zhao, H., Cho, E. S., et al. (2007). Smad4 is required to regulate the fate of cranial neural crest cells. *Developmental Biology*, *312*(1), 435–447.

Kogerman, P., Grimm, T., Kogerman, L., Krause, D., Unden, A. B., Sandstedt, B., et al. (1999). Mammalian suppressor-of-fused modulates nuclear-cytoplasmic shuttling of Gli-1. *Nature Cell Biology*, *1*(5), 312–319.

Komori, T. (2003). Requisite roles of Runx2 and Cbfb in skeletal development. *Journal of Bone and Mineral Metabolism*, *21*(4), 193–197.

Komori, T., Yagi, H., Nomura, S., Yamaguchi, A., Sasaki, K., Deguchi, K., et al. (1997). Targeted disruption of Cbfa1 results in a complete lack of bone formation owing to maturational arrest of osteoblasts. *Cell*, *89*(5), 755–764.

Kondo, S., Schutte, B. C., Richardson, R. J., Bjork, B. C., Knight, A. S., Watanabe, Y., et al. (2002). Mutations in IRF6 cause Van der Woude and popliteal pterygium syndromes. *Nature Genetics*, *32*(2), 285–289.

Kondoh, H., & Kamachi, Y. (2010). SOX-partner code for cell specification: Regulatory target selection and underlying molecular mechanisms. *The International Journal of Biochemistry & Cell Biology*, *42*(3), 391–399.

Kumar, D., Shadrach, J. L., Wagers, A. J., & Lassar, A. B. (2009). Id3 is a direct transcriptional target of Pax7 in quiescent satellite cells. *Molecular Biology of the Cell*, *20*(14), 3170–3177.

Lamouille, S., Xu, J., & Derynck, R. (2014). Molecular mechanisms of epithelial-mesenchymal transition. *Nature Reviews. Molecular Cell Biology*, *15*(3), 178–196.

Lan, F., Bayliss, P. E., Rinn, J. L., Whetstine, J. R., Wang, J. K., Chen, S., et al. (2007). A histone H3 lysine 27 demethylase regulates animal posterior development. *Nature*, *449*(7163), 689–694.

Lander, R., Nordin, K., & LaBonne, C. (2011). The F-box protein Ppa is a common regulator of core EMT factors Twist, Snail, Slug, and Sip1. *The Journal of Cell Biology*, *194*(1), 17–25.

Landin Malt, A., Cesario, J. M., Tang, Z., Brown, S., & Jeong, J. (2014). Identification of a face enhancer reveals direct regulation of LIM homeobox 8 (Lhx8) by wingless-int (WNT)/beta-catenin signaling. *The Journal of Biological Chemistry*, *289*(44), 30289–30301.

Lasorella, A., Benezra, R., & Iavarone, A. (2014). The ID proteins: Master regulators of cancer stem cells and tumour aggressiveness. *Nature Reviews. Cancer*, *14*(2), 77–91.

Laugier-Anfossi, F., & Villard, L. (2000). Molecular characterization of a new human T-box gene (TBX22) located in xq21.1 encoding a protein containing a truncated T-domain. *Gene*, *255*(2), 289–296.

Lederer, D., Grisart, B., Digilio, M. C., Benoit, V., Crespin, M., Ghariani, S. C., et al. (2012). Deletion of KDM6A, a histone demethylase interacting with MLL2, in three patients with Kabuki syndrome. *American Journal of Human Genetics*, *90*(1), 119–124.

Lefebvre, V., Li, P., & de Crombrugghe, B. (1998). A new long form of Sox5 (L-Sox5), Sox6 and Sox9 are coexpressed in chondrogenesis and cooperatively activate the type II collagen gene. *The EMBO Journal*, *17*(19), 5718–5733.

Lenton, K., James, A. W., Manu, A., Brugmann, S. A., Birker, D., Nelson, E. R., et al. (2011). Indian hedgehog positively regulates calvarial ossification and modulates bone morphogenetic protein signaling. *Genesis*, *49*(10), 784–796.

Li, J., Huang, X., Xu, X., Mayo, J., Bringas, P., Jr., Jiang, R., et al. (2011). SMAD4-mediated WNT signaling controls the fate of cranial neural crest cells during tooth morphogenesis. *Development*, *138*(10), 1977–1989.

Li, B., Samanta, A., Song, X., Iacono, K. T., Bembas, K., Tao, R., et al. (2007). FOXP3 interactions with histone acetyltransferase and class II histone deacetylases are required for repression. *Proceedings of the National Academy of Sciences of the United States of America*, *104*(11), 4571–4576.

Lien, W. H., & Fuchs, E. (2014). Wnt some lose some: Transcriptional governance of stem cells by Wnt/beta-catenin signaling. *Genes & Development*, *28*(14), 1517–1532.

Lim, K. E., Park, N. R., Che, X., Han, M. S., Jeong, J. H., Kim, S. Y., et al. (2015). Core binding factor beta of osteoblasts maintains cortical bone mass via stabilization of Runx2 in mice. *Journal of Bone and Mineral Research*, *30*(4), 715–722.
Lin, C. R., Kioussi, C., O'Connell, S., Briata, P., Szeto, D., Liu, F., et al. (1999). Pitx2 regulates lung asymmetry, cardiac positioning and pituitary and tooth morphogenesis. *Nature*, *401*(6750), 279–282.
Lin, Q., Schwarz, J., Bucana, C., & Olson, E. N. (1997). Control of mouse cardiac morphogenesis and myogenesis by transcription factor MEF2C. *Science*, *276*(5317), 1404–1407.
Lindgren, A. M., Hoyos, T., Talkowski, M. E., Hanscom, C., Blumenthal, I., Chiang, C., et al. (2013). Haploinsufficiency of KDM6A is associated with severe psychomotor retardation, global growth restriction, seizures and cleft palate. *Human Genetics*, *132*(5), 537–552.
Little, R. D., Carulli, J. P., Del Mastro, R. G., Dupuis, J., Osborne, M., Folz, C., et al. (2002). A mutation in the LDL receptor-related protein 5 gene results in the autosomal dominant high-bone-mass trait. *American Journal of Human Genetics*, *70*(1), 11–19.
Liu, W., Lan, Y., Pauws, E., Meester-Smoor, M. A., Stanier, P., Zwarthoff, E. C., et al. (2008). The Mn1 transcription factor acts upstream of Tbx22 and preferentially regulates posterior palate growth in mice. *Development*, *135*(23), 3959–3968.
Liu, Z., Tang, Y., Qiu, T., Cao, X., & Clemens, T. L. (2006). A dishevelled-1/Smad1 interaction couples WNT and bone morphogenetic protein signaling pathways in uncommitted bone marrow stromal cells. *The Journal of Biological Chemistry*, *281*(25), 17156–17163.
Liu, J. A., Wu, M. H., Yan, C. H., Chau, B. K., So, H., Ng, A., et al. (2013). Phosphorylation of Sox9 is required for neural crest delamination and is regulated downstream of BMP and canonical Wnt signaling. *Proceedings of the National Academy of Sciences of the United States of America*, *110*(8), 2882–2887.
Lo Iacono, N., Mantero, S., Chiarelli, A., Garcia, E., Mills, A. A., Morasso, M. I., et al. (2008). Regulation of Dlx5 and Dlx6 gene expression by p63 is involved in EEC and SHFM congenital limb defects. *Development*, *135*(7), 1377–1388.
Lopes, M., Goupille, O., Saint Cloment, C., Lallemand, Y., Cumano, A., & Robert, B. (2011). Msx genes define a population of mural cell precursors required for head blood vessel maturation. *Development*, *138*(14), 3055–3066.
Lu, J. R., Bassel-Duby, R., Hawkins, A., Chang, P., Valdez, R., Wu, H., et al. (2002). Control of facial muscle development by MyoR and capsulin. *Science*, *298*(5602), 2378–2381.
Mallo, M., & Alonso, C. R. (2013). The regulation of Hox gene expression during animal development. *Development*, *140*(19), 3951–3963.
Mao, J., Maye, P., Kogerman, P., Tejedor, F. J., Toftgard, R., Xie, W., et al. (2002). Regulation of Gli1 transcriptional activity in the nucleus by Dyrk1. *The Journal of Biological Chemistry*, *277*(38), 35156–35161.
McKinsey, T. A., Zhang, C. L., & Olson, E. N. (2001). Control of muscle development by dueling HATs and HDACs. *Current Opinion in Genetics & Development*, *11*(5), 497–504.
McKinsey, T. A., Zhang, C. L., & Olson, E. N. (2002). MEF2: A calcium-dependent regulator of cell division, differentiation and death. *Trends in Biochemical Sciences*, *27*(1), 40–47.
McMahon, A. P., Ingham, P. W., & Tabin, C. J. (2003). Developmental roles and clinical significance of hedgehog signaling. *Current Topics in Developmental Biology*, *53*, 1–114.
Medici, D., Hay, E. D., & Olsen, B. R. (2008). Snail and Slug promote epithelial-mesenchymal transition through beta-catenin-T-cell factor-4-dependent expression of transforming growth factor-beta3. *Molecular Biology of the Cell*, *19*(11), 4875–4887.
Merrill, A. E., Bochukova, E. G., Brugger, S. M., Ishii, M., Pilz, D. T., Wall, S. A., et al. (2006). Cell mixing at a neural crest-mesoderm boundary and deficient ephrin-

Eph signaling in the pathogenesis of craniosynostosis. *Human Molecular Genetics, 15*(8), 1319–1328.

Milet, C., Maczkowiak, F., Roche, D. D., & Monsoro-Burq, A. H. (2013). Pax3 and Zic1 drive induction and differentiation of multipotent, migratory, and functional neural crest in Xenopus embryos. *Proceedings of the National Academy of Sciences of the United States of America, 110*(14), 5528–5533.

Miraoui, H., & Marie, P. J. (2010). Pivotal role of Twist in skeletal biology and pathology. *Gene, 468*(1–2), 1–7.

Mishina, Y., & Snider, T. N. (2014). Neural crest cell signaling pathways critical to cranial bone development and pathology. *Experimental Cell Research, 325*(2), 138–147.

Mitsiadis, T. A., & Drouin, J. (2008). Deletion of the Pitx1 genomic locus affects mandibular tooth morphogenesis and expression of the Barx1 and Tbx1 genes. *Developmental Biology, 313*(2), 887–896.

Miyake, N., Koshimizu, E., Okamoto, N., Mizuno, S., Ogata, T., Nagai, T., et al. (2013). MLL2 and KDM6A mutations in patients with Kabuki syndrome. *American Journal of Medical Genetics Part A, 161A*(9), 2234–2243.

Molofsky, A. V., Pardal, R., Iwashita, T., Park, I. K., Clarke, M. F., & Morrison, S. J. (2003). Bmi-1 dependence distinguishes neural stem cell self-renewal from progenitor proliferation. *Nature, 425*(6961), 962–967.

Moncaut, N., Cross, J. W., Siligan, C., Keith, A., Taylor, K., Rigby, P. W. J., et al. (2012). Musculin and TCF21 coordinate the maintenance of myogenic regulatory factor expression levels during mouse craniofacial development. *Development, 139*(5), 958–967.

Moncaut, N., Rigby, P. W., & Carvajal, J. J. (2013). Dial M(RF) for myogenesis. *The FEBS Journal, 280*(17), 3980–3990.

Morello, R., Bertin, T. K., Schlaubitz, S., Shaw, C. A., Kakuru, S., Munivez, E., et al. (2008). Brachy-syndactyly caused by loss of Sfrp2 function. *Journal of Cellular Physiology, 217*(1), 127–137.

Mori-Akiyama, Y., Akiyama, H., Rowitch, D. H., & de Crombrugghe, B. (2003). Sox9 is required for determination of the chondrogenic cell lineage in the cranial neural crest. *Proceedings of the National Academy of Sciences of the United States of America, 100*(16), 9360–9365.

Murakami, S., Nifuji, A., & Noda, M. (1997). Expression of Indian hedgehog in osteoblasts and its posttranscriptional regulation by transforming growth factor-beta. *Endocrinology, 138*(5), 1972–1978.

Murdoch, B., DelConte, C., & Garcia-Castro, M. I. (2012). Pax7 lineage contributions to the mammalian neural crest. *PLoS One*, 7(7), e41089.

Murray, S. S., Glackin, C. A., Winters, K. A., Gazit, D., Kahn, A. J., & Murray, E. J. (1992). Expression of helix-loop-helix regulatory genes during differentiation of mouse osteoblastic cells. *Journal of Bone and Mineral Research*, 7(10), 1131–1138.

Murray, S. A., & Gridley, T. (2006). Snail family genes are required for left-right asymmetry determination, but not neural crest formation, in mice. *Proceedings of the National Academy of Sciences of the United States of America, 103*(27), 10300–10304.

Murray, S. A., Oram, K. F., & Gridley, T. (2007). Multiple functions of Snail family genes during palate development in mice. *Development, 134*(9), 1789–1797.

Nabeshima, Y., Hanaoka, K., Hayasaka, M., Esumi, E., Li, S. W., Nonaka, I., et al. (1993). Myogenin gene disruption results in perinatal lethality because of severe muscle defect. *Nature, 364*(6437), 532–535.

Nakanishi, R., Akiyama, H., Kimura, H., Otsuki, B., Shimizu, M., Tsuboyama, T., et al. (2008). Osteoblast-targeted expression of Sfrp4 in mice results in low bone mass. *Journal of Bone and Mineral Research, 23*(2), 271–277.

Nawshad, A., Medici, D., Liu, C. C., & Hay, E. D. (2007). TGFbeta3 inhibits E-cadherin gene expression in palate medial-edge epithelial cells through a Smad2-Smad4-LEF1 transcription complex. *Journal of Cell Science, 120*(Pt 9), 1646–1653.

Neubuser, A., Peters, H., Balling, R., & Martin, G. R. (1997). Antagonistic interactions between FGF and BMP signaling pathways: A mechanism for positioning the sites of tooth formation. *Cell, 90*(2), 247–255.

Nie, X., Deng, C. X., Wang, Q., & Jiao, K. (2008). Disruption of Smad4 in neural crest cells leads to mid-gestation death with pharyngeal arch, craniofacial and cardiac defects. *Developmental Biology, 316*(2), 417–430.

Nissen, R. M., Yan, J., Amsterdam, A., Hopkins, N., & Burgess, S. M. (2003). Zebrafish foxi one modulates cellular responses to Fgf signaling required for the integrity of ear and jaw patterning. *Development, 130*(11), 2543–2554.

Ogawa, T., Kapadia, H., Feng, J. Q., Raghow, R., Peters, H., & D'Souza, R. N. (2006). Functional consequences of interactions between Pax9 and Msx1 genes in normal and abnormal tooth development. *The Journal of Biological Chemistry, 281*(27), 18363–18369.

Ohtani, N., Zebedee, Z., Huot, T. J., Stinson, J. A., Sugimoto, M., Ohashi, Y., et al. (2001). Opposing effects of Ets and Id proteins on p16INK4a expression during cellular senescence. *Nature, 409*(6823), 1067–1070.

Otto, F., Thornell, A. P., Crompton, T., Denzel, A., Gilmour, K. C., Rosewell, I. R., et al. (1997). Cbfa1, a candidate gene for cleidocranial dysplasia syndrome, is essential for osteoblast differentiation and bone development. *Cell, 89*(5), 765–771.

Palma, V., Lim, D. A., Dahmane, N., Sanchez, P., Brionne, T. C., Herzberg, C. D., et al. (2005). Sonic hedgehog controls stem cell behavior in the postnatal and adult brain. *Development, 132*(2), 335–344.

Parada, C., Li, J., Iwata, J., Suzuki, A., & Chai, Y. (2013). CTGF mediates Smad-dependent transforming growth factor beta signaling to regulate mesenchymal cell proliferation during palate development. *Molecular and Cellular Biology, 33*(17), 3482–3493.

Park, S. J., Kim, R. J., & Nam, J. S. (2011). Inhibitor of DNA-binding 4 contributes to the maintenance and expansion of cancer stem cells in 4T1 mouse mammary cancer cell line. *Laboratory Animal Research, 27*(4), 333–338.

Peirano, R. I., & Wegner, M. (2000). The glial transcription factor Sox10 binds to DNA both as monomer and dimer with different functional consequences. *Nucleic Acids Research, 28*(16), 3047–3055.

Peters, H., Neubuser, A., Kratochwil, K., & Balling, R. (1998). Pax9-deficient mice lack pharyngeal pouch derivatives and teeth and exhibit craniofacial and limb abnormalities. *Genes & Development, 12*(17), 2735–2747.

Plouhinec, J. L., Roche, D. D., Pegoraro, C., Figueiredo, A. L., Maczkowiak, F., Brunet, L. J., et al. (2014). Pax3 and Zic1 trigger the early neural crest gene regulatory network by the direct activation of multiple key neural crest specifiers. *Developmental Biology, 386*(2), 461–472.

Qin, Q., Xu, Y., He, T., Qin, C., & Xu, J. (2012). Normal and disease-related biological functions of Twist1 and underlying molecular mechanisms. *Cell Research, 22*(1), 90–106.

Qiu, M., Bulfone, A., Ghattas, I., Meneses, J. J., Christensen, L., Sharpe, P. T., et al. (1997). Role of the Dlx homeobox genes in proximodistal patterning of the branchial arches: Mutations of Dlx-1, Dlx-2, and Dlx-1 and -2 alter morphogenesis of proximal skeletal and soft tissue structures derived from the first and second arches. *Developmental Biology, 185*(2), 165–184.

Qiu, M., Bulfone, A., Martinez, S., Meneses, J. J., Shimamura, K., Pedersen, R. A., et al. (1995). Null mutation of Dlx-2 results in abnormal morphogenesis of proximal first

and second branchial arch derivatives and abnormal differentiation in the forebrain. *Genes & Development, 9*(20), 2523–2538.
Reardon, W., & Winter, R. M. (1994). Saethre-chotzen syndrome. *Journal of Medical Genetics, 31*(5), 393–396.
Resende, T. P., Ferreira, M., Teillet, M. A., Tavares, A. T., Andrade, R. P., & Palmeirim, I. (2010). Sonic hedgehog in temporal control of somite formation. *Proceedings of the National Academy of Sciences of the United States of America, 107*(29), 12907–12912.
Rezsohazy, R. (2014). Non-transcriptional interactions of Hox proteins: Inventory, facts, and future directions. *Developmental Dynamics, 243*(1), 117–131.
Rice, D. P. C., Aberg, T., Chan, Y. S., Tang, Z. Q., Kettunen, P. J., Pakarinen, L., et al. (2000). Integration of FGF and TWIST in calvarial bone and suture development. *Development, 127*(9), 1845–1855.
Rice, R., Rice, D. P. C., Olsen, B. R., & Thesleff, I. (2003). Progression of calvarial bone development requires Foxc1 regulation of Msx2 and Alx4. *Developmental Biology, 262*(1), 75–87.
Rice, R., Rice, D. P., & Thesleff, I. (2005). Foxc1 integrates Fgf and Bmp signalling independently of twist or noggin during calvarial bone development. *Developmental Dynamics, 233*(3), 847–852.
Richardson, R. J., Dixon, J., Jiang, R., & Dixon, M. J. (2009). Integration of IRF6 and Jagged2 signalling is essential for controlling palatal adhesion and fusion competence. *Human Molecular Genetics, 18*(14), 2632–2642.
Richardson, R. J., Dixon, J., Malhotra, S., Hardman, M. J., Knowles, L., Boot-Handford, R. P., et al. (2006). Irf6 is a key determinant of the keratinocyte proliferation-differentiation switch. *Nature Genetics, 38*(11), 1329–1334.
Rinon, A., Molchadsky, A., Nathan, E., Yovel, G., Rotter, V., Sarig, R., et al. (2011). p53 coordinates cranial neural crest cell growth and epithelial-mesenchymal transition/delamination processes. *Development, 138*(9), 1827–1838.
Roberts, E. C., Deed, R. W., Inoue, T., Norton, J. D., & Sharrocks, A. D. (2001). Id helix-loop-helix proteins antagonize pax transcription factor activity by inhibiting DNA binding. *Molecular and Cellular Biology, 21*(2), 524–533.
Robledo, R. F., Rajan, L., Li, X., & Lufkin, T. (2002). The Dlx5 and Dlx6 homeobox genes are essential for craniofacial, axial, and appendicular skeletal development. *Genes & Development, 16*(9), 1089–1101.
Roy, K., de la Serna, I. L., & Imbalzano, A. N. (2002). The myogenic basic helix-loop-helix family of transcription factors shows similar requirements for SWI/SNF chromatin remodeling enzymes during muscle differentiation in culture. *Journal of Biological Chemistry, 277*(37), 33818–33824.
Ruiz i Altaba, A. (1997). Catching a Gli-mpse of Hedgehog. *Cell, 90*(2), 193–196.
Ruiz i Altaba, A. (1998). Combinatorial Gli gene function in floor plate and neuronal inductions by Sonic hedgehog. *Development, 125*(12), 2203–2212.
Ruiz i Altaba, A. (1999). Gli proteins encode context-dependent positive and negative functions: Implications for development and disease. *Development, 126*(14), 3205–3216.
Ruzinova, M. B., & Benezra, R. (2003). Id proteins in development, cell cycle and cancer. *Trends in Cell Biology, 13*(8), 410–418.
Saadi, I., Das, P., Zhao, M., Raj, L., Ruspita, I., Xia, Y., et al. (2013). Msx1 and Tbx2 antagonistically regulate Bmp4 expression during the bud-to-cap stage transition in tooth development. *Development, 140*(13), 2697–2702.
Sakai, D., Suzuki, T., Osumi, N., & Wakamatsu, Y. (2006). Cooperative action of Sox9, Snail2 and PKA signaling in early neural crest development. *Development, 133*(7), 1323–1333.

Sambasivan, R., Gayraud-Morel, B., Dumas, G., Cimper, C., Paisant, S., Kelly, R. G., et al. (2009). Distinct regulatory cascades govern extraocular and pharyngeal arch muscle progenitor cell fates. *Developmental Cell, 16*(6), 810–821.

Sato, T., Sasai, N., & Sasai, Y. (2005). Neural crest determination by co-activation of Pax3 and Zic1 genes in Xenopus ectoderm. *Development, 132*(10), 2355–2363.

Scapoli, L., Palmieri, A., Martinelli, M., Pezzetti, F., Carinci, P., Tognon, M., et al. (2005). Strong evidence of linkage disequilibrium between polymorphisms at the IRF6 locus and nonsyndromic cleft lip with or without cleft palate, in an Italian population. *American Journal of Human Genetics, 76*(1), 180–183.

Schwarz, D., Varum, S., Zemke, M., Scholer, A., Baggiolini, A., Draganova, K., et al. (2014). Ezh2 is required for neural crest-derived cartilage and bone formation. *Development, 141*(4), 867–877.

Seo, J. H., Jin, Y. H., Jeong, H. M., Kim, Y. J., Jeong, H. G., Yeo, C. Y., et al. (2009). Calmodulin-dependent kinase II regulates Dlx5 during osteoblast differentiation. *Biochemical and Biophysical Research Communications, 384*(1), 100–104.

Sharp, T., Wang, J., Li, X., Cao, H., Gao, S., Moreno, M., et al. (2014). A pituitary homeobox 2 (Pitx2):microRNA-200a-3p:beta-catenin pathway converts mesenchymal cells to amelogenin-expressing dental epithelial cells. *The Journal of Biological Chemistry, 289*(39), 27327–27341.

Shih, H. P., Gross, M. K., & Kioussi, C. (2007). Cranial muscle defects of Pitx2 mutants result from specification defects in the first branchial arch. *Proceedings of the National Academy of Sciences of the United States of America, 104*(14), 5907–5912.

Smith, R. S., Zabaleta, A., Kume, T., Savinova, O. V., Kidson, S. H., Martin, J. E., et al. (2000). Haploinsufficiency of the transcription factors FOXC1 and FOXC2 results in aberrant ocular development. *Human Molecular Genetics, 9*(7), 1021–1032.

Smits, P., Li, P., Mandel, J., Zhang, Z., Deng, J. M., Behringer, R. R., et al. (2001). The transcription factors L-Sox5 and Sox6 are essential for cartilage formation. *Developmental Cell, 1*(2), 277–290.

Sock, E., Pagon, R. A., Keymolen, K., Lissens, W., Wegner, M., & Scherer, G. (2003). Loss of DNA-dependent dimerization of the transcription factor SOX9 as a cause for campomelic dysplasia. *Human Molecular Genetics, 12*(12), 1439–1447.

Soo, K., O'Rourke, M. P., Khoo, P. L., Steiner, K. A., Wong, N., Behringer, R. R., et al. (2002). Twist function is required for the morphogenesis of the cephalic neural tube and the differentiation of the cranial neural crest cells in the mouse embryo. *Developmental Biology, 247*(2), 251–270.

St Amand, T. R., Zhang, Y., Semina, E. V., Zhao, X., Hu, Y., Nguyen, L., et al. (2000). Antagonistic signals between BMP4 and FGF8 define the expression of Pitx1 and Pitx2 in mouse tooth-forming anlage. *Developmental Biology, 217*(2), 323–332.

Stecca, B., & Ruiz, I. A. A. (2010). Context-dependent regulation of the GLI code in cancer by HEDGEHOG and non-HEDGEHOG signals. *Journal of Molecular Cell Biology, 2*(2), 84–95.

Stolt, C. C., & Wegner, M. (2010). SoxE function in vertebrate nervous system development. *The International Journal of Biochemistry & Cell Biology, 42*(3), 437–440.

Strobl-Mazzulla, P. H., & Bronner, M. E. (2012). A PHD12-Snail2 repressive complex epigenetically mediates neural crest epithelial-to-mesenchymal transition. *The Journal of Cell Biology, 198*(6), 999–1010.

Suphapeetiporn, K., Tongkobpetch, S., Siriwan, P., & Shotelersuk, V. (2007). TBX22 mutations are a frequent cause of non-syndromic cleft palate in the Thai population. *Clinical Genetics, 72*(5), 478–483.

Tanaka, J. L., Ono, E., Filho, E. M., Castilho, J. C., Moraes, L. C., & Moraes, M. E. (2006). Cleidocranial dysplasia: Importance of radiographic images in diagnosis of the condition. *Journal of Oral Science, 48*(3), 161–166.

Taneyhill, L. A., Coles, E. G., & Bronner-Fraser, M. (2007). Snail2 directly represses cadherin6B during epithelial-to-mesenchymal transitions of the neural crest. *Development, 134*(8), 1481–1490.

Taniguchi, T., Ogasawara, K., Takaoka, A., & Tanaka, N. (2001). IRF family of transcription factors as regulators of host defense. *Annual Review of Immunology, 19*, 623–655.

Thomason, H. A., Zhou, H., Kouwenhoven, E. N., Dotto, G. P., Restivo, G., Nguyen, B. C., et al. (2010). Cooperation between the transcription factors p63 and IRF6 is essential to prevent cleft palate in mice. *The Journal of Clinical Investigation, 120*(5), 1561–1569.

Tian, F., Wu, M., Deng, L., Zhu, G., Ma, J., Gao, B., et al. (2014). Core binding factor beta (Cbfbeta) controls the balance of chondrocyte proliferation and differentiation by upregulating Indian hedgehog (Ihh) expression and inhibiting parathyroid hormone-related protein receptor (PPR) expression in postnatal cartilage and bone formation. *Journal of Bone and Mineral Research, 29*(7), 1564–1574.

Tien, C. L., Jones, A., Wang, H., Gerigk, M., Nozell, S., & Chang, C. (2015). Snail2/Slug cooperates with Polycomb repressive complex 2 (PRC2) to regulate neural crest development. *Development, 142*(4), 722–731.

Tokuzawa, Y., Yagi, K., Yamashita, Y., Nakachi, Y., Nikaido, I., Bono, H., et al. (2010). Id4, a new candidate gene for senile osteoporosis, acts as a molecular switch promoting osteoblast differentiation. *PLoS Genetics, 6*(7), e1001019.

Topol, L., Chen, W., Song, H., Day, T. F., & Yang, Y. (2009). Sox9 inhibits Wnt signaling by promoting beta-catenin phosphorylation in the nucleus. *The Journal of Biological Chemistry, 284*(5), 3323–3333.

Trainor, P. A., & Krumlauf, R. (2001). Hox genes, neural crest cells and branchial arch patterning. *Current Opinion in Cell Biology, 13*(6), 698–705.

Trinh, B. Q., Barengo, N., & Naora, H. (2011). Homeodomain protein DLX4 counteracts key transcriptional control mechanisms of the TGF-beta cytostatic program and blocks the antiproliferative effect of TGF-beta. *Oncogene, 30*(24), 2718–2729.

Tuteja, G., & Kaestner, K. H. (2007). SnapShot: Forkhead transcription factors I. *Cell, 130*(6), 1160.

Tyurina, O. V., Guner, B., Popova, E., Feng, J. C., Schier, A. F., Kohtz, J. D., et al. (2005). Zebrafish Gli3 functions as both an activator and a repressor in Hedgehog signaling. *Developmental Biology, 277*(2), 537–556.

Ueta, C., Iwamoto, M., Kanatani, N., Yoshida, C., Liu, Y., Enomoto-Iwamoto, M., et al. (2001). Skeletal malformations caused by overexpression of Cbfa1 or its dominant negative form in chondrocytes. *Journal of Cell Biology, 153*(1), 87–99.

Vainio, S., Karavanova, I., Jowett, A., & Thesleff, I. (1993). Identification of BMP-4 as a signal mediating secondary induction between epithelial and mesenchymal tissues during early tooth development. *Cell, 75*(1), 45–58.

van der Horst, A., de Vries-Smits, A. M., Brenkman, A. B., van Triest, M. H., van den Broek, N., Colland, F., et al. (2006). FOXO4 transcriptional activity is regulated by monoubiquitination and USP7/HAUSP. *Nature Cell Biology, 8*(10), 1064–1073.

Venza, M., Visalli, M., Venza, I., Torino, C., Tripodo, B., Melita, R., et al. (2009). Altered binding of MYF-5 to FOXE1 promoter in non-syndromic and CHARGE-associated cleft palate. *Journal of Oral Pathology & Medicine, 38*(1), 18–23.

Vernon, A. E., & LaBonne, C. (2006). Slug stability is dynamically regulated during neural crest development by the F-box protein Ppa. *Development, 133*(17), 3359–3370.

Verzi, M. P., Agarwal, P., Brown, C., McCulley, D. J., Schwarz, J. J., & Black, B. L. (2007). The transcription factor MEF2C is required for craniofacial development. *Developmental Cell, 12*(4), 645–652.

Verzi, M. P., Anderson, J. P., Dodou, E., Kelly, K. K., Greene, S. B., North, B. J., et al. (2002). N-twist, an evolutionarily conserved bHLH protein expressed in the developing CNS, functions as a transcriptional inhibitor. *Developmental Biology, 249*(1), 174–190.

von Scheven, G., Bothe, I., Ahmed, M. U., Alvares, L. E., & Dietrich, S. (2006). Protein and genomic organisation of vertebrate MyoR and Capsulin genes and their expression during avian development. *Gene Expression Patterns, 6*(4), 383–393.

Wagner, T., Wirth, J., Meyer, J., Zabel, B., Held, M., Zimmer, J., et al. (1994). Autosomal sex reversal and campomelic dysplasia are caused by mutations in and around the SRY-related gene SOX9. *Cell, 79*(6), 1111–1120.

Wang, J., & Abate-Shen, C. (2012). The MSX1 homeoprotein recruits G9a methyltransferase to repressed target genes in myoblast cells. *PLoS One*, 7(5), e37647.

Wang, Q., Fang, W. H., Krupinski, J., Kumar, S., Slevin, M., & Kumar, P. (2008). Pax genes in embryogenesis and oncogenesis. *Journal of Cellular and Molecular Medicine, 12*(6A), 2281–2294.

Wang, Y., Groppe, J. C., Wu, J., Ogawa, T., Mues, G., D'Souza, R. N., et al. (2009). Pathogenic mechanisms of tooth agenesis linked to paired domain mutations in human PAX9. *Human Molecular Genetics, 18*(15), 2863–2874.

Wang, Y., Li, Y. P., Paulson, C., Shao, J. Z., Zhang, X., Wu, M., et al. (2014). Wnt and the Wnt signaling pathway in bone development and disease. *Frontiers in Bioscience (Landmark Edition), 19*, 379–407.

Wegner, M. (2008). A matter of identity: Transcriptional control in oligodendrocytes. *Journal of Molecular Neuroscience, 35*(1), 3–12.

Wissmuller, S., Kosian, T., Wolf, M., Finzsch, M., & Wegner, M. (2006). The high-mobility-group domain of Sox proteins interacts with DNA-binding domains of many transcription factors. *Nucleic Acids Research, 34*(6), 1735–1744.

Wu, M., Li, C., Zhu, G., Wang, Y., Jules, J., Lu, Y., et al. (2014). Deletion of core-binding factor beta (Cbfbeta) in mesenchymal progenitor cells provides new insights into Cbfbeta/Runxs complex function in cartilage and bone development. *Bone, 65*, 49–59.

Yamagata, K., Daitoku, H., Takahashi, Y., Namiki, K., Hisatake, K., Kako, K., et al. (2008). Arginine methylation of FOXO transcription factors inhibits their phosphorylation by Akt. *Molecular Cell, 32*(2), 221–231.

Yan, J., Xu, L., Crawford, G., Wang, Z., & Burgess, S. M. (2006). The forkhead transcription factor FoxI1 remains bound to condensed mitotic chromosomes and stably remodels chromatin structure. *Molecular and Cellular Biology, 26*(1), 155–168.

Yang, M. H., Hsu, D. S., Wang, H. W., Wang, H. J., Lan, H. Y., Yang, W. H., et al. (2010). Bmi1 is essential in Twist1-induced epithelial-mesenchymal transition. *Nature Cell Biology, 12*(10), 982–992.

Yang, J., Mani, S. A., Donaher, J. L., Ramaswamy, S., Itzykson, R. A., Come, C., et al. (2004). Twist, a master regulator of morphogenesis, plays an essential role in tumor metastasis. *Cell, 117*(7), 927–939.

Yang, G., Yuan, G., Li, X., Liu, P., Chen, Z., & Fan, M. (2014). BMP-2 induction of Dlx3 expression is mediated by p38/Smad5 signaling pathway in osteoblastic MC3T3-E1 cells. *Journal of Cellular Physiology, 229*(7), 943–954.

Yang, G., Yuan, G., Ye, W., Cho, K. W., & Chen, Y. (2014). An atypical canonical bone morphogenetic protein (BMP) signaling pathway regulates Msh homeobox 1 (Msx1) expression during odontogenesis. *The Journal of Biological Chemistry, 289*(45), 31492–31502.

Yao, W., Cheng, Z., Shahnazari, M., Dai, W., Johnson, M. L., & Lane, N. E. (2010). Overexpression of secreted frizzled-related protein 1 inhibits bone formation and attenuates parathyroid hormone bone anabolic effects. *Journal of Bone and Mineral Research, 25*(2), 190–199.

Yates, P. R., Atherton, G. T., Deed, R. W., Norton, J. D., & Sharrocks, A. D. (1999). Id helix-loop-helix proteins inhibit nucleoprotein complex formation by the TCF ETS-domain transcription factors. *The EMBO Journal, 18*(4), 968–976.

Ye, L., Fan, Z., Yu, B., Chang, J., Al Hezaimi, K., Zhou, X., et al. (2012). Histone demethylases KDM4B and KDM6B promotes osteogenic differentiation of human MSCs. *Cell Stem Cell, 11*(1), 50–61.

Yoon, W. J., Cho, Y. D., Cho, K. H., Woo, K. M., Baek, J. H., Cho, J. Y., et al. (2008). The Boston-type craniosynostosis mutation MSX2 (P148H) results in enhanced susceptibility of MSX2 to ubiquitin-dependent degradation. *The Journal of Biological Chemistry, 283*(47), 32751–32761.

Yoshida, J., Horiuchi, A., Kikuchi, N., Hayashi, A., Osada, R., Ohira, S., et al. (2009). Changes in the expression of E-cadherin repressors, Snail, Slug, SIP1, and Twist, in the development and progression of ovarian carcinoma: The important role of Snail in ovarian tumorigenesis and progression. *Medical Molecular Morphology, 42*(2), 82–91.

Yoshida, T., Phylactou, L. A., Uney, J. B., Ishikawa, I., Eto, K., & Iseki, S. (2005). Twist is required for establishment of the mouse coronal suture. *Journal of Anatomy, 206*(5), 437–444.

Yoshida, C. A., Yamamoto, H., Fujita, T., Furuichi, T., Ito, K., Inoue, K., et al. (2004). Runx2 and Runx3 are essential for chondrocyte maturation, and Runx2 regulates limb growth through induction of Indian hedgehog. *Genes & Development, 18*(8), 952–963.

Zabeen, B., Mohsin, F., Taher, A., Khan, M., Azad, K., & Nahar, N. (2008). Cleidocranial dysplasia. *Mymensingh Medical Journal, 17*(1), 82–84.

Zalc, A., Rattenbach, R., Aurade, F., Cadot, B., & Relaix, F. (2015). Pax3 and Pax7 play essential safeguard functions against environmental stress-induced birth defects. *Developmental Cell, 33*(1), 56–66.

Zhang, Z., Gutierrez, D., Li, X., Bidlack, F., Cao, H., Wang, J., et al. (2013). The LIM homeodomain transcription factor LHX6: A transcriptional repressor that interacts with pituitary homeobox 2 (PITX2) to regulate odontogenesis. *The Journal of Biological Chemistry, 288*(4), 2485–2500.

Zhang, H., Hu, G., Wang, H., Sciavolino, P., Iler, N., Shen, M. M., et al. (1997). Heterodimerization of Msx and Dlx homeoproteins results in functional antagonism. *Molecular and Cellular Biology, 17*(5), 2920–2932.

Zhao, Y., Guo, Y. J., Tomac, A. C., Taylor, N. R., Grinberg, A., Lee, E. J., et al. (1999). Isolated cleft palate in mice with a targeted mutation of the LIM homeobox gene lhx8. *Proceedings of the National Academy of Sciences of the United States of America, 96*(26), 15002–15006.

PART 3

Disease Models, Human Genetics, Genomics and Dynamic Imaging

CHAPTER SIXTEEN

Mouse Models of Rare Craniofacial Disorders

Annita Achilleos*, Paul A. Trainor*,†,1
*Stowers Institute for Medical Research, Kansas City, Missouri, USA
†Department of Anatomy & Cell Biology, University of Kansas Medical Center, Kansas City, Kansas, USA
[1]Corresponding author: e-mail address: pat@stowers.org

Contents

Abstract

A rare disease is defined as a condition that affects less than 1 in 2000 individuals. Currently more than 7000 rare diseases have been documented, and most are thought to be of genetic origin. Rare diseases primarily affect children, and congenital craniofacial syndromes and disorders constitute a significant proportion of rare diseases, with over 700 having been described to date. Modeling craniofacial disorders in animal models has been instrumental in uncovering the etiology and pathogenesis of numerous conditions and in some cases has even led to potential therapeutic avenues for their prevention. In this chapter, we focus primarily on two general classes of rare disorders, ribosomopathies and ciliopathies, and the surprising finding that the disruption of fundamental, global processes can result in tissue-

Current Topics in Developmental Biology, Volume 115
ISSN 0070-2153
http://dx.doi.org/10.1016/bs.ctdb.2015.07.011

specific craniofacial defects. In addition, we discuss recent advances in understanding the pathogenesis of an extremely rare and specific craniofacial condition known as syngnathia, based on the first mouse models for this condition. Approximately 1% of all babies are born with a minor or major developmental anomaly, and individuals suffering from rare diseases deserve the same quality of treatment and care and attention to their disease as other patients.

1. INTRODUCTION

Craniofacial development is an intricate process that requires several distinct signaling pathways and elaborate morphogenetic movements encompassing each of the ectoderm, mesoderm, and endoderm germ layers. In addition, a transient, migratory, and multipotent cell population called neural crest cells (NCC) that can be considered the "fourth" germ layer also contributes significantly to craniofacial development. In fact, NCC will give rise to the majority of the facial mesenchyme and subsequently the craniofacial skeleton.

NCC arise during early embryonic development at the border of the neural and nonneural ectoderm. Although derived primarily from the neuroepithelium, the nonneural or surface ectoderm is also able to generate small numbers of NCC. After induction, NCC undergo an epithelial-to-mesenchymal transition, delaminate from the neural plate or tube, and migrate throughout nearly the entire embryo. Upon reaching their destinations, NCC differentiate into a broad range of cell and tissue types, including cranial and spinal ganglia, enteric neurons, melanocytes, endocrine and paraendocrine cells, connective tissue, and craniofacial bone and cartilage (Bronner & LeDouarin, 2012; Santagati & Rijli, 2003). NCC are generally divided into two primary groups, cranial and trunk NCC, based on the axial levels from which they are derived in the embryo. The major difference between these two populations is the ability of the cranial NCC to differentiate into bone, cartilage, and connective tissue, whereas trunk NCC generally lack this ability.

Cranial NCC arise from the diencephalon, mesencephalon, and rhombencephalon and generate the majority of the bone and cartilage of the face and skull (Fig. 1). The most anterior cranial NCC populate the frontonasal region, which will give rise to the frontal and nasal bones, whereas posterior cranial NCC populate the pharyngeal arches (PAs), which will give rise to the upper and lower jaw, middle ear, and structures in the neck. The PAs comprise a reiterated series of outgrowths on the lateral side of the head that

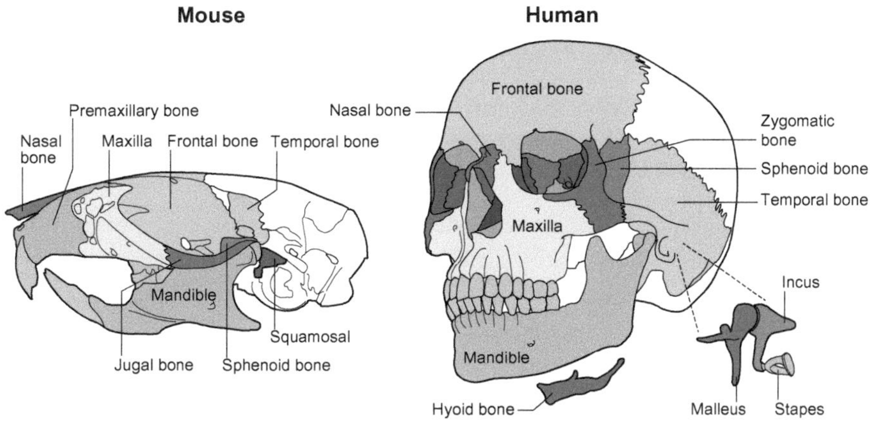

Figure 1 Contribution of neural crest cells to the craniofacial skeleton. Cranial neural crest cells gives rise to the majority of the craniofacial skeleton. All the colored bones in the mouse (left) and human (right) skull are derived from neural crest cells. (A color version of this figure is located at the end of this chapter.)

are highly conserved throughout vertebrate evolution (Frisdal & Trainor, 2014). Mammals exhibit five pairs of PAs that develop rostrocaudally in a serial manner. Each PA is composed externally of ectoderm and lined internally by endoderm, enveloping a mesenchymal core composed of NCC and mesoderm. The juxtaposition of the endoderm and ectoderm forms an internal cleft and an external pouch, respectively, that separates each arch from its neighbor. Complex reciprocal signaling interactions between the ectoderm, endoderm, mesoderm, and NCC ensure proper development of the craniofacial skeleton (Frisdal & Trainor, 2014).

Considering the multiple and distinct stages of NCC development and their contribution to the craniofacial skeleton, it is not surprising that many congenital craniofacial abnormalities, which constitute one out of three birth defects, are due to defects in NCC development. Such abnormalities are collectively referred to as neurocristopathies. Perturbation of distinct stages of NCC development results in different craniofacial deficiencies and syndromes. For example, Treacher Collins syndrome (TCS) arises due to defects in the generation and proliferation of NCC, whereas syngnathia is associated with defective differentiation of NCC.

Here, we describe rare craniofacial neurocristopathies and efforts to understand their genetic and cellular basis as well as the pathogenesis of these disorders using mouse models. Interestingly, many of these tissue-specific disorders are caused by mutations in genes involved in global cellular processes such as ribosome and cilium biogenesis. Termed ribosomopathies and ciliopathies, respectively, these congenital disorders have changed the way we think about the composition and function of ribosomes and cilia, as these global processes have surprisingly tissue-specific roles and effects.

2. RIBOSOMOPATHIES

The ribosome is a large ribonucleoprotein complex responsible for translating mRNAs into proteins in all living organisms. The mature eukaryotic 80S ribosome is composed of two subunits, the small (40S) and the large (60S) subunit, each of which consists of ribosomal proteins (RPs) and ribosomal RNAs (rRNAs). The 40S ribosome, comprised of the 18S rRNA and 33 RPs, decodes the mRNA, whereas the 60S ribosome, comprised of the 5S, 5.8S, and 28S rRNAs together with 46 RPs, catalyzes the formation of peptide bonds (Lafontaine & Tollervey, 2001). The process of generating ribosomes, termed ribosome biogenesis, is elaborate and utilizes all three RNA Polymerases (Pol I, II, III). The four rRNAs are transcribed by RNA Pol I (18S, 5.8S, 28S) and Pol III (5S), whereas about 80 RPs, accessory proteins, and approximately 70 small nucleolar RNAs (snoRNA) are transcribed by Pol II. Additionally, more than 200 proteins are involved in the production, processing, and assembling of the rRNA alone (Kressler, Hurt, & Bassler, 2010; Lafontaine & Tollervey, 2001).

Transcription of rDNA genes is the first step in ribosome biogenesis and is considered to be a rate-limiting step (Fig. 2). Transcription of the 47S precursor rRNA, which includes the 18S, 5.8S, and 28S, by Pol I takes place in the nucleolus, whereas transcription of the 5S rRNA by Pol III happens in the nucleus. Altogether, rDNA transcription accounts for about 60% of overall cellular transcription (Laferte et al., 2006). Subsequently, the 47S rRNA is processed by covalent modifications, bound by RPs, and cleaved several times to produce the individual 5.8S, 18S, and 28S rRNAs, which will then—along with the 5S rRNA—assemble into their respective ribosomal subunits together with the appropriate RPs. These subunits are then exported from the nucleus into the cytoplasm to form the mature and active 80S ribosome.

Given the importance of the ribosome in protein synthesis, which is a basal, global process vital for cell survival and proliferation, it is remarkable that mutations in genes disrupting the process of ribosome biogenesis result in congenital disorders with tissue-specific defects. Studies of these unique ribosomopathy conditions in animal systems, and specifically in mammalian models such as mice, have been instrumental in unraveling tissue-specific roles for ribosomal genes during embryogenesis.

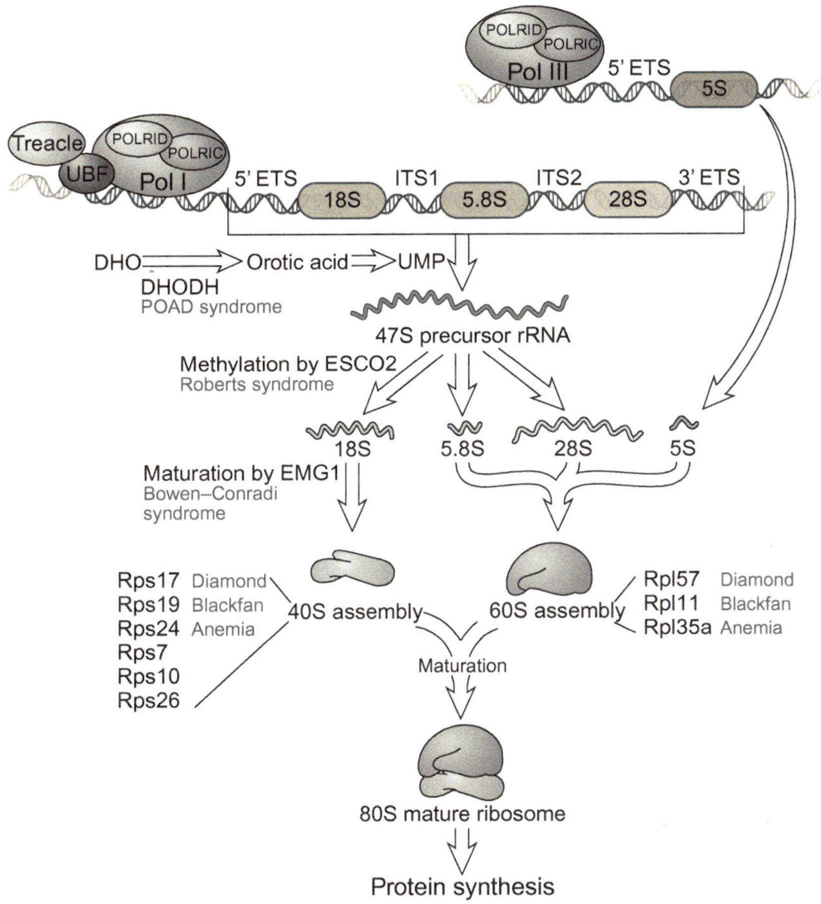

Figure 2 The ribosome biogenesis pathway and ribosomopathies. Schematic figure summarizing the major stages of ribosome biogenesis. The genes and the corresponding ribosomopathies are depicted. Note that disruption of ribosome biogenesis at different stages leads to different syndromes. (A color version of this figure is located at the end of this chapter.)

2.1 Treacher Collins Syndrome

TCS (OMIM 154500) is a congenital birth defect also known as mandibulofacial dysostosis or Franceschetti–Zwahlen–Klein syndrome. It occurs with an incidence of about 1 in 50,000 human live births and was described by George Andreas Berry in 1889, followed in more detail by the ophthalmologist Treacher Collins in 1900 (Kadakia, Helman, Badhey, Saman, & Ducic, 2014; Sakai & Trainor, 2009). Later in 1940, ophthalmologists Franceschetti and Klein further delineated and termed the condition mandibulofacial dysostosis (Franceschetti & Klein, 1949).

2.1.1 Clinical Features

TCS originates during the fifth to eighth week of gestation. It affects the proper formation of the facial prominences and first and second PAs and consequently their derivatives in the head and neck (Fig. 3B; Kadakia et al., 2014). The malformations and the degree of severity are highly variable in TCS patients; however, a characteristic feature is general hypoplasia of the facial bones: mainly the maxilla, the mandible, and the zygomatic complex. Other major clinical features include downward slanting of the palpebral fissures with colobomas of the lower eyelids and absence of the medial third of the lower eyelashes due to clefting of the inferolateral orbital, giving TCS patients characteristic downslanting eyes (Trainor & Andrews, 2013). Atresia of the external auditory canals and deformities of the middle ear including anomalies in the auditory ossicles, together with frequent fusion of the rudimentary malleus and incus, partial absence of the stapes and oval window, or even complete absence of the middle ear and epitympanic space can lead to conductive hearing loss (Trainor, Dixon, &

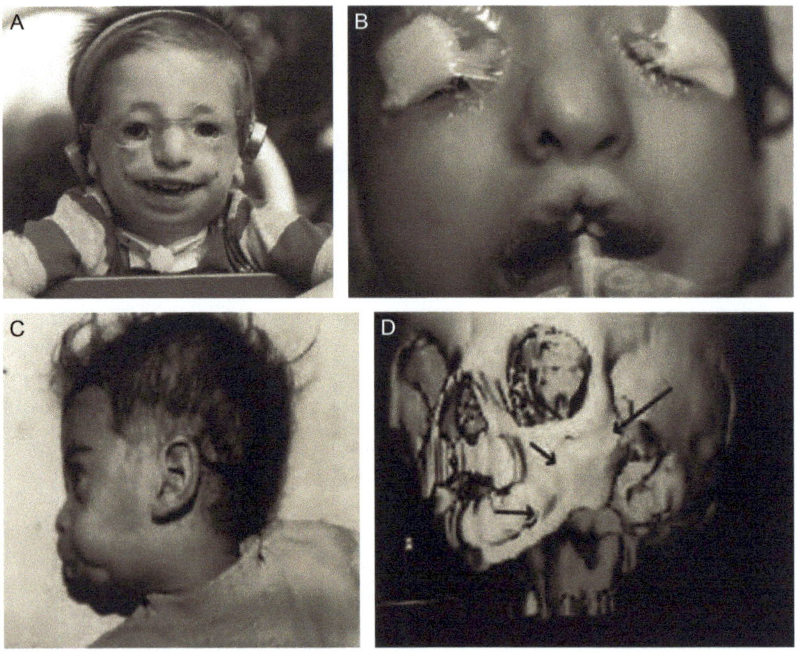

Figure 3 Examples of rare craniofacial disorders. Images of patients with (A) Treacher Collins syndrome, (B) oral-facial-digital syndrome type 1, and (C) syngnathia. (D) A 3D CT scan image showing a bilateral fusion of the mandible to the maxilla, an underlying feature of syngnathia. *(B) Adapted from Gonzalez, Castro, Nieto, and Bouzan (2014).* (A color version of this figure is located at the end of this chapter.)

Dixon, 2009). Moreover many patients exhibit cleft or high-arched palate, dental malocclusion with anterior open bite due to retrognathia, and airway dysfunction such as tracheostoma or choanal stenosis/atresia. Together, these anomalies lead to speech and language difficulties, visual impairment, breathing difficulty, and obstructive sleep apnea (Kadakia et al., 2014).

2.1.2 Genetics

Mutations in three genes have been identified in association with TCS: *TCOF1*, *POLR1C*, and *POLR1D*. About 60% of all mutations are thought to arise *de novo*. The remainder is inherited predominantly in an autosomal dominant fashion, but recessive inheritance has also been recorded. Mutations in *TCOF1* are autosomal dominant and account for about 80% of patients with TCS. Mutations in *POLR1C* are autosomal recessive and mutations in *POLR1D* are predominantly autosomal dominant but can also be recessive (Dauwerse et al., 2011; Schaefer et al., 2014; Trainor et al., 2009; Treacher Collins Syndrome Collaborative Group, 1996). Mutations in *POLR1C* and *POLR1D* account for about 2% of the patients sequenced to date, which suggests that mutations in additional genes may also be causative for TCS.

Interestingly, *TCOF1*, *POLR1C*, and *POLR1D* are part of the ribosomal DNA (rDNA) transcription machinery (Fig. 2). *TCOF1* encodes the nucleolar phosphoprotein Treacle that physically interacts with upstream binding factor (UBF), a transcription factor required for the initiation of transcription by Pol I (Russell & Zomerdijk, 2006; Valdez, Henning, So, Dixon, & Dixon, 2004). Not much is known about *POLR1C* and *POLR1D* other than that they are subunits of Pol I and Pol III, which are responsible for transcription for all the rRNAs and that they interact strongly with each other and are essential for Pol I and Pol III assembly (Larkin & Guilfoyle, 1997; Mann, Buhler, Treich, & Sentenac, 1987; Yao, Yamamoto, Nishi, Nogi, & Muramatsu, 1996). Mutations in *TCOF1*, *POLR1C*, and *POLR1D* together with their roles in ribosome biogenesis are therefore consistent with the description of TCS as a ribosomopathy.

2.1.3 Mouse Models and Mechanism

The *Tcof1*$^{+/-}$ mouse is the first and only TCS mouse model to have been described, and it has revealed much about the cellular function of *Tcof1* and the pathogenesis of TCS (Dixon et al., 2006; Jones et al., 2008). *Tcof1* haploinsufficient mice mimic the craniofacial deformities observed in TCS patients, including general hypoplasia of the craniofacial skeleton, with misshapen and hypoplastic maxilla and palatine bones, occasional cleft palate, and choanal atresia (Fig. 4B; Dixon et al., 2006). At the cellular level, Treacle, the protein encoded by *Tcof1*, is required for proper ribosome biogenesis. Haploinsufficiency of *Tcof1* causes a reduction in ribosome production, which results in nucleolar stress. Nucleolar stress then leads to stabilization of p53 in the neuroepithelium, the birthplace of NCC. Consequently, elevated cell death is observed in these NCC precursors, resulting in fewer migrating and therefore fewer differentiating NCC (Dixon et al., 2006). A deficiency in the number of NCC is the primary cause of craniofacial deformities observed in the *Tcof1*$^{+/-}$ mouse model of TCS and by extrapolation in human patients.

Spatiotemporal characterization of *Tcof1* expression in the developing embryo revealed broad yet dynamic activity in the neuroepithelium and in migrating and differentiating NCC. This pattern coincides with the onset of NCC specification and persists throughout their migration and differentiation consistent with the phenotypes observed in *Tcof1* haploinsufficient embryos (Dixon et al., 2006). Furthermore, analysis of Treacle in a cell culture system revealed colocalization with UBF, suggesting its possible function within the rDNA transcription machinery (Valdez et al., 2004). Indeed, upon downregulation of *Tcof1 in vitro*, rDNA transcription is greatly reduced, consistent with the *in vivo* reduction in ribosome biogenesis observed in *Tcof1*$^{+/-}$ embryos (Valdez et al., 2004). Treacle has also been

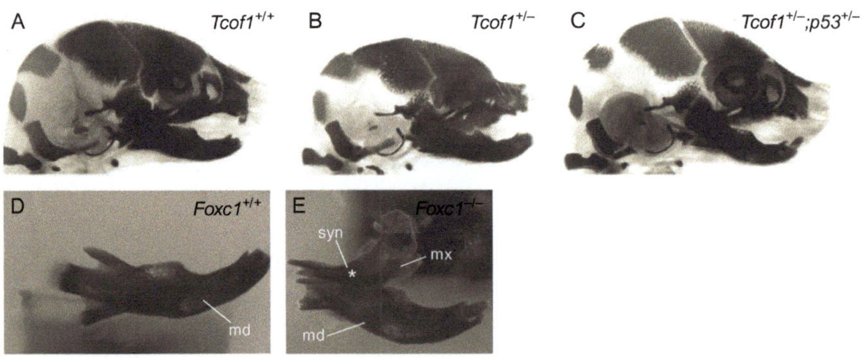

Figure 4 Mouse models of rare craniofacial disorders. Alizarin red (bone) and alcian blue (cartilage) skeletal staining. (A) Wild-type E18.5 mouse embryo; (B) *Tcof1*$^{+/-}$ littermate that resembles Treacher Collins syndrome (TCS). (C) *Tcof1*$^{+/-}$;*p53*$^{+/-}$ embryo with normal craniofacial morphology. (D) The mandible of a wild-type E18.5 mouse embryo. (E) Mandible of an E18.5 *Foxc1*$^{-/-}$ embryo fused to the maxilla, which is consistent with syngnathia. *Panel (C): Adapted from Jones et al. (2008).* (A color version of this figure is located at the end of this chapter.)

associated with the early stages of pre-RNA processing. Specifically, Treacle was shown to interact with Nop56-associated pre-ribosomal ribonucleoprotein (pre-rRNPs) complexes, which methylate 2′-*O*-ribose moieties on pre-rRNA (Hayano et al., 2003). Taken together, these data support a role for Treacle in ribosome biogenesis and defines TCS as a ribosomopathy.

Nucleolar stress caused by insufficient ribosome biogenesis leads to activation and stabilization of p53 and p53-dependent cell death (Rubbi & Milner, 2003). *Tcof1*$^{+/-}$ mouse embryos exhibit over twofold higher levels of p53 than their control littermates. Inhibiting p53 by genetic or pharmacological means prevents the craniofacial defects in *Tcof1*$^{+/-}$ embryos. p53 inhibition suppresses cell death in the neuroepithelium, facilitating the proper formation and survival of NCC (Fig. 4C; Jones et al., 2008). These results support a role for p53 downstream of perturbed ribosome biogenesis and implicate p53 in the potential prevention of TCS. Although inactivation of a tumor suppressor is not a viable preventative therapy, these results showcase the power of genetic mouse models in uncovering the cellular mechanisms of congenital disease and their utility in designing potential preventative therapy.

Interestingly, the rescue of *Tcof1*$^{+/-}$ embryos via *p53* loss of function correlated with suppression of neuroepithelial cell death and the restoration of normal numbers of NCC, but failed to reverse the ribosome biogenesis defects (Jones et al., 2008). This finding implies that *Tcof1* may play roles in neural progenitor cells and NCC processes other than ribosome biogenesis. *In vivo* data have shown that during metaphase, Treacle localizes to the centrosomes of neuroepithelial cells and is important for proper spindle orientation and chromosome segregation (Sakai, Dixon, Dixon, & Trainor, 2012). Additionally, Treacle has been implicated in the regulation of oxidative stress and the DNA damage response (Ciccia et al., 2014; Duan, Kelsen, Clarkson, Ji, & Merali, 2010; Larsen et al., 2014). Treacle is required for the translocation of NBS1, a central regulator of DNA damage response, into the nucleoli upon DNA damage. This in turn triggers pan-nuclear silencing of rDNA transcription by Pol I in order to maintain genomic integrity after DNA damage (Ciccia et al., 2014; Larsen et al., 2014). Taken together, these data suggest a protective role of *Tcof1* in oxidative stress and the DNA damage response. It has been speculated that in TCS patients, Treacle fails to globally shut down rDNA transcription upon DNA damage, which leads to genome instability and cell death (Ciccia et al., 2014; Larsen et al., 2014). It would be interesting to study the DNA damage response and the oxidative stress pathways in *Tcof1*$^{+/-}$ mice in order to gain a better understanding of the function of Treacle and the underlying pathogenesis of TCS.

Furthermore, it will be important to study *POLR1C* and *POLR1D* mouse models to determine whether their functions overlap with those of *TCOF1* in the pathogenesis of TCS. However, although the majority of traditional knockout mouse models have been helpful in understanding the cellular function of genes, they do not necessarily mimic the exact pathophysiology of human disease. Human diseases are often not caused by null alleles, but by hypomorphic, dominant negative or dominant positive alleles. Therefore, the function of these gene mutations in humans can be quite different physiologically from total loss of function or null mutations in mouse models. Thus, it will be invaluable to mimic specific mutations that TCS patients carry in *TCOF1*, *POLR1C*, and *POLR1D* by using the TALEN or CRISPR system for genome editing (Doudna & Charpentier, 2014; Sommer, Peters, Baumgart, & Beyer, 2015).

2.2 Diamond–Blackfan Anemia

Diamond–Blackfan anemia (DBA; OMIM 105650) is a congenital disorder that occurs in about 1 per 200,000 live births and is considered one of the founding ribosomopathies (Narla & Ebert, 2010). It was originally characterized as an anemia in infancy and childhood before being later defined as a congenital hypoplastic anemia (Lipton & Ellis, 2009).

2.2.1 Clinical Features

DBA is usually diagnosed during the first year of life primarily due to symptoms such as pallor and lethargy. Patients exhibit anemia (deficiency of red blood cells), macrocytosis (enlarged erythrocytes), reticulocytopenia, and a selective decrease or absence of erythroid precursors in an otherwise cellularly normal bone marrow (Narla & Ebert, 2010). Additionally, approximately half of the patients also exhibit craniofacial abnormalities, including cleft lip and/or palate as well as microtia, or other physical abnormalities like thumb deformities and/or cardiac defects (Gazda et al., 2008; Lipton & Ellis, 2009; Ross & Zarbalis, 2014; Trainor & Merrill, 2014).

2.2.2 Genetics

About half of the cases of DBA exhibit autosomal dominant inheritance, whereas the remainder arise *de novo* (Gazda et al., 2008). No matter the nature of the mutation, all DBA patients are heterozygous, suggesting that the homozygous condition is embryonic lethal (Nakhoul et al., 2014).

Mutations in *RPS19*, which encodes a RP, were the first to be identified in association with DBA and they occur in 25% of known DBA cases (Gazda et al., 2004; Nakhoul et al., 2014). Another 10–15 small and large ribosomal subunit RPs have also been implicated in DBA and, together, mutations in RPs account for about 50–70% of DBA patients (Gerrard et al., 2013; Nakhoul et al., 2014). Interestingly, mutations in different RPs result in distinct clinical manifestations with specific phenotypes. For example, mutations in *RPL5* are more likely to be associated with cleft lip and/or palate, whereas mutations in *RPL11* correspond to a higher incidence of thumb abnormalities. In contrast, mutations in *RPS19* have a low occurrence of either (Gazda et al., 2008). Nevertheless, the phenotypes characteristic of DBA are variable even within families, similar to TCS, and therefore environmental factors may also play a role in the manifestation of DBA.

Recently, mutations in the hematopoietic transcription factor *GATA1* were identified in two DBA pedigrees, making *GATA1* the first non-ribosomal gene to be implicated in DBA (Sankaran et al., 2012). However, specialized ribosomes have previously been shown to selectively control translation of specific proteins, possibly explaining why mutations in RPs can cause tissue-specific phenotypes (Kondrashov et al., 2011). Mutations in RP genes may lead to a selective decrease in GATA1 translation, making Gata1 a potentially common factor in the pathogenesis of DBA (Ludwig et al., 2014).

2.2.3 Mouse Models and Mechanism

Several potential mouse models of DBA have been generated, but to date none fully recapitulate the human condition. Several RP mutant mice have been generated that carry deletions, missense, or splicing mutations derived through targeted gene deletion, forward genetic mutagenesis screens, or have arisen spontaneously (Matsson et al., 2004; Terzian & Box, 2013). Homozygosity has typically led to lethality prior to implantation, whereas heterozygosity has had no effect (McGowan & Mason, 2011). However, *dark skin* (*Dsk*) mice with dominant negative and missense mutations in *Rps19* exhibit growth retardation and mild anemia (Devlin, Dacosta, Mohandas, Elliott, & Bodine, 2010; McGowan et al., 2008). *Dsk* mice also display a white belly spot, which is suggestive of defects in NCC-derived melanocyte development.

Despite the absence of a complete DBA mouse model, much has been learned from mice and zebrafish carrying mutations in RPs, as well as from primary cells from human patients. A common theme from these studies is that defects in ribosome biogenesis can lead to a decrease in the 40S, 60S, or 80S ribosomal subunits depending on the RP mutated (Narla & Ebert, 2010). Consequently, this decrease results in a global or a selective suppression of protein synthesis, which can lead to nucleolar stress. During nucleolar stress, certain RPs (including RPL11, RPL5, and RPL23) bind to and sequester MDM2, which normally ubiquitinates p53, targeting it for degradation. Without MDM2 function, p53 accumulates and activates the cell death program (Rubbi & Milner, 2003). Similar to the prevention of craniofacial anomalies in *Tcof1*$^{+/-}$; *p53*$^{+/-}$ mice, inhibition of p53 prevents anemia and developmental anomalies in *rpl11* zebrafish morphants, *Rps19* mutant zebrafish, and *Rps19* mutant mice, suggesting a similar molecular mechanism underlying the pathogenesis of distinct ribosomopathies that is p53 dependent (Chakraborty, Uechi, Higa, Torihara, & Kenmochi, 2009; Danilova, Sakamoto, & Lin, 2008; Jaako et al., 2011; McGowan et al., 2008).

Another pathway that has been explored for potential treatment and/or prevention of DBA is the mTOR pathway, an important regulator of ribosome biogenesis and consequently cell growth and proliferation. Activation of the mTOR pathway by amino acids such as L-leucine leads to phosphorylation of the RNA Pol I initiation complex, which in turns results in the stimulation of rDNA transcription (Hannan et al., 2003; Mayer, Zhao, Yuan, & Grummt, 2004). Dietary supplementation with L-leucine rescues the anemia in an *Rps19* mutant mouse model and the craniofacial abnormalities in *rps14* and *rps19* morphant zebrafish (Jaako et al., 2012; Payne et al., 2012). Remarkably, dietary supplementation with L-leucine successfully treated the anemia in some DBA patients (Cmejlova et al., 2006; Pospisilova, Cmejlova, Hak, Adam, & Cmejla, 2007). However, mutations in RPs can cause a reduction in the assembly of one of the ribosomal subunits that can then cause a global reduction in protein translation or a more selective reduction in specific proteins. Therefore, although L-leucine supplementation is exciting as a novel treatment approach, elevating the levels of rDNA transcription by activating the mTOR pathway will not necessarily "fix" proper assembly of the ribosome.

2.3 Bowen–Conradi Syndrome

Bowen–Conradi syndrome (BCS; OMIM 211180) is an autosomal recessive disorder that occurs almost exclusively within the Hutterite population of the Canadian Prairies at a frequency of 1 in 355 live births and causes death within the first year of life (Lowry et al., 2003). It has been predicted that 1 in 10 is a carrier within the Hutterite population.

2.3.1 Clinical Features

BCS is characterized by pre- and postnatal growth retardation, with a failure to thrive, and is currently the only ribosomopathy characterized by severe neurological impairment from birth (Armistead et al., 2015; Lowry et al., 2003). BCS patients also exhibit micrognathia, prominent nose with a lack of glabellar angle, joint anomalies such as camptodactyly, as well as microcephaly and severe psychomotor delay (Lowry et al., 2003). These features overlap with cerebro-oculo-facial-skeletal syndrome and trisomy 18 making diagnosis challenging.

2.3.2 Genetics

In 2009, the identification of c.400A/G, p.D86G mutation in *EMG1* was reported as causative of BCS in the Hutterite population (Armistead et al., 2009). EMG1 (essential for mitotic growth 1), which is also known as Nep1 in yeast, is part of the small subunit processome, a ribonucleoprotein complex essential for maturation of the 18S rRNA and assembly of the 40S ribosome (Sondalle & Baserga, 2014). EMG1 is highly conserved throughout archaea and eukaryotes, and it is believed to act as a pseudouridine methyltransferase during processing of the 18S rRNA (Wurm et al., 2010).

2.3.3 Mouse Models and Mechanism

Emg1$^{-/-}$ mouse embryos fail to develop into blastocysts and die prior to implantation (Wu, Sandhu, Patel, Triggs-Raine, & Ding, 2010). This severity is not surprising because the mouse mutation results in null alleles of *Emg1*. In contrast, the human D86G substitution does not cause a complete deficiency, but rather primarily affects the function of the protein (Armistead et al., 2009; Wu et al., 2010). EMG1 functions as a dimer or a multimer, and the D86G substitution increases the affinity of EMG1 monomers resulting in a 10-fold increase in dimers (Armistead et al., 2009). Interestingly, in yeast, the equivalent D86G substitution does not affect its role as a methyltransferase (Meyer et al., 2011). Therefore, it has been suggested that the elevated dimerization of mutant EMG1 prevents its localization to the nucleolus and thus the D86G mutation results in BCS due to a defect in the assembly of the 40S ribosome and not in rRNA modification (Meyer et al., 2011).

Recently, a new mouse model was generated for BCS by knocking-in the D86G substitution in an effort to better mimic the human condition (Armistead et al., 2015). Homozygous mice (*Emg*$^{G/G}$) are smaller than their siblings, exhibit neural tube closure defects, and die between E8.5 and E12.5. *Emg*$^{G/G}$ embryos exhibit decreased cell proliferation, and this phenotype was also detected in brain tissue and bone marrow autopsies of BCS patients. Furthermore, derivation of *Emg*$^{G/G}$ mouse fibroblasts showed a high number of binucleated cells pointing to a defect in mitosis that could be the cause of growth arrest in BCS (Armistead et al., 2015). Interestingly, mitotic arrest and ribosome biogenesis have also previously been linked in other ribosomopathies such as TCS and Shwachman–Diamond syndrome. Thus, further study of this relationship in the context of these diseases will be important for a better understanding of the pathogenesis of ribosomopathies (Austin et al., 2008; Sakai et al., 2012).

2.4 Postaxial Acrofacial Dysostosis

Postaxial acrofacial dysostosis (POADS; OMIM 263750), also known as Miller, Genée–Wiedemann, and Wildervanck–Smith syndrome, is an autosomal recessive disorder with an occurrence of 1 in 1,000,000 live births (Wieczorek, 2013).

2.4.1 Clinical Features

The craniofacial deformities observed in POADS are almost identical to the characteristic features of TCS and include micrognathia, orofacial clefts, malar hypoplasia, coloboma of the lower eyelid, and cup-shaped ears. Unlike TCS, however, POADS patients also exhibit postaxial limb defects that include absence of the fourth or fifth digits, or both rays of the hands and feet with or without ulnar and fibular hypoplasia (Trainor & Andrews, 2013).

2.4.2 Genetics

POADS, the first genetic disorder for which the causative gene was identified via whole exome sequencing, correlates with mutations in *dihydroorotate dehydrogenase* (*DHODH*), a gene associated with the mitochondrial electron transport chain and *de novo* pyrimidine biosynthesis pathway (Ng et al., 2010). Fourteen different mutations in the coding region of *DHODH* have been identified to date and, surprisingly, all affected individuals carry compound heterozygous mutations in *DHODH*, which is atypical of an autosomal dominant disorder (Rainger et al., 2012).

2.4.3 Mouse Models and Mechanism

DHODH is one of three enzymes that catalyze the six enzymatic reactions needed for *de novo* synthesis of pyrimidine. DHODH catalyzes the conversion of dihydroorotate to orotic acid, which is then converted to uridine monophosphate, the RNA nucleotide essential for ribosome biogenesis (Brosnan & Brosnan, 2007; Minet, Dufour, & Lacroute, 1992). There are currently no mouse models for POADS, but inhibition of DHODH by the immunosuppressant leflunomide in zebrafish completely abolished NCC, resulting in a reduction of their derivatives as well as a disruption in the jaw cartilage (White et al., 2011). Moreover, microarray analysis of leflunomide-treated zebrafish embryos showed downregulation of NCC genes. Leflunomide also caused a decrease in the self-renewal capacity of rat NC stem cells. These data suggest defects in NCC development are a potential cause of the pathogenesis of POADS.

Interestingly, suppression of *DHODH* leads to p53-dependent cell death and, independent of this role, *DHODH* has also been linked to oxidative stress, illustrating similarities in the pathogenesis of TCS and POADS (Khutornenko, Dalina, Chernyak, Chumakov, & Evstafieva, 2014; White et al., 2011). Generating mouse models for POADS that mimic the mutations identified in humans are necessary to provide a better understanding of this disease and the cellular and tissue abnormalities associated with this condition.

2.5 Roberts Syndrome

Roberts syndrome (RBS; OMIM 268300) is another rare genetic disorder that is inherited in an autosomal recessive manner and currently only about 150 affected individuals have been reported.

2.5.1 Clinical Features

RBS is characterized by slow pre- and postnatal growth, mental retardation, craniofacial defects including cleft lip and palate, hypertelorism, hypoplastic nasal alae, micrognathia, and ear abnormalities as well as limb defects that include shortened arm and leg bones (Trainor & Merrill, 2014).

2.5.2 Genetics

RBS is caused by mutations in *ESCO2* (establishment of cohesion 1 homolog 2). The yeast homolog, *Eco1*, is an acetyltransferase crucial in the assembly of cohesin (Vega et al., 2005). Cohesin is a multiprotein complex that holds sister chromatids together from S phase until the start of mitosis, helping to ensure genomic integrity (Haarhuis, Elbatsh, & Rowland, 2014). The acetyltransferase activity of Eco1 is required for the establishment of cohesin–DNA interactions during S phase (Skibbens, Corson, Koshland, & Hieter, 1999; Toth et al., 1999). Mutations in *Eco1* result in sensitivity to DNA damage agents that cause double-strand breaks, which is an emerging theme in ribosomopathies (Bose et al., 2012; Lu et al., 2010).

2.5.3 Mouse Models and Mechanism

Esco2 morphant zebrafish exhibit smaller heads and eyes, abnormal pigmentation, reduced and deformed craniofacial cartilage, as well as defects in pectoral fin development and cardiac problems (Monnich, Kuriger, Print, & Horsfield, 2011), all of which resemble RBS. In contrast, *Esco2* homozygous mouse embryos die by the eight-cell stage due to problems in chromatid cohesion. However, conditional knockout of *Esco2* specifically in the neuroepithelium of the dorsal telencephalon (*Emx1-CRE;Esco2flx/flx*) resulted in viable mice with microcephaly, a deformity found in some RBS patients (Whelan et al., 2012). At midgestation, *Emx1-CRE;Esco2flx/flx* embryos are characterized by agenesis of the neocortical and hippocampal neuroepithelium due to cell death. Given the craniofacial deformities that patients exhibit, it would be interesting to conditionally knock out *Esco2* from NCC in mice and examine whether it recapitulates the human condition. Moreover, given the early embryonic requirements for *Esco2*, it will be important to control deletion of *Esco2* temporally to determine if *Esco2* has other tissue-specific roles later in development. Nonetheless, analysis of RBS mutations in yeast and human patient cell lines revealed defects in ribosome biogenesis, rRNA production, and protein translation as well as a disruption in nucleolar morphology (Bose et al., 2012). It is clear that RBS can be categorized as a ribosomopathy; however, additional work using the *Esco2* conditional mouse model is needed for a more detailed understanding of RBS.

The identification and characterization of ribosomopathies in humans and their subsequent study in mouse and zebrafish models has revealed new and surprising roles for the molecular machines called ribosomes. The discovery that mutations in genes involved in ribosome biogenesis are not necessarily embryonic lethal, but instead cause tissue-specific craniofacial defects, has changed our view of the roles and functions of global cellular processes during development and disease. Mouse models have helped reveal the etiology and pathogenesis of ribosomopathies and have also been instrumental in furthering our understanding of other rare craniofacial disorders that arise within ciliopathies or disorders of cilia development and function.

3. CILIOPATHIES

Primary cilia are finger-like projections that extend from the surface of nearly every vertebrate cell in order to receive and process signaling cues. Each cilium is composed of an axoneme, a microtubule-based structure; the basal body (centrosomes) that connects the axoneme to the cell body; and the specialized ciliary membrane that covers the axoneme and hosts numerous signaling pathway receptors (Fig. 5; Chang, Schock, Attia, Stottmann, & Brugmann, 2015; Sorokin, 1962). Another critical component is the transition zone, which is found at the base of the axoneme and the distal basal body. The transition zone accommodates Y-shaped protein structures that connect the axonemal microtubules to the ciliary membrane and also serves as the ciliary gate required for establishing the cilium as its own compartment separated from the rest of the cell body (Dawe, Farr, & Gull, 2007). The primary cilium is absolutely required for the transduction of important signaling pathways including the Hedgehog (Hh), platelet-derived growth factor (PDGF) alpha, polycystin, and Wnt pathways (Pedersen & Rosenbaum, 2008).

Ciliogenesis, the formation of the cilium, is an elaborate process that starts with differentiation of the mother centriole into a basal body followed by its docking at the apical membrane. The basal body then provides a scaffold for axonemal microtubules to polymerize giving rise to the axoneme and thus extending the cilium into the extracellular space. This extension requires intraflagellar transport (IFT) machinery for the transport of tubulin and other building blocks of the axoneme. IFT is also critical for the function of the cilium because this microtubule motor-based process shuttles and localizes proteins including signal transduction components to the mature cilia. This transport is bidirectional along the axoneme with IFT proteins that travel to the distal end of the microtubules being referred to as anterograde, whereas proximal transport toward the cell body is referred to as retrograde (Reiter, Blacque, & Leroux, 2012).

Mutations in genes encoding proteins involved in the formation, structure, maintenance, or function of primary cilia can lead to ciliopathies (Fig. 5; Badano, Mitsuma, Beales, & Katsanis, 2006). Many ciliopathies exhibit craniofacial malformations among other deformities, suggesting a

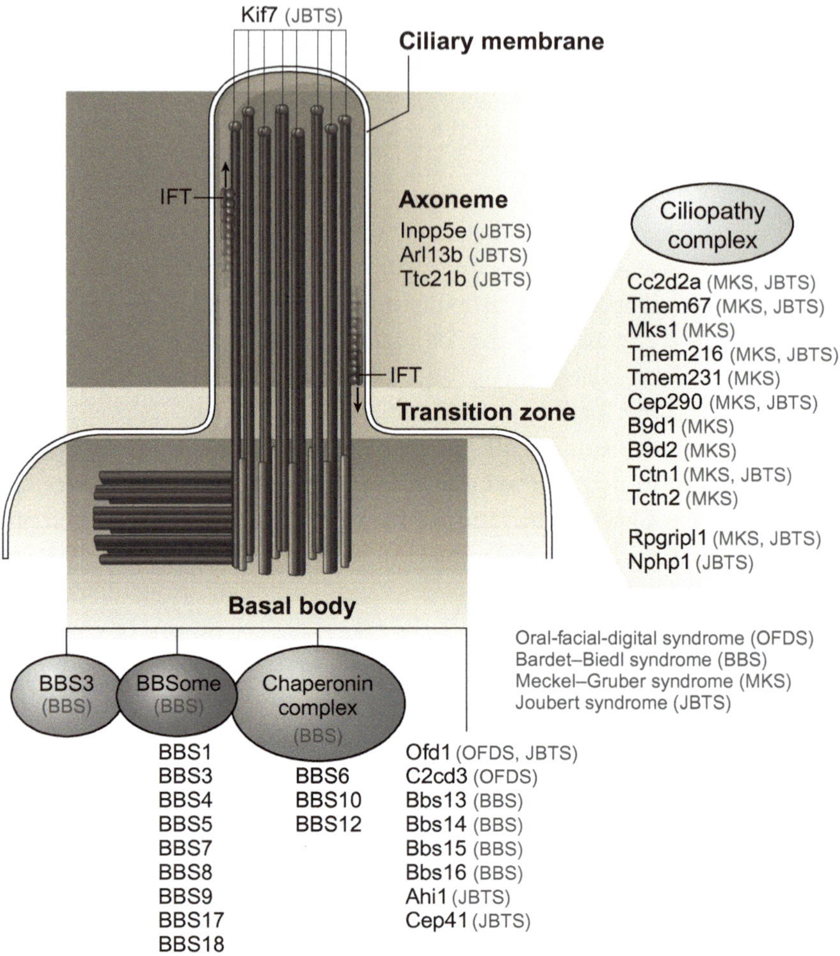

Figure 5 The primary cilium and ciliopathies. Schematic representation of a primary cilium detailing the main compartments of the cilium as well as the products of the genes and their localization and involvement in specific ciliopathies. (A color version of this figure is located at the end of this chapter.)

role for cilia in NCC development. The following section focuses on the role of cilia in craniofacial structures and rare craniofacial disorders based on mouse models of ciliopathies.

3.1 Oral-Facial-Digital Syndrome 1

Oral-facial-digital syndrome 1 (OFDS1; OMIM 311200) is an X-linked dominant disorder that affects 1 in 50,000–250,000 newborns and is the most common subtype among 13 different oral-facial-digital conditions.

All OFDS1 patients are female, as male carriers die during the first or second trimester of pregnancy. Although primarily an inherited disorder, 30% of the cases are sporadic (Gurrieri, Franco, Toriello, & Neri, 2007; Wettke-Schafer & Kantner, 1983).

3.1.1 Clinical Features

OFDS1 is characterized by craniofacial anomalies, digit deformities, central nervous system abnormalities, and polycystic kidney disease (Thauvin-Robinet et al., 2006). Craniofacial malformations occur in over 87% of reported cases and include facial asymmetry, hypertelorism, micrognathia, broadened nasal ridges, cleft palate, high-arched palate, lingual hamartomas, hypodontia, and hyperplastic buccal fernula that can lead to clefting (Brugmann, Cordero, & Helms, 2010; Gurrieri et al., 2007; Thauvin-Robinet et al., 2006).

3.1.2 Genetics

OFD1 is the only gene to be associated with OFDS1 thus far, although it still does not account for most of the OFDS1 and other OFD subtypes (Ferrante et al., 2001; Rakkolainen, Ala-Mello, Kristo, Orpana, & Jarvela, 2002; Romio et al., 2003). *OFD1* is widely expressed from early stages of development in all tissues affected in OFDS1 in both humans and mice (de Conciliis et al., 1998; Ferrante et al., 2003; Romio et al., 2004, 2003). *OFD1* encodes a centrosomal protein that localizes at the basal body of primary cilia. Numerous mutations and deletions have been identified in *OFD1* with most of them resulting in truncated protein products that are nonfunctional (Ferrante et al., 2001; Thauvin-Robinet et al., 2006).

Similar to the ribosomopathy disorders described earlier, there is wide phenotypic variability even within the same family. In this case, the variability has been attributed to somatic mosaicism due to random X inactivation, but environmental factors may also play a role (Franco & Ballabio, 2006).

3.1.3 Mouse Models and Mechanism

Ofd1 knockout mice recapitulate the main features of OFDS including craniofacial and limb abnormalities (Ferrante et al., 2006). The phenotypes of *Ofd1* mice are more severe than the human condition, probably due to differences in the pattern of X inactivation between humans and mice. Cilia in *Ofd1* mutant male embryos are absent from the embryonic node, consistent with a failure of left–right axis specification. Additionally, *Ofd1* mutant embryos exhibit abnormal patterning of the neural tube associated with an absence of Shh and considerable reduction in expression of Shh downstream target genes including Ptc1 and Gli1 (Ferrante et al., 2006). The neural tube phenotype resembles that of some IFT mutants, suggesting a genetic interaction between *Ofd1* and IFT genes (Ferrante et al., 2006). IFT is required for cilia formation and maintenance, and IFT proteins are also involved in the Shh pathway via regulation of Gli protein function (Liu, Wang, & Niswander, 2005). Interestingly, Ofd1 has been hypothesized to functionally interact with the Gli proteins, and, consistent with this idea, loss of *Ofd1* disrupts the ciliary processing of Gli3 resulting in phenotypes that resemble *Gli3* mutants (Brugmann, Cordero, et al., 2010; Toriello, 2009).

Recently, a link between FGF-related syndromes and ciliopathies was observed in the high-arched palate phenotype of *Fuzzy* (*Fuz*) mutant mice (Tabler et al., 2013). *Fuz*, a planar cell polarity effector essential for targeted membrane trafficking, regulates ciliogenesis, although it is not structurally or physically part of the cilium (Gray et al., 2009; Park, Haigo, & Wallingford, 2006). It has been suggested that the high-arched phenotype in *Fuz* mice is due to expanded *Fgf8* expression, which results in abnormal NCC migration into the facial prominences and an enlarged maxillary process. Strikingly, *Ofd1* mutant mice also exhibit *Fgf8* expansion, enlarged maxillary processes, and a high arch palate (Tabler et al., 2013).

Recently, a mutation in *C2CD3* was identified in a 4-year-old male that presented with canonical OFDS1 in addition to microcephaly, micropenis, and severe intellectual disability, potentially representing a new OFD subtype (Thauvin-Robinet et al., 2014). C2CD3 colocalizes with OFD1 at the distal end of centrioles, which initiate the assembly of cilia. *Hearty* mutant mice that carry a mutation in *C2cd3* exhibit neural tube defects, abnormal left–right asymmetry axis determination, and polydactyly (Hoover et al., 2008). C2cd3, similar to Ofd1, is required for ciliogenesis and is also an essential regulator of the intracellular transduction of Hedgehog signaling. These observations add to accumulating data that OFDS should be classified as a ciliopathy.

3.2 Bardet–Biedl Syndrome

Bardet–Biedl syndrome (BBS; OMIM 209900) is an autosomal recessive condition with a general occurrence of 1 in 140,000 newborns, but up to 1 in 17,000 newborns in some isolated regions (Klysik, 2008).

3.2.1 Clinical Features

BBS is pleiotropic and is characterized by retinal degeneration, obesity, polydactyly, learning difficulties, and renal and genital anomalies (Fig. 3A; M'Hamdi, Ouertani, & Chaabouni-Bouhamed, 2014). A subgroup of BBS patients also exhibit craniofacial deformities including hypertelorism, deep set eyes, downward slanting palpebral fissures, high-arched palate, migrognathia, a flat nasal bridge with anteverted nares and prominent nasolabial folds, a long philtrum, and a thin upper lip. Interestingly, a small percentage also displays Hirschsprung disease, which is a neurocristopathy affecting the enteric nervous system (Beales, Elcioglu, Woolf, Parker, & Flinter, 1999).

3.2.2 Genetics

To date, mutations in 18 genes have been identified that account for 70–80% of the BBS cases, and they have been classified as *BBS1-18*. All of the BBS proteins localize to the centrosome/basal bodies or ciliary axoneme, and seven of them form a complex termed the BBSome (BSS1, 2, 4, 5, 7, 8, and 9). Interestingly, mutations in specific *BBS* genes are more prevalent than others in different geographic regions (Brugmann, Cordero, et al., 2010; M'Hamdi et al., 2014; Nachury et al., 2007).

3.2.3 Mouse Models and Mechanism

BBS proteins can associate with cofactors to promote trafficking of vesicles to the cilium, but their function still remains undetermined. One hypothesis is that BBS proteins indirectly affect cilia by bringing additional proteins to the centrosome, the failure of which leads to centrosome/basal body abnormalities and ciliary dysfunction. Alternatively, BBS proteins may have a role within the cilium, to promote cohesion between IFT subcomplexes (Brugmann, Cordero, et al., 2010; Nachury et al., 2007).

Mouse models have begun to shed some light on the function of BBS proteins. *Bbs2*, *Bbs4*, and *Bbs6* knockout mice exhibit some of the human phenotypes including retinal regeneration, but their cilia and basal body structures are not defective, suggesting a role in IFT, and not in the assembly of cilia (Fath et al., 2005; Mykytyn et al., 2004; Nishimura et al., 2004). Further studies have shown that the BBSome functions to promote cilia membrane biogenesis and serves as a coat complex for ciliary membrane protein trafficking (Jin et al., 2010; Nachury et al., 2007).

Surprisingly, craniofacial development has not been extensively studied in these BBS mouse models, perhaps because the phenotypes are subtle. Three-dimensional surface modeling of *Bbs4*$^{-/-}$ and *Bbs6*$^{-/-}$ adult mice, however, revealed a larger mid-face width to height ratio and a shorter snout, which was attributed to premaxillary and maxillary hypoplasia (Tobin et al., 2008). Furthermore, *bbs* morphant zebrafish exhibit craniofacial malformations including shortening of the anterior neurocranium, migrognathia, and cyclopia, phenotypes that were attributed to an inability of the NCC to migrate into the facial prominences. These morphant zebrafish also exhibited Hirschsprung disease, which results from the inability of the NCC to form a complete enteric nervous system. Interestingly, disruption of Wnt and Shh signaling, two signaling pathways regulated by cilia, may underlie these NCC defects (Tobin et al., 2008).

More work clearly needs to be done using the currently available mouse models to understand the relationship between primary cilia and CNC. However, new models also need to be generated that mimic the severe craniofacial defects characteristic of the human mutations. Moreover, given the number of the different BBS genes mutated, generating compound heterozygous or homozygous mice will help uncover any level of redundancy that exists among BBS genes.

3.3 Meckel–Gruber Syndrome

Meckel–Gruber syndrome (MKS; OMIM 249000) is an autosomal recessive disorder that occurs in about 1 in 140,000 live births and results in perinatal lethality. It occurs at a higher frequency in more isolated populations and is especially prevalent in Finns (Szymanska, Hartill, & Johnson, 2014).

3.3.1 *Clinical Features*

MKS is characterized by polycystic kidneys, cardiac defects, polydactyly, and numerous craniofacial abnormalities including occipital encephalocele, microcephaly, sloping forehead, cleft lip/palate, macrostomia, micrognathia, microphthalmia, and tongue defects (Brugmann, Cordero, et al., 2010; Szymanska et al., 2014). Lethality occurs either *in utero* or shortly after birth usually due to pulmonary hypoplasia (Szymanska et al., 2014).

3.3.2 *Genetics*

There are 11 subtypes of MKS and mutations in 11 genes have been identified to date. The first MKS locus to be identified was *MKS1* and subsequently mutations in 10 more genes have been identified: *TMEM216*,

TMEM67, *CEP290*, *RPGRIP1L*, *CC2D2A*, *NPHP3*, *TCTN2*, *B9D1*, *B9D2*, and TMEM231 (Szymanska et al., 2014).

3.3.3 Mouse Models and Mechanism

Several mouse models have been generated to study the different MKS subtypes and much has been learned regarding the role of ciliogenesis in the pathogenesis of MKS.

3.3.4 *Mks1*

Two independent chemical mutagenesis screens recovered mice that carried mutations in *Mks1* (Cui et al., 2011; Weatherbee, Niswander, & Anderson, 2009). MKS1 is a B9-domain containing protein that localizes to the basal body of mammalian cells. MKS1 regulates ciliogenesis by interacting with meckelin and two other B9-domain containing proteins, B9D1 and B9D2, mutations in which have also been identified in MKS patients (Dawe, Smith, et al., 2007; Kyttala et al., 2006; Williams, Winkelbauer, Schafer, Michaud, & Yoder, 2008). *Mks1* mutant mice *kerouac* (*krc*)/$Mks1^{krc}$ and $Mks1^{del64-323}$ both display several characteristic MKS deformities including polydactyly, cystic kidneys, and craniofacial defects such as exencephaly, reduced ossification of craniofacial bones, and cleft palate. Both mutants also display a dramatic reduction in the number of primary cilia in many but not all tissues, which may explain the tissue-specific phenotypes observed in both mouse and human MKS. Moreover, these mouse models revealed that *Mks1* functions in the dorsal–ventral patterning of the neural tube and the anterior–posterior patterning of the limb bud by regulating the Shh pathway (Cui et al., 2011; Weatherbee et al., 2009). A more recently generated mouse model, $Mks1^{tm1a(EUCOMM)Wtsi}$, exhibits similar defects to Mks^{krc} and $Mks^{del64-323}$ mice (Wheway et al., 2013). Interestingly, in addition to perturbation of Shh signaling, the Wnt and mTOR pathways were also upregulated in association with ectopic cell proliferation. Upregulation of the mTOR pathway may explain the formation of cysts in the kidney; however, further studies are needed to determine how MKS1 alters the Wnt1 and mTOR pathways.

3.3.5 *Tmem67*

MKS subtype 3 (MKS3) is caused by mutations in *TMEM67*, which encodes meckelin, an orphan transmembrane receptor found in the renal epithelium and the primary cilium (Smith et al., 2006). Two mouse models for MKS3 have been described: the *bpck* mouse that carries a spontaneous deletion including the *Tmem67* gene and the *Tmem67*$^{tm1(Dgen/H)}$ mouse that carries a targeted allele of *Tmem67* (Cook et al., 2009; Garcia-Gonzalo et al., 2011). Both models exhibit phenotypes that resemble MKS3 including polycystic kidneys and hydrocephalus, but do not show other classic MKS phenotypic characteristics such as polydactyly and encephalocele. Recently however, congenic *Tmem67*$^{tm1(Dgen/H)}$ embryos revealed some of the MKS craniofacial deformities including exencephaly, prosencephalon dysgenesis, and encephalocele (Abdelhamed et al., 2013). All mutants exhibit dysmorphic and fewer or absent primary cilia in a tissue-specific distribution that is consistent with their phenotypes. The congenic *Tmem67*$^{tm1(Dgen/H)}$ mice also revealed deregulation of the Shh and Wnt pathways in the developing brain. Therefore, it was proposed that TMEM67 regulates the trafficking of Gli proteins and Disheveled in the cilium (Abdelhamed et al., 2013). The congenic mutants revealed a degree of phenotypic severity consistent with human MKS and, together with the other *Tmem67* mutant mice, could serve as models to understand the variability in human patients carrying the same genetic mutations (Abdelhamed et al., 2013).

3.3.6 *Rpgrip1l*

RPGRIP1L contains several protein–protein interaction domains and is localized in the basal body centrosome complex as well as the transition zone along with nephrocystins 1 and 4 (Arts et al., 2007; Delous et al., 2007; Sang et al., 2011; Vierkotten, Dildrop, Peters, Wang, & Ruther, 2007). Mutations in *RPGRIP1L* are associated with MKS subtype 5. *Rpgrip1l*$^{-/-}$ (*Ftm*) mice die neonatally and exhibit phenotypes observed in human patients with MKS including exencephaly, polydactyly, laterality defects, microphthalmia, cleft lip, and hypoplastic lower jaw (Delous et al., 2007; Vierkotten et al., 2007). *Ftm*$^{-/-}$ embryos, although able to assemble cilia, have fewer cilia, which are unable to transduce the Hh signaling properly and thus manifest Hh-deficiency related deformities (Vierkotten et al., 2007).

3.3.7 *Cc2d2a*

Cc2d2a also contains protein–protein interaction domains and localizes to the basal body (Garcia-Gonzalo et al., 2011; Gorden et al., 2008; Tallila, Jakkula, Peltonen, Salonen, & Kestila, 2008). Mutations in *CC2D2A* are associated with MKS subtype 6. *Cc2d2a*$^{-/-}$ mouse embryos die just prior to birth and exhibit pleiotropic phenotypes resembling MKS including microphthalmia, exencephaly, polydactyly, and defects in multiple organs as well as *situs inversus* (Garcia-Gonzalo et al., 2011; Veleri et al., 2014).

Cc2d2a is part of a complex that includes multiple ciliopathy-associated proteins. This complex localizes to the transition zone, includes Mks1, Tmem216, Tmem67, Tmem231, Cep290, B9d1, Tctn1, and Tctn2, and regulates ciliogenesis and ciliary membrane composition (Chih et al., 2012; Garcia-Gonzalo et al., 2011). Cdc2d2a was shown to localize to the subdistal appendages in the mother centriole and its loss inhibits the initiation of cilia biogenesis from the basal body. As a consequence, *Cc2d2a*$^{-/-}$ embryos have fewer and defective cilia, which leads to defects in Shh signaling, helping to account for the pleiotropic phenotypes (Veleri et al., 2014).

3.3.8 *Nphp3*

Nephrocystin-3 (Nphp3) interacts with Nphp1 and Nphp2, two proteins involved in nephronophthisis, a group of autosomal recessive cystic kidney diseases, that localize to the ciliary base (Olbrich et al., 2003). Mutations in *NPHP3* are associated with MKS subtype 7. Not much is known about Nphp3 other than it contains protein–protein interacting domains and is expressed during mouse embryonic development in the node, neural tissue, kidney tubules, retina, respiratory epithelium, biliary tract, and liver (Bergmann et al., 2008; Olbrich et al., 2003). Two mouse models exist bearing mutations in *Nephrocystin-3* (*Nphp3*). *Nphp3*^{pcy} is a spontaneous hypomorphic mutation, and *Nphp3*^{ko} is a targeted null mutation (Bergmann et al., 2008; Olbrich et al., 2003; Omran et al., 2001; Takahashi et al., 1986). *Nphp3*$^{pcy/pcy}$ mice exhibit adolescent nephronophthisis with no other apparent deformities, whereas *Nphp3*$^{ko/ko}$ display MKS-like symptoms including midgestational embryonic lethality and defective left–right asymmetry (Bergmann et al., 2008). Interestingly, no craniofacial defects have been described in these mutants.

3.3.9 *Tctn2*

Tectonic2 (*Tctn2*) encodes a transmembrane protein that is part of the ciliopathy complex with Cc2d2a that localizes to the transition zone (Garcia-Gonzalo et al., 2011). Mutations in *TCTN2* are associated with MKS subtype 8. *Tctn2*$^{-/-}$ embryos exhibit neural tube closure defects, exencephaly, microphthalmia, cleft palate, polydactyly, and left–right asymmetry defects (Sang et al., 2011). Similar to other ciliopathy genes, *Tctn2* is required for ciliogenesis and, consequently, for proper Hh signaling transduction.

3.3.10 B9d1 *and* Tmem231

B9d1 and *Tmem231* knockout mouse embryos exhibit similar lethality during late gestation and MKS-like phenotypes including microphthalmia and syndactyly (Chih et al., 2012). Consistent with these phenotypes is a loss of cilia that results in misregulation of the Shh signaling pathway. B9d1 and Tmem231 are part of the ciliopathy complex in the transition zone required for ciliogenesis and ciliary membrane composition (Chih et al., 2012; Garcia-Gonzalo et al., 2011). Mutations in *B9D1* are associated with MKS subtype 9.

3.3.11 *B9d2*

The *Stumpy* mutant mouse carries a conditional deletion of the *B9d2* gene (Town et al., 2008). Similar to other ciliopathy mouse mutants, *Stumpy* mice exhibit perinatal lethality, hydrocephaly, polycystic kidneys, and abnormalities in the olfactory bulb, the hippocampus, and the cerebellum. They have a dramatic decrease in primary cilia and an absence of ciliary axonemes in affected tissues, which then result in deregulation of the Shh signaling pathway, in association with aforementioned phenotypes (Breunig et al., 2008; Town et al., 2008). B9d2 is also part of the ciliopathy complex that localizes to the transition zone of the cilium (Chih et al., 2012) and mutations in *B9D2* are associated with MKS subtype 10.

3.4 Joubert Syndrome

Joubert syndrome (JBTS; OMIM 213300) is an autosomal recessive neurodevelopmental condition with 10 subtypes and an occurrence of 1:100,000 (Szymanska et al., 2014).

3.4.1 *Clinical Features*

JBTS patients exhibit a characteristic brain abnormality called the molar tooth sign that can be detected by MRI and is the result of abnormal development of the cerebellar vermis and the brain stem. Common features of JBTS include developmental delay, ataxia, hyperpnea, hypotonia, sleep apnea, abnormal eye and tongue movements, renal/liver deformities, and polydactyly. JBTS patients also have craniofacial malformations including a long face and frontonasal prominence, bitemporal narrowing, ptosis of the eyelid, protruding nasal bridge and tip, prognathism, eyebrow deformities, trapezoid-shaped mouth, lower lip eversion, and thick ear lobes (Maria, Boltshauser, Palmer, & Tran, 1999). Based on the mouse models described below, the genes implicated in JBTS are involved in signaling, rather than the actual structure of the cilium, which is likely the reason JBTS has milder defects compared to other ciliopathies.

3.4.2 Genetics

There are currently 22 genes associated with JBTS: *INPP5E, TMEM216, AHI1, NPHP1, CEP290, TMEM67, RPGRIP1L, ARL13B, CC2D2A, OFD1, TTC21B, KIF7, TCTN1, TMEM231, TMEM237, CEP41, TMEM138, C5orf42, TCTN3, ZNF423, CSPP1*, and PDE6D. Interestingly, *TMEM67, RPGRIP1L, CC2D2A, TMEM216*, and *CEP290* are allelic to MKS (Szymanska et al., 2014).

3.4.3 Mouse Models and Mechanism

Several mouse models of JBTS have been generated revealing the cellular basis of the disease as well as the role of different JBTS associated genes in ciliogenesis.

3.4.4 Inpp5E

Inpp5E encodes lipid phosphatase 5, which localizes to the ciliary axoneme and hydrolyzes the 5-phosphate of PtdIns(3,4,5)P3 and PtdIns(4,5)P2 (Astle, Horan, Ooms, & Mitchell, 2007; Bielas et al., 2009; Jacoby et al., 2009). *Innp5E*$^{\Delta/\Delta}$ mice die soon after birth and exhibit polydactyly, anophthalmia, polycystic kidneys, cleft palate, anencephaly, exencephaly, and delayed ossification of metacarpals and phalanges (Jacoby et al., 2009). *Inpp5E*$^{\Delta/\Delta}$ embryos have fewer primary cilia, with altered morphology, whereas *Inpp5E*$^{\Delta/\Delta}$ mouse embryonic fibroblasts are able to assemble cilia but fail to maintain them. It has been hypothesized that cilium stability is regulated by growth factor receptors such as PDGFRα that regulate ciliary PtdIns(3,4,5)P3 production and cilia disassembly. Therefore, in the absence of Inpp5E, cilia disassemble rapidly due to the accumulation of PtdIns(3,4,5) P3 (Jacoby et al., 2009). The phenotypes of the *Inpp5E*$^{\Delta/\Delta}$ embryos are more severe than those of JBTS human patients, which may again reflect variable severity of the causative mutations in humans versus the null mouse (Plotnikova et al., 2015).

3.4.5 Ahi1

Ahi1 encodes the signaling protein Jouberin (Jbn), which localizes to the basal body of the cilium and was the first gene to be identified in JBT (Dixon-Salazar et al., 2004; Ferland et al., 2004). *Ahi1*$^{-/-}$null mice survive until birth with no apparent embryonic deformities except a smaller cerebellum and underdeveloped vermis that resembles human JBT. These mice are smaller, with slightly reduced brain size, and have polycystic kidneys and higher mortality (Lancaster et al., 2011, 2009). *Ahi1*$^{-/-}$ mice also display rapid loss of photoreceptors in the retina (Louie et al., 2010). Examination of primary cilia revealed unchanged number, length, and morphology, suggesting a role for Jbn in signaling rather than ciliogenesis. Indeed, Jbn was shown to modulate Wnt signaling downstream of β-catenin and has been hypothesized to facilitate nuclear translocation of β-catenin.

3.4.6 *Arl13b*

Arl13b is a small GTPase of the Arf/Arl family that localizes to the ciliary axoneme. A mutation in *Arl13b* was found in the *hennin* (hnn) mutant mouse, as part of an ENU screen, and has subsequently been used to study JBTS (Garcia-Garcia et al., 2005). *hnn* embryos exhibit an open neural tube, randomized heart looping, abnormal eyes, polydactyly, and midgestational lethality (Caspary, Larkins, & Anderson, 2007). *hnn* embryos have short cilia with defective axonemes that perturb the Gli gradient and thus the Shh signaling necessary for the spatial organization of the ventral neural tube. Studies in zebrafish showed that Arl13b functions in ciliary membrane extension to regulate ciliary length (Lu et al., 2015). Moreover, conditional *Arl13b* mice revealed the presence of guidance cue receptors essential for interneuronal migration on the interneuronal primary cilia, whose concentration and dynamics are controlled by Arl13b (Higginbotham et al., 2012; Su, Bay, Mariani, Hillman, & Caspary, 2012).

3.4.7 *Ttc21b*

The *alien* (*aln*) mouse mutant was identified in a chemical mutagenesis screen affecting late embryogenesis and carries a mutation in the *Ttc21b* gene (Herron et al., 2002; Tran et al., 2008). *Ttc21b*$^{aln/aln}$ embryos exhibit polydactyly, delayed eye and forebrain development, and neural tube defects. Ttc2b localizes to the ciliary axoneme and in its absence the axonemes have an abnormal morphology and accumulate IFT proteins due to impairment of IFT. *Ttc21b*$^{aln/aln}$ embryos exhibit activated Shh signaling in association with disrupted IFT (Tran et al., 2008). The *aln* mutation was particularly useful in uncoupling the roles of anterograde and retrograde transport in Shh signaling and it was suggested, although not shown, that anterograde IFT is required for Gli activation, which is then modulated by retrograde IFT.

3.4.8 *Kif7*

There are currently two mouse models with mutations in *Kif7*: a targeted *Kif7* knockout and the ENU-induced *matariki* (*maki*) allele. Similar to other JBTS mouse models, $Kif7^{-/-}$ and $Kif7^{maki}$ embryos exhibit polydactyly, exencephaly, and microphthalmia, phenotypes consistent with defects in the Hh pathway (Endoh-Yamagami et al., 2009; Liem, He, Ocbina, & Anderson, 2009). $Kif7^{-/-}$ embryos display patterning defects in the neural tube and limb buds due to deregulation of the Shh signaling. Kif7, a kinesin family member, localizes to the base of the cilia in the absence of Shh and accumulates at the distal tip of the primary cilia upon activation of the Shh pathway. Based on studies of the $Kif7^{-/-}$ embryos, *Kif7* regulates Shh signaling by controlling the efficient localization of Gli3 to cilia and processing Gli3 to its repressor form (Endoh-Yamagami et al., 2009). Studies of $Kif7^{maki}$ embryos revealed a role for Kif7 downstream of Smoothened and upstream of Gli2 controlling Shh signaling in both a positive and negative manner (He, Subramanian, et al., 2014; Liem et al., 2009). It was recently demonstrated that Kif7 does not traffic Gli proteins into the cilia, but instead controls cilium architecture by modulating microtubule dynamics at the tip and creating a single cilium tip compartment where Gli proteins can be regulated (He, Subramanian, et al., 2014).

3.4.9 *Tctn1*

Tectonic1 (*Tctn1*) mouse mutants die during late gestation and display holoprosencephaly, a failure to form the floor plate and to pattern the neural tube properly, as well as defects in left–right asymmetry, all phenotypes typically associated with defective Shh signaling (Mitchell et al., 2001; Reiter & Skarnes, 2006). Tctn1 is part of the ciliopathy complex that localizes to the transition zone of the cilium and regulates ciliogenesis and ciliary membrane composition (Garcia-Gonzalo et al., 2011). Epistasis experiments revealed that Tctn1 modulates Shh signaling downstream of Smoothened (Reiter & Skarnes, 2006). Examination of the node and neural tube shows a dramatic reduction in the number of cilia (Garcia-Gonzalo et al., 2011). In contrast, cilia number in the limb buds is not affected, but these cilia fail to localize membrane-associated proteins such as Smoothened and Arl13b.

3.4.10 *Cep41*

Cep41 localizes to the basal body and regulates ciliary entry of the polyglutamylase enzyme TTLL6, which modulates tubulin glutamylation that is necessary for axonemal formation (Lee et al., 2012). $Cep41^{Gt/Gt}$ embryos have a malformed hindbrain, exencephaly, brain hemorrhage, cardiac deformities, and embryonic lethality. Similar to other ciliopathy mouse mutants, *Cep41*$^{Gt/Gt}$ embryos exhibit phenotypic variability, suggesting extragenic phenotypic modifiers. The cause of the phenotypes, as well as the architecture of the cilia in these mutants, has not been examined yet; hence, further studies are needed to better understand the role of ciliary glutamylation in mammals. However, studies in primary cells from human patients suggest that CEP41 is not required for ciliogenesis (Lee et al., 2012).

3.4.11 The Role of Cilia in NCC and Craniofacial Development

Much has been learned about the role of cilia in NCC and craniofacial development from an additional array of mutant mice exhibiting dysfunctional cilia. The currently favored hypothesis is that primary cilia are not absolutely required for proper NCC specification (Brugmann, Allen, et al., 2010; Chang et al., 2015). Primary cilia do, however, play an important role in NCC migration, because the receptors of several chemotactic molecules essential for NCC migration localize to primary cilia (Schneider et al., 2005; Theveneau & Mayor, 2012). In support of this idea, NCC in zebrafish morphants and mouse embryonic fibroblasts lacking cilia display aberrant migration patterns (Chang et al., 2015). Furthermore, the importance of primary cilia for proliferation of cranial NCC has also been demonstrated. NCC-specific excision of *Kif3a*, which encodes a ciliary microtubule motor, causes an increase in proliferation that leads to midfacial expansion (Brugmann, Allen, et al., 2010).

Thus, ciliary proteins can have cell- and tissue-specific effects that are consistent with the ciliopathies described here. Many of the phenotypic anomalies are common to many ciliopathies, but there are also ciliopathy-specific defects in individual tissues. Moreover, the degree of severity varies from ciliopathy to ciliopathy, which further supports a tissue-specific role for ciliary proteins that was previously not appreciated.

Thus, similar to ribosomopathies, mouse models of ciliopathies have revealed tissue-specific roles for a global cellular structure, in this case, the cilium. Future research is now aimed at understanding how cilia-associated genes regulate signaling pathways in a tissue-specific manner. In contrast to ribosomopathies and ciliopathies that encompass rare disorders of craniofacial development but with other pleiotropic phenotypes, there are a number of rare but specific craniofacial anomalies such as syngnathia, for which mouse models have been instrumental recently in revealing their etiology and pathogenesis.

4. SYNGNATHIA

Syngnathia (OMIM 119550), the bony fusion of the upper and lower jaw, is a rare congenital disorder with only about 60 cases reported in the literature to date (Fig. 3C and D; Inman, Purcell, Kume, & Trainor, 2013).

4.1 Clinical Features

Syngnathia causes restriction of the mouth opening affecting feeding, breathing, and communication. Currently syngnathia is classified into four types: Type 1a, simple anterior syngnathia by bony fusion of the alveolar ridge only, with no other craniofacial congenital anomalies; Type 1b, complex anterior syngnathia by bony fusion of the alveolar ridges associated with other craniofacial defects; Type 2a, simple zygomatic-mandibular syngnathia by bony fusion of the mandible to zygoma; and Type 2b, complex zygomatic-mandibular syngnathia by bony fusion of the mandible to the zygoma associated with cleft palate and/or temporomandibular joint (TMJ) ankylosis (Laster, Temkin, Zarfin, & Kushnir, 2001). Syngnathia can also be associated with syndromes such as Van der Woude and cleft palate lateral alveolar synechiae syndrome. In one rare case, syngnathia was associated with maxillary duplication, nasal cleft, and other noncraniofacial defects including situs inversus, left choanal stenosis, and dextrocardia (Patel, Porras, & Lypka, 2015). Furthermore, in another rare case, a newborn presented with both syngnathia and craniosynostosis (Ajike et al., 2008). Depending on when the fusion occurs during development, patients can display complete maxillomandibular-zygomatic fusion, or bony fusion of the maxillary ridge to the mandibular ridge.

4.2 Genetics

To date, specific mutations in humans have not yet been identified in association with syngnathia. However, inheritance has been reported in a few cases. In one case, the patient was born to consanguineous parents and had a sibling with a similar defect (Ugurlu et al., 1999), consistent with an autosomal recessive mode of inheritance. Another case of a newborn with syngnathia who was also born of consanguineous parents further supports autosomal recessive inheritance (Villanueva-Garcia, Contreras-Masse, Villa-Guillen, Ramon-Garcia, & Murguia-De Sierra, 2009).

4.3 Mouse Models and Mechanism

Although no causative loci have been identified in human patients with syngnathia, three mutant mouse models have been described exhibiting syngnathia: *Foxc1*$^{-/-}$, *Fgf8*$^{\Delta 2,3/neo}$, and *Wnt1Cre;pMes-Bmp4* (He, Hu, et al., 2014; Inman et al., 2013). *Foxc1* is a member of the forkhead box winged helix transcription factor family and in mice *Foxc1* plays a role in meningeal, calvarial, ocular, somitic, and renal development (Carlsson & Mahlapuu, 2002; Hannenhalli & Kaestner, 2009). *Foxc1*$^{-/-}$-null mouse embryos have defects in maxillary and mandibular structures, along with bony syngnathia and agenesis of the TMJ which is the functional jaw joint essential for jaw articulation (Fig. 4D and E; Inman et al., 2013). More specifically, syngnathia is present in the form of bilateral fusion of the maxillary zygomatic complex to the dentary bone. *Foxc1*$^{-/-}$ embryos additionally exhibit defects in palate formation with a slightly higher arch and malformed rugae, as well as facial muscle patterning abnormalities. These phenotypes are consistent with defects in the development of the NCC-derived mesenchyme of PA1. Indeed, *Foxc1*$^{-/-}$ embryos display defects in the osteogenic differentiation of the NCC.

Fgf and Bmp signaling pathways play critical roles in PA1 patterning and development. *Fgf8* expression in PA1 is reduced in the absence of *Foxc1* and severely hypomorphic *Fgf8*$^{null/neo}$ mice also exhibit syngnathia (Inman et al., 2013). Moreover, *Foxc1* genetically interacts with *Fgf8* such that decreasing *Fgf8* dosage results in increasingly severe syngnathic phenotypes. Downstream of *Focx1-Fgf8* signaling, localization of *Dlx2* and *Dlx5*, two members of the *Dlx* transcription factor family that are necessary in the mesenchyme of PA1 for the proper pattern, and morphology of the jaw are altered in *Foxc1*$^{-/-}$ embryos. Taken together, syngnathia and TMJ agenesis in *Foxc1*$^{-/-}$ embryos are both due to perturbed Fgf8 signaling in PA1. This defect leads to alterations in *Dlx2* and *Dlx5* activity, resulting in regionalized patterning changes in the maxillary and mandibular mesenchyme that ultimately cause defects in development of the upper and lower jaw (Inman et al., 2013).

The third mouse model of syngnathia, *Wnt1Cre;pMes-Bmp4* mice, results from transgenic overexpression of *Bmp4* in NCC (He, Hu, et al., 2014). Bone morphogenetic proteins (BMPs) are secreted proteins associated with the extracellular matrix and regulate several developmental processes including bone and cartilage formation (Brazil, Church, Surae, Godson, & Martin, 2015). *Wnt1Cre;pMes-Bmp4* mice exhibit bony syngnathia and TMJ agenesis in addition to several other craniofacial deformities including a shortened snout, open eyelids, absence of external ears, and cleft palate (He, Hu, et al., 2014). Further examination revealed enhanced osteoblast differentiation that could explain the jaw fusion. Similar to *Foxc1*$^{-/-}$ mutant embryos, expression of patterning genes such as *Dlx2*, *Hand2*, and *Msx1* is altered in PA1, likely due to reduction of *Fgf8* expression in the oral ectoderm.

Much has been learned from these mouse models regarding the molecular mechanisms underlying the pathogenesis of syngnathia. However, the spatiotemporal requirement for *Foxc1* and *Bmp4* in musculoskeletal development still needs to be examined. Recently, a *Foxc1* conditional allele was generated that could potentially address this issue (Seo et al., 2012). In addition, it will be informative to determine which genes *Foxc1* regulates in the cranial mesenchyme and the oral ectoderm of PA1 using an unbiased approach like RNA sequencing to further our understanding of jaw development and the pathogenesis of jaw disorders. The rarity of syngnathia and especially isolated syngnathia make it challenging for human geneticists to isolate responsible loci. However, analysis of mutations in *Foxc1*, *Bmp4*, and *Fgf8* and their downstream targets provide a promising starting point.

5. CONCLUSIONS AND PERSPECTIVES

During the past 15 years, completion of the first human genome sequence followed by rapid advances in sequencing has accelerated the discovery of the underlying etiology of human genetic disorders. In parallel, mouse models of human congenital disorders have been an instrumental source of knowledge and discovery and resulted in extraordinary findings regarding the cellular basis and pathogenesis of rare craniofacial conditions. Ribosomopathies, the class of diseases caused by disruption in ribosome biogenesis, serve as a classic example of a global cellular process with tissue-specific effects and functions. Mutations in different ribosomal genes manifest in distinct phenotypes and are classified as unique human syndromes, which is extraordinary given how critical ribosomes are to the basic functioning of every cell. Even more surprising is the fact that humans exhibiting defective ribosome biogenesis survive with malformations restricted to certain parts of their bodies, with the craniofacial skeleton being an important example. The existence of specialized ribosomes could help account for the spatiotemporal specificity of the deformities observed in ribosomopathies. There are more than 50 RPs and 4 different rRNAs that make up the ribosome, and it was recently suggested that the composition of ribosomes in different tissues could result from different combinations of RPs (Xue & Barna, 2012). Knocking-out specific RPs results in tissue-specific developmental defects, demonstrating that not all ribosomes are the same (Kondrashov et al., 2011; Xue et al., 2015). Moreover, RPs translate specific mRNAs adding yet another level of gene expression regulation not previously appreciated. Therefore, mutations in RPs associated with DBA could elicit their tissue-specific effects by failing to translate their target mRNAs. Although current DBA mouse models do not represent the human condition very well, it would be interesting to determine the ribosome profiles in these mutants and explore whether translation of only specific mRNAs is affected.

TCS is a bit more challenging to explain because *Tcof1* and presumably *Polr1c* and *Polr1d* play a primary role during the transcription of rRNAs, which is the first step in ribosome biogenesis. One would expect that failure to transcribe rRNA would be incompatible with life. However, RNA Pol I could also be "specialized" and its composition, like the ribosome, may also be tissue specific. Another possibility, although less likely, could be that the rRNA composition of ribosomes, similar to the RP composition, is distinct in each tissue. Treacle is part of the Pol I transcription initiation machinery, but not Pol III. Thus, presumably transcription of only the 5.8S, 18S, and 28S is affected when Treacle is disrupted, and not 5S which is transcribed by Pol III. This disruption is likely to change the rRNA composition of the ribosome. A simpler hypothesis is that NCC are simply more sensitive to protein translation levels than other tissues. The generation of mice carrying mutations in *Polr1c* and *Polr1d* will be helpful in elucidating the role of rRNA transcription in ribosomopathies and craniofacial development.

Although more pleiotropic, ciliopathies fall into the same category of disruption of a global process with tissue-specific effects, because cilia are found in nearly every cell in the human body. Significant discoveries have been made in the last decade regarding cilia and the cellular basis of ciliopathies using mouse models, however much remains to be learned regarding their composition, the signaling pathways they help transduce between cells, and their uniqueness that translates into tissue specificity. Moreover, the role of cilia in the NCC needs to be further investigated to pinpoint at which stage(s) of CNC development each ciliary protein or process is needed. This analysis is now possible with the availability of several mouse models of human ciliopathies and other ciliopathic mouse mutants, many of which carry conditional alleles that can be used in a tissue-specific or a temporal manner.

It is also remarkable that some rare syndromes are caused by mutations in the same genes and are thus allelic. Meckel–Gruber (MKS) and Joubert

(JBTS) syndromes are a classic example of allelic syndromes and several mouse models of these conditions exist. The use of CRISPR/Cas9 technology to generate mouse mutants carrying the exact mutations that human patients have, will be helpful in elucidating the distinct effects that different mutations in the same gene have on cilia and ultimately in the cellular pathogenesis of these ciliopathies.

In summary, the use of mouse models has been invaluable in decoding the cellular mechanisms and pathogenesis of rare craniofacial disorders in humans including ribosomopathies, ciliopathies, and the rarest of the congenital conditions like syngnathia. The advantage of mouse models in their application to understanding human disease resides in their comparative mammalian anatomy and physiology, together with a sophisticated array of genetic tools for exploring gene function in a precise spatiotemporal manner. Much is still to be learned, but the future looks promising with improved phenotyping in humans and new gene editing approaches for mouse models.

ACKNOWLEDGMENTS

We are indebted to Mark Miller for artistic illustrations in Figs. 1, 2, and 5, and Dr Kimberly Inman for images of wild-type and *Foxc1*$^{-/-}$ embryos in Fig. 4. We are also very grateful to the Children's Craniofacial Association for providing Fig. 3A. Research in the Trainor laboratory is supported by the Stowers Institute for Medical Research and the National Institute for Dental and Craniofacial Research (DE 016082).

REFERENCES

Abdelhamed, Z. A., Wheway, G., Szymanska, K., Natarajan, S., Toomes, C., Inglehearn, C., et al. (2013). Variable expressivity of ciliopathy neurological phenotypes that encompass Meckel-Gruber syndrome and Joubert syndrome is caused by complex de-regulated ciliogenesis, Shh and Wnt signalling defects. *Human Molecular Genetics*, *22*(7), 1358–1372. http://dx.doi.org/10.1093/hmg/dds546.

Ajike, S. O., Adeosun, O. O., Adebayo, E. T., Anyiam, J. O., Jalo, I., & Chom, N. D. (2008). Congenital bilateral fusion of the maxillomandibular alveolar processes with craniosynostosis: Report of a rare case. *Nigerian Journal of Clinical Practice*, *11*(1), 77–80.

Armistead, J., Khatkar, S., Meyer, B., Mark, B. L., Patel, N., Coghlan, G., et al. (2009). Mutation of a gene essential for ribosome biogenesis, EMG1, causes Bowen-Conradi syndrome. *American Journal of Human Genetics*, *84*(6), 728–739. http://dx.doi.org/10.1016/j.ajhg.2009.04.017.

Armistead, J., Patel, N., Wu, X., Hemming, R., Chowdhury, B., Basra, G. S., et al. (2015). Growth arrest in the ribosomopathy, Bowen-Conradi syndrome, is due to dramatically reduced cell proliferation and a defect in mitotic progression. *Biochimica et Biophysica Acta*, *1852*(5), 1029–1037. http://dx.doi.org/10.1016/j.bbadis.2015.02.007.

Arts, H. H., Doherty, D., van Beersum, S. E., Parisi, M. A., Letteboer, S. J., Gorden, N. T., et al. (2007). Mutations in the gene encoding the basal body protein RPGRIP1L, a nephrocystin-4 interactor, cause Joubert syndrome. *Nature Genetics*, *39*(7), 882–888. http://dx.doi.org/10.1038/ng2069.

Astle, M. V., Horan, K. A., Ooms, L. M., & Mitchell, C. A. (2007). The inositol polyphosphate 5-phosphatases: Traffic controllers, waistline watchers and tumour suppressors? *Biochemical Society Symposium, 74,* 161–181. http://dx.doi.org/10.1042/BSS0740161.

Austin, K. M., Gupta, M. L., Jr., Coats, S. A., Tulpule, A., Mostoslavsky, G., Balazs, A. B., et al. (2008). Mitotic spindle destabilization and genomic instability in Shwachman-Diamond syndrome. *Journal of Clinical Investigation, 118*(4), 1511–1518. http://dx.doi.org/10.1172/JCI33764.

Badano, J. L., Mitsuma, N., Beales, P. L., & Katsanis, N. (2006). The ciliopathies: An emerging class of human genetic disorders. *Annual Review of Genomics and Human Genetics, 7,* 125–148. http://dx.doi.org/10.1146/annurev.genom.7.080505.115610.

Beales, P. L., Elcioglu, N., Woolf, A. S., Parker, D., & Flinter, F. A. (1999). New criteria for improved diagnosis of Bardet-Biedl syndrome: Results of a population survey. *Journal of Medical Genetics, 36*(6), 437–446.

Bergmann, C., Fliegauf, M., Bruchle, N. O., Frank, V., Olbrich, H., Kirschner, J., et al. (2008). Loss of nephrocystin-3 function can cause embryonic lethality, Meckel-Gruber-like syndrome, situs inversus, and renal-hepatic-pancreatic dysplasia. *American Journal of Human Genetics, 82*(4), 959–970. http://dx.doi.org/10.1016/j.ajhg.2008.02.017.

Bielas, S. L., Silhavy, J. L., Brancati, F., Kisseleva, M. V., Al-Gazali, L., Sztriha, L., et al. (2009). Mutations in INPP5E, encoding inositol polyphosphate-5-phosphatase E, link phosphatidyl inositol signaling to the ciliopathies. *Nature Genetics, 41*(9), 1032–1036. http://dx.doi.org/10.1038/ng.423.

Bose, T., Lee, K. K., Lu, S., Xu, B., Harris, B., Slaughter, B., et al. (2012). Cohesin proteins promote ribosomal RNA production and protein translation in yeast and human cells. *PLoS Genetics, 8*(6), e1002749. http://dx.doi.org/10.1371/journal.pgen.1002749.

Brazil, D. P., Church, R. H., Surae, S., Godson, C., & Martin, F. (2015). BMP signalling: Agony and antagony in the family. *Trends in Cell Biology, 25*(5), 249–264. http://dx.doi.org/10.1016/j.tcb.2014.12.004.

Breunig, J. J., Sarkisian, M. R., Arellano, J. I., Morozov, Y. M., Ayoub, A. E., Sojitra, S., et al. (2008). Primary cilia regulate hippocampal neurogenesis by mediating sonic hedgehog signaling. *Proceedings of the National Academy of Sciences of the United States of America, 105*(35), 13127–13132. http://dx.doi.org/10.1073/pnas.0804558105.

Bronner, M. E., & LeDouarin, N. M. (2012). Development and evolution of the neural crest: An overview. *Developmental Biology, 366*(1), 2–9. http://dx.doi.org/10.1016/j.ydbio.2011.12.042.

Brosnan, M. E., & Brosnan, J. T. (2007). Orotic acid excretion and arginine metabolism. *The Journal of Nutrition, 137*(6 Suppl. 2), 1656S–1661S.

Brugmann, S. A., Allen, N. C., James, A. W., Mekonnen, Z., Madan, E., & Helms, J. A. (2010). A primary cilia-dependent etiology for midline facial disorders. *Human Molecular Genetics, 19*(8), 1577–1592. http://dx.doi.org/10.1093/hmg/ddq030.

Brugmann, S. A., Cordero, D. R., & Helms, J. A. (2010). Craniofacial ciliopathies: A new classification for craniofacial disorders. *American Journal of Medical Genetics. Part A, 152A*(12), 2995–3006. http://dx.doi.org/10.1002/ajmg.a.33727.

Carlsson, P., & Mahlapuu, M. (2002). Forkhead transcription factors: Key players in development and metabolism. *Developmental Biology, 250*(1), 1–23.

Caspary, T., Larkins, C. E., & Anderson, K. V. (2007). The graded response to Sonic Hedgehog depends on cilia architecture. *Developmental Cell, 12*(5), 767–778. http://dx.doi.org/10.1016/j.devcel.2007.03.004.

Chakraborty, A., Uechi, T., Higa, S., Torihara, H., & Kenmochi, N. (2009). Loss of ribosomal protein L11 affects zebrafish embryonic development through a p53-dependent apoptotic response. *PLoS One, 4*(1), e4152. http://dx.doi.org/10.1371/journal.pone.0004152.

Chang, C. F., Schock, E. N., Attia, A. C., Stottmann, R. W., & Brugmann, S. A. (2015). The ciliary baton: Orchestrating neural crest cell development. *Current Topics in Developmental Biology*, *111*, 97–134. http://dx.doi.org/10.1016/bs.ctdb.2014.11.004.

Chih, B., Liu, P., Chinn, Y., Chalouni, C., Komuves, L. G., Hass, P. E., et al. (2012). A ciliopathy complex at the transition zone protects the cilia as a privileged membrane domain. *Nature Cell Biology*, *14*(1), 61–72. http://dx.doi.org/10.1038/ncb2410.

Ciccia, A., Huang, J. W., Izhar, L., Sowa, M. E., Harper, J. W., & Elledge, S. J. (2014). Treacher Collins syndrome TCOF1 protein cooperates with NBS1 in the DNA damage response. *Proceedings of the National Academy of Sciences of the United States of America*, *111*(52), 18631–18636. http://dx.doi.org/10.1073/pnas.1422488112.

Cmejlova, J., Dolezalova, L., Pospisilova, D., Petrtylova, K., Petrak, J., & Cmejla, R. (2006). Translational efficiency in patients with Diamond-Blackfan anemia. *Haematologica*, *91*(11), 1456–1464.

Cook, S. A., Collin, G. B., Bronson, R. T., Naggert, J. K., Liu, D. P., Akeson, E. C., et al. (2009). A mouse model for Meckel syndrome type 3. *Journal of the American Society of Nephrology*, *20*(4), 753–764. http://dx.doi.org/10.1681/ASN.2008040412.

Cui, C., Chatterjee, B., Francis, D., Yu, Q., SanAgustin, J. T., Francis, R., et al. (2011). Disruption of Mks1 localization to the mother centriole causes cilia defects and developmental malformations in Meckel-Gruber syndrome. *Disease Models & Mechanisms*, *4*(1), 43–56. http://dx.doi.org/10.1242/dmm.006262.

Danilova, N., Sakamoto, K. M., & Lin, S. (2008). Ribosomal protein S19 deficiency in zebrafish leads to developmental abnormalities and defective erythropoiesis through activation of p53 protein family. *Blood*, *112*(13), 5228–5237. http://dx.doi.org/10.1182/blood-2008-01-132290.

Dauwerse, J. G., Dixon, J., Seland, S., Ruivenkamp, C. A., van Haeringen, A., Hoefsloot, L. H., et al. (2011). Mutations in genes encoding subunits of RNA polymerases I and III cause Treacher Collins syndrome. *Nature Genetics*, *43*(1), 20–22. http://dx.doi.org/10.1038/ng.724.

Dawe, H. R., Farr, H., & Gull, K. (2007). Centriole/basal body morphogenesis and migration during ciliogenesis in animal cells. *Journal of Cell Science*, *120*(Pt. 1), 7–15. http://dx.doi.org/10.1242/jcs.03305.

Dawe, H. R., Smith, U. M., Cullinane, A. R., Gerrelli, D., Cox, P., Badano, J. L., et al. (2007). The Meckel-Gruber Syndrome proteins MKS1 and meckelin interact and are required for primary cilium formation. *Human Molecular Genetics*, *16*(2), 173–186. http://dx.doi.org/10.1093/hmg/ddl459.

de Conciliis, L., Marchitiello, A., Wapenaar, M. C., Borsani, G., Giglio, S., Mariani, M., et al. (1998). Characterization of Cxorf5 (71-7A), a novel human cDNA mapping to Xp22 and encoding a protein containing coiled-coil alpha-helical domains. *Genomics*, *51*(2), 243–250. http://dx.doi.org/10.1006/geno.1998.5348.

Delous, M., Baala, L., Salomon, R., Laclef, C., Vierkotten, J., Tory, K., et al. (2007). The ciliary gene RPGRIP1L is mutated in cerebello-oculo-renal syndrome (Joubert syndrome type B) and Meckel syndrome. *Nature Genetics*, *39*(7), 875–881. http://dx.doi.org/10.1038/ng2039.

Devlin, E. E., Dacosta, L., Mohandas, N., Elliott, G., & Bodine, D. M. (2010). A transgenic mouse model demonstrates a dominant negative effect of a point mutation in the RPS19 gene associated with Diamond-Blackfan anemia. *Blood*, *116*(15), 2826–2835. http://dx.doi.org/10.1182/blood-2010-03-275776.

Dixon, J., Jones, N. C., Sandell, L. L., Jayasinghe, S. M., Crane, J., Rey, J. P., et al. (2006). Tcof1/Treacle is required for neural crest cell formation and proliferation deficiencies that cause craniofacial abnormalities. *Proceedings of the National Academy of Sciences of the United States of America*, *103*(36), 13403–13408. http://dx.doi.org/10.1073/pnas.0603730103.

Dixon-Salazar, T., Silhavy, J. L., Marsh, S. E., Louie, C. M., Scott, L. C., Gururaj, A., et al. (2004). Mutations in the AHI1 gene, encoding jouberin, cause Joubert syndrome with cortical polymicrogyria. *American Journal of Human Genetics*, *75*(6), 979–987. http://dx.doi.org/10.1086/425985.

Doudna, J. A., & Charpentier, E. (2014). Genome editing. The new frontier of genome engineering with CRISPR-Cas9. *Science*, *346*(6213), 1258096. http://dx.doi.org/10.1126/science.1258096.

Duan, X., Kelsen, S. G., Clarkson, A. B., Jr., Ji, R., & Merali, S. (2010). SILAC analysis of oxidative stress-mediated proteins in human pneumocytes: New role for treacle. *Proteomics*, *10*(11), 2165–2174. http://dx.doi.org/10.1002/pmic.201000020.

Endoh-Yamagami, S., Evangelista, M., Wilson, D., Wen, X., Theunissen, J. W., Phamluong, K., et al. (2009). The mammalian Cos2 homolog Kif7 plays an essential role in modulating Hh signal transduction during development. *Current Biology*, *19*(15), 1320–1326. http://dx.doi.org/10.1016/j.cub.2009.06.046.

Fath, M. A., Mullins, R. F., Searby, C., Nishimura, D. Y., Wei, J., Rahmouni, K., et al. (2005). Mkks-null mice have a phenotype resembling Bardet-Biedl syndrome. *Human Molecular Genetics*, *14*(9), 1109–1118. http://dx.doi.org/10.1093/hmg/ddi123.

Ferland, R. J., Eyaid, W., Collura, R. V., Tully, L. D., Hill, R. S., Al-Nouri, D., et al. (2004). Abnormal cerebellar development and axonal decussation due to mutations in AHI1 in Joubert syndrome. *Nature Genetics*, *36*(9), 1008–1013. http://dx.doi.org/10.1038/ng1419.

Ferrante, M. I., Barra, A., Truong, J. P., Banfi, S., Disteche, C. M., & Franco, B. (2003). Characterization of the OFD1/Ofd1 genes on the human and mouse sex chromosomes and exclusion of Ofd1 for the Xpl mouse mutant. *Genomics*, *81*(6), 560–569.

Ferrante, M. I., Giorgio, G., Feather, S. A., Bulfone, A., Wright, V., Ghiani, M., et al. (2001). Identification of the gene for oral-facial-digital type I syndrome. *American Journal of Human Genetics*, *68*(3), 569–576.

Ferrante, M. I., Zullo, A., Barra, A., Bimonte, S., Messaddeq, N., Studer, M., et al. (2006). Oral-facial-digital type I protein is required for primary cilia formation and left-right axis specification. *Nature Genetics*, *38*(1), 112–117. http://dx.doi.org/10.1038/ng1684.

Franceschetti, A., & Klein, D. (1949). The mandibulofacial dysostosis: A new hereditary syndrome. *Acta Ophthalmologica*, *27*(2), 143–224.

Franco, B., & Ballabio, A. (2006). X-inactivation and human disease: X-linked dominant male-lethal disorders. *Current Opinion in Genetics & Development*, *16*(3), 254–259. http://dx.doi.org/10.1016/j.gde.2006.04.012.

Frisdal, A., & Trainor, P. A. (2014). Development and evolution of the pharyngeal apparatus. *Wiley Interdisciplinary Reviews. Developmental Biology*, *3*(6), 403–418. http://dx.doi.org/10.1002/wdev.147.

Garcia-Garcia, M. J., Eggenschwiler, J. T., Caspary, T., Alcorn, H. L., Wyler, M. R., Huangfu, D., et al. (2005). Analysis of mouse embryonic patterning and morphogenesis by forward genetics. *Proceedings of the National Academy of Sciences of the United States of America*, *102*(17), 5913–5919. http://dx.doi.org/10.1073/pnas.0501071102.

Garcia-Gonzalo, F. R., Corbit, K. C., Sirerol-Piquer, M. S., Ramaswami, G., Otto, E. A., Noriega, T. R., et al. (2011). A transition zone complex regulates mammalian ciliogenesis and ciliary membrane composition. *Nature Genetics*, *43*(8), 776–784. http://dx.doi.org/10.1038/ng.891.

Gazda, H. T., Sheen, M. R., Vlachos, A., Choesmel, V., O'Donohue, M. F., Schneider, H., et al. (2008). Ribosomal protein L5 and L11 mutations are associated with cleft palate and abnormal thumbs in Diamond-Blackfan anemia patients. *American Journal of Human Genetics*, *83*(6), 769–780. http://dx.doi.org/10.1016/j.ajhg.2008.11.004.

Gazda, H. T., Zhong, R., Long, L., Niewiadomska, E., Lipton, J. M., Ploszynska, A., et al. (2004). RNA and protein evidence for haplo-insufficiency in Diamond-Blackfan anaemia patients with RPS19 mutations. *British Journal of Haematology*, *127*(1), 105–113. http://dx.doi.org/10.1111/j.1365-2141.2004.05152.x.

Gerrard, G., Valganon, M., Foong, H. E., Kasperaviciute, D., Iskander, D., Game, L., et al. (2013). Target enrichment and high-throughput sequencing of 80 ribosomal protein genes to identify mutations associated with Diamond-Blackfan anaemia. *British Journal of Haematology, 162*(4), 530–536. http://dx.doi.org/10.1111/bjh.12397.
Gonzalez, M. G., Castro, M. P., Nieto, D. V., & Bouzan, J. C. (2014). Oral-facial-digital syndrome type I: Surgical approach and a case report. *Journal of Plastic, Reconstructive & Aesthetic Surgery, 67*(3), 396–398. http://dx.doi.org/10.1016/j.bjps.2013.06.044.
Gorden, N. T., Arts, H. H., Parisi, M. A., Coene, K. L., Letteboer, S. J., van Beersum, S. E., et al. (2008). CC2D2A is mutated in Joubert syndrome and interacts with the ciliopathy-associated basal body protein CEP290. *American Journal of Human Genetics, 83*(5), 559–571. http://dx.doi.org/10.1016/j.ajhg.2008.10.002.
Gray, R. S., Abitua, P. B., Wlodarczyk, B. J., Szabo-Rogers, H. L., Blanchard, O., Lee, I., et al. (2009). The planar cell polarity effector Fuz is essential for targeted membrane trafficking, ciliogenesis and mouse embryonic development. *Nature Cell Biology, 11*(10), 1225–1232. http://dx.doi.org/10.1038/ncb1966.
Gurrieri, F., Franco, B., Toriello, H., & Neri, G. (2007). Oral-facial-digital syndromes: Review and diagnostic guidelines. *American Journal of Medical Genetics. Part A, 143A*(24), 3314–3323. http://dx.doi.org/10.1002/ajmg.a.32032.
Haarhuis, J. H., Elbatsh, A. M., & Rowland, B. D. (2014). Cohesin and its regulation: On the logic of X-shaped chromosomes. *Developmental Cell, 31*(1), 7–18. http://dx.doi.org/10.1016/j.devcel.2014.09.010.
Hannan, K. M., Brandenburger, Y., Jenkins, A., Sharkey, K., Cavanaugh, A., Rothblum, L., et al. (2003). mTOR-dependent regulation of ribosomal gene transcription requires S6K1 and is mediated by phosphorylation of the carboxy-terminal activation domain of the nucleolar transcription factor UBF. *Molecular and Cellular Biology, 23*(23), 8862–8877.
Hannenhalli, S., & Kaestner, K. H. (2009). The evolution of Fox genes and their role in development and disease. *Nature Reviews. Genetics, 10*(4), 233–240. http://dx.doi.org/10.1038/nrg2523.
Hayano, T., Yanagida, M., Yamauchi, Y., Shinkawa, T., Isobe, T., & Takahashi, N. (2003). Proteomic analysis of human Nop56p-associated pre-ribosomal ribonucleoprotein complexes. Possible link between Nop56p and the nucleolar protein treacle responsible for Treacher Collins syndrome. *The Journal of Biological Chemistry, 278*(36), 34309–34319.
He, F., Hu, X., Xiong, W., Li, L., Lin, L., Shen, B., et al. (2014). Directed Bmp4 expression in neural crest cells generates a genetic model for the rare human bony syngnathia birth defect. *Developmental Biology, 391*(2), 170–181. http://dx.doi.org/10.1016/j.ydbio.2014.04.013.
He, M., Subramanian, R., Bangs, F., Omelchenko, T., Liem, K. F., Jr., Kapoor, T. M., et al. (2014). The kinesin-4 protein Kif7 regulates mammalian Hedgehog signalling by organizing the cilium tip compartment. *Nature Cell Biology, 16*(7), 663–672. http://dx.doi.org/10.1038/ncb2988.
Herron, B. J., Lu, W., Rao, C., Liu, S., Peters, H., Bronson, R. T., et al. (2002). Efficient generation and mapping of recessive developmental mutations using ENU mutagenesis. *Nature Genetics, 30*(2), 185–189. http://dx.doi.org/10.1038/ng812.
Higginbotham, H., Eom, T. Y., Mariani, L. E., Bachleda, A., Hirt, J., Gukassyan, V., et al. (2012). Arl13b in primary cilia regulates the migration and placement of interneurons in the developing cerebral cortex. *Developmental Cell, 23*(5), 925–938. http://dx.doi.org/10.1016/j.devcel.2012.09.019.
Hoover, A. N., Wynkoop, A., Zeng, H., Jia, J., Niswander, L. A., & Liu, A. (2008). C2cd3 is required for cilia formation and Hedgehog signaling in mouse. *Development, 135*(24), 4049–4058. http://dx.doi.org/10.1242/dev.029835.
Inman, K. E., Purcell, P., Kume, T., & Trainor, P. A. (2013). Interaction between Foxc1 and Fgf8 during mammalian jaw patterning and in the pathogenesis of syngnathia. *PLoS Genetics, 9*(12), e1003949. http://dx.doi.org/10.1371/journal.pgen.1003949.

Jaako, P., Debnath, S., Olsson, K., Bryder, D., Flygare, J., & Karlsson, S. (2012). Dietary L-leucine improves the anemia in a mouse model for Diamond-Blackfan anemia. *Blood*, *120*(11), 2225–2228. http://dx.doi.org/10.1182/blood-2012-05-431437.

Jaako, P., Flygare, J., Olsson, K., Quere, R., Ehinger, M., Henson, A., et al. (2011). Mice with ribosomal protein S19 deficiency develop bone marrow failure and symptoms like patients with Diamond-Blackfan anemia. *Blood*, *118*(23), 6087–6096. http://dx.doi.org/10.1182/blood-2011-08-371963.

Jacoby, M., Cox, J. J., Gayral, S., Hampshire, D. J., Ayub, M., Blockmans, M., et al. (2009). INPP5E mutations cause primary cilium signaling defects, ciliary instability and ciliopathies in human and mouse. *Nature Genetics*, *41*(9), 1027–1031. http://dx.doi.org/10.1038/ng.427.

Jin, H., White, S. R., Shida, T., Schulz, S., Aguiar, M., Gygi, S. P., et al. (2010). The conserved Bardet-Biedl syndrome proteins assemble a coat that traffics membrane proteins to cilia. *Cell*, *141*(7), 1208–1219. http://dx.doi.org/10.1016/j.cell.2010.05.015.

Jones, N. C., Lynn, M. L., Gaudenz, K., Sakai, D., Aoto, K., Rey, J. P., et al. (2008). Prevention of the neurocristopathy Treacher Collins syndrome through inhibition of p53 function. *Nature Medicine*, *14*(2), 125–133. http://dx.doi.org/10.1038/nm1725.

Kadakia, S., Helman, S. N., Badhey, A. K., Saman, M., & Ducic, Y. (2014). Treacher Collins syndrome: The genetics of a craniofacial disease. *International Journal of Pediatric Otorhinolaryngology*, *78*(6), 893–898. http://dx.doi.org/10.1016/j.ijporl.2014.03.006.

Khutornenko, A. A., Dalina, A. A., Chernyak, B. V., Chumakov, P. M., & Evstafieva, A. G. (2014). The role of dihydroorotate dehydrogenase in apoptosis induction in response to inhibition of the mitochondrial respiratory chain complex III. *Acta Naturae*, *6*(1), 69–75.

Klysik, M. (2008). Ciliary syndromes and treatment. *Pathology, Research and Practice*, *204*(2), 77–88. http://dx.doi.org/10.1016/j.prp.2007.10.013.

Kondrashov, N., Pusic, A., Stumpf, C. R., Shimizu, K., Hsieh, A. C., Xue, S., et al. (2011). Ribosome-mediated specificity in Hox mRNA translation and vertebrate tissue patterning. *Cell*, *145*(3), 383–397. http://dx.doi.org/10.1016/j.cell.2011.03.028.

Kressler, D., Hurt, E., & Bassler, J. (2010). Driving ribosome assembly. *Biochimica et Biophysica Acta*, *1803*(6), 673–683. http://dx.doi.org/10.1016/j.bbamcr.2009.10.009.

Kyttala, M., Tallila, J., Salonen, R., Kopra, O., Kohlschmidt, N., Paavola-Sakki, P., et al. (2006). MKS1, encoding a component of the flagellar apparatus basal body proteome, is mutated in Meckel syndrome. *Nature Genetics*, *38*(2), 155–157. http://dx.doi.org/10.1038/ng1714.

Laferte, A., Favry, E., Sentenac, A., Riva, M., Carles, C., & Chedin, S. (2006). The transcriptional activity of RNA polymerase I is a key determinant for the level of all ribosome components. *Genes and Development*, *20*(15), 2030–2040. http://dx.doi.org/10.1101/gad.386106.

Lafontaine, D. L., & Tollervey, D. (2001). The function and synthesis of ribosomes. *Nature Reviews. Molecular Cell Biology*, *2*(7), 514–520. http://dx.doi.org/10.1038/35080045.

Lancaster, M. A., Gopal, D. J., Kim, J., Saleem, S. N., Silhavy, J. L., Louie, C. M., et al. (2011). Defective Wnt-dependent cerebellar midline fusion in a mouse model of Joubert syndrome. *Nature Medicine*, *17*(6), 726–731. http://dx.doi.org/10.1038/nm.2380.

Lancaster, M. A., Louie, C. M., Silhavy, J. L., Sintasath, L., Decambre, M., Nigam, S. K., et al. (2009). Impaired Wnt-beta-catenin signaling disrupts adult renal homeostasis and leads to cystic kidney ciliopathy. *Nature Medicine*, *15*(9), 1046–1054. http://dx.doi.org/10.1038/nm.2010.

Larkin, R. M., & Guilfoyle, T. J. (1997). Reconstitution of yeast and Arabidopsis RNA polymerase alpha-like subunit heterodimers. *The Journal of Biological Chemistry*, *272*(19), 12824–12830.

Larsen, D. H., Hari, F., Clapperton, J. A., Gwerder, M., Gutsche, K., Altmeyer, M., et al. (2014). The NBS1-Treacle complex controls ribosomal RNA transcription in response to DNA damage. *Nature Cell Biology*, *16*(8), 792–803. http://dx.doi.org/10.1038/ncb3007.
Laster, Z., Temkin, D., Zarfin, Y., & Kushnir, A. (2001). Complete bony fusion of the mandible to the zygomatic complex and maxillary tuberosity: Case report and review. *International Journal of Oral and Maxillofacial Surgery*, *30*(1), 75–79. http://dx.doi.org/10.1054/ijom.2000.0009.
Lee, J. E., Silhavy, J. L., Zaki, M. S., Schroth, J., Bielas, S. L., Marsh, S. E., et al. (2012). CEP41 is mutated in Joubert syndrome and is required for tubulin glutamylation at the cilium. *Nature Genetics*, *44*(2), 193–199. http://dx.doi.org/10.1038/ng.1078.
Liem, K. F., Jr., He, M., Ocbina, P. J., & Anderson, K. V. (2009). Mouse Kif7/Costal2 is a cilia-associated protein that regulates Sonic hedgehog signaling. *Proceedings of the National Academy of Sciences of the United States of America*, *106*(32), 13377–13382. http://dx.doi.org/10.1073/pnas.0906944106.
Lipton, J. M., & Ellis, S. R. (2009). Diamond-Blackfan anemia: Diagnosis, treatment, and molecular pathogenesis. *Hematology/Oncology Clinics of North America*, *23*(2), 261–282. http://dx.doi.org/10.1016/j.hoc.2009.01.004.
Liu, A., Wang, B., & Niswander, L. A. (2005). Mouse intraflagellar transport proteins regulate both the activator and repressor functions of Gli transcription factors. *Development*, *132*(13), 3103–3111. http://dx.doi.org/10.1242/dev.01894.
Louie, C. M., Caridi, G., Lopes, V. S., Brancati, F., Kispert, A., Lancaster, M. A., et al. (2010). AHI1 is required for photoreceptor outer segment development and is a modifier for retinal degeneration in nephronophthisis. *Nature Genetics*, *42*(2), 175–180. http://dx.doi.org/10.1038/ng.519.
Lowry, R. B., Innes, A. M., Bernier, F. P., McLeod, D. R., Greenberg, C. R., Chudley, A. E., et al. (2003). Bowen-Conradi syndrome: A clinical and genetic study. *American Journal of Medical Genetics. Part A*, *120A*(3), 423–428. http://dx.doi.org/10.1002/ajmg.a.20059.
Lu, S., Goering, M., Gard, S., Xiong, B., McNairn, A. J., Jaspersen, S. L., et al. (2010). Eco1 is important for DNA damage repair in S. cerevisiae. *Cell Cycle*, *9*(16), 3315–3327. http://dx.doi.org/10.4161/cc.9.16.12673.
Lu, H., Toh, M. T., Narasimhan, V., Thamilselvam, S. K., Choksi, S. P., & Roy, S. (2015). A function for the Joubert syndrome protein Arl13b in ciliary membrane extension and ciliary length regulation. *Developmental Biology*, *397*(2), 225–236. http://dx.doi.org/10.1016/j.ydbio.2014.11.009.
Ludwig, L. S., Gazda, H. T., Eng, J. C., Eichhorn, S. W., Thiru, P., Ghazvinian, R., et al. (2014). Altered translation of GATA1 in Diamond-Blackfan anemia. *Nature Medicine*, *20*(7), 748–753. http://dx.doi.org/10.1038/nm.3557.
Mann, C., Buhler, J. M., Treich, I., & Sentenac, A. (1987). RPC40, a unique gene for a subunit shared between yeast RNA polymerases A and C. *Cell*, *48*(4), 627–637.
Maria, B. L., Boltshauser, E., Palmer, S. C., & Tran, T. X. (1999). Clinical features and revised diagnostic criteria in Joubert syndrome. *Journal of Child Neurology*, *14*(9), 583–590. discussion 590–581.
Matsson, H., Davey, E. J., Draptchinskaia, N., Hamaguchi, I., Ooka, A., Leveen, P., et al. (2004). Targeted disruption of the ribosomal protein S19 gene is lethal prior to implantation. *Molecular and Cellular Biology*, *24*(9), 4032–4037.
Mayer, C., Zhao, J., Yuan, X., & Grummt, I. (2004). mTOR-dependent activation of the transcription factor TIF-IA links rRNA synthesis to nutrient availability. *Genes & Development*, *18*(4), 423–434. http://dx.doi.org/10.1101/gad.285504.

McGowan, K. A., Li, J. Z., Park, C. Y., Beaudry, V., Tabor, H. K., Sabnis, A. J., et al. (2008). Ribosomal mutations cause p53-mediated dark skin and pleiotropic effects. *Nature Genetics*, *40*(8), 963–970. http://dx.doi.org/10.1038/ng.188.

McGowan, K. A., & Mason, P. J. (2011). Animal models of Diamond Blackfan anemia. *Seminars in Hematology*, *48*(2), 106–116. http://dx.doi.org/10.1053/j.seminhematol.2011.02.001.

Meyer, B., Wurm, J. P., Kotter, P., Leisegang, M. S., Schilling, V., Buchhaupt, M., et al. (2011). The Bowen-Conradi syndrome protein Nep1 (Emg1) has a dual role in eukaryotic ribosome biogenesis, as an essential assembly factor and in the methylation of Psi1191 in yeast 18S rRNA. *Nucleic Acids Research*, *39*(4), 1526–1537. http://dx.doi.org/10.1093/nar/gkq931.

M'Hamdi, O., Ouertani, I., & Chaabouni-Bouhamed, H. (2014). Update on the genetics of Bardet-Biedl syndrome. *Molecular Syndromology*, *5*(2), 51–56. http://dx.doi.org/10.1159/000357054.

Minet, M., Dufour, M. E., & Lacroute, F. (1992). Cloning and sequencing of a human cDNA coding for dihydroorotate dehydrogenase by complementation of the corresponding yeast mutant. *Gene*, *121*(2), 393–396.

Mitchell, K. J., Pinson, K. I., Kelly, O. G., Brennan, J., Zupicich, J., Scherz, P., et al. (2001). Functional analysis of secreted and transmembrane proteins critical to mouse development. *Nature Genetics*, *28*(3), 241–249. http://dx.doi.org/10.1038/90074.

Monnich, M., Kuriger, Z., Print, C. G., & Horsfield, J. A. (2011). A zebrafish model of Roberts syndrome reveals that Esco2 depletion interferes with development by disrupting the cell cycle. *PLoS One*, *6*(5), e20051. http://dx.doi.org/10.1371/journal.pone.0020051.

Mykytyn, K., Mullins, R. F., Andrews, M., Chiang, A. P., Swiderski, R. E., Yang, B., et al. (2004). Bardet-Biedl syndrome type 4 (BBS4)-null mice implicate Bbs4 in flagella formation but not global cilia assembly. *Proceedings of the National Academy of Sciences of the United States of America*, *101*(23), 8664–8669. http://dx.doi.org/10.1073/pnas.0402354101.

Nachury, M. V., Loktev, A. V., Zhang, Q., Westlake, C. J., Peranen, J., Merdes, A., et al. (2007). A core complex of BBS proteins cooperates with the GTPase Rab8 to promote ciliary membrane biogenesis. *Cell*, *129*(6), 1201–1213. http://dx.doi.org/10.1016/j.cell.2007.03.053.

Nakhoul, H., Ke, J., Zhou, X., Liao, W., Zeng, S. X., & Lu, H. (2014). Ribosomopathies: Mechanisms of disease. *Clinical Medicine Insights. Blood Disorders*, 7, 7–16. http://dx.doi.org/10.4137/CMBD.S16952.

Narla, A., & Ebert, B. L. (2010). Ribosomopathies: Human disorders of ribosome dysfunction. *Blood*, *115*(16), 3196–3205. http://dx.doi.org/10.1182/blood-2009-10-178129.

Ng, S. B., Buckingham, K. J., Lee, C., Bigham, A. W., Tabor, H. K., Dent, K. M., et al. (2010). Exome sequencing identifies the cause of a Mendelian disorder. *Nature Genetics*, *42*(1), 30–35. http://dx.doi.org/10.1038/ng.499.

Nishimura, D. Y., Fath, M., Mullins, R. F., Searby, C., Andrews, M., Davis, R., et al. (2004). Bbs2-null mice have neurosensory deficits, a defect in social dominance, and retinopathy associated with mislocalization of rhodopsin. *Proceedings of the National Academy of Sciences of the United States of America*, *101*(47), 16588–16593. http://dx.doi.org/10.1073/pnas.0405496101.

Olbrich, H., Fliegauf, M., Hoefele, J., Kispert, A., Otto, E., Volz, A., et al. (2003). Mutations in a novel gene, NPHP3, cause adolescent nephronophthisis, tapeto-retinal degeneration and hepatic fibrosis. *Nature Genetics*, *34*(4), 455–459. http://dx.doi.org/10.1038/ng1216.

Omran, H., Haffner, K., Burth, S., Fernandez, C., Fargier, B., Villaquiran, A., et al. (2001). Human adolescent nephronophthisis: Gene locus synteny with polycystic kidney disease in pcy mice. *Journal of the American Society of Nephrology*, *12*(1), 107–113.

Park, T. J., Haigo, S. L., & Wallingford, J. B. (2006). Ciliogenesis defects in embryos lacking inturned or fuzzy function are associated with failure of planar cell polarity and Hedgehog signaling. *Nature Genetics*, *38*(3), 303–311. http://dx.doi.org/10.1038/ng1753.

Patel, S. D., Porras, S., & Lypka, M. (2015). Journey to chew: A case of maxillary duplication and bony syngnathia. *Journal of Cranio-Maxillo-Facial Surgery*, *43*(1), 57–61. http://dx.doi.org/10.1016/j.jcms.2014.10.010.

Payne, E. M., Virgilio, M., Narla, A., Sun, H., Levine, M., Paw, B. H., et al. (2012). L-Leucine improves the anemia and developmental defects associated with Diamond-Blackfan anemia and del(5q) MDS by activating the mTOR pathway. *Blood*, *120*(11), 2214–2224. http://dx.doi.org/10.1182/blood-2011-10-382986.

Pedersen, L. B., & Rosenbaum, J. L. (2008). Intraflagellar transport (IFT) role in ciliary assembly, resorption and signalling. *Current Topics in Developmental Biology*, *85*, 23–61. http://dx.doi.org/10.1016/S0070-2153(08)00802-8.

Plotnikova, O. V., Seo, S., Cottle, D. L., Conduit, S., Hakim, S., Dyson, J. M., et al. (2015). INPP5E interacts with AURKA, linking phosphoinositide signaling to primary cilium stability. *Journal of Cell Science*, *128*(2), 364–372. http://dx.doi.org/10.1242/jcs.161323.

Pospisilova, D., Cmejlova, J., Hak, J., Adam, T., & Cmejla, R. (2007). Successful treatment of a Diamond-Blackfan anemia patient with amino acid leucine. *Haematologica*, *92*(5), e66–e67.

Rainger, J., Bengani, H., Campbell, L., Anderson, E., Sokhi, K., Lam, W., et al. (2012). Miller (Genee-Wiedemann) syndrome represents a clinically and biochemically distinct subgroup of postaxial acrofacial dysostosis associated with partial deficiency of DHODH. *Human Molecular Genetics*, *21*(18), 3969–3983. http://dx.doi.org/10.1093/hmg/dds218.

Rakkolainen, A., Ala-Mello, S., Kristo, P., Orpana, A., & Jarvela, I. (2002). Four novel mutations in the OFD1 (Cxorf5) gene in Finnish patients with oral-facial-digital syndrome 1. *Journal of Medical Genetics*, *39*(4), 292–296.

Reiter, J. F., Blacque, O. E., & Leroux, M. R. (2012). The base of the cilium: Roles for transition fibres and the transition zone in ciliary formation, maintenance and compartmentalization. *EMBO Reports*, *13*(7), 608–618. http://dx.doi.org/10.1038/embor.2012.73.

Reiter, J. F., & Skarnes, W. C. (2006). Tectonic, a novel regulator of the Hedgehog pathway required for both activation and inhibition. *Genes & Development*, *20*(1), 22–27. http://dx.doi.org/10.1101/gad.1363606.

Romio, L., Fry, A. M., Winyard, P. J., Malcolm, S., Woolf, A. S., & Feather, S. A. (2004). OFD1 is a centrosomal/basal body protein expressed during mesenchymal-epithelial transition in human nephrogenesis. *Journal of the American Society of Nephrology*, *15*(10), 2556–2568. http://dx.doi.org/10.1097/01.ASN.0000140220.46477.5C.

Romio, L., Wright, V., Price, K., Winyard, P. J., Donnai, D., Porteous, M. E., et al. (2003). OFD1, the gene mutated in oral-facial-digital syndrome type 1, is expressed in the metanephros and in human embryonic renal mesenchymal cells. *Journal of the American Society of Nephrology*, *14*(3), 680–689.

Ross, A. P., & Zarbalis, K. S. (2014). The emerging roles of ribosome biogenesis in craniofacial development. *Frontiers in Physiology*, *5*, 26. http://dx.doi.org/10.3389/fphys. 2014.00026.

Rubbi, C. P., & Milner, J. (2003). Disruption of the nucleolus mediates stabilization of p53 in response to DNA damage and other stresses. *The EMBO Journal*, *22*(22), 6068–6077. http://dx.doi.org/10.1093/emboj/cdg579.

Russell, J., & Zomerdijk, J. C. (2006). The RNA polymerase I transcription machinery. *Biochemical Society Symposium*, *73*, 203–216.

Sakai, D., Dixon, J., Dixon, M. J., & Trainor, P. A. (2012). Mammalian neurogenesis requires Treacle-Plk1 for precise control of spindle orientation, mitotic progression, and maintenance of neural progenitor cells. *PLoS Genetics*, *8*(3), e1002566. http://dx.doi.org/10.1371/journal.pgen.1002566.

Sakai, D., & Trainor, P. A. (2009). Treacher Collins syndrome: Unmasking the role of Tcof1/treacle. *The International Journal of Biochemistry & Cell Biology*, *41*(6), 1229–1232. http://dx.doi.org/10.1016/j.biocel.2008.10.026.

Sang, L., Miller, J. J., Corbit, K. C., Giles, R. H., Brauer, M. J., Otto, E. A., et al. (2011). Mapping the NPHP-JBTS-MKS protein network reveals ciliopathy disease genes and pathways. *Cell*, *145*(4), 513–528. http://dx.doi.org/10.1016/j.cell.2011.04.019.

Sankaran, V. G., Ghazvinian, R., Do, R., Thiru, P., Vergilio, J. A., Beggs, A. H., et al. (2012). Exome sequencing identifies GATA1 mutations resulting in Diamond-Blackfan anemia. *The Journal of Clinical Investigation*, *122*(7), 2439–2443. http://dx.doi.org/10.1172/JCI63597.

Santagati, F., & Rijli, F. M. (2003). Cranial neural crest and the building of the vertebrate head. *Nature Reviews. Neuroscience*, *4*(10), 806–818. http://dx.doi.org/10.1038/nrn1221.

Schaefer, E., Collet, C., Genevieve, D., Vincent, M., Lohmann, D. R., Sanchez, E., et al. (2014). Autosomal recessive POLR1D mutation with decrease of TCOF1 mRNA is responsible for Treacher Collins syndrome. *Genetics in Medicine*, *16*(9), 720–724. http://dx.doi.org/10.1038/gim.2014.12.

Schneider, L., Clement, C. A., Teilmann, S. C., Pazour, G. J., Hoffmann, E. K., Satir, P., et al. (2005). PDGFRalphaalpha signaling is regulated through the primary cilium in fibroblasts. *Current Biology*, *15*(20), 1861–1866. http://dx.doi.org/10.1016/j.cub. 2005.09.012.

Seo, S., Singh, H. P., Lacal, P. M., Sasman, A., Fatima, A., Liu, T., et al. (2012). Forkhead box transcription factor FoxC1 preserves corneal transparency by regulating vascular growth. *Proceedings of the National Academy of Sciences of the United States of America*, *109*(6), 2015–2020. http://dx.doi.org/10.1073/pnas.1109540109.

Skibbens, R. V., Corson, L. B., Koshland, D., & Hieter, P. (1999). Ctf7p is essential for sister chromatid cohesion and links mitotic chromosome structure to the DNA replication machinery. *Genes & Development*, *13*(3), 307–319.

Smith, U. M., Consugar, M., Tee, L. J., McKee, B. M., Maina, E. N., Whelan, S., et al. (2006). The transmembrane protein meckelin (MKS3) is mutated in Meckel-Gruber syndrome and the wpk rat. *Nature Genetics*, *38*(2), 191–196. http://dx.doi.org/10.1038/ng1713.

Sommer, D., Peters, A. E., Baumgart, A. K., & Beyer, M. (2015). TALEN-mediated genome engineering to generate targeted mice. *Chromosome Research*, *23*, 43–55. http://dx.doi.org/10.1007/s10577-014-9457-1.

Sondalle, S. B., & Baserga, S. J. (2014). Human diseases of the SSU processome. *Biochimica et Biophysica Acta*, *1842*(6), 758–764. http://dx.doi.org/10.1016/j.bbadis.2013.11.004.

Sorokin, S. (1962). Centrioles and the formation of rudimentary cilia by fibroblasts and smooth muscle cells. *The Journal of Cell Biology*, *15*, 363–377.

Su, C. Y., Bay, S. N., Mariani, L. E., Hillman, M. J., & Caspary, T. (2012). Temporal deletion of Arl13b reveals that a mispatterned neural tube corrects cell fate over time. *Development*, *139*(21), 4062–4071. http://dx.doi.org/10.1242/dev.082321.

Szymanska, K., Hartill, V. L., & Johnson, C. A. (2014). Unraveling the genetics of Joubert and Meckel-Gruber syndromes. *Journal of Pediatric Genetics*, *3*(2), 65–78. http://dx.doi.org/10.3233/PGE-14090.

Tabler, J. M., Barrell, W. B., Szabo-Rogers, H. L., Healy, C., Yeung, Y., Perdiguero, E. G., et al. (2013). Fuz mutant mice reveal shared mechanisms between ciliopathies and FGF-related syndromes. *Developmental Cell*, *25*(6), 623–635. http://dx.doi.org/10.1016/j.devcel.2013.05.021.

Takahashi, H., Ueyama, Y., Hibino, T., Kuwahara, Y., Suzuki, S., Hioki, K., et al. (1986). A new mouse model of genetically transmitted polycystic kidney disease. *The Journal of Urology*, *135*(6), 1280–1283.

Tallila, J., Jakkula, E., Peltonen, L., Salonen, R., & Kestila, M. (2008). Identification of CC2D2A as a Meckel syndrome gene adds an important piece to the ciliopathy puzzle. *American Journal of Human Genetics*, *82*(6), 1361–1367. http://dx.doi.org/10.1016/j.ajhg.2008.05.004.

Terzian, T., & Box, N. (2013). Genetics of ribosomal proteins: "Curiouser and curiouser" *PLoS Genetics, 9*(1), e1003300. http://dx.doi.org/10.1371/journal.pgen.1003300.

Thauvin-Robinet, C., Cossee, M., Cormier-Daire, V., Van Maldergem, L., Toutain, A., Alembik, Y., et al. (2006). Clinical, molecular, and genotype-phenotype correlation studies from 25 cases of oral-facial-digital syndrome type 1: A French and Belgian collaborative study. *Journal of Medical Genetics, 43*(1), 54–61. http://dx.doi.org/10.1136/jmg.2004.027672.

Thauvin-Robinet, C., Lee, J. S., Lopez, E., Herranz-Perez, V., Shida, T., Franco, B., et al. (2014). The oral-facial-digital syndrome gene C2CD3 encodes a positive regulator of centriole elongation. *Nature Genetics, 46*(8), 905–911. http://dx.doi.org/10.1038/ng.3031.

Theveneau, E., & Mayor, R. (2012). Neural crest migration: Interplay between chemorepellents, chemoattractants, contact inhibition, epithelial-mesenchymal transition, and collective cell migration. *Wiley Interdisciplinary Reviews. Developmental Biology, 1*(3), 435–445. http://dx.doi.org/10.1002/wdev.28.

Tobin, J. L., Di Franco, M., Eichers, E., May-Simera, H., Garcia, M., Yan, J., et al. (2008). Inhibition of neural crest migration underlies craniofacial dysmorphology and Hirschsprung's disease in Bardet-Biedl syndrome. *Proceedings of the National Academy of Sciences of the United States of America, 105*(18), 6714–6719. http://dx.doi.org/10.1073/pnas.0707057105.

Toriello, H. V. (2009). Are the oral-facial-digital syndromes ciliopathies? *American Journal of Medical Genetics. Part A, 149A*(5), 1089–1095. http://dx.doi.org/10.1002/ajmg.a.32799.

Toth, A., Ciosk, R., Uhlmann, F., Galova, M., Schleiffer, A., & Nasmyth, K. (1999). Yeast cohesin complex requires a conserved protein, Eco1p(Ctf7), to establish cohesion between sister chromatids during DNA replication. *Genes & Development, 13*(3), 320–333.

Town, T., Breunig, J. J., Sarkisian, M. R., Spilianakis, C., Ayoub, A. E., Liu, X., et al. (2008). The stumpy gene is required for mammalian ciliogenesis. *Proceedings of the National Academy of Sciences of the United States of America, 105*(8), 2853–2858. http://dx.doi.org/10.1073/pnas.0712385105.

Trainor, P. A., & Andrews, B. T. (2013). Facial dysostoses: Etiology, pathogenesis and management. *American Journal of Medical Genetics. Part C, Seminars in Medical Genetics, 163C*(4), 283–294. http://dx.doi.org/10.1002/ajmg.c.31375.

Trainor, P. A., Dixon, J., & Dixon, M. J. (2009). Treacher Collins syndrome: Etiology, pathogenesis and prevention. *European Journal of Human Genetics, 17*(3), 275–283. http://dx.doi.org/10.1038/ejhg.2008.221.

Trainor, P. A., & Merrill, A. E. (2014). Ribosome biogenesis in skeletal development and the pathogenesis of skeletal disorders. *Biochimica et Biophysica Acta, 1842*(6), 769–778. http://dx.doi.org/10.1016/j.bbadis.2013.11.010.

Tran, P. V., Haycraft, C. J., Besschetnova, T. Y., Turbe-Doan, A., Stottmann, R. W., Herron, B. J., et al. (2008). THM1 negatively modulates mouse sonic hedgehog signal transduction and affects retrograde intraflagellar transport in cilia. *Nature Genetics, 40*(4), 403–410. http://dx.doi.org/10.1038/ng.105.

Treacher Collins Syndrome Collaborative Group. (1996). Positional cloning of a gene involved in the pathogenesis of Treacher Collins syndrome. *Nature Genetics, 12*(2), 130–136. http://dx.doi.org/10.1038/ng0296-130.

Ugurlu, K., Turan, T., Urganci, N., Gozu, A., Gunay, Y., & Bas, L. (1999). Fusion of maxillary and mandibular alveolar process together with a median mandibular cleft: A rare congenital anomaly. *Journal of Cranio-Maxillo-Facial Surgery, 27*(2), 105–108.

Valdez, B. C., Henning, D., So, R. B., Dixon, J., & Dixon, M. J. (2004). The Treacher Collins syndrome (TCOF1) gene product is involved in ribosomal DNA gene transcription by interacting with upstream binding factor. *Proceedings of the National Academy of Sciences of the United States of America, 101*(29), 10709–10714.

Vega, H., Waisfisz, Q., Gordillo, M., Sakai, N., Yanagihara, I., Yamada, M., et al. (2005). Roberts syndrome is caused by mutations in ESCO2, a human homolog of yeast ECO1 that is essential for the establishment of sister chromatid cohesion. *Nature Genetics*, *37*(5), 468–470. http://dx.doi.org/10.1038/ng1548.

Veleri, S., Manjunath, S. H., Fariss, R. N., May-Simera, H., Brooks, M., Foskett, T. A., et al. (2014). Ciliopathy-associated gene Cc2d2a promotes assembly of subdistal appendages on the mother centriole during cilia biogenesis. *Nature Communications*, *5*, 4207. http://dx.doi.org/10.1038/ncomms5207.

Vierkotten, J., Dildrop, R., Peters, T., Wang, B., & Ruther, U. (2007). Ftm is a novel basal body protein of cilia involved in Shh signalling. *Development*, *134*(14), 2569–2577. http://dx.doi.org/10.1242/dev.003715.

Villanueva-Garcia, D., Contreras-Masse, G., Villa-Guillen, M., Ramon-Garcia, G., & Murguia-De Sierra, T. (2009). Syngnathism in an infant born to consanguineous parents. *American Journal of Medical Genetics. Part A*, *149A*(10), 2303–2305. http://dx.doi.org/10.1002/ajmg.a.32734.

Weatherbee, S. D., Niswander, L. A., & Anderson, K. V. (2009). A mouse model for Meckel syndrome reveals Mks1 is required for ciliogenesis and Hedgehog signaling. *Human Molecular Genetics*, *18*(23), 4565–4575. http://dx.doi.org/10.1093/hmg/ddp422.

Wettke-Schafer, R., & Kantner, G. (1983). X-linked dominant inherited diseases with lethality in hemizygous males. *Human Genetics*, *64*(1), 1–23.

Whelan, G., Kreidl, E., Wutz, G., Egner, A., Peters, J. M., & Eichele, G. (2012). Cohesin acetyltransferase Esco2 is a cell viability factor and is required for cohesion in pericentric heterochromatin. *The EMBO Journal*, *31*(1), 71–82. http://dx.doi.org/10.1038/emboj.2011.381.

Wheway, G., Abdelhamed, Z., Natarajan, S., Toomes, C., Inglehearn, C., & Johnson, C. A. (2013). Aberrant Wnt signalling and cellular over-proliferation in a novel mouse model of Meckel-Gruber syndrome. *Developmental Biology*, *377*(1), 55–66. http://dx.doi.org/10.1016/j.ydbio.2013.02.015.

White, R. M., Cech, J., Ratanasirintrawoot, S., Lin, C. Y., Rahl, P. B., Burke, C. J., et al. (2011). DHODH modulates transcriptional elongation in the neural crest and melanoma. *Nature*, *471*(7339), 518–522. http://dx.doi.org/10.1038/nature09882.

Wieczorek, D. (2013). Human facial dysostoses. *Clinical Genetics*, *83*(6), 499–510. http://dx.doi.org/10.1111/cge.12123.

Williams, C. L., Winkelbauer, M. E., Schafer, J. C., Michaud, E. J., & Yoder, B. K. (2008). Functional redundancy of the B9 proteins and nephrocystins in Caenorhabditis elegans ciliogenesis. *Molecular Biology of the Cell*, *19*(5), 2154–2168. http://dx.doi.org/10.1091/mbc.E07-10-1070.

Wu, X., Sandhu, S., Patel, N., Triggs-Raine, B., & Ding, H. (2010). EMG1 is essential for mouse pre-implantation embryo development. *BMC Developmental Biology*, *10*, 99. http://dx.doi.org/10.1186/1471-213X-10-99.

Wurm, J. P., Meyer, B., Bahr, U., Held, M., Frolow, O., Kotter, P., et al. (2010). The ribosome assembly factor Nep1 responsible for Bowen-Conradi syndrome is a pseudouridine-N1-specific methyltransferase. *Nucleic Acids Research*, *38*(7), 2387–2398. http://dx.doi.org/10.1093/nar/gkp1189.

Xue, S., & Barna, M. (2012). Specialized ribosomes: A new frontier in gene regulation and organismal biology. *Nature Reviews. Molecular Cell Biology*, *13*(6), 355–369. http://dx.doi.org/10.1038/nrm3359.

Xue, S., Tian, S., Fuji, K., Kladwang, W., Das, R., & Barna, M. (2015). RNA regulons in Hox5′ UTRs confer ribosome specificity to gene regulation. *Nature*, *517*(7532), 33–38.

Yao, Y., Yamamoto, K., Nishi, Y., Nogi, Y., & Muramatsu, M. (1996). Mouse RNA polymerase I 16-kDa subunit able to associate with 40-kDa subunit is a homolog of yeast AC19 subunit of RNA polymerases I and III. *The Journal of Biological Chemistry*, *271*(51), 32881–32885.

CHAPTER SEVENTEEN

From Bench to Bedside and Back: Improving Diagnosis and Treatment of Craniofacial Malformations Utilizing Animal Models

Alice F. Goodwin[*,†], **Rebecca Kim**[*,†], **Jeffrey O. Bush**[*,‡,§,1], **Ophir D. Klein**[*,†,¶,§,1]

*Program in Craniofacial Biology, University of California San Francisco, San Francisco, California, USA
†Department of Orofacial Sciences, University of California San Francisco, San Francisco, California, USA
‡Department of Cell and Tissue Biology, University of California San Francisco, San Francisco, California, USA
¶Department of Pediatrics, University of California San Francisco, San Francisco, California, USA
§Institute for Human Genetics, University of California San Francisco, San Francisco, California, USA
[1]Corresponding authors: e-mail address: jeffrey.bush@ucsf.edu; ophir.klein@ucsf.edu

Contents

Abstract

Craniofacial anomalies are among the most common birth defects and are associated with increased mortality and, in many cases, the need for lifelong treatment. Over the past few decades, dramatic advances in the surgical and medical care of these patients have led to marked improvements in patient outcomes. However, none of the treatments currently in clinical use address the underlying molecular causes of these disorders. Fortunately, the field of craniofacial developmental biology provides a strong foundation for improved diagnosis and for therapies that target the genetic causes

Current Topics in Developmental Biology, Volume 115
ISSN 0070-2153
http://dx.doi.org/10.1016/bs.ctdb.2015.07.003

of birth defects. In this chapter, we discuss recent advances in our understanding of the embryology of craniofacial conditions, and we focus on the use of animal models to guide rational therapies anchored in genetics and biochemistry.

1. MODELS TO UNCOVER GENETICS OF CLEFT LIP AND PALATE

Cleft lip with or without cleft palate (CL/P) is the most common congenital anomaly of craniofacial development, affecting approximately 1 in 700 live births (Dixon, Marazita, Beaty, & Murray, 2011; Iwata, Parada, & Chai, 2011; Jugessur, Farlie, & Kilpatrick, 2009). It is estimated that 70% of CL/P cases are isolated and nonsyndromic and that 30% occur as part of one of more than 500 Mendelian syndromes (Dixon et al., 2011). Over the past few decades, mouse models have proved an invaluable tool in understanding the etiology of CL/P (Table 1; Bush & Jiang, 2012; Gritli-Linde, 2008; Jiang, Bush, & Lidral, 2006). These models have led to exponential growth in our knowledge of the molecular and cellular basis for clefts and have contributed greatly to the understanding of general developmental mechanisms. To date, the principle clinical benefit of information gained from these studies has been to improve diagnostics and estimates of recurrence risks (Dixon et al., 2011; Grosen et al., 2010). Looking toward the future, our improved understanding of the developmental processes underlying cleft pathogenesis increasingly suggests opportunities for developing new preventative and therapeutic approaches.

From an embryologic perspective, formation of the lip and palate has developmental similarities in that both require the coordinated growth and fusion of embryonic prominences. However, these are temporally distinct events that involve different morphogenetic programs. The development of the lip occurs during the 4th through 6th week of human development and involves fusions of the maxillary, medial nasal, and lateral nasal processes, whereas secondary palate development occurs between the 6th and 12th week and involves the fusion of the maxillary-derived secondary palatal shelves (Bush & Jiang, 2012; Jiang et al., 2006). Consistent with this temporal asynchrony, cleft lip and cleft palate can occur separately or together; cleft lip is defined as a gap between the philtrum and the lateral upper lip, and cleft palate is defined as a gap in the secondary palate (roof of the mouth). Although the lip and palate form through separate developmental programs, many genetic factors play a role in both processes.

Table 1 Clefting Genes and Candidate Loci with Associated Mouse Models of Orofacial Clefting

Human Disease	Gene/Locus	Mouse Phenotype
Nonsyndromic CL/P	*MSX1*	Cleft palate (Satokata & Maas, 1994)
Nonsyndromic CL/P	*TGFβ3*	Cleft palate (Kaartinen et al., 1995; Proetzel et al., 1995)
Nonsyndromic CL/P	*IRF6*	CP (Ingraham et al., 2006; Richardson et al., 2006)
Nonsyndromic CL/P	*8q24 locus*	Deletion of 8q24 region in mice results in CL/P possibly by loss of *Myc* regulation (Uslu et al., 2014)
Nonsyndromic CL/P	*PDGFC*	CP (Ding et al., 2004)
Nonsyndromic CL/P	*VAX1*	CP (Bertuzzi, Hindges, Mui, O'Leary, & Lemke, 1999)
Nonsyndromic CL/P	*ARHGAP29* (*ABCA4 locus*)	CP (Leslie et al., 2012)
Nonsyndromic CL/P	*BMP4*	Conditional ablation in the epithelium results in CL (Liu et al., 2005)
Nonsyndromic CL/P	*FGFR2*	Conditional ablation in the epithelium results in CP (Hosokawa et al., 2009; Rice et al., 2004)
Nonsyndromic CL/P	*MYH9*	Conditional ablation in the epithelium results in fusion defects; compound homozygous loss of Myh9; Myh10 results in CP (Kim et al., 2015)
Syndromic CL/P	**Gene/Locus**	**Mouse Phenotype**
Van Der Woude syndrome (OMIM#119300)	*IRF6*	CP (Ingraham et al., 2006; Richardson et al., 2006)
Ankyloblepharon–ectodermal defects–cleft lip/palate (AEC) (OMIM#106260); Ectrodactyly, ectodermal dysplasia, and cleft lip/palate(EEC) (OMIM#129900)	*P63*	Knock-in of human allele results in autosomal dominant CP and phenocopies AEC (Ferone et al., 2012), and homozygous null displays CL/P and phenocopies EEC (Moretti et al., 2010; Thomason, Dixon, & Dixon, 2008)

Continued

Table 1 Clefting Genes and Candidate Loci with Associated Mouse Models of Orofacial Clefting—cont'd

Syndromic CL/P	Gene/Locus	Mouse Phenotype
Loeys–Dietz syndrome (OMIM#609192)	*TGFBR1*, *TGFBR2*, *SMAD3*, *TGFB2*	Conditional ablation of TgfBR1 or TgfBR2 in palate epithelium or NCC results in CP (Dudas et al., 2006; Ito et al., 2003; Iwata et al., 2012; Xu et al., 2006)
Stickler type 1 (OMIM#108300)	*COL2A1*	CP (Li, Prockop, et al., 1995)
Stickler type 2 (OMIM#604841)	*COL11A1*	CP (Li, Lacerda, et al., 1995; Seegmiller, Fraser, & Sheldon, 1971)
Smith–Lemli–Opitz syndrome (OMIM#270400)	*DHCR7*	CP phenocopies SLOS (Wassif et al., 2001)
Treacher Collins (OMIM#154500)	*TCOF1*	CP phenocopies Treacher Collins (Dixon et al., 2006; Jones et al., 2008)
Craniofrontonasal syndrome (CFNS) (OMIM#304110)	*EFNB1*	CP phenocopies CFNS (Bush & Soriano, 2010; Compagni, Logan, Klein, & Adams, 2003)
Pierre Robin syndrome (OMIM#261800)	*SOX9*	CP (Bi et al., 2001)
Andersen Syndrome (OMIM#170390)	*KCNJ2*	CP (Zaritsky, Eckman, Wellman, Nelson, & Schwarz, 2000)
Cleft palate with or without ankyloglossia, X-linked (CPX) (OMIM#303400)	*TBX22*	CP phenocopies CPX (Pauws et al., 2009)
DiGeorge syndrome (OMIM#188400)	*TBX1*	CP phenocopies DiGeorge (Jerome & Papaioannou, 2001)
Branchiooculofacial syndrome (OMIM#113620)	*TFAP2a*	Chimeras and hypomorphic embryos have CL/P (Green et al., 2015; Nottoli, Hagopian-Donaldson, Zhang, Perkins, & Williams, 1998), ablation from NCC results in CP (Brewer, Feng, Huang, Sullivan, & Williams, 2004), deletion from midface results in midline cleft

Table 1 Clefting Genes and Candidate Loci with Associated Mouse Models of Orofacial Clefting—cont'd

Syndromic CL/P	Gene/Locus	Mouse Phenotype
		(Nelson & Williams, 2004), hypomorphic allele results in CL/P
Greig cephalopolysyndactyly syndrome (OMIM#175700)	GLI3	CP (Huang, Goudy, Ketova, Litingtung, & Chiang, 2008)
Holoprosencephaly 9 (OMIM#610829)	GLI2	CP (Mo et al., 1997)
Coloboma, heart anomaly, choanal atresia, retardation, genital and ear anomalies (CHARGE) (OMIM#214800)	CHD7	CP phenocopies CHARGE (Sperry et al., 2014)
Hypogonadotropic hypogonadism with or without anosmia (Kallman syndrome) (OMIM#147950)	FGFR1	CP (Trokovic, Trokovic, Mai, & Partanen, 2003) or CP with midline cleft (Wang et al., 2013)
Crouzon syndrome(OMIM#101200)	FGFR2	CP, phenocopies Crouzon (Eswarakumar, Horowitz, Locklin, Morriss-Kay, & Lonai, 2004)
Apert syndrome (OMIM#101200)	FGFR2	CP, phenocopies Apert (Martinez-Abadias et al., 2013)
Otopalatodigital syndrome (OMIM#311300)	FLNA	CP (Hart et al., 2006)
Lymphedema-distichiasis syndrome (OMIM#153400)	FOXC2	CP (Iida et al., 1997; Winnier, Hargett, & Hogan, 1997)
Hypothyroidism, athyroidal, with spiky hair and cleft Palate (OMIM#241850)	FOXE1	CP, models hereditary thyroid dysgenesis and cleft palate (De Felice et al., 1998)
Glass syndrome (OMIM#612313)	*SATB2*	CP phenocopies 2q32-q33 deletion syndrome (Britanova et al., 2006)
Saethre–Chotzen syndrome (OMIM#101400)	TWIST1	Ablation in mandibular NCCs results in posterior CP (Zhang et al., 2012)

Complete tables of human clefting genes and loci, including those without mouse models can be found in Dixon et al. (2011).

Prior to the advent of genome-wide methodologies, candidate genes for linkage and mutation analysis were selected largely on the basis of phenotypes arising from gene-targeted mutations in mice. For example, targeted mutation of the *Msx1* homeobox gene resulted in syndromic cleft secondary palate and tooth agenesis in homozygous mutant mice and led to the evaluation of its human ortholog as a human CL/P candidate (Satokata & Maas, 1994). Linkage disequilibrium was found between *MSX1* and human CL/P, and ultimately, mutations in *MSX1* were identified in patients with syndromic cleft palate with tooth agenesis as well as in patients with nonsyndromic CL/P; in total, mutations in *MSX1* may contribute to around 2% of nonsyndromic CL/P cases (Jezewski et al., 2003; Lidral et al., 1998). Integration of mouse genetics, *ex vivo*, and molecular techniques has advanced our understanding of the developmental basis for *Msx1* involvement in CL/P. *Msx1* controls cell proliferation in the anterior secondary palate by maintaining mesenchymal expression of *Bmp4* in the palatal mesenchyme (Alappat, Zhang, & Chen, 2003; Parada & Chai, 2012; Zhang et al., 2002). Whereas mutations in *MSX1* have been identified in humans with cleft lip, *Msx1*$^{-/-}$ mice do not exhibit cleft lip, but rather cleft palate only. The related MSX2 might partially compensate for MSX1 loss in mice, because mice lacking MSX1 and MSX2 exhibit bilateral CL/P as part of dramatic craniofacial dysmorphogenesis (Ishii et al., 2005). Interestingly, *Msx1*$^{-/-}$ and *Pax9*$^{-/-}$ mice exhibit cleft lip with variable penetrance, possibly due to a shared role of MSX1 and PAX9 in the regulation of *Bmp4* or other factors critical for lip morphogenesis (Liu et al., 2005; Nakatomi et al., 2010; Zhou et al., 2013). On the other hand, compound loss of *Dlx5*$^{-/-}$ and *Msx1*$^{-/-}$ in mice restored cell proliferation and rescued the cleft palate phenotype (Han et al., 2009). These results exemplify the complexity of the genetic networks controlling lip and palate development and support the likely importance of genetic interaction in human CL/P. Indeed, PAX9 has been confirmed as a contributor to human syndromic and nonsyndromic CL/P (Ichikawa et al., 2006; Schuffenhauer et al., 1999). Mouse genetics studies designed to elucidate genetic pathways have therefore been highly useful in identifying candidate genes for human genetic analysis.

The TGF*β3* gene has been considered as a candidate human clefting gene based on the cleft palate phenotype in *Tgfβ3*$^{-/-}$ homozygous mice (Kaartinen et al., 1995; Proetzel et al., 1995). Though no loss-of-function mutations have been identified in *TGFβ3* in nonsyndromic cleft patients, association with the *TGFβ3* genomic locus has been demonstrated (Lidral et al., 1998). Interestingly, treatment of mouse *Tgfβ3*$^{-/-}$ homozygous palatal shelves with exogenous TGFβ3 in explant culture resulted in a rescue of palate fusion, even when treated with a low dose for a relatively short time (Taya, O'Kane, & Ferguson, 1999). Further, exogenous TGFβ3 induced fusion of explanted lip primordia in CL/Fr, a mouse line that exhibits high susceptibility to cleft lip with unknown genetic etiology (Juriloff & Fraser, 1980; Kohama, Nonaka, Hosokawa, Shum, & Ohishi, 2002; Muraoka et al., 2005). These studies suggest the possibility of modulation of TGFβ signaling as a potential therapeutic approach for nonsyndromic CL/P.

Notably, mutations in other TGFβ pathway components including *TGFB2*, *SMAD3*, *TGFβR1*, or *TGFβR2* cause syndromic cleft palate as part of autosomal dominant Loeys–Dietz syndrome (LDS; Loeys et al., 2005). The majority of LDS cases are caused by mutations in *TGFβR1* or *TGFβR2* and result in impaired kinase activity without altered expression or localization of the receptor, though the net effect is an apparent increase in TGFβ signaling (Loeys et al., 2005; Van Laer, Dietz, & Loeys, 2014). Marfan syndrome (MFS) has significant clinical overlap with LDS, but does not include clefting phenotypes and is also thought to involve overactive TGFβ signaling (Dietz et al., 1991; Habashi et al., 2006). In mice, TGFβ-neutralizing antibodies or the angiotensin II type 1 receptor (AT1) blocker losartan was used to decrease TGFβ signaling, which led to successful treatment of aortic aneurism. On this basis, several trials have been conducted in MFS patients to determine whether suppressing TGFβ signaling can reduce aortic aneurism (Bowen & Connolly, 2014; Chiu et al., 2013; Groenink et al., 2013; Habashi et al., 2006; Lacro et al., 2014). It will be interesting to determine whether similar improvements can be obtained for craniofacial phenotypes in LDS mouse models. However, prenatal use of such a strategy for LDS is likely to be complicated by embryonic consequences of blockade of TGFβ signaling and by timing of diagnosis, which raises an important general point: many of the genes involved in clefting have pleiotropic roles in the embryo, and it is not trivial to devise treatments that would lead to blockade of the pathway in one organ but not others.

During secondary palate fusion in mice, signaling by TgFβR2 is required for the expression of the gene encoding the IRF6 transcription factor, and this regulation is required for normal palate development (Iwata et al., 2013). Human mutations in *IRF6* cause Van der Woude syndrome (VWS), an autosomal dominant ectodermal dysplasia that causes CL/P with lip pits (Kondo et al., 2002). Linkage between *IRF6* and nonsyndromic CL/P has been shown in a number of studies, supporting its importance in human clefting (Dixon et al., 2011; Zucchero et al., 2004). Mice harboring mutations orthologous to those causing VWS phenocopy the disease and exhibit defects in epithelial differentiation that result in oral adhesions and cleft palate (Ingraham et al., 2006; Richardson et al., 2006; Stottmann, Bjork, Doyle, & Beier, 2010). Interestingly, expression of *Irf6* is regulated by the transcription factor P63 in mice, and mutations in *P63* in humans cause ectrodactyly–ectodermal dysplasia clefting (EEC) and ankyloblepharon ectodermal dysplasia clefting (AEC). Further, mice harboring compound heterozygous mutations in *Irf6* and *p63* exhibit fully penetrant cleft palate, connecting these ectodermal dysplasias to VWS through a molecular mechanism.

More recently, several genome-wide association studies (GWAS) have identified a number of additional loci that are likely to be involved in nonsyndromic cleft lip and palate (Beaty et al., 2010, 2011; Birnbaum et al., 2009; Grant et al., 2009; Mangold et al., 2010). In addition to confirmation of *IRF6* as a key susceptibility locus, four previously unidentified loci, including candidate genes with previously unknown involvement in craniofacial development, were identified (Mangold, Ludwig, & Nothen, 2011). Interestingly, all of these studies identified a locus at 8q24 as having significant involvement in nonsyndromic CL/P. Subsequent work in mice has indicated that this region includes a remote *cis*-acting enhancer for the *Myc* transcription factor (Uslu et al., 2014). Deletion of this region resulted in a dramatic decrease in expression of *Myc* specifically in the developing lip and a low frequency of CL/P, suggesting that regulatory mutations in *MYC* may contribute to human clefting linked to the 8q24 locus (Uslu et al., 2014). Estimation of the contribution of the four loci identified in these studies (IRF6, 8q24, 17q22, and 10q25.3) suggests that these loci might account for approximately 55% of variation in CL/P (Marazita, 2012). These data may challenge the traditional view of CL/P as a highly multifactorial disorder, raising new hopes for designing therapeutics if mechanistic insights for these loci can be obtained.

Gene–environment interactions have long been known to be a significant contributor to orofacial clefting in humans, and maternal smoking and alcohol consumption, as well as nutritional deficiency, have been implicated (Dixon et al., 2011). The extent of this contribution is not clear, however, and few susceptibility loci have been identified consistently. Recently, a GWAS approach has been used to identify gene–environment interactions underlying nonsyndromic cleft palate. Several loci, which harbor genes with previously unrecognized roles in orofacial clefting, were shown to have significant interaction with maternal smoking or alcohol use, whereas these loci did not reach significance if considered independently of smoking or alcohol use (Beaty et al., 2011). Mouse teratogen studies have shown that environmental influences can contribute to orofacial clefting, but very few studies have brought modern molecular and cell biological approaches to bear on understanding the mechanisms of these interactions. Such studies will be critical for understanding the mechanisms of action for these genes and determining if preventative approaches can reduce the frequency of clefting in humans.

2. TREACHER COLLINS: PROOF OF CONCEPT OF A NONSURGICAL THERAPEUTIC FOR A CRANIOFACIAL SYNDROME

Treacher Collins syndrome (TCS) is a congenital disorder that affects craniofacial development. Mutations in *treacle (TCOF1)* were the first identified cause of TCS, and since then additional genetic causes have been identified as well (Dixon, Trainor, & Dixon, 2007). Over 120 mutations in *TCOF1*, half of which occur spontaneously, have been identified, and most result in a premature stop codon, suggesting that haploinsufficiency of *TCOF1* underlies the pathogenesis of TCS (Dixon et al., 2007).

Individuals with TCS exhibit a spectrum of craniofacial anomalies, including hypoplasia of the facial bones, particularly the mandible and zygomatic complex; downward slanting of the palpebral fissures, usually with colobomas of the lower eyelids and a paucity of lashes medial to the defect; abnormalities of the external ears and atresia of external auditory canals, resulting in hearing loss; and choanal atresia. Cleft palate and absence of the zygomatic arch may occur in severe cases (Edwards et al., 1996; Fazen, Elmore, & Nadler, 1967; Phelps, Poswillo, & Lloyd, 1981; Rovin, Dachi, Borenstein, & Cotter, 1964). There is high phenotypic variability among TCS individuals, and some individuals are retrospectively diagnosed after their more severely affected siblings or children are born.

Tcof1 expression in mouse embryos coincides with the formation and migration of neural crest cells, suggesting a role in their development (Dixon et al., 2006). Studies of neural crest cells in *Tcof1*$^{+/-}$ embryos revealed that, although the migration pattern remained unchanged, there was a reduction in the number of migrating neural crest cells compared to wild-type embryos (Dixon et al., 2006). Consistent with this observation, there was a significant increase in apoptosis and reduction in proliferation of neuroepithelial and neural crest cells in *Tcof1*$^{+/-}$ embryos. Interestingly, the half-life and cytoplasmic concentration of P53 were increased in *Tcof1*$^{+/-}$ embryos, and proapoptotic genes that are transcriptional targets of P53 showed increased expression levels as well (Jones et al., 2008).

Genetically ablating one or both alleles of *Trp53*, which encodes P53, in *Tcof1*$^{+/-}$ embryos resulted in dosage-dependent rescue of the TCS-like phenotypes (Jones et al., 2008). All *Tcof1*$^{+/-}$;*Trp53*$^{-/-}$ pups survived beyond weaning with a fully rescued craniofacial phenotype. However, some of the pups developed tumors, which are commonly observed in *Trp53*$^{-/-}$ mice (reviewed in Levine, 1997). Chemical inhibition of P53 with pifithrin-α at the onset of neural crest cell migration led to a significant decrease in neuroepithelial apoptosis and subsequent alleviation of TCS-related phenotypes, suggesting the exciting possibility that small molecules administered *in utero* could treat TCS (Jones et al., 2008).

The successful rescue experiments in the TCS mouse model provide a conceptual advance in terms of medical treatment of birth defects resulting from ribosome abnormalities. This work also suggests that inhibition of P53-mediated apoptosis is a potential strategy for prenatal treatment of TCS. However, the strong correlation between P53 inhibition and tumorigenesis indicates that targeting P53 itself is not likely to be a practical therapeutic. Instead, investigating the TCOF effectors that are more specifically involved in TCS pathogenesis would be the logical next step to develop a safe and effective therapeutic for TCS.

3. RASopathies: UNDERSTANDING AND DEVELOPING TREATMENT FOR SYNDROMES OF THE RAS PATHWAY

The RASopathies are a group of syndromes characterized by dysregulation of signaling through the RAS pathway and include neurofibromatosis type 1 (NF1), Noonan syndrome (NS), NS with multiple lentigines, capillary malformation–AV malformation syndrome, Legius syndrome, Costello syndrome (CS), and cardio-facio-cutaneous syndrome (CFC; Fig. 1; Tidyman & Rauen, 2009). The RASopathies are caused by mutations in genes that encode components of the RAS pathway, and these conditions have both unique and overlapping characteristics.

The *RAS* genes, *HRAS*, *NRAS*, and *KRAS*, encode small, monomeric GTPases that are activated by upstream regulators, including receptor tyrosine kinases, G-protein-coupled receptors, and integrins. Once activated, RAS-GTP signals through multiple effector pathways, including

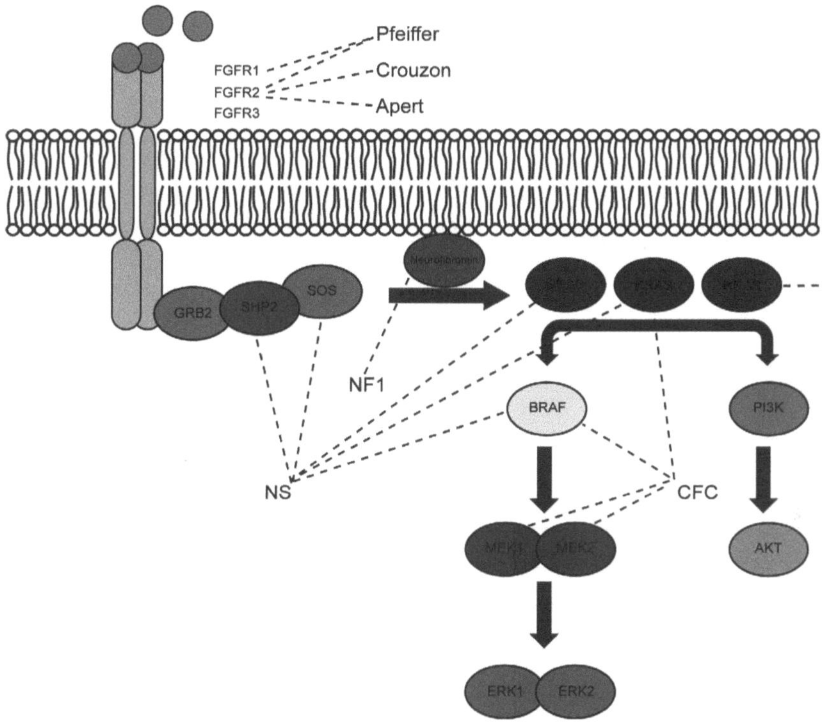

Figure 1 Syndromes of the RAS pathway. Schematic of the RAS signaling pathway with dashed lines connecting the syndrome with the protein in the pathway encoded by the causative mutated gene. Syndromes described in the text. NS, Noonan syndrome; NF1, neurofibromatosis 1; CFC, cardio-facio-cutaneous syndrome; CS, Costello syndrome. (A color version of this figure is located at the end of this chapter.)

RAF/MEK/ERK, PI3K/AKT, TIAM1/Rac, and RALGDS/Ral. Mouse models have advanced our understanding of the *in vivo* role of the RAS signaling pathway as well as how it is regulated, and animal models have also pointed to therapeutic targets for the RASopathies. Here, we focus on three of the RASopathies, NS, CS, and CFC, all of which have important craniofacial phenotypes.

NS typically presents with proportional short stature, facial dysmorphia, including a typical "triangular facies" with dolicocephaly, prominent forehead, pointed chin and hypertelorism, as well as blood and cardiovascular abnormalities (Fig. 2A and B; Collins & Turner, 1973; Noonan, 1968; Zenker, 2009). In approximately 50% of NS cases, mutations in the PTPN11 gene, which encodes the SHP2 phosphatase, cause pathogenesis by increasing phosphatase activity (Tartaglia et al., 2001); other genes

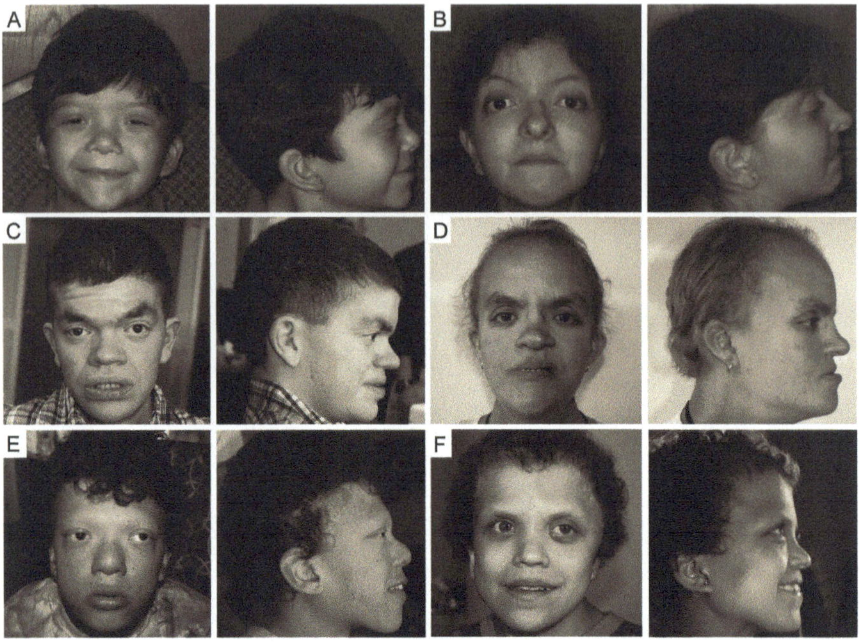

Figure 2 Craniofacial features of the RASopathies. The RASopathies are caused by mutations in genes encoding different proteins in the RAS pathway and have both unique and overlapping craniofacial characteristics. (A) Frontal and profile photographs of a 7-year-old male and (B) 21-year-old female with Noonan syndrome (NS), (C) a 20-year-old male and (D) 23-year-old female with Costello syndrome (CS), and (E) a 15-year-old male and (F) 15-year-old female with cardio-facio-cutaneous syndrome (CFC). (A color version of this figure is located at the end of this chapter.)

known to cause NS include *SOS1* (~10%) (Roberts et al., 2007; Tartaglia et al., 2007), *RAF1* (3–5%) (Pandit et al., 2007; Razzaque et al., 2007), *KRAS* (1–2%) (Schubbert et al., 2006; Zenker et al., 2007), *NRAS* (<1%) (Cirstea et al., 2010), and *SHOC2* (<1%) (Cordeddu et al., 2009). Several mouse models with mutations in the genes responsible for NS have revealed the different contributions of specific genes to the syndrome pathology and shed light on the complexity of the RAS pathway. Mice expressing the NS-associated mutation $Ptpn11^{D61G}$ ($Ptpn11^{D61G/+}$) have decreased body weight and length, craniofacial dysmorphia including decreased skull length, increased intercanthal distance and wide, blunt snout, myeloproliferative disease (MPD), and a range of cardiac defects (Araki et al., 2004). These mice do not develop hypertrophic cardiomyopathy (HCM), reflecting the fact that only 10% of NS individuals with *PTPN11* mutations develop HCM. However, approximately 95% of patients carrying *RAF1* mutations that cause increased kinase activity develop HCM, and knock-in mice expressing the kinase-activating NS mutation *Raf1*^{L613V} have short stature, facial dysmorphia including short nose and wide-set eyes, and these mice develop HCM (Wu et al., 2011). Mice with the *Sos1*^{E846K}-activating mutation display increased embryonic and perinatal mortality, short stature, craniofacial dysmorphia with short skull length, increased intercanthal distance with blunt snout due to depression of the frontal bone, and severe cardiac hypertrophy (Chen et al., 2010). Another strain of mice carrying an endogenous *K-Ras*^{V14I} germline mutation, the most frequent *KRAS* mutation in NS individuals, display many of the phenotypic abnormalities observed in NS, including small size, craniofacial dysmorphism with increased skull width and height but decreased length, wide separation between the eyes and blunt snout, and cardiac defects (Hernandez-Porras et al., 2014). Moreover, these mice develop fatal MPD, a disease similar to a leukemia seen in patients with NS. Although these NS mouse models share craniofacial characteristics similar to the triangular facies of NS individuals, the cardiac conditions vary depending upon the gene mutation, emphasizing the specific effects of distinct gene mutations.

CS is characterized by craniofacial malformations including macrocephaly, bitemporal narrowing, convex facial profile, full cheeks and large mouth, dermatologic anomalies, cardiac defects, musculoskeletal abnormalities, growth delay, and cognitive deficits (Fig. 2C and D; Goodwin, Oberoi, et al., 2014; Rauen, 2007). Nearly all individuals with CS have a heterozygous, *de novo* germline mutation in *HRAS* that results in a constitutively active protein (Aoki et al., 2005; Estep, Tidyman, Teitell, Cotter, & Rauen, 2006). The majority of the mutations encode a HRAS^{G12S} protein. There are no mouse models expressing HRAS^{G12S}; however, mice expressing *HRAS*^{G12V}, the mutation most commonly expressed in HRAS tumors (Schubbert, Bollag, & Shannon, 2007), phenocopy and provide an effective model of CS. A mouse model with a germline G12V mutation within the endogenous H-*Ras* locus (*Hras*$^{G12V/geo}$) had several characteristics of CS, including craniofacial dysmorphia due to depression of the anterior frontal bone, shortening and depression of the nasal bridge and premaxillary bone, and choanal atresia, and HCM including cardiac enlargement and hypertrophic cardiomyocytes (Schuhmacher et al., 2008). Another CS mouse model using a targeted conditional activating *Hras*^{G12V} mutation (*CC/FR-Hras*^{G12V}) displayed many traits associated with CS, including failure to thrive and decreased growth rate, relative macrocephaly, deviated septum, and myocardial fibrosis (Chen et al., 2009). These mice also exhibited an enamel defect characterized by hyperproliferative and disorganized ameloblasts similar to the hypoplastic enamel phenotype observed in CS individuals (Goodwin, Tidyman, et al., 2014). Although these mouse models have provided insight into the features of CS, more work is necessary to understand the underlying pathogenesis of CS.

CFC is a multiple congenital anomaly disorder characterized by craniofacial dysmorphia including macrocephaly, bitemporal narrowing, convex facial profile and hypoplastic supraorbital ridges, ectodermal abnormalities, congenital heart defects, growth delays, and neurocognitive deficits (Fig. 2E and F; Goodwin et al., 2013; Pierpont et al., 2014). CFC is caused by heterozygous, activating germline mutations in *KRAS*, *BRAF*, *MAP2K1* (*MEK1*), or *MAP2K2* (*MEK2*), all components of the RAS/MAPK pathway (Niihori et al., 2006; Rodriguez-Viciana et al., 2006). BRAF mutations have been found in human cancers, and the *BRAF*$^{\mathrm{V600E}}$ mutation is the most commonly detected (Davies et al., 2002); the kinase activity of this mutation is 10- to 50-fold higher than that of other *BRAF* mutations responsible for CFC (Wan et al., 2004). The *B-Raf*$^{+/LSLV600E}$ mouse model expresses the constitutively active *B-Raf*$^{\mathrm{V600E}}$ allele but at 5–10% the level of the wild-type allele to model the BRAF activity in CFC individuals more closely. This model phenocopies features of CFC, including reduced size and body weight, more rounded and shorter heads due to changes in the shape of the frontal and parietal bones, cataracts, neurological defects including hyperactivity and seizures, and cardiomegaly due to an increase in the number of cardiomyocytes (Urosevic et al., 2011).

Additional mouse models of high- or intermediate-activity *BRAF* mutations have been developed. A conditional knock-in mouse for the intermediate-activity mutant *Braf*$^{\mathrm{L597V}}$ phenocopies features of CFC including short stature, blunt nose, cardiac hypertrophy, and predisposition to develop benign tumors including papillomas and intestinal polyps (Andreadi et al., 2012). Another model expressing the *Braf*$^{\mathrm{Q241R}}$ mutation, which is the most frequent mutation in the CFC population, manifests several features of CFC, including heart defects and craniofacial anomalies such as mandibular hypoplasia, before dying embryonically (Inoue et al., 2014). Also, CFC zebrafish models expressing a kinase-activating *BRAF*$^{\mathrm{Q257R}}$ allele or kinase-inactivating *BRAF*$^{\mathrm{G596V}}$ allele develop craniofacial anomalies (Anastasaki, Estep, Marais, Rauen, & Patton, 2009). These CFC models have elucidated the biological effects of specific BRAF mutations and revealed the importance of BRAF activity levels in CFC.

Drugs developed to target the RAS pathway in cancer have enormous potential for treatment of developmental syndromes caused by its dysregulation (Rauen et al., 2011, 2015). Utilizing animal models, researchers have begun testing the potential of RAS pathway inhibitors in treatment of the RASopathies. Treatment of NS mice with the MEK1/2 inhibitor PD0325901 restored their weight and length with partial improvement of the heart defects (Chen et al., 2010; Wu et al., 2011). Additionally, when NS mice were treated *in utero* with PD0325901, the craniofacial defects in the mice improved (Wu et al., 2011), and *in utero* treatment with U0126, an inhibitor of MEK1/2, resulted in complete rescue of the development and growth of the neural crest-derived bones of the face (Nakamura et al., 2007). In CS mice, the enamel defect was rescued in adult mice treated with PD0325901; however, treatment with the PI3K inhibitor GDC0941 had no effect on the disrupted ameloblasts or enamel defect, indicating that signaling through MEK is critical in enamel development (Goodwin, Tidyman, et al., 2014). Treatment of the CFC zebrafish model with low doses of PD0325901 at early stages of development ameliorated craniofacial defects (Anastasaki, Rauen, & Patton, 2012). Treatment of the $Braf^{Q241R/+}$ embryos with PD0325091 rescued embryonic lethality and ameliorated edema and craniofacial defects, and furthermore, treatment in combination with a histone 3 demethylase inhibitor, GSK-J4, additionally rescued enlarged cardiac valves, suggesting the potential of combined therapies in treating the RASopathies (Inoue et al., 2014). Thus, drugs targeting the RAS pathway have potential in the treatment of RASopathies, but the complexity of RAS dysregulation in these syndromes means that further work will be required to determine optimal timing and dosage of drugs with specific targets.

4. CRANIOSYNOSTOSIS: PURSUING GENETIC AND PHARMACEUTICAL ALTERNATIVES TO SURGICAL TREATMENT

Craniosynostosis is the premature fusion of calvarial sutures, and multiple craniosynostosis syndromes, including Apert, Pfeiffer, and Crouzon, are caused by mutations in fibroblast growth factor receptors (FGFRs). Several mutations in FGFR1 and FGFR2 result in phenotypes that vary in severity yet share some characteristics. These mutations all result in gain of function of FGFR signaling, and a variety of specific molecular mechanisms have been shown to achieve this net effect. Mouse models have been invaluable in understanding how altered FGFR signaling contributes to the pathogenesis of these syndromes.

Apert syndrome is defined by craniosynostosis involving coronal sutures, midface hypoplasia, and bony and/or cutaneous syndactyly of the hands and feet (Bonaventure & El Ghouzzi, 2003; Wilkie, Oldridge, Tang, & Maxson, 2001). Individuals with Apert syndrome also have varying degrees of malformations of the nervous, respiratory, cardiovascular, and genitourinary systems. Furthermore, they have distinct craniofacial characteristics including increased incidence of cleft palate, narrow, high-arched palate and hypoplastic maxilla, hypodontia, delayed eruption, and dental crowding (Kreiborg & Cohen, 1992; Letra et al., 2007). Apert syndrome is caused by heterozygous, gain-of-function mutations affecting FGFR2, and more than 98% of individuals diagnosed with Apert syndrome carry a Ser252Trp (FGFR2 S252W; 66%) or Pro253Arg (FGFR2 P253R; 32%) mutation (Park et al., 1995; Wilkie et al., 1995). The altered amino acids lie in the linker region between the second and third extracellular immunoglobulin-like domains (IgII and IgIII) that mediate FGF ligand binding. The region encoding IgIII undergoes alternative splicing to form two well-described isoforms, FGFR2b and FGFR2c, which are expressed in the epithelium and mesenchyme, respectively, and have differing affinities for specific FGF ligands. Both mutations have been shown to increase FGFR2 ligand-binding affinity (Anderson, Burns, Enriquez-Harris, Wilkie, & Heath, 1998; Yu, Herr, Waksman, & Ornitz, 2000); the Ser252Trp mutation increases the affinity of FGFR2 for a subset of FGFs, whereas the Pro253Arg mutation increases affinity of FGFR2 for FGF indiscriminately (Ibrahimi et al., 2001).

Mouse models of the two most common Apert syndrome mutations in FGFR2 have shed light on the cellular mechanisms by which these mutations lead to syndromic phenotypes. The Fgfr2-Ser250Trp mutation resulted in mice that exhibit craniosynostosis, which is most pronounced in the coronal sutures, and skull malformations (Chen, Li, Li, Engel, & Deng, 2003; Wang et al., 2005). Increased Bax expression and apoptosis in the suture mesenchyme (Chen et al., 2003), as well as increased proliferation of chondrocytes and osteoblasts (Wang et al., 2005), have been suggested to contribute to craniosynostosis pathology in other mouse models of Apert syndrome.

When the P253R mutation was inserted in the *Fgfr2* gene using a knock-in approach, the mice had premature closure of the coronal suture, as well as retarded growth of the synchondroses of the cranial base and growth plates of long bones, due to abnormalities in both osteogenesis and chondrogenesis (Wang et al., 2010; Yin et al., 2008). Of note, Fgfr2S250T mice did not develop syndactyly, whereas in rare instances Fgfr2P253R mice did. These data correlate with genotype–phenotype studies showing that the Pro253Arg mutation causes more severe syndactyly, whereas cleft palate is more common in Ser252Trp individuals (Slaney et al., 1996).

Furthermore, mutations causing ectopic expression of FGFR2 splice variants can result in increased ligand-dependent activation of the receptor. *De novo Alu* insertions within or upstream of alternatively spliced exon 9 (IIIc) of *FGFR2* (Oldridge et al., 1999) resulted in ectopic expression of FGFR2b in mesenchymal tissues of the limb bud that normally express FGFR2c, suggesting a possible mechanism for the syndactyly phenotype in Apert syndrome. Similarly, excision of a single copy of *FgfR2-IIIc* in mice resulted in a gain-of-function mutation causing precocious ossification of the coronal sutures, zygomatic arch joints, and the sternebrae, as well as major defects in the kidney, lung, and lacrimal glands (Hajihosseini, Wilson, De Moerlooze, & Dickson, 2001). In this model, *FgfR2-IIIb* expression is substantially elevated in calvarial sutures and zygomatic joints, suggesting that cells in the sutures would respond to a broader array of FGF ligands and undergo differentiation prematurely, resulting in premature suture ossification. Thus, Apert syndrome appears to circumvent the biochemical and developmental regulatory mechanisms that are normally imposed by tissue-specific alternative splicing of *Fgfr2*, resulting in ectopic ligand-dependent receptor activation (Yu & Ornitz, 2001).

Similar to Apert syndrome, Pfeiffer syndrome is characterized by premature fusion of several cranial sutures, widely spaced eyes, a small nose, and midface hypoplasia (Pfeiffer, 1964; Vogels & Fryns, 2006). Additionally, individuals with Pfeiffer syndrome have hand and foot anomalies, including broad thumbs, cutaneous syndactly, shortened fingers, and medially deviated broad toes; however, unlike Apert syndrome, syndactyly is not a feature of Pfeiffer syndrome. Pfeiffer syndrome is caused by mutations in both FGFR1 and FGFR2, including a cytosine to guanine conversion in exon 5 of FGFR1 that results in a proline to arginine substitution in the extracellular domain (Muenke et al., 1994; Schell et al., 1995). When the Pro250Arg mutation was introduced into *Fgfr1*, mimicking the human mutation, the mice exhibited craniosynostosis, a dome-shaped skull, and a shortened snout (Zhou et al., 2000).

The phenotype of Crouzon syndrome is similar to that of Pfeiffer syndrome and includes craniosynostosis with abnormal skull shape and prominent eyes secondary to early fusion of the cranial sutures, without digital malformations like syndactyly (Crouzon, 1912). More than 30 different point mutations in the extracellular domain of FGFR2, most destroying or creating a cysteine residue, have been identified in individuals with Crouzon syndrome (Reardon et al., 1994; Wilkie et al., 1995). When FGFR2/Neu chimeras were generated with these mutations and expressed in NIH 3T3 cell cultures, FGFR2 was constitutively active due to aberrant intermolecular disulfide bonds (Galvin, Hart, Meyer, Webster, & Donoghue, 1996). These data suggest that the underlying pathology of Crouzon syndrome is the creation of unpaired cysteine residues, which facilitate the formation of intermolecular disulphide bonds, causing ligand-independent dimerization, phosphorylation, and signaling. The common Pfeiffer/Crouzon gain-of-function mutation Cys342Tyr was introduced into *Fgfr2c*, and the *Fgfr2c*$^{C342Y/+}$ heterozygote mice were characterized by a shortened face, protruding eyes, and premature fusion of cranial sutures (Eswarakumar et al., 2004), due to ligand-independent activation of FGFR2 via disulfide bond dimerization (Eswarakumar et al., 2006). Notably, when a mutation inhibiting FRS2 binding was introduced together with the C342Y mutation, the phenotype was rescued, indicating that the FRS2 adaptor is a key mediator of FGFR2 signaling in craniosynostosis pathogenesis and suggesting a potential point of therapeutic intervention (Eswarakumar et al., 2006).

Currently, treatment of craniosynostosis is almost exclusively surgical and consists of removing fused sutures with osteotomies and reconstructing the skull (Panchal & Uttchin, 2003). More recently, surgeons have utilized distraction osteogenesis, in which osteotomies are performed and force is applied by an external or internal device to separate the skull fragments (Gasparini, Di Rocco, Tamburrini, & Pelo, 2012). Additionally, specific synostoses may be treated with minimally invasive endoscopic repairs that result in reduced blood loss and earlier hospital discharge (Jimenez & Barone, 1998). Due to the morbidity associated with multiple surgeries to treat craniosynostosis, pharmacologic approaches are an attractive option to consider. Utilizing mouse models to understand the underlying pathogenesis of craniosynostosis associated with FGFR mutations has pinpointed downstream effectors that may serve as targets in treatment of FGFR-related craniosynostosis syndromes. Treatment of calvarial explants from Crouzon-like *Fgfr2c*$^{C342Y/+}$ mice with the small-molecule FGFR inhibitors PLX052 or PD173074 prevented premature suture fusion (Eswarakumar et al., 2006; Perlyn, Morriss-Kay, Darvann, Tenenbaum, & Ornitz, 2006). Similarly, treatment of cultured calvaria from Apert-like P253R mice with the ERK1/2 inhibitor PD98059 partially alleviated the coronal suture fusion (Yin et al., 2008). *In vivo*, mice with the Fgfr2S252W mutation expressing a small hairpin RNA targeting *Fgfr2*^{S252W} mRNA had patent cranial sutures and normal skull development (Shukla, Coumoul, Wang, Kim, & Deng, 2007). Furthermore, increased phosphorylation of ERK1/2 was observed in *Fgfr2*^{S252W} mice, and treatment of these mice *in utero* at E18.5 and postnatally with U0126, an inhibitor of MEK1/2, resulted in rescue of the craniosynostosis phenotype (Shukla et al., 2007). Thus, utilization of mouse models has elucidated potential downstream targets of FGFR signaling for pharmacological treatment of craniosynostosis. However, resynostosis has been noted after withdrawal of inhibitors, and further investigation is necessary to develop drugs with specific targets to maximize efficacy and minimize side effects.

5. XLHED: DEVELOPING TREATMENT BASED ON KNOWLEDGE GAINED FROM MOUSE AND CANINE MODELS

In 1848, an early case study reported a male subject with an "almost complete absence of hair; the teeth being not more than four in number; the delicate structure of the skin; and the absence of sensible perspiration and tears" (Thurnam, 1848). In 1875, Charles Darwin made similar observations on a family of males in India (Darwin, 1875). These men suffered from a hereditary condition that disrupts the development of ectodermal structures and their appendages, termed ectodermal dysplasia (ED; Smith, 1929).

Over 150 syndromes have been categorized as EDs, and the most prevalent subgroup is comprised of the hypohidrotic ectodermal dysplasias (HEDs), which affect more than 1 in 17,000 live births (Itin & Fistarol, 2004; Pinheiro & Freire-Maia, 1994). Clinical features of HED include sparseness of scalp and body hair (hypotrichosis); reduced sweating (hypohidrosis) and tear production; and malformation and reduced number of teeth (hypodontia), with an average of nine permanent teeth (Fig. 3; Clarke, 1987; Kobielak et al., 2001; Lexner, Bardow, Hertz, Nielsen, & Kreiborg, 2007; Solomon & Keuer, 1980). In addition, many HED patients exhibit abnormalities in facial appearance, such as prominent forehead, depressed nasal bridge, hypoplastic nose, prominent supraorbital ridges, and thick lips (Goodwin, Larson, et al., 2014; Gunduz Arslan, Devecioglu Kama, Ozer, & Yavuz, 2007; Johnson et al., 2002; Lexner et al., 2007). These

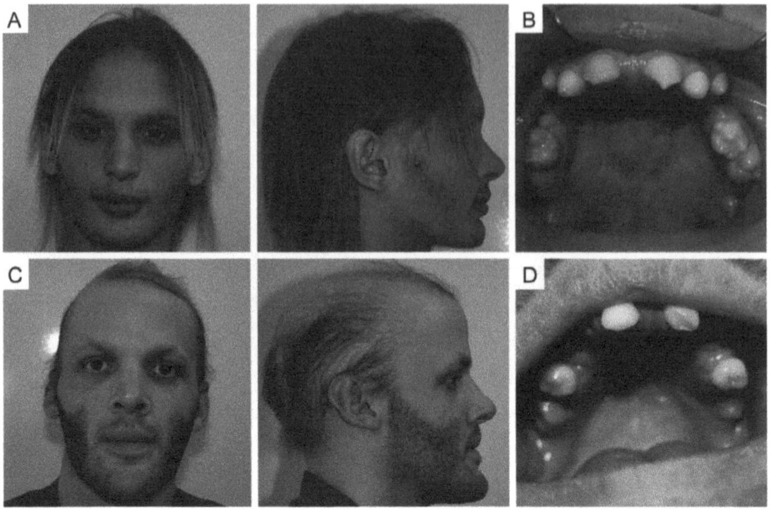

Figure 3 Craniofacial and dental characteristics of XLHED. (A) 13-year-old male with facial characteristics of XLHED including short face with proportionally longer chin and midface, narrow and pointed nose, narrow mouth with full, rounded lower lip and (B) dental features such as missing teeth and conical-shaped incisors and molars with abnormal cuspal morphology. (C) A 21-year-old male with midface hypoplasia and prognathic mandible and chin and short philtrum, and (D) missing teeth, typical of XLHED. (A color version of this figure is located at the end of this chapter.)

clinical phenotypes serve as the initial diagnostic features for most affected individuals with HED. Early diagnosis is crucial because affected children can develop recurrent hyperpyrexia due to impaired sweating and temperature control ability, which can result in febrile seizures; some authors have suggested that neurologic damage can result and have proposed a mortality rate of about 30% during the first 2 years of life (Clarke, 1987; Salisbury & Stothers, 1981).

HED individuals carry mutations in one or more of the core genes in the ectodysplasin pathway: ectodysplasin ligand (*EDA*), ectodysplasin receptor (*EDAR*), or EDAR-associated death domain adaptor protein (*EDARADD*; reviewed in Sadier, Viriot, Pantalacci, & Laudet 2014). EDA is a type II transmembrane tumor necrosis factor superfamily member that is proteolytically processed into a soluble trimeric form to bind the receptor, EDAR. Upon activation, EDAR recruits EDARADD via its intracellular death domain region. In turn, EDARADD transduces the signal to activate the NF-κB pathway, which mediates the transcriptional activation of the target genes involved in hair, sweat gland, and tooth development (Lefebvre,

Fliniaux, Schneider, & Mikkola, 2012). In rare cases, HED can be associated with immune deficiency caused by dysregulation of NF-κB (Doffinger et al., 2001; Jain et al., 2001; Zonana et al., 2000). EDA has several splicing isoforms; EDA-A1 and -A2 interact with EDAR and EDA2R (formerly known as XEDAR), respectively.

Mutations in *EDAR* and *EDARADD* are inherited in an autosomal dominant or recessive fashion, whereas *EDA* mutations are inherited in an X-linked manner. Phenotypes caused by mutations at different points along the EDA pathway are clinically similar, with autosomal dominant forms of *EDAR* and *EDARADD* exhibiting milder features. Here, we will focus on X-linked HED (XLHED), which is caused by mutations in *EDA* and is the most common form of HED.

Tabby mice that exhibit clinical features of XLHED were described in the mid-1900s, but it was not until three decades later that the *Tabby* gene was confirmed to be orthologous to human *EDA* through synteny mapping and comparison of predicted amino sequence (Bayes et al., 1998; Falconer, 1952; Ferguson et al., 1997; Srivastava et al., 1997). *Tabby* mice carry spontaneous loss-of-function mutations in *Eda*, resulting in phenotypes comparable to those observed in XLHED individuals: hypodontia with incomplete penetrance; reduced size and cusps of the molars; reduction in four pelage hair types (awl, auchene, guard, zigzag) to a single awl-like hair type; missing hair on tail and retroauricular regions; missing sweat glands on footpads; and reduction in size of various glands, including lacrimal and salivary glands (Cui et al., 2003; Gruneberg, 1965, 1971; Pispa et al., 1999; Srivastava et al., 1997). In contrast, overexpression of the *Eda-A1* transgene in developing murine ectoderm resulted in supernumerary mammary glands and teeth, hypertrophy of sebaceous glands, and abnormal composition and structure of pelage hairs (Mustonen et al., 2003; Srivastava et al., 2001). These mouse models with dysregulation of the EDA pathway suggest the importance of the EDA pathway in fine-tuning the size, number, and morphology of ectodermal organs.

Controlled expression of a tetracycline-regulated *Eda-A1* transgene during *Tabby* embryogenesis revealed that ectodysplasin signaling is required for hair development within defined spatiotemporal windows (Cui, Kunisada, Esibizione, Douglass, & Schlessinger, 2009). *Tabby* phenotypes can also be ameliorated by administration of recombinant Fc:EDA1, which contains the EDAR-binding domain linked to the IgG1 Fc domain to allow trafficking across the placental barrier and trimerization of the ligand (Fig. 4; Gaide & Schneider, 2003). As with transgenic rescue, Fc:EDA1 efficacy was timing

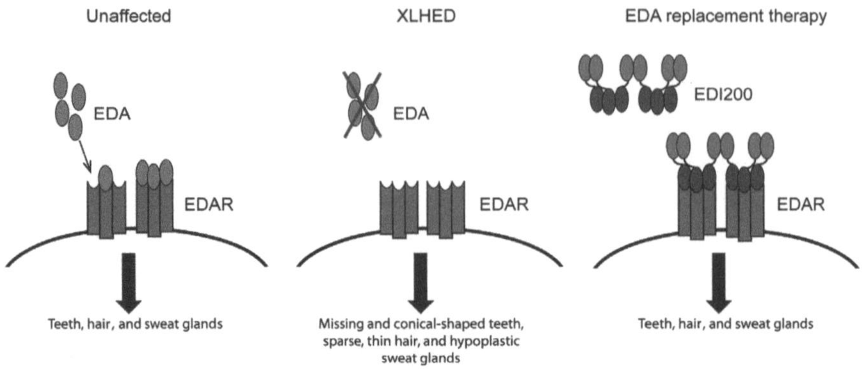

Figure 4 Schematic of recombinant EDA therapy. In unaffected individuals, EDA binds EDAR, activating downstream signaling that results in tooth, hair, and sweat gland development. XLHED individuals are missing EDA, which results in missing and conical-shaped teeth, sparse, thin hair, and hypoplastic sweat glands. EDI200 therapy replaces missing EDA in XLHED and rescues tooth, hair, and sweat gland development. (A color version of this figure is located at the end of this chapter.)

dependent; fetal, but not neonatal, administration rescued tooth morphology, as well as some hair and gland phenotypes. Interestingly, whereas transgenic expression of Eda-A1 was able to rescue the number but not the morphology of molars, Fc:EDA1 treatment restored morphology but not molar tooth number (Gaide & Schneider, 2003).

Subsequently, a colony of dogs with XLHED-like features was characterized and found to carry a spontaneous mutation in *Eda* that resulted in truncated EDA-A1 and -A2 (Casal, Jezyk, Greek, Goldschmidt, & Patterson, 1997). EDA replacement experiments were then performed in the canine model, but because immunoglobulins are not transferred transplacentally in dogs, Fc:EDA1 was administered postnatally via intravenous injection. The treatment effectively rescued the sweat glands, as observed in *Tabby* mice. Although hair growth was not improved, the number and shape of permanent teeth were restored in most of the treated dogs (Casal et al., 2007). Furthermore, susceptibility to eye and respiratory infections was also reduced through restoration of lacrimal as well as tracheal and bronchial glands. These results collectively demonstrate that short-term treatment with recombinant EDA during a therapeutic window can dramatically improve the XLHED phenotype in murine and canine models. Importantly, the treatment did not cause any noticeable side effects during or after the injections, showing great promise for the potential utility of such a therapeutic intervention in patients with XLHED. Currently, clinical trials are underway to administer recombinant EDA to XLHED-affected males in the immediate postnatal period (Fig. 4), constituting the first attempt to treat a structural birth defect in humans with a targeted drug therapy.

6. CONCLUDING THOUGHTS

The conditions that we have reviewed in this chapter, including orofacial clefting, the RASopathies, craniosynostosis syndromes, and XLHED, illustrate the exciting recent advances in our understanding of the molecular underpinnings of craniofacial conditions. These disorders lie at various points along a spectrum that has at one end basic discovery science and at the other end clinical trials of targeted therapies. The studies described above have led to an increased understanding of the etiology of these conditions and have revealed the pathways and processes underlying craniofacial development as well as general principles that are applicable to our broader understanding of organogenesis. These examples provide hope that, as the field of craniofacial biology continues to move forward, a shift will ensue from surgical correction toward medical treatment and even prevention. The barriers to such a goal include the timing of diagnosis, developmental pleiotropy of signaling pathways, and the unpredictable consequences of specific gene mutations in a given individual. Overcoming these barriers will require continued intensive study of the basic mechanisms that underlie these diseases.

ACKNOWLEDGMENTS

The authors were funded by the following NIDCR grants: U01-DE024440 and R01-DE021420 to O.D.K., R01-DE023337 to J.O.B., and F30-DE025160 to R.K.

REFERENCES

Alappat, S., Zhang, Z. Y., & Chen, Y. P. (2003). Msx homeobox gene family and craniofacial development. *Cell Research*, *13*, 429–442.

Anastasaki, C., Estep, A. L., Marais, R., Rauen, K. A., & Patton, E. E. (2009). Kinase-activating and kinase-impaired cardio-facio-cutaneous syndrome alleles have activity during zebrafish development and are sensitive to small molecule inhibitors. *Human Molecular Genetics*, *18*, 2543–2554.

Anastasaki, C., Rauen, K. A., & Patton, E. E. (2012). Continual low-level MEK inhibition ameliorates cardio-facio-cutaneous phenotypes in zebrafish. *Disease Models & Mechanisms*, *5*, 546–552.

Anderson, J., Burns, H. D., Enriquez-Harris, P., Wilkie, A. O., & Heath, J. K. (1998). Apert syndrome mutations in fibroblast growth factor receptor 2 exhibit increased affinity for FGF ligand. *Human Molecular Genetics*, *7*, 1475–1483.

Andreadi, C., Cheung, L. K., Giblett, S., Patel, B., Jin, H., Mercer, K., et al. (2012). The intermediate-activity (L597V)BRAF mutant acts as an epistatic modifier of oncogenic RAS by enhancing signaling through the RAF/MEK/ERK pathway. *Genes & Development, 26*, 1945–1958.

Aoki, Y., Niihori, T., Kawame, H., Kurosawa, K., Ohashi, H., Tanaka, Y., et al. (2005). Germline mutations in HRAS proto-oncogene cause Costello syndrome. *Nature Genetics, 37*, 1038–1040.

Araki, T., Mohi, M. G., Ismat, F. A., Bronson, R. T., Williams, I. R., Kutok, J. L., et al. (2004). Mouse model of Noonan syndrome reveals cell type- and gene dosage-dependent effects of Ptpn11 mutation. *Nature Medicine, 10*, 849–857.

Bayes, M., Hartung, A. J., Ezer, S., Pispa, J., Thesleff, I., Srivastava, A. K., et al. (1998). The anhidrotic ectodermal dysplasia gene (EDA) undergoes alternative splicing and encodes ectodysplasin-A with deletion mutations in collagenous repeats. *Human Molecular Genetics, 7*, 1661–1669.

Beaty, T. H., Murray, J. C., Marazita, M. L., Munger, R. G., Ruczinski, I., Hetmanski, J. B., et al. (2010). A genome-wide association study of cleft lip with and without cleft palate identifies risk variants near MAFB and ABCA4. *Nature Genetics, 42*, 525–529.

Beaty, T. H., Ruczinski, I., Murray, J. C., Marazita, M. L., Munger, R. G., Hetmanski, J. B., et al. (2011). Evidence for gene-environment interaction in a genome wide study of nonsyndromic cleft palate. *Genetic Epidemiology, 35*, 469–478.

Bertuzzi, S., Hindges, R., Mui, S. H., O'Leary, D. D., & Lemke, G. (1999). The homeodomain protein vax1 is required for axon guidance and major tract formation in the developing forebrain. *Genes & Development, 13*, 3092–3105.

Bi, W., Huang, W., Whitworth, D. J., Deng, J. M., Zhang, Z., Behringer, R. R., et al. (2001). Haploinsufficiency of Sox9 results in defective cartilage primordia and premature skeletal mineralization. *Proceedings of the National Academy of Sciences of the United States of America, 98*, 6698–6703.

Birnbaum, S., Ludwig, K. U., Reutter, H., Herms, S., Steffens, M., Rubini, M., et al. (2009). Key susceptibility locus for nonsyndromic cleft lip with or without cleft palate on chromosome 8q24. *Nature Genetics, 41*, 473–477.

Bonaventure, J., & El Ghouzzi, V. (2003). Molecular and cellular bases of syndromic craniosynostoses. *Expert Reviews in Molecular Medicine, 5*, 1–17.

Bowen, J. M., & Connolly, H. M. (2014). Of Marfan's syndrome, mice, and medications. *The New England Journal of Medicine, 371*, 2127–2128.

Brewer, S., Feng, W., Huang, J., Sullivan, S., & Williams, T. (2004). Wnt1-Cre-mediated deletion of AP-2alpha causes multiple neural crest-related defects. *Developmental Biology, 267*, 135–152.

Britanova, O., Depew, M. J., Schwark, M., Thomas, B. L., Miletich, I., Sharpe, P., et al. (2006). Satb2 haploinsufficiency phenocopies 2q32-q33 deletions, whereas loss suggests a fundamental role in the coordination of jaw development. *American Journal of Human Genetics, 79*, 668–678.

Bush, J. O., & Jiang, R. (2012). Palatogenesis: Morphogenetic and molecular mechanisms of secondary palate development. *Development, 139*, 231–243.

Bush, J. O., & Soriano, P. (2010). Ephrin-B1 forward signaling regulates craniofacial morphogenesis by controlling cell proliferation across Eph-ephrin boundaries. *Genes & Development, 24*, 2068–2080.

Casal, M. L., Jezyk, P. F., Greek, J. M., Goldschmidt, M. H., & Patterson, D. F. (1997). X-linked ectodermal dysplasia in the dog. *The Journal of Heredity, 88*, 513–517.

Casal, M. L., Lewis, J. R., Mauldin, E. A., Tardivel, A., Ingold, K., Favre, M., et al. (2007). Significant correction of disease after postnatal administration of recombinant ectodysplasin A in canine X-linked ectodermal dysplasia. *American Journal of Human Genetics, 81*, 1050–1056.

Chen, L., Li, D., Li, C., Engel, A., & Deng, C. X. (2003). A Ser252Trp [corrected] substitution in mouse fibroblast growth factor receptor 2 (Fgfr2) results in craniosynostosis. *Bone*, *33*, 169–178.

Chen, X., Mitsutake, N., LaPerle, K., Akeno, N., Zanzonico, P., Longo, V. A., et al. (2009). Endogenous expression of Hras(G12V) induces developmental defects and neoplasms with copy number imbalances of the oncogene. *Proceedings of the National Academy of Sciences of the United States of America*, *106*, 7979–7984.

Chen, P. C., Wakimoto, H., Conner, D., Araki, T., Yuan, T., Roberts, A., et al. (2010). Activation of multiple signaling pathways causes developmental defects in mice with a Noonan syndrome-associated Sos1 mutation. *The Journal of Clinical Investigation*, *120*, 4353–4365.

Chiu, H. H., Wu, M. H., Wang, J. K., Lu, C. W., Chiu, S. N., Chen, C. A., et al. (2013). Losartan added to beta-blockade therapy for aortic root dilation in Marfan syndrome: A randomized, open-label pilot study. *Mayo Clinic Proceedings*, *88*, 271–276.

Cirstea, I. C., Kutsche, K., Dvorsky, R., Gremer, L., Carta, C., Horn, D., et al. (2010). A restricted spectrum of NRAS mutations causes Noonan syndrome. *Nature Genetics*, *42*, 27–29.

Clarke, A. (1987). Hypohidrotic ectodermal dysplasia. *Journal of Medical Genetics*, *24*, 659–663.

Collins, E., & Turner, G. (1973). The Noonan syndrome—A review of the clinical and genetic features of 27 cases. *The Journal of Pediatrics*, *83*, 941–950.

Compagni, A., Logan, M., Klein, R., & Adams, R. H. (2003). Control of skeletal patterning by ephrinB1-EphB interactions. *Developmental Cell*, *5*, 217–230.

Cordeddu, V., Di Schiavi, E., Pennacchio, L. A., Ma'ayan, A., Sarkozy, A., Fodale, V., et al. (2009). Mutation of SHOC2 promotes aberrant protein N-myristoylation and causes Noonan-like syndrome with loose anagen hair. *Nature Genetics*, *41*, 1022–1026.

Crouzon, O. (1912). Dysostose cranio-faciale hereditaire. *Bulletins et mémoires de la Société médicale des hôpitaux de Paris*, *33*, 545–555.

Cui, C. Y., Durmowicz, M., Ottolenghi, C., Hashimoto, T., Griggs, B., Srivastava, A. K., et al. (2003). Inducible mEDA-A1 transgene mediates sebaceous gland hyperplasia and differential formation of two types of mouse hair follicles. *Human Molecular Genetics*, *12*, 2931–2940.

Cui, C. Y., Kunisada, M., Esibizione, D., Douglass, E. G., & Schlessinger, D. (2009). Analysis of the temporal requirement for eda in hair and sweat gland development. *Journal of Investigative Dermatology*, *129*(4), 984–993.

Darwin, C. (1875). *Insectivorous Plants. The variations of animals and plants under domestication: Vol. 2* (p. 319). London: John Murray.

Davies, H., Bignell, G. R., Cox, C., Stephens, P., Edkins, S., Clegg, S., et al. (2002). Mutations of the BRAF gene in human cancer. *Nature*, *417*, 949–954.

De Felice, M., Ovitt, C., Biffali, E., Rodriguez-Mallon, A., Arra, C., Anastassiadis, K., et al. (1998). A mouse model for hereditary thyroid dysgenesis and cleft palate. *Nature Genetics*, *19*, 395–398.

Dietz, H. C., Cutting, G. R., Pyeritz, R. E., Maslen, C. L., Sakai, L. Y., Corson, G. M., et al. (1991). Marfan syndrome caused by a recurrent de novo missense mutation in the fibrillin gene. *Nature*, *352*, 337–339.

Ding, H., Wu, X., Bostrom, H., Kim, I., Wong, N., Tsoi, B., et al. (2004). A specific requirement for PDGF-C in palate formation and PDGFR-alpha signaling. *Nature Genetics*, *36*, 1111–1116.

Dixon, J., Jones, N. C., Sandell, L. L., Jayasinghe, S. M., Crane, J., Rey, J. P., et al. (2006). Tcof1/Treacle is required for neural crest cell formation and proliferation deficiencies that cause craniofacial abnormalities. *Proceedings of the National Academy of Sciences of the United States of America*, *103*, 13403–13408.

Dixon, M. J., Marazita, M. L., Beaty, T. H., & Murray, J. C. (2011). Cleft lip and palate: Understanding genetic and environmental influences. *Nature Reviews. Genetics*, *12*, 167–178.

Dixon, J., Trainor, P., & Dixon, M. J. (2007). Treacher Collins syndrome. *Orthodontics & Craniofacial Research*, *10*, 88–95.

Doffinger, R., Smahi, A., Bessia, C., Geissmann, F., Feinberg, J., Durandy, A., et al. (2001). X-linked anhidrotic ectodermal dysplasia with immunodeficiency is caused by impaired NF-kappaB signaling. *Nature Genetics*, *27*, 277–285.

Dudas, M., Kim, J., Li, W. Y., Nagy, A., Larsson, J., Karlsson, S., et al. (2006). Epithelial and ectomesenchymal role of the type I TGF-beta receptor ALK5 during facial morphogenesis and palatal fusion. *Developmental Biology*, *296*, 298–314.

Edwards, S. J., Fowlie, A., Cust, M. P., Liu, D. T., Young, I. D., & Dixon, M. J. (1996). Prenatal diagnosis in Treacher Collins syndrome using combined linkage analysis and ultrasound imaging. *Journal of Medical Genetics*, *33*, 603–606.

Estep, A. L., Tidyman, W. E., Teitell, M. A., Cotter, P. D., & Rauen, K. A. (2006). HRAS mutations in Costello syndrome: Detection of constitutional activating mutations in codon 12 and 13 and loss of wild-type allele in malignancy. *American Journal of Medical Genetics. Part A*, *140*, 8–16.

Eswarakumar, V. P., Horowitz, M. C., Locklin, R., Morriss-Kay, G. M., & Lonai, P. (2004). A gain-of-function mutation of Fgfr2c demonstrates the roles of this receptor variant in osteogenesis. *Proceedings of the National Academy of Sciences of the United States of America*, *101*, 12555–12560.

Eswarakumar, V. P., Ozcan, F., Lew, E. D., Bae, J. H., Tome, F., Booth, C. J., et al. (2006). Attenuation of signaling pathways stimulated by pathologically activated FGF-receptor 2 mutants prevents craniosynostosis. *Proceedings of the National Academy of Sciences of the United States of America*, *103*, 18603–18608.

Falconer, D. S. (1952). A totally sex-linked gene in the house mouse. *Nature*, *169*, 664–665.

Fazen, L. E., Elmore, J., & Nadler, H. L. (1967). Mandibulofacial dysostosis (Treacher-Collins syndrome). *American Journal of Diseases of Children*, *113*, 405–410.

Ferguson, B. M., Brockdorff, N., Formstone, E., Ngyuen, T., Kronmiller, J. E., & Zonana, J. (1997). Cloning of Tabby, the murine homolog of the human EDA gene: Evidence for a membrane-associated protein with a short collagenous domain. *Human Molecular Genetics*, *6*, 1589–1594.

Ferone, G., Thomason, H. A., Antonini, D., De Rosa, L., Hu, B., Gemei, M., et al. (2012). Mutant p63 causes defective expansion of ectodermal progenitor cells and impaired FGF signalling in AEC syndrome. *EMBO Molecular Medicine*, *4*, 192–205.

Gaide, O., & Schneider, P. (2003). Permanent correction of an inherited ectodermal dysplasia with recombinant EDA. *Nature Medicine*, *9*, 614–618.

Galvin, B. D., Hart, K. C., Meyer, A. N., Webster, M. K., & Donoghue, D. J. (1996). Constitutive receptor activation by Crouzon syndrome mutations in fibroblast growth factor receptor (FGFR)2 and FGFR2/Neu chimeras. *Proceedings of the National Academy of Sciences of the United States of America*, *93*, 7894–7899.

Gasparini, G., Di Rocco, C., Tamburrini, G., & Pelo, S. (2012). External craniofacial osteodistraction in complex craniosynostoses. *Child's Nervous System: ChNS: Official Journal of the International Society for Pediatric Neurosurgery*, *28*, 1565–1570.

Goodwin, A. F., Larson, J. R., Jones, K. B., Liberton, D. K., Landan, M., Wang, Z., et al. (2014). Craniofacial morphometric analysis of individuals with X-linked hypohidrotic ectodermal dysplasia. *Molecular Genetics & Genomic Medicine*, *2*, 422–429.

Goodwin, A. F., Oberoi, S., Landan, M., Charles, C., Groth, J., Martinez, A., et al. (2013). Craniofacial and dental development in cardio-facio-cutaneous syndrome: The importance of Ras signaling homeostasis. *Clinical Genetics*, *83*, 539–544.

Goodwin, A. F., Oberoi, S., Landan, M., Charles, C., Massie, J. C., Fairley, C., et al. (2014). Craniofacial and dental development in Costello syndrome. *American Journal of Medical Genetics. Part A*, *164A*, 1425–1430.

Goodwin, A. F., Tidyman, W. E., Jheon, A. H., Sharir, A., Zheng, X., Charles, C., et al. (2014). Abnormal Ras signaling in Costello syndrome (CS) negatively regulates enamel formation. *Human Molecular Genetics*, *23*, 682–692.

Grant, S. F., Wang, K., Zhang, H., Glaberson, W., Annaiah, K., Kim, C. E., et al. (2009). A genome-wide association study identifies a locus for nonsyndromic cleft lip with or without cleft palate on 8q24. *The Journal of Pediatrics*, *155*, 909–913.

Green, R. M., Feng, W., Phang, T., Fish, J. L., Li, H., Spritz, R. A., et al. (2015). Tfap2a-dependent changes in mouse facial morphology result in clefting that can be ameliorated by a reduction in Fgf8 gene dosage. *Disease Models & Mechanisms*, *8*, 31–43.

Gritli-Linde, A. (2008). The etiopathogenesis of cleft lip and cleft palate: Usefulness and caveats of mouse models. *Current Topics in Developmental Biology*, *84*, 37–138.

Groenink, M., den Hartog, A. W., Franken, R., Radonic, T., de Waard, V., Timmermans, J., et al. (2013). Losartan reduces aortic dilatation rate in adults with Marfan syndrome: A randomized controlled trial. *European Heart Journal*, *34*, 3491–3500.

Grosen, D., Bille, C., Pedersen, J. K., Skytthe, A., Murray, J. C., & Christensen, K. (2010). Recurrence risk for offspring of twins discordant for oral cleft: A population-based cohort study of the Danish 1936-2004 cleft twin cohort. *American Journal of Medical Genetics. Part A*, *152A*, 2468–2474.

Gruneberg, H. (1965). Genes and genotypes affecting the teeth of the mouse. *Journal of Embryology and Experimental Morphology*, *14*, 137–159.

Gruneberg, H. (1971). The tabby syndrome in the mouse. *Proceedings of the Royal Society of London, Series B: Biological Sciences*, *179*, 139–156.

Gunduz Arslan, S., Devecioglu Kama, J., Ozer, T., & Yavuz, I. (2007). Craniofacial and upper airway cephalometrics in hypohidrotic ectodermal dysplasia. *Dento Maxillo Facial Radiology*, *36*, 478–483.

Habashi, J. P., Judge, D. P., Holm, T. M., Cohn, R. D., Loeys, B. L., Cooper, T. K., et al. (2006). Losartan, an AT1 antagonist, prevents aortic aneurysm in a mouse model of Marfan syndrome. *Science*, *312*, 117–121.

Hajihosseini, M. K., Wilson, S., De Moerlooze, L., & Dickson, C. (2001). A splicing switch and gain-of-function mutation in FgfR2-IIIc hemizygotes causes Apert/Pfeiffer-syndrome-like phenotypes. *Proceedings of the National Academy of Sciences of the United States of America*, *98*, 3855–3860.

Han, J., Mayo, J., Xu, X., Li, J., Bringas, P., Jr., Maas, R. L., et al. (2009). Indirect modulation of Shh signaling by Dlx5 affects the oral-nasal patterning of palate and rescues cleft palate in Msx1-null mice. *Development*, *136*, 4225–4233.

Hart, A. W., Morgan, J. E., Schneider, J., West, K., McKie, L., Bhattacharya, S., et al. (2006). Cardiac malformations and midline skeletal defects in mice lacking filamin A. *Human Molecular Genetics*, *15*, 2457–2467.

Hernandez-Porras, I., Fabbiano, S., Schuhmacher, A. J., Aicher, A., Canamero, M., Camara, J. A., et al. (2014). K-RasV14I recapitulates Noonan syndrome in mice. *Proceedings of the National Academy of Sciences of the United States of America*, *111*, 16395–16400.

Hosokawa, R., Deng, X., Takamori, K., Xu, X., Urata, M., Bringas, P., Jr., et al. (2009). Epithelial-specific requirement of FGFR2 signaling during tooth and palate development. *Journal of Experimental Zoology. Part B, Molecular and Developmental Evolution*, *312B*, 343–350.

Huang, X., Goudy, S. L., Ketova, T., Litingtung, Y., & Chiang, C. (2008). Gli3-deficient mice exhibit cleft palate associated with abnormal tongue development. *Developmental Dynamics: An Official Publication of the American Association of Anatomists*, *237*, 3079–3087.

Ibrahimi, O. A., Eliseenkova, A. V., Plotnikov, A. N., Yu, K., Ornitz, D. M., & Mohammadi, M. (2001). Structural basis for fibroblast growth factor receptor 2 activation in Apert syndrome. *Proceedings of the National Academy of Sciences of the United States of America, 98*, 7182–7187.

Ichikawa, E., Watanabe, A., Nakano, Y., Akita, S., Hirano, A., Kinoshita, A., et al. (2006). PAX9 and TGFB3 are linked to susceptibility to nonsyndromic cleft lip with or without cleft palate in the Japanese: Population-based and family-based candidate gene analyses. *Journal of Human Genetics, 51*, 38–46.

Iida, K., Koseki, H., Kakinuma, H., Kato, N., Mizutani-Koseki, Y., Ohuchi, H., et al. (1997). Essential roles of the winged helix transcription factor MFH-1 in aortic arch patterning and skeletogenesis. *Development, 124*, 4627–4638.

Ingraham, C. R., Kinoshita, A., Kondo, S., Yang, B., Sajan, S., Trout, K. J., et al. (2006). Abnormal skin, limb and craniofacial morphogenesis in mice deficient for interferon regulatory factor 6 (Irf6). *Nature Genetics, 38*, 1335–1340.

Inoue, S., Moriya, M., Watanabe, Y., Miyagawa-Tomita, S., Niihori, T., Oba, D., et al. (2014). New BRAF knockin mice provide a pathogenetic mechanism of developmental defects and a therapeutic approach in cardio-facio-cutaneous syndrome. *Human Molecular Genetics, 23*, 6553–6566.

Ishii, M., Han, J., Yen, H. Y., Sucov, H. M., Chai, Y., & Maxson, R. E., Jr. (2005). Combined deficiencies of Msx1 and Msx2 cause impaired patterning and survival of the cranial neural crest. *Development, 132*, 4937–4950.

Itin, P. H., & Fistarol, S. K. (2004). Ectodermal dysplasias. *American Journal of Medical Genetics. Part C, Seminars in Medical Genetics, 131C*, 45–51.

Ito, Y., Yeo, J. Y., Chytil, A., Han, J., Bringas, P., Jr., Nakajima, A., et al. (2003). Conditional inactivation of Tgfbr2 in cranial neural crest causes cleft palate and calvaria defects. *Development, 130*, 5269–5280.

Iwata, J., Hacia, J. G., Suzuki, A., Sanchez-Lara, P. A., Urata, M., & Chai, Y. (2012). Modulation of noncanonical TGF-beta signaling prevents cleft palate in Tgfbr2 mutant mice. *The Journal of Clinical Investigation, 122*, 873–885.

Iwata, J., Parada, C., & Chai, Y. (2011). The mechanism of TGF-beta signaling during palate development. *Oral Diseases, 17*, 733–744.

Iwata, J., Suzuki, A., Pelikan, R. C., Ho, T. V., Sanchez-Lara, P. A., Urata, M., et al. (2013). Smad4-Irf6 genetic interaction and TGFbeta-mediated IRF6 signaling cascade are crucial for palatal fusion in mice. *Development, 140*, 1220–1230.

Jain, A., Ma, C. A., Liu, S., Brown, M., Cohen, J., & Strober, W. (2001). Specific missense mutations in NEMO result in hyper-IgM syndrome with hypohydrotic ectodermal dysplasia. *Nature Immunology, 2*, 223–228.

Jerome, L. A., & Papaioannou, V. E. (2001). DiGeorge syndrome phenotype in mice mutant for the T-box gene, Tbx1. *Nature Genetics, 27*, 286–291.

Jezewski, P. A., Vieira, A. R., Nishimura, C., Ludwig, B., Johnson, M., O'Brien, S. E., et al. (2003). Complete sequencing shows a role for MSX1 in non-syndromic cleft lip and palate. *Journal of Medical Genetics, 40*, 399–407.

Jiang, R., Bush, J. O., & Lidral, A. C. (2006). Development of the upper lip: Morphogenetic and molecular mechanisms. *Developmental Dynamics: An Official Publication of the American Association of Anatomists, 235*, 1152–1166.

Jimenez, D. F., & Barone, C. M. (1998). Endoscopic craniectomy for early surgical correction of sagittal craniosynostosis. *Journal of Neurosurgery, 88*, 77–81.

Johnson, E. L., Roberts, M. W., Guckes, A. D., Bailey, L. J., Phillips, C. L., & Wright, J. T. (2002). Analysis of craniofacial development in children with hypohidrotic ectodermal dysplasia. *American Journal of Medical Genetics, 112*, 327–334.

Jones, N. C., Lynn, M. L., Gaudenz, K., Sakai, D., Aoto, K., Rey, J. P., et al. (2008). Prevention of the neurocristopathy Treacher Collins syndrome through inhibition of p53 function. *Nature Medicine, 14*, 125–133.

Jugessur, A., Farlie, P. G., & Kilpatrick, N. (2009). The genetics of isolated orofacial clefts: From genotypes to subphenotypes. *Oral Diseases*, *15*, 437–453.

Juriloff, D. M., & Fraser, F. C. (1980). Genetic maternal effects on cleft lip frequency in A/J and CL/Fr mice. *Teratology*, *21*, 167–175.

Kaartinen, V., Voncken, J. W., Shuler, C., Warburton, D., Bu, D., Heisterkamp, N., et al. (1995). Abnormal lung development and cleft palate in mice lacking TGF-beta 3 indicates defects of epithelial-mesenchymal interaction. *Nature Genetics*, *11*, 415–421.

Kim, S., Lewis, A. E., Singh, V., Ma, X., Adelstein, R., & Bush, J. O. (2015). Convergence and extrusion are required for normal fusion of the mammalian secondary palate. *PLoS Biology*, *13*(4), e1002122. http://dx.doi.org/10.1371/journal.pbio.1002122.

Kobielak, K., Kobielak, A., Roszkiewicz, J., Wierzba, J., Limon, J., & Trzeciak, W. H. (2001). Mutations in the EDA gene in three unrelated families reveal no apparent correlation between phenotype and genotype in the patients with an X-linked anhidrotic ectodermal dysplasia. *American Journal of Medical Genetics*, *100*, 191–197.

Kohama, K., Nonaka, K., Hosokawa, R., Shum, L., & Ohishi, M. (2002). TGF-beta-3 promotes scarless repair of cleft lip in mouse fetuses. *Journal of Dental Research*, *81*, 688–694.

Kondo, S., Schutte, B. C., Richardson, R. J., Bjork, B. C., Knight, A. S., Watanabe, Y., et al. (2002). Mutations in IRF6 cause Van der Woude and popliteal pterygium syndromes. *Nature Genetics*, *32*, 285–289.

Kreiborg, S., & Cohen, M. M., Jr. (1992). The oral manifestations of Apert syndrome. *Journal of Craniofacial Genetics and Developmental Biology*, *12*, 41–48.

Lacro, R. V., Dietz, H. C., Sleeper, L. A., Yetman, A. T., Bradley, T. J., Colan, S. D., et al. (2014). Atenolol versus losartan in children and young adults with Marfan's syndrome. *The New England Journal of Medicine*, *371*, 2061–2071.

Lefebvre, S., Fliniaux, I., Schneider, P., & Mikkola, M. L. (2012). Identification of ectodysplasin target genes reveals the involvement of chemokines in hair development. *The Journal of Investigative Dermatology*, *132*, 1094–1102.

Leslie, E. J., Mansilla, M. A., Biggs, L. C., Schuette, K., Bullard, S., Cooper, M., et al. (2012). Expression and mutation analyses implicate ARHGAP29 as the etiologic gene for the cleft lip with or without cleft palate locus identified by genome-wide association on chromosome 1p22. *Birth Defects Research. Part A, Clinical and Molecular Teratology*, *94*, 934–942.

Letra, A., de Almeida, A. L., Kaizer, R., Esper, L. A., Sgarbosa, S., & Granjeiro, J. M. (2007). Intraoral features of Apert's syndrome. *Oral Surgery, Oral Medicine, Oral Pathology, Oral Radiology, and Endodontics*, *103*, e38–e41.

Levine, A. J. (1997). p53, the cellular gatekeeper for growth and division. *Cell*, *88*, 323–331.

Lexner, M. O., Bardow, A., Hertz, J. M., Nielsen, L. A., & Kreiborg, S. (2007). Anomalies of tooth formation in hypohidrotic ectodermal dysplasia. *International Journal of Paediatric Dentistry/The British Paedodontic Society [and] the International Association of Dentistry for Children*, *17*, 10–18.

Li, Y., Lacerda, D. A., Warman, M. L., Beier, D. R., Yoshioka, H., Ninomiya, Y., et al. (1995). A fibrillar collagen gene, Col11a1, is essential for skeletal morphogenesis. *Cell*, *80*, 423–430.

Li, S. W., Prockop, D. J., Helminen, H., Fassler, R., Lapvetelainen, T., Kiraly, K., et al. (1995). Transgenic mice with targeted inactivation of the Col2 alpha 1 gene for collagen II develop a skeleton with membranous and periosteal bone but no endochondral bone. *Genes & Development*, *9*, 2821–2830.

Lidral, A. C., Romitti, P. A., Basart, A. M., Doetschman, T., Leysens, N. J., Daack-Hirsch, S., et al. (1998). Association of MSX1 and TGFB3 with nonsyndromic clefting in humans. *American Journal of Human Genetics*, *63*, 557–568.

Liu, W., Sun, X., Braut, A., Mishina, Y., Behringer, R. R., Mina, M., et al. (2005). Distinct functions for Bmp signaling in lip and palate fusion in mice. *Development*, *132*, 1453–1461.

Loeys, B. L., Chen, J., Neptune, E. R., Judge, D. P., Podowski, M., Holm, T., et al. (2005). A syndrome of altered cardiovascular, craniofacial, neurocognitive and skeletal development caused by mutations in TGFBR1 or TGFBR2. *Nature Genetics*, *37*, 275–281.

Mangold, E., Ludwig, K. U., Birnbaum, S., Baluardo, C., Ferrian, M., Herms, S., et al. (2010). Genome-wide association study identifies two susceptibility loci for nonsyndromic cleft lip with or without cleft palate. *Nature Genetics*, *42*, 24–26.

Mangold, E., Ludwig, K. U., & Nothen, M. M. (2011). Breakthroughs in the genetics of orofacial clefting. *Trends in Molecular Medicine*, *17*, 725–733.

Marazita, M. L. (2012). The evolution of human genetic studies of cleft lip and cleft palate. *Annual Review of Genomics and Human Genetics*, *13*, 263–283.

Martinez-Abadias, N., Holmes, G., Pankratz, T., Wang, Y., Zhou, X., Jabs, E. W., et al. (2013). From shape to cells: Mouse models reveal mechanisms altering palate development in Apert syndrome. *Disease Models & Mechanisms*, *6*, 768–779.

Mo, R., Freer, A. M., Zinyk, D. L., Crackower, M. A., Michaud, J., Heng, H. H., et al. (1997). Specific and redundant functions of Gli2 and Gli3 zinc finger genes in skeletal patterning and development. *Development*, *124*, 113–123.

Moretti, F., Marinari, B., Lo Iacono, N., Botti, E., Giunta, A., Spallone, G., et al. (2010). A regulatory feedback loop involving p63 and IRF6 links the pathogenesis of 2 genetically different human ectodermal dysplasias. *The Journal of Clinical Investigation*, *120*, 1570–1577.

Muenke, M., Schell, U., Hehr, A., Robin, N. H., Losken, H. W., Schinzel, A., et al. (1994). A common mutation in the fibroblast growth factor receptor 1 gene in Pfeiffer syndrome. *Nature Genetics*, *8*, 269–274.

Muraoka, N., Shum, L., Fukumoto, S., Nomura, T., Ohishi, M., & Nonaka, K. (2005). Transforming growth factor-beta3 promotes mesenchymal cell proliferation and angiogenesis mediated by the enhancement of cyclin D1, Flk-1, and CD31 gene expression during CL/Fr mouse lip fusion. *Birth Defects Research. Part A, Clinical and Molecular Teratology*, *73*, 956–965.

Mustonen, T., Pispa, J., Mikkola, M. L., Pummila, M., Kangas, A. T., Pakkasjarvi, L., et al. (2003). Stimulation of ectodermal organ development by Ectodysplasin-A1. *Developmental Biology*, *259*, 123–136.

Nakamura, T., Colbert, M., Krenz, M., Molkentin, J. D., Hahn, H. S., Dorn, G. W., 2nd., et al. (2007). Mediating ERK 1/2 signaling rescues congenital heart defects in a mouse model of Noonan syndrome. *The Journal of Clinical Investigation*, *117*, 2123–2132.

Nakatomi, M., Wang, X. P., Key, D., Lund, J. J., Turbe-Doan, A., Kist, R., et al. (2010). Genetic interactions between Pax9 and Msx1 regulate lip development and several stages of tooth morphogenesis. *Developmental Biology*, *340*, 438–449.

Nelson, D. K., & Williams, T. (2004). Frontonasal process-specific disruption of AP-2alpha results in postnatal midfacial hypoplasia, vascular anomalies, and nasal cavity defects. *Developmental Biology*, *267*, 72–92.

Niihori, T., Aoki, Y., Narumi, Y., Neri, G., Cave, H., Verloes, A., et al. (2006). Germline KRAS and BRAF mutations in cardio-facio-cutaneous syndrome. *Nature Genetics*, *38*, 294–296.

Noonan, J. A. (1968). Hypertelorism with Turner phenotype. A new syndrome with associated congenital heart disease. *American Journal of Diseases of Children*, *116*, 373–380.

Nottoli, T., Hagopian-Donaldson, S., Zhang, J., Perkins, A., & Williams, T. (1998). AP-2-null cells disrupt morphogenesis of the eye, face, and limbs in chimeric mice. *Proceedings of the National Academy of Sciences of the United States of America*, *95*, 13714–13719.

Oldridge, M., Zackai, E. H., McDonald-McGinn, D. M., Iseki, S., Morriss-Kay, G. M., Twigg, S. R., et al. (1999). De novo alu-element insertions in FGFR2 identify a distinct pathological basis for Apert syndrome. *American Journal of Human Genetics*, *64*, 446–461.

Panchal, J., & Uttchin, V. (2003). Management of craniosynostosis. *Plastic and Reconstructive Surgery*, *111*, 2032–2048. quiz 2049.
Pandit, B., Sarkozy, A., Pennacchio, L. A., Carta, C., Oishi, K., Martinelli, S., et al. (2007). Gain-of-function RAF1 mutations cause Noonan and LEOPARD syndromes with hypertrophic cardiomyopathy. *Nature Genetics*, *39*, 1007–1012.
Parada, C., & Chai, Y. (2012). Roles of BMP signaling pathway in lip and palate development. *Frontiers of Oral Biology*, *16*, 60–70.
Park, W. J., Theda, C., Maestri, N. E., Meyers, G. A., Fryburg, J. S., Dufresne, C., et al. (1995). Analysis of phenotypic features and FGFR2 mutations in Apert syndrome. *American Journal of Human Genetics*, *57*, 321–328.
Pauws, E., Hoshino, A., Bentley, L., Prajapati, S., Keller, C., Hammond, P., et al. (2009). Tbx22null mice have a submucous cleft palate due to reduced palatal bone formation and also display ankyloglossia and choanal atresia phenotypes. *Human Molecular Genetics*, *18*, 4171–4179.
Perlyn, C. A., Morriss-Kay, G., Darvann, T., Tenenbaum, M., & Ornitz, D. M. (2006). A model for the pharmacological treatment of crouzon syndrome. *Neurosurgery*, *59*, 210–215. discussion 210–215.
Pfeiffer, R. A. (1964). Dominant hereditary acrocephalosyndactylia. *Zeitschrift für Kinderheilkunde*, *90*, 301–320.
Phelps, P. D., Poswillo, D., & Lloyd, G. A. (1981). The ear deformities in mandibulofacial dysostosis (Treacher Collins syndrome). *Clinical Otolaryngology and Allied Sciences*, *6*, 15–28.
Pierpont, M. E., Magoulas, P. L., Adi, S., Kavamura, M. I., Neri, G., Noonan, J., et al. (2014). Cardio-facio-cutaneous syndrome: Clinical features, diagnosis, and management guidelines. *Pediatrics*, *134*, e1149–e1162.
Pinheiro, M., & Freire-Maia, N. (1994). Ectodermal dysplasias: A clinical classification and a causal review. *American Journal of Medical Genetics*, *53*, 153–162.
Pispa, J., Jung, H. S., Jernvall, J., Kettunen, P., Mustonen, T., Tabata, M. J., et al. (1999). Cusp patterning defect in Tabby mouse teeth and its partial rescue by FGF. *Developmental Biology*, *216*, 521–534.
Proetzel, G., Pawlowski, S. A., Wiles, M. V., Yin, M., Boivin, G. P., Howles, P. N., et al. (1995). Transforming growth factor-beta 3 is required for secondary palate fusion. *Nature Genetics*, *11*, 409–414.
Rauen, K. A. (2007). HRAS and the Costello syndrome. *Clinical Genetics*, *71*, 101–108.
Rauen, K. A., Banerjee, A., Bishop, W. R., Lauchle, J. O., McCormick, F., McMahon, M., et al. (2011). Costello and cardio-facio-cutaneous syndromes: Moving toward clinical trials in RASopathies. *American Journal of Medical Genetics. Part C, Seminars in Medical Genetics*, *157C*, 136–146.
Rauen, K. A., Huson, S. M., Burkitt-Wright, E., Evans, D. G., Farschtschi, S., Ferner, R. E., et al. (2015). Recent developments in neurofibromatoses and RASopathies: Management, diagnosis and current and future therapeutic avenues. *American Journal of Medical Genetics Part A*, *167A*, 1–10.
Razzaque, M. A., Nishizawa, T., Komoike, Y., Yagi, H., Furutani, M., Amo, R., et al. (2007). Germline gain-of-function mutations in RAF1 cause Noonan syndrome. *Nature Genetics*, *39*, 1013–1017.
Reardon, W., Winter, R. M., Rutland, P., Pulleyn, L. J., Jones, B. M., & Malcolm, S. (1994). Mutations in the fibroblast growth factor receptor 2 gene cause Crouzon syndrome. *Nature Genetics*, *8*, 98–103.
Rice, R., Spencer-Dene, B., Connor, E. C., Gritli-Linde, A., McMahon, A. P., Dickson, C., et al. (2004). Disruption of Fgf10/Fgfr2b-coordinated epithelial-mesenchymal interactions causes cleft palate. *The Journal of Clinical Investigation*, *113*, 1692–1700.

Richardson, R. J., Dixon, J., Malhotra, S., Hardman, M. J., Knowles, L., Boot-Handford,-R. P., et al. (2006). Irf6 is a key determinant of the keratinocyte proliferation-differentiation switch. *Nature Genetics*, *38*, 1329–1334.

Roberts, A. E., Araki, T., Swanson, K. D., Montgomery, K. T., Schiripo, T. A., Joshi, V. A., et al. (2007). Germline gain-of-function mutations in SOS1 cause Noonan syndrome. *Nature Genetics*, *39*, 70–74.

Rodriguez-Viciana, P., Tetsu, O., Tidyman, W. E., Estep, A. L., Conger, B. A., Cruz, M. S., et al. (2006). Germline mutations in genes within the MAPK pathway cause cardio-facio-cutaneous syndrome. *Science*, *311*, 1287–1290.

Rovin, S., Dachi, S. F., Borenstein, D. B., & Cotter, W. B. (1964). Mandibulofacial dysostosis, a familial study of five generations. *The Journal of Pediatrics*, *65*, 215–221.

Sadier, A., Viriot, L., Pantalacci, S., & Laudet, V. (2014). The ectodysplasin pathway: From diseases to adaptations. *Trends in Genetics*, *30*(1), 24–31.

Salisbury, D. M., & Stothers, J. K. (1981). Hypohidrotic ectodermal dysplasia and sudden infant death. *Lancet*, *1*, 153–154.

Satokata, I., & Maas, R. (1994). Msx1 deficient mice exhibit cleft palate and abnormalities of craniofacial and tooth development. *Nature Genetics*, *6*, 348–356.

Schell, U., Hehr, A., Feldman, G. J., Robin, N. H., Zackai, E. H., de Die-Smulders, C., et al. (1995). Mutations in FGFR1 and FGFR2 cause familial and sporadic Pfeiffer syndrome. *Human Molecular Genetics*, *4*, 323–328.

Schubbert, S., Bollag, G., & Shannon, K. (2007). Deregulated Ras signaling in developmental disorders: New tricks for an old dog. *Current Opinion in Genetics & Development*, *17*, 15–22.

Schubbert, S., Zenker, M., Rowe, S. L., Boll, S., Klein, C., Bollag, G., et al. (2006). Germline KRAS mutations cause Noonan syndrome. *Nature Genetics*, *38*, 331–336.

Schuffenhauer, S., Leifheit, H. J., Lichtner, P., Peters, H., Murken, J., & Emmerich, P. (1999). De novo deletion (14)(q11.2q13) including PAX9: Clinical and molecular findings. *Journal of Medical Genetics*, *36*, 233–236.

Schuhmacher, A. J., Guerra, C., Sauzeau, V., Canamero, M., Bustelo, X. R., & Barbacid, M. (2008). A mouse model for Costello syndrome reveals an Ang II-mediated hypertensive condition. *The Journal of Clinical Investigation*, *118*, 2169–2179.

Seegmiller, R., Fraser, F. C., & Sheldon, H. (1971). A new chondrodystrophic mutant in mice. Electron microscopy of normal and abnormal chondrogenesis. *The Journal of Cell Biology*, *48*, 580–593.

Shukla, V., Coumoul, X., Wang, R. H., Kim, H. S., & Deng, C. X. (2007). RNA interference and inhibition of MEK-ERK signaling prevent abnormal skeletal phenotypes in a mouse model of craniosynostosis. *Nature Genetics*, *39*, 1145–1150.

Slaney, S. F., Oldridge, M., Hurst, J. A., Moriss-Kay, G. M., Hall, C. M., Poole, M. D., et al. (1996). Differential effects of FGFR2 mutations on syndactyly and cleft palate in Apert syndrome. *American Journal of Human Genetics*, *58*, 923–932.

Smith, J. (1929). Hereditary ectodermal dysplasia. *Archives of Disease in Childhood*, *4*, 215–226.

Solomon, L. M., & Keuer, E. J. (1980). The ectodermal dysplasias. Problems of classification and some newer syndromes. *Archives of Dermatology*, *116*, 1295–1299.

Sperry, E. D., Hurd, E. A., Durham, M. A., Reamer, E. N., Stein, A. B., & Martin, D. M. (2014). The chromatin remodeling protein CHD7, mutated in CHARGE syndrome, is necessary for proper craniofacial and tracheal development. *Developmental Dynamics: An Official Publication of the American Association of Anatomists*, *243*, 1055–1066.

Srivastava, A. K., Durmowicz, M. C., Hartung, A. J., Hudson, J., Ouzts, L. V., Donovan, D. M., et al. (2001). Ectodysplasin-A1 is sufficient to rescue both hair growth and sweat glands in Tabby mice. *Human Molecular Genetics*, *10*, 2973–2981.

Srivastava, A. K., Pispa, J., Hartung, A. J., Du, Y., Ezer, S., Jenks, T., et al. (1997). The Tabby phenotype is caused by mutation in a mouse homologue of the EDA gene that reveals

novel mouse and human exons and encodes a protein (ectodysplasin-A) with collagenous domains. *Proceedings of the National Academy of Sciences of the United States of America, 94*, 13069–13074.
Stottmann, R. W., Bjork, B. C., Doyle, J. B., & Beier, D. R. (2010). Identification of a Van der Woude syndrome mutation in the cleft palate 1 mutant mouse. *Genesis, 48*, 303–308.
Tartaglia, M., Mehler, E. L., Goldberg, R., Zampino, G., Brunner, H. G., Kremer, H., et al. (2001). Mutations in PTPN11, encoding the protein tyrosine phosphatase SHP-2, cause Noonan syndrome. *Nature Genetics, 29*, 465–468.
Tartaglia, M., Pennacchio, L. A., Zhao, C., Yadav, K. K., Fodale, V., Sarkozy, A., et al. (2007). Gain-of-function SOS1 mutations cause a distinctive form of Noonan syndrome. *Nature Genetics, 39*, 75–79.
Taya, Y., O'Kane, S., & Ferguson, M. W. (1999). Pathogenesis of cleft palate in TGF-beta3 knockout mice. *Development, 126*, 3869–3879.
Thomason, H. A., Dixon, M. J., & Dixon, J. (2008). Facial clefting in Tp63 deficient mice results from altered Bmp4, Fgf8 and Shh signaling. *Developmental Biology, 321*, 273–282.
Thurnam, J. (1848). Two cases in which the skin, hair and teeth were very imperfectly developed. *Medico-Chirurgical Transactions, 31*, 71–82.
Tidyman, W. E., & Rauen, K. A. (2009). The RASopathies: Developmental syndromes of Ras/MAPK pathway dysregulation. *Current Opinion in Genetics & Development, 19*, 230–236.
Trokovic, N., Trokovic, R., Mai, P., & Partanen, J. (2003). Fgfr1 regulates patterning of the pharyngeal region. *Genes & Development, 17*, 141–153.
Urosevic, J., Sauzeau, V., Soto-Montenegro, M. L., Reig, S., Desco, M., Wright, E. M., et al. (2011). Constitutive activation of B-Raf in the mouse germ line provides a model for human cardio-facio-cutaneous syndrome. *Proceedings of the National Academy of Sciences of the United States of America, 108*, 5015–5020.
Uslu, V. V., Petretich, M., Ruf, S., Langenfeld, K., Fonseca, N. A., Marioni, J. C., et al. (2014). Long-range enhancers regulating Myc expression are required for normal facial morphogenesis. *Nature Genetics, 46*, 753–758.
Van Laer, L., Dietz, H., & Loeys, B. (2014). Loeys-Dietz syndrome. *Advances in Experimental Medicine and Biology, 802*, 95–105.
Vogels, A., & Fryns, J. P. (2006). Pfeiffer syndrome. *Orphanet Journal of Rare Diseases, 1*, 19.
Wan, P. T., Garnett, M. J., Roe, S. M., Lee, S., Niculescu-Duvaz, D., Good, V. M., et al. (2004). Mechanism of activation of the RAF-ERK signaling pathway by oncogenic mutations of B-RAF. *Cell, 116*, 855–867.
Wang, C., Chang, J. Y., Yang, C., Huang, Y., Liu, J., You, P., et al. (2013). Type 1 fibroblast growth factor receptor in cranial neural crest cell-derived mesenchyme is required for palatogenesis. *The Journal of Biological Chemistry, 288*, 22174–22183.
Wang, Y., Sun, M., Uhlhorn, V. L., Zhou, X., Peter, I., Martinez-Abadias, N., et al. (2010). Activation of p38 MAPK pathway in the skull abnormalities of Apert syndrome Fgfr2 (+P253R) mice. *BMC Developmental Biology, 10*, 22.
Wang, Y., Xiao, R., Yang, F., Karim, B. O., Iacovelli, A. J., Cai, J., et al. (2005). Abnormalities in cartilage and bone development in the Apert syndrome FGFR2(+/S252W) mouse. *Development, 132*, 3537–3548.
Wassif, C. A., Zhu, P., Kratz, L., Krakowiak, P. A., Battaile, K. P., Weight, F. F., et al. (2001). Biochemical, phenotypic and neurophysiological characterization of a genetic mouse model of RSH/Smith–Lemli–Opitz syndrome. *Human Molecular Genetics, 10*, 555–564.
Wilkie, A. O., Oldridge, M., Tang, Z., & Maxson, R. E., Jr. (2001). Craniosynostosis and related limb anomalies. *Novartis Foundation Symposium, 232*, 122–133. discussion 133–143.

Wilkie, A. O., Slaney, S. F., Oldridge, M., Poole, M. D., Ashworth, G. J., Hockley, A. D., et al. (1995). Apert syndrome results from localized mutations of FGFR2 and is allelic with Crouzon syndrome. *Nature Genetics, 9*, 165–172.

Winnier, G. E., Hargett, L., & Hogan, B. L. (1997). The winged helix transcription factor MFH1 is required for proliferation and patterning of paraxial mesoderm in the mouse embryo. *Genes & Development, 11*, 926–940.

Wu, X., Simpson, J., Hong, J. H., Kim, K. H., Thavarajah, N. K., Backx, P. H., et al. (2011). MEK-ERK pathway modulation ameliorates disease phenotypes in a mouse model of Noonan syndrome associated with the Raf1(L613V) mutation. *The Journal of Clinical Investigation, 121*, 1009–1025.

Xu, X., Han, J., Ito, Y., Bringas, P., Jr., Urata, M. M., & Chai, Y. (2006). Cell autonomous requirement for Tgfbr2 in the disappearance of medial edge epithelium during palatal fusion. *Developmental Biology, 297*, 238–248.

Yin, L., Du, X., Li, C., Xu, X., Chen, Z., Su, N., et al. (2008). A Pro253Arg mutation in fibroblast growth factor receptor 2 (Fgfr2) causes skeleton malformation mimicking human Apert syndrome by affecting both chondrogenesis and osteogenesis. *Bone, 42*, 631–643.

Yu, K., Herr, A. B., Waksman, G., & Ornitz, D. M. (2000). Loss of fibroblast growth factor receptor 2 ligand-binding specificity in Apert syndrome. *Proceedings of the National Academy of Sciences of the United States of America, 97*, 14536–14541.

Yu, K., & Ornitz, D. M. (2001). Uncoupling fibroblast growth factor receptor 2 ligand binding specificity leads to Apert syndrome-like phenotypes. *Proceedings of the National Academy of Sciences of the United States of America, 98*, 3641–3643.

Zaritsky, J. J., Eckman, D. M., Wellman, G. C., Nelson, M. T., & Schwarz, T. L. (2000). Targeted disruption of Kir2.1 and Kir2.2 genes reveals the essential role of the inwardly rectifying K(+) current in K(+)-mediated vasodilation. *Circulation Research, 87*, 160–166.

Zenker, M. (2009). Genetic and pathogenetic aspects of Noonan syndrome and related disorders. *Hormone Research, 72*(Suppl. 2), 57–63.

Zenker, M., Lehmann, K., Schulz, A. L., Barth, H., Hansmann, D., Koenig, R., et al. (2007). Expansion of the genotypic and phenotypic spectrum in patients with KRAS germline mutations. *Journal of Medical Genetics, 44*, 131–135.

Zhang, Y., Blackwell, E. L., McKnight, M. T., Knutsen, G. R., Vu, W. T., & Ruest, L. B. (2012). Specific inactivation of Twist1 in the mandibular arch neural crest cells affects the development of the ramus and reveals interactions with hand2. *Developmental Dynamics: An Official Publication of the American Association of Anatomists, 241*, 924–940.

Zhang, Z., Song, Y., Zhao, X., Zhang, X., Fermin, C., & Chen, Y. (2002). Rescue of cleft palate in Msx1-deficient mice by transgenic Bmp4 reveals a network of BMP and Shh signaling in the regulation of mammalian palatogenesis. *Development, 129*, 4135–4146.

Zhou, Q., Li, M. J., Zhu, W. L., Guo, J. Z., Wang, Y., Li, Y., et al. (2013). Association between interferon regulatory factor 6 gene polymorphisms and nonsyndromic cleft lip with or without cleft palate in a Chinese population. *The Cleft Palate-Craniofacial Journal, 50*, 570–576.

Zhou, Y. X., Xu, X., Chen, L., Li, C., Brodie, S. G., & Deng, C. X. (2000). A Pro250Arg substitution in mouse Fgfr1 causes increased expression of Cbfa1 and premature fusion of calvarial sutures. *Human Molecular Genetics, 9*, 2001–2008.

Zonana, J., Elder, M. E., Schneider, L. C., Orlow, S. J., Moss, C., Golabi, M., et al. (2000). A novel X-linked disorder of immune deficiency and hypohidrotic ectodermal dysplasia is allelic to incontinentia pigmenti and due to mutations in IKK-gamma (NEMO). *American Journal of Human Genetics, 67*, 1555–1562.

Zucchero, T. M., Cooper, M. E., Maher, B. S., Daack-Hirsch, S., Nepomuceno, B., Ribeiro, L., et al. (2004). Interferon regulatory factor 6 (IRF6) gene variants and the risk of isolated cleft lip or palate. *The New England Journal of Medicine, 351*, 769–780.

CHAPTER EIGHTEEN

Signaling Pathways in Craniofacial Development: Insights from Rare Skeletal Disorders

Cynthia L. Neben*, Amy E. Merrill*,†,1
*Department of Biochemistry and Molecular Biology, Keck School of Medicine, University of Southern California, Los Angeles, California, USA
†Center for Craniofacial Molecular Biology, Ostrow School of Dentistry, University of Southern California, Los Angeles, California, USA
[1]Corresponding author: e-mail address: amerrill@usc.edu

Contents

Abstract

In the developing embryo, signaling pathways define how the mesenchymal precursors of bone form individual skeletal elements with the proper size, shape, orientation, and integration. Disruptions to these signaling processes lead to a large variety of congenital conditions categorized as skeletal dysplasias. While individually these skeletal disorders are rare, collectively they represent a significant cause of disability in the United States. Here, we discuss how the study of these rare events in human development reveal novel and unexpected insights into signaling mechanisms that regulate normal skeletal development and, in the process, advance novel molecular-based therapies for treatment of rare and common bone diseases alike.

Current Topics in Developmental Biology, Volume 115
ISSN 0070-2153
http://dx.doi.org/10.1016/bs.ctdb.2015.09.005

1. RARE SKELETAL DISORDERS AND THEIR CLASSIFICATION

Heritable disorders of the human skeletal system, known as skeletal dysplasias, are one of the largest classes of birth defects with over 450 clinically recognized conditions (Warman et al., 2011). These disorders arise from a variety of errors in the complex processes regulating skeletal development and exhibit a wide range of abnormalities from mild joint irregularities to severe dwarfism with prenatal lethality. While each skeletal dysplasia is relatively rare, their collective incidence rate of 1 in 5000 live births represents a significant health concern (Krakow & Rimoin, 2010). Despite this, these conditions remain difficult to diagnose because clinical manifestations and severity can be heterogeneous within a single disorder and overlapping between distinct disorders.

The skeletal dysplasias can be broadly divided into three classes: the osteodysplasias, which have abnormalities in bone density and mineralization; the chondrodysplasias, which exhibit irregularities in cartilage growth and organization; and the dysostoses, which have malformations of individual bones or a group of bones. These general categories are neither absolute nor mutually exclusive, and the distinctions between them become further blurred as the disease mechanisms of skeletal dysplasias become clearer. Currently, more detailed classifications are used clinically to aid in disease recognition and inform genetic risk assessment, prognosis, and treatment.

The Nosology and Classification of Genetic Skeletal Disorders integrates our ever-growing body of knowledge of the genetic, molecular, and biochemical features with clinical, radiographic, and pathological findings (Warman et al., 2011). In its most current form, it organizes the skeletal dysplasias into 40 distinct groups (Warman et al., 2011). Many groups are based on a common underlying gene or pathway, as in the *FGFR3* chondrodysplasias. Macroscopic bone features characterize other groups, such as in the bent bone dysplasias. Changes in bone density delineate groups for increased versus decreased bone density. Various groups are based on radiographic abnormalities that localize to specific bone structures (epiphysis, metaphysis, or diaphysis) and segments (rhizo, meso, or acro) that are indicated in their namesakes using descriptors constructed from Greek stems. For example, the metaphyseal dysplasias are characterized by radiographic abnormalities in the growth plate (metaphysis) while the acromesomelic dysplasias show abnormal shortening of the bones in the hands/feet (acro) and forearm/forelegs (meso). Multiple groups are dedicated to the dysostoses, including groups for cleidocranial dysplasia and craniosynostosis syndromes.

Beyond aiding in disease recognition, classification systems serve as a catalog of genes regulating human skeletal development and homeostasis. Of the known skeletal dysplasias, 70% are associated with one or more of at least 230 different genes that code for a variety of protein products such as those involved in mRNA processing, ribosome biosynthesis, extracellular matrix, metabolism, proliferation, transcriptional regulation, and signal transduction (Warman et al., 2011). Mutations in the same gene can cause multiple conditions with wide phenotypic variability; similarly, mutations in different genes can result in disorders with overlapping clinical features. While disease-causing genes have been identified for many disorders, there are a significant number of conditions where the molecular basis of disease is unknown. Historically, disease gene identification relied on labor-intensive approaches that involved candidate gene analysis and linkage mapping. These techniques were less successful in studies with a small patient pool or locus heterogeneity. With the advent of next-generation DNA sequencing, disease-causing genes that evaded previous detection are being identified at an accelerating rate using whole-genome and whole-exome approaches.

The discoveries made in rare skeletal disorders have uncovered novel genes that regulate skeletal development, revealed new functions for already identified genes, and linked genes into common molecular networks. Here, we highlight the impact of these studies on our understanding of the signaling pathways regulating skeletal development, with a particular emphasis on the craniofacial complex. We also discuss how this knowledge, which in many cases could not have been predicted, is advancing development of novel molecular-based therapeutics for both rare and common diseases of bone.

2. BUILDING THE SKELETON

The complex and elegant morphology of the human skeletal system arises through a stepwise series of developmental events, where each event depends on that before it. The first event observable in skeletal development is aggregation of previously dispersed mesenchymal cells, derived from mesoderm and neural crest, into condensations whose shape and position prefigure the future skeletal elements. Following the process of condensation, signaling pathways initiate mesenchymal differentiation to build bone with or without a cartilage precursor.

In bone arising from cartilage, known as endochondral bone, differentiation of the mesenchymal condensation into chondrocytes depends initially upon the SOX trio, a trifecta of SOX5, SOX6, and SOX9 SRY-related transcription factors (Bi, Deng, Zhang, Behringer, & de Crombrugghe, 1999; Smits et al., 2001). In humans, haploinsufficiency for *SOX9* causes campomelic dysplasia, a severe chondrodysplasia affecting formation of all endochondral bones and named after the characteristic angulated long bones (Foster et al., 1994; Wagner et al., 1994). Once the cartilage anlage is established, cells at the periphery form the perichondrium, which houses osteoprogenitor cells and regulates subsequent bone growth by sending signals to the underlying chondrocytes. (Kronenberg, 2007). Subsequently, those chondrocytes at the midpoint of the bone shaft mature into hypertrophic chondrocytes and secrete matrix rich in proteoglycans and collagens. Mouse studies show that this chondrocyte maturation process is regulated by Mef2c, Mef2d, Runx2, and Runx3 (Arnold et al., 2007; Inada et al., 1999; Kim, Otto, Zabel, & Mundlos, 1999). Simultaneously, perichondrial cells adjacent to the zone of hypertrophy begin expressing Runx2; these cells form the periosteum and differentiate into osteoblasts. *RUNX2* haploinsufficiency in cleidocranial dysplasia, a disorder with deficient and delayed bone ossification, revealed the essential role of RUNX2 in osteoblast differentiation (Mundlos et al., 1997; Otto et al., 1997). Newly-formed osteoblasts from the periosteum remain at the periphery to build the bone collar and travel along invading blood vessels to fill the matrix lacunae left void by apoptosed hypertrophic chondrocytes. Subsequent longitudinal growth of endochondral bones occurs interstitially at the growth plate, which contains stratified layers of columnar chondrocytes that progress sequentially through zones of proliferation, prehypertrophy, and hypertrophy to terminate at the mineralized front of the newly forming bone.

On the other hand, bone created without a cartilage precursor, known as intramembranous bone, arises from direct osteogenic differentiation nucleated within the condensation. This process is initiated by Runx2 and subsequently supported by its downstream transcriptional target SP7/Osterix (Nakashima et al., 2002). *SP7* is essential for both endochondral and intramembranous bone formation, and frameshift mutations in this gene cause recessive osteogenesis imperfecta, a disorder of extremely low bone mass (Lapunzina et al., 2010). Growth of intramembranous bone is tightly regulated at the osteogenic fronts of the bone in order to maintain a self-renewing population of skeletal progenitors at the advancing surface. Transcriptional regulators such as Msx2 and Twist maintain the proper balance between proliferation and differentiation in skeletogenic mesenchyme. This is best described in the calvarial sutures, where there are organizing centers for appositional growth of intramembranous bone (Bialek et al., 2004; Ishii et al., 2003; Merrill et al., 2006). These mouse studies are supported by human genetics. Mutations in *MSX2* that increase protein function cause premature fusion of the suture, or craniosynostosis, whereas, mutations that reduce MSX2 function lead to ossification defects within the calvarium (Jabs et al., 1993; Wilkie et al., 2000). On the other hand, truncating mutations in *TWIST* also cause craniosynostosis (Howard et al., 1997; el Ghouzzi et al., 1997).

3. SIGNALING PATHWAYS IN SKELETAL DEVELOPMENT AND THEIR ASSOCIATED SKELETAL DISORDERS

Of the many important signaling molecules regulating bone formation, most fall into a few developmental pathways grouped by their response to the following signaling ligand families: Hedgehog (HH), Parathyroid hormone-related protein (PTHrP), WNT, NOTCH, Transforming growth factor β (TGFβ), Bone morphogenetic protein (BMP), and Fibroblast growth factor (FGF). Most ligands are secreted into the extracellular space and establish a gradient across a field of cells, eliciting distinct responses depending on their concentration. Other ligands are tethered to the plasma membrane through covalent hydrophobic modifications or endogenous transmembrane domains and, under certain stimuli, are released into the extracellular space via enzymatic cleavage. Nevertheless, all ligands provide positional information to those cells expressing the corresponding receptors. These receptors relay messages across the plasma membrane by launching signal transduction cascades through a series of cleavages and/or post-translational modification steps that ultimately alter gene expression in the nucleus. Depending on the duration and intensity of the signal, these events can initiate different cellular programs for proliferation, growth, and differentiation.

4. HEDGEHOG

Hedgehog signaling is central to growth, patterning, and morphogenesis of many tissues within the developing embryo, and thus its misregulation leads to multiple pathologies. The hedgehog ligand was originally discovered in a *Drosophila* mutagenesis screen and named after its corresponding mutant whose larval cuticle was covered in a lawn of spikey denticles (Nusslein-Volhard & Wieschaus, 1980). Subsequently, three mammalian homologous of Hedgehog were identified: Sonic Hedgehog (SHH), Indian Hedgehog (IHH), and Desert Hedgehog (DHH). All Hedgehog ligands are secreted factors synthesized as large precursors that undergo autocatalytic cleavage into N- and C-terminal fragments (Hall, Porter, Beachy, & Leahy, 1995; Porter et al., 1996). Following the split, the C-terminal fragment is rapidly degraded by the proteasome, and the N-terminal domain is covalently modified with lipid moieties and transported to the cell surface, where it associates with the extracellular side of the plasma membrane (Lewis & Eisen, 2001; Zeng et al., 2001). With the help of Dispatched and Scube, active Hedgehog ligands are released from the producing cell's membrane and diffuse through the extracellular space as a morphogen, creating a concentration gradient that elicits distinct cellular responses (Burke et al., 1999; Ma et al., 2002; Tukachinsky, Kuzmickas, Jao, Liu, & Salic, 2012). The distinct cellular response to Hedgehog depends on the target cell type and its discrete concentration threshold. Target cells detect and respond to Hedgehog ligands through the Patched receptor, of which there are two homologues in mammals (PTCH1 and PTCH2). The Patched receptor, unlike many signaling receptors, is repressed by ligand binding. In the absence of Hedgehog ligands, PTCH suppresses Smoothened (SMO), a G protein-coupled receptor, by triggering its degradation and regulating its intracellular trafficking. Hedgehog binding to PTCH relieves SMO inhibition, allowing SMO to accumulate in the primary cilium and mediate downstream signaling cascades that dislodge GLI transcription factors (GLI 1, 2, and 3) from a repressive complex with KIF7 and Suppressor of Fused (SUFU). GLI then enters the nucleus and activates expression of its target genes.

In vertebrates, SHH and IHH appear to be the principal ligands regulating skeletal development. SHH acts at early stages to pattern skeletogenic mesenchyme of the axial and appendicular skeleton. SHH from the ventral floor plate of the neural tube and notochord regulates patterning of the axial skeleton by defining the region within the somite that becomes sclerotome, the skeletogenic mesenchyme from which the intervertebral discs, vertebrae, and ribs are derived (Chiang et al., 1996; Fan et al., 1995; Fan & Tessier-Lavigne, 1994; Johnson, Laufer, Riddle, & Tabin, 1994; McMahon et al., 1998). In the craniofacial complex, SHH from the frontonasal ectoderm zone regulates proximal–distal and dorsal–ventral patterning as well as directing outgrowth and differentiation of the cranial neural crest cell-derived skeletal structures that arise from the facial primordia (Abzhanov & Tabin, 2004; Hu, Marcucio, & Helms, 2003; Jeong, Mao, Tenzen, Kottmann, & McMahon, 2004). In the appendicular skeleton, SHH from the zone of polarizing activity regulates limb bud outgrowth and anterior–posterior polarity, which is critical for digit patterning (Riddle, Johnson, Laufer, & Tabin, 1993).

Misregulation of Hedgehog signaling results in several human conditions that are associated with a wide-spectrum of limb and facial skeletal anomalies. The most common disorder linked to the Hedgehog pathway is holoprosencephaly (HPE), which occurs in 1/250 gestations with a reduced birth prevalence of 1/10,000 due to intrauterine lethality (Leoncini et al., 2008; Matsunaga & Shiota, 1977). HPE has varying degrees of severity that range from a single central maxillary incisor and hypertelorism to premaxillary agenesis with a single-nostril nose to, on the extreme end, a blind-ending proboscis above a single median eye due to impaired midline cleavage of the embryonic forebrain (Muenke, Gurrieri, et al., 1994). *SHH* was the first disease gene to be identified in HPE, and thus the disorder is largely associated with a deficiency in SHH signaling (Belloni et al., 1996; Roessler et al., 1996).

Additional disease loci have since been mapped and identified for HPE. In some cases, the gene defects are known to reduce SHH signaling, including missense mutations in *PTCH1* that enhance its repressive activity and truncation mutations in *GLI2* (Ming et al., 2002; Roessler et al., 2003). In other cases, gene defects identified in HPE such as *CDON*, *SIX3*, *ZIC2*, and *TGIF* did not at first have an obvious link to SHH signaling, and their association with HPE strongly suggested a mechanistic link (Bae et al., 2011; Brown et al., 1998; Gripp et al., 2000; Wallis et al., 1999). Evidence now shows this to be the case (Mercier et al., 2011). For example, identification of *CDON* mutations in HPE was first to indicate that this PTCH1 coreceptor is required to promote SHH signaling during neural tube patterning, as well as in craniofacial and vertebral development (Allen, Tenzen, & McMahon, 2007). *SIX3* mutations in HPE led to the discovery that this homeobox-containing transcription factor directly activates *SHH* transcription in the anterior diencephalon (Jeong et al., 2008). *ZIC2* mutations in HPE assisted in identifying the necessity for this zinc finger transcription factor in development of the prechordal plate, a SHH-expressing organizing center that is required for forebrain midline morphogenesis (Warr et al., 2008). By showing that SIX3 and ZIC2 lie upstream of

SHH signaling, these studies in return provide an explanation for why the most severe forms of HPE are associated with mutations in these two genes.

HPE is also an occasional finding in Pallister–Hall syndrome, which is characterized by a spectrum of anomalies ranging from mild facial dysmorphism and polydactyly to agnathia, short limbs, dislocated hips, and neonatal lethality. This disorder is caused by dominant mutations in *GLI3*, a downstream modulator of SHH signaling (Kang, Graham, Olney, & Biesecker, 1997). Dominant mutations in *GLI3* are also responsible for Grieg cephalopolysyndactyly syndrome, another disorder with markedly variable expressivity that is characterized by polydactyly, calvarial defects, and in some cases, craniosynostosis (Vortkamp, Gessler, & Grzeschik, 1991). The *GLI3* mutations found in these disorders, along with their associated clinical phenotype, identified GLI3 as the human homolog of the *Drosophila* hedgehog transcriptional effector Cubitus interruptes (Ci) and uncovered its bifunctional nature in mammals (Shin, Kogerman, Lindstrom, Toftgard, & Biesecker, 1999). In the presence of SHH, full-length GLI3 functions as a transcriptional activator, and in its absence, GLI3 is cleaved to produce a repressor (Aza-Blanc, Lin, Ruiz i Altaba, & Kornberg, 2000). Truncating mutations in the middle of *GLI3* generally produce a constitutive repressive protein in Pallister–Hall syndrome, while deletions or missense mutations in other parts of *GLI3* disrupt DNA binding, leading to haploinsufficiency in Grieg cephalopolysyndactyly syndrome (Demurger et al., 2015; Johnston et al., 2005, 2010). These correlations between the location and type of mutations helped to clarify GLI3 as both an activator and repressor in SHH signaling.

After major events in skeletal patterning are completed by SHH, IHH regulates cell-fate specification during intramembranous and endochondral bone formation. In intramembranous bone of the skull, IHH is expressed at the osteogenic front where it regulates osteoblast differentiation (Lenton et al., 2011; Rice et al., 2010). In endochondral bone of the skull and limb, IHH is expressed by the prehypertrophic chondrocytes in the growth plate, where it promotes chondrocyte proliferation in the neighboring proliferative zone, triggers the onset of osteoblast differentiation in the surrounding perichondrium/periosteum, and delays chondrocyte hypertrophy in part through a negative-feedback loop with PTHrP (Lanske et al., 1996; Vortkamp et al., 1996). Recessive missense mutations in *IHH* cause Acrocapitofemoral dysplasia, which is characterized by short stature and cone-shaped epiphyses (Hellemans et al., 2003). While this disorder is not clinically associated with dysmorphic craniofacial features, mouse experiments have demonstrated Ihh involvement in craniofacial development: *Ihh* knockout mice have calvaria with reduced expanse and thickness in addition to widened sutures (Abzhanov, Rodda, McMahon, & Tabin, 2007; Lenton et al., 2011; St-Jacques, Hammerschmidt, & McMahon, 1999). This inconsistency strongly suggests that the human *IHH* mutations do not result in strict loss-of-function and instead alter domains of IHH that are critical to its signaling capacity and range (Gao et al., 2009).

Gene mutations that increase Hedgehog signaling are associated with the cancer-predisposing condition basal cell nevus syndrome, also known as Gorlin syndrome. This familial disorder presents with basal cell tumors, jaw cysts, general overgrowth, as well as skeletal abnormalities in the face, calvaria, ribs, hands, and vertebrae. The cause of basal cell nevus syndrome is linked to dominant mutations in negative regulators for SHH signaling including *PTCH1*, *PTCH2*, and *SUFU* (Fan et al., 2008; Hahn et al., 1996; Smyth et al., 1999; Johnson et al., 1996; Pastorino et al., 2009). In the case of *PTCH1*, hemizygous inactivation is associated with developmental abnormalities while homozygous inactivation, caused by a somatic second-hit mutation, results in neoplasm (Hahn et al., 1996). These discoveries showed that SHH signaling is a dosage-sensitive pathway linking differentiation and growth control in skeletal developmental and cellular transformation.

5. PARATHYROID HORMONE-RELATED PROTEIN

PTHrP was first discovered as the cause of hypercalcemia of malignancy, which results when tumors ectopically secrete PTHrP into circulation (Suva et al., 1987). The normal physiological role of PTHrP was uncovered using mouse genetic studies, where it was shown to regulate endochondral bone through the Parathyroid hormone receptor 1 (Pthr1). In mice, genetic inactivation of *Pthrp* and *Pthr1* produces an overlapping phenotype of reduced chondrocyte proliferation and accelerated formation of endochondral bones, including those in the chondrocranium (Karaplis et al., 1994; Lanske et al., 1996). Subsequent studies showed that PTHrP produced by the perichondrial cells and epiphyseal chondrocytes acts as a paracrine factor in the growth plate by promoting proliferation of columnar chondrocytes and delaying their hypertrophy (Amizuka, Warshawsky, Henderson, Goltzman, & Karaplis, 1994; Lee et al., 1996). When PTHrP levels are low, chondrocytes exit the cell cycle and become hypertrophic (St-Jacques et al., 1999; Vortkamp et al., 1996). The role of PTHrP in the growth plate is intimately associated with IHH signaling through a feedback loop: IHH in the prehypertrophic chondrocytes stimulates PTHrP in the perichondrial cells and epiphyseal chondrocytes to delay their hypertrophy.

Chondrodysplasia, Blomstrand type is caused by inactivating mutations in *PTHR1*, which presents with advanced endochondral bone maturation and increase bone density similar to that seen in the *Pthr1* knockout mouse (Jobert et al., 1998). Metaphyseal chondrodysplasia, Murk Jansen type, is caused by activating mutations in *PTHR1* and exhibits a reciprocal phenotype to that of Blomstrand type. These dominant missense mutations result in constitutive, ligand-independent activation of the receptor and cause hypercalcemia, severe short stature, short bowed limbs, enlarged joints, micrognathia, frontonasal hyperplasia, and sclerosis in the cranial bones (Schipani, Kruse, & Juppner, 1995). Overexpression of Pthrp or expression of a constitutively active Pthr1 in mice results in growth plates almost exclusively made of proliferating chondrocytes (Schipani et al., 1997; Weir et al., 1996). These disorders showed that PTHR1 is the major mediator for the endocrine function of PTHrP and Parathyroid hormone (PTH) in mineral homeostasis, while also playing a paracrine/autocrine role in the metaphyseal growth plate.

Binding of PTHrP to PTHR1 activates two major G protein signal cascades: adenylate cyclase and phospholipase C. Adenylate cyclase stimulation increases production of cAMP, leading to Protein kinase A (PKA)-mediated activation of transcription factors such as Sox9 and downregulates expression of the cyclin-dependent kinase inhibitor p57 (Huang, Chung, Kronenberg, & de Crombrugghe, 2001; MacLean et al., 2004). Activation of phospholipase C, on the other hand, induces accumulation of IP3 which increases intracellular calcium and DAG to induce Protein kinase C (PKC)-mediated activation of the transcription factor complex AP1 (Ionescu et al., 2001). The rare human disorder acrodysostosis suggests that PTHrP-mediated activation of the adenylate cyclase pathway is critical for its paracrine/autocrine role in endochondral bone formation. Mutations in the cAMP-dependent regulatory subunit of PKA *PRKAR1A* or the phosphodiesterase *PDE4D* that degrades cAMP cause nasal and maxillary hypoplasia, short stature, epiphyseal abnormalities, severe brachydactyly, advanced skeletal maturation, and resistance to parathyroid hormone treatment (Lee et al., 2012; Linglart et al., 2011; Michot et al., 2012). The knowledge gained from these rare disorders has advanced our understanding of PTHrP signaling in bone and provided rationale for the use of intermittent PTH/PTHrP anabolic therapy to treat common disorders of low bone mass including osteoporosis (Esbrit & Alcaraz, 2013).

6. WNT

WNTs, the human homologues of *Drosophila wingless*, are a highly conserved family of cysteine-rich secreted proteins that control key aspects of skeletal development. In vertebrates, WNTs act at short-range on closely situated cells expressing various WNT receptors and WNT coreceptors. With more than 19 WNTs, 15 different receptors/coreceptors, and an ever-growing list of regulatory proteins and intracellular components, the WNT signaling network has unparalleled complexity. While the combinatorial code for WNT signaling is not yet completely clear in the skeleton, some generalizations can be made. The seven-transmembrane-containing FRIZZLED proteins act as the common WNT receptors, and the alternative use of LRP or ROR coreceptors determine if intracellular signaling is dependent (canonical) or independent (noncanonical) of the transcriptional co-activator β-catenin. In canonical signaling, β-catenin is targeted for proteasomal degradation in the absence of WNT by a destruction complex composed of Axin, APC, WTX, and the kinases GSK3 and CK1γ. Upon WNT binding to the FRIZZLED-LRP receptor complex, Disheveled (DVL) scaffold proteins dock on the cytoplasmic tail of FRIZZLED, oligomerize, and recruit the destruction complex to the cytoplasmic tail of LRP (Clevers & Nusse, 2012; Li et al., 2012). These events protect cytoplasmic β-catenin from phosphorylation, allowing its stabilization and nuclear translocation. Once in the nucleus, β-catenin interacts with TCF/LEF to activate WNT target genes controlling proliferation and differentiation (Behrens et al., 1996). In noncanonical signaling, WNTs trigger homodimerization of ROR and formation of a ternary receptor complex with FRIZZLED. This is followed by recruitment of DVL to FRIZZLED and activation of the small GTPases RHO and RAC. Subsequent downstream activation of ROCK and JNK leads to cytoskeletal rearrangements that control cell polarity and motility.

Human genetic studies have pinpointed the importance of individual WNTs, WNT receptors, and coreceptors in regulating development of the craniofacial complex and limb. Early in development, canonical and noncanonical WNT signaling regulate craniofacial patterning as well as proximo-distal outgrowth and dorso-ventral patterning in the limb (Brugmann et al., 2007; Parr & McMahon, 1994). The importance of *WNT3* is strikingly clear in tetra-amelia where loss-of-function mutations cause severe craniofacial abnormalities and absence of all four limbs

(Niemann et al., 2004). Loss-of-function mutations in the noncanonical ligand *WNT7A* are also associated with craniofacial abnormalities and limb aplasia/hypoplasia in Fuhrmann syndrome and Al-Awadi/Raas-Rothschild syndrome (Woods et al., 2006). Mutations that reduce function of the noncanonical coreceptor ROR2 cause recessive Robinow syndrome, a severe skeletal dysplasia with mesomelic shortening of the long bones, dysmorphic facies, and segmental defect in the spine (Afzal et al., 2000; van Bokhoven et al., 2000). That Robinow syndrome shares phenotypic features with *Wnt5a* null mice led to the identification of *WNT5A* mutations in a dominant form of Robinow syndrome (Person et al., 2010; Yamaguchi, Bradley, McMahon, & Jones, 1999). Robinow syndrome and omodysplasia have substantial phenotypic overlap, including short limbs and facial dysmorphism (Venditti et al., 2002). Recently, it was shown that omodysplasia is caused by nonsense mutations in the WNT5A receptor *FRIZZLED2* (Saal et al., 2015; Sato, Yamamoto, Sakane, Koyama, & Kikuchi, 2010). These studies, along with elegant molecular studies, demonstrate that WNT5A is the principal ligand for ROR2 and FRIZZLED2 in human craniofacial and limb development (Mikels & Nusse, 2006).

Increased canonical WNT signaling also adversely affects skeletal development. Cenani–Lenz syndactyly syndrome, characterized by severe shortening of the limbs, syndactyly, as well as mild facial dysmorphism, results from *LRP4* missense mutations that block the coreceptors antagonistic effects on LRP6-mediated activation of β-catenin signaling (Li et al., 2010). Together these disorders show that craniofacial and limb patterning are sensitive to changes in the levels of canonical and noncanonical WNT signaling.

The WNT pathway also regulates lineage determination in skeletogenic mesenchyme, a general process affecting development of all skeletal elements. For example, both canonical and noncanonical signaling enhance early and terminal differentiation of osteoblasts and repress chondrocyte differentiation and maturation within the developing bone and joint space (Akiyama et al., 2004; Day, Guo, Garrett-Beal, & Yang, 2005; Spater et al., 2006; Yang, Topol, Lee, & Wu, 2003). A role for WNT signaling in the skeleton was first demonstrated by a series of human genetic studies of rare skeletal disorders with severely altered bone mass. Homozygous loss-of-function mutations in the WNT coreceptor *LRP5* were identified in the low bone mass disorder osteoporosis-pseudoglioma (Ai, Holmen, Van Hul, Williams, & Warman, 2005; Gong et al., 2001). Shortly thereafter, heterozygous gain-of-function mutations in *LRP5* were found in the high bone mass disorders Van Buchem disease, osteosclerosis, and osteopetrosis (Boyden et al., 2002; Little et al., 2002; Van Wesenbeeck et al., 2003). Striking similarities between these high bone mass disorders and sclerosteosis, a disorder recognized a year earlier to be caused by nonsense mutations in a novel gene called *SCLEROSTIN* (*SOST*), led to the discovery that SOST is a WNT antagonist (Balemans et al., 2001; Brunkow et al., 2001; Semenov, Tamai, & He, 2005). Later it was shown that the high bone mass *LRP5* mutations increased coreceptor activity by disrupting SOST inhibitor binding (Balemans et al., 2007). Together these studies demonstrate that increased WNT signaling leads to high bone mass and decreased WNT signaling causes low bone mass.

More recent genetic studies have further confirmed the relationship between WNT signaling levels and bone mass. Dominant missense mutations that reduce secretion of SOST, therefore increasing WNT signaling, were identified in craniodiaphyseal dysplasia, a condition with severe thickening of the craniofacial bones (Kim et al., 2011). Loss-of-function in WTX, a member of the β-catenin destruction complex causes excess bone hardening in osteopathia striata with cranial sclerosis (Jenkins et al., 2009). Furthermore, missense mutations in the coreceptor *LRP4* that impair the inhibitory action of SOST causes skeletal overgrowth in Sclerosteosis 2 (Leupin et al., 2011). Conversely, reduced WNT signaling caused by missense mutations in the coreceptor *LRP6* were found in a family with osteoporosis, coronary artery disease, and metabolic syndrome (Mani et al., 2007). By identifying WNT as a means to control bone mass, these studies in rare skeletal disorders have led to the development of therapeutic monoclonal antibodies that target WNT pathway members such as SOST to improve bone mass in the treatment of common diseases such as osteoporosis (Li et al., 2009; McClung et al., 2013).

7. NOTCH

NOTCH signaling is an evolutionarily conserved pathway that regulates cell-fate determination via short-range cell-to-cell communication during development. The pathway was first described in *Drosophila* when a mutation of the *Notch* gene resulted in small notches in the tips of their wings (Wharton, Johansen, Xu, & Artavanis-Tsakonas, 1985). Since then, its impressive complexity and functional diversity has been attributed to a small number of core components. In mammals, these core components include four NOTCH transmembrane receptors (NOTCH 1–4) that interact extracellularly with five canonical NOTCH transmembrane ligands (JAGGED1, JAGGED2, and Delta-like (DLL) 1, 3, 4) expressed on a neighboring cell (Andersson, Sandberg, & Lendahl, 2011). NOTCH ligand–receptor interactions trigger a series of proteolytic cleavage steps by γ-secretase and ADAM metalloproteases to release the NOTCH intracellular domain (NICD). NICD interacts with the DNA-binding protein CLS and translocates to the nucleus to regulate downstream target genes (Andersson et al., 2011). NOTCH signaling specificity is derived from various receptor–ligand combinations, posttranslational modifications of NOTCH receptors by Fringe family members such as LUNATIC FRINGE, and an ever-growing list of auxiliary proteins (Hicks et al., 2000).

NOTCH signaling plays a key role in morphogenesis during development by establishing and maintaining cellular boundaries that subdivide an originally homogeneous tissue field into distinct cell populations. In vertebrates, NOTCH signaling regulates development of the somites, which give rise to the vertebrae and ribs. During somitogenesis, NOTCH signaling first mediates mesodermal segmentation into somites by regulating an oscillating protein gradient called the segmentation clock; once the somites are formed, NOTCH establishes the anterior–posterior polarity of each somite (Feller, Schneider, Schuster-Gossler, & Gossler, 2008; Ferjentsik et al., 2009). It was the discovery of *JAGGED1* loss-of-function mutations in Alagille syndrome, which is characterized by short distal phalanges, characteristic facies, and abnormalities in vertebral body shape and segmentation, that first identified a role for NOTCH signaling in skeletal development (Li et al., 1997; Oda et al., 1997). Detailed studies of these *JAGGED1* mutations showed that glycosylation of JAGGED1 is critical for its intracellular trafficking to the cell membrane (Morrissette, Colliton, & Spinner, 2001). Alagille syndrome has also been associated with truncating mutations in *NOTCH2*, suggesting that it is the critical NOTCH receptor in bone (McDaniell et al., 2006). Interestingly, a later clinical report identified craniosynostosis as a feature of Alagille syndrome, hinting for the first time that NOTCH signaling regulates suture development (Kamath, Bason, Piccoli, Krantz, & Spinner, 2003). That NOTCH regulates formation of the suture has since been confirmed in mice: *Jagged1* expression in the mesodermal compartment of the developing coronal suture is necessary to maintain a cellular boundary between osteogenic and nonosteogenic cells (Yen, Ting, & Maxson, 2010). This study demonstrates that loss of the cellular boundary mediated by NOTCH signaling is likely the cause of abnormal suture fusion in Alagille syndrome.

Spondylocostal dysostosis (SCDO) was first linked to recessive loss-of-function mutations in *DLL3* on the basis of similarity between the disorder and the *Dll3* mutant mouse (Bulman et al., 2000; Kusumi et al., 1998). SCDO is a heterogeneous group of disorders with multiple vertebral segmentation defects including hemivertebrae, as well as posterior fusion and absence of ribs. Craniofacial findings include broad forehead, wide nasal bridge, enlarged posterior skull, and in some cases, under-ossified lacunae in the calvarial vault. Other forms of the disorder have also been linked to mutations in genes that code for critical components of the NOTCH pathway including LUNATIC FRINGE, MESP2, and the NOTCH pathway effector HES7 (Sparrow et al., 2006; Sparrow, Guillen-Navarro, Fatkin, & Dunwoodie, 2008; Whittock et al., 2004). These genes are essential for the segmentation clock and suggest the developmental origins of the skeletal defects in SCDO results from delayed and/or irregular somite formation.

Increased NOTCH signaling is also associated with abnormal skeletal development. Truncating mutations in *NOTCH2* that inhibit proteosomal degradation of the NICD, leading to persistent NOTCH signaling, are responsible for Hajdu–Cheney syndrome. This is a dominant disorder with short stature, decreased bone density, persistent calvarial sutures, micrognathia, and joint laxity (Isidor et al., 2011; Simpson et al., 2011). This syndrome strongly supports a role for NOTCH signaling in both skeletal development and bone homeostasis and is consistent with a role for NOTCH signaling in osteoblastogenesis and osteoclastogenesis (Fukushima et al., 2008; Zanotti et al., 2008). A regulatory role for NOTCH in bone mass in this disorder identifies this pathway as a potential target for therapeutic intervention of osteoporosis.

8. TRANSFORMING GROWTH FACTOR BETA SUPERFAMILY

The TGFβ superfamily is a group of over 30 structurally related secreted proteins that regulate virtually all aspects of development. Among the members of the TGFβ superfamily are TGFβs, Activins, Nodals, Growth differentiation factors (GDFs), and BMPs. These factors transmit signals across the plasma membrane through a heteromeric complex of Type I and Type II serine/threonine kinase receptors. Various combinations of Type I and Type II receptors selectively mediate signals through specific members of the TGFβ family of ligands. For example, BMPs activate signaling through combinations of one of the Type I receptors BMPR1A, BMPR1B, or ACVR1 and one of the Type II receptors BMPR2, ACVR2A, or ACVR2B. Upon ligand binding, the Type II receptor phosphorylates the Type I receptor, which in turn initiates intracellular signaling through phosphorylation of the receptor-regulated SMADS (R-SMADs). R-SMADs mediating BMP signaling are SMAD1, SMAD5, and SMAD8, and those for TGFβ signaling are SMAD2 and SMAD3. Activated R-SMADs interact with the co-mediator SMAD4 and together translocate to the nucleus where they act as transcription factors to regulate gene expression. Although the TGFβ superfamily pathways use overlapping components, signaling specificity is created by distinct spatiotemporal expression patterns of the pathway members and their antagonists. Two inhibitory SMADS (I-SMADS), SMAD6 and SMAD7, act as intracellular antagonists of the pathway by interacting with Type I receptors and blocking phosphorylation and activation of R-SMADs. Extracellular antagonists, including Noggin, Chordin, and BMPER, halt signaling by binding the ligands and blocking their interactions with specific cell-surface receptors. These and other regulatory mechanisms, including proteosome-mediated degradation of R-SMADs, fine-tune temporal and spatial activity as well as the amplitude of canonical signaling. Receptors of the TGFβ superfamily can also signal through non-SMAD pathways that include various branches of the MAP kinase pathway. For example, TGFβs and BMPs can activate ERK, JNK, and P38 (Huang, Yuan, Tu, Zou, & Li, 2014; Lee et al., 2007; Mulder, 2000; Yamaguchi et al., 1995).

BMPs were originally identified as inducers of bone and cartilage (Urist, 1965; Wozney et al., 1988). Since, mouse genetics have attributed canonical and noncanonical BMP signaling to multiple events during endochondral ossification, such as patterning, cell-fate determination, and differentiation. In particular, BMP ligands and receptors regulate skeletal patterning by establishing the size of the skeletal condensation initially through recruitment and later during distal outgrowth (Merino et al., 1999; Pizette, Abate-Shen, & Niswander, 2001). During histogenesis, BMPs promote differentiation and maturation of both chondrocytes and osteoblasts (Hatakeyama, Tuan, & Shum, 2004; Yoon et al., 2005). The GDFs, a subclass of BMPs, regulate chondrocyte differentiation, bone growth, and joint morphogenesis in the limbs, skull, and axial skeleton (Storm & Kingsley, 1999).

TGFβ was originally identified as a secreted factor capable of inducing fibroblast growth, collagen production, and cartilage formation (Roberts et al., 1986; Seyedin et al., 1987). TGFβ signaling plays a critical role in early development, and loss of its components leads to early embryonic lethality; therefore, mouse models that delete or overexpress TGFβ pathway components using a tissue-specific approach have been particularly useful in elucidating the role of TGFβ signaling in skeletal development. These studies have shown that TGFβ regulates patterning of the axial, appendicular, and craniofacial skeleton, with different members participating in the development of distinct skeletal elements (Baffi et al., 2004; Ito et al., 2003; Sanford et al., 1997). Like BMP, TGFβ stimulates the growth and differentiation of osteoprogenitor cells. Later TGFβ regulates bone homeostasis by acting as a central coordinator of bone remodeling and production by coupling the activity of bone-resorbing osteoclasts and bone-forming osteoblasts (Tang et al., 2009).

While the role of TGFβ superfamily in skeletal development was initially defined by phenotypes observed in genetically altered mice, the importance of this pathway in humans was first confirmed by the identification of human skeletal disorders with altered BMP signaling. *NOGGIN* mutations that abrogate protein function were found in multiple synostoses syndrome, a condition with facial dysmorphism, craniosynostosis, and multiple joint fusions within the digits, elbow, and spine (Gong et al., 1999). This finding demonstrated that a gradient of BMP, regulated by NOGGIN, is critical for establishing a specific subset of fibrous and synovial joints. Later, it was found that *GDF5* mutations that increase ligand activity were also responsible for multiple synostoses syndrome (Dawson et al., 2006). While increased GDF5 signaling can lead to joint loss, reduced GDF5 signaling caused by inactivating mutations results in an allelic series of chondrodysplasias with abnormalities in the bones and joints of the mesomelic and acromelic limb segments, ranging from severe shortening and joint dislocations to missing bones (Faiyaz-Ul-Haque, Ahmad, Wahab, et al., 2002; Faiyaz-Ul-Haque, Ahmad, Zaidi, et al., 2002; Thomas et al., 1997, 1996). Together these data suggest that there is a joint-specific threshold necessary for GDF5 signaling and that NOGGIN is required for appropriate antagonism of GDF5 to establish and maintain joint boundaries. GDF6, a close relative of GDF5, also regulates joint formation. Loss-of-function mutations in *GDF6* cause Klippel–Feil syndrome which presents with bone fusions in the wrists and ankles, as well as in the cervical and upper thoracic vertebrae (Tassabehji et al., 2008). Craniofacial findings associated with this disorder include microtia and laryngeal cartilage malformations. While the sites of joint fusions in Klippel–Feil syndrome and multiple synostoses syndrome do not completely overlap, the cause of the fusions appears common: there is a failure to segment the skeletal condensation into separate bones and establish the joint (Settle et al., 2003). Mechanistically, this may be explained if GDF5 and GDF6 regulate c-Jun, a transcription factor necessary for joint formation, through BMP receptor-mediated activation of JNK (Kan & Tabin, 2013).

Misregulation of BMP signaling has also been found in diaphanospondylodysostosis, a disorder caused by *BMPER* inactivation and characterized by vertebral segmentation defects, craniofacial anomalies, and incompletely differentiated mesenchyme in the ribs and vertebrae (Funari et al., 2010). BMPER was originally identified as a BMP antagonist, and only later found to also be a BMP agonist (Moser et al., 2003; Rentzsch, Zhang, Kramer, Sebald, & Hammerschmidt, 2006). Diaphanospondylodysostosis has similar features to *Bmp4* and *Bmp7* knockout mice, supporting a role for BMPER as a agonist for these particular ligands in the skeleton. The activity of BMPER as a BMP agonist or antagonist is concentration dependent: low BMPER concentrations increase SMAD1/SMAD5 phosphorylation, and high BMPER levels increases ERK1/ERK2 phosphorylation (Heinke et al., 2008). This suggests that BMPER regulates the balance between canonical and noncanonical BMP signaling in skeletal development and the pathogenesis of diaphanospondylodysostosis.

The bone-inducing properties of BMP have been long known and harnessed clinically for use in open and nonunion fractures of long bones and with spinal fusion surgery (Boden, Kang, Sandhu, & Heller, 2002; Govender et al., 2002). However, a key role for the BMP Type I receptor *ACVR1* in BMP-mediated osteoinduction was determined with the identification of mutations that constitutive activated *ACVR1* in fibrodysplasia ossificans progressive, a dominant disorder of progressive ectopic ossification of skeletal muscle, fascia, tendons, and ligaments (Shore et al., 2006).

Enhanced TGFβ signaling is also a major contributor to the pathology of multiple skeletal dysplasias. TGFβ is synthesized as a precursor protein, secreted, and stored in the extracellular matrix as part of a latency complex with its pro-domain. For TGFβ to become biologically active, it must be released from its latent state by a variety of factors. Mutations within the latency-associated peptide-coding region of *TGFβ1* prevent formation of the latent complex and cause a constitutive activation in Camurati–Engelmann (Kinoshita et al., 2000; Saito et al., 2001). Camurati–Engelmann disease presents with cortical thickening of the long bone diaphysis, sclerotic changes in the skull base, and bone overgrowth in the ear, thus supporting a role for TGFβ1 in regulating bone mass. Similarly, elevated signaling caused by *SMAD4* mutations that prolong stability and concurrently increase R-SMAD activation cause Myhre syndrome with dysmorphic facial features, thickened calvaria, short stature, and abnormal vertebrae and joint stiffness (Caputo et al., 2012; Le Goff & Cormier-Daire, 2012).

Myhre syndrome has clinical overlap with at least two *Fibrillin* disorders, geleophysic dysplasia and acromicric dysplasia. Fibrillin forms extracellular microfirils and can negatively regulate TGFβ signaling by sequestering the TGFβ latent complex or positively regulate signaling by concentrating TGFβ at sites of intended function (Arteaga-Solis et al., 2001). Mutations in *Fibrillin* that increase the TGFβ bioavailability play a primary pathogenic role in geleophysic dysplasia, acromicric dysplasia, and also Marfan syndrome (Chaudhry et al., 2007; Dietz, Loeys, Carta, & Ramirez, 2005; Kaartinen & Warburton, 2003; Le Goff et al., 2011). Skeletal findings in Marfan syndrome include skeletal overgrowth, early onset osteoarthritis, inequality of limb length, retrognathia, and highly arched palate. Identification of mutations in *TGFβR1* and *TGFβR2* in the Marfan-like condition Loeys–Dietz syndrome first indicated a role for Fibrillin in regulating TGFβ signaling (Loeys et al., 2005). Skeletal findings in Loeys–Dietz syndrome include abnormal spine curvature, craniosynostosis, bent digits, cleft palate, micrognathia, and craniofacial dysmorphism. Later Loeys–Dietz syndrome became associated with mutations in *SMAD3*, *TGFβ2*, and *TGFβ3* (Lindsay et al., 2012; Rienhoff et al., 2013; van de Laar et al., 2011). Despite the phenotypic suggestion of enhanced TGFβ signaling in Loeys–Dietz syndrome, the disorder is caused by heterozygous loss-of-function mutations in positive effectors of TGFβ signaling components. This paradox was reconciled by the finding that compensatory autocrine or paracrine mechanisms respond to a relative loss of TGFβ signaling by increasing signaling through noncanonical (SMAD-independent) signaling routes (Iwata et al., 2012; Lindsay et al., 2012). The insightful work on Marfan and Loeys–Dietz syndromes has led to the development of several anti-TGFβ therapies, such as losartan and TGFβ-neutralizing antibodies, to treat these and other disorders with increased TGFβ signaling (Akhurst & Hata, 2012; Habashi et al., 2006; Piccolo et al., 2014).

The importance of extracellular matrix proteins regulating TGFβ is also shown by human mutations in *BMP1*. BMP1, while not technically a member of the TGFβ superfamily, regulates bone mass by cleaving procollagens using its activity as a metalloproteinase. Homozygous missense mutations that hinder the ability of BMP1 to promote assembly of type I collagen were found in a subtype of osteogenesis imperfecta, a disorder of bone fragility and low bone mass (Martinez-Glez et al., 2012; Valencia et al., 2014). Type I collagen is the most abundant component of the bone's extracellular matrix, and most cases of osteogenesis imperfect are caused by mutations in the *COLLAGEN 1A* genes. The extracellular matrix of bone is an important reservoir for signaling factors, and it was recently shown that defective type I collagen biosynthesis in osteogenesis imperfecta increases the bioavailability of TGFβ, leading to excessive TGFβ signaling (Grafe et al., 2014). This finding that increased TGFβ signaling is a primary mechanism in the pathogenesis of osteogenesis imperfecta provides new rationale to treat this disorder with anti-TGFβ therapies.

9. FIBROBLAST GROWTH FACTOR

FGF was initially discovered and named for its ability to stimulate fibroblast proliferation (Armelin, 1973; Gospodarowicz & Moran, 1975). FGFs have since been recognized as fundamental regulators of development and a critical components of embryonic stem cell cultures to maintain pluripotency. The FGF family is comprised of 22 ligands, most of which are secreted and act as paracrine factors to transduce signals through four distinct FGF receptor (FGFR) tyrosine kinases. Select FGFs have intracellular activities or act as endocrine factors. FGF activity and specificity is regulated in part by tissue-specific alternative mRNA splicing of FGFRs, which produces receptors with distinct ligand binding properties. Ligand binding induces the formation of an activated receptor complex consisting of two FGFs, a FGFR dimer, and cell-surface heparin sulfate proteoglycans. Activated receptors autophosphorylate intracellular tyrosine residues and subsequently phosphorylate adapter proteins juxtamembrane to launch signal transduction cascades including Ras-MAP kinase, PI3 kinase/AKT, PLC-γ/PKC, and STAT1/p21 pathways (Ornitz & Itoh, 2015).

FGF signaling induces different responses in different cell types. Yet, correlations between the biological response and either receptor–ligand combinations, specific signal transduction cascades activated, or differential signaling kinetics do not completely explain this phenomenon. Evidence shows that FGF signaling elicits cell-specific effects through differential regulation of gene transcription (Dailey, Ambrosetti, Mansukhani, & Basilico, 2005). Even so, FGF signaling does not have pathway-specific transcription factors and instead acts through ubiquitously expressed transcription factors utilized by a number of other signaling pathways. Thus, many of the molecular details behind the specificity of the transcriptional response to FGF signaling are still to be uncovered.

Given the multifunctional nature of FGF signaling, it is not surprising that this pathway plays key roles in skeletal development including cell-fate determination, proliferation, patterning, and morphogenesis. Early in development, FGF signaling is essential for induction and patterning of neural crest cells and mesoderm, which are the embryonic sources of skeletal progenitors (Kimelman & Kirschner, 1987; Villanueva, Glavic, Ruiz, & Mayor, 2002). In the craniofacial complex, FGF plays a role in polarity and symmetry of the facial primordia as well as patterning of the craniofacial bones (Kubota & Ito, 2000; Tucker, Yamada, Grigoriou, Pachnis, & Sharpe, 1999). In the limb bud, FGF is essential for proximo-distal patterning and outgrowth (Boulet, Moon, Arenkiel, & Capecchi, 2004; Sun, Mariani, & Martin, 2002; Xu, Weinstein, Li, & Deng, 1999). After the major events in skeletal patterning are complete, FGF signaling regulates cell-fate specification during intramembranous and endochondral bone formation. During intramembranous bone formation, *Fgfr1* and *Fgfr2* are expressed in the condensing mesenchyme, and FGFR activation promotes proliferation and differentiation in osteogenic cells, most notably at the cranial sutures (Iseki, Wilkie, & Morriss-Kay, 1999; Rice et al., 2000). In endochondral bone ossification, proliferating chondrocytes express *Fgfr3* whereas hypertrophic chondrocytes express *Fgfr1*. *Fgfr2* is expressed at high levels in the condensing mesenchyme and later in the perichondrium and periosteum (Ornitz & Itoh, 2015). In the growth plate, FGF signaling inhibits chondrocyte proliferation and maturation while it promotes preosteoblast proliferation, differentiation, and function (Wang, Green, Zhao, & Ornitz, 2001; Yu, Herr, Waksman, & Ornitz, 2000).

Over 20 skeletal dysplasias result from misregulation of FGF signaling, and examining their disease mechanisms has significantly advanced our knowledge of the pathway's role in bone. The finding of dominant missense *FGFR3* mutations in achondroplasia, the most common genetic form of dwarfism, first distinguished a role for FGF signaling in the skeleton (Rousseau et al., 1994; Shiang et al., 1994). It was later shown that the *FGFR3* mutations in this form of rhizomelic short-limb dwarfism cause ligand-independent constitutive activation and subsequently reduce chondrocyte proliferation (Henderson et al., 2000). Phenotypic similarities between achondroplasia and Thanatophoric dysplasia, a neonatal lethal short-limbed dwarfism with variable presentation, led to identification of additional disease-causing *FGFR3* mutations (Rousseau et al., 1996; Tavormina, Rimoin, et al., 1995; Tavormina, Shiang, et al., 1995). Thanatophoric dysplasia subtypes are correlated with receptor domains affected: straight femurs with severe cloverleaf skull deformity in type I are associated with missense mutations near the auto-phosphorylation site that cause ligand-insensitive constitutive activation, whereas curved femurs and variable degrees of cloverleaf skull in type 2 are associated with a cysteine substitution in the extracellular domain that cause ectopic disulfide bonds between receptor monomers and ligand-sensitive hyperactivation of the receptor (Naski, Wang, Xu, & Ornitz, 1996; Wilcox et al., 1998). Thus, the severity of the chondrodysplasia is linked with the degree of FGFR3 activation. Interestingly, the kinase domain residue altered in thanatophoric dysplasia type I is also associated with the *FGFR3* missense mutation in severe achondroplasia with developmental delay and acanthosis nigricans (SADDAN) (Bellus et al., 1999; Tavormina, Shiang, et al., 1995). While all the *FGFR3* missense mutations in these disorders strongly induce ERK activation, STAT1 activation is restricted to mutations in the kinase domain activation loop. Together this data suggests that FGFR3 signaling through MAPK activation is largely responsible for the receptor's role in chondrocyte growth control and that abnormal activation of STAT1 is a feature unique to the most severe cases (Krejci et al., 2008).

While cloverleaf skull deformity had been described in the *FGFR3* chondrodysplasias, a role for the receptor in intramembranous ossification and suture development was not appreciated until the identification of *FGFR3* mutations in the craniosynostosis disorders Crouzon syndrome with acanthosis nigricans and Muenke syndrome (Meyer et al., 2012; Muenke et al., 1997). This was later confirmed in mouse models for both Muenke syndrome and achondroplasia, where it was found that Fgfr3 is expressed in the midsuture mesenchyme and that *Fgfr3* activating mutations can lead to suture fusion and ossification defects within the intramembranous bones of the calvaria (Di Rocco et al., 2014; Twigg et al., 2009).

Craniosynostosis is the most common birth defect linked to FGF signaling, with an incidence rate of 1 in 2000 live births (Kimonis, Gold, Hoffman, Panchal, & Boyadjiev, 2007). *FGFR2* was the first disease gene to be associated with dominant craniosynostosis syndromes when missense mutations increasing receptor activity were identified in Crouzon and Jackson–Weiss syndromes (Jabs et al., 1994; Reardon et al., 1994). The craniofacial and limb abnormalities in these disorders underlie the critical role of FGFR2 in skeletal development, and FGFR2 expression in the growth centers for both endochondral and intramembranous bone suggested the receptor's importance in the pathogenesis of other forms of craniosynostosis as well. Indeed, Apert, Beare–Stevenson cutis gyrate, Saethre–Chotzen, and Antley–Bixler syndromes are now all associated with *FGFR2* mutations that increase receptor activity (Chun, Siegel-Bartelt, Chitayat, Phillips, & Ray, 1998; Paznekas et al., 1998; Przylepa et al., 1996; Wilkie et al., 1995). These *FGFR2* and *FGFR3* mutations set a precedent for FGF signaling in bone, and subsequently, *FGFR1* mutations were identified in Pfeiffer, Jackson–Weiss, and Hartsfield syndromes (Muenke, Schell, et al., 1994; Roscioli et al., 2000; Simonis et al., 2013). Activating mutations in *FGFR1* and *FGFR2* are associated with a spectrum of disorders with a wide range of abnormalities in the craniofacial complex and hand/foot. While these disorders are phenotypically related, the mechanisms of receptor activation are distinct and can involve ligand-independent dimerization, constitutive receptor activation, prolonged receptor signaling, or reduced specificity in ligand binding (Anderson, Burns, Enriquez-Harris, Wilkie, & Heath, 1998; Mangasarian, Li, Mansukhani, & Basilico, 1997; Meyers et al., 1996; Robertson et al., 1998; Yu et al., 2000). Correlating phenotype to genotype is difficult as clinically different phenotypes can be caused by distinct mutations in the same amino acid. Nevertheless, like *FGFR3* disorders, the degree to which mutations activate FGFR1 or FGFR2 and their downstream cascades likely regulates phenotypic severity. In fact, complex craniosynostoses in humans and mice can be caused by reduced function of the ERK1/2 inhibitor *ERF*, which encodes an inhibitory ETS transcription factor (Twigg et al., 2013). This work suggests that FGF signaling through Ras–MAPK–ERK transduction cascade regulates osteogenesis in the cranial sutures.

In general, mutations in *FGFR1* and *FGFR2* cause syndromes involving craniosynostosis, whereas mutations in *FGFR3* are largely associated with the chondrodysplasia syndromes. However, there are crossover disorders that share characteristics of both craniosynostosis and chondrodysplasia syndromes. Osteoglophonic dysplasia is characterized by craniosynostosis as well as rhizomelic dwarfism and results from activating mutations in *FGFR1* (White et al., 2005). This finding revealed a crucial role for FGFR1 as a negative regulator of long bone growth, which was previously unrecognized. Bent bone dysplasia syndrome (BBDS), a dominant disorder with craniosynostosis and bent long bones in the lower extremities, may also represent a crossover disorder. BBDS results from missense mutations in the transmembrane domain of *FGFR2* that shift the subcellular distribution of the receptor from the membrane to the nucleolus, where it activates ribosomal DNA transcription by limiting RUNX2-mediated transcriptional repression (Merrill et al., 2012; Neben et al., 2014). Nuclear localization of FGFR1–3 and select FGFs have been reported in many cell types under normal and pathological conditions. There is growing evidence that strongly suggests nuclear FGFRs regulate gene expression (Chioni & Grose, 2012; Degnin, Laederich, & Horton, 2011; Gatius et al., 2011; Han, Xiao, & Quarles, 2015; Schmahl, Kim, Colvin, Ornitz, & Capel, 2004; Stachowiak, Fang, Myers, Dunham, & Stachowiak, 2003). The unique skeletal defects in BBDS that have not been previously observed in FGFR disorders, such as poor mineralization of the calvaria, hypoplastic clavicles, and pubis, as well as a thickened periosteum/perichondrium, reveal a noncanonical nuclear role of FGFR2 in regulating the binary choice of osteoprogenitor cell to proliferate or differentiate via ribosome biogenesis (Neben et al., 2014).

Skeletal development is also sensitive to decreases in FGF signaling. Phenotypic overlap between *Fgfr3* null mice and camptodactyly, tall stature, and hearing-loss (CASTHL) syndrome guided the discovery of *FGFR3* missense mutations in the kinase domain that reduce receptor function (Toydemir et al., 2006). This suggests that stature can be regulated in a dose-dependent manner by signaling through FGFR3. Dominant mutations in *FGFR2* and *FGFR3* that localize to the tyrosine kinase domain and reduce receptor function are found in Lacrimoauriculodentodigital (LADD) syndrome (Rohmann et al., 2006). Skeletal findings in LADD syndrome include cup-shaped pinnas, micrognathia, and variable digit defects. Heterozygous nonsense or missense mutations of *FGF10* are also associated with LADD syndrome (Milunsky, Zhao, Maher, Colby, & Everman, 2006). This disorder suggests that FGFR2 and FGFR3 have overlapping and yet nonredundant functions in digit patterning and formation of elastic cartilage.

Missense mutations in *FGF9* that reduce binding to FGFRs cause multiple synostosis syndrome type 3 with fusion of the nasal bone and the frontal processes of the maxilla and fusions of proximal interphalangeal, carpal–tarsal, and humeroradial joints (Wu et al., 2009). The progressive joint fusions in this disorder result from ectopic cartilage formation in the joint space, making them strikingly similar to those in multiple synostosis syndromes type 1 and 2, which are caused by mutations in the BMP pathway members *NOGGIN* and *GDF5*, respectively (Dawson et al., 2006; Gong et al., 1999). Such a phenotypic overlap suggests FGF9 plays a role in joint specification by defining the boundary of articulated skeletal elements through BMP signaling.

10. FUTURE DIRECTIONS

The aforementioned studies of rare events in human development have revealed much about signaling pathways themselves and their key roles in regulating histogenesis and morphogenesis in skeletal development. Although the craniofacial findings in many skeletal dysplasias may not be the primary distinguishing feature clinically, the abnormalities are profound (Table 1), and this is important from both a clinical and mechanistic perspective. Reexamination of the craniofacial defects in these disorders reveals that distinct phenotypes can arise from mutations in genes that lie in the same signaling pathway. Similarly, overlapping phenotypes can occur from mutations in genes that function in distinct signaling pathways. This indicates that the individual activities of these signaling pathways are neither linear nor independent from one another in the craniofacial skeleton. In fact, many of these pathways crosstalk, have combinatorial effects, and converge on relatively few transcriptional effectors that regulate the stepwise transition of a self-renewing progenitor cell into a terminally differentiated chondrocyte or osteoblast (Fig. 1). Furthermore, mutations in many of these transcriptional effectors themselves have been connected to skeletal dysplasias with craniofacial abnormalities, creating a pathway-based phenotypic link and exposing the web of interconnectedness between transcription factors and their upstream regulators.

Genetic analyses of rare human birth defects, along with their models in organisms such as the mouse, will continue to uncover new revelations about signaling pathways in skeletal development. One poorly understood aspect of skeletal development is the signaling mechanisms that integrate bone with their surrounding soft tissue, including tendons and ligaments. The surfaces of bones are adorned with projections and ridges of diverse shapes and sizes that mark points of soft tissue attachment. As such, these projections and ridges are the mineralized end of a graded transitional tissue called the enthesis that allows transmission of forces to and from the skeleton. While it is known that the enthesis develops from a distinct pool of progenitor cells that are added on to the primary bone template, many questions remain about the signals that regulate specification and their eventual segregation into tenocytes/ligamentocytes or chondrocytes (Blitz et al., 2009; Blitz, Sharir, Akiyama, & Zelzer, 2013; Sugimoto et al., 2013). We suggest

Table 1 Craniofacial Findings in the Skeletal Dysplasias

Pathway	Disorder (MIM)	Gene/s	Craniofacial Findings	Citation
BMP	Diaphanospondylodysostosis (608022)	*BMPER*	Short nose, depressed nasal bridge, hypertelorism, low-set ears	Funari et al. (2010)
	Fibrodysplasia ossificans progressive (135100)	*ACVR1*	Malformations of the TMJ and mandibular hypoplasia prior to the onset of heterotypic ossification	Shore et al. (2006)
	Klippel–Feil syndrome (118100)	*GDF6*	Macrocephaly, facial asymmetry, cleft palate	Tassabehji et al. (2008)
	Microphthalmia, syndromic 6 (607932)	*BMP4*	Brachycephaly, retrognathia, fleshy low-set angular ears	Reis et al. (2011)
	Multiple synostoses type 1(186500) and type 2 (610017)	*NOGGIN*, *GDF5*	Fusion of the frontal process of the maxilla and nasal bones, broad hemicylindrical nose, osteosclerotic deafness	Gong et al. (1999) and Dawson et al., 2006
	Myhre syndrome (139210)	*SMAD4*	Thickened calvaria, maxillary hypoplasia, prognathism	Le Goff and Cormier-Daire (2012) and Caputo et al. (2012)
	Osteogenesis imperfecta type XIII (614856)	*BMP1*	Calvarial wormian bones, wide-open anterior fontanel	Martinez-Glez et al. (2012)
FGF	Achondroplasia (100800)	*FGFR3*	Macrocephaly, frontal bossing, midfacial hypoplasia	Shiang et al. (1994) and Rousseau et al. (1994)
	Antley–Bixler syndrome (207410)	*FGFR2*	Trapezoidocephaly synostosis, midfacial hypoplasia	Chun et al. (1998)

Apert syndrome (101200)	*FGFR2*	Craniosynostosis, midfacial hypoplasia	Wilkie et al. (1995)
Beare–Stevenson cutis gyrata syndrome (123790)	*FGFR2*	Craniosynostosis, midface hypoplasia, high nasal bridge, high-arched and narrow palate, ear defects	Przylepa et al. (1996)
Bent bone dysplasia syndrome (614592)	*FGFR2*	Poorly mineralized calvaria, craniosynostosis, midface hypoplasia, micrognathia, low-set ears, prenatal teeth	Merrill et al. (2012)
CATSHL syndrome (610474)	*FGFR3*	Microcephaly, high arched palate, hearing loss	Toydemir et al. (2006)
Craniosynostosis (600775)	*ERF*	Multiple suture synostosis, macrocephaly, midface hypoplasia	Twigg et al. (2013)
Crouzon syndrome (123500) with acanthosis nigricans (612247)	*FGFR2*, *FGFR3*	Craniosynostosis, hypoplastic maxilla, prognathism, hypertelorism, parrot-beaked nose	Reardon et al. (1994) and Meyers et al. (1995)
Deafness with LAMM (610706)	*FGF3*	Microtia, microdontia	Tekin et al. (2007)
Hartsfield syndrome (615465)	*FGFR1*	Triad of holoprosencephaly, cleft lip and palate	Simonis et al. (2013)
Hypogonadotropic hypogonadism (612702)	*FGF8*	Cleft palate	Falardeau et al. (2008)
Jackson–Weiss syndrome (123150)	*FGFR1*, *FGFR2*	Craniosynostosis, midfacial hypoplasia	Jabs et al. (1994)

Continued

Table 1 Craniofacial Findings in the Skeletal Dysplasias—cont'd

Pathway	Disorder (MIM)	Gene/s	Craniofacial Findings	Citation
	LADD syndrome (149730)	*FGFR2*, *FGFR3*, *FGF10*	Broad anterior fontanel, high forehead, abnormal ear shape, small and peg-shaped lateral incisors, mild enamel dysplasia	Rohmann et al. (2006) and Milunsky et al. (2006)
	Muenke syndrome (602849)	*FGFR3*	Uni- or bicoronal synostosis, macrocephaly, midfacial hypoplasia	Muenke et al. (1997)
	Multiple synostosis syndrome type 3 (612961)	*FGF9*	Facial dysmorphism, broad hemicylindrical nose, conductive hearing loss	Wu et al. (2009)
	Osteoglophonic dysplasia (166250)	*FGFR1*	Depressed nasal bridge, frontal bossing, prognathism	White et al. (2005)
	Pfeiffer syndrome (101600)	*FGFR1*, *FGFR2*	Craniosynostosis, midface deficiency, cloverleaf skull	Muenke, Schell, et al. (1994), and Schell et al. (1995)
	SADDAN (616482)	*FGFR3*	Frontal bossing, large anterior fontanel, depressed nasal bridge	Tavormina et al. (1999)
	Saethre–Chotzen syndrome (101400)	*FGFR2*, *TWIST1*	Mild acrocephaly, facial asymmetry; thin, long, pointed nose; unusually shaped ear with prominent crus	Howard et al. (1997), el Ghouzzi et al. (1997), and Paznekas et al. (1998)
	Scaphocephaly (609579)	*FGFR2*	Macrocephaly, scaphocephaly, maxillary retrusion, hypertelorism	McGillivray et al. (2005)
	Thanatophoric dysplasia (187600)	*FGFR3*	Macrocrania, cloverleaf skull, frontal bossing	Reardon et al. (1994)
	Trigonocephaly 1 (190440)	*FGFR1*	Triangular shape of the head, microcephaly, craniosynostosis, ear abnormalities	Kress, Petersen, Collmann, and Grimm (2000)

HH	Acrocapitofemoral dysplasia (607778)	*IHH*	None identified	Hellemans et al. (2003)
	Basal cell nevus syndrome (109400)	*PTCH2*, *PTCH1*, *SUFU*	Marfanoid habitus, frontal bossing with large occipitofrontal circumference enlarged jaw, cleft palate	Johnson et al. (1996), Hahn et al. (1996), Smyth et al. (1999), and Pastorino et al. (2009)
	Greig cephalopolysyndactyly syndrome (175700)	*GLI3*	Scaphocephaly, frontal bossing, hypertelorism	Vortkamp et al. (1991)
	Holoprosencephaly type 2 (157170), type 3 (142945), type 4 (142946), type 5 (609637), type 7 (610828), type 9 (610829), type 11 (614226)	*SIX3*, *SHH*, *TGIF*, *ZIC2*, *PTCH1*, *GLI*, *CDON*	Diminished frontonasal angle, bilateral cleft lip/palate, hypoplastic premaxilla, malocclusion, hypoplastic nose, large ears, hypotelorism	Wallis et al. (1999), Roessler et al. (1996, 2003), Brown et al. (1998), Ming et al. (2002), and Bae et al. (2011)
	Pallister–Hall syndrome (146510)	*GLI3*	Micrognathia, cleft palate, short nose with flattened bridge	Kang et al. (1997)
	Solitary median maxillary central incisor (147250)	*SHH*	Single (unpaired) deciduous and permanent maxillary central incisors	Nanni et al. (2001)
NOTCH	Alagille syndrome (118450)	*JAG1*, *NOTCH2*	Prominent forehead, triangular faces, small anteriorly pointed chin, deep-set eyes	Li et al. (1997) and Oda et al. (1997)
	Hajdu–Cheney syndrome (600275)	*NOTCH2*	Midface hypoplasia, cleft palate, widely spaced and downward-slanting eyes	Simpson et al. (2011) and Isidor et al. (2011)
	Spondylocostal dysostosis types 1–4 (277300, 608681, 609813, 613686)	*DLL3*, *MESP2*, *LFNG*, *HES7*	Enlarged posterior skull, broad forehead, wide nasal bridge	Bulman et al. (2000), Whittock et al. (2004), and Sparrow et al. (2006)

Continued

Table 1 Craniofacial Findings in the Skeletal Dysplasias—cont'd

Pathway	Disorder (MIM)	Gene/s	Craniofacial Findings	Citation
PTHrP	Acrodysostosis 1 (101800)	*PRKAR1A*	Nasal and maxillary hypoplasia, prognathism	Linglart et al. (2011) and Lee et al. (2012)
	Chondrodysplasia, Blomstrand type (215045)	*PTHR1*	Sclerosis in the cranial bones, prominent upper face, small mandible	Jobert et al. (1998)
	Metaphyseal chondrodysplasia, Murk Jansen type (156400)	*PTHR1*	Small mandible, frontonasal hyperplasia, sclerosis of the cranial bones	Schipani et al. (1995)
TGFβ	Camurati–Engelmann disease (131300)	*TGFβ1*	Sclerotic bone disease including skull base; skull hyperostosis to severe thickening	Kinoshita et al. (2000)
	Loeys–Dietz syndrome types 1–5 (609192, 610168, 613795, 614816, 615582)	*TGFβR1*, *TGFβR2*, *SMAD3*, *TGFβ2*, *TGFβ3*	Skeletal overgrowth, craniosynostosis, dolichocephaly, frontal bossing, cleft palate, malocclusion, hypertelorism	Loeys et al. (2005), Mizuguchi et al. (2004), van de Laar et al. (2011), Lindsay et al. (2012), and Rienhoff et al. (2013)
WNT	Al-Awadi/Raas-Rothschild syndrome (276820)	*WNT7A*	Parieto-occipital skull defects, very dysplastic, large ears, a high and narrow palate	Woods et al. (2006)
	Cenani–Lenz syndactyly syndrome (212780)	*LRP4*	Prominent forehead, micrognathia, hypertelorism, down-slanting palpebral fissures	Li et al. (2010)
	Craniodiaphyseal dysplasia (122860)	*SOST*	Leonine facies, progressive stenosis of craniofacial foramina	Kim et al. (2011)

Donnai–Barrow syndrome (222448)	*LRP2*	Prominent brow, short nose, hypertelorism	Kantarci et al. (2007)
Endosteal hyperostosis (144750)	*LRP5*	Macrocephaly, sclerosis of the cranial vault and mandible, widened and deepened mandible, hypertelorism, nasal obstruction	Van Wesenbeeck et al. (2003)
Fuhrmann syndrome (228930)	*WNT7A*	Cleft lip and palate	Woods et al. (2006)
Omodysplasia 1 (258315)	*FRIZZLED2*	Round face, frontal bossing, small chin, bilateral cleft lip and cleft palate, short nose with a depressed broad bridge	Saal et al. (2015)
Osteopathia striata with cranial sclerosis (300373)	*WTX*	Cranial sclerosis macrocephaly, frontal bossing, microretrognathia, cleft palate, broad nasal bridge, hypertelorism	Jenkins et al. (2009)
Osteopetrosis, autosomal dominant (607634)	*LRP5*	Osteosclerosis in the cranial vault	Van Wesenbeeck et al. (2003)
Osteoporosis-pseudoglioma syndrome (259770)	*LRP5*	Increased bone density	Gong et al. (2001) and Ai et al. (2005)
Osteosclerosis (144750)	*LRP5*	Increased bone density	Van Wesenbeeck et al. (2003)
Robinow syndrome, recessive (268310), dominant (180700)	*ROR2*, *WNT5A*	Frontal bossing, hypertelorism, broad nose; fetal face, frontal bossing, depressed nasal bridge	Afzal et al. (2000) and Person et al. (2010)

Continued

Table 1 Craniofacial Findings in the Skeletal Dysplasias—cont'd

Pathway	Disorder (MIM)	Gene/s	Craniofacial Findings	Citation
	Sclerosteosis 1 (269500), Sclerosteosis 2 (614305)	*SOST*, *LRP4*	type 1: Sclerosis of the skull, square mandible; type 2: frontal bossing, midface hypoplasia, enlarged sclerotic mandible, prognathism, malocclusion	Brunkow et al. (2001) and Leupin et al. (2011)
	Tetra-amelia syndrome (273395)	*WNT3*	Absent frontal bones, malformed mouth, and absent nose and ears	Niemann et al. (2004)
	Van Buchem disease, type 1 (239100), type 2 (607636)	*SOST*, *LRP5*	Osteosclerosis of the skull and mandible; enlarged mandible	Brunkow et al. (2001) and Van Wesenbeeck et al. (2003)

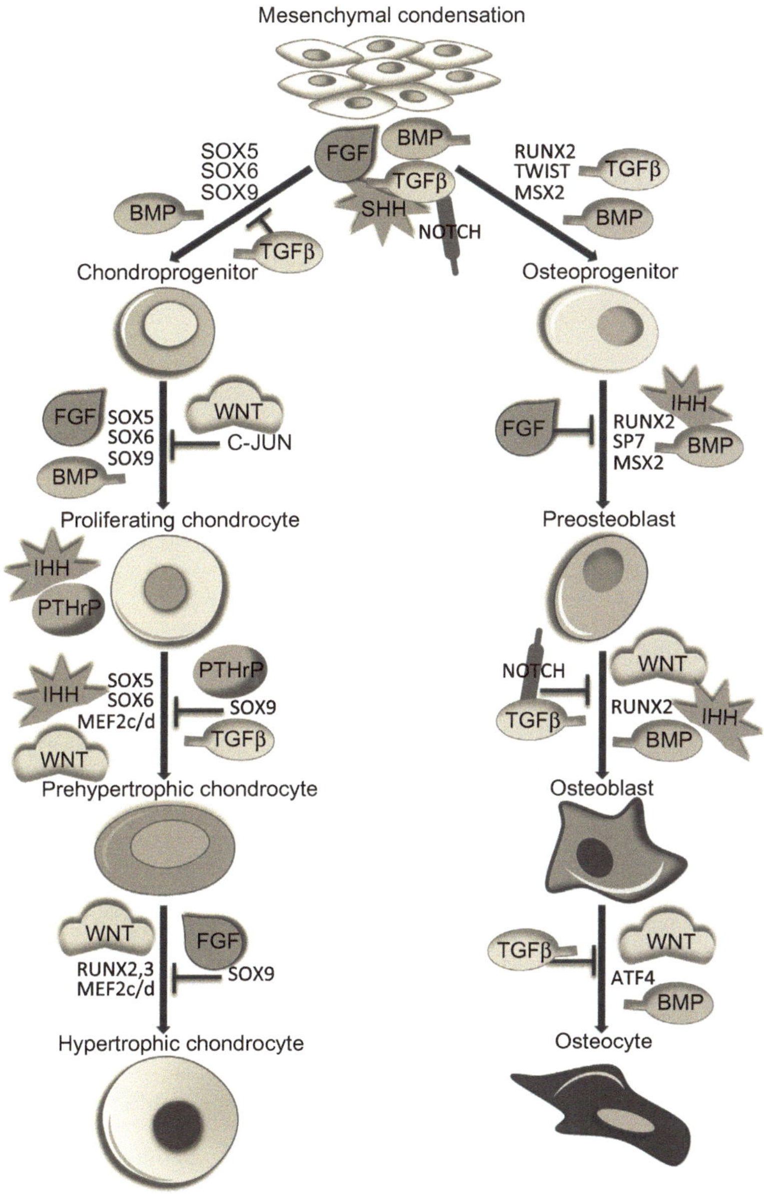

Figure 1 Signaling pathways have combinatorial effects in skeletal development. The key pathways regulating skeletal development include Hedgehog (SHH, IHH), Parathyroid hormone-related protein (PTHrP), WNT, NOTCH, Transforming growth factor β (TGFβ), Bone morphogenetic protein (BMP), and Fibroblast growth factor (FGF). These pathways have combinatorial effects and converge on relatively few transcriptional effectors during the stepwise transition from a self-renewing progenitor cell into a terminally differentiated chondrocyte or osteoblast. (A color version of this figure is located at the end of this chapter.)

that closer examination of these transitional tissues in skeletal dysplasias that exhibit joint laxity, contractures, dislocations, fusions, and/or abnormal bone angulation will provide new insights into the signaling mechanisms that regulate developmental and functional integration of individual bones into the musculoskeletal system.

ACKNOWLEDGMENTS

The authors are supported by the following grants: NIDCR #1R01DE025222 and March of Dimes Research Grant #6-FY15-233 to A.E.M.; and NICHD #5T32HD060549 to C.L.N. We would like to thank Joanna Salva and Ryan Roberts for helpful comments on the chapter.

REFERENCES

Abzhanov, A., Rodda, S. J., McMahon, A. P., & Tabin, C. J. (2007). Regulation of skeletogenic differentiation in cranial dermal bone. *Development*, *134*, 3133–3144.

Abzhanov, A., & Tabin, C. J. (2004). Shh and Fgf8 act synergistically to drive cartilage outgrowth during cranial development. *Developmental Biology*, *273*, 134–148.

Afzal, A. R., Rajab, A., Fenske, C. D., Oldridge, M., Elanko, N., Ternes-Pereira, E., et al. (2000). Recessive Robinow syndrome, allelic to dominant brachydactyly type B, is caused by mutation of ROR2. *Nature Genetics*, *25*, 419–422.

Ai, M., Holmen, S. L., Van Hul, W., Williams, B. O., & Warman, M. L. (2005). Reduced affinity to and inhibition by DKK1 form a common mechanism by which high bone mass-associated missense mutations in LRP5 affect canonical Wnt signaling. *Molecular and Cellular Biology*, *25*, 4946–4955.

Akhurst, R. J., & Hata, A. (2012). Targeting the TGFbeta signalling pathway in disease. *Nature Reviews Drug Discovery*, *11*, 790–811.

Akiyama, H., Lyons, J. P., Mori-Akiyama, Y., Yang, X., Zhang, R., Zhang, Z., et al. (2004). Interactions between Sox9 and beta-catenin control chondrocyte differentiation. *Genes and Development*, *18*, 1072–1087.

Allen, B. L., Tenzen, T., & McMahon, A. P. (2007). The Hedgehog-binding proteins Gas1 and Cdo cooperate to positively regulate Shh signaling during mouse development. *Genes and Development*, *21*, 1244–1257.

Amizuka, N., Warshawsky, H., Henderson, J. E., Goltzman, D., & Karaplis, A. C. (1994). Parathyroid hormone-related peptide-depleted mice show abnormal epiphyseal cartilage development and altered endochondral bone formation. *The Journal of Cell Biology*, *126*, 1611–1623.

Anderson, J., Burns, H. D., Enriquez-Harris, P., Wilkie, A. O., & Heath, J. K. (1998). Apert syndrome mutations in fibroblast growth factor receptor 2 exhibit increased affinity for FGF ligand. *Human Molecular Genetics*, 7, 1475–1483.

Andersson, E. R., Sandberg, R., & Lendahl, U. (2011). Notch signaling: Simplicity in design, versatility in function. *Development*, *138*, 3593–3612.

Armelin, H. A. (1973). Pituitary extracts and steroid hormones in the control of 3T3 cell growth. *Proceedings of the National Academy of Sciences of the United States of America*, *70*, 2702–2706.

Arnold, M. A., Kim, Y., Czubryt, M. P., Phan, D., McAnally, J., Qi, X., et al. (2007). MEF2C transcription factor controls chondrocyte hypertrophy and bone development. *Developmental Cell*, *12*, 377–389.

Arteaga-Solis, E., Gayraud, B., Lee, S. Y., Shum, L., Sakai, L., & Ramirez, F. (2001). Regulation of limb patterning by extracellular microfibrils. *The Journal of Cell Biology*, *154*, 275–281.

Aza-Blanc, P., Lin, H. Y., Ruiz i Altaba, A., & Kornberg, T. B. (2000). Expression of the vertebrate Gli proteins in Drosophila reveals a distribution of activator and repressor activities. *Development, 127*, 4293–4301.
Bae, G. U., Domene, S., Roessler, E., Schachter, K., Kang, J. S., Muenke, M., et al. (2011). Mutations in CDON, encoding a hedgehog receptor, result in holoprosencephaly and defective interactions with other hedgehog receptors. *The American Journal of Human Genetics, 89*, 231–240.
Baffi, M. O., Slattery, E., Sohn, P., Moses, H. L., Chytil, A., & Serra, R. (2004). Conditional deletion of the TGF-beta type II receptor in Col2a expressing cells results in defects in the axial skeleton without alterations in chondrocyte differentiation or embryonic development of long bones. *Developmental Biology, 276*, 124–142.
Balemans, W., Devogelaer, J. P., Cleiren, E., Piters, E., Caussin, E., & Van Hul, W. (2007). Novel LRP5 missense mutation in a patient with a high bone mass phenotype results in decreased DKK1-mediated inhibition of Wnt signaling. *Journal of Bone and Mineral Research, 22*, 708–716.
Balemans, W., Ebeling, M., Patel, N., Van Hul, E., Olson, P., Dioszegi, M., et al. (2001). Increased bone density in sclerosteosis is due to the deficiency of a novel secreted protein (SOST). *Human Molecular Genetics, 10*, 537–543.
Behrens, J., von Kries, J. P., Kuhl, M., Bruhn, L., Wedlich, D., Grosschedl, R., et al. (1996). Functional interaction of beta-catenin with the transcription factor LEF-1. *Nature, 382*, 638–642.
Belloni, E., Muenke, M., Roessler, E., Traverso, G., Siegel-Bartelt, J., Frumkin, A., et al. (1996). Identification of Sonic hedgehog as a candidate gene responsible for holoprosencephaly. *Nature Genetics, 14*, 353–356.
Bellus, G. A., Bamshad, M. J., Przylepa, K. A., Dorst, J., Lee, R. R., Hurko, O., et al. (1999). Severe achondroplasia with developmental delay and acanthosis nigricans (SADDAN): Phenotypic analysis of a new skeletal dysplasia caused by a Lys650Met mutation in fibroblast growth factor receptor 3. *American Journal of Medical Genetics, 85*, 53–65.
Bi, W., Deng, J. M., Zhang, Z., Behringer, R. R., & de Crombrugghe, B. (1999). Sox9 is required for cartilage formation. *Nature Genetics, 22*, 85–89.
Bialek, P., Kern, B., Yang, X., Schrock, M., Sosic, D., Hong, N., et al. (2004). A twist code determines the onset of osteoblast differentiation. *Developmental Cell, 6*, 423–435.
Blitz, E., Sharir, A., Akiyama, H., & Zelzer, E. (2013). Tendon-bone attachment unit is formed modularly by a distinct pool of Scx- and Sox9-positive progenitors. *Development (Cambridge, England), 140*, 2680–2690.
Blitz, E., Viukov, S., Sharir, A., Shwartz, Y., Galloway, J. L., Pryce, B. A., et al. (2009). Bone ridge patterning during musculoskeletal assembly is mediated through SCX regulation of Bmp4 at the tendon-skeleton junction. *Developmental Cell, 17*, 861–873.
Boden, S. D., Kang, J., Sandhu, H., & Heller, J. G. (2002). Use of recombinant human bone morphogenetic protein-2 to achieve posterolateral lumbar spine fusion in humans: a prospective, randomized clinical pilot trial: 2002 Volvo Award in clinical studies. *Spine (Phila Pa 1976), 27*, 2662–2673.
Boulet, A. M., Moon, A. M., Arenkiel, B. R., & Capecchi, M. R. (2004). The roles of Fgf4 and Fgf8 in limb bud initiation and outgrowth. *Developmental Biology, 273*, 361–372.
Boyden, L. M., Mao, J., Belsky, J., Mitzner, L., Farhi, A., Mitnick, M. A., et al. (2002). High bone density due to a mutation in LDL-receptor-related protein 5. *The New England Journal of Medicine, 346*, 1513–1521.
Brown, S. A., Warburton, D., Brown, L. Y., Yu, C. Y., Roeder, E. R., Stengel-Rutkowski, S., et al. (1998). Holoprosencephaly due to mutations in ZIC2, a homologue of Drosophila odd-paired. *Nature Genetics, 20*, 180–183.

Brugmann, S. A., Goodnough, L. H., Gregorieff, A., Leucht, P., ten Berge, D., Fuerer, C., et al. (2007). Wnt signaling mediates regional specification in the vertebrate face. *Development, 134*, 3283–3295.

Brunkow, M. E., Gardner, J. C., Van Ness, J., Paeper, B. W., Kovacevich, B. R., Proll, S., et al. (2001). Bone dysplasia sclerosteosis results from loss of the SOST gene product, a novel cystine knot-containing protein. *The American Journal of Human Genetics, 68*, 577–589.

Bulman, M. P., Kusumi, K., Frayling, T. M., McKeown, C., Garrett, C., Lander, E. S., et al. (2000). Mutations in the human delta homologue, DLL3, cause axial skeletal defects in spondylocostal dysostosis. *Nature Genetics, 24*, 438–441.

Burke, R., Nellen, D., Bellotto, M., Hafen, E., Senti, K. A., Dickson, B. J., et al. (1999). Dispatched, a novel sterol-sensing domain protein dedicated to the release of cholesterol-modified hedgehog from signaling cells. *Cell, 99*, 803–815.

Caputo, V., Cianetti, L., Niceta, M., Carta, C., Ciolfi, A., Bocchinfuso, G., et al. (2012). A restricted spectrum of mutations in the SMAD4 tumor-suppressor gene underlies Myhre syndrome. *The American Journal of Human Genetics, 90*, 161–169.

Chaudhry, S. S., Cain, S. A., Morgan, A., Dallas, S. L., Shuttleworth, C. A., & Kielty, C. M. (2007). Fibrillin-1 regulates the bioavailability of TGFbeta1. *The Journal of Cell Biology, 176*, 355–367.

Chiang, C., Litingtung, Y., Lee, E., Young, K. E., Corden, J. L., Westphal, H., et al. (1996). Cyclopia and defective axial patterning in mice lacking Sonic hedgehog gene function. *Nature, 383*, 407–413.

Chioni, A. M., & Grose, R. (2012). FGFR1 cleavage and nuclear translocation regulates breast cancer cell behavior. *The Journal of Cell Biology, 197*, 801–817.

Chun, K., Siegel-Bartelt, J., Chitayat, D., Phillips, J., & Ray, P. N. (1998). FGFR2 mutation associated with clinical manifestations consistent with Antley-Bixler syndrome. *American Journal of Medical Genetics, 77*, 219–224.

Clevers, H., & Nusse, R. (2012). Wnt/beta-catenin signaling and disease. *Cell, 149*, 1192–1205.

Dailey, L., Ambrosetti, D., Mansukhani, A., & Basilico, C. (2005). Mechanisms underlying differential responses to FGF signaling. *Cytokine and Growth Factor Reviews, 16*, 233–247.

Dawson, K., Seeman, P., Sebald, E., King, L., Edwards, M., Williams, J., 3rd, et al. (2006). GDF5 is a second locus for multiple-synostosis syndrome. *The American Journal of Human Genetics, 78*, 708–712.

Day, T. F., Guo, X., Garrett-Beal, L., & Yang, Y. (2005). Wnt/beta-catenin signaling in mesenchymal progenitors controls osteoblast and chondrocyte differentiation during vertebrate skeletogenesis. *Developmental Cell, 8*, 739–750.

Degnin, C. R., Laederich, M. B., & Horton, W. A. (2011). Ligand activation leads to regulated intramembrane proteolysis of fibroblast growth factor receptor 3. *Molecular Biology of the Cell, 22*, 3861–3873.

Demurger, F., Ichkou, A., Mougou-Zerelli, S., Le Merrer, M., Goudefroye, G., Delezoide, A. L., et al. (2015). New insights into genotype-phenotype correlation for GLI3 mutations. *European Journal of Human Genetics, 23*, 92–102.

Di Rocco, F., Biosse Duplan, M., Heuze, Y., Kaci, N., Komla-Ebri, D., Munnich, A., et al. (2014). FGFR3 mutation causes abnormal membranous ossification in achondroplasia. *Human Molecular Genetics, 23*, 2914–2925.

Dietz, H. C., Loeys, B., Carta, L., & Ramirez, F. (2005). Recent progress towards a molecular understanding of Marfan syndrome. *American Journal of Medical Genetics Part C: Seminars in Medical Genetics, 139C*, 4–9.

el Ghouzzi, V., Le Merrer, M., Perrin-Schmitt, F., Lajeunie, E., Benit, P., Renier, D., et al. (1997). Mutations of the TWIST gene in the Saethre-Chotzen syndrome. *Nature Genetics, 15*, 42–46.

Esbrit, P., & Alcaraz, M. J. (2013). Current perspectives on parathyroid hormone (PTH) and PTH-related protein (PTHrP) as bone anabolic therapies. *Biochemical Pharmacology*, *85*, 1417–1423.

Faiyaz-Ul-Haque, M., Ahmad, W., Wahab, A., Haque, S., Azim, A. C., Zaidi, S. H., et al. (2002a). Frameshift mutation in the cartilage-derived morphogenetic protein 1 (CDMP1) gene and severe acromesomelic chondrodysplasia resembling Grebe-type chondrodysplasia. *American Journal of Medical Genetics*, *111*, 31–37.

Faiyaz-Ul-Haque, M., Ahmad, W., Zaidi, S. H., Haque, S., Teebi, A. S., Ahmad, M., et al. (2002b). Mutation in the cartilage-derived morphogenetic protein-1 (CDMP1) gene in a kindred affected with fibular hypoplasia and complex brachydactyly (DuPan syndrome). *Clinical Genetics*, *61*, 454–458.

Fan, Z., Li, J., Du, J., Zhang, H., Shen, Y., Wang, C. Y., et al. (2008). A missense mutation in PTCH2 underlies dominantly inherited NBCCS in a Chinese family. *Journal of Medical Genetics*, *45*, 303–308.

Fan, C. M., Porter, J. A., Chiang, C., Chang, D. T., Beachy, P. A., & Tessier-Lavigne, M. (1995). Long-range sclerotome induction by sonic hedgehog: Direct role of the amino-terminal cleavage product and modulation by the cyclic AMP signaling pathway. *Cell*, *81*, 457–465.

Fan, C. M., & Tessier-Lavigne, M. (1994). Patterning of mammalian somites by surface ectoderm and notochord: Evidence for sclerotome induction by a hedgehog homolog. *Cell*, *79*, 1175–1186.

Falardeau, J., Chung, W. C., Beenken, A., Raivio, T., Plummer, L., Sidis, Y., et al. (2008). Decreased FGF8 signaling causes deficiency of gonadotropin-releasing hormone in humans and mice. *The Journal of Clinical Investigation*, *118*, 2822–2831.

Feller, J., Schneider, A., Schuster-Gossler, K., & Gossler, A. (2008). Noncyclic Notch activity in the presomitic mesoderm demonstrates uncoupling of somite compartmentalization and boundary formation. *Genes and Development*, *22*, 2166–2171.

Ferjentsik, Z., Hayashi, S., Dale, J. K., Bessho, Y., Herreman, A., De Strooper, B., et al. (2009). Notch is a critical component of the mouse somitogenesis oscillator and is essential for the formation of the somites. *PLoS Genetics*, *5*, e1000662.

Foster, J. W., Dominguez-Steglich, M. A., Guioli, S., Kwok, C., Weller, P. A., Stevanovic, M., et al. (1994). Campomelic dysplasia and autosomal sex reversal caused by mutations in an SRY-related gene. *Nature*, *372*, 525–530.

Fukushima, H., Nakao, A., Okamoto, F., Shin, M., Kajiya, H., Sakano, S., et al. (2008). The association of Notch2 and NF-kappaB accelerates RANKL-induced osteoclastogenesis. *Molecular and Cellular Biology*, *28*, 6402–6412.

Funari, V. A., Krakow, D., Nevarez, L., Chen, Z., Funari, T. L., Vatanavicharn, N., et al. (2010). BMPER mutation in diaphanospondylodysostosis identified by ancestral autozygosity mapping and targeted high-throughput sequencing. *The American Journal of Human Genetics*, *87*, 532–537.

Gao, B., Hu, J., Stricker, S., Cheung, M., Ma, G., Law, K. F., et al. (2009). A mutation in Ihh that causes digit abnormalities alters its signalling capacity and range. *Nature*, *458*, 1196–1200.

Gatius, S., Velasco, A., Azueta, A., Santacana, M., Pallares, J., Valls, J., et al. (2011). FGFR2 alterations in endometrial carcinoma. *Modern Pathology*, *24*, 1500–1510.

Gong, Y., Krakow, D., Marcelino, J., Wilkin, D., Chitayat, D., Babul-Hirji, R., et al. (1999). Heterozygous mutations in the gene encoding noggin affect human joint morphogenesis. *Nature Genetics*, *21*, 302–304.

Gong, Y., Slee, R. B., Fukai, N., Rawadi, G., Roman-Roman, S., Reginato, A. M., et al. (2001). LDL receptor-related protein 5 (LRP5) affects bone accrual and eye development. *Cell*, *107*, 513–523.

Gospodarowicz, D., & Moran, J. S. (1975). Mitogenic effect of fibroblast growth factor on early passage cultures of human and murine fibroblasts. *The Journal of Cell Biology*, *66*, 451–457.

Govender, S., Csimma, C., Genant, H. K., Valentin-Opran, A., Amit, Y., Arbel, R., et al. (2002). Recombinant human bone morphogenetic protein-2 for treatment of open tibial fractures: A prospective, controlled, randomized study of four hundred and fifty patients. *The Journal of Bone and Joint Surgery. American Volume*, *84-A*, 2123–2134.

Grafe, I., Yang, T., Alexander, S., Homan, E. P., Lietman, C., Jiang, M. M., et al. (2014). Excessive transforming growth factor-beta signaling is a common mechanism in osteogenesis imperfecta. *Nature Medicine*, *20*, 670–675.

Gripp, K. W., Wotton, D., Edwards, M. C., Roessler, E., Ades, L., Meinecke, P., et al. (2000). Mutations in TGIF cause holoprosencephaly and link NODAL signalling to human neural axis determination. *Nature Genetics*, *25*, 205–208.

Habashi, J. P., Judge, D. P., Holm, T. M., Cohn, R. D., Loeys, B. L., Cooper, T. K., et al. (2006). Losartan, an AT1 antagonist, prevents aortic aneurysm in a mouse model of Marfan syndrome. *Science*, *312*, 117–121.

Hahn, H., Wicking, C., Zaphiropoulous, P. G., Gailani, M. R., Shanley, S., Chidambaram, A., et al. (1996). Mutations of the human homolog of Drosophila patched in the nevoid basal cell carcinoma syndrome. *Cell*, *85*, 841–851.

Hall, T. M., Porter, J. A., Beachy, P. A., & Leahy, D. J. (1995). A potential catalytic site revealed by the 1.7-A crystal structure of the amino-terminal signalling domain of Sonic hedgehog. *Nature*, *378*, 212–216.

Han, X., Xiao, Z., & Quarles, L. D. (2015). Membrane and integrative nuclear fibroblastic growth factor receptor (FGFR) regulation of FGF-23. *The Journal of Biological Chemistry*, *290*, 20101.

Hatakeyama, Y., Tuan, R. S., & Shum, L. (2004). Distinct functions of BMP4 and GDF5 in the regulation of chondrogenesis. *Journal of Cellular Biochemistry*, *91*, 1204–1217.

Heinke, J., Wehofsits, L., Zhou, Q., Zoeller, C., Baar, K. M., Helbing, T., et al. (2008). BMPER is an endothelial cell regulator and controls bone morphogenetic protein-4-dependent angiogenesis. *Circulation Research*, *103*, 804–812.

Hellemans, J., Coucke, P. J., Giedion, A., De Paepe, A., Kramer, P., Beemer, F., et al. (2003). Homozygous mutations in IHH cause acrocapitofemoral dysplasia, an autosomal recessive disorder with cone-shaped epiphyses in hands and hips. *The American Journal of Human Genetics*, 72, 1040–1046.

Henderson, J. E., Naski, M. C., Aarts, M. M., Wang, D., Cheng, L., Goltzman, D., et al. (2000). Expression of FGFR3 with the G380R achondroplasia mutation inhibits proliferation and maturation of CFK2 chondrocytic cells. *Journal of Bone and Mineral Research*, *15*, 155–165.

Hicks, C., Johnston, S. H., diSibio, G., Collazo, A., Vogt, T. F., & Weinmaster, G. (2000). Fringe differentially modulates Jagged1 and Delta1 signalling through Notch1 and Notch2. *Nature Cell Biology*, *2*, 515–520.

Howard, T. D., Paznekas, W. A., Green, E. D., Chiang, L. C., Ma, N., Ortiz de Luna, R. I., et al. (1997). Mutations in TWIST, a basic helix-loop-helix transcription factor, in Saethre-Chotzen syndrome. *Nature Genetics*, *15*, 36–41.

Hu, D., Marcucio, R. S., & Helms, J. A. (2003). A zone of frontonasal ectoderm regulates patterning and growth in the face. *Development*, *130*, 1749–1758.

Huang, W., Chung, U. I., Kronenberg, H. M., & de Crombrugghe, B. (2001). The chondrogenic transcription factor Sox9 is a target of signaling by the parathyroid hormone-related peptide in the growth plate of endochondral bones. *Proceedings of the National Academy of Sciences of the United States of America*, *98*, 160–165.

Huang, R. L., Yuan, Y., Tu, J., Zou, G. M., & Li, Q. (2014). Opposing TNF-alpha/IL-1beta- and BMP-2-activated MAPK signaling pathways converge on Runx2 to regulate BMP-2-induced osteoblastic differentiation. *Cell Death and Disease*, *5*, e1187.

Inada, M., Yasui, T., Nomura, S., Miyake, S., Deguchi, K., Himeno, M., et al. (1999). Maturational disturbance of chondrocytes in Cbfa1-deficient mice. *Developmental Dynamics*, *214*, 279–290.

Ionescu, A. M., Schwarz, E. M., Vinson, C., Puzas, J. E., Rosier, R., Reynolds, P. R., et al. (2001). PTHrP modulates chondrocyte differentiation through AP-1 and CREB signaling. *The Journal of Biological Chemistry*, *276*, 11639–11647.

Iseki, S., Wilkie, A. O., & Morriss-Kay, G. M. (1999). Fgfr1 and Fgfr2 have distinct differentiation- and proliferation-related roles in the developing mouse skull vault. *Development*, *126*, 5611–5620.

Ishii, M., Merrill, A. E., Chan, Y. S., Gitelman, I., Rice, D. P., Sucov, H. M., et al. (2003). Msx2 and Twist cooperatively control the development of the neural crest-derived skeletogenic mesenchyme of the murine skull vault. *Development*, *130*, 6131–6142.

Isidor, B., Lindenbaum, P., Pichon, O., Bezieau, S., Dina, C., Jacquemont, S., et al. (2011). Truncating mutations in the last exon of NOTCH2 cause a rare skeletal disorder with osteoporosis. *Nature Genetics*, *43*, 306–308.

Ito, Y., Yeo, J. Y., Chytil, A., Han, J., Bringas, P., Jr., Nakajima, A., et al. (2003). Conditional inactivation of Tgfbr2 in cranial neural crest causes cleft palate and calvaria defects. *Development*, *130*, 5269–5280.

Iwata, J., Hacia, J. G., Suzuki, A., Sanchez-Lara, P. A., Urata, M., & Chai, Y. (2012). Modulation of noncanonical TGF-beta signaling prevents cleft palate in Tgfbr2 mutant mice. *The Journal of Clinical Investigation*, *122*, 873–885.

Jabs, E. W., Li, X., Scott, A. F., Meyers, G., Chen, W., Eccles, M., et al. (1994). Jackson-Weiss and Crouzon syndromes are allelic with mutations in fibroblast growth factor receptor 2. *Nature Genetics*, *8*, 275–279.

Jabs, E. W., Muller, U., Li, X., Ma, L., Luo, W., Haworth, I. S., et al. (1993). A mutation in the homeodomain of the human MSX2 gene in a family affected with autosomal dominant craniosynostosis. *Cell*, *75*, 443–450.

Jenkins, Z. A., van Kogelenberg, M., Morgan, T., Jeffs, A., Fukuzawa, R., Pearl, E., et al. (2009). Germline mutations in WTX cause a sclerosing skeletal dysplasia but do not predispose to tumorigenesis. *Nature Genetics*, *41*, 95–100.

Jeong, Y., Leskow, F. C., El-Jaick, K., Roessler, E., Muenke, M., Yocum, A., et al. (2008). Regulation of a remote Shh forebrain enhancer by the Six3 homeoprotein. *Nature Genetics*, *40*, 1348–1353.

Jeong, J., Mao, J., Tenzen, T., Kottmann, A. H., & McMahon, A. P. (2004). Hedgehog signaling in the neural crest cells regulates the patterning and growth of facial primordia. *Genes and Development*, *18*, 937–951.

Jobert, A. S., Zhang, P., Couvineau, A., Bonaventure, J., Roume, J., Le Merrer, M., et al. (1998). Absence of functional receptors for parathyroid hormone and parathyroid hormone-related peptide in Blomstrand chondrodysplasia. *The Journal of Clinical Investigation*, *102*, 34–40.

Johnson, R. L., Laufer, E., Riddle, R. D., & Tabin, C. (1994). Ectopic expression of Sonic hedgehog alters dorsal-ventral patterning of somites. *Cell*, *79*, 1165–1173.

Johnson, R. L., Rothman, A. L., Xie, J., Goodrich, L. V., Bare, J. W., Bonifas, J. M., et al. (1996). Human homolog of patched, a candidate gene for the basal cell nevus syndrome. *Science*, *272*, 1668–1671.

Johnston, J. J., Olivos-Glander, I., Killoran, C., Elson, E., Turner, J. T., Peters, K. F., et al. (2005). Molecular and clinical analyses of Greig cephalopolysyndactyly and

Pallister-Hall syndromes: Robust phenotype prediction from the type and position of GLI3 mutations. *The American Journal of Human Genetics, 76*, 609–622.
Johnston, J. J., Sapp, J. C., Turner, J. T., Amor, D., Aftimos, S., Aleck, K. A., et al. (2010). Molecular analysis expands the spectrum of phenotypes associated with GLI3 mutations. *Human Mutation, 31*, 1142–1154.
Kaartinen, V., & Warburton, D. (2003). Fibrillin controls TGF-beta activation. *Nature Genetics, 33*, 331–332.
Kamath, B. M., Bason, L., Piccoli, D. A., Krantz, I. D., & Spinner, N. B. (2003). Consequences of JAG1 mutations. *Journal of Medical Genetics, 40*, 891–895.
Kan, A., & Tabin, C. J. (2013). c-Jun is required for the specification of joint cell fates. *Genes & Development, 27*, 514–524.
Kang, S., Graham, J. M., Jr., Olney, A. H., & Biesecker, L. G. (1997). GLI3 frameshift mutations cause autosomal dominant Pallister-Hall syndrome. *Nature Genetics, 15*, 266–268.
Kantarci, S., Al-Gazali, L., Hill, R. S., Donnai, D., Black, G. C., Bieth, E., et al. (2007). Mutations in LRP2, which encodes the multiligand receptor megalin, cause Donnai-Barrow and facio-oculo-acoustico-renal syndromes. *Nature Genetics, 39*, 957–959.
Karaplis, A. C., Luz, A., Glowacki, J., Bronson, R. T., Tybulewicz, V. L., Kronenberg, H. M., et al. (1994). Lethal skeletal dysplasia from targeted disruption of the parathyroid hormone-related peptide gene. *Genes and Development, 8*, 277–289.
Kim, S. J., Bieganski, T., Sohn, Y. B., Kozlowski, K., Semenov, M., Okamoto, N., et al. (2011). Identification of signal peptide domain SOST mutations in autosomal dominant craniodiaphyseal dysplasia. *Human Genetics, 129*, 497–502.
Kim, I. S., Otto, F., Zabel, B., & Mundlos, S. (1999). Regulation of chondrocyte differentiation by Cbfa1. *Mechanisms of Development, 80*, 159–170.
Kimelman, D., & Kirschner, M. (1987). Synergistic induction of mesoderm by FGF and TGF-beta and the identification of an mRNA coding for FGF in the early Xenopus embryo. *Cell, 51*, 869–877.
Kimonis, V., Gold, J. A., Hoffman, T. L., Panchal, J., & Boyadjiev, S. A. (2007). Genetics of craniosynostosis. *Seminars in Pediatric Neurology, 14*, 150–161.
Kinoshita, A., Saito, T., Tomita, H., Makita, Y., Yoshida, K., Ghadami, M., et al. (2000). Domain-specific mutations in TGFB1 result in Camurati-Engelmann disease. *Nature Genetics, 26*, 19–20.
Krakow, D., & Rimoin, D. L. (2010). The skeletal dysplasias. *Genetics in Medicine, 12*, 327–341.
Krejci, P., Salazar, L., Kashiwada, T. A., Chlebova, K., Salasova, A., Thompson, L. M., et al. (2008). Analysis of STAT1 activation by six FGFR3 mutants associated with skeletal dysplasia undermines dominant role of STAT1 in FGFR3 signaling in cartilage. *PLoS One, 3*, e3961.
Kress, W., Petersen, B., Collmann, H., & Grimm, T. (2000). An unusual FGFR1 mutation (fibroblast growth factor receptor 1 mutation) in a girl with non-syndromic trigonocephaly. *Cytogenetics and Cell Genetics, 91*, 138–140.
Kronenberg, H. M. (2007). The role of the perichondrium in fetal bone development. *Annals of the New York Academy of Sciences, 1116*, 59–64.
Kubota, Y., & Ito, K. (2000). Chemotactic migration of mesencephalic neural crest cells in the mouse. *Developmental Dynamics, 217*, 170–179.
Kusumi, K., Sun, E. S., Kerrebrock, A. W., Bronson, R. T., Chi, D. C., Bulotsky, M. S., et al. (1998). The mouse pudgy mutation disrupts Delta homologue Dll3 and initiation of early somite boundaries. *Nature Genetics, 19*, 274–278.
Lanske, B., Karaplis, A. C., Lee, K., Luz, A., Vortkamp, A., Pirro, A., et al. (1996). PTH/PTHrP receptor in early development and Indian hedgehog-regulated bone growth. *Science, 273*, 663–666.

Lapunzina, P., Aglan, M., Temtamy, S., Caparros-Martin, J. A., Valencia, M., Leton, R., et al. (2010). Identification of a frameshift mutation in Osterix in a patient with recessive osteogenesis imperfecta. *The American Journal of Human Genetics*, *87*, 110–114.

Le Goff, C., & Cormier-Daire, V. (2012). From tall to short: The role of TGFbeta signaling in growth and its disorders. *American Journal of Medical Genetics Part C: Seminars in Medical Genetics*, *160C*, 145–153.

Le Goff, C., Mahaut, C., Wang, L. W., Allali, S., Abhyankar, A., Jensen, S., et al. (2011). Mutations in the TGFbeta binding-protein-like domain 5 of FBN1 are responsible for acromicric and geleophysic dysplasias. *The American Journal of Human Genetics*, *89*, 7–14.

Lee, H., Graham, J. M., Jr., Rimoin, D. L., Lachman, R. S., Krejci, P., Tompson, S. W., et al. (2012). Exome sequencing identifies PDE4D mutations in acrodysostosis. *American Journal of Human Genetics*, *90*, 746–751.

Lee, K., Lanske, B., Karaplis, A. C., Deeds, J. D., Kohno, H., Nissenson, R. A., et al. (1996). Parathyroid hormone-related peptide delays terminal differentiation of chondrocytes during endochondral bone development. *Endocrinology*, *137*, 5109–5118.

Lee, M. K., Pardoux, C., Hall, M. C., Lee, P. S., Warburton, D., Qing, J., et al. (2007). TGF-beta activates Erk MAP kinase signalling through direct phosphorylation of ShcA. *The EMBO Journal*, *26*, 3957–3967.

Lenton, K., James, A. W., Manu, A., Brugmann, S. A., Birker, D., Nelson, E. R., et al. (2011). Indian hedgehog positively regulates calvarial ossification and modulates bone morphogenetic protein signaling. *Genesis*, *49*, 784–796.

Leoncini, E., Baranello, G., Orioli, I. M., Anneren, G., Bakker, M., Bianchi, F., et al. (2008). Frequency of holoprosencephaly in the International Clearinghouse Birth Defects Surveillance Systems: Searching for population variations. *Birth Defects Research, Part A: Clinical and Molecular Teratology*, *82*, 585–591.

Leupin, O., Piters, E., Halleux, C., Hu, S., Kramer, I., Morvan, F., et al. (2011). Bone overgrowth-associated mutations in the LRP4 gene impair sclerostin facilitator function. *The Journal of Biological Chemistry*, *286*, 19489–19500.

Lewis, K. E., & Eisen, J. S. (2001). Hedgehog signaling is required for primary motoneuron induction in zebrafish. *Development*, *128*, 3485–3495.

Li, L., Krantz, I. D., Deng, Y., Genin, A., Banta, A. B., Collins, C. C., et al. (1997). Alagille syndrome is caused by mutations in human Jagged1, which encodes a ligand for Notch1. *Nature Genetics*, *16*, 243–251.

Li, V. S., Ng, S. S., Boersema, P. J., Low, T. Y., Karthaus, W. R., Gerlach, J. P., et al. (2012). Wnt signaling through inhibition of beta-catenin degradation in an intact Axin1 complex. *Cell*, *149*, 1245–1256.

Li, X., Ominsky, M. S., Warmington, K. S., Morony, S., Gong, J., Cao, J., et al. (2009). Sclerostin antibody treatment increases bone formation, bone mass, and bone strength in a rat model of postmenopausal osteoporosis. *Journal of Bone and Mineral Research*, *24*, 578–588.

Li, Y., Pawlik, B., Elcioglu, N., Aglan, M., Kayserili, H., Yigit, G., et al. (2010). LRP4 mutations alter Wnt/beta-catenin signaling and cause limb and kidney malformations in Cenani-Lenz syndrome. *The American Journal of Human Genetics*, *86*, 696–706.

Lindsay, M. E., Schepers, D., Bolar, N. A., Doyle, J. J., Gallo, E., Fert-Bober, J., et al. (2012). Loss-of-function mutations in TGFB2 cause a syndromic presentation of thoracic aortic aneurysm. *Nature Genetics*, *44*, 922–927.

Linglart, A., Menguy, C., Couvineau, A., Auzan, C., Gunes, Y., Cancel, M., et al. (2011). Recurrent PRKAR1A mutation in acrodysostosis with hormone resistance. *The New England Journal of Medicine*, *364*, 2218–2226.

Little, R. D., Carulli, J. P., Del Mastro, R. G., Dupuis, J., Osborne, M., Folz, C., et al. (2002). A mutation in the LDL receptor-related protein 5 gene results in the autosomal dominant high-bone-mass trait. *The American Journal of Human Genetics*, *70*, 11–19.

Loeys, B. L., Chen, J., Neptune, E. R., Judge, D. P., Podowski, M., Holm, T., et al. (2005). A syndrome of altered cardiovascular, craniofacial, neurocognitive and skeletal development caused by mutations in TGFBR1 or TGFBR2. *Nature Genetics*, *37*, 275–281.

Ma, Y., Erkner, A., Gong, R., Yao, S., Taipale, J., Basler, K., et al. (2002). Hedgehog-mediated patterning of the mammalian embryo requires transporter-like function of dispatched. *Cell*, *111*, 63–75.

MacLean, H. E., Guo, J., Knight, M. C., Zhang, P., Cobrinik, D., & Kronenberg, H. M. (2004). The cyclin-dependent kinase inhibitor p57(Kip2) mediates proliferative actions of PTHrP in chondrocytes. *The Journal of Clinical Investigation*, *113*, 1334–1343.

Mangasarian, K., Li, Y., Mansukhani, A., & Basilico, C. (1997). Mutation associated with Crouzon syndrome causes ligand-independent dimerization and activation of FGF receptor-2. *Journal of Cellular Physiology*, *172*, 117–125.

Mani, A., Radhakrishnan, J., Wang, H., Mani, A., Mani, M. A., Nelson-Williams, C., et al. (2007). LRP6 mutation in a family with early coronary disease and metabolic risk factors. *Science*, *315*, 1278–1282.

Martinez-Glez, V., Valencia, M., Caparros-Martin, J. A., Aglan, M., Temtamy, S., Tenorio, J., et al. (2012). Identification of a mutation causing deficient BMP1/mTLD proteolytic activity in autosomal recessive osteogenesis imperfecta. *Human Mutation*, *33*, 343–350.

Matsunaga, E., & Shiota, K. (1977). Holoprosencephaly in human embryos: Epidemiologic studies of 150 cases. *Teratology*, *16*, 261–272.

McClung, M. R., Lewiecki, E. M., Geller, M. L., Bolognese, M. A., Peacock, M., Weinstein, R. L., et al. (2013). Effect of denosumab on bone mineral density and biochemical markers of bone turnover: 8-year results of a phase 2 clinical trial. *Osteoporosis International*, *24*, 227–235.

McDaniell, R., Warthen, D. M., Sanchez-Lara, P. A., Pai, A., Krantz, I. D., Piccoli, D. A., et al. (2006). NOTCH2 mutations cause Alagille syndrome, a heterogeneous disorder of the notch signaling pathway. *The American Journal of Human Genetics*, *79*, 169–173.

McGillivray, G., Savarirayan, R., Cox, T. C., Stojkoski, C., McNeil, R., Bankier, A., et al. (2005). Familial scaphocephaly syndrome caused by a novel mutation in the FGFR2 tyrosine kinase domain. *Journal of Medical Genetics*, *42*, 656–662.

McMahon, J. A., Takada, S., Zimmerman, L. B., Fan, C. M., Harland, R. M., & McMahon, A. P. (1998). Noggin-mediated antagonism of BMP signaling is required for growth and patterning of the neural tube and somite. *Genes and Development*, *12*, 1438–1452.

Mercier, S., Dubourg, C., Garcelon, N., Campillo-Gimenez, B., Gicquel, I., Belleguic, M., et al. (2011). New findings for phenotype-genotype correlations in a large European series of holoprosencephaly cases. *Journal of Medical Genetics*, *48*, 752–760.

Merino, R., Rodriguez-Leon, J., Macias, D., Ganan, Y., Economides, A. N., & Hurle, J. M. (1999). The BMP antagonist Gremlin regulates outgrowth, chondrogenesis and programmed cell death in the developing limb. *Development*, *126*, 5515–5522.

Merrill, A. E., Bochukova, E. G., Brugger, S. M., Ishii, M., Pilz, D. T., Wall, S. A., et al. (2006). Cell mixing at a neural crest-mesoderm boundary and deficient ephrin-Eph signaling in the pathogenesis of craniosynostosis. *Human Molecular Genetics*, *15*, 1319–1328.

Merrill, A. E., Sarukhanov, A., Krejci, P., Idoni, B., Camacho, N., Estrada, K. D., et al. (2012). Bent bone dysplasia-FGFR2 type, a distinct skeletal disorder, has deficient canonical FGF signaling. *The American Journal of Human Genetics*, *90*, 550–557.

Meyer, M., Muller, A. K., Yang, J., Moik, D., Ponzio, G., Ornitz, D. M., et al. (2012). FGF receptors 1 and 2 are key regulators of keratinocyte migration in vitro and in wounded skin. *Journal of Cell Science*, *125*, 5690–5701.

Meyers, G. A., Day, D., Goldberg, R., Daentl, D. L., Przylepa, K. A., Abrams, L. J., et al. (1996). FGFR2 exon IIIa and IIIc mutations in Crouzon, Jackson-Weiss, and Pfeiffer syndromes: Evidence for missense changes, insertions, and a deletion due to alternative RNA splicing. *The American Journal of Human Genetics*, *58*, 491–498.

Michot, C., Le Goff, C., Goldenberg, A., Abhyankar, A., Klein, C., Kinning, E., et al. (2012). Exome sequencing identifies PDE4D mutations as another cause of acrodysostosis. *The American Journal of Human Genetics*, *90*, 740–745.

Mikels, A. J., & Nusse, R. (2006). Purified Wnt5a protein activates or inhibits beta-catenin-TCF signaling depending on receptor context. *PLoS Biology*, *4*, e115.

Milunsky, J. M., Zhao, G., Maher, T. A., Colby, R., & Everman, D. B. (2006). LADD syndrome is caused by FGF10 mutations. *Clinical Genetics*, *69*, 349–354.

Ming, J. E., Kaupas, M. E., Roessler, E., Brunner, H. G., Golabi, M., Tekin, M., et al. (2002). Mutations in PATCHED-1, the receptor for SONIC HEDGEHOG, are associated with holoprosencephaly. *Human Genetics*, *110*, 297–301.

Mizuguchi, T., Collod-Beroud, G., Akiyama, T., Abifadel, M., Harada, N., Morisaki, T., et al. (2004). Heterozygous TGFBR2 mutations in Marfan syndrome. *Nature Genetics*, *36*, 855–860.

Morrissette, J. D., Colliton, R. P., & Spinner, N. B. (2001). Defective intracellular transport and processing of JAG1 missense mutations in Alagille syndrome. *Human Molecular Genetics*, *10*, 405–413.

Moser, M., Binder, O., Wu, Y., Aitsebaomo, J., Ren, R., Bode, C., et al. (2003). BMPER, a novel endothelial cell precursor-derived protein, antagonizes bone morphogenetic protein signaling and endothelial cell differentiation. *Molecular and Cellular Biology*, *23*, 5664–5679.

Muenke, M., Gripp, K. W., McDonald-McGinn, D. M., Gaudenz, K., Whitaker, L. A., Bartlett, S. P., et al. (1997). A unique point mutation in the fibroblast growth factor receptor 3 gene (FGFR3) defines a new craniosynostosis syndrome. *The American Journal of Human Genetics*, *60*, 555–564.

Muenke, M., Gurrieri, F., Bay, C., Yi, D. H., Collins, A. L., Johnson, V. P., et al. (1994). Linkage of a human brain malformation, familial holoprosencephaly, to chromosome 7 and evidence for genetic heterogeneity. *Proceedings of the National Academy of Sciences of the United States of America*, *91*, 8102–8106.

Muenke, M., Schell, U., Hehr, A., Robin, N. H., Losken, H. W., Schinzel, A., et al. (1994). A common mutation in the fibroblast growth factor receptor 1 gene in Pfeiffer syndrome. *Nature Genetics*, *8*, 269–274.

Mulder, K. M. (2000). Role of Ras and Mapks in TGFbeta signaling. *Cytokine and Growth Factor Reviews*, *11*, 23–35.

Mundlos, S., Otto, F., Mundlos, C., Mulliken, J. B., Aylsworth, A. S., Albright, S., et al. (1997). Mutations involving the transcription factor CBFA1 cause cleidocranial dysplasia. *Cell*, *89*, 773–779.

Nakashima, K., Zhou, X., Kunkel, G., Zhang, Z., Deng, J. M., Behringer, R. R., et al. (2002). The novel zinc finger-containing transcription factor osterix is required for osteoblast differentiation and bone formation. *Cell*, *108*, 17–29.

Nanni, L., Ming, J. E., Du, Y., Hall, R. K., Aldred, M., Bankier, A., et al. (2001). SHH mutation is associated with solitary median maxillary central incisor: A study of 13 patients and review of the literature. *American Journal of Medical Genetics*, *102*, 1–10.

Naski, M. C., Wang, Q., Xu, J., & Ornitz, D. M. (1996). Graded activation of fibroblast growth factor receptor 3 by mutations causing achondroplasia and thanatophoric dysplasia. *Nature Genetics*, *13*, 233–237.

Neben, C. L., Idoni, B., Salva, J. E., Tuzon, C. T., Rice, J. C., Krakow, D., et al. (2014). Bent bone dysplasia syndrome reveals nucleolar activity for FGFR2 in ribosomal DNA transcription. *Human Molecular Genetics*, *23*, 5659–5671.

Niemann, S., Zhao, C., Pascu, F., Stahl, U., Aulepp, U., Niswander, L., et al. (2004). Homozygous WNT3 mutation causes tetra-amelia in a large consanguineous family. *The American Journal of Human Genetics*, *74*, 558–563.

Nusslein-Volhard, C., & Wieschaus, E. (1980). Mutations affecting segment number and polarity in Drosophila. *Nature*, *287*, 795–801.

Oda, T., Elkahloun, A. G., Pike, B. L., Okajima, K., Krantz, I. D., Genin, A., et al. (1997). Mutations in the human Jagged1 gene are responsible for Alagille syndrome. *Nature Genetics*, *16*, 235–242.

Ornitz, D. M., & Itoh, N. (2015). The fibroblast growth factor signaling pathway. *Wiley Interdisciplinary Reviews. Developmental Biology*, *4*, 215–266.

Otto, F., Thornell, A. P., Crompton, T., Denzel, A., Gilmour, K. C., Rosewell, I. R., et al. (1997). Cbfa1, a candidate gene for cleidocranial dysplasia syndrome, is essential for osteoblast differentiation and bone development. *Cell*, *89*, 765–771.

Parr, B. A., & McMahon, A. P. (1994). Wnt genes and vertebrate development. *Current Opinion in Genetics and Development*, *4*, 523–528.

Pastorino, L., Ghiorzo, P., Nasti, S., Battistuzzi, L., Cusano, R., Marzocchi, C., et al. (2009). Identification of a SUFU germline mutation in a family with Gorlin syndrome. *American Journal of Medical Genetics Part A*, *149A*, 1539–1543.

Paznekas, W. A., Cunningham, M. L., Howard, T. D., Korf, B. R., Lipson, M. H., Grix, A. W., et al. (1998). Genetic heterogeneity of Saethre-Chotzen syndrome, due to TWIST and FGFR mutations. *The American Journal of Human Genetics*, *62*, 1370–1380.

Person, A. D., Beiraghi, S., Sieben, C. M., Hermanson, S., Neumann, A. N., Robu, M. E., et al. (2010). WNT5A mutations in patients with autosomal dominant Robinow syndrome. *Developmental Dynamics: An Official Publication of the American Association of Anatomists*, *239*, 327–337.

Piccolo, P., Mithbaokar, P., Sabatino, V., Tolmie, J., Melis, D., Schiaffino, M. C., et al. (2014). SMAD4 mutations causing Myhre syndrome result in disorganization of extracellular matrix improved by losartan. *European Journal of Human Genetics*, *22*, 988–994.

Pizette, S., Abate-Shen, C., & Niswander, L. (2001). BMP controls proximodistal outgrowth, via induction of the apical ectodermal ridge, and dorsoventral patterning in the vertebrate limb. *Development*, *128*, 4463–4474.

Porter, J. A., Ekker, S. C., Park, W. J., von Kessler, D. P., Young, K. E., Chen, C. H., et al. (1996). Hedgehog patterning activity: Role of a lipophilic modification mediated by the carboxy-terminal autoprocessing domain. *Cell*, *86*, 21–34.

Przylepa, K. A., Paznekas, W., Zhang, M., Golabi, M., Bias, W., Bamshad, M. J., et al. (1996). Fibroblast growth factor receptor 2 mutations in Beare-Stevenson cutis gyrata syndrome. *Nature Genetics*, *13*, 492–494.

Reardon, W., Winter, R. M., Rutland, P., Pulleyn, L. J., Jones, B. M., & Malcolm, S. (1994). Mutations in the fibroblast growth factor receptor 2 gene cause Crouzon syndrome. *Nature Genetics*, *8*, 98–103.

Reis, L. M., Tyler, R. C., Schilter, K. F., Abdul-Rahman, O., Innis, J. W., Kozel, B. A., et al. (2011). BMP4 loss-of-function mutations in developmental eye disorders including SHORT syndrome. *Human Genetics*, *130*, 495–504.

Rentzsch, F., Zhang, J., Kramer, C., Sebald, W., & Hammerschmidt, M. (2006). Crossveinless 2 is an essential positive feedback regulator of Bmp signaling during zebrafish gastrulation. *Development*, *133*, 801–811.

Rice, D. P., Aberg, T., Chan, Y., Tang, Z., Kettunen, P. J., Pakarinen, L., et al. (2000). Integration of FGF and TWIST in calvarial bone and suture development. *Development*, *127*, 1845–1855.
Rice, D. P., Connor, E. C., Veltmaat, J. M., Lana-Elola, E., Veistinen, L., Tanimoto, Y., et al. (2010). Gli3Xt-J/Xt-J mice exhibit lambdoid suture craniosynostosis which results from altered osteoprogenitor proliferation and differentiation. *Human Molecular Genetics*, *19*, 3457–3467.
Riddle, R. D., Johnson, R. L., Laufer, E., & Tabin, C. (1993). Sonic hedgehog mediates the polarizing activity of the ZPA. *Cell*, *75*, 1401–1416.
Rienhoff, H. Y., Jr., Yeo, C. Y., Morissette, R., Khrebtukova, I., Melnick, J., Luo, S., et al. (2013). A mutation in TGFB3 associated with a syndrome of low muscle mass, growth retardation, distal arthrogryposis and clinical features overlapping with Marfan and Loeys-Dietz syndrome. *American Journal of Medical Genetics Part A*, *161A*, 2040–2046.
Roberts, A. B., Sporn, M. B., Assoian, R. K., Smith, J. M., Roche, N. S., Wakefield, L. M., et al. (1986). Transforming growth factor type beta: Rapid induction of fibrosis and angiogenesis in vivo and stimulation of collagen formation in vitro. *Proceedings of the National Academy of Sciences of the United States of America*, *83*, 4167–4171.
Robertson, S. C., Meyer, A. N., Hart, K. C., Galvin, B. D., Webster, M. K., & Donoghue, D. J. (1998). Activating mutations in the extracellular domain of the fibroblast growth factor receptor 2 function by disruption of the disulfide bond in the third immunoglobulin-like domain. *Proceedings of the National Academy of Sciences of the United States of America*, *95*, 4567–4572.
Roessler, E., Belloni, E., Gaudenz, K., Jay, P., Berta, P., Scherer, S. W., et al. (1996). Mutations in the human Sonic Hedgehog gene cause holoprosencephaly. *Nature Genetics*, *14*, 357–360.
Roessler, E., Du, Y. Z., Mullor, J. L., Casas, E., Allen, W. P., Gillessen-Kaesbach, G., et al. (2003). Loss-of-function mutations in the human GLI2 gene are associated with pituitary anomalies and holoprosencephaly-like features. *Proceedings of the National Academy of Sciences of the United States of America*, *100*, 13424–13429.
Rohmann, E., Brunner, H. G., Kayserili, H., Uyguner, O., Nurnberg, G., Lew, E. D., et al. (2006). Mutations in different components of FGF signaling in LADD syndrome. *Nature Genetics*, *38*, 414–417.
Roscioli, T., Flanagan, S., Kumar, P., Masel, J., Gattas, M., Hyland, V. J., et al. (2000). Clinical findings in a patient with FGFR1 P252R mutation and comparison with the literature. *American Journal of Medical Genetics*, *93*, 22–28.
Rousseau, F., Bonaventure, J., Legeai-Mallet, L., Pelet, A., Rozet, J. M., Maroteaux, P., et al. (1994). Mutations in the gene encoding fibroblast growth factor receptor-3 in achondroplasia. *Nature*, *371*, 252–254.
Rousseau, F., el Ghouzzi, V., Delezoide, A. L., Legeai-Mallet, L., Le Merrer, M., Munnich, A., et al. (1996). Missense FGFR3 mutations create cysteine residues in thanatophoric dwarfism type I (TD1). *Human Molecular Genetics*, *5*, 509–512.
Saal, H. M., Prows, C. A., Guerreiro, I., Donlin, M., Knudson, L., Sund, K. L., et al. (2015). A mutation in FRIZZLED2 impairs Wnt signaling and causes autosomal dominant omodysplasia. *Human Molecular Genetics*, *24*, 3399–3409.
Saito, T., Kinoshita, A., Yoshiura, K., Makita, Y., Wakui, K., Honke, K., et al. (2001). Domain-specific mutations of a transforming growth factor (TGF)-beta 1 latency-associated peptide cause Camurati-Engelmann disease because of the formation of a constitutively active form of TGF-beta 1. *The Journal of Biological Chemistry*, *276*, 11469–11472.
Sanford, L. P., Ormsby, I., Gittenberger-de Groot, A. C., Sariola, H., Friedman, R., Boivin, G. P., et al. (1997). TGFbeta2 knockout mice have multiple developmental

defects that are non-overlapping with other TGFbeta knockout phenotypes. *Development*, *124*, 2659–2670.

Sato, A., Yamamoto, H., Sakane, H., Koyama, H., & Kikuchi, A. (2010). Wnt5a regulates distinct signalling pathways by binding to Frizzled2. *The EMBO Journal*, *29*, 41–54.

Schell, U., Hehr, A., Feldman, G. J., Robin, N. H., Zackai, E. H., de Die-Smulders, C., et al. (1995). Mutations in FGFR1 and FGFR2 cause familial and sporadic Pfeiffer syndrome. *Human Molecular Genetics*, *4*, 323–328.

Schipani, E., Jensen, G. S., Pincus, J., Nissenson, R. A., Gardella, T. J., & Juppner, H. (1997). Constitutive activation of the cyclic adenosine 3',5'-monophosphate signaling pathway by parathyroid hormone (PTH)/PTH-related peptide receptors mutated at the two loci for Jansen's metaphyseal chondrodysplasia. *Molecular Endocrinology*, *11*, 851–858.

Schipani, E., Kruse, K., & Juppner, H. (1995). A constitutively active mutant PTH-PTHrP receptor in Jansen-type metaphyseal chondrodysplasia. *Science*, *268*, 98–100.

Schmahl, J., Kim, Y., Colvin, J. S., Ornitz, D. M., & Capel, B. (2004). Fgf9 induces proliferation and nuclear localization of FGFR2 in Sertoli precursors during male sex determination. *Development*, *131*, 3627–3636.

Semenov, M., Tamai, K., & He, X. (2005). SOST is a ligand for LRP5/LRP6 and a Wnt signaling inhibitor. *The Journal of Biological Chemistry*, *280*, 26770–26775.

Settle, S. H., Jr., Rountree, R. B., Sinha, A., Thacker, A., Higgins, K., & Kingsley, D. M. (2003). Multiple joint and skeletal patterning defects caused by single and double mutations in the mouse Gdf6 and Gdf5 genes. *Developmental Biology*, *254*, 116–130.

Seyedin, S. M., Segarini, P. R., Rosen, D. M., Thompson, A. Y., Bentz, H., & Graycar, J. (1987). Cartilage-inducing factor-B is a unique protein structurally and functionally related to transforming growth factor-beta. *The Journal of Biological Chemistry*, *262*, 1946–1949.

Shiang, R., Thompson, L. M., Zhu, Y. Z., Church, D. M., Fielder, T. J., Bocian, M., et al. (1994). Mutations in the transmembrane domain of FGFR3 cause the most common genetic form of dwarfism, achondroplasia. *Cell*, *78*, 335–342.

Shin, S. H., Kogerman, P., Lindstrom, E., Toftgard, R., & Biesecker, L. G. (1999). GLI3 mutations in human disorders mimic Drosophila cubitus interruptus protein functions and localization. *Proceedings of the National Academy of Sciences of the United States of America*, *96*, 2880–2884.

Shore, E. M., Xu, M., Feldman, G. J., Fenstermacher, D. A., Cho, T. J., Choi, I. H., et al. (2006). A recurrent mutation in the BMP type I receptor ACVR1 causes inherited and sporadic fibrodysplasia ossificans progressiva. *Nature Genetics*, *38*, 525–527.

Simonis, N., Migeotte, I., Lambert, N., Perazzolo, C., de Silva, D. C., Dimitrov, B., et al. (2013). FGFR1 mutations cause Hartsfield syndrome, the unique association of holoprosencephaly and ectrodactyly. *Journal of Medical Genetics*, *50*, 585–592.

Simpson, M. A., Irving, M. D., Asilmaz, E., Gray, M. J., Dafou, D., Elmslie, F. V., et al. (2011). Mutations in NOTCH2 cause Hajdu-Cheney syndrome, a disorder of severe and progressive bone loss. *Nature Genetics*, *43*, 303–305.

Smits, P., Li, P., Mandel, J., Zhang, Z., Deng, J. M., Behringer, R. R., et al. (2001). The transcription factors L-Sox5 and Sox6 are essential for cartilage formation. *Developmental Cell*, *1*, 277–290.

Smyth, I., Narang, M. A., Evans, T., Heimann, C., Nakamura, Y., Chenevix-Trench, G., et al. (1999). Isolation and characterization of human patched 2 (PTCH2), a putative tumour suppressor gene inbasal cell carcinoma and medulloblastoma on chromosome 1p32. *Human Molecular Genetics*, *8*, 291–297.

Sparrow, D. B., Chapman, G., Wouters, M. A., Whittock, N. V., Ellard, S., Fatkin, D., et al. (2006). Mutation of the LUNATIC FRINGE gene in humans causes spondylocostal dysostosis with a severe vertebral phenotype. *The American Journal of Human Genetics*, *78*, 28–37.

Sparrow, D. B., Guillen-Navarro, E., Fatkin, D., & Dunwoodie, S. L. (2008). Mutation of Hairy-and-Enhancer-of-Split-7 in humans causes spondylocostal dysostosis. *Human Molecular Genetics*, *17*, 3761–3766.

Spater, D., Hill, T. P., O'Sullivan, R. J., Gruber, M., Conner, D. A., & Hartmann, C. (2006). Wnt9a signaling is required for joint integrity and regulation of Ihh during chondrogenesis. *Development*, *133*, 3039–3049.

Stachowiak, E. K., Fang, X., Myers, J., Dunham, S., & Stachowiak, M. K. (2003). cAMP-induced differentiation of human neuronal progenitor cells is mediated by nuclear fibroblast growth factor receptor-1 (FGFR1). *Journal of Neurochemistry*, *84*, 1296–1312.

St-Jacques, B., Hammerschmidt, M., & McMahon, A. P. (1999). Indian hedgehog signaling regulates proliferation and differentiation of chondrocytes and is essential for bone formation. *Genes and Development*, *13*, 2072–2086.

Storm, E. E., & Kingsley, D. M. (1999). GDF5 coordinates bone and joint formation during digit development. *Developmental Biology*, *209*, 11–27.

Sugimoto, Y., Takimoto, A., Akiyama, H., Kist, R., Scherer, G., Nakamura, T., et al. (2013). Scx+/Sox9+ progenitors contribute to the establishment of the junction between cartilage and tendon/ligament. *Development (Cambridge, England)*, *140*, 2280–2288.

Sun, X., Mariani, F. V., & Martin, G. R. (2002). Functions of FGF signalling from the apical ectodermal ridge in limb development. *Nature*, *418*, 501–508.

Suva, L. J., Winslow, G. A., Wettenhall, R. E., Hammonds, R. G., Moseley, J. M., Diefenbach-Jagger, H., et al. (1987). A parathyroid hormone-related protein implicated in malignant hypercalcemia: Cloning and expression. *Science*, *237*, 893–896.

Tang, Y., Wu, X., Lei, W., Pang, L., Wan, C., Shi, Z., et al. (2009). TGF-beta1-induced migration of bone mesenchymal stem cells couples bone resorption with formation. *Nature Medicine*, *15*, 757–765.

Tassabehji, M., Fang, Z. M., Hilton, E. N., McGaughran, J., Zhao, Z., de Bock, C. E., et al. (2008). Mutations in GDF6 are associated with vertebral segmentation defects in Klippel-Feil syndrome. *Human Mutation*, *29*, 1017–1027.

Tavormina, P. L., Bellus, G. A., Webster, M. K., Bamshad, M. J., Fraley, A. E., McIntosh, I., et al. (1999). A novel skeletal dysplasia with developmental delay and acanthosis nigricans is caused by a Lys650Met mutation in the fibroblast growth factor receptor 3 gene. *American Journal of Human Genetics*, *64*, 722–731.

Tavormina, P. L., Rimoin, D. L., Cohn, D. H., Zhu, Y. Z., Shiang, R., & Wasmuth, J. J. (1995). Another mutation that results in the substitution of an unpaired cysteine residue in the extracellular domain of FGFR3 in thanatophoric dysplasia type I. *Human Molecular Genetics*, *4*, 2175–2177.

Tavormina, P. L., Shiang, R., Thompson, L. M., Zhu, Y. Z., Wilkin, D. J., Lachman, R. S., et al. (1995). Thanatophoric dysplasia (types I and II) caused by distinct mutations in fibroblast growth factor receptor 3. *Nature Genetics*, *9*, 321–328.

Tekin, M., Hismi, B. O., Fitoz, S., Ozdag, H., Cengiz, F. B., Sirmaci, A., et al. (2007). Homozygous mutations in fibroblast growth factor 3 are associated with a new form of syndromic deafness characterized by inner ear agenesis, microtia, and microdontia. *American Journal of Human Genetics*, *80*, 338–344.

Thomas, J. T., Kilpatrick, M. W., Lin, K., Erlacher, L., Lembessis, P., Costa, T., et al. (1997). Disruption of human limb morphogenesis by a dominant negative mutation in CDMP1. *Nature Genetics*, *17*, 58–64.

Thomas, J. T., Lin, K., Nandedkar, M., Camargo, M., Cervenka, J., & Luyten, F. P. (1996). A human chondrodysplasia due to a mutation in a TGF-beta superfamily member. *Nature Genetics*, *12*, 315–317.

Toydemir, R. M., Brassington, A. E., Bayrak-Toydemir, P., Krakowiak, P. A., Jorde, L. B., Whitby, F. G., et al. (2006). A novel mutation in FGFR3 causes camptodactyly, tall stature, and hearing loss (CATSHL) syndrome. *The American Journal of Human Genetics, 79*, 935–941.

Tucker, A. S., Yamada, G., Grigoriou, M., Pachnis, V., & Sharpe, P. T. (1999). Fgf-8 determines rostral-caudal polarity in the first branchial arch. *Development, 126*, 51–61.

Tukachinsky, H., Kuzmickas, R. P., Jao, C. Y., Liu, J., & Salic, A. (2012). Dispatched and scube mediate the efficient secretion of the cholesterol-modified hedgehog ligand. *Cell Reports, 2*, 308–320.

Twigg, S. R., Healy, C., Babbs, C., Sharpe, J. A., Wood, W. G., Sharpe, P. T., et al. (2009). Skeletal analysis of the Fgfr3(P244R) mouse, a genetic model for the Muenke craniosynostosis syndrome. *Developmental Dynamics, 238*, 331–342.

Twigg, S. R., Vorgia, E., McGowan, S. J., Peraki, I., Fenwick, A. L., Sharma, V. P., et al. (2013). Reduced dosage of ERF causes complex craniosynostosis in humans and mice and links ERK1/2 signaling to regulation of osteogenesis. *Nature Genetics, 45*, 308–313.

Urist, M. R. (1965). Bone: Formation by autoinduction. *Science, 150*, 893–899.

Valencia, M., Caparros-Martin, J. A., Sirerol-Piquer, M. S., Garcia-Verdugo, J. M., Martinez-Glez, V., Lapunzina, P., et al. (2014). Report of a newly indentified patient with mutations in BMP1 and underlying pathogenetic aspects. *American Journal of Medical Genetics Part A, 164A*, 1143–1150.

van Bokhoven, H., Celli, J., Kayserili, H., van Beusekom, E., Balci, S., Brussel, W., et al. (2000). Mutation of the gene encoding the ROR2 tyrosine kinase causes autosomal recessive Robinow syndrome. *Nature Genetics, 25*, 423–426.

van de Laar, I. M., Oldenburg, R. A., Pals, G., Roos-Hesselink, J. W., de Graaf, B. M., Verhagen, J. M., et al. (2011). Mutations in SMAD3 cause a syndromic form of aortic aneurysms and dissections with early-onset osteoarthritis. *Nature Genetics, 43*, 121–126.

Van Wesenbeeck, L., Cleiren, E., Gram, J., Beals, R. K., Benichou, O., Scopelliti, D., et al. (2003). Six novel missense mutations in the LDL receptor-related protein 5 (LRP5) gene in different conditions with an increased bone density. *The American Journal of Human Genetics, 72*, 763–771.

Venditti, C. P., Farmer, J., Russell, K. L., Friedrich, C. A., Alter, C., Canning, D., et al. (2002). Omodysplasia: An affected mother and son. *American Journal of Medical Genetics, 111*, 169–177.

Villanueva, S., Glavic, A., Ruiz, P., & Mayor, R. (2002). Posteriorization by FGF, Wnt, and retinoic acid is required for neural crest induction. *Developmental Biology, 241*, 289–301.

Vortkamp, A., Gessler, M., & Grzeschik, K. H. (1991). GLI3 zinc-finger gene interrupted by translocations in Greig syndrome families. *Nature, 352*, 539–540.

Vortkamp, A., Lee, K., Lanske, B., Segre, G. V., Kronenberg, H. M., & Tabin, C. J. (1996). Regulation of rate of cartilage differentiation by Indian hedgehog and PTH-related protein. *Science, 273*, 613–622.

Wagner, T., Wirth, J., Meyer, J., Zabel, B., Held, M., Zimmer, J., et al. (1994). Autosomal sex reversal and campomelic dysplasia are caused by mutations in and around the SRY-related gene SOX9. *Cell, 79*, 1111–1120.

Wallis, D. E., Roessler, E., Hehr, U., Nanni, L., Wiltshire, T., Richieri-Costa, A., et al. (1999). Mutations in the homeodomain of the human SIX3 gene cause holoprosencephaly. *Nature Genetics, 22*, 196–198.

Wang, Q., Green, R. P., Zhao, G., & Ornitz, D. M. (2001). Differential regulation of endochondral bone growth and joint development by FGFR1 and FGFR3 tyrosine kinase domains. *Development, 128*, 3867–3876.

Warman, M. L., Cormier-Daire, V., Hall, C., Krakow, D., Lachman, R., LeMerrer, M., et al. (2011). Nosology and classification of genetic skeletal disorders: 2010 revision. *American Journal of Medical Genetics Part A, 155A*, 943–968.

Warr, N., Powles-Glover, N., Chappell, A., Robson, J., Norris, D., & Arkell, R. M. (2008). Zic2-associated holoprosencephaly is caused by a transient defect in the organizer region during gastrulation. *Human Molecular Genetics*, *17*, 2986–2996.

Weir, E. C., Philbrick, W. M., Amling, M., Neff, L. A., Baron, R., & Broadus, A. E. (1996). Targeted overexpression of parathyroid hormone-related peptide in chondrocytes causes chondrodysplasia and delayed endochondral bone formation. *Proceedings of the National Academy of Sciences of the United States of America*, *93*, 10240–10245.

Wharton, K. A., Johansen, K. M., Xu, T., & Artavanis-Tsakonas, S. (1985). Nucleotide sequence from the neurogenic locus notch implies a gene product that shares homology with proteins containing EGF-like repeats. *Cell*, *43*, 567–581.

White, K. E., Cabral, J. M., Davis, S. I., Fishburn, T., Evans, W. E., Ichikawa, S., et al. (2005). Mutations that cause osteoglophonic dysplasia define novel roles for FGFR1 in bone elongation. *The American Journal of Human Genetics*, *76*, 361–367.

Whittock, N. V., Sparrow, D. B., Wouters, M. A., Sillence, D., Ellard, S., Dunwoodie, S. L., et al. (2004). Mutated MESP2 causes spondylocostal dysostosis in humans. *The American Journal of Human Genetics*, *74*, 1249–1254.

Wilcox, W. R., Tavormina, P. L., Krakow, D., Kitoh, H., Lachman, R. S., Wasmuth, J. J., et al. (1998). Molecular, radiologic, and histopathologic correlations in thanatophoric dysplasia. *American Journal of Medical Genetics*, *78*, 274–281.

Wilkie, A. O., Slaney, S. F., Oldridge, M., Poole, M. D., Ashworth, G. J., Hockley, A. D., et al. (1995). Apert syndrome results from localized mutations of FGFR2 and is allelic with Crouzon syndrome. *Nature Genetics*, *9*, 165–172.

Wilkie, A. O., Tang, Z., Elanko, N., Walsh, S., Twigg, S. R., Hurst, J. A., et al. (2000). Functional haploinsufficiency of the human homeobox gene MSX2 causes defects in skull ossification. *Nature Genetics*, *24*, 387–390.

Woods, C. G., Stricker, S., Seemann, P., Stern, R., Cox, J., Sherridan, E., et al. (2006). Mutations in WNT7A cause a range of limb malformations, including Fuhrmann syndrome and Al-Awadi/Raas-Rothschild/Schinzel phocomelia syndrome. *The American Journal of Human Genetics*, *79*, 402–408.

Wozney, J. M., Rosen, V., Celeste, A. J., Mitsock, L. M., Whitters, M. J., Kriz, R. W., et al. (1988). Novel regulators of bone formation: Molecular clones and activities. *Science*, *242*, 1528–1534.

Wu, X. L., Gu, M. M., Huang, L., Liu, X. S., Zhang, H. X., Ding, X. Y., et al. (2009). Multiple synostoses syndrome is due to a missense mutation in exon 2 of FGF9 gene. *The American Journal of Human Genetics*, *85*, 53–63.

Xu, X., Weinstein, M., Li, C., & Deng, C. (1999). Fibroblast growth factor receptors (FGFRs) and their roles in limb development. *Cell and Tissue Research*, *296*, 33–43.

Yamaguchi, T. P., Bradley, A., McMahon, A. P., & Jones, S. (1999). A Wnt5a pathway underlies outgrowth of multiple structures in the vertebrate embryo. *Development*, *126*, 1211–1223.

Yamaguchi, K., Shirakabe, K., Shibuya, H., Irie, K., Oishi, I., Ueno, N., et al. (1995). Identification of a member of the MAPKKK family as a potential mediator of TGF-beta signal transduction. *Science*, *270*, 2008–2011.

Yang, Y., Topol, L., Lee, H., & Wu, J. (2003). Wnt5a and Wnt5b exhibit distinct activities in coordinating chondrocyte proliferation and differentiation. *Development*, *130*, 1003–1015.

Yen, H. Y., Ting, M. C., & Maxson, R. E. (2010). Jagged1 functions downstream of Twist1 in the specification of the coronal suture and the formation of a boundary between osteogenic and non-osteogenic cells. *Developmental Biology*, *347*, 258–270.

Yoon, B. S., Ovchinnikov, D. A., Yoshii, I., Mishina, Y., Behringer, R. R., & Lyons, K. M. (2005). Bmpr1a and Bmpr1b have overlapping functions and are essential for chondrogenesis in vivo. *Proceedings of the National Academy of Sciences of the United States of America, 102*, 5062–5067.

Yu, K., Herr, A. B., Waksman, G., & Ornitz, D. M. (2000). Loss of fibroblast growth factor receptor 2 ligand-binding specificity in Apert syndrome. *Proceedings of the National Academy of Sciences of the United States of America, 97*, 14536–14541.

Zanotti, S., Smerdel-Ramoya, A., Stadmeyer, L., Durant, D., Radtke, F., & Canalis, E. (2008). Notch inhibits osteoblast differentiation and causes osteopenia. *Endocrinology, 149*, 3890–3899.

Zeng, X., Goetz, J. A., Suber, L. M., Scott, W. J., Jr., Schreiner, C. M., & Robbins, D. J. (2001). A freely diffusible form of Sonic hedgehog mediates long-range signalling. *Nature, 411*, 716–720.

CHAPTER NINETEEN

Clinical and Genomic Approaches for the Diagnosis of Craniofacial Disorders

Pedro A. Sanchez-Lara[1]
Director of Craniofacial Genetics, Children's Hospital Los Angeles, Los Angeles, California, USA
Assistant Professor of Clinical Pediatrics, University of Southern California, Los Angeles, California, USA
Assistant Professor of Clinical Pathology, University of Southern California, Los Angeles, California, USA
Research Instructor Center for Craniofacial Molecular Biology, Herman Ostrow School of Dentistry, University of Southern California, Los Angeles, California, USA
[1]Corresponding author: e-mail address: pasanchez@chla.usc.edu

Contents

Abstract

With the rapid development of readily accessible molecular diagnostic tools, a growing number of patients and families with craniofacial anomalies will have access to a confirmed molecular diagnosis. This chapter provides an overview to current clinical and molecular resources and approaches used by diagnostician today.

Clarifying the underlying cause of a congenital defect is necessary to provide proper counseling, identify carrier/risk status of family members, inform prognosis and direct appropriate management, treatments, and surveillance recommendations. The use of molecular testing has evolved to confirm a suspected clinical diagnosis, establish a diagnosis in an unclear condition and end a diagnostic odyssey for many children with underlying syndromes, but the use of these techniques to understand common nonsyndromic malformations like clefts and craniosynostosis is still an active area of research that will contribute to clinical care in the future.

Current Topics in Developmental Biology, Volume 115
ISSN 0070-2153
http://dx.doi.org/10.1016/bs.ctdb.2015.09.004

1. INTRODUCTION

A diagnostic evaluation for a craniofacial disorder is based on the knowledge from embryology, dysmorphology, clinical genetics, and pediatrics. With the rapid development of readily accessible molecular diagnostic tools, patients, and families will have access to a confirmed molecular diagnosis, shortening their diagnostic odyssey, and facilitating their clinical care and genetic counseling. As a consequence, families and medical providers are turning to the clinical and basic science research communities for recommendations and targeted therapies to improve outcomes and prognoses.

Clarifying the underlying cause of a congenital defect is necessary to provide proper counseling, identify carrier/risk status of family members, inform prognosis and direct appropriate management, treatments, and surveillance recommendations. The use of molecular testing has evolved to confirm a suspected clinical diagnosis, establish a diagnosis in an unclear condition and end a diagnostic odyssey for many children with underlying syndromes, but the use of these techniques to understand common non-syndromic malformations like clefts and craniosynostosis is still an active area of research that will contribute to clinical care in the future. It is not the intention of this chapter to encompass all elements of a diagnostic evaluation as one can refer to a growing number of texts, articles, and databases regarding the diagnostic approach to a dysmorphic child (Aase, 1990; Epstein, Erickson, & Wynshaw-Boris, 2008; Ferretti, 2006; Graham & Sanchez-Lara, 2014; Hall, Allanson, Gripp, & Slavotinek, 2007; Johns Hopkins University, n.d.; Jones, Jones, & Campo, 2013; London Medical Databases, n.d.). Rather, this chapter will provide an overview to some of the current molecular resources and approaches used today.

2. CLINICAL EVALUATION AND DYSMORPHOLOGY

The field of dysmorphology has played a major role in the prenatal and postnatal clinical diagnostic evaluation of patients with craniofacial malformations. A comprehensive diagnostic consultation includes a pre-evaluation review of photographs and prior records including data from prior molecular, cytogenetic and biochemical tests, and other procedures; a thorough pregnancy, birth, family, medical, surgical, and developmental history; and detailed review of systems. It is particularly important to not overlook the prenatal exposure history and timing for common recognized teratogens (Abbott, 2010), and to be able to recognize a teratogenic syndrome and be able to reassure parents regarding common exposures to medications regarding medications, which are not teratogenic. The importance of common maternal conditions (e.g., obesity, diabetes, alcoholism), which increase the risk for certain malformations, can help to prevent future malformations. One of the most valuable tools in an evaluation is a thorough history and pedigree analysis, which can alert the clinician of potential Mendelian disorders with incomplete penetrance and variable expressivity. The physical examination should always include standard growth measurements plotted appropriately and adjusted for prematurity and sex. One should also follow a prescribed examination format whenever possible to avoid overlooking clues in the proband and other potentially affected relatives. A differential diagnosis can then be formulated and additional data may be collected to narrow this list of conditions or confirm a suspected diagnosis. Follow-up data can include imaging, additional subspecialty consultations, examining photographs at a younger age, imaging studies, or laboratory studies (molecular, cytogenetic, or biochemical testing). It is also often helpful to just reevaluate the patient after a short time if the child is very young or records are not complete.

An individual with "dysmorphic features" has a pattern of morphologic abnormalities that is outside the normal range of development either by descriptive observation or measurement (Hall et al., 2007). Clinicians may refer to such a collection of major and minor anomalies in an individual as a "syndromic" based solely on their experience. Most care providers for dysmorphic patients develop skills in recognizing the "gestault" of common patterns of malformation, but for less frequent malformation syndromes, molecular diagnosis can be quite helpful. Recognizing the pattern of major and minor anomalies through a thorough history and examination is critical for an accurate diagnosis.

A *syndrome* can be defined as a particular set of anomalies occurring together with some frequency. For example patients with Treacher Collins syndrome (Fig. 1B) often have malar hypoplasia, lower lid colobomas, micrognathia, and occasionally cleft palate or microtia, but there are many nuances in diagnosing different syndromes associated with mandibulofacial dysostosis, and the accuracy of a clinical diagnosis can hinge on a single feature such as the presence or absence of hand abnormalities or microcephaly. Multiple structural defects can also result from a *sequence* in which a single primary defect in morphogenesis produces multiple abnormalities through a cascading process of secondary and tertiary problems in morphogenesis.

Isolated structural defects can arise through any of four basic mechanistic categories: *deformations, malformations, disruptions,* and *dysplasias* (Fig. 1; Table 1).

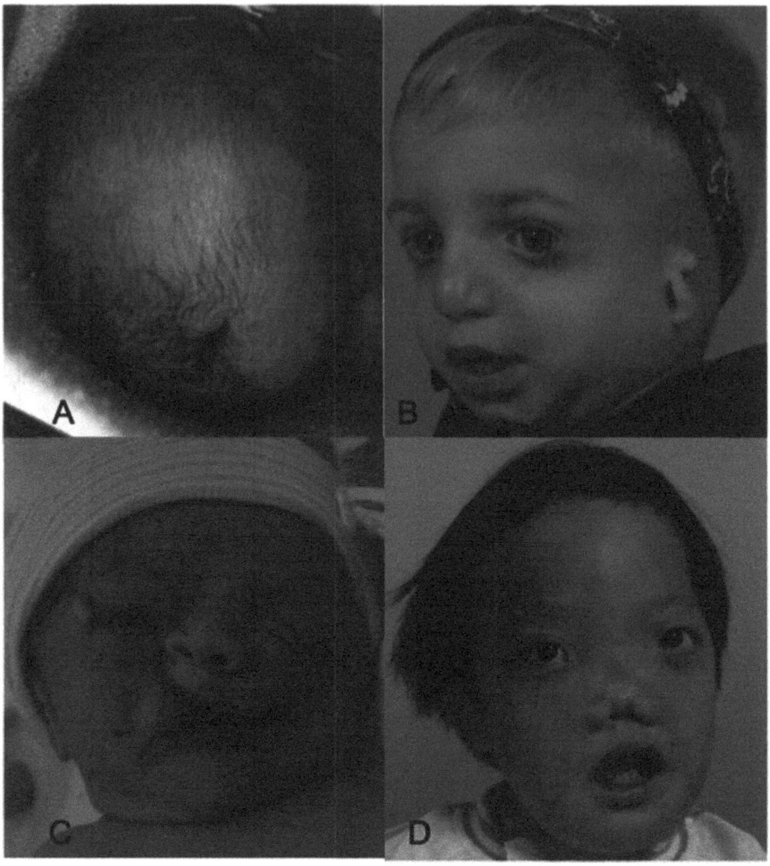

Figure 1 Examples of a deformation, malformation, disruption, and dysplasia. (A) Deformation example: top view of a child's head with plagiocephaly and left-sided flattening of the occiput (Note the anterior position of the left ear as a secondary finding). (B) Malformation example: child with Treacher Collins syndrome (lower lid colobomas, microtia/anotia, micrognathia, wearing a BAHA for conductive hearing loss). (C) Disruption example: multiple facial defects secondary to amniotic band sequence. (D) Dysplasia example: child with craniometaphyseal dysplasia, a progressive bony dysplasia with thickening of the craniofacial bones. (A color version of this figure is located at the end of this chapter.)

Deformational abnormalities are produced by intrauterine mechanical forces that distort otherwise normal fetal structures. Deformational anomalies often occur late in gestation as the fetus outgrows the size of the uterus and fetal constraint becomes more apparent. It is thought that after an abnormal mechanical stress is removed, the defect may begin to correct spontaneously, or correct with gentle bracing to counter the initial deforming forces. Intrauterine constraint can be secondary to factors specific to the mother (small pelvic outlet, primigravida, bicornuate uterus, or uterine fibroids),

Table 1 Morphogenesis of Congenital Anomalies

Deformations	Response of normal tissue to a mechanical force
Malformations	Primary error in tissue morphogenesis
Disruptions	Breakdown of previously normal tissues
Dysplasias	Abnormal gross structure because of abnormal tissue architecture

to the individual pregnancy (multiple gestation or intra-abdominal pregnancy), or to the individual fetus (oligohydramnios, malformations such as bilateral renal agenesis or abnormal fetal presentations, e.g., breech presentation) (see Graham & Sanchez, 2016).

In contrast, *malformations* denote a primary problem in the morphogenesis of a tissue (Opitz, 2000). An isolated malformation can arise as a localized error in embryonic development with a multifactorial etiology and varied recurrence risk or higher birth prevalence among certain populations. A multiple malformation syndrome results in a primary developmental anomaly of two or more systems in which all of the structural defects are secondary to a single underlying etiology, which can be environmental (teratogenic), genetic (e.g., single-nucleotide variant) or chromosomal. *Disruptions* are structural defects caused by the active destruction and breakdown of normal tissues (e.g., amniotic band syndrome/amnion rupture sequence (Iqbal, Derderian, Cheng, Lee, & Hirose, 2015)). A fourth category of problems, termed *dysplasias*, results in abnormal structure because the tissues from which individual structures are formed are abnormal; for example, the defective cartilage and bone in various skeletal dysplasias (Lachman & Taybi, 2007).

It is often helpful to prepare a differential diagnosis for major malformations while tracking and comparing their associated *minor anomalies*. *Major malformations* are disturbances of embryogenesis and these defects affect medically important structures such as the palate, eyes, brain, heart, and kidneys. Most major malformations occur as isolated defects although they can often be essential to meeting the diagnostic criteria for a recognizable syndrome. *Minor anomalies* and *common variants* (prevalence >4%) are considered subtle developmental disturbances and are of no serious medical or cosmetic consequence but may be rather common in the general population. Minor anomalies and common variants often occur in structures that are embryologically complex such often as the face, ears, hands, and feet; the distinction between these categories is one of degree and prevalence (Aase, 1990; Merks, van Karnebeek, Caron, & Hennekam, 2003). Mendelian disorders are typically very rare and have highly penetrant phenotypes. Outside of pharmacogenetics, the adoption and clinical utility of rare and more common genetic variants with low penetrance is yet to be determined (Fig. 2).

The use of standardized nomenclature and of phenotype descriptors is essential to both clinical and molecular diagnostics. Over the course of several years, an international group of dysmorphology experts set out to define this work by dividing into in six working groups, each focused on a particular area of the human body. In 2005 and 2006, this international working group met face-to-face developed a standardized set of definitions and terms now known as the Elements of Morphology which describe the physical variations now used in human phenotypic analyses. The Elements of Morphology, resulted in six articles proposing consensus definitions for almost 400 phenotypic variations of the head and face; periorbital region; ear, nose, and philtrum; mouth and lips; and hands and feet (Carey, Allanson, Hennekam, & Biesecker, 2012). Today, integration of clinical and phenotypic data is critical to filtering genomic results and ultimately clinical molecular reporting. The Human Phenotype Ontology (HPO) project (http://www.human-phenotype-ontology.org) has provided an expanded structure and consists of a comprehensive set of over 10,000 terms describing human phenotypic abnormalities and several thousand subclass relations between the various HPO classes (Kohler et al., 2014).

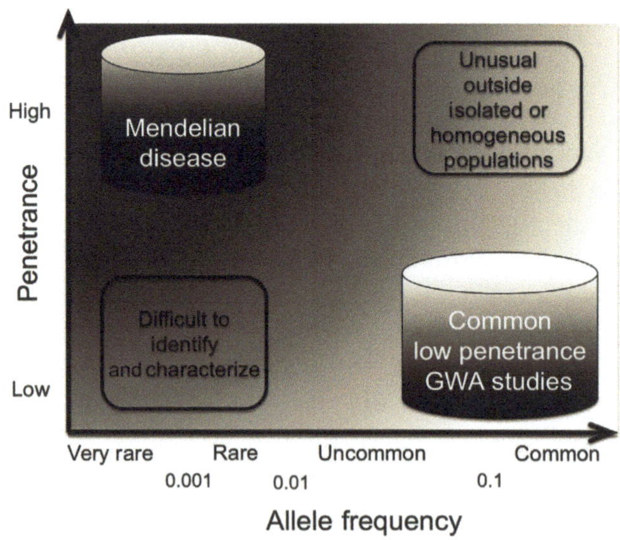

Figure 2 Illustration of categories/conditions with various allele frequencies (*x*-axis) and penetrance (*y*-axis). Currently only highly penetrant and rare conditions are assessed in a clinical setting. Rare and less penetrant variants are difficult to identify and characterize for clinical utility. *(Figure adapted from McCarthy et al., 2009)*

The types of biological samples used to screen for genetic abnormalities have also expanded with the recent advances in molecular technology. Prenatal testing can now begin before implantation with preimplantation genetic diagnosis or after fertilization with invasive procedures such as chorionic villous sampling and amniocentesis (Rechitsky et al., 2015). More recently, there has been a growing shift away from invasive obstetric procedures. More pregnant women are choosing noninvasive prenatal screening (NIPS) of cell-free DNA from the maternal blood (Manegold-Brauer et al., 2015; Willems et al., 2014). NIPS has been successful at detecting aneuploidy and some microdeletion syndromes although confirmatory testing is still recommended (Gregg et al., 2013). Postnatally, blood or saliva is typically used for germline mutation screening but buccal/cheek swabs, skin biopsies, or other tissue testing can be used when suspecting an underling mosaic condition (Reddy & Mak, 2001). Genetic counseling concerning a fairly straightforward autosomal dominant Mendelian disorder can become quite complicated when considering the spectrum of presentations, potential mosaicism, and molecular test results and outcomes (Fig. 3).

Father's phenotype	+++	++	+	–	–	–	–
Father's blood test	+++	+/–	+/–	–	–	–	–
Father's Rec. risk	50%	? 1–50%	? 1–50%	? 1–50%	General Pop. risk	General Pop. risk	General Pop. risk
Sons blood test	+++	+++	+++	+++	+++	+/–	–
Sons phenotype	+++	+++	+++	+++	+++	++	+
Sons Rec. risk	50%	50%	50%	50%	50%	? 1–50%	General Pop. risk

Figure 3 Hypothetical example of an autosomal dominant condition and potential phenotype/genotype spectrum of results when mosaicism is considered. A mutation inherited from the father (A–D); a *de novo* mutation in the male child (E) and post-fertilization somatic mosaicism in the child (F–G). The father's blood DNA test result could be abnormal (+++) (A) potentially abnormal, mosaic, or normal (+/–) (B and C), or normal in (D–G) (–). Figure (D) demonstrates paternal gonadal mosaicism. Figure (E) shows the male child with a *de novo* mutation and (F–G) illustrates decreasing amounts of segmental mosaicism. Affected tissue is illustrated within the colored area (Rec., recurrence; Pop., population). (A color version of this figure is located at the end of this chapter.)

3. CHROMOSOMAL ABERRATIONS

In 1956, Tjio and Levan were the first to accurately describe that humans have 46 chromosomes. This discovery was technically the first very low-resolution whole-genome analysis and opened the field of clinical cytogenetics. As standard-staining techniques such as G-Banding and other visualization methods evolved, the technical ability to detect aneuploidy (e.g., Trisomy 13), large structural rearrangements and some defined microdeletion syndromes became common in clinical practice. Chromosomal aberrations can be associated with up to half of early spontaneous abortions, 6% of stillbirths, around 5% of couples with a history of two or more miscarriages, and approximately 0.5% of newborns (Patil, Rao, & Majumdar, 2014). G-banding chromosome analysis performed at the 350–500 band level does not typically detect segments smaller than 5–10 Mb (Gardner, Sutherland, & Shaffer, 2012). Florescent *in situ* hybridization (FISH) techniques and subtelomere FISH, which bind to specific regions of the chromosome by matching complementary sequences, have improved the ability to detect small cryptic deletions and rearrangements and has a resolution as low as 35 kb (Fig. 4C) (Bejjani & Shaffer, 2008). One primary limitation of G-Banding and FISH apart from the resolution is the need for fixed intact or cultured cells.

4. EVOLUTION OF COPY NUMBER DETECTION

After the first draft of the human genome was published, almost any genomic region could be fluorescently labeled and available commercially for diagnostic purposes. Subtelomeric FISH became one of the most helpful tools to screen chromosomes for cryptic terminal deletions or duplications related to unbalanced translocations (Durmaz et al., 2015). The next wave of technology came from multiplexing thousands of genomic regions on chromosome array chips to efficiently screen the whole genome for cryptic copy number anomalies. As array density and resolution evolved, there was a transition from lower resolution bacterial artificial chromosome arrays to comparative genomic hybridization analyses, oligo and single nucleotide polymorphism (SNP) arrays. The latter two are now often combined and these are the most common array types in use today. Arrays containing SNPs have an advantage over other types of arrays because they can detect copy neutral loss of heterozygosity flagging potential consanguinity and recessive

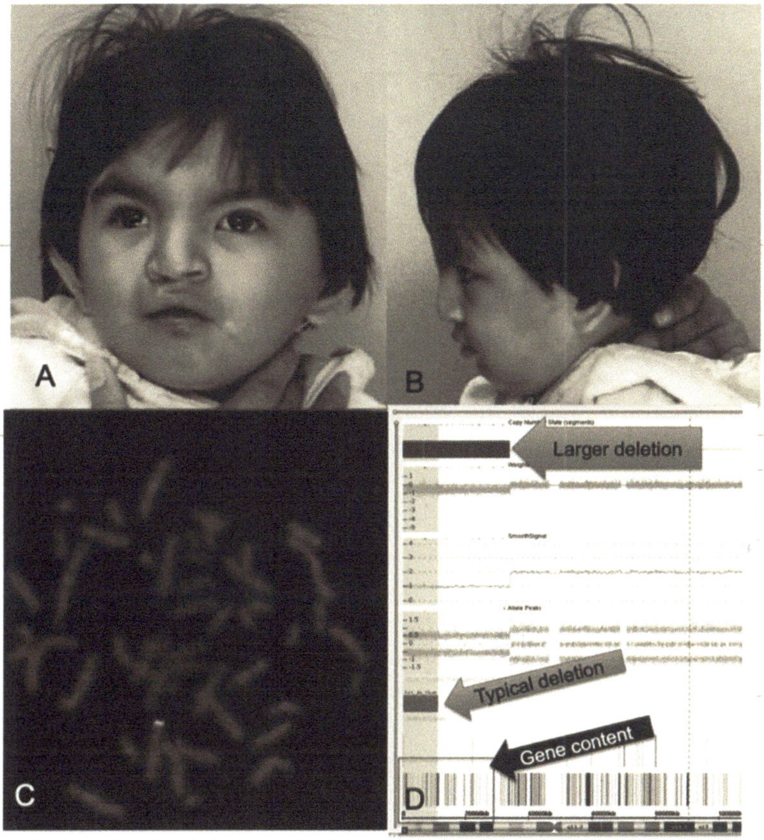

Figure 4 (A and B) Female patient with cleft lip and palate secondary to Cri Du Chat syndrome (5p minus microdeletion syndrome). (C) Florescence *in situ* hybridization (FISH) green probe for 5p confirmed presence of a single copy. (D) Chromosome microarray detected a 33.7 Mb deletion significantly larger than the typical 5p minus deletion. (A color version of this figure is located at the end of this chapter.)

candidate genes, and can flag individual chromosomes raising concern for uniparental disomy (Alkuraya, 2012). Arrays have significantly advanced our understanding of chromosomal disorders and facilitated the ability to compare an individual's genomic breakpoints with gene content (Fig. 4). A chromosomal microarray cannot detect balanced chromosomal rearrangements such as translocations and inversions, nor can it assess positional information for insertions. Other molecular techniques such as quantitative real-time polymerase chain reaction (qPCR) or multiplex ligation-dependent probe amplification are used to detect, quantify, and confirm chromosomal copy number changes but these cannot screen a large number of regions across the whole genome. Exon-specific oligo arrays are a more cost-effective way to screen a large number of genes for copy number changes. There are now several copy number detecting bioinformatic algorithms using next-generation sequencing (NGS) (Abel & Duncavage, 2013; Legault, Girard, Lemieux Perreault, Rouleau, & Dube, 2015; Xie & Tammi, 2009). These NGS-based copy number techniques are beginning to transition into the clinical setting where they will likely replace current chromosomal microarray techniques. Extensive validation and optimization are still needed to increase the diagnostic sensitivity and reduce the number of false positives that require follow-up confirmatory testing.

The diagnostic yield of a karyotype or chromosomal microarray analysis used to screen craniofacial malformations varies by type and category. For example, in some studies, the diagnostic yield for isolated nonsyndromic single-suture craniosynostosis cases is very low or near zero. Yet in syndromic craniosynostosis, the yield ranges from 6.7% to 28%. The vast majority (85%) of craniosynostosis cases with an abnormal karyotype or array are typically midline (metopic and sagittal) sutures (Johnson & Wilkie, 2011; Lattanzi et al., 2012; Wilkie et al., 2010). Karyotype and chromosome microarray studies in individuals with oral clefts also have varying degrees of diagnostic yield depending on whether if the clefts were detected prenatally or postnatally. Maarse et al. summarized a comprehensive review of prenatal and postnatal chromosomal and microarray studies (Maarse et al., 2012). Of 407 fetuses with oral clefts, cleft lip and palate had the highest prevalence of associated anomalies (54%, range 39.1–66%). There were 23 cases of cleft lip without cleft palate, and three of these had associated anomalies while only one had a chromosomal defect. Studies that grouped both cleft lip and/or cleft palate had a lower prevalence of associated anomalies (29.9%, ranging 17.2–57.1%). The prevalence of chromosomal defects in cleft cases with associated anomalies was 50% (74/146) while it was 0.9% in cases with clefts that were formerly presumed to be isolated. Of 28,953 postnatally assessed infants, almost all chromosomal abnormalities were found in association with additional anomalies. Cleft palate was the category most frequently associated with other anomalies (45.9%, range 22.2–78.3%). The prevalence of associated anomalies in cleft lip cases was approximately 10%. One study of isolated cleft lip cases found a chromosomal defect in 1.8% (2/110) cases (both having a 22q11.2 deletion). Overall, the diagnostic yield of screening for chromosomal defects in cases of cleft lip with or without cleft palate was 9.5% range (0.5–12.6%) (Maarse et al., 2012).

5. SINGLE-CANDIDATE GENE AND PANEL-BASED TESTING

PCR generates of thousands to millions of copies of a particular DNA sequence (Saiki et al., 1985). Efforts to develop tools utilize PCR led to other molecular techniques such as restriction fragment length polymorphism (RFLP), single-strand confirmation polymorphism (SSCP), and improvement in targeted classical chain-termination sequencing-based methods (Sanger sequencing). Up until the development of NGS, SSCP, and RFLP were the most widely used techniques for mutation screening but were not able to detect every mutation (Durmaz et al., 2015).

Most recognizable craniofacial syndromes are monogenic Mendelian disorders, but even within the same condition there is often *allelic heterogeneity* with most individuals having different mutations in the same gene. The exceptions to this are the recognizable craniosynostosis syndromes involving the *FGFR* genes (*FGFR1, FGFR2, FGFR3*) (Aleck, 2004). In craniofacial medicine, there are a growing number of clinically indistinguishable or overlapping phenotypes that may be caused by mutations in different genes (*locus heterogeneity*). Examples include rasopathies, cohesinopathies, mandibulofacial dysostoses, and Stickler syndrome (Alzahrani, Al Hazzaa, Tayeb, & Alkuraya, 2015; Bezniakow, Gos, & Obersztyn, 2014; Parenti et al., 2015; Robin, Moran, & Ala-Kokko, 1993; Vincent et al., 2015). When using a major malformation as the only search criterion, the number of genes involved can range from just a few to more than a hundred. For many genetically heterogeneous craniofacial disorders, the full complement of causal genes is yet to be established (Beaty et al., 2013; Wilkie et al., 2010).

Determining genetic causality for a particular disease and establishing a molecular diagnosis in clinical practice can be challenging. In recent years, exome and genome sequencing have increased the rate of gene discovery for single-gene disorders among patients with suspected, but previously undiagnosed, genetic disorders. Although exome and genome sequencing are becoming more readily available, the value of molecular diagnosis should be viewed from a clinical perspective as similar to other diagnostic tests. The decision to proceed with molecular testing must integrate many factors specific to clinical status of the affected individual, such as probability of diagnostic yield and the patient's/families personal preference.

The ability to confirm a clinical diagnosis and stop a diagnostic odyssey has historically occurred without any molecular testing, yet the growing number of clinically indistinguishable craniofacial conditions that have mutations in different genes (*locus heterogeneity*) has spread uncertainty in the clinical setting and prompted more confirmatory molecular testing. A craniofacial condition such as Treacher Collins syndrome, which had initially been reported to be monogenic and autosomal dominant, has subsequently been found to be multigenic with autosomal dominant (*TCOF1, POLR1D*) and recessive forms (*POLR1C)* (Vincent et al., 2015). Stickler syndrome is another common condition with a growing number of genes related to an overlapping phenotype with autosomal dominant (*COL2A1, COL11A1, COL11A2, VCAN*) and autosomal recessive inheritance (*COL9A1, COL9A2, COL9A3, LOXL3*) (Acke et al., 2014; Alzahrani et al., 2015).

As of today, Sanger sequencing remains the gold standard molecular diagnostic tool used to screen DNA for unknown point mutations in defined genes; this may change as the confidence and quality of newer technologies improve (Neveling et al., 2013). Up until recently, some larger genes had remained inaccessible to clinical testing because the older methods were too burdensome on laboratory staff, or the condition was too rare for a test to be commercially viable.

6. NEXT-GENERATION SEQUENCING

In 2001, a new era of rapid gene identification emerged after the first draft of the human genome sequence was released. Large families with multiple affected members could easily have their condition identified and mapped using a whole-genome linkage approach. With the introduction of massively parallel sequencing in 2005, the technology also known as *NGS* or "Next Gen" became a cost-effective tool with advantages in detecting genetic variations ranging from single nucleotides to larger structural rearrangements. *Whole-exome sequencing* (WES) was rapidly developed to perform NGS on exon-enriched DNA. Companion software was developed to code the sequences and quickly identify protein-coding mutations. Using WES in Mendelian disorders has expedited identification of causative genes from a small number of individuals within a family (Zhang, 2014). In 2010, Miller syndrome (a condition with severe micrognathia, cleft lip and/or palate, hypoplasia or aplasia of the posterior elements of the limbs, coloboma of the eyelids, and supernumerary nipples) became the first condition for which WES facilitated the discovery of the underlying causative gene (Ng et al., 2010).

Broad molecular screening techniques have traditionally been reserved for disorders that are difficult to diagnose because of their nonspecific presentation, which is too subtle or has too many overlapping phenotypes to confidently determine which gene to sequence based on clinical features alone. Yet with the dramatic drop in NGS sequencing costs, many centers have begun considering a targeted-gene panel as their first approach.

Limitations of NGS for gene panels or WES include uneven efficiency in exon pull down or capture leading to missing exons or gaps in coverage. Structural variants such as large deletions, duplications, and rearrangements are not routinely detected or assessed outside of a research setting. Nucleotide triplet-repeat disorders cannot be detected nor quantified. Noncoding regulatory sequence mutations in the promoter, intronic sequences, and deep intronic splice site mutations will also not be detected. Signs of uniparental disomy or epigenetic changes are also not detected.

7. WHOLE-EXOME SEQUENCING AND WHOLE-GENOME SEQUENCING

WES has the advantages of speed and efficiency relative to Sanger sequencing of multiple genes, delivering results in weeks to months. Candidate variants can be significantly reduced in number (from an initial 20,000 to 30,000 per individual exome) using clinically validated bioinformatic pipelines. Targeted Sanger sequencing is often used to confirm the presence or absence of the potential variants in the individual being diagnosed, and in affected and unaffected family members. Outside of the fairly recognizable conditions and clearly pathogenic mutations, there remains a need to standardize bioinformatic pipelines as well as expand the lists of clinically relevant genes with associated phenotypes. Despite this, there are a growing number of centers that are jumping directly to WES when the diagnosis is unclear, but a genetic etiology is clearly suspected (Pengelly et al., 2014).

An average human exome contains between 20,000 and 25,000 coding variants, of which ~10,000 variants are nonsynonymous, and can potentially consist of the genetic factors that cause a Mendelian disease (Gonzaga-Jauregui, Lupski, & Gibbs, 2012). An individual with phenotypic manifestations would likely carry 40 to 100 deleterious variants simultaneously.

Despite the promise of ending the diagnostic odyssey, several limitations of NGS impede the complete explanation of an underlying genetic diagnosis. First, although WES has made great progress in the diagnosis of monogenetic diseases, the human genome is highly variable, and rare variants outside coding exons can also confer disease. Approximately 85% of such diseases are caused by the coding exons and therefore by default >15% lie outside the exons, including single nucleotide variations (SNVs) and copy number variants (CNVs), in the noncoding, conserved, or regulatory regions. The latter data cannot be assayed with the use of WES but may be assessed by using whole-genome sequencing (WGS) (Gonzaga-Jauregui et al., 2012)

Even if the causative variant lies within the exome, WES relying on NGS technologies is not without challenges and limitations. In the human exome, about 1 in every 1,000 nucleotides will vary. Bioinformatic filtering can eliminate most variants that are presumed irrelevant to the condition of concern. The final steps of variant filtering involve manual research and review. This requires expertise, including knowledge of the patient's phenotype and family history, current knowledge, and exploration of published variant-disease associations, and judgment, all of which can contribute to variability in the final variant interpretation.

Reproducibility at the level of final manual variant interpretation has not been fully characterized; comparisons of exome capture platforms using detection of known medically relevant variants as the sequencing result for assessment indicate small differences in detection, gaps in capture platforms, and regions of poor coverage that are unlikely to benefit from improved capture or greater depth of sequencing. Tools to improve the computational and variant analysis pipelines are available and more are under development. Like Sanger sequencing, current standard review of NGS data cannot detect large deletions or duplications of DNA or nucleotide repeats that can cause disease, although this barrier may soon be surmounted. Error rates due to uneven sequencing coverage, gaps in exon capture prior to sequencing, difficulties with narrowing the large initial number of variants to manageable numbers without losing likely candidate mutations, poorly annotated variant databases, lack of standardized procedures, and identifying mutations in unrelated genes or unknown genes are all issues. Detailed guidance from regulatory or professional organizations is under development, and the variability contributed by the different platforms and procedures used by clinical laboratories offering exome sequencing as a clinical service is currently unknown.

There are also unresolved ethical questions that arise as a result of genetic testing. Some of these questions concern the reporting of incidental findings, such as identifying medically relevant mutations in genes unrelated to the diagnostic question and sex chromosome abnormalities. Although the cost of data storage is decreasing, there remains a question about what to do with remaining data and who has ownership of and rights to this information. There is also a large amount of ancillary data with thousands of variants of uncertain significance detected some of which may be reinterpreted at a future date as being pathogenic and disease-associated. Further social and ethical issues can arise when family studies are performed such as when there is a discovery of nonpaternity or of incidental off-target actionable variants in parents.

WGS enables the identification of the majority of variant types including single-nucleotide variants (SNVs), insertions and deletions both in coding regions and noncoding regions, and also has the advantage of detecting copy number variations (CNVs) and structural variations.

WES is gradually being optimized to identify mutations in increasing proportions of the protein-coding exome, but WGS is currently touted as a superior alternative, especially as cost decreases. The distribution of quality parameters for single-nucleotide variants (SNVs) and insertions/deletions has been shown to be more uniform for WGS than for WES. Therefore WGS may be slightly more efficient than WES for detecting mutations in the targeted exome (Belkadi et al., 2015).

8. FUTURE PERSPECTIVES

As the cost of WES and WGS drops, this type of targeted approach to genetically heterogeneous conditions will become more useful and accessible to clinical practice. Large datasets of both genotypic and phenotypic information will eventually be mined to discover and characterize the clinical significance of low penetrant variants. Future research will also explore how environmental factors interact with the genome, and contribute to abnormal craniofacial development and human health.

REFERENCES

Aase, J. M. (1990). *Diagnostic dysmorphology*. New York: Plenum Medical Book Co. pp. xiii, 299 p.

Abbott, B. D. (2010). The etiology of cleft palate: A 50-year search for mechanistic and molecular understanding. *Birth Defects Research. Part B, Developmental and Reproductive Toxicology*, *89*, 266–274.

Abel, H. J., & Duncavage, E. J. (2013). Detection of structural DNA variation from next generation sequencing data: A review of informatic approaches. *Cancer Genetics*, *206*, 432–440.

Acke, F. R., et al. (2014). Novel pathogenic COL11A1/COL11A2 variants in Stickler syndrome detected by targeted NGS and exome sequencing. *Molecular Genetics and Metabolism*, *113*, 230–235.

Aleck, K. (2004). Craniosynostosis syndromes in the genomic era. *Seminars in Pediatric Neurology*, *11*, 256–261.

Alkuraya, F. S. (2012). Discovery of rare homozygous mutations from studies of consanguineous pedigrees. *Current Protocols in Human Genetics*. Chapter 6, Unit 6.12.

Alzahrani, F., Al Hazzaa, S. A., Tayeb, H., & Alkuraya, F. S. (2015). LOXL3, encoding lysyl oxidase-like 3, is mutated in a family with autosomal recessive Stickler syndrome. *Human Genetics*, *134*, 451–453.

Beaty, T. H., et al. (2013). Confirming genes influencing risk to cleft lip with/without cleft palate in a case-parent trio study. *Human Genetics*, *132*, 771–781.

Bejjani, B. A., & Shaffer, L. G. (2008). Clinical utility of contemporary molecular cytogenetics. *Annual Review of Genomics and Human Genetics*, *9*, 71–86.

Belkadi, A., et al. (2015). Whole-genome sequencing is more powerful than whole-exome sequencing for detecting exome variants. *Proceedings of the National Academy of Sciences of the United States of America*, *112*, 5473–5478.

Bezniakow, N., Gos, M., & Obersztyn, E. (2014). The RASopathies as an example of RAS/MAPK pathway disturbances—Clinical presentation and molecular pathogenesis of selected syndromes. *Developmental Period Medicine*, *18*, 285–296.

Carey, J. C., Allanson, J. E., Hennekam, R. C., & Biesecker, L. G. (2012). Standard terminology for phenotypic variations: The elements of morphology project, its current progress, and future directions. *Human Mutation*, *33*, 781–786.

Durmaz, A. A., et al. (2015). Evolution of genetic techniques: Past, present, and beyond. *BioMed Research International*, *2015*, 461524.

Epstein, C. J., Erickson, R. P., & Wynshaw-Boris, A. J. (2008). *Inborn errors of development: The molecular basis of clinical disorders of morphogenesis. Oxford monographs on medical genetics* (2nd ed.). Oxford; New York: Oxford University Press. pp. xl, 1617 p.

Ferretti, P. (2006). *Embryos, genes, and birth defects* (2nd ed.). Chichester, England; Hoboken, NJ: Wiley. pp. xiv, 547 p.

Gardner, R. J. M., Sutherland, G. R., & Shaffer, L. G. (2012). *Chromosome abnormalities and genetic counseling. Oxford monographs on medical genetics* (4th ed.). Oxford; New York: Oxford University Press. pp. xiv, 634 p.

Gonzaga-Jauregui, C., Lupski, J. R., & Gibbs, R. A. (2012). Human genome sequencing in health and disease. *Annual Review of Medicine*, *63*, 35–61.

Graham, J. M., & Sanchez-Lara, P. A., Jr. (2016). *Smith's recognizable patterns of human deformation* (4th ed.). Philadelphia: Elsevier.

Gregg, A. R., et al. (2013). ACMG statement on noninvasive prenatal screening for fetal aneuploidy. *Genetics in Medicine*, *15*, 395–398.

Hall, J. G., Allanson, J. E., Gripp, K. W., & Slavotinek, A. (2007). *Handbook of physical measurements* (2nd ed.). Oxford; New York: Oxford University Press. pp. xii, 507 p.

Iqbal, C. W., Derderian, S. C., Cheng, Y., Lee, H., & Hirose, S. (2015). Amniotic band syndrome: A single-institutional experience. *Fetal Diagnosis and Therapy*, *37*, 1–5.

Johns Hopkins University and National Center for Biotechnology Information Online Mendelian Inheritance in Man. www.ncbi.nlm.nih.gov/omim.

Johnson, D., & Wilkie, A. O. (2011). Craniosynostosis. *European Journal of Human Genetics*, *19*, 369–376.

Jones, K. L., Jones, M. C., & Campo, M. d. (2013). *Smith's recognizable patterns of human malformation* (7th ed.). Philadelphia, PA: Elsevier Saunders. pp. xiii, 998 p.

Kohler, S., et al. (2014). The Human Phenotype Ontology project: Linking molecular biology and disease through phenotype data. *Nucleic Acids Research*, *42*, D966–D974.

Lachman, R. S., & Taybi, H. (2007). *Taybi and Lachman's radiology of syndromes, metabolic disorders, and skeletal dysplasias* (5th ed.). Philadelphia: Mosby Elsevier. pp. xxiii, 1365 p.

Lattanzi, W., et al. (2012). Genetic basis of single-suture synostoses: Genes, chromosomes and clinical implications. *Child's Nervous System*, *28*, 1301–1310.

Legault, M. A., Girard, S., Lemieux Perreault, L. P., Rouleau, G. A., & Dube, M. P. (2015). Comparison of sequencing based CNV discovery methods using monozygotic twin quartets. *PLoS One*, *10*, e0122287.

London Medical Databases Winter–Baraitser Dysmorphology Database. www.lmdatabases.com.
Maarse, W., et al. (2012). A systematic review of associated structural and chromosomal defects in oral clefts: When is prenatal genetic analysis indicated? *Journal of Medical Genetics, 49*, 490–498.
Manegold-Brauer, G., et al. (2015). Uptake of non-invasive prenatal testing (NIPT) and impact on invasive procedures in a tertiary referral center. *Archives of Gynecology and Obstetrics, 292*, 543–548.
McCarthy, M. I., Abecasis, G. R., Cardon, L. R., Goldstein, D. B., Little, J., Ioannidis, J. P., et al. (2008). Genome-wide association studies for complex traits: consensus, uncertainty and challenges. *Nature Reviews Genetics, 9*(5), 356–369.
Merks, J. H., van Karnebeek, C. D., Caron, H. N., & Hennekam, R. C. (2003). Phenotypic abnormalities: Terminology and classification. *American Journal of Medical Genetics. Part A, 123a*, 211–230.
Neveling, K., et al. (2013). A post-hoc comparison of the utility of sanger sequencing and exome sequencing for the diagnosis of heterogeneous diseases. *Human Mutation, 34*, 1721–1726.
Ng, S. B., et al. (2010). Exome sequencing identifies MLL2 mutations as a cause of Kabuki syndrome. *Nature Genetics, 42*, 790–793.
Opitz, J. M. (2000). Heterogeneity and minor anomalies. *American Journal of Medical Genetics, 92*, 373–375.
Parenti, I., et al. (2015). Broadening of cohesinopathies: Exome sequencing identifies mutations in ANKRD11 in two patients with Cornelia de Lange-overlapping phenotype. *Clinical Genetics*. Advanced online publication. http://dx.doi.org/10.1111/cge.12564.
Patil, S., Rao, R. S., & Majumdar, B. (2014). Chromosomal and multifactorial genetic disorders with oral manifestations. *Journal of International Oral Health, 6*, 118–125.
Pengelly, R. J., et al. (2015). Resolving clinical diagnoses for syndromic cleft lip and/or palate phenotypes using whole-exome sequencing. *Clinical Genetics, 88*, 441–449.
Rechitsky, S., et al. (2015). First systematic experience of preimplantation genetic diagnosis for single-gene disorders, and/or preimplantation human leukocyte antigen typing, combined with 24-chromosome aneuploidy testing. *Fertility and Sterility, 103*, 503–512.
Reddy, K. S., & Mak, L. (2001). Mosaic unbalanced structural abnormalities confirmed using FISH on buccal mucosal cells. *Annales de Génétique, 44*, 37–40.
Robin, N. H., Moran, R. T., & Ala-Kokko, L. (1993–2015). Stickler syndrome. 2000 Jun 9 [Updated 2014 Nov 26]. In R. A. Pagon, M. P. Adam, & H. H. Ardinger, et al.(Eds.), *GeneReviews® [Internet]*. Seattle (WA): University of Washington, Seattle. Available from: http://www.ncbi.nlm.nih.gov/books/NBK1302/.
Saiki, R. K., et al. (1985). Enzymatic amplification of beta-globin genomic sequences and restriction site analysis for diagnosis of sickle cell anemia. *Science, 230*, 1350–1354.
Vincent, M., et al. (2015). Treacher Collins syndrome: A clinical and molecular study based on a large series of patients. *Genetics in Medicine*. Advanced online publication. http://dx.doi.org/10.1038/gim.2015.29.
Wilkie, A. O., et al. (2010). Prevalence and complication of single-gene and chromosomal disorders in craniosynostosis. *Pediatrics, 126*, e391–e400.
Willems, P. J., et al. (2014). The first 3,000 Non-Invasive Prenatal Tests (NIPT) with the Harmony test in Belgium and the Netherlands. *Facts, Views & Vision in ObGyn, 6*, 7–12.
Xie, C., & Tammi, M. T. (2009). CNV-seq, a new method to detect copy number variation using high-throughput sequencing. *BMC Bioinformatics, 10*, 80.
Zhang, X. (2014). Exome sequencing greatly expedites the progressive research of Mendelian diseases. *Frontiers of Medicine, 8*, 42–57.

CHAPTER TWENTY

Morphometrics, 3D Imaging, and Craniofacial Development

Benedikt Hallgrimsson*[,1], Christopher J. Percival*, Rebecca Green*, Nathan M. Young[†], Washington Mio[‡], Ralph Marcucio[†]

*Department of Cell Biology and Anatomy, Alberta Children's Hospital Research Institute, and McCaig Bone and Joint Institute, University of Calgary, Calgary, Alberta, Canada

[†]Department of Orthopaedic Surgery, San Francisco General Hospital, Orthopaedic Trauma Institute, University of California San Francisco, San Francisco, California, USA

[‡]Department of Mathematics, Florida State University, Tallahassee, Florida, USA

[1]Corresponding author: e-mail address: bhallgri@ucalgary.ca

Contents

Abstract

Recent studies have shown how volumetric imaging and morphometrics can add significantly to our understanding of morphogenesis, the developmental basis for variation, and the etiology of structural birth defects. On the other hand, the complex questions and diverse imaging data in developmental biology present morphometrics with more complex challenges than applications in virtually any other field. Meeting these challenges is necessary in order to understand the mechanistic basis for variation

Current Topics in Developmental Biology, Volume 115
ISSN 0070-2153
http://dx.doi.org/10.1016/bs.ctdb.2015.09.003

in complex morphologies. This chapter reviews the methods and theory that enable the application of modern landmark-based morphometrics to developmental biology and craniofacial development, in particular. We discuss the theoretical foundations of morphometrics as applied to development and review the basic approaches to the quantification of morphology. Focusing on geometric morphometrics, we discuss the principal statistical methods for quantifying and comparing morphological variation and covariation structure within and among groups. Finally, we discuss the future directions for morphometrics in developmental biology that will be required for approaches that enable quantitative integration across the genotype–phenotype map.

1. INTRODUCTION

Answering the question of how developmental mechanisms result in morphogenesis is a key goal of developmental biology. The study of molecular mechanisms underlying morphogenesis has become increasingly quantitative and integrative with the development of genomic technologies. Yet, the quantitative study of morphology has been fairly peripheral to the mechanistic study of morphogenesis (Hallgrímsson, Boughner, Turinsky, Logan, & Sensen, 2009). In recent years, quantitative analysis of morphology has become more prevalent in the study of morphogenesis. New imaging techniques provide increasingly accessible and higher throughput imaging of embryonic morphology. Combined with novel quantitative approaches, this allows for analyses and visualizations that are increasingly intuitive and accessible to developmental biologists and geneticists. Further, studies of the developmental basis for morphogenesis involve increasingly complex analyses of multiple genetic factors or treatments. Finally, there is increased interest in the relationship between the determinants of normal variation and the genetics of structural birth defects, especially in the field of craniofacial biology (Cooper & Albertson, 2008; Hallgrímsson et al., 2009; Heuze et al., 2014; Houle, 2010; Houle, Govindaraju, & Omholt, 2010; Young, Chong, Hu, Hallgrimsson, & Marcucio, 2010). Such studies demand more refined quantitative phenotypic assessment than might have sufficed in the past. This chapter reviews the use of 3D imaging and morphometrics for the study of craniofacial development.

Developmental biology is focused on revealing the processes and the interactions among processes that result in embryogenesis and growth (Love, 2014). Processes and mechanisms underlying organismal development are the focus of study for developmental biologists, and the field assembles approaches and techniques that tackle this basic question and its various components (Burian, 2005). As such, processes and mechanisms are the focus of study for developmental biologists. Although morphology is, ultimately, the phenomenon that the field seeks to explain, it is far distant from most developmental biology research that seeks to reveal molecular and cellular mechanisms. Accordingly, phenotypic variation often takes the role of a predicted outcome of an experimental perturbation, and an observation showing morphological difference represents an abnormality that points toward a mechanism of interest. For these reasons, developmental biologists have not generally been interested in phenotypic variation *per se*. Phenotypic outcomes are usually seen as discrete. Mutants are described as *having a phenotype*, and variation in experimental outcomes at the phenotypic level is more often a nuisance than an object of study as it is thought to obscure the phenotype of interest.

In contrast, phenotypic variation has a central epistemological position within evolutionary biology. Although its coherence can be debated, evolutionary biology has a unifying theory (Sober, 1994), within which selection acts on phenotypic variation and evolution occurs through particulate inheritance of genes. Therefore, the quantitative study of morphological variation has been much more prominent within evolutionary biology than within developmental biology (Bookstein et al., 1985). Much of the theory of geometric morphometrics (GMM), for example, was developed to answer evolutionary questions. Within this chapter, we discuss some of the ways that the resulting morphometric concepts and methods can be applied to answer questions of development. However, development also presents huge challenges to morphometrics. Chief among these is the need to integrate quantification across levels from genetic to cellular to morphological. In particular, the need to quantitatively integrate molecular and morphological imaging is a major challenge. Current efforts are being made to meet these challenges. Further, the quantification of morphology across developmental stages often spans ranges of morphological variation that challenge existing techniques, particularly when some attempt is made to identify and preserve homologies, defined as the biological correspondence of structures or location across individuals and ontogeny. Finally, innovation in imaging techniques has led to increasingly rich image datasets that challenge both existing theory and computational power. We discuss these challenges, their potential solutions, and progress on their development (below).

2. MORPHOMETRICS AND MORPHOSPACES

Morphometrics is the quantification and statistical analysis of form. Form is the combination of size and shape of a geometric object in an arbitrary orientation and location. Shape is what remains of the geometry of such an object once you standardize for size. Note that in morphometrics, unlike in common parlance, form and shape do not mean the same thing.

A key concept underlying much of the quantitative analysis of form is the idea that morphology can be mapped in a systematic way, often within a "morphospace." Morphospaces are maps showing how shapes are defined by quantitative traits. Although the basic idea arguably derives from the deformation grids of Thompson (1942/1961, 1917), the idea of a quantitative morphospace was first operationalized by Raup (1966). While the concept of a phenotypic distance between specimens or groups of specimens, based on the measures of multiple traits, was used in numerical taxonomy at that time (Sneath & Sokal, 1973; Sokal & Rohlf, 1995b), Raup created the visual metaphor of the morphospace. He created a model composed of three parameters that could be used to describe the potential diversity of shell form (Fig. 1A), his phenotype of interest. He then plotted the evolutionary diversity of coiled shell animals within the three-dimensional morphospace defined by these parameters and found that large areas of potential morphospace were not occupied. This raised the interesting question of why there are no naturally occurring shells that represent all possible regions of morphospace. The proposed explanations included shell function and constraints imposed by developmental processes. In this way, Raup's morphospace concept was foundational for evolutionary developmental biology because early concepts of developmental constraint were used to explain gaps in the space of possible morphologies (Alberch, 1982; Maynard Smith et al., 1985).

Raup's elegantly simple morphospace provides a surprisingly complete description of shell coiling, but it belies the arbitrariness of such spaces more generally. Morphospaces are defined by their axes. Often such axes are multivariate summaries of form data as in Fig. 1B and C. In very few circumstances, it is possible to construct a morphospace based on developmental parameters, such as the expression or diffusion gradients of particular morphogens, because developmental systems are rarely understood at that level of granularity. This means that while it is possible to plot measured morphology and quantify morphological variation in accordance with the axes

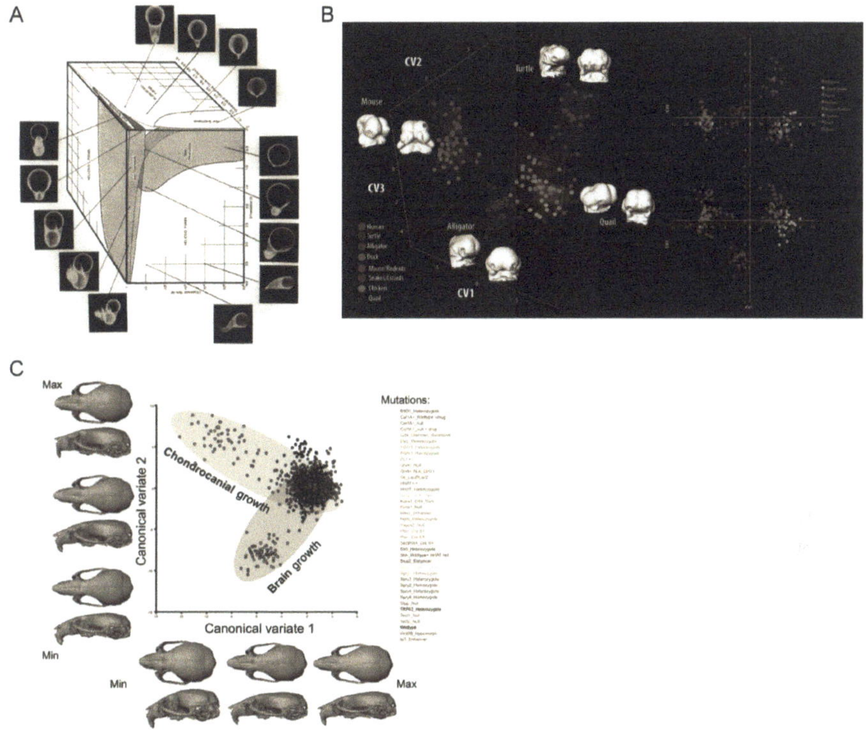

Figure 1 (A) Raup's morphospace for shell coiling (Raup, 1966). (B) Morphology for amniote embryos, including humans, is shown in a space constructed using principal components analysis from 3D landmark data (Young et al., 2014). (C) Analysis of mouse mutants with in a canonical variates analysis-based space. This plots shows craniofacial effects for mutations affecting chondrocranial growth and brain size forming two distinct common axes of covariation. (A color version of this figure is located at the end of this chapter.)

(i.e., chosen parameters) of a morphospace, it is rarely possible to capture the all options for potential variation within a plot or single shape space. Depending on how a morphospace is defined, developmental changes can occur in ways that are not captured. Further, for axes defined on the basis of covariation structure, it must be borne in mind that developmental perturbations can alter covariation structure in ways that can be hard to predict (Hallgrimsson et al., 2009; Jamniczky & Hallgrimsson, 2009; Mitteroecker, 2009). Morphospaces are generally arbitrary with respect to development, as they are not usually defined by direct developmental parameters. Morphospaces are, nonetheless, essential tools in that they can be used to quantitatively compare and situate morphologies relative to each other. As phenomics gains traction and intersects more meaningfully with developmental biology, careful thought will need to be devoted to the construction of the morphospaces within which phenomic analyses are conducted.

As with all morphometric tools, care must be taken to interpret morphospaces within the context in which they are constructed. To be useful, morphospaces must have a few key properties. The first is that relative locations and distances in such spaces must have biological meaning (Mitteroecker & Huttegger, 2009). This means that forms that are similar should cluster together and those that are dissimilar should be far apart and that the distances among them should be proportional to some underlying biological difference. This is critical if a measure of dispersion of a sample within a space is to have any relation to the distance between the mean of two samples or if the distances among various mutant phenotypes are to be compared. Second, directions within the morphospace should have biological meaning. Without biological meaning, it becomes impossible to use the morphospace to predict morphologies based on a continuous relationship or determine whether a group of related mutations or treatments produce effects in the same direction. Closely related to this is the requirement of colinearity, meaning that parallel trajectories in morphospace should represent comparable axes of phenotypic change. For example, in two samples that differ in mean shape, one should be able to compare the effect of a continuous variable such as age, size, or some treatment under the expectation that these effects would produce a displacement in the same direction for both shapes even though they are at different locations in morphospace. Finally, the axes of a morphospace should be independent and consist of commensurate units. For a more fulsome discussion of these properties, see Mitteroecker and Huttegger (2009).

Often, however, morphospaces fail to meet these criteria (Mitteroecker & Huttegger, 2009), making the data contained within them difficult to interpret. Raup's shell coiling space, for instance, uses the parameters of a mathematical model of coiling to create the axes of the space. The problem with this is that there is not a natural scale relationship among the axes. Further, the axes are not all independent as some involve overlapping sets of input parameters. The lack of a natural scale means that angles or directions in Raup's morphospace are arbitrary. When two phenotypic measures are dependent on one another, then using both measures can lead to a morphospace within which there are regions that cannot be occupied simply because some combination of values is not possible, rather than because of some biological, developmental, or evolutionary constraint. Spaces defined by ratios among traits such as limb proportions or size ratios within structures often produce such artifacts.

3. APPROACHES TO MORPHOMETRICS

There are several approaches to doing morphometrics. All of these approaches have advantages and limitations. There is no single correct approach to morphometrics that applies to all problems. Approaches are more or less correct or appropriate only in the context of the question being asked. For this reason, it is critical to understand what different morphometric techniques do and their respective limitations. Much effort has been devoted in morphometrics to debates over the use of one approach or another. Too often such debates are framed within the assumption that there is a single correct approach to the quantification of form. Under this assumption, such debates have sometimes taken on a religious fervor. However, the quantification of form is not an end in itself. Rather, it is a means of answering biological questions. All methods will result in something less than a perfect quantification of the phenomena of interest. One must always ask whether the method used generates useful quantifications and there are occasions when using more than one approach and interpreting the differences among them leads to greater insight into the biological question at hand.

In the age of omics, debates over methods have practical implications because of the increasing need to compare and integrate data or results across studies. This is a good reason to favor some methods over others. On the other hand, maintaining a diversity of approaches in a field can lead to innovation that might otherwise be missed. Balancing these two considerations is important both for reviewers of grants and papers and those who design "omic" resources involving morphometric data.

3.1 Landmark-Based Methods

Traditional morphometrics relied on the phenotypic measurements such as linear distances, angles, weights, and areas. Most modern morphometric approaches, however, are based on analysis of landmarks. Not all landmarks are equally useful. Bookstein classified landmarks into three types based on their information content and utility (Bookstein, 1991). Type 1 landmarks represent discrete identifiable points that usually occur at the intersection of distinct anatomical structures. Intersections of sutures in bones, foramina, and vascular branching points are examples of such landmarks. Type 2 landmarks represent points of maximal curvature along definable features. The tip of facial prominence or the inflection point of a cleft would be examples of such points. Type 3 landmarks are defined along extremes that are often defined by other points. These types of landmarks contain progressively less biological information. Type 3 landmarks, specifically, are deficient because they rely on information from other landmarks.

Semi-landmarks are a special case of Type 3 landmarks that are constructed by distributing points across a surface defined by other landmarks (Gunz, Mitteroecker, & Bookstein, 2005). Usually, semi-landmarks are placed in an equidistant grid across a surface and then they are "slid" to optimize their position relative to the average shape for the sample. This sliding step is important as it places the landmarks in positions where they correspond better to each other across individuals (Gunz & Mitteroecker, 2013). Skipping this step results in artifacts caused by the arbitrary assumption of equidistance among landmarks (Gunz & Mitteroecker, 2013).

Mention that some methods for dense landmark quantification of surfaces are semi-landmark techniques taken to the extreme?

3.2 Geometric Morphometrics

GMM is the most established body of morphometric theory for handling landmark-based data. Paraphrasing Rohlf's definition of the field (http://life.bio.sunysb.edu/morph/), GMM is the multivariate statistical analysis of form as represented by Cartesian coordinate data. There are many excellent introductions to GMM (Mitteroecker & Gunz, 2009) and its use in developmental biology (Cooper & Albertson, 2008; Hallgrímsson et al., 2009). We will not attempt a thorough review here. GMM methods rely on superimposition of landmark coordinate data in order to place individuals into a common morphospace. In the earliest example of this, Bookstein proposed using two fixed landmarks to register a larger landmark set. After scaling and superimposition of these two landmarks, the locations of the remaining landmarks represent variation in shape between specimens (Bookstein, 1986, 1991). Currently, the most commonly used form of superimposition is the generalized Procrustes analysis (GPA). (Mitteroecker & Gunz, 2009). Developed in its more general form for multivariate statistics by Gower (1975), GPA was adapted to morphometric data by Rohlf and Slice (Rohlf & Bookstein, 1990; Rohlf & Slice, 1990).

GPA places multiple individual specimens into the same shape space by scaling, translating, and rotating the landmark coordinates (Fig. 2). To scale

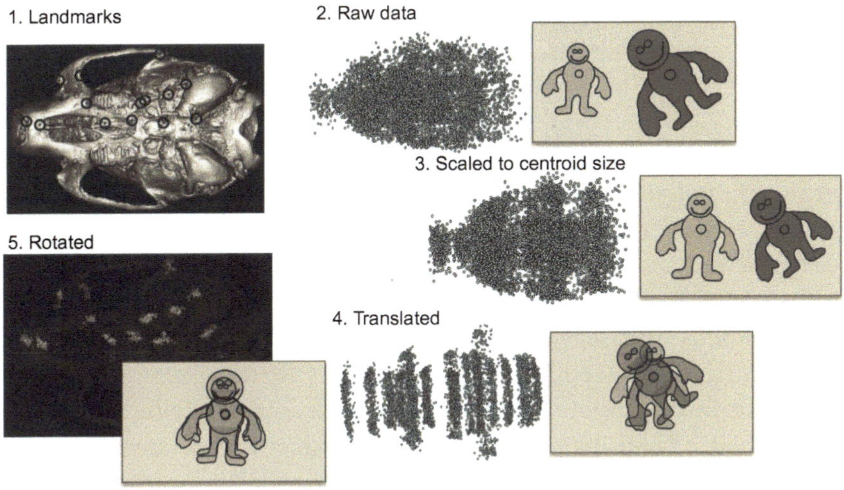

Figure 2 General procrustes superimposition. (1) A 2D view of a 3D adult mouse landmark set. (2) The full scatter for the raw landmarks in this dataset. In this particular datasets, the two sides of each specimen are treated as separate individuals. (3) The scatter for all individuals (left and right sides) after scaling the LM coordinates for centroid size. (4) The scatter after translation so as to center the centroids of each individual. (5). The distribution of Procrustes residuals after rotation and, where appropriate, reflection superimposed on a ventral view of a mouse. (A color version of this figure is located at the end of this chapter.)

the landmark coordinates, GPA uses the centroid size (CS), a measure of scale representing the overall spread of a specimen's landmark configuration from its centroid. CS is calculated as the square root of the summed squared deviations from the center (arithmetic mean) of each landmark configuration (Rohlf & Slice, 1990). The arithmetic mean of each landmark configuration is the centroid. Each specimen's landmark configuration is then translated so they share a centroid. Finally, each landmark configuration is rotated around the common centroid such that the deviation of each landmark from the average position for that landmark is minimized. This involves an iterative process. In the first step, the configurations are fitted to an arbitrary rotational angle for which each specimen is rotated to the angle that minimizes the sum of the squared distances from each landmark to its mean. From this first least-squares fit step, an average landmark configuration is generated and the least-squares fit is repeated. The GPA process does this repeatedly until further steps produce no improvement in the fit (Dryden & Mardia, 1998; Mitteroecker & Gunz, 2009; Rohlf & Slice, 1990). The result of the superimposition process is a set of modified landmark coordinates, termed Procrustes coordinates, for each specimen. These coordinates preserve the geometric relationship among a specimen's landmarks and express each specimen as a set of departures from each landmark mean. Preservation of geometry throughout the analysis generates much of the power of geometric morphometric methods.

Although Procrustes superimposed data correspond to a definable morphospace, that space is not linear (Kendall, 1984). For this reason, analyses are often conducted on data in which the Procrustes coordinates have been projected to a tangent space that is approximately linear (Dryden & Mardia, 1998; Rohlf, 1999). Most commonly, Procrustes coordinates are transformed to partial warp scores. Partial warp scores are based on Bookstein's thin-plate spline method which uses a bending energy metaphor to calculate the continuous deformation of surfaces (Bookstein, 1989, 1991). This method is fundamental in computer graphics and image informatics, and is widely used in engineering. Whether to use Procrustes coordinates or partial warp scores does not matter for most morphometric analyses as they tend to involve fairly small amounts of shape variation (Rohlf, 1999). In developmental biology, however, analyses often span large shape distances produced by morphogenesis where this choice may well matter for the results of an analysis. In general, it is critical for users of morphometric methods to know and understand the precise form of superimposition to which their data has been subjected, particularly when using comprehensive morphometric analysis software instead of custom-made programs.

There are several important advantages of geometric morphometric methods over more traditional morphometrics that rely principally on linear distances, ratios, or angles. The fact that GMM methods preserve geometry throughout an analysis means that variation within and between groups or the variation that corresponds to covariates of interest can be visualized as displacements of individual landmarks. These displacements can be used to generate 2- or 3-dimensional deformations of an anatomical object, showing how the object changes across the morphospace. Figure 3 shows examples of such analyses applied to microCT scan data for mouse embryos. Here, wireframes show the 3D deformations that correspond to ontogenetic variation in craniofacial shape. A related advantage is that the Procrustes superimposition-based methods generate useful morphospaces in which multiple groups can be meaningfully compared. Such spaces have all of the desired properties listed above (Mitteroecker & Huttegger, 2009).

Variation in size and shape can be successfully quantified using GMM methods because they are based on a rigorous definition of shape

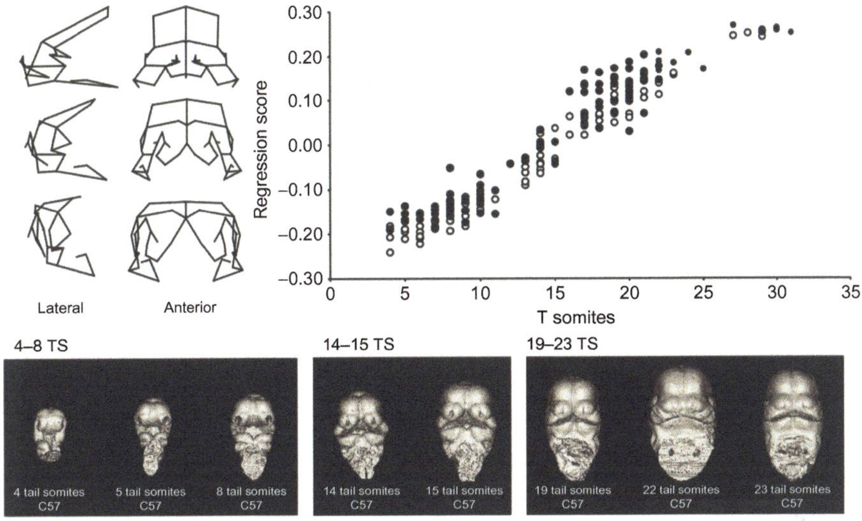

Figure 3 Multivariate regression on Procrustes superimposed landmark data allows calculation of the shape changes that correspond to that regression. Here, tail somite stage is regressed on landmark coordinates that capture craniofacial shape variation during face formation. The wireframes show the deformations that correspond to the regression in frontal and lateral views, while the scatter plot shows the regression scores plotted against tail somite stage.

(Kendall's shape space and its linearizations). Shape variation can be decomposed into components that relate to biological factors of interest and that can be visualized as deformations of landmark configurations and surfaces, a key advantage in terms of biological relevance. Figure 3B shows an example of this in which the shape variation related to stage and to size are quantified separately within a sample of mouse embryos and this is explained in more detail in the next section.

As with all good methods, GMM approaches also have limitations. The most important of these is that superimposition distributes some fraction of variation across landmarks (Marquez, Cabeen, Woods, & Houle, 2012). This is known as the Pinocchio effect (Fig. 4). The Pinocchio effect is simply a consequence of the GMM definition of shape. Once Pinocchio's nose has elongated, the shape of his profile has changed and those changes are accurately described by the small differences in landmark positions across the head as well as the point of the nose even if only the nose has actually changed. However, as Klingenberg (2013) points out, from a GMM perspective there is no *a priori* reason why this difference in shape should be described in

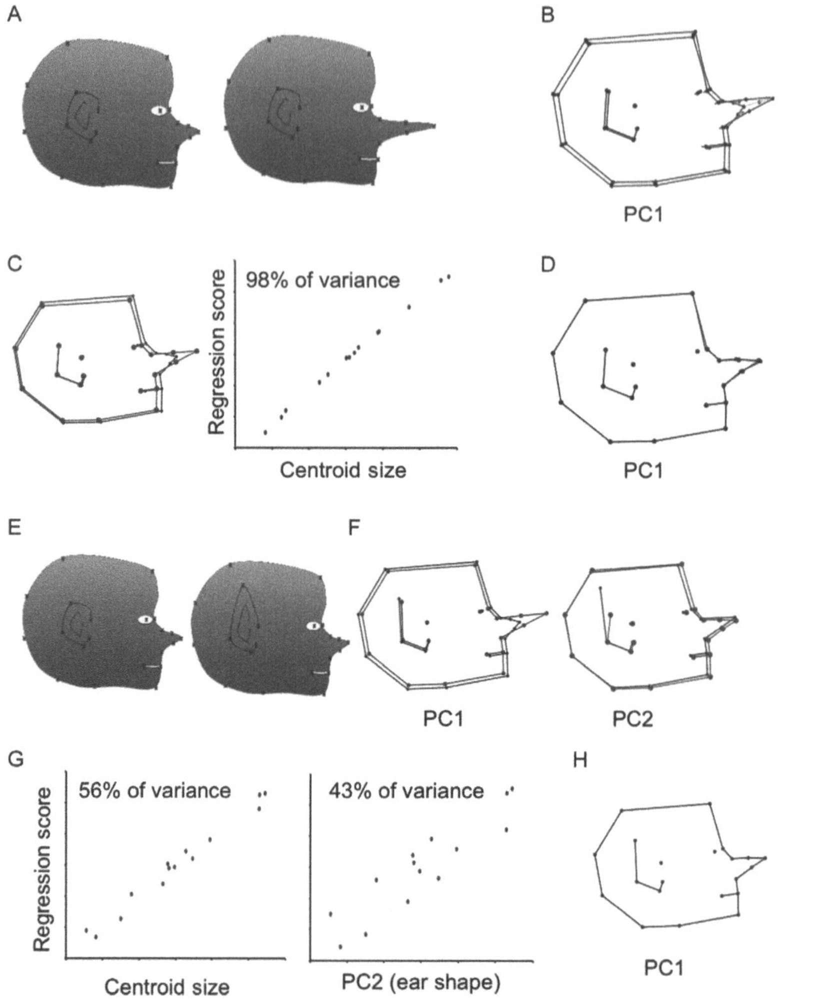

Figure 4 Illustration of the Pinocchio effect. In (A), a dataset is generated in which a profile of Pinocchio's head varies only in that the nose gets longer. This only displaces the three landmarks on the nose. However, a PCA based on a Procrustes fit of those data shows small displacements of landmarks that were not actually displaced although the largest displacements involve the nose (B). However, the increase in nose length is perfectly correlated with centroid size and so regressing on CS completely removes both the nose length variation and its associated artifacts (C, D). In (D), shape is standardized to the sample average and so shows an intermediate length nose. If we add an additional shape component, ear shape (E) that is uncorrelated with nose length, the Pinocchio effect is complicated slightly. Here, variation is spread across two PCs (F), one which captures nose length and another that captures ear length. These effects and all their associated changes can be removed from the data as shown in (G) and (H), either by regressing out centroid size and PC2 or PC1 and PC2. The artifactual shape changes, however, are preserved in the shape variation associated with these regressions. (A color version of this figure is located at the end of this chapter.)

terms of the nose moving forward rather than the head moving backward. The profile has simple changed in shape. Further, as the Pinocchio's nose grows, CS increases. In a longitudinal analysis of Pinocchio profiles that capture the growth of the nose, the combination of CS and shape using GMM does accurately portray what is happening to Pinocchio (Fig. 4). Even if you add a second localized shape change such as elongation of the ears, it is still theoretically possible to quantify and remove the artifactual variation. This can be hard in practice, however. The Pinocchio effect does not mean that Procrustes superimposed data contain no information about local variation. However, the fact that variation gets distributed to some degree across landmarks means that care must be taken to ensure that what appears to be local variation is not driven by variation somewhere else in a landmark configuration, particularly when dealing with subtle changes in shape.

Various methods have been proposed to minimize the distorting effects of variation at particular landmarks. Rohlf and Slice (1990), for example, proposed a resistant fit superimposition for such cases, while Marquez et al. (2012) suggest an interpolation-based method. There is, however, little agreement on the use of such methods (Klingenberg, 2013) and this remains an active area of investigation (Marquez et al., 2012). Although the choice of superimposition method may in practice default to GPA (Richtsmeier, 2005), this need not be so. If one suspects that particular landmarks are introducing large amounts of variation that obscures other signals of interest, it is perfectly reasonable to try an alternative superimposition, such as the resistant fit version (Rohlf & Slice, 1990), to determine whether this significantly alters the results of an analysis.

Scaling is necessary for superimposition of each specimen's landmark coordinates into the same shape space. This makes direct comparisons of form across ontogeny impossible. Since variation in size and the associated allometric (age-shape) changes are critical components of phenotypic change across development, researchers may find it useful to add CS or a component of age to a GMM analysis as a covariate (refer to size-shape space analysis and geomorph) or to use a different set of morphometric methods. In general, size and size-related change in shape are important components of any analysis. These are rarely features of shape variation that should simply be removed. Rather, these aspects of form variation need to be quantified carefully and analyzed either separately or conjointly with shape variation (see below).

3.3 Euclidean Distance Matrix Analysis

The other major body of landmark-based morphometric methods is Euclidean distance matrix analysis or EDMA (Lele & Richtsmeier, 2001). EDMA is a method of morphometric analysis that avoids superimposition and its drawbacks altogether. Instead of comparing variation of superimposed landmark coordinates, EDMA techniques first represent each specimen as a matrix of linear distances between all possible pairs of landmarks (Lele & Richtsmeier, 1991), which can be scaled by a chosen scaling factor (Lele & Cole, 1996). Morphological differences between groups can be pinpointed to specific linear distances on an object through pairwise comparisons of mean form or shape matrices, followed by bootstrapping to estimate the significance of these differences (Fig. 5) (Lele & Richtsmeier, 1991). The ability to conclusively identify specific regions of an object that differ between two samples is not always possible with commonly used multivariate data exploration techniques, including principal components analysis (PCA), where a given landmark might be highly weighted on multiple principal components (PCs) and each PC is associated with many moderately weighted landmarks of potential interest. Additionally, scaling phenotypic values by a measure of size is not automatically performed for EDMA analysis, allowing for comparisons of ontogenetic growth trajectories

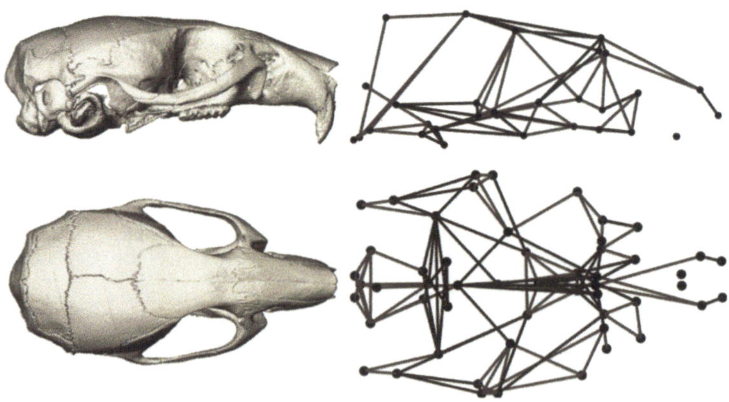

Figure 5 The strongest (>5% difference in means) significantly ($\alpha = 0.05$) different linear dimensions identified in a EDMA SHAPE comparison of skulls from 129 and C57 mouse strains, viewed from the lateral (top) and superior (bottom) aspects. Measurements were standardized by centroid size before analysis. Red lines are dimensions for which C57 is relatively longer than 129, while blue lines are relatively longer in 129 than C57. (A color version of this figure is located at the end of this chapter.)

between groups (Richtsmeier, Cheverud, Danahey, Corner, & Lele, 1993; Richtsmeier, Corner, Grausz, Cheverud, & Danahey, 1993), without having to reintroduce a previously removed measure of scale. Growth analysis methods can also be applied to other situations, including comparisons of the morphological effects of a specific mutation on two different genetic backgrounds.

The major advantage of EDMA is that it avoids the superimposition step yet allows analysis of morphological variation throughout a morphologically complex object. This is important if an analysis depends critically on the ability to precisely localize variation within a structure. The major disadvantage of EDMA is that since it is based on the analysis of linear distances, it does not deal overtly with shape. Methods exist within EDMA to quantify allometry and ontogenetic change in form but they are more difficult to implement within a complex multivariate analysis. Further development of EDMA methods and software may well address these limitations. It is not necessarily the case that one must make a choice between using EDMA or GMM. Similar results from multiple, relatively independent methods of analysis can provide stronger support for subsequent interpretations and conclusions (Martínez-Abadías et al., 2010). Although not currently as popular as GMM methods, EDMA methods continue to be adapted to answer new research questions (Hill et al., 2013; Motch Perrine et al., 2014).

3.4 Image Analysis-Based Methods

Finally, there is a growing body of methods that are designed to extract morphometric information directly from 2D and 3D image data, without the use of sparse landmarks. Some rely on automated landmarking of a few key features, then evenly distributing landmarks across the remainder of the surface (Heimann & Meinzer, 2009). Others rely on analysis of deformation data on a voxel-by-voxel basis (Chakravarty et al., 2011; Heimann & Meinzer, 2009; Joshi, Srivastava, & Mio, 2005; Srivastava, Joshi, Mio, & Liu, 2005; Wong, Maezawa, Lerch, & Henkelman, 2014). Most applications of such methods have been in neuroimaging but there are examples of applications to craniofacial morphology (Chakravarty et al., 2011; Chinthapalli et al., 2012; Hopman, Merks, Suttie, Hennekam, & Hammond, 2014) and embryonic development (Kristensen, Parsons, Hallgrimsson, & Boyd, 2008). For voxel-based methods, thresholding and registration can produce a large amount of error that must be minimized or considered in analyses (Bookstein, 2001). Voxel-based and dense landmark-based methods involve a superimposition step of some sort and so share the pitfalls of GMM approaches. On the other hand, statistical shape models and voxel-based morphometry make use of much more of the image data that is contained in a 2D (e.g., photograph) or 3D image (e.g., microCT or optical projection tomography image) than approaches based on sparse landmark sets. With further methods development, such approaches may be adaptable to the full range of applications currently available through GMM methods.

4. LANDMARKS AND HOMOLOGY

An assumption of most morphometric methods is that landmarks are homologous across individuals, samples, and within individuals across ages or stages in the case of longitudinal data (Bookstein, 1991; Oxnard & O'Higgins, 2009). A particular landmark should effectively correspond to the same point across these situations, but what does landmark homology mean in practice? Leaving aside the debate over homology as representations of discrete, biologically defined, developmental entities versus resemblance caused by continuity of information (Hall, 2007; Jamniczky, 2008; Roth, 1984, 1991; Van Valen, 1982; Wagner, 2007), homology is a difficult problem for morphometrics (Klingenberg, 2008). This is particularly the case for

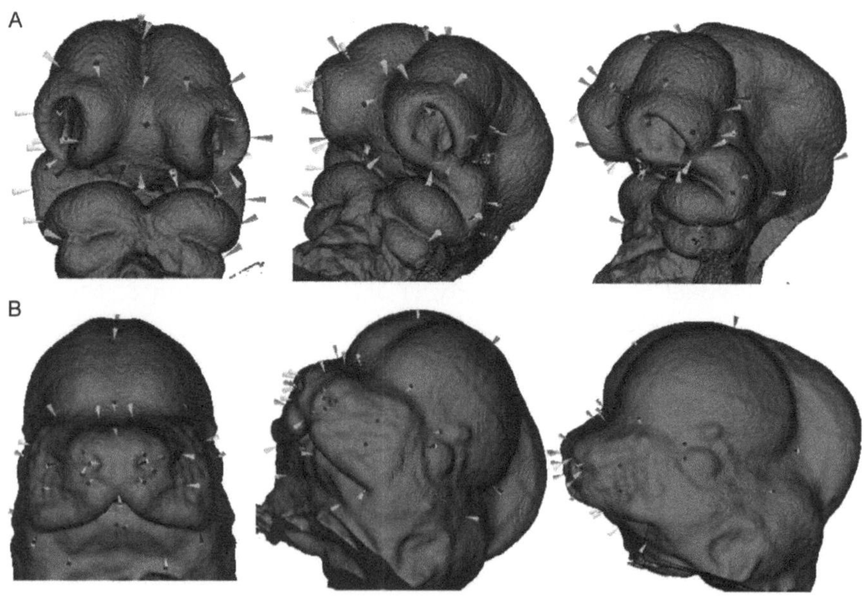

Figure 6 Standard embryo craniofacial landmark set (Percival et al., 2014). (A color version of this figure is located at the end of this chapter.)

applications of morphometrics to morphogenesis in embryos (Percival, Green, Marcucio, & Hallgrimsson, 2014). Figure 6 shows a standard landmark set that we use for mouse embryonic craniofacial morphology from E10 to E12.5. While these landmarks are chosen with care, the meaning of homology even within this ontogenetic range is sometimes unclear. The point defined at the caudal extreme of the cleft between the maxillary prominence and the lateral nasal prominence is a good example (Fig. 7). This location gives rise eventually to lacrimal duct and so it should be a good homologous point across stages. Yet, as morphogenesis proceeds, this initially obvious anatomical feature moves in relation to the position of the eye, while descendants of the cells found at this landmark at E9.5 are probably buried as the prominences fuse and the whisker field develops by E12.5. Since our landmarking scheme is intended to track the growth and movement of the facial prominences, we define this landmark such that it remains at the border between cells derived from the maxillary and lateral nasal prominences rather than maintaining a standard geometric relationship with other cranial features like the eye (Percival et al., 2014). In this case, we have made a conscious choice to attempt to follow the specific cell population rather a point defined by its geometric relationship with other features. Unfortunately, this may not always be possible, particularly across a wide range of developmental stages.

The homology issue becomes much more difficult when one needs to assemble samples across wider ranges of developmental stages. Here there are two issues. The first is that structures come and go. For the quantitative study of face formation, we have used soft tissue landmarks that are mostly

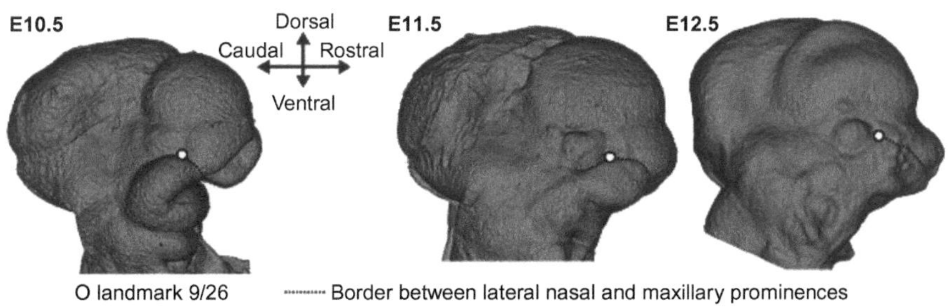

Figure 7 Right lateral view of the surface craniofacial morphology of E10.5, E11.5, and E12.5 mouse specimens in a standard orientation. As previously defined (Percival et al., 2014), the homologous location of a landmark between the maxillary and the lateral nasal prominences is shown. We defined this landmark such that it remains at the border between these cell populations rather than maintaining a standard geometric relationship with other cranial features like the eye.

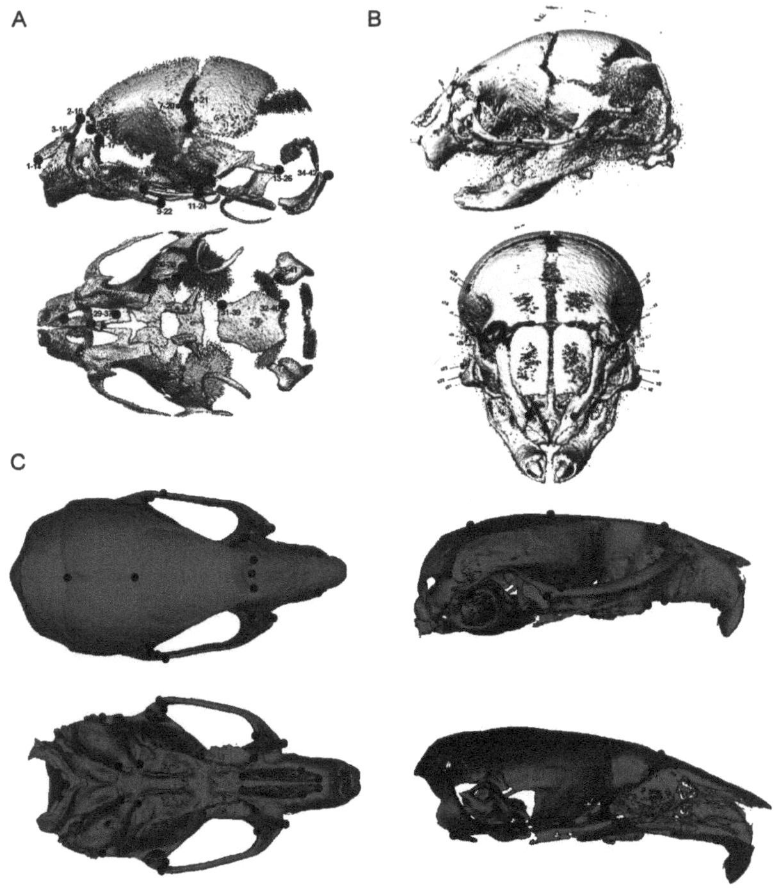

Figure 8 (A) Late fetal (E17.5) (Gonzalez, Lotto, & Hallgrimsson, 2014), (B) neonatal (Boughner et al., 2008), and (C) adult mouse landmark sets used in recent studies by our group. (A color version of this figure is located at the end of this chapter.)

located on the epithelial surface. For later fetal craniofacial development, while it is possible to rely on surface features, one is usually interested in other structures such as the skeletal features that are forming at that time. For analysis of postnatal morphology, one may be interested in features of the skull or the brain. Combining such analyses in a coherent way is an unresolved challenge. In practice, we have approached this issue by defining standard landmark sets that are particular to a range of stages and structures. Figure 8 shows standard craniofacial landmark sets for late-stage fetuses, neonates, and adult mice.

These issues underline the importance of taking care in the definition of landmark sets. Poorly or improperly defined landmarks can lead to false conclusions. Phenomic analyses that rely on large datasets are particularly vulnerable to drift in landmark position. Reliance on manual landmarking is a significant barrier to the use of image data for phenomic analyses. Manual landmarks vary among and can drift over time for landmarkers. Landmarking protocols are also exceedingly difficult to standardize across labs (Percival et al., 2014). For this reason, development of automated methods that can handle the complexity of real datasets is crucial. Methods exist for 3D landmarking from image data (Brett & Taylor, 1999; Douglas, 2004) and there are examples of their application in particular contexts (Chakravarty et al., 2011; Subburaj, Ravi, & Agarwal, 2009). Due to the complexity and noisiness of most image datasets, however, the majority of morphometric work still relies heavily on manual landmarking.

5. QUANTIFYING VARIATION

5.1 Comparing Shape and Size Among Groups

Morphometric analyses that address questions in developmental biology seek to quantify variation within samples and to compare variation among samples. Morphometric data based on 3D landmarks are multivariate. The statistical techniques for describing variation within groups and comparing that across groups rely heavily on matrix algebra. A fundamental element of any morphometric analysis is the variance–covariance matrix (VCM). Such matrices consist of the set of landmark by coordinate variances along their diagonal and all of the pairwise covariances in the off-diagonal cells of the matrix. An intuitive way to express the variation in such matrices is to use a multivariate data reduction technique such as PCA. A PCA uses the VCM to create new variables that correspond to progressively smaller proportions of the total variance in the sample. Imagine a bivariate dataset in which x and y are partially correlated (Fig. 9). In this case, one could construct a new variable that captures the covariation among x and y. This would be the first PC. Every individual point in that dataset can be projected onto this variable, resulting in a set of PC scores for each individual. In PCA, the second component is defined in the direction that is orthogonal to the first component but contains the second largest amount of covariance. In a 2D dataset, there is only one such option (Fig. 9A), but in a multidimensional dataset, PC2 describes the second largest direction of variation. The variance of all phenotypic measures explained by a PC is its eigenvalue

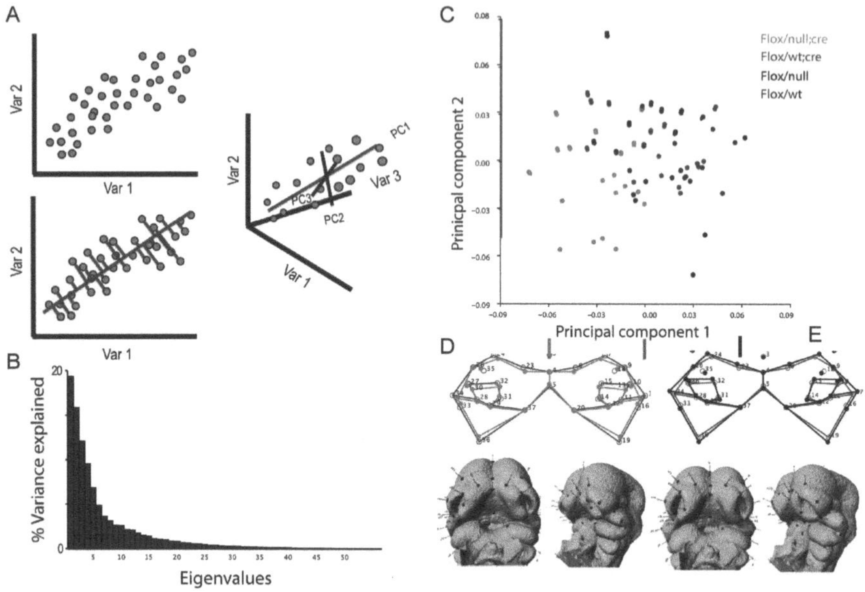

Figure 9 Principal components analysis in geometric morphometrics. (A) Explanation of PCA (from Zelditch, 2004). The 2D plots illustrate PC1 and the projection of points (individuals) onto PC1 to create PC scores. The 3D plot shows how the scatter among three variables would correspond to principal components. (B–D) Example of the use of principal components analysis to investigate a the effect of a mutation on craniofacial shape in mice (Green et al., 2015). (A color version of this figure is located at the end of this chapter.)

while its direction is its eigenvector. Within GMM, the eigenvector of each PC is its orientation with respect to the original variables. These orientations allow the visualization of the shape transformations that correspond to variation along each PC. Figure 9B–E illustrates this using PCA plots for a mouse embryo craniofacial dataset.

Although PCA is a useful way to approach a multivariate morphometric data, it is important to note that the assumption of orthogonality is one made for convenience and not because it corresponds to biology in any way. Real covariance structures can be very complex, determined by the variances of multiple interacting and overlaid developmental factors (Hallgrimsson et al., 2009; Mitteroecker, 2009). This can easily create situations where a biological effect gets distributed across multiple PCs in real datasets. The orthogonality assumption also means that every PC is uncorrelated with every other; each PC is dependent on the definition of the PCs that precede it. PCs capture axes of covariance in the VCM, but they estimate those axes with error (Zelditch, 2004). If the identity of a particular PC is important to an interpretation of biological differences, resampling approaches can be used to determine whether that PC is significantly different from its neighbors (Zelditch, 2004).

PCA plots are a useful first step to take when examining variation among groups such as genotypes or treatments. Subtle differences among groups, however, may not be readily evident on such plots or the differences may be distributed across several PCs. Canonical variates analysis (CVA) orients that data along axes that maximally distinguish groups that are defined *a priori* (Marcus, Corti, Loy, Naylor, & Slice, 1996). CVA is best used as a data exploration technique rather than as a definitive test of the hypothesis that two groups are different. For landmark data, the large numbers of variables and the covariance structure or Procrustes superimposed landmark data produce a situation where CVA may separate groups that actually have the same mean shape when the number of variables is close to the number of specimens (Mitteroecker & Gunz, 2009). To put this more simply, use of CVA is not advised when the number of specimens under study is comparable to the number of variables in the analysis (e.g., number of landmarks multiplied by number of dimensions), particularly when there are a large number of groups to separate. For the same reason, the Mahalanobis permutation test should not be used to test for differences in mean shape when sample sizes are small. In most cases, tests of significance for differences among groups should rely on permutation tests that are not affected by the ratio of variables to sample size, such as the Procrustes distance permutation test (Gunz & Mitteroecker, 2013; Gunz et al., 2005; Mitteroecker & Gunz, 2009). The Procrustes distance is also the basis for the parametric Goodall's *F*-test, which performs well under most GMM conditions (Rohlf, 2000). Finally, there are also EDMA-based tests for matrices of linear distances that can be used to test for differences in form before and after scaling for size (Lele & Richtsmeier, 1995, 2001).

For many analyses, one may wish to quantify the phenotypic variance of a sample and compare it to others. For example, mutations can affect both the phenotypic mean as well as the variance about the mean, a phenomenon known as canalization (Wagner, Booth, & Bagheri-Chaichian, 1997). For landmark data, phenotypic variance can be quantified in several ways. The distribution of shape variation at the individual level can be quantified using the Procrustes distance of each individual from the mean. The variance of these distances can also serve as a sample variance for shape (Zelditch, Lundrigan, & Garland, 2004). This measure is equivalent to the trace of the VCM, or the sum of the diagonal elements in the VCM.

5.2 Comparing Covariation Structures

Although covariation structures may seem arcane to most developmental biologists, this is actually fundamental to most advanced morphometrics. The rich datasets that we obtain from 3D images require that we apply data reduction techniques in order to extract meaning from them. Such data reductions techniques (e.g., PCA) are all based on analysis of covariance structure. To compare relationships among complex datasets, we also need to use methods that relate different covariance structures to one another. To study how brain morphology affects the face during face formation, for example, we have used such methods (Parsons et al., 2011). Similarly, if we want to relate variation in cellular dynamics to measures of morphogenesis over time, such analyses are also based on quantification and comparisons of covariance structures. For example, one might wish to determine how a mutation influences morphogenesis via spatially structured changes in gene expression, and cellular dynamics (proliferation, hypertrophy, polarity, etc.). To test this, one could obtain external shape data along with data on cellular parameters and gene expression for multiple individuals. The resulting dataset would be large and heterogeneous. Relating such datasets to each other requires sophisticated multivariate approaches, most of which rely on analysis of covariance structure.

In most datasets, PCA reveals a salient feature of morphological variation: multivariate morphological variation tends to be structured such that the majority of the variance is captured by the first few PCs. This can be visualized using a histogram of the eigenvalues from a PCA (Fig. 9B). This phenomenon, morphological integration (Olson & Miller, 1958), reflects the influence of developmental processes that vary during development and produce patterns of coordinated variation among phenotypic traits (Cheverud, 1996b; Hallgrimsson et al., 2009; Wagner, 1990). The strength of morphological integration can be compared using the tendency for variation to concentrate in the first few PCs. The more variance is biased toward the first few PCs, the higher the variance of the eigenvalues (VE) for each PC. Therefore, the VE can be used to measure the strength of integration (Wagner, 1990). Variation tends to concentrate in the first few PCs and decreases across each subsequent PC. An alteration in the rate of this decrease demonstrates integration. VE can be derived from a covariance matrix and scaled to the mean eigenvalue or simply derived from the correlation matrix. VE estimates integration more reliably than the older and more intuitive metric of simply using correlation coefficients (Fisher-z

transformed for normality), particularly in the case of heterogeneous matrices such as those we usually encounter in biology (Pavlicev, Cheverud, & Wagner, 2009). VE is dependent on the number of traits and, for this reason, Pavlicev et al. recommend using the relative eigenvalue variance (Pavlicev et al., 2009). Here, VE is scaled to the maximum eigenvalue of a correlation matrix. In some datasets, even with scaling to the mean eigenvalue, there is a persistent correlation between VE and overall variance with the absolute scale of the eigenvalues as demonstrated by resampling the VE and the sample variance (Hallgrimsson et al., 2009; Young, Wagner, & Hallgrimsson, 2010). This artifact is usually small but it is a concern when comparing samples that vary dramatically in phenotypic variance. VE can be compared among groups by resampling.

Covariation structure can be quantified and compared in a multitude of ways. See Steppan, Phillips, and Houle (2002) for a high-level review. The simplest and most direct way to compare correlation of covariance matrices is by using the matrix correlation, which is simply the correlation among the corresponding cells in two matrices. Matrix correlations can also be adjusted for matrix repeatability (Cheverud, 1996a). To determine the significance of a matrix correlation, one can use a Mantel's test which permutes the matrices by randomly shuffling their cells, calculating a matrix correlation at each iteration (Sokal & Rohlf, 1995a). Klingenberg modified the Mantel's test for landmark data by shuffling the coordinates for each landmark as a unit rather than treating them as independent cells (Klingenberg & McIntyre, 1998). Random skewers is a method that provides for a more flexible analysis of the differences and similarities in covariance structure (Cheverud, 1996a). This method has been used successfully for GMM data (Jamniczky & Hallgrimsson, 2009).

Covariation among parts of a structure defined by landmarks or among distinct structures defined by landmarks can be analyzed by means of partial least squares (Rohlf & Corti, 2000). This approach can provide evidence for interaction between structures during development. This is useful, for instance, if one suspects that a perturbation that directly affects one structure influences the development of another. This approach is preferable to multivariate approaches that have been frequently used in the past, including canonical correlation analysis (Rohlf & Corti, 2000). Importantly, the blocks of landmarks that are compared must have independent Procrustes fits so as to remove the possibility that variance from one set of landmarks affects the other. Many studies apply and develop the application of these methods to landmark data (Bastir & Rosas, 2005; Bookstein et al., 2003; Klingenberg, Badyaev, Sowry, & Beckwith, 2001; Klingenberg, Mebus, & Auffray, 2003;

Mitteroecker, 2009; Mitteroecker & Bookstein, 2007; Mitteroecker, Gunz, Neubauer, & Muller, 2012) including analysis of embryonic craniofacial structures (Parsons et al., 2011). Bookstein refers to partial least squares as singular warp analysis when applied in GMM (Bookstein et al., 2003).

5.3 Allometry—The Shape Correlates of Size and Stage

Allometry is a salient feature of the covariance structures of most complex morphologies. Traits are said to exhibit allometric variation when they do not scale isometrically to some measure of size (Gould, 1966). Shape and size are inextricably linked in most organisms. Allometry is also a prominent feature of shape variation for complex morphological traits such as craniofacial shape (Fig. 10A). For example, allometry represents roughly a third of the total shape variation in baboon skulls and in mice (Frost, Marcus, Bookstein, Reddy, & Delson, 2003; Hallgrimsson et al., 2009). In GMM, Procrustes superimposition scales the landmark sets that describe a structure to CS, leaving the shape component that is correlated with size. This component represents allometric variation or the tendency for structures of different sizes to be differently shaped. Allometry is often divided into ontogenetic, static, or evolutionary (Klingenberg & Zimmermann, 1992). Ontogenetic allometry refers to shape changes with ontogenetic stage or age. This is the most important type of allometry for most studies of development. Static allometry refers to the shape correlates of size independently of age. Evolutionary allometry refers to the shape correlates of size among species. The relationships among these three types of allometry have been a significant area of inquiry within evolutionary biology (Lande, 1979).

In some cases, most of the allometric variation is captured by the first component in a PCA (Klingenberg, 1998). In others, allometry can be more complex. In such cases, multivariate regression can be used to regress shape on CS, age, or other relevant measure (Monteiro, 1999). Figure 10B shows this in a sample of mouse embryos that vary in age and CS. Here, regression of shape on stage quantifies ontogenetic allometry, while regression of shape on CS quantifies static allometry. This analysis was based on multiple multivariate regression. As there is colinearity between the effects of stage and size, it is difficult to estimate the independent shape effects of each. These regression-based methods can be used more generally to quantify or standardize for the effect of other covariates.

Allometry can complicate comparisons among genotypes or treatments when they vary in size as well as shape. In such cases, it can be necessary to

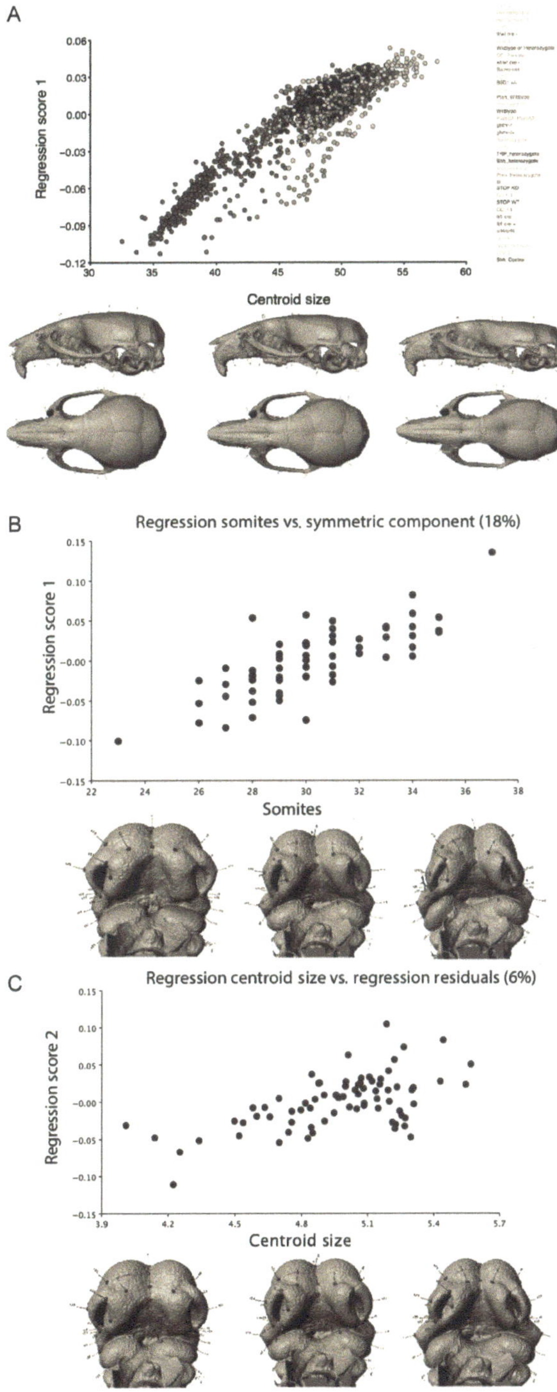

Figure 10 Allometry in adult mice and in mouse embryos (E1.0–12.5). (A) The allometric component of shape variation in the parental strains and F1 crosses for the collaborative cross mice. (B, C) The results of multiple multivariate regression of shape on both tail somite stage (B) and size (C). (B) shows the component of shape variation that is related to tail somite stage, while (C) shows the static allometry component that is perpendicular to the variation in (C). A challenge in such analyses is the colinearity among the effects of stage and size. (A color version of this figure is located at the end of this chapter.)

determine what the static allometric shape variation is independent of the treatment or mutation of interest. As an example of this, we have used mice that have reduced growth due to a mutation in the growth hormone-releasing hormone receptor to quantify the allometric component of a mutational effect (Boughner et al., 2008). This is a common problem as many mutations affect size with most resulting in a reduction in size. A further complication is that many mutations may be associated with a reduction in developmental rate. For these reasons, quantification of both ontogenetic and static allometry is a critical component of the phenotypic analysis of any mutant. A useful tool in such analyses is to do a combined analysis of shape and size, or form space, by adding CS back into a morphometric analysis (Mitteroecker & Bookstein, 2009; Smith et al., 2015).

6. 3D IMAGING FOR MORPHOMETRICS

A variety of imaging techniques exist for generating data for morphometric analysis, and morphometric techniques are broadly applicable across these techniques (Norris et al., 2013). A full review of these methods is beyond the scope of this chapter. Instead, we will briefly describe the three imaging techniques that have been most widely used to create 3D image data for embryos and describe their chief advantages and limitations.

6.1 Computed Microtomography

Computed microtomography (μCT) is the technique that has most commonly been used in the quantitative study of craniofacial morphogenesis (Billington et al., 2015; Boughner et al., 2008; Chong et al., 2012; Green et al., 2015; Hu, Young, Li, et al., 2015; Parsons et al., 2008; Percival et al., 2014; Schmidt et al., 2010; Young, Chong, et al., 2010; Young et al., 2014). Although μCT was initially used mostly for visualizing skeletal tissues, it has proven to be an effective modality for 3D volumetric imaging of embryonic soft tissues (Boughner et al., 2008; Metscher, 2009; Parsons et al., 2008; Wong et al., 2014). The major issue for μCT imaging of embryos is the distortions that are produced by fixation or desiccation (Schmidt et al., 2010). Scanning in a liquid medium is also problematic because of the low contrast between embryonic tissues and water in a CT scan. We have experimented with a variety of techniques and have found that a 4% formaldehyde + 5% biological grade glutaraldehyde-buffered fix produces scans with the least amount of fixation-related shrinkage and shape distortion (Schmidt et al., 2010) (Fig. 11). Embryos are scanned in air but

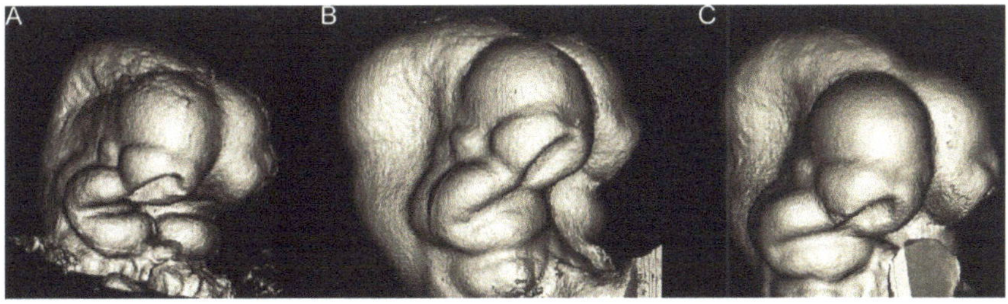

Figure 11 MicroCT scans of embryos fixed using different protocols shown at the same scale, illustrating the effect of different fixation and scanning procedures on morphology. (A) 4% Formaldehyde + Bouin's. (B) 4% Formaldehyde + 1% glutaraldehyde with iothalamate meglumine for contrast. (C) 4% Formaldehyde + 5% glutaraldehyde plus contrast agent (Schmidt et al., 2010).

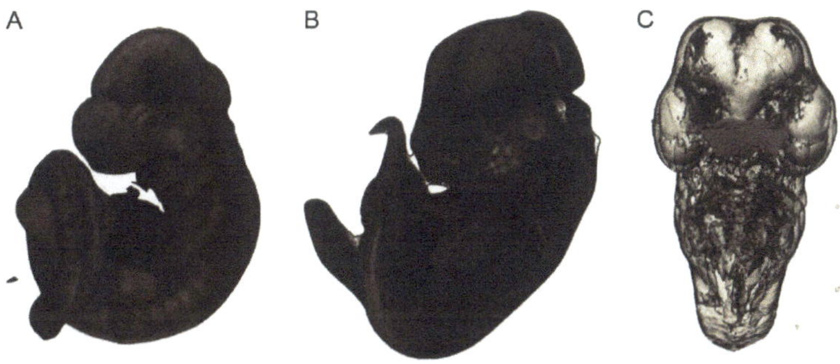

Figure 12 Examples of optical projection tomography images. (A) E10.5 mouse embryo stained with Sytox Green. (B) E12.5 mouse embryo stained with Sytox Green (blue) and Ser10 phosphohistone H3 (green). (C) Hamburger Hamilton Stage 23 chick embryo showing Shh expression in the frontonasal ectodermal zone (FEZ) highlighted in red. (A color version of this figure is located at the end of this chapter.)

carefully positioned for scanning so as to avoid compression or damage that can alter facial morphology.

6.2 Optical Projection Tomography

Optical projection tomography (OPT) is a form of optical computed tomography scanning that is based on visible light rather than X-rays (Sharpe, 2009; Sharpe et al., 2002). The principal advantage of this technique is one can use optical molecular markers, such as fluorescent tags, to image different tissues or regions expressing particular proteins, protein modifications, or mRNAs. Current instrumentation allows visualization in several UV or white light channels, allowing simultaneous visualization of several markers as well as generation of 3D morphology in a single scan session (Fig. 12). This combination of molecular and morphological imaging makes OPT a very powerful technique for morphometrics in developmental biology. The principal drawback of the technique is that the need to partially clear tissues increases processing artifacts that affect morphology. Further, processing and imaging is more time consuming than microCT. Finally, the fact that OPT relies on lenses to focus images limits the depth of focus and thus the size of tissue that can be imaged in this way (Sharpe, 2009).

6.3 High-Resolution Magnetic Resonance Imaging

High-resolution magnetic resonance imaging (μMRI) has been used to image embryos for over two decades (Smith, 1999; Smith, Effmann, & Johnson, 1992; Smith, Johnson, Groman, & Linney, 1994; Smith, Linney, Huff, & Johnson, 1996). μMRI offers tremendous advantages over other volumetric imaging modalities in terms of soft tissue contrast (Baghdadi et al., 2011). Advances in molecular imaging based on MRI also have tremendous potential to advance the abilities of this imaging technique. μMRI can be performed in unfixed tissues in a liquid medium, which means that specimen distortion is minimal. The drawbacks of μMRI are the availability and cost of scanning time and the specialized expertise required to optimize scanning protocols. Further, the spatial resolution and signal to noise are determined by the strength of the magnet, and so high-resolution imaging studies of embryos require powerful and expensive magnets. Even so, μMRI scanning resolutions are generally between 20 and 40 μm, compared to 5 μm or higher for μCT. Although the soft tissue differentiation is much better than with μCT, the large voxel size results in significant pixilation error for small embryos. At higher resolutions, scan times on a per specimen basis can be prohibitively long (Powell & Wilson, 2012). For these reasons, morphometric analysis of craniofacial morphogenesis has not to date employed μMRI. Instead, efforts have focused on atlas construction (Petiet et al., 2008) or automated organism-wide detection of abnormalities (Nieman, Wong, & Henkelman, 2011).

7. FUTURE DIRECTIONS—INTEGRATING MOLECULAR AND ANATOMICAL IMAGING

Recent studies have used morphometrics and 3D imaging to quantify variation in craniofacial morphogenesis (Billington et al., 2015; Green et al.,

2015; Hu, Young, Xu, et al., 2015; Smith et al., 2015; Young et al., 2014). These studies have shown that morphometric analyses can significantly add to mechanistic studies of development and thus that morphometrics and 3D imaging of embryos have significant potential for developmental biology. However, these current studies fall far short of realizing the full potential of the quantitative study of variation.

The questions of developmental biology are multilayered and increasingly complex. The broader opportunity for the combination of imaging and morphometrics lies in the challenge of quantitative integration across the genotype–phenotype map, that is, the development of techniques and statistical methods that quantify morphology in combination with molecular- and cellular-level data. We are now in the very early stages of realizing this goal using a combination of innovative imaging technology, molecular methods, and quantitative methods. OPT imaging allows us to step toward this goal. Using OPT we have been able to, for example, relate variation in the 3D shape of a gene expression domain to morphogenesis (Hu, Young, Xu, et al., 2015; Mio et al., 2015). OPT allowed the simultaneous imaging of a 3D gene expression pattern and external embryonic morphology. Quantifying the shape of the gene expression pattern required a novel morphometric method for the analysis of fuzzy shapes. In collaboration with Washington Mio, we solved this problem by applying spherical harmonics analysis to the Shh gene expression domains in chick embryos (Fig. 13A). OPT is currently limited in that it can only image expression for three genes in the same embryos.

Combining data on cellular dynamics and morphology offers tremendous potential to study mechanisms of morphogenesis and their relation to structural birth defects. This is also a difficult problem. MicroCT and OPT are both potential avenues toward this goal. It is possible to successfully perform whole mount immunohistochemistry on mouse embryos to obtain both scans and sections (Fig. 13B). This is very labor intensive, however, and the partial digestion of the embryo that is required for the whole mount immunohistochemistry step greatly complicates the μCT scanning. OPT offers the ability to obtain such data simultaneously from the same embryo. The challenge here is spatial resolution which at a nominal level of 5 μm is currently insufficient to make out signals in individual cells. However, spatial variation in staining intensity may be provide useful cellular-level data (Fig. 13C).

At a methodological level, developmental biology presents morphometrics with a massive challenge. 3D imaging modalities offer rich datasets and

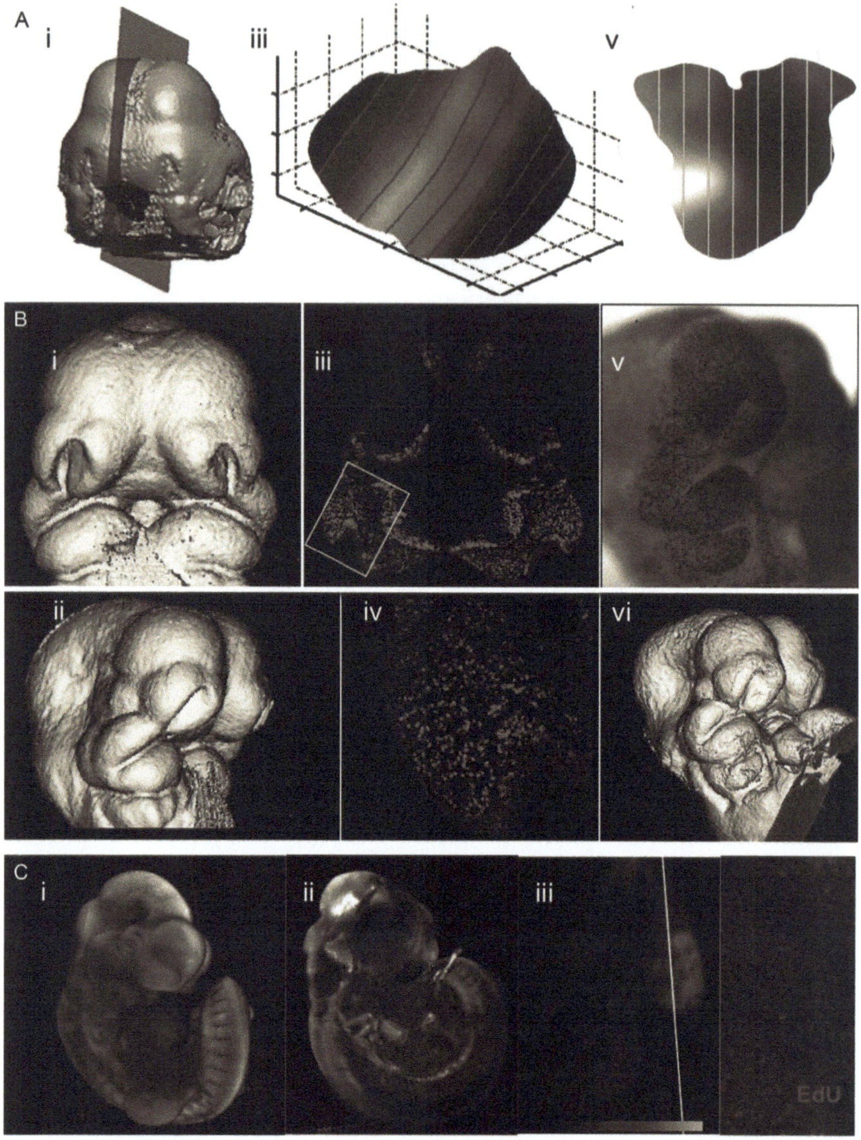

Figure 13 (A) Analysis of shape of gene expression. (i) OPT scan of whole mount *in situ* showing Shh expression (red). The green plane indicates the slices in (i) and (ii). (ii) Section of FEZ parallel to sagittal plane. (iii) Surface spline extracted from raw FEZ. (B) MicroCT renderings and cell proliferation data from the same specimens. (i and ii) 3D reconstruction of μCT taken after processing but before sectioning. (iii and iv) Hoechst 33342 staining to visualize cell nuclei (blue) with cells in S phase visualized using EdU + Alexa Fluor® 488 labeling (green) at 5 × and 200 ×. (v) Specimen processed whole mount for anti-PHH3 primary antibody to identify M-phase cells. (vi) MicroCT rendering of the same specimen after treatment. (C) Images derived from OPT imaging of EdU-stained embryo counterstained with Sytox Green. (B) Volume view of Sytox Green channel showing regional difference in cell density. (C) Volume view of EdU channel showing regional difference in cell proliferation. (D) Heat map of virtual section of lateral nasal prominence of (C) showing local proliferation differences. (E) EdU incorporation as viewed in a traditional confocal section EdU stain (red) and DAPI (blue). (A color version of this figure is located at the end of this chapter.)

we lack methods to make effective use of the information contained in these images. Conducting statistical tests on quantifications derived from the full complexity of such image sets will likely run into computational power issues. Integrating this information across modalities and across levels poses an additional challenge in terms of both methods and computational power. These challenges are well worth investing in. Integrating 3D imaging based on a firm foundation of morphometric theory will be key if developmental biology is to move toward quantitative frameworks that allow for predictive modeling of morphogenesis (Boehm et al., 2010). Personalized or precision medicine is unlikely to make significant inroads for structural birth defects based on genomics alone (Hallgrimsson, Mio, Marcucio, & Spritz, 2014). Developmental processes are simply too complex for prediction to bypass quantitative understanding the role of development in the genotype–phenotype map. That can only be realized through significant investment in imaging and morphometrics as applied to development.

ACKNOWLEDGMENTS

This work is supported by NSERC grant #238992-12 to B.H., NIH-NIDCR grants 1R01DE021708 and 1U01DE024440 to B.H. and R.M., and 1R01DE021708 to Rich Spritz and B.H. as well as the Canadian Foundation for Innovation, Alberta Innovates Health Solutions, and the University of Calgary.

REFERENCES

Alberch, P. (1982). Developmental constraints in evolutionary processes. In J. T. Bonner (Ed.), *Development in evolution* (pp. 313–332). Berlin and New York: Springer-Verlag.

Baghdadi, L., Zamyadi, M., Sled, J. G., Schneider, J. E., Bhattacharya, S., Henkelman, R. M., et al. (2011). Semi-automatic segmentation of multiple mouse embryos in MR images. *BMC Bioinformatics*, *12*, 237.

Bastir, M., & Rosas, A. (2005). Hierarchical nature of morphological integration and modularity in the human posterior face. *American Journal of Physical Anthropology*, *128*(1), 26–34.

Billington, C. J., Jr., Schmidt, B., Marcucio, R. S., Hallgrimsson, B., Gopalakrishnan, R., & Petryk, A. (2015). Impact of retinoic acid exposure on midfacial shape variation and manifestation of holoprosencephaly in Twsg1 mutant mice. *Disease Models & Mechanisms*, *8*(2), 139–146.

Boehm, B., Westerberg, H., Lesnicar-Pucko, G., Raja, S., Rautschka, M., Cotterell, J., et al. (2010). The role of spatially controlled cell proliferation in limb bud morphogenesis. *PLoS Biology*, *8*(7), e1000420.

Bookstein, F. L. (1986). Size and shape spaces for landmark data in two dimensions. *Statistical Science*, *1*(2), 181–222.

Bookstein, F. L. (1989). Principal warps: Thin-plate splines and the decomposition of deformations. *IEEE Transactions on Pattern Analysis and Machine Intelligence*, *11*(6), 567–585.

Bookstein, F. L. (1991). *Morphometric tools for landmark data*. Cambridge: Cambridge University Press.

Bookstein, F. L. (2001). "Voxel-based morphometry" should not be used with imperfectly registered images. *NeuroImage, 14*(6), 1454–1462.
Bookstein, B., Chernoff, B., Elder, R., Humphries, J., Smith, G., & Strauss, R. (1985). *Morphometrics in evolutionary biology*. Philadelphia: The Academy of Natural Sciences of Philadelphia.
Bookstein, F. L., Gunz, P., Mitteroecker, P., Prossinger, H., Schaefer, K., & Seidler, H. (2003). Cranial integration in Homo: Singular warps analysis of the midsagittal plane in ontogeny and evolution. *Journal of Human Evolution, 44*(2), 167–187.
Boughner, J. C., Wat, S., Diewert, V. M., Young, N. M., Browder, L. W., & Hallgrimsson, B. (2008). Short-faced mice and developmental interactions between the brain and the face. *Journal of Anatomy, 213*(6), 646–662.
Brett, A. D., & Taylor, C. J. (1999). A framework for automated landmark generation for automated 3D statistical model construction. *Information processing in medical imaging* (pp. 376-381). Berlin: Springer.
Burian, R. M. (2005). *The epistemology of development, evolution, and genetics*. Cambridge: Cambridge University Press.
Chakravarty, M. M., Aleong, R., Leonard, G., Perron, M., Pike, G. B., Richer, L., et al. (2011). Automated analysis of craniofacial morphology using magnetic resonance images. *PLoS ONE, 6*(5), e20241.
Cheverud, J. (1996a). Quantitative genetic analysis of cranial morphology in the cotton-top (Saguinus oedipus) and saddle-back (S. fuscicollis) tamarins. *Journal of Evolutionary Biology, 9*(1), 5–42.
Cheverud, J. M. (1996b). Developmental integration and the evolution of pleiotropy. *American Zoologist, 36*, 44–50.
Chinthapalli, K., Bartolini, E., Novy, J., Suttie, M., Marini, C., Falchi, M., et al. (2012). Atypical face shape and genomic structural variants in epilepsy. *Brain, 135*(Pt. 10), 3101–3114.
Chong, H. J., Young, N. M., Hu, D., Jeong, J., McMahon, A. P., Hallgrimsson, B., et al. (2012). Signaling by SHH rescues facial defects following blockade in the brain. *Developmental Dynamics, 241*(2), 247–256.
Cooper, W. J., & Albertson, R. C. (2008). Quantification and variation in experimental studies of morphogenesis. *Developmental Biology, 321*(2), 295–302.
Douglas, T. S. (2004). Image processing for craniofacial landmark identification and measurement: A review of photogrammetry and cephalometry. *Computerized Medical Imaging and Graphics, 28*(7), 401–409.
Dryden, I. L., & Mardia, K. V. (1998). *Statistical shape analysis*. Chichester: John Wiley & Sons.
Frost, S. R., Marcus, L. F., Bookstein, F. L., Reddy, D. P., & Delson, E. (2003). Cranial allometry, phylogeography, and systematics of large-bodied papionins (primates: Cercopithecinae) inferred from geometric morphometric analysis of landmark data. *The Anatomical Record, 275A*(2), 1048–1072.
Gonzalez, P. N., Lotto, F. P., & Hallgrimsson, B. (2014). Canalization and developmental instability of the fetal skull in a mouse model of maternal nutritional stress. *American Journal of Physical Anthropology, 154*(4), 544–553.
Gould, S. J. (1966). Allometry and size in ontogeny and phylogeny. *Biological Reviews, 41*(4), 587–640.
Gower, J. C. (1975). Generalized procrustes analysis. *Psychometrika, 40*(1), 33–51.
Green, R. M., Feng, W., Phang, T., Fish, J. L., Li, H., Spritz, R. A., et al. (2015). Tfap2a-dependent changes in mouse facial morphology result in clefting that can be ameliorated by a reduction in Fgf8 gene dosage. *Disease Models & Mechanisms, 8*(1), 31–43.
Gunz, P., & Mitteroecker, P. (2013). Semilandmarks: A method for quantifying curves and surfaces. *Hystrix—The Italian Journal of Mammalogy, 24*(1), 103–109.

Gunz, P., Mitteroecker, P., & Bookstein, F. L. (2005). Semilandmarks in three dimensions. In D. E. Slice (Ed.), *Modern morphometrics in physical anthropology* (pp. 73–98). New York: Kluwer Academic/Plenum Publishers.
Hall, B. K. (2007). Homoplasy and homology: Dichotomy or continuum? *Journal of Human Evolution, 52*(5), 473–479.
Hallgrímsson, B., Boughner, J. C., Turinsky, A., Logan, C., & Sensen, C. W. (2009). Geometric morphometrics and the study of development. In B. Hallgrimsson & C. W. Sensen (Eds.), *Advanced imaging in biology and medicine: Technology, software environments, applications*. Berlin: Springer-Verlag.
Hallgrimsson, B., Jamniczky, H., Young, N. M., Rolian, C., Parsons, T. E., Boughner, J. C., et al. (2009). Deciphering the palimpsest: Studying the relationship between morphological integration and phenotypic covariation. *Evolutionary Biology, 36*(4), 355–376.
Hallgrimsson, B., Mio, W., Marcucio, R. S., & Spritz, R. (2014). Let's face it—Complex traits are just not that simple. *PLoS Genetics, 10*(11), e1004724.
Heimann, T., & Meinzer, H.-P. (2009). Statistical shape models for 3D medical image segmentation: A review. *Medical Image Analysis, 13*(4), 543–563.
Heuze, Y., Martinez-Abadias, N., Stella, J. M., Arnaud, E., Collet, C., Garcia Fructuoso, G., et al. (2014). Quantification of facial skeletal shape variation in fibroblast growth factor receptor-related craniosynostosis syndromes. *Birth Defects Research. Part A, Clinical and Molecular Teratology, 100*(4), 250–259.
Hill, C. A., Martinez-Abadias, N., Motch, S. M., Austin, J. R., Wang, Y., Jabs, E. W., et al. (2013). Postnatal brain and skull growth in an Apert syndrome mouse model. *American Journal of Medical Genetics. Part A, 161A*(4), 745–757.
Hopman, S. M., Merks, J. H., Suttie, M., Hennekam, R. C., & Hammond, P. (2014). Face shape differs in phylogenetically related populations. *European Journal of Human Genetics, 22*(11), 1268–1271.
Houle, D. (2010). Colloquium papers: Numbering the hairs on our heads: The shared challenge and promise of phenomics. *Proceedings of the National Academy of Sciences of the United States of America, 107*(Suppl. 1), 1793–1799.
Houle, D., Govindaraju, D. R., & Omholt, S. (2010). Phenomics: The next challenge. *Nature Reviews. Genetics, 11*(12), 855–866.
Hu, D., Young, N. M., Li, X., Xu, Y., Hallgrimsson, B., & Marcucio, R. S. (2015). A dynamic Shh expression pattern, regulated by SHH and BMP signaling, coordinates fusion of primordia in the amniote face. *Development, 142*(3), 567–574.
Hu, D., Young, N., Xu, Q., Jamniczky, H., Green, R., Mio, W., et al. (2015). Signals from the brain induce variation in avian facial shape. *Developmental Dynamics, 244*(9), 1133–1143.
Jamniczky, H. (2008). Phenotypic integration patterns support an account of homology as a manifestation of evolvability. *Evolutionary Biology, 35*(4), 312–316.
Jamniczky, H. A., & Hallgrimsson, B. (2009). A comparison of covariance structure in wild and laboratory muroid crania. *Evolution, 63*(6), 1540–1556.
Joshi, S. H., Srivastava, A., & Mio, W. (2005). Elastic shape models for interpolations of curves in image sequences. *Information Processing in Medical Imaging Conference, 19*, 541–552.
Kendall, D. G. (1984). Shape-manifolds, procrustean metrics and complex projective spaces. *Bulletin of the London Mathematical Society, 16*, 81–121.
Klingenberg, C. P. (1998). Heterochrony and allometry: The analysis of evolutionary change in ontogeny. *Biological Reviews of the Cambridge Philosophical Society, 73*(1), 79–123.
Klingenberg, C. P. (2008). Novelty and "homology-free" morphometrics: What's in a name? *Evolutionary Biology, 35*(3), 186–190.
Klingenberg, C. P. (2013). Visualizations in geometric morphometrics: How to read and how to make graphs showing shape changes. *Hystrix—The Italian Journal of Mammalogy, 24*(1), 15–24.

Klingenberg, C. P., Badyaev, A. V., Sowry, S. M., & Beckwith, N. J. (2001). Inferring developmental modularity from morphological integration: Analysis of individual variation and asymmetry in bumblebee wings. *The American Naturalist, 157*(1), 11–23.
Klingenberg, C. P., & McIntyre, G. S. (1998). Geometric morphometrics of developmental instability: Analyzing patterns of fluctuating asymmetry with Procrustes methods. *Evolution, 52*(5), 1363–1375.
Klingenberg, C. P., Mebus, K., & Auffray, J. C. (2003). Developmental integration in a complex morphological structure: How distinct are the modules in the mouse mandible? *Evolution & Development, 5*(5), 522–531.
Klingenberg, C. P., & Zimmermann, M. (1992). Static, ontogenetic, and evolutionary allometry: A multivariate comparison in nine species of water striders. *The American Naturalist, 140*(4), 601–620.
Kristensen, E., Parsons, T. E., Hallgrimsson, B., & Boyd, S. K. (2008). A novel 3-D image-based morphological method for phenotypic analysis. *IEEE Transactions on Biomedical Engineering, 55*(12), 2826–2831.
Lande, R. (1979). Quantitative genetic analysis of multivariate evolution: Applied to brain: body size allometry. *Evolution, 33*, 203–215.
Lele, S., & Cole, T. M., III. (1996). A new test for shape differences when variance-covariance matrices are unequal. *Journal of Human Evolution, 31*(3), 193–212.
Lele, S., & Richtsmeier, J. T. (1991). Euclidean distance matrix analysis: A coordinate-free approach for comparing biological shapes using landmark data. *American Journal of Physical Anthropology, 86*(3), 415–427.
Lele, S., & Richtsmeier, J. T. (1995). Euclidean distance matrix analysis: Confidence intervals for form and growth differences. *American Journal of Physical Anthropology, 98*(1), 73–86.
Lele, S., & Richtsmeier, J. T. (2001). *An invariant approach to the statistical analysis of shapes.* Boca Raton: Chapman & Hall.
Love, A. C. (2014). The erotetic organization of developmental biology. In A. Minelli & T. Pradeu (Eds.), *Towards a theory of development* (pp. 33–55). Oxford: Oxford University Press.
Marcus, L. F., Corti, M., Loy, A., Naylor, G. J., & Slice, D. E. (1996). *Advances in morphometrics.* New York: Plenum Press.
Martínez-Abadías, N., Percival, C., Aldridge, K., et al. (2010). Beyond the closed suture in Apert mouse models: evidence of primary effects of FGFR2 signaling on facial shape at P0. *Developmental Dynamics, 239*, 3058–3071.
Marquez, E. J., Cabeen, R., Woods, R. P., & Houle, D. (2012). The measurement of local variation in shape. *Evolutionary Biology, 39*(3), 419–439.
Maynard Smith, J., Burian, R., Kauffman, S., Alberch, P., Campbell, J., Goodwin, B., et al. (1985). Developmental constraints and evolution. *The Quarterly Review of Biology, 60*(3), 265–287.
Metscher, B. D. (2009). MicroCT for developmental biology: A versatile tool for high-contrast 3D imaging at histological resolutions. *Developmental Dynamics, 238*(3), 632–640.
Mio, W., Xu, Q., Jamniczky, H., Hu, D., Green, R. M., Marcucio, R. S., et al. (2015). Correlations between the morphology of sonic hedgehog expression domains and embryonic craniofacial shape. *Evolutionary Biology, 42*(3), 379–386.
Mitteroecker, P. (2009). The developmental basis of variational modularity: Insights from quantitative genetics, morphometrics, and developmental biology. *Evolutionary Biology, 36*(4), 377–385.
Mitteroecker, P., & Bookstein, F. (2007). The conceptual and statistical relationship between modularity and morphological integration. *Systematic Biology, 56*(5), 818–836.
Mitteroecker, P., & Bookstein, F. (2009). The ontogenetic trajectory of the phenotypic covariance matrix, with examples from craniofacial shape in rats and humans. *Evolution, 63*(3), 727–737.

Mitteroecker, P., & Gunz, P. (2009). Advances in geometric morphometrics. *Evolutionary Biology*, *36*(2), 235–247.
Mitteroecker, P., Gunz, P., Neubauer, S., & Muller, G. (2012). How to explore morphological integration in human evolution and development? *Evolutionary Biology*, *39*(4), 536–553.
Mitteroecker, P., & Huttegger, S. M. (2009). The concept of morphospaces in evolutionary and developmental biology: Mathematics and metaphors. *Biological Theory*, *4*(1), 54–67.
Monteiro, L. R. (1999). Multivariate regression models and geometric morphometrics: The search for causal factors in the analysis of shape. *Systematic Biology*, *48*(1), 192–199.
Motch Perrine, S. M., Cole, T. M., 3rd., Martinez-Abadias, N., Aldridge, K., Jabs, E. W., & Richtsmeier, J. T. (2014). Craniofacial divergence by distinct prenatal growth patterns in Fgfr2 mutant mice. *BMC Developmental Biology*, *14*, 8.
Nieman, B. J., Wong, M. D., & Henkelman, R. M. (2011). Genes into geometry: Imaging for mouse development in 3D. *Current Opinion in Genetics & Development*, *21*(5), 638–646.
Norris, F. C., Wong, M. D., Greene, N. D., Scambler, P. J., Weaver, T., Weninger, W. J., et al. (2013). A coming of age: Advanced imaging technologies for characterising the developing mouse. *Trends in Genetics*, *29*(12), 700–711.
Olson, E. C., & Miller, R. A. (1958). *Morphological integration*. Chicago: University of Chicago Press.
Oxnard, C., & O'Higgins, P. (2009). Biology clearly needs morphometrics. Does morphometrics need biology? *Biological Theory*, *4*(1), 84–97.
Parsons, T. E., Kristensen, E., Hornung, L., Diewert, V. M., Boyd, S. K., German, R. Z., et al. (2008). Phenotypic variability and craniofacial dysmorphology: Increased shape variance in a mouse model for cleft lip. *Journal of Anatomy*, *212*(2), 135–143.
Parsons, T. E., Schmidt, E. J., Boughner, J. C., Jamniczky, H. A., Marcucio, R. S., & Hallgrimsson, B. (2011). Epigenetic integration of the developing brain and face. *Developmental Dynamics*, *240*(10), 2233–2244.
Pavlicev, M., Cheverud, J., & Wagner, G. P. (2009). Measuring morphological integration using eigenvalue variance. *Evolutionary Biology*, *36*, 157–170.
Percival, C. J., Green, R., Marcucio, R., & Hallgrimsson, B. (2014). Surface landmark quantification of embryonic mouse craniofacial morphogenesis. *BMC Developmental Biology*, *14*, 31.
Petiet, A. E., Kaufman, M. H., Goddeeris, M. M., Brandenburg, J., Elmore, S. A., & Johnson, G. A. (2008). High-resolution magnetic resonance histology of the embryonic and neonatal mouse: A 4D atlas and morphologic database. *Proceedings of the National Academy of Sciences*, *105*(34), 12331–12336.
Powell, K., & Wilson, D. (2012). 3-Dimensional imaging modalities for phenotyping genetically engineered mice. *Veterinary Pathology Online*, *49*(1), 106–115.
Raup, D. M. (1966). Geometric analysis of shell coiling. *Journal of Paleontology*, *40*, 1178–1190.
Richtsmeier, J. T. (2005). An invariant approach to the study of fluctuating asymmetry. In D. E. Slice (Ed.), *Modern morphometrics in physical anthropology* (pp. 187–212). New York: Kluwer/Plenum Publishers.
Richtsmeier, J. T., Cheverud, J., Danahey, S., Corner, B., & Lele, S. (1993). Sexual dimorphism of ontogeny in the crab-eating macaque (Macaca fascicularis). *Journal of Human Evolution*, *25*(1), 1–30.
Richtsmeier, J. T., Corner, B. D., Grausz, H. M., Cheverud, J. M., & Danahey, S. E. (1993). The role of postnatal growth pattern in the production of facial morphology. *Systematic Biology*, *42*(3), 307–330.
Rohlf, F. J. (1999). Shape statistics: Procrustes superimposition and tangent spaces. *Journal of Classification*, *16*, 197–223.
Rohlf, F. J. (2000). Statistical power comparisons among alternative morphometric methods. *American Journal of Physical Anthropology*, *111*(4), 463–478.

Rohlf, F. J., & Bookstein, F. L. (1990). In *Proceedings of the Michigan morphometrics workshop: Vol. vii* (p. 380). Ann Arbor, MI: University of Michigan Museum of Zoology.
Rohlf, F. J., & Corti, M. (2000). Use of two-block partial least-squares to study covariation in shape. *Systematic Biology, 49*(4), 740–753.
Rohlf, F. J., & Slice, D. E. (1990). Extensions of the Procrustes method for the optical superimposition of landmarks. *Systematic Zoology, 39*(1), 40–59.
Roth, V. L. (1984). On homology. *Biological Journal of the Linnean Society, 22*, 13–29.
Roth, V. L. (1991). Homology and hierarchies: Problems solved and unresolved. *Journal of Evolutionary Biology, 4*, 167–194.
Schmidt, E. J., Parsons, T. E., Jamniczky, H. A., Gitelman, J., Trpkov, C., Boughner, J. C., et al. (2010). Micro-computed tomography-based phenotypic approaches in embryology: Procedural artifacts in assessments of embryonic craniofacial growth and development. *BMC Developmental Biology, 10*, 18.
Sharpe, J. (2009). Optical projection tomography. In C. W. Sensen & B. Hallgrimsson (Eds.), *Advanced imaging in biology and medicine* (pp. 199–224). Berlin: Springer.
Sharpe, J., Ahlgren, U., Perry, P., Hill, B., Ross, A., Hecksher-Sorensen, J., et al. (2002). Optical projection tomography as a tool for 3D microscopy and gene expression studies. *Science, 296*(5567), 541–545.
Smith, B. R. (1999). Visualizing human embryos. *Scientific American, 280*(3), 76–81.
Smith, B. R., Effmann, E. L., & Johnson, G. A. (1992). MR microscopy of chick embryo vasculature. *Journal of Magnetic Resonance Imaging, 2*(2), 237–240.
Smith, B. R., Johnson, G. A., Groman, E. V., & Linney, E. (1994). Magnetic resonance microscopy of mouse embryos. *Proceedings of the National Academy of Sciences of the United States of America, 91*(9), 3530–3533.
Smith, B. R., Linney, E., Huff, D. S., & Johnson, G. A. (1996). Magnetic resonance microscopy of embryos. *Computerized Medical Imaging and Graphics, 20*(6), 483–490.
Smith, F. J., Percival, C. J., Young, N. M., Hu, D., Schneider, R. A., Marcucio, R. S., et al. (2015). Divergence of craniofacial developmental trajectories among avian embryos. *Developmental Dynamics, 244*(9), 1158–1167.
Sneath, P. H. A, & Sokal, R. R. (1973). *Numerical taxonomy; the principles and practice of numerical classification* (xv, 573 p.). San Francisco: W. H. Freeman.
Sober, E. (1994). *Conceptual issues in evolutionary biology*. Cambridge, MA: MIT Press.
Sokal, R. R., & Rohlf, F. J. (1995a). *Biometry*. New York: W.H. Freeman and Company.
Sokal, R. R., & Rohlf, F. J. (1995b). *Biometry: The principles and practice of statistics in biological research*. San Francisco: W.H. Freeman and Company.
Srivastava, A., Joshi, S. H., Mio, W., & Liu, X. (2005). Statistical shape analysis: Clustering, learning, and testing. *IEEE Transactions on Pattern Analysis and Machine Intelligence, 27*(4), 590–602.
Steppan, S. J., Phillips, P. C., & Houle, D. (2002). Comparative quantitative genetics: Evolution of the G matrix. *Trends in Ecology & Evolution, 17*(7), 320–327.
Subburaj, K., Ravi, B., & Agarwal, M. (2009). Automated identification of anatomical landmarks on 3D bone models reconstructed from CT scan images. *Computerized Medical Imaging and Graphics, 33*(5), 359–368.
Thompson, A. (1942/1961). *On growth and form*. Cambridge: Cambridge University Press.
Thompson, D. A. W. (1917). In *On growth and form: Vol. xv* (p.793). Cambridge, England: Cambridge University Press.
Van Valen, L. M. (1982). Homology and causes. *Journal of Morphology, 173*(3), 305–312.
Wagner, G. P. (1990). A comparative study of morphological integration in *Apis mellifera* (Insecta, Hymenoptera). *Zeitschrift Fur Zoologische Systematik Und Evolutionsforschung, 28*, 48–61.
Wagner, G. P. (2007). The developmental genetics of homology. *Nature Reviews. Genetics, 8*(6), 473–479.

Wagner, G. P., Booth, G., & Bagheri-Chaichian, H. (1997). A population genetic theory of canalization. *Evolution, 51*(2), 329–347.
Wong, M. D., Maezawa, Y., Lerch, J. P., & Henkelman, R. M. (2014). Automated pipeline for anatomical phenotyping of mouse embryos using micro-CT. *Development, 141*(12), 2533–2541.
Young, N. M., Chong, H. J., Hu, D., Hallgrimsson, B., & Marcucio, R. S. (2010). Quantitative analyses link modulation of sonic hedgehog signaling to continuous variation in facial growth and shape. *Development, 137*(20), 3405–3409.
Young, N. M., Hu, D., Lainoff, A. J., Smith, F. J., Diaz, R., Tucker, A. S., et al. (2014). Embryonic bauplans and the developmental origins of facial diversity and constraint. *Development, 141*(5), 1059–1063.
Young, N. M., Wagner, G. P., & Hallgrimsson, B. (2010). Development and the evolvability of human limbs. *Proceedings of the National Academy of Sciences of the United States of America, 107*(8), 3400–3405.
Zelditch, M. L. (2004). *Geometric morphometrics for biologists: A primer.* Boston, MA: Elsevier Academic Press.
Zelditch, M. L., Lundrigan, B. L., & Garland, T. (2004). Developmental regulation of skull morphology. I. Ontogenetic dynamics of variance. *Evolution & Development, 6*(3), 194–206.

CHAPTER TWENTY-ONE

Imaging the Cell and Molecular Dynamics of Craniofacial Development

Challenges and New Opportunities in Imaging Developmental Tissue Patterning

Le A. Trinh, Scott E. Fraser[1]

Molecular and Computational Biology, University of Southern California, Los Angeles, California, USA

[1]Corresponding author: e-mail address: sfraser@provost.usc.edu

Contents

Abstract

The development of the vertebrate head requires cell–cell and tissue–tissue interactions between derivatives of the three germ layers to coordinate morphogenetic movements in four dimensions (4D: *x, y, z, t*). The high spatial and temporal resolution offered by optical microscopy has made it the main imaging modularity for capturing the molecular and cellular dynamics of developmental processes. In this chapter, we highlight the challenges and new opportunities provided by emerging technologies that enable dynamic, high-information-content imaging of craniofacial development. We discuss the challenges of varying spatial and temporal scales encountered from the biological and technological perspectives. We identify molecular and fluorescence imaging technology that can provide solutions to some of the challenges. Application of the techniques described within this chapter combined with considerations of the biological and technical challenges will aid in formulating the best image-based studies to extend our understanding of the genetic and environmental influences underlying craniofacial anomalies.

Current Topics in Developmental Biology, Volume 115
ISSN 0070-2153
http://dx.doi.org/10.1016/bs.ctdb.2015.09.002

1. INTRODUCTION

Embryonic processes are driven by the action of gene regulatory networks (GRNs) that control cell–cell interactions, differentiation, divisions, and coordinated movements. These concerted outputs of the GRN execute the patterning and the morphogenetic changes that generate the features of the head and face or any organ within the embryo. Thus, understanding how the GRN translates into the cellular behaviors that coordinate the cell and tissue rearrangements giving rise to a functioning organ has been both fascinating and a major challenge to scientists for generations. The ability to directly observe morphogenetic processes and cell behaviors that contribute to coordinated tissue rearrangements has been a long-standing goal for developmental biologists. However, the dynamic imaging of key events in embryogenesis has been hampered by the inherent nature of developmental processes, as they take place over multiple spatial and temporal scales. To resolve cell behaviors in their *in vivo* context, it is necessary to image at cellular resolution ($\sim$1–10 μm scale), as the cells are migrating, dividing, or undergoing differentiation, within tissues that are patterning at a much larger scale ($\sim$1 mm scale). In addition to this mismatch in scale, there is a similar mismatch in timing: cellular processes such as rearrangements, division, and migration can occur in seconds to minutes; in contrast, the entire morphogenetic event can take place over days to weeks, depending on the animal model system being studied. The ability to capture these broad spatial and temporal scales is further complicated by the need to understand the multicomponent molecular networks that contribute to the dynamic cell and tissue behavior.

Imaging studies of craniofacial development encounter these same multiple, opposing needs that have provided long-standing challenges to dynamic imaging of developmental events. The entire process of craniofacial development, from the initial specification and migration of neural crest cells to the eventual formation of the bony skeleton structures, ranges from several days in animal models such as zebrafish to over a week in mouse and avian embryos (Neuhauss et al., 1996; Rossant & Tam, 2002; Wilkie & Morriss-Kay, 2001; Yelick, Driever, Neuhauss, & Stashenko, 1996; Yelick & Schilling, 2002). Depending on the developmental stages and particular model organisms studied, there are distinct challenges to achieving both cellular and tissue resolution due to the spatial and temporal scales over which the cellular processes occur, the morphological structures that can obscure the imaging window, and the composition of the tissue. These challenges and their potential solutions are the major issues that will be presented in the following materials.

Adding to challenges of dynamic imaging is the growing recognition that the formation of structures like the face is not the result of a single mechanism acting in isolation; instead such structures derive from interactions among many distinct components, at several levels of regulation—genes, regulatory networks, and cells—in varying contexts. Such interacting systems exhibit properties of nonlinear dynamics and emergent behavior that cannot easily be inferred from studies of the individual components in isolation. Thus, researchers requires the means to follow multiple cellular and molecular events in the same preparation. We require multiplex and multimodal imaging tools, able to detect multiple gene expression patterns in a single specimen, and to perform quantitative studies of the molecular components within the live embryo so that the interplay between gene networks can be addressed.

In this chapter, we discuss the challenges mentioned above and highlight some of the emerging technological solutions to these challenges. We discuss the issues related to imaging craniofacial development in terms of both the biological and the instrumental temporal and spatial scales. We highlight current imaging approaches and consider new opportunities in imaging to provide higher spatial and molecular content to imaging data. Finally, we reflect on the potential for these emerging technologies to provide high-resolution spatial, temporal, and molecular content that will provide new insights into normal and abnormal craniofacial development.

2. CHALLENGE OF TEMPORAL AND SPATIAL SCALES IN IMAGING OF CRANIOFACIAL DEVELOPMENT

2.1 Biological Challenges to Dynamic Imaging

The development of craniofacial structures can be divided into a number of discrete stages. Initially, neural crest cells are induced from the border of the neural plate and epidermis (Milet & Monsoro-Burq, 2012; Moury & Jacobson, 1990; Selleck & Bronner-Fraser, 1995). Upon induction, the neural crest cells migrate in stereotypical streams to intermingle with endoderm-derived pharyngeal arches (Kulesa & Fraser, 2000), also known as the branchial arches, which develop as a series of bilateral endodermal outpockets from the pharynx (Graham & Smith, 2001). Regional proliferation of neural crest cells within the pharyngeal arches helps create the facial prominences (Graham, Okabe, & Quinlan, 2005). Along with migration and proliferation of neural crest cells, the cranial placodes develop from the regional thickening and differentiation of the ectoderm to become intermingled with neural crest cells and give rise to the cranial ganglia (Baker & Bronner-Fraser, 2001; Schlosser, 2006). The facial prominences eventually fuse to form the upper and lower palates and jaw features (Bush & Jiang, 2012). Finally, directional growth, condensation, and differentiation of the osteoblasts together shape the skeletonized facial structures (Kimmel, Miller, & Moens, 2001; Knight & Schilling, 2006). The molecular controls regulating these various steps in craniofacial developed have been reviewed extensively (Betancur, Bronner-Fraser, & Sauka-Spengler, 2010; Liu, Rooker, & Helms, 2010; Sauka-Spengler & Bronner-Fraser, 2008; Simoes-Costa & Bronner, 2015; Szabo-Rogers, Smithers, Yakob, & Liu, 2010; Wilkie & Morriss-Kay, 2001). In the following sections, we discuss the challenges presented by each of these stages for dynamic imaging experiments.

The biological challenges encountered in imaging of the developmental processes are both stage- and scale-dependent. Much of our understanding of cell movements and tissue rearrangements in the formation of the face has focused on the early stages, namely neural crest induction and migration, due to the accessibility of these stages to dynamic imaging. The induction of neural crest cells takes place during gastrulation and extends into early organogenesis, a relatively short developmental time window, taking less than 12 h in model organisms such as the mouse, chick, and zebrafish (Aybar & Mayor, 2002; Basch, Selleck, & Bronner-Fraser, 2000; Selleck & Bronner-Fraser, 1995; Serbedzija, Bronner-Fraser, & Fraser, 1992). Additionally, the presumptive neural plate and epidermis, where the neural crest cells form in the ectodermal germ layer, are the most superficial tissue layer and easiest to image. As they migrate, the neural crest cells remain relatively superficial and are accessible to microscopy through both dorsal and lateral views. The rapid development and suitability for imaging makes these stages of craniofacial development accessible to multiple optical imaging modalities, including epifluorescence microscopy and confocal laser scanning microscopy (CLSM) (Andersen, Asuri, Clay, & Halloran, 2010; Clay & Halloran, 2010; Collazo, Bronner-Fraser, & Fraser, 1993; Hochgreb-Hagele & Bronner, 2013; Kulesa & Fraser, 1998). Vital dye tracing experiments in mouse, avian, and *Xenopus* models offer information on cell lineages and can provide dynamic cell motion including complete cell trajectories and velocities (Collazo et al., 1993; Kulesa & Fraser, 1998; Serbedzija et al., 1992). Higher resolution cellular behaviors, such as the formation of cell protrusions, directional blebbing of the cell soma, and epithelial-to-mesenchymal transitions through the loss of cell–cell adhesion have been observed using both vital dyes and genetically encoded fluorescent reporters that specifically label cell membranes, junctions, and nuclei (Ahlstrom & Erickson, 2009; Berndt, Clay, Langenberg, & Halloran, 2008; Kasemeier-Kulesa, Kulesa, & Lefcort, 2005).

While the early processes of neural crest specification and migration are relatively easy to image using vital stains and genetically encoded reporters to label cells and capture their motion with optical microscopy, this is not the case for later stages of craniofacial development; these stages occur in embryos that have undergone significant growth and have often developed more complex morphological structures that can obscure the cells and tissue of interest. For example, during primary palate fusion, when the maxillary processes fuse with the frontonasal prominence to form the upper jaw, the superior view is obscured by the nasal and head processes while the inferior view is inaccessible due to the presence of the lower jaw. Thus, the very anatomical structures that are critical for the formation of facial features also present many of the challenges from a dynamic imaging perspective in most model organisms. Additionally, as the embryo grows, the specimen becomes thicker and less accessible to light microscopy, as signal deteriorates due to photon scattering from the inhomogeneity of the tissue composition. Moreover, the *in utero* development of mouse embryos limits the ability to observe morphogenetic events *in vivo*. For these reasons, much of our understanding of cell and tissue motion in the later stages of craniofacial development comes from inferred motion using developmental time series of 3D volumetric images of fluorescent optical projection tomography, magnetic resonance imaging (MRI), and high-resolution sonographs (Abramyan, Thivichon-Prince, & Richman, 2015; Cerny et al., 2004; Eames et al., 2013; Hu, Marcucio, & Helms, 2003; Lee, Bedard, Buchtova, Fu, & Richman, 2004; McGonnell, Clarke, & Tickle, 1998; Roelfsema, Grijseels, Hop, & Wladimiroff, 2007). These 3D volumetric approaches allow imaging of specimens much larger than is possible using CLSM; however, they are unable to show detailed dynamic cell behaviors and lack the cellular and subcellular resolution offered by CLSM.

To overcome the challenges of imaging large specimen and model organisms that develop *in utero*, tissue explants and embryo cultures can be employed. Adapting specimens for culturing and *ex vivo* imaging is decidedly more difficult (Garcia, Udan, Hadjantonakis, & Dickinson, 2011; Jones et al., 2002; Piliszek, Kwon, & Hadjantonakis, 2011). Culture systems typically require a rich supply of nutrients in complex media and an efficient means of nutrient transfer through the thick specimen (Piliszek et al., 2011; Rivera-Perez, Jones, & Tam, 2010). The constant motion of the culture medium is often required for nutrient transfer and gas exchange in culture systems to enable growth of explanted tissues or embryos (Tam, 1998). This constant motion provides an added technical challenge for imaging, as it can move the specimen, making it difficult to remain focused on an area of interest. This adds to the challenges inherent in imaging a region of interest within regions of rapid tissue growth in mammalian embryos. For example, cranial structures in the mouse embryo increase six times in size (~1–6 mm in biparietal diameter) between 8.5 and 16.0 days *post coitum* (dpc) (Mu, Slevin, Qu, McCormick, & Adamson, 2008). Even with perfect culturing systems, imaging the later stages of craniofacial development remains challenging due to the inherent technological limitations of optical imaging that are discussed in the following sections.

In those stages of craniofacial development that are relatively accessible to dynamic imaging, the biological process being studied and imaging parameters must be carefully considered, as there are trade-offs in capturing both the broad tissue arrangements and the detailed cellular behaviors that drive morphogenesis. A large field of view is necessary to capture complete cell trajectories within the context of embryonic structures that the cells migrate through. To image the multiple streams of neural crest cell migrations, fields of view on the order of 1 mm × 2 mm are needed for avian embryos (Kulesa & Fraser, 2000). The larger field of view is often achieved with lower magnification objectives (10 × magnification) that have lower numerical apertures (NA ~ 0.3) leading to lower spatial resolution of the image. Borders between closely juxtaposed neighboring cells and fine structures such as cell protrusions become more difficult to detect with lower spatial resolution (Kulesa & Gammill, 2010). To capture dynamic cell behaviors such as cell protrusions and directional blebbing, higher spatial and temporal resolution are required. Cell protrusions and blebs are often transient behavior that last for as little as a few seconds, extending a few microns in length and with submicron diameters (Ahlstrom & Erickson, 2009; Berndt et al., 2008; Kasemeier-Kulesa et al., 2005). In order to distinguish such a feature in time and space, the sampling spatial and temporal frequency must be at least double the speed of the process and half the size of the feature (Murphy & Davidson, 2013). For example, to capture cell blebbing, Berndt and colleagues imaged cells every 5 s, as each protrusion lasted 30–60 s (Berndt et al., 2008). To distinguish a cell protrusion that is 0.5 μm in diameter, the acquisition pixel size must be at most 0.25 μm to fulfill the Nyquist theorem of sampling. However, increasing resolution requires an increase in the number of pixels collected, resulting in larger digital image files. Increasing the pixel resolution by two-fold results in the files sizes that are four-fold larger for 2D (x, y) image data, and eight-fold for 3D (x, y, z) image data. Therefore, it is important to carefully match the resolution with the biological criteria of the experiment and the anatomy of the specimen. The different length scales in time and space that is needed to image cellular and developmental dynamics are mapped in relations to the capabilities of imaging modalities in Fig. 1.

2.2 Technological Challenges to Dynamic Imaging

Advanced techniques in optical microscopy have been the primary modality for the imaging of dynamic cell behaviors and tissue morphogenesis in development. While optical techniques are well positioned to assist in providing information on the positions, motions, and behavior of cells and tissues, the ability to push the envelope to achieve higher temporal, spatial, and spectral resolution over extended periods of time is hampered by the technical challenges intrinsic to optical microscopy. The ideal microscope for the dynamic imaging of 3D developmental processes, such as craniofacial development, would combine several, somewhat contradictory capabilities: thin optical sectioning, high signal-to-noise ratio, large field of view, high throughput, and low photobleaching. In this section, we discuss the approaches in optical microscopy for imaging of cell and tissue dynamics and their capabilities for achieving fast 3D volumetric imaging over time.

2.2.1 Confocal Laser Scanning Microscopy

CLSM has been the workhorse for dynamic imaging of early stages of neural crest induction and migration as it provides high lateral (x–y) and axial (z) resolution. When used correctly, the lateral resolution of CLSM is approximately 200 nm (Pawley, 2006), depending upon the numerical aperture of objective and the wavelength of illumination. Axial resolution is achieved through a confocal pinhole that rejects emitted fluorescence from above and below the focal plane to eliminate out-of-focus light that would otherwise blur the image (Fig. 2A; Pawley, 2006). The maximum lateral and axial resolution is dictated by the diffraction limit of light. However, there is a tradeoff between thin optical sectioning achieved by reducing the pinhole size (increasing axial resolution) and decreasing signal that reaches the detectors. This loss in light collection cannot be compensated for by increasing the

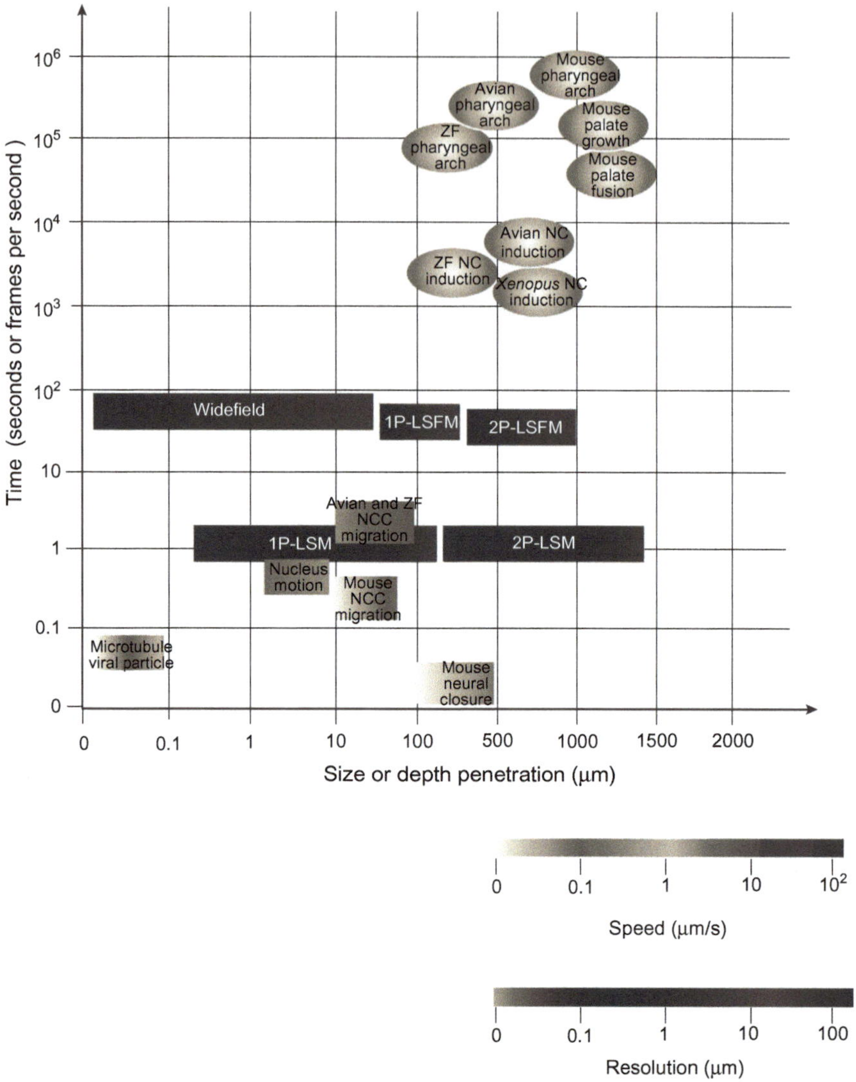

Figure 1 Varying biological and instrumental spatial and temporal scales encountered in dynamic imaging of development. Plot of biological processes and optical microscopy techniques with *y*-axis representing time or scan rates and *x*-axis representing size or depth penetration. For biological processes, time is mapped based on the length of time a particular process takes to develop (gray ovals) or speed of the delineated biological processes calculated as microns/seconds (rectangular scale bar). For optical techniques, scan rate is calculated for the number of 512×512 pixel frames per second. Scale bar for lateral (*x*–*y*) resolution microscopy techniques is color-coded in teal to magenta (lower right corner). Scale bar for speed is color-coded in light blue to red

exciting laser light, as an increase in laser power will result in increased photobleaching and phototoxicity. In dynamic imaging experiments, cellular processes and tissue morphogenesis are greatly dependent on the health of the tissue. Therefore, it is critical to detect fluorescent labels from the specimen using conditions that do not cause even subtle photodamage. Increasing the detection aperture of the CLSM can allow more photons to be detected; however, this sacrifices the depth resolution by increasing the thickness of the optical sectioning of the CLSM. In some settings, this drawback can become an advantage, as opening the confocal aperture increases the depth of field of each optical section, better capturing some cellular dynamics and reducing the number of optical sections needed to capture cell trajectories in a three-dimensional specimen (Kulesa & Fraser, 2000). Thus, it is important to define the goals of the imaging study, so that optimal compromises can be made between speed, resolution, and laser power.

Although CLSM provides high spatial resolution, it is not suited for deep tissue imaging needed to follow the morphogenetic events in later stages of craniofacial development. Tissue size and heterogeneity result in increased scattering of both the exciting light and the emitted light between the focal plane and the objective lens, resulting in an exponential decay in image quality with depth. For example, during the growth phase of craniofacial development in mouse embryos, the frontal prominence enlarges in size by fourfold (Young, Wat, Diewert, Browder, & Hallgrimsson, 2007), becoming over 1 mm in width and depth. In contrast, the optical performance of the CLSM is best suited to samples from 50 to 100 μm in depth (Centonze & White, 1998; Pawley, 2006). Thus, despite dramatic technical improvements in microscopes and microscope objectives, it remains challenging for CLSM to meet the demands of dynamic imaging of thick samples with complex tissue composition.

Figure 1—Cont'd (lower right corner). Numbers for biological processes plotted were exacted from the following references; diameter of microtubule (Faisal, White, & Laughlin, 2005), motion of viral particle on microtubule (Arhel et al., 2006), diameter of nucleus and subnuclear motion (Gorisch, Lichter, & Rippe, 2005; Li & Xie, 2011), rate of neural crest migration in mouse, avian, and zebrafish (ZF) (Hochgreb-Hagele & Bronner, 2013; Kawakami, Umeda, Nakagata, Takeo, & Yamamura, 2011; Kulesa & Fraser, 2000), rate of neural tube closure (Pyrgaki, Trainor, Hadjantonakis, & Niswander, 2010), avian, *Xenopus*, and zebrafish neural crest induction (Aybar & Mayor, 2002), avian, mouse, and zebrafish pharyngeal arch formation (Szabo-Rogers et al., 2010; Yelick & Schilling, 2002), and mouse palate growth and fusion (Bush & Jiang, 2012). (A color version of this figure is located at the end of this chapter.)

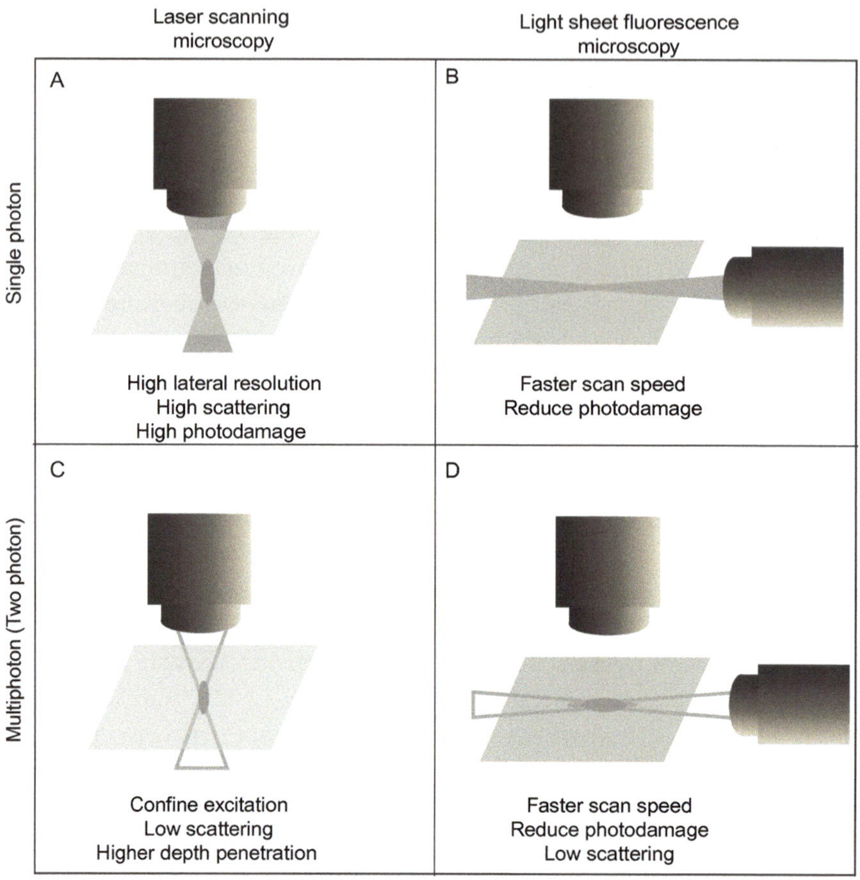

Figure 2 Approaches in optical microscopy for capturing cell and tissue dynamics. Geometry of laser scanning microscopy (LSM) (A and C) and light-sheet fluorescence microscopy (LSFM) (B and D) with single (A and B) or multiphoton (C and D) excitation. (A) Single-photon LSM provides high lateral (*x–y*) and axial (*z*) resolution through a confocal pinhole that rejects light from above and below the focal plane; however, out-of-focus excitation results in high light scattering and photodamage of the sample. (B) LSFM uses a sheet-shaped beam to selectively excite the plane of interest. Fluorescence is collected with a widefield camera perpendicular to the excitation beam resulting in faster scanning rates and reduce photodamage. (C) Nonlinear excitation of fluorescence in multiphoton LSM confines the excitation to the region of interest. Confined excitation results in reduce photodamage while near-infra-red wavelengths used for excitation allows for lower light scattering and higher depth penetration. (D) Multiphoton LSFM excitation in light-sheet mode provides the same advantages as LSM with an increase in scanning rate. (A color version of this figure is located at the end of this chapter.)

2.2.2 Multiphoton Laser Scanning Microscopy

Multiphoton or two-photon LSM (TPLSM) extends the depth penetration of high-resolution optical imaging by increasing the wavelength of the exciting light, relying on the nonlinear excitation of fluorescence only at the focal plane to create an optical section in the tissue (Fig. 2C). In TPLSM, short pulses of laser light in the infra-red (IR) wavelengths (750–1200 nm) replace the typical visible light lasers used in CLSM. At the focus of the objective lens, these lower energy photons become so concentrated that through simultaneous absorption more than one photon can excite the fluorescent dye, resulting in the emission of a single photon with a much shorter wavelength (dependent on the fluorophore, typically in the visible range). No confocal aperture is required to block the light from above and below the focal plane, as the light is not sufficiently concentrated to result in excitation away from the focus, increasing the light collection efficiency. Because light scattering is dramatically decreased for longer excitation wavelengths, studies using TPLSM have reported to resolve fine structures such as axons and neuronal cell bodies at depths of more than 1 mm (Dombeck, Harvey, Tian, Looger, & Tank, 2010; Kobat, Horton, & Xu, 2011). In addition to the increased depth penetration of TPLSM, photobleaching and phototoxicity are significantly decreased, as there is little excitation outside of the focal plane; in contrast, CLSM excites fluorescence throughout the depth of the sample (Fig. 2).

Multiphoton imaging typically uses a single wavelength from a pulse laser to excite a fluorophore. Therefore, imaging of multiple fluorophores is sometimes difficult to achieve because excitation of multiple fluorophores can require tuning of the laser to different excitation wavelengths. The recent development of turn-key ultrafast light sources that combine the typical Titanium–Sapphire (Ti:S) laser with an optical parametric oscillator (OPO) increases the color range of fluorescent labels that can be used in TPLSM (Mahou et al., 2012). The OPO converts an input laser wavelength into laser outputs of longer wavelength, extending the system so it can excite spectrally distinct fluorophores. Interestingly, in addition to the two wavelengths made possible by the use of the OPO, a third excitation can be provided by the spatiotemporal combination of these two wavelengths. Thus, a femtosecond laser (Ti:S) coupled to an OPO can extend the number of distinct fluorophores that can be selectively excited in a sample to three, when an appropriate combination of fluorophores is selected. Each of the fluorophores must be spectrally separated with carefully chosen emission filters and an appropriate number of detectors to excite multiple fluorophores simultaneously (Mahou et al., 2012). The ability to image multiple fluorophores increases the informational content of dynamic imaging experiments.

TPLSM, similar to CLSM, uses raster scanning of the laser beam to generate an image. In raster scanning, the acquisition speed is limited by the galvanometer mirrors that drive the deflection of the laser beam. Galvanometers scan at rates of several microseconds per pixel, translating into scan rates of 0.5–2 s, depending on the x–y dimensions of the image. Adding the z-dimension to capture an entire 3D volume can thus require time scales of minutes. To overcome such slow acquisition times, resonant scanning mirrors capable of rates of 30–60 frames per second have been implemented in LSM systems (Pawley, 2006). However, as the scanning speed increases, the number of photons detected per pixel decreases. Therefore, there is a constant challenge to meet the opposing needs of faster, deeper, and better resolution in dynamic imaging.

2.2.3 Light-Sheet Fluorescence Microscopy

Recent refinements in light-sheet fluorescence microscopy (LSFM) offers a potential answer to the challenges of imaging large and highly scattering 3D samples (Huisken, Swoger, Del Bene, Wittbrodt, & Stelzer, 2004; Verveer et al., 2007). Unlike CLSM, where the excitation light passes through the entire sample (and the confocal aperture rejects the light from planes other than the focal plane), LSFM employs a sheet-shaped excitation beam, projected in from the side of the sample, to selectively excite only the plane of interest (Fig. 2B and D; Greger, Swoger, & Stelzer, 2007). Because there is little excitation above or below the light sheet, the fluorescence can be collected without rejecting any of the emitted light with a confocal aperture and with reduced photobleaching/toxicity (up to ~100-fold), extending the time over which the specimen can be imaged (Truong, Supatto, Koos, Choi, & Fraser, 2011). By collecting the signal in parallel from all the pixels of an optical section with a widefield camera, it becomes possible to reach imaging speeds more than 100-fold faster than raster scanning systems. However, LSFM depends critically on the quality of the light sheet, as any scattering of the exciting light will sacrifice the thinness of the excited domain. The magnitude and the effects of scattering on the light sheet can be minimized by employing a pulsed IR laser to achieve two-photon excitation (Truong et al., 2011), as the IR light is scattered ~16-fold less than visible light so the light sheet will not be as broadened by light scattering; furthermore, since the excitation depends on the square of the instantaneous intensity, it is unlikely that randomly scattered light could create excitation outside of the intended light sheet. These are the same advantages of TPLSM, avoiding the losses introduced by a confocal aperture and making TPLSM the technique of choice for intravital 3D imaging (Gobel, Kampa, & Helmchen, 2007; Zipfel, Williams, & Webb, 2003), due to its greater imaging depth and reduced out-of-plane excitation (Centonze & White, 1998; Gilbert, Hoffman, Capitano, & So, 2000). Similarly, when used in light-sheet mode, two-photon excitation increases the imaging depth, reduces the out-of-plane excitation due to scattering, and reduces photobleaching (Truong et al., 2011), making 2p-LSFM a powerful approach for imaging 3D specimens (Fig. 2D). Such advantages are particularly needed to capture the later stages of craniofacial development when structures of the head become too thick to image with sufficient temporal and spatial resolution or when the tissues become too light scattering.

2.3 Challenge of Higher Content Molecular Imaging

The need for higher content molecular imaging has emerged in conjunction with the understanding that developmental processes are driven by multiple gene networks that are in constant interplay to shape cells, tissues, and embryonic structures. For example, dynamic imaging of early stages of neural crest cell migration has shown that there are significant heterogeneities in cell behavior (Kasemeier-Kulesa et al., 2005; Kulesa, Teddy, Stark, Smith, & McLennan, 2008; Nishiyama et al., 2012), while gene profiling of individual neural crest cells indicates that there are distinct molecular signatures for subpopulations of neural crest cells even during the migration process (McLennan et al., 2015). A major challenge for dynamic imaging is to be able to combine the information of dynamic cell behavior offered by current optical imaging with molecular readouts of gene expression, cellular metabolism, and signaling events to better understand the interplay between complex gene network interactions and morphogenetic events. In this section, we discuss recent advances in molecular and imaging techniques that can be applied to craniofacial development to meet the challenge of obtaining high-information-content image data.

2.3.1 Multiplex Detection of Gene Expression

Spatially, information on gene expression that is important for craniofacial development is traditionally assessed by *in situ* hybridization (ISH) to detect mRNA expression in a morphological context. With standard ISH approaches, it is challenging to simultaneously map the expression of multiple target mRNAs within a single intact embryo (Barroso-Chinea et al., 2007; Denkers, Garcia-Villalba, Rodesch, Nielson, & Mauch, 2004; Harland, 1991; Speel, Hopman, & Komminoth, 1999). For this reason, reports of gene expression overlap in developmental studies are inferred though multiple samples or staining of serial sections for multiple mRNAs of interest. Due to variability between specimens, accurate mapping of spatial relationships between the regulatory loci of different genes requires multiplexed experiments in which multiple mRNAs are detected within a single biological sample.

Multiplex ISH hairpin chain reaction (HCR) is an mRNA detection approach that has been developed to meet the challenge of resolving multiple gene expression patterns in biological samples at subcellular resolution (Choi et al., 2010). The approach uses sequence-specific DNA or RNA probes coupled to a hairpin-based amplification system that allows for the multiplex fluorescent detection of mRNAs of interest. The sequence-specific probes contain defined initiator sequences that can trigger the polymerization of specific hairpins conjugated with fluorescent dyes. The hairpins are metastable in the absence of initiators; the introduction of a single initiator species selectively triggers the cognate hairpins to polymerize off of the initiator sequence, leading to an amplified fluorescent signal per molecule of mRNA (Choi et al., 2010). Multiple pairs of hairpins have been developed that are orthogonal to other pairs, allowing for the simultaneous amplification and detection of five gene targets within a single biological sample (Choi, Beck, & Pierce, 2014; Choi et al., 2010). By using this technique, McLennan and colleagues demonstrated that the migratory neural crest streams exhibit spatially distinct gene patterns, with cells at the migratory front expressing a discrete molecular signature (McLennan et al., 2015). In addition to multiplexing, *in situ* HCR permits mRNAs to be imaged at subcellular resolution, as the HCR amplification polymers remain tethered to their initiating probes, yielding sharp signal localization. This capability provides the potential to use *in situ* HCR to study the timing of gene transcription, as nascent transcripts can be detected within the nucleus of whole embryos (Ruf-Zamojski, Trivedi, Fraser, & Trinh, 2015), allowing the cells actively transcribing a gene to be identified. Those cells that are no longer actively transcribing will have signal in their cytoplasm but not their nuclei. Furthermore, alternative splice variants can be distinguished from each other, as the sequence-specific probes are ~50 bp in length, providing higher molecular content information when probes are carefully designed to target specific regions of transcripts.

Branched DNA (bDNA) is an alternative ISH technique that holds the potential for multiplex detection of mRNA in biological samples. It relies on multistep amplification of tree-like DNA structures that have fluorescent dyes conjugated to DNA molecules (Battich, Stoeger, & Pelkmans, 2013; Collins et al., 1997). Primary probes of 20–30 bp bind to specific targets and are recognized by subsequent steps of hybridization of preamplifiers and fluorescently labeled DNA molecules that bind to the preamplifier to form tree-like structures of DNA complexes. Originally developed for *in vitro* detection of viral transcripts in cell lysates, bDNA has been adapted for *in vivo* quantification of transcripts (Battich et al., 2013; Cao et al., 1995; Collins et al., 1997; Kern et al., 1996). Similar to *in situ* HCR, bDNA leads to fluorescent signal amplification and discrete subcellular localization of transcripts (Battich et al., 2013; Wang et al., 2012).

While both bDNA and *in situ* HCR hold the potential to provide high-content image data for gene expression, they both are currently only applicable to fixed samples and can detect at most five genes within a single sample. This precludes dynamic imaging of gene expression changes; however, dynamic information can be inferred from a single sample though analysis of cytoplasmic and nascent transcripts. Additionally, new orthogonal amplifiers can be developed to increase the number of genes that can be simultaneously detected using these techniques. However, with increasing numbers of amplifier systems, both bDNA and *in situ* HCR will encounter the issue of spectral bandwidth for detection of multiple fluorophores. Fluorescent dyes conjugated to these amplification systems have broad and overlapping emission spectra that cannot be easily separated using standard emission filters (Patterson, Day, & Piston, 2001; Shaner, Steinbach, & Tsien, 2005). For spectral separation of multicolor fluorescence imaging, there are some emerging advanced microscopy techniques, discussed in the following sections, that enable separation of closely overlapping emission signals and permit higher content information for molecular imaging.

2.3.2 Hyperspectral Multiplexing Imaging

Dynamic imaging of craniofacial developmental processes, as with the majority of live tissue and whole embryo imaging approaches, relies on vital stains and fluorescent proteins to discretely label populations of cells, subcellular compartments, and/or proteins. As new fluorophores are developed, the number of labels that could be used for multicolor fluorescence imaging can dramatically increase the informational content of image data. However, separating the signal from each dye is made challenging by their fluorescence spectrum overlap with other fluorophores and with endogenous fluorescence signals. Hyperspectral imaging provides the potential to distinguish multiple fluorophores, as well as a means to assess biochemical interactions in biological samples in a label-free manner that extends beyond conventional fluorescent protein tagging and morphological imaging.

Hyperspectral imaging takes advantage of wavelengths emitted by or reflected from a sample, adding a spectral dimension to imaging data (Cutrale, Salih, & Gratton, 2013; Favreau et al., 2014). In hyperspectral imaging, a complete electromagnetic spectrum in the visible range is collected for each pixel within an image (Fig. 3). The spectral information can come from exogenous labels such as fluorescent proteins or from endogenous metabolites that exhibit a fluorescent signature. For exogenous labels, the high spectral information allows distinction of fluorescent proteins or dyes that have overlapping emission spectra. More importantly, signaling molecules and cellular metabolites such as retinoid acid, folic acid, and NADH have spectral signatures that can be distinguished using hyperspectral imaging (Palero et al., 2011; Rocheleau, Head, & Piston, 2004). Additionally, a number of studies using hyperspectral imaging have shown that different tissues have distinct spectral properties (Holzer et al., 2011; Lu et al., 2014; Panasyuk et al., 2007; Park et al., 2009; Zuzak et al., 2011). Tumors and intestinal tissue have been successfully distinguished by hyperspectral imaging, permitting recognition of different domains, or cell types without the need for exogenous labels (Akbari, Uto, Kosugi, Kojima, & Tanaka, 2011; Best et al., 2013; McCormack et al., 2014). Thus, hyperspectral imaging can provide a new dimension to dynamic imaging that extends beyond traditional labels with fluorescent probes and has great potential for new discoveries in craniofacial imaging. In particular, the ability to detect endogenous molecules such as retinoids and folic acid, both important regulators of craniofacial development, provides the motivation to dramatically increase the use of hyperspectral imaging in both developmental studies and therapeutic interventions.

A challenge in hyperspectral imaging is the analysis of the large spectral data sets that result from acquiring a spectrum for every pixel within an image. The result of one two-dimensional (2D) image acquisition is a three-dimensional (3D) cube of data with spatial (x and y) and spectral (λ) information. As a consequence, a z-stack will result in a 4D (x, y, z, λ) dataset and a time-lapse will result in a 5D (x, y, z, λ, t) hypercube. Principal Component Analysis (PCA), Support Vector Machine (SVM), and Spectral Phasor (SP) methods have been employed for segmenting hyperspectral data

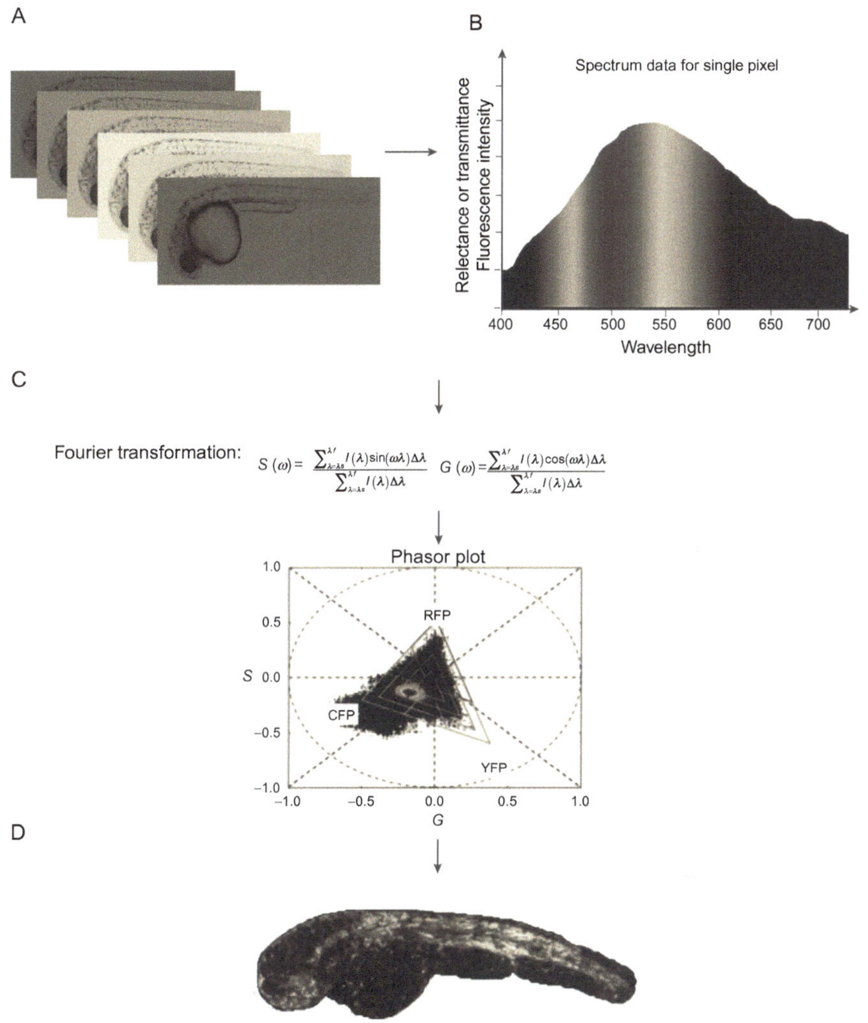

Figure 3 Hyperspectral imaging workflow of highspectral imaging and data analysis using Spectral Phasor approach. (A) In hyperspectral imaging, samples are illuminated with a single wavelength and (B) transmitted or reflected fluorescence intensity is collected for the entire visible spectrum at each pixel. (C) Fourier transform of the multidimension data (*x*, *y*, *z*, λ, and/or *t*) condenses the data into a coordinate *S* and *G* on the Phasor plot. *S* and *G* coordinates depend on the intensity, $I(\lambda)$, the wavelength, λ, and Harmonic number, ω (Cutrale et al., 2013). (D) Each *S* and *G* coordinate on the Phasor plot can be mapped back to corresponding pixels in the image data. *Image of Spectral Phasor plot and corresponding image data courtesy of F. Cutrale.* (A color version of this figure is located at the end of this chapter.)

(Cutrale et al., 2013; Garcia-Allende et al., 2008; Gowen, O'Donnell, Cullen, Downey, & Frias, 2007; Kong, Zhang, Liu, Nie, & He, 2013; Rodarmel & Shan, 2002). Hyperspectral image datasets are dramatically larger than a simple grayscale dataset (*n* times larger, where *n* is the number of spectral channels acquired). This means that a typical 32-channel hyperspectral dataset will be 32 times larger than a common grayscale dataset, which could make it much more cumbersome to display, process, and store. PCA and SVM approaches are computationally expensive, and the resulting segmentation or classification approaches often fail when the data is complicated. Additionally, PCA relies on a linear transformation and the maximum variance to reduce data into two dimensions. This can lead to changes between datasets if there is any variation between samples. In contrast, the SP method offers an efficient representation of the spectral data as vectors in Fourier space, thereby condensing each spectrum in the x, y, z, t space into a single point in a 2D Phasor plot without data loss (Cutrale et al., 2013). SP is based on the fast Sine and Cosine Fourier transform to represent each point in the Phasor plot based on the pixel intensity and wavelength captured from the sample (Fig. 3). SP does not alter the data through fitting, but instead provides a general reference system in the Fourier domain that allows a user to explore the data. Furthermore, SP is computationally faster than PCA and SVM, making it more readily accessible to many research laboratories. Software tools for each of these approaches are becoming available in the public domain, making it possible for a variety of users to deploy these advanced tools without major expense.

2.3.3 Fluorescence Lifetime Imaging

In addition to distinct spectral properties, all fluorescent molecules have characteristic excited-state lifetimes (the time between when the dye absorbs a photon and when it emits a photon) that can be measured and captured using fluorescence lifetime imaging microscopy (FLIM). FLIM measures the time required for a population of excited fluorophores to return to the ground state through fluorescence and other non-radiative processes (Digman, Caiolfa, Zamai, & Gratton, 2008). Fluorescence lifetime is a parameter largely independent of the fluorescence intensity and fluorophore concentration, a unique advantage of fluorescence imaging. This key advantage allows for imaging of fluorescence signal in microenvironments where concentrations may be low or in which the emission spectra could be altered. The fluorescence lifetime of a molecule can be sensitive to a great variety of internal factors defined by the fluorophore structure and external factors that include temperature, polarity, and the presence of fluorescence quenchers (Digman et al., 2008). Although this could be viewed as a shortcoming, it presents powerful opportunities to turn the fluorescent signal into a reporter of local environment. The sensitivity to fluorescence quenchers such as oxygen is the basis for the measurement of redox states in cells and tissues using FLIM. The endogenous fluorophore NADH is quenched when it is oxidized to NAD+, while FAD is fluorescent and FADH is not (Lakowicz, Szmacinski, Nowaczyk, & Johnson, 1992). Hence, measuring the fluorescence lifetimes of NADH and FAD can be used to determine the oxygen concentrations and redox states of cells (Chance, Schoener, Oshino, Itshak, & Nakase, 1979). In fact, NADH FLIM signatures have been associated with cell proliferation and the differentiation state of cells in the intestine and elsewhere, and characteristic difference have been used to identify neural and other stem cells (Quinn et al., 2013; Stringari, Nourse, Flanagan, & Gratton, 2012). These studies suggest that the time dimension of the fluorescence signal, that is almost always ignored, can provide a powerful imaging parameter in developmental studies.

In addition to measuring endogenous fluorescence, FLIM can be used to assess protein interaction in Förster resonance energy transfer (FRET) experiments. FRET involves the direct transfer of the excited state from a donor fluorophore to an acceptor fluorophore (Stryer, 1978). The transfer of energy does not involve the emission and absorption of a photon, and can only occur when the two fluorophores are aligned and in close proximity (~10 nm). The steep dependence on molecular distance has allowed FRET to be used as a molecular yardstick and has made it an important molecular imaging technique for studies of protein interactions *in vivo* (Roy, Hohng, & Ha, 2008). FRET experiments require that the proteins of interest are labeled with donor or acceptor fluorophores; the amount of FRET between the two labeled proteins can be measured through fluorescence intensity or fluorescence lifetime. In intensity-based FRET experiments, the losses of donor fluorescence intensity and/or an increase in acceptor fluorescence intensity are measured and used to calculate the amount of exchange (Miyawaki, 2003). While ratiometric detection of FRET is relatively easy to implement on many CLSM systems, it is limited by several issues, including donor emission bleed-through and direct excitation of the acceptor; furthermore, differences in protein concentration between acceptor and donor have to be carefully controlled. The use of FLIM detection for FRET experiments overcomes many of these challenges, as a single measurement—the fluorescence lifetime of the donor—is needed for the detection of FRET

(Sun, Day, & Periasamy, 2011). Additionally, as FLIM is independent of the concentration of the fluorophore, FLIM detection is far less sensitive to variations in protein concentrations between cells, and a wider range of samples can be compared to provide informative readouts of molecular interactions.

A number of FRET probes have been generated to sense signaling events, cell adhesion, and molecular tensions (Borghi et al., 2012; Itoh et al., 2002; Kraynov et al., 2000; Mochizuki et al., 2001; Seth, Otomo, Yin, & Rosen, 2003). Many of these probes label genetic pathways and cellular processes that have been implicated in craniofacial development. For example, FRET probes have been generated to sense the activation of Rac1 and RhoA (Yoshizaki et al., 2003). These probes have been applied to *in vitro* and *in vivo* analysis of neural crest cell migration in *Xenopus* and zebrafish. A FRET probe for RhoA activation has been used to demonstrate that directional migration of neural crest cells is controlled by the planar cell polarity pathway mediated by Wnt signaling at sites of cell–cell contacts via RhoA activation (Carmona-Fontaine et al., 2008). Furthermore, analyzing the activity of the small GTPases Cdc42, Rac1, and RhoA activity by FRET showed that syndecan-4 inhibits Rac1 at the trailing edge of migratory neural crest cells (Matthews et al., 2008). By using FRET probes for these small GTPases in combination with genetic perturbations and dynamic imaging, a model emerges for neural crest cell migration that is dependent on the balance of RhoA activation and Rac1 inhibition at the leading and trailing edges. More importantly, these examples demonstrate that FRET-based molecular probes can provide detail about molecular mechanisms when applied to molecular imaging of processes in craniofacial development.

2.3.4 Fluorescence Correlation Spectroscopy

Another fluorescence imaging approach that holds great potential for studying molecular dynamics of protein *in vivo* is fluorescence correlation spectroscopy (FCS). FCS is a fluctuation-based imaging technique that provides information on concentrations, mobility, and interactions of macromolecules (Haustein & Schwille, 2007). In contrast to most fluorescence imaging modalities, FCS extracts information from the fluorescence intensity fluctuations in a sample: as labeled molecules pass in and out of a small confocal volume (<1 femtoliter), the signal fluctuates depending on the number of molecules. Because the technique is based on fluctuations, it requires fine spatial resolution, achieved by focusing a laser beam with a high NA objective to create a diffraction-limited spot, and a small confocal aperture to collect data from a minimum-sized volume element. Comparison of fluorescence intensity fluctuations on the time-scale of milliseconds to seconds by autocorrelation allows local concentrations, mobility coefficients, or characteristic rate constants of inter- or intramolecular reactions of fluorescently labeled biomolecules to be determined. Fluorescence cross-correlation spectroscopy (FCCS) is a dual-color variant of FCS for studying the molecular binding (including affinity and stoichiometry), enzyme kinetics, and dynamic co-localization of two distinctly labeled proteins (Bacia, Kim, & Schwille, 2006).

FCS was originally applied to study the binding kinetics of fluorescent dyes to DNA and the diffusion of dyes in lipid bilayers (Magde, Elson, & Webb, 1972, 1974; Meyer & Schindler, 1988). Advances in instrumentation, image analysis, and biological labels have recently moved FCS from *in vitro* applications in which macromolecules are measured in homogenous solution to *in vivo* measurements of protein concentration and diffusion rates in cells and whole embryos (Nowak, Machate, Yu, Gupta, & Brand, 2011; Ries, Yu, Burkhardt, Brand, & Schwille, 2009; Yu et al., 2009). Quantitative results in FCS rely on the size of the confocal volume that can be determined experimentally. Conventional FCS uses a static detection volume and assumes the volume is a Gaussian. However, *in vivo* the confocal volume is difficult to determine due to tissue absorption and scattering that can change the depth of the detection volume. Additionally, cell movements, changes in shape of a tissue as an organism undergoes growth, and morphogenesis can lead to constant changes in the detection volume. To account for these challenges, corresponding FCS techniques that modify the way in which fluorescence fluctuations are sampled have been developed, such as two-focus and scanning fluorescence correlation spectroscopy (SFCS) and raster imaging correlation spectroscopy (RICS) (Dertinger et al., 2007; Digman et al., 2005; Ries, Chiantia, & Schwille, 2009; Ries & Schwille, 2006; Ries, Yu, et al., 2009). Two-focus FCS uses two laterally shifted and overlapping laser foci at a fixed and known distance, while SFCS applies continuous wave excitation to analyze membrane dynamics by changing the position of the detection volume to span vertically with respect to the membrane. Both of these approaches require signification modifications to the LSM and have not been broadly used. In contrast, RICS takes advantage of the raster scanning of LSM to measure fluorescence intensity across a region of interest one pixel at a time. Because each pixel is collected at a different time, there is both spatial and temporal information included in each pixel that can be analyzed by cross- and autocorrelations. The pixel dwell time or sampling rate of the raster scan can be adjusted from microseconds to milliseconds to allow measurements of both fast and slow molecular dynamics. The main advantage of RICS over other modifications to FCS is that it can be performed using a commercially available LSM, making it accessible to many researchers.

While these technical improvements have expanded the application of FCS to cells and whole embryos, FCS is still challenging due to the inherent need for macromolecules to emit a fluorescent signature and be expressed at physiological levels. Most macromolecules, in particular proteins, do not possess intrinsic fluorescence signatures. Therefore, fluorescent fusion proteins have to be generated to perform FCS experiments. However, for relevant dynamics, the fluorescent fusion proteins need to be titrated to match endogenous levels. For these reasons, the application of FCS to *in vivo* studies has been limited. Only recently have these techniques been applied *in vivo* to vertebrate animal models. Experiments in zebrafish have used FCS and FCCS to determine whether Fgf8 acts as a morphogen during development and for quantifying interactions between Fgf receptors and ligands (Nowak et al., 2011; Yu et al., 2009). In these experiments, the fluorescently labeled proteins had to be carefully calibrated to match endogenous levels. With the generation of endogenously fluorescent tag proteins through gene trapping (Trinh et al., 2011), such technical challenges would be minimized and hold the potential to enable wider use of FCS approaches to assess protein dynamics and measure protein concentrations in various developmental processes.

3. PERSPECTIVE

Understanding cell–cell interactions, how neighbor relationships change over time, and the cellular basis for morphogenetic processes driving formation of craniofacial features remains a challenge. Emerging imaging techniques are well positioned to assist in this challenge. The combination of advanced fluorescence imaging techniques and molecular approaches presents new opportunities for researchers to obtain high-content information on the cellular and molecular dynamics of craniofacial development. To fully realize the potential for imaging to contribute to our understanding of developmental processes, these emerging techniques must move out of specialty laboratories and into more available facilities. To do so, modifications to existing hardware need to be developed that push the imaging speed and depth capability of the microscopes. In addition, more diverse and efficient labeling approaches are needed to permit experiments that capture the molecular complexity of gene networks as they function within cell and tissues. Finally, methods that allow high-throughput labeling and imaging must be developed to speed the development of new *in vivo* analyses of network activity and function. The future of extrapolating molecular imaging tools out of cell culture and into intact, developing embryos holds great promise for better defining the cellular molecular basis of craniofacial development.

REFERENCES

Abramyan, J., Thivichon-Prince, B., & Richman, J. M. (2015). Diversity in primary palate ontogeny of amniotes revealed with 3D imaging. *Journal of Anatomy*, *226*(5), 420–433. http://dx.doi.org/10.1111/joa.12291. PMID:25904546.

Ahlstrom, J. D., & Erickson, C. A. (2009). The neural crest epithelial-mesenchymal transition in 4D: A 'tail' of multiple non-obligatory cellular mechanisms. *Development*, *136*(11), 1801–1812. http://dx.doi.org/10.1242/dev.034785. PMID:19429784. PMCID: PMCPMC2680107.

Akbari, H., Uto, K., Kosugi, Y., Kojima, K., & Tanaka, N. (2011). Cancer detection using infrared hyperspectral imaging. *Cancer Science*, *102*(4), 852–857. http://dx.doi.org/10.1111/j.1349-7006.2011.01849.x. PMID:21205093.

Andersen, E., Asuri, N., Clay, M., & Halloran, M. (2010). Live imaging of cell motility and actin cytoskeleton of individual neurons and neural crest cells in zebrafish embryos. *Journal of Visualized Experiments*, *36*, 1726–1729. http://dx.doi.org/10.3791/1726.. PMID:20130524. PMCID:PMCPMC2818709.

Arhel, N., Genovesio, A., Kim, K. A., Miko, S., Perret, E., Olivo-Marin, J. C., et al. (2006). Quantitative four-dimensional tracking of cytoplasmic and nuclear HIV-1 complexes. *Nature Methods*, *3*(10), 817–824. http://dx.doi.org/10.1038/nmeth928. PMID:16990814.

Aybar, M. J., & Mayor, R. (2002). Early induction of neural crest cells: Lessons learned from frog, fish and chick. *Current Opinion in Genetics & Development*, *12*(4), 452–458. PMID:12100892.

Bacia, K., Kim, S. A., & Schwille, P. (2006). Fluorescence cross-correlation spectroscopy in living cells. *Nature Methods*, *3*(2), 83–89. http://dx.doi.org/10.1038/nmeth822. PMID:16432516. Epub 2006/01/25; [pii]1038/nmeth822.

Baker, C. V., & Bronner-Fraser, M. (2001). Vertebrate cranial placodes I. Embryonic induction. *Developmental Biology*, *232*(1), 1–61. http://dx.doi.org/10.1006/dbio.2001.0156. PMID:11254347.

Barroso-Chinea, P., Aymerich, M. S., Castle, M. M., Perez-Manso, M., Tunon, T., Erro, E., et al. (2007). Detection of two different mRNAs in a single section by dual in situ hybridization: A comparison between colorimetric and fluorescent detection. *Journal of Neuroscience Methods*, *162*(1–2), 119–128. http://dx.doi.org/10.1016/J.Jneumeth.2006.12.017. PMID:ISI:000246429000015.

Basch, M. L., Selleck, M. A., & Bronner-Fraser, M. (2000). Timing and competence of neural crest formation. *Developmental Neuroscience*, *22*(3), 217–227. http://dx.doi.org/10.1159/000017444. PMID:10894985.

Battich, N., Stoeger, T., & Pelkmans, L. (2013). Image-based transcriptomics in thousands of single human cells at single-molecule resolution. *Nature Methods*, *10*(11), 1127–1133. http://dx.doi.org/10.1038/nmeth.2657. PMID:24097269.

Berndt, J. D., Clay, M. R., Langenberg, T., & Halloran, M. C. (2008). Rho-kinase and myosin II affect dynamic neural crest cell behaviors during epithelial to mesenchymal transition in vivo. *Developmental Biology, 324*(2), 236–244. http://dx.doi.org/10.1016/j.ydbio.2008.09.013. PMID:18926812. PMCID:PMCPMC2650264.

Best, S. L., Thapa, A., Jackson, N., Olweny, E., Holzer, M., Park, S., et al. (2013). Renal oxygenation measurement during partial nephrectomy using hyperspectral imaging may predict acute postoperative renal function. *Journal of Endourology, 27*(8), 1037–1040. http://dx.doi.org/10.1089/end.2012.0683. PMID:23544949.

Betancur, P., Bronner-Fraser, M., & Sauka-Spengler, T. (2010). Assembling neural crest regulatory circuits into a gene regulatory network. *Annual Review of Cell and Developmental Biology, 26*, 581–603. http://dx.doi.org/10.1146/annurev.cellbio.042308.113245. PMID:19575671. PMCID:PMCPMC4040144.

Borghi, N., Sorokina, M., Shcherbakova, O. G., Weis, W. I., Pruitt, B. L., Nelson, W. J., et al. (2012). E-cadherin is under constitutive actomyosin-generated tension that is increased at cell-cell contacts upon externally applied stretch. *Proceedings of the National Academy of Sciences of the United States of America, 109*(31), 12568–12573. http://dx.doi.org/10.1073/pnas.1204390109. PMID:22802638. PMCID:PMCPMC3411997.

Bush, J. O., & Jiang, R. (2012). Palatogenesis: Morphogenetic and molecular mechanisms of secondary palate development. *Development, 139*(2), 231–243. http://dx.doi.org/10.1242/dev.067082. PMID:22186724. PMCID:PMCPMC3243091.

Cao, Y., Ho, D. D., Todd, J., Kokka, R., Urdea, M., Lifson, J. D., et al. (1995). Clinical evaluation of branched DNA signal amplification for quantifying HIV type 1 in human plasma. *AIDS Research and Human Retroviruses, 11*(3), 353–361. PMID:7786581.

Carmona-Fontaine, C., Matthews, H. K., Kuriyama, S., Moreno, M., Dunn, G. A., Parsons, M., et al. (2008). Contact inhibition of locomotion in vivo controls neural crest directional migration. *Nature, 456*(7224), 957–961. http://dx.doi.org/10.1038/nature07441. PMID:19078960. PMCID:PMCPMC2635562.

Centonze, V. E., & White, J. G. (1998). Multiphoton excitation provides optical sections from deeper within scattering specimens than confocal imaging. *Biophysical Journal, 75*(4), 2015–2024. http://dx.doi.org/10.1016/S0006-3495(98)77643-X. PMID:9746543. PMCID:PMCPMC1299873.

Cerny, R., Lwigale, P., Ericsson, R., Meulemans, D., Epperlein, H. H., & Bronner-Fraser, M. (2004). Developmental origins and evolution of jaws: New interpretation of "maxillary" and "mandibular". *Developmental Biology, 276*(1), 225–236. http://dx.doi.org/10.1016/j.ydbio.2004.08.046. PMID:15531376.

Chance, B., Schoener, B., Oshino, R., Itshak, F., & Nakase, Y. (1979). Oxidation-reduction ratio studies of mitochondria in freeze-trapped samples. NADH and flavoprotein fluorescence signals. *The Journal of Biological Chemistry, 254*(11), 4764–4771. PMID:220260.

Choi, H. M., Beck, V. A., & Pierce, N. A. (2014). Next-generation in situ hybridization chain reaction: Higher gain, lower cost, greater durability. *ACS Nano, 8*(5), 4284–4294. http://dx.doi.org/10.1021/nn405717p. PMID:24712299. PMCID:PMCPMC4046802.

Choi, H. M., Chang, J. Y., Trinh le, A., Padilla, J. E., Fraser, S. E., & Pierce, N. A. (2010). Programmable in situ amplification for multiplexed imaging of mRNA expression. *Nature Biotechnology, 28*(11), 1208–1212. http://dx.doi.org/10.1038/nbt.1692. PMID:21037591. PMCID:PMCPMC3058322.

Clay, M. R., & Halloran, M. C. (2010). Control of neural crest cell behavior and migration: Insights from live imaging. *Cell Adhesion & Migration, 4*(4), 586–594. http://dx.doi.org/10.4161/cam.4.4.12902. PMID:20671421. PMCID:PMCPMC3011273.

Collazo, A., Bronner-Fraser, M., & Fraser, S. E. (1993). Vital dye labelling of *Xenopus laevis* trunk neural crest reveals multipotency and novel pathways of migration. *Development, 118*(2), 363–376. PMID:7693414.

Collins, M. L., Irvine, B., Tyner, D., Fine, E., Zayati, C., Chang, C., et al. (1997). A branched DNA signal amplification assay for quantification of nucleic acid targets below 100 molecules/ml. *Nucleic Acids Research*, *25*(15), 2979–2984. PMID:9224596. PMCIDPMCPMC146852.

Cutrale, F., Salih, A., & Gratton, E. (2013). Spectral Phasor approach for fingerprinting of photo-activatable fluorescent proteins Dronpa, Kaede and KikGR. *Methods and Applications in Fluorescence*, *1*(3), 35001. http://dx.doi.org/10.1088/2050-6120/1/3/035001. PMID:24040513. PMCID:PMCPMC3771857.

Denkers, N., Garcia-Villalba, P., Rodesch, C. K., Nielson, K. R., & Mauch, T. J. (2004). FISHing for chick genes: Triple-label whole-mount fluorescence in situ hybridization detects simultaneous and overlapping gene expression in avian embryos. *Developmental Dynamics*, *229*(3), 651–657. http://dx.doi.org/10.1002/Dvdy.20005. PMID:ISI:000189359600026.

Dertinger, T., Pacheco, V., von der Hocht, I., Hartmann, R., Gregor, I., & Enderlein, J. (2007). Two-focus fluorescence correlation spectroscopy: A new tool for accurate and absolute diffusion measurements. *Chemphyschem*, *8*(3), 433–443. http://dx.doi.org/10.1002/cphc.200600638. PMID:17269116.

Digman, M. A., Brown, C. M., Sengupta, P., Wiseman, P. W., Horwitz, A. R., & Gratton, E. (2005). Measuring fast dynamics in solutions and cells with a laser scanning microscope. *Biophysical Journal*, *89*(2), 1317–1327. http://dx.doi.org/10.1529/biophysj.105.062836. PMID:15908582. PMCID:PMCPMC1366616.

Digman, M. A., Caiolfa, V. R., Zamai, M., & Gratton, E. (2008). The phasor approach to fluorescence lifetime imaging analysis. *Biophysical Journal*, *94*(2), L14–L16. http://dx.doi.org/10.1529/biophysj.107.120154. PMID:17981902. PMCID:PMCPMC2157251.

Dombeck, D. A., Harvey, C. D., Tian, L., Looger, L. L., & Tank, D. W. (2010). Functional imaging of hippocampal place cells at cellular resolution during virtual navigation. *Nature Neuroscience*, *13*(11), 1433–1440. http://dx.doi.org/10.1038/nn.2648. PMID:20890294. PMCID:PMCPMC2967725.

Eames, B. F., DeLaurier, A., Ullmann, B., Huycke, T. R., Nichols, J. T., Dowd, J., et al. (2013). FishFace: Interactive atlas of zebrafish craniofacial development at cellular resolution. *BMC Developmental Biology*, *13*, 23. http://dx.doi.org/10.1186/1471-213X-13-23. PMID:23714426. PMCID:PMCPMC3698193.

Faisal, A. A., White, J. A., & Laughlin, S. B. (2005). Ion-channel noise places limits on the miniaturization of the brain's wiring. *Current Biology*, *15*(12), 1143–1149. http://dx.doi.org/10.1016/j.cub.2005.05.056. PMID:15964281.

Favreau, P. F., Hernandez, C., Heaster, T., Alvarez, D. F., Rich, T. C., Prabhat, P., et al. (2014). Excitation-scanning hyperspectral imaging microscope. *Journal of Biomedical Optics*, *19*(4), 046010. http://dx.doi.org/10.1117/1.JBO.19.4.046010. PMID:24727909. PMCID:PMCPMC3983524.

Garcia, M. D., Udan, R. S., Hadjantonakis, A. K., & Dickinson, M. E. (2011). Preparation of postimplantation mouse embryos for imaging. pdb prot5594. *Cold Spring Harbor Protocols*, *2011*(4), 394–396. http://dx.doi.org/10.1101/pdb.prot5594.. PMID:21460043.

Garcia-Allende, P. B., Anabitarte, F., Conde, O. M., Mirapeix, J., Madruga, F. J., & Lopez-Higuera, J. M. (2008). Support vector machines in hyperspectral imaging spectroscopy with application to material identification. *SPIE Proceedings*, *6966*, 69661V-1–69661V-11.

Gilbert, R. J., Hoffman, M., Capitano, A., & So, P. T. (2000). Imaging of three-dimensional epithelial architecture and function in cultured CaCo2a monolayers with two-photon excitation microscopy. *Microscopy Research and Technique*, *51*(2), 204–210. http://dx.doi.org/10.1002/1097-0029(20001015)51:2<204::AID-JEMT11>3.0.CO;2-W. PMID:11054870.

Gobel, W., Kampa, B. M., & Helmchen, F. (2007). Imaging cellular network dynamics in three dimensions using fast 3D laser scanning. *Nature Methods*, *4*(1), 73–79. http://dx.doi.org/10.1038/nmeth989. PMID:17143280.

Gorisch, S. M., Lichter, P., & Rippe, K. (2005). Mobility of multi-subunit complexes in the nucleus: Accessibility and dynamics of chromatin subcompartments. *Histochemistry and Cell Biology, 123*(3), 217–228. http://dx.doi.org/10.1007/s00418-005-0752-y. PMID:15830242.

Gowen, A. A., O'Donnell, C. P., Cullen, P. J., Downey, G., & Frias, J. M. (2007). Hyperspectral imaging—An emerging process analytical tool for food quality and safety control. *Trends in Food Science & Technology, 18*, 590–598.

Graham, A., Okabe, M., & Quinlan, R. (2005). The role of the endoderm in the development and evolution of the pharyngeal arches. *Journal of Anatomy, 207*(5), 479–487. http://dx.doi.org/10.1111/j.1469-7580.2005.00472.x. PMID:16313389. PMCID:PMCPMC1571564.

Graham, A., & Smith, A. (2001). Patterning the pharyngeal arches. *Bioessays, 23*(1), 54–61. http://dx.doi.org/10.1002/1521-1878(200101)23:1<54::AID-BIES1007>3.0.CO;2-5. PMID:11135309.

Greger, K., Swoger, J., & Stelzer, E. H. (2007). Basic building units and properties of a fluorescence single plane illumination microscope. *The Review of Scientific Instruments, 78*(2), 023705. PMID:17578115.

Harland, R. M. (1991). In situ hybridization: An improved whole-mount method for *Xenopus* embryos. *Methods in Cell Biology, 36*, 685–695.

Haustein, E., & Schwille, P. (2007). Fluorescence correlation spectroscopy: Novel variations of an established technique. *Annual Review of Biophysics and Biomolecular Structure, 36*, 151–169. http://dx.doi.org/10.1146/annurev.biophys.36.040306.132612. PMID:17477838.

Hochgreb-Hagele, T., & Bronner, M. E. (2013). A novel FoxD3 gene trap line reveals neural crest precursor movement and a role for FoxD3 in their specification. *Developmental Biology, 374*(1), 1–11. http://dx.doi.org/10.1016/j.ydbio.2012.11.035. PMID:23228892. PMCID:PMCPMC3553214.

Holzer, M. S., Best, S. L., Jackson, N., Thapa, A., Raj, G. V., Cadeddu, J. A., et al. (2011). Assessment of renal oxygenation during partial nephrectomy using hyperspectral imaging. *The Journal of Urology, 186*(2), 400–404. http://dx.doi.org/10.1016/j.juro.2011.03.162. PMID:21679994.

Hu, D., Marcucio, R. S., & Helms, J. A. (2003). A zone of frontonasal ectoderm regulates patterning and growth in the face. *Development, 130*(9), 1749–1758. PMID:12642481.

Huisken, J., Swoger, J., Del Bene, F., Wittbrodt, J., & Stelzer, E. H. (2004). Optical sectioning deep inside live embryos by selective plane illumination microscopy. *Science, 305*(5686), 1007–1009. http://dx.doi.org/10.1126/science.1100035. PMID: 15310904.

Itoh, R. E., Kurokawa, K., Ohba, Y., Yoshizaki, H., Mochizuki, N., & Matsuda, M. (2002). Activation of rac and cdc42 video imaged by fluorescent resonance energy transfer-based single-molecule probes in the membrane of living cells. *Molecular and Cellular Biology, 22*(18), 6582–6591. PMID:12192056. PMCID:PMCPMC135619.

Jones, E. A., Crotty, D., Kulesa, P. M., Waters, C. W., Baron, M. H., Fraser, S. E., et al. (2002). Dynamic in vivo imaging of postimplantation mammalian embryos using whole embryo culture. *Genesis, 34*(4), 228–235. http://dx.doi.org/10.1002/gene.10162. PMID:12434332.

Kasemeier-Kulesa, J. C., Kulesa, P. M., & Lefcort, F. (2005). Imaging neural crest cell dynamics during formation of dorsal root ganglia and sympathetic ganglia. *Development, 132*(2), 235–245. http://dx.doi.org/10.1242/dev.01553. PMID:15590743.

Kawakami, M., Umeda, M., Nakagata, N., Takeo, T., & Yamamura, K. (2011). Novel migrating mouse neural crest cell assay system utilizing P0-Cre/EGFP fluorescent time-lapse imaging. *BMC Developmental Biology, 11*, 68. http://dx.doi.org/10.1186/1471-213X-11-68. PMID:22070366. PMCID:PMCPMC3224755.

Kern, D., Collins, M., Fultz, T., Detmer, J., Hamren, S., Peterkin, J. J., et al. (1996). An enhanced-sensitivity branched-DNA assay for quantification of human immunodeficiency virus type 1 RNA in plasma. *Journal of Clinical Microbiology*, *34*(12), 3196–3202. PMID:8940471. PMCID,PMCPMC229482.

Kimmel, C. B., Miller, C. T., & Moens, C. B. (2001). Specification and morphogenesis of the zebrafish larval head skeleton. *Developmental Biology*, *233*(2), 239–257. http://dx.doi.org/10.1006/dbio.2001.0201. PMID:11336493.

Knight, R. D., & Schilling, T. F. (2006). Cranial neural crest and development of the head skeleton. *Advances in Experimental Medicine and Biology*, *589*, 120–133. http://dx.doi.org/10.1007/978-0-387-46954-6_7. PMID:17076278.

Kobat, D., Horton, N. G., & Xu, C. (2011). In vivo two-photon microscopy to 1.6-mm depth in mouse cortex. *Journal of Biomedical Optics*, *16*(10), 106014. http://dx.doi.org/10.1117/1.3646209. PMID:22029361.

Kong, W., Zhang, C., Liu, F., Nie, P., & He, Y. (2013). Rice seed cultivar identification using near-infrared hyperspectral imaging and multivariate data analysis. *Sensors*, *13*, 8916–8927.

Kraynov, V. S., Chamberlain, C., Bokoch, G. M., Schwartz, M. A., Slabaugh, S., & Hahn, K. M. (2000). Localized Rac activation dynamics visualized in living cells. *Science*, *290*(5490), 333–337. PMID:11030651.

Kulesa, P. M., & Fraser, S. E. (1998). Neural crest cell dynamics revealed by time-lapse video microscopy of whole embryo chick explant cultures. *Developmental Biology*, *204*(2), 327–344. http://dx.doi.org/10.1006/dbio.1998.9082. PMID:9882474.

Kulesa, P. M., & Fraser, S. E. (2000). In ovo time-lapse analysis of chick hindbrain neural crest cell migration shows cell interactions during migration to the branchial arches. *Development*, *127*(6), 1161–1172. PMID:10683170.

Kulesa, P. M., & Gammill, L. S. (2010). Neural crest migration: Patterns, phases and signals. *Developmental Biology*, *344*(2), 566–568. http://dx.doi.org/10.1016/j.ydbio.2010.05.005. PMID:20478296. PMCID:PMCPMC2936914.

Kulesa, P. M., Teddy, J. M., Stark, D. A., Smith, S. E., & McLennan, R. (2008). Neural crest invasion is a spatially-ordered progression into the head with higher cell proliferation at the migratory front as revealed by the photoactivatable protein, KikGR. *Developmental Biology*, *316*(2), 275–287. http://dx.doi.org/10.1016/j.ydbio.2008.01.029. PMID:18328476. PMCID:PMCPMC3501347.

Lakowicz, J. R., Szmacinski, H., Nowaczyk, K., & Johnson, M. L. (1992). Fluorescence lifetime imaging of free and protein-bound NADH. *Proceedings of the National Academy of Sciences of the United States of America*, *89*(4), 1271–1275. PMID:1741380. PMCID, PMCPMC48431.

Lee, S. H., Bedard, O., Buchtova, M., Fu, K., & Richman, J. M. (2004). A new origin for the maxillary jaw. *Developmental Biology*, *276*(1), 207–224. http://dx.doi.org/10.1016/j.ydbio.2004.08.045. PMID:15531375.

Li, G. W., & Xie, X. S. (2011). Central dogma at the single-molecule level in living cells. *Nature*, *475*(7356), 308–315. http://dx.doi.org/10.1038/nature10315. PMID:21776076. PMCID:PMCPMC3600414.

Liu, B., Rooker, S. M., & Helms, J. A. (2010). Molecular control of facial morphology. *Seminars in Cell & Developmental Biology*, *21*(3), 309–313. http://dx.doi.org/10.1016/j.semcdb.2009.09.002. PMID:19747977. PMCID:PMCPMC3410822.

Lu, G., Halig, L., Wang, D., Qin, X., Chen, Z. G., & Fei, B. (2014). Spectral-spatial classification for noninvasive cancer detection using hyperspectral imaging. *Journal of Biomedical Optics*, *19*(10), 106004. http://dx.doi.org/10.1117/1.JBO.19.10.106004. PMID:25277147. PMCID:PMCPMC4183763.

Magde, D., Elson, E., & Webb, W. W. (1972). Thermodynamic fluctuations in a reacting system-measurement by fluorescence correlation spectroscopy. *Physical Review Letters*, *29*, 705–708.

Magde, D., Elson, E. L., & Webb, W. W. (1974). Fluorescence correlation spectroscopy. II. An experimental realization. *Biopolymers*, *13*(1), 29–61. http://dx.doi.org/10.1002/bip.1974.360130103. PMID:4818131.

Mahou, P., Zimmerley, M., Loulier, K., Matho, K. S., Labroille, G., Morin, X., et al. (2012). Multicolor two-photon tissue imaging by wavelength mixing. *Nature Methods*, *9*(8), 815–818. http://dx.doi.org/10.1038/nmeth.2098. PMID:22772730.

Matthews, H. K., Marchant, L., Carmona-Fontaine, C., Kuriyama, S., Larrain, J., Holt, M. R., et al. (2008). Directional migration of neural crest cells in vivo is regulated by Syndecan-4/Rac1 and non-canonical Wnt signaling/RhoA. *Development*, *135*(10), 1771–1780. http://dx.doi.org/10.1242/dev.017350. PMID:18403410.

McCormack, D. R., Walsh, A. J., Sit, W., Arteaga, C. L., Chen, J., Cook, R. S., et al. (2014). In vivo hyperspectral imaging of microvessel response to trastuzumab treatment in breast cancer xenografts. *Biomedical Optics Express*, *5*(7), 2247–2261. http://dx.doi.org/10.1364/BOE.5.002247. PMID:25071962. PMCID:PMCPMC4102362.

McGonnell, I. M., Clarke, J. D., & Tickle, C. (1998). Fate map of the developing chick face: Analysis of expansion of facial primordia and establishment of the primary palate. *Developmental Dynamics*, *212*(1), 102–118. http://dx.doi.org/10.1002/(SICI)1097-0177(199805)212:1<102::AID-AJA10>3.0.CO;2-9. PMID:9603428.

McLennan, R., Schumacher, L. J., Morrison, J. A., Teddy, J. M., Ridenour, D. A., Box, A. C., et al. (2015). Neural crest migration is driven by a few trailblazer cells with a unique molecular signature narrowly confined to the invasive front. *Development*, *142*(11), 2014–2025. http://dx.doi.org/10.1242/dev.117507. PMID:25977364.

Meyer, T., & Schindler, H. (1988). Particle counting by fluorescence correlation spectroscopy. Simultaneous measurement of aggregation and diffusion of molecules in solutions and in membranes. *Biophysical Journal*, *54*(6), 983–993. http://dx.doi.org/10.1016/S0006-3495(88)83036-4. PMID:3233275. PMCIDPMCPMC1330411.

Milet, C., & Monsoro-Burq, A. H. (2012). Neural crest induction at the neural plate border in vertebrates. *Developmental Biology*, *366*(1), 22–33. http://dx.doi.org/10.1016/j.ydbio.2012.01.013. PMID:22305800.

Miyawaki, A. (2003). Visualization of the spatial and temporal dynamics of intracellular signaling. *Developmental Cell*, *4*(3), 295–305. PMID:12636912.

Mochizuki, N., Yamashita, S., Kurokawa, K., Ohba, Y., Nagai, T., Miyawaki, A., et al. (2001). Spatio-temporal images of growth-factor-induced activation of Ras and Rap1. *Nature*, *411*(6841), 1065–1068. http://dx.doi.org/10.1038/35082594. PMID:11429608.

Moury, J. D., & Jacobson, A. G. (1990). The origins of neural crest cells in the axolotl. *Developmental Biology*, *141*(2), 243–253. PMID:2210034.

Mu, J., Slevin, J. C., Qu, D., McCormick, S., & Adamson, S. L. (2008). In vivo quantification of embryonic and placental growth during gestation in mice using micro-ultrasound. *Reproductive Biology and Endocrinology*, *6*, 34. http://dx.doi.org/10.1186/1477-7827-6-34. PMID:18700008. PMCID:PMCPMC2527569.

Murphy, D. B., & Davidson, M. W. (2013). *Fundamentals of light microscopy and electronic imaging* (2nd ed.). Hoboken, NJ: Wiley-Blackwell xiii, 538 pp.

Neuhauss, S. C., Solnica-Krezel, L., Schier, A. F., Zwartkruis, F., Stemple, D. L., Malicki, J., et al. (1996). Mutations affecting craniofacial development in zebrafish. *Development*, *123*, 357–367. PMID:9007255.

Nishiyama, C., Uesaka, T., Manabe, T., Yonekura, Y., Nagasawa, T., Newgreen, D. F., et al. (2012). Trans-mesenteric neural crest cells are the principal source of the colonic enteric nervous system. *Nature Neuroscience*, *15*(9), 1211–1218. http://dx.doi.org/10.1038/nn.3184. PMID:22902718.

Nowak, M., Machate, A., Yu, S. R., Gupta, M., & Brand, M. (2011). Interpretation of the FGF8 morphogen gradient is regulated by endocytic trafficking. *Nature Cell Biology*, *13*(2), 153–158. http://dx.doi.org/10.1038/ncb2155. PMID:21258372.

Palero, J. A., Bader, A. N., de Bruijn, H. S., der Ploeg van den Heuvel, A., Sterenborg, H. J., & Gerritsen, H. C. (2011). In vivo monitoring of protein-bound and free NADH during ischemia by nonlinear spectral imaging microscopy. *Biomedical Optics Express*, *2*(5), 1030–1039. http://dx.doi.org/10.1364/BOE.2.001030. PMID:21559117. PMCID: PMCPMC3087562.

Panasyuk, S. V., Yang, S., Faller, D. V., Ngo, D., Lew, R. A., Freeman, J. E., et al. (2007). Medical hyperspectral imaging to facilitate residual tumor identification during surgery. *Cancer Biology & Therapy*, *6*(3), 439–446. PMID:17374984.

Park, S. O., Wankhede, M., Lee, Y. J., Choi, E. J., Fliess, N., Choe, S. W., et al. (2009). Real-time imaging of de novo arteriovenous malformation in a mouse model of hereditary hemorrhagic telangiectasia. *The Journal of Clinical Investigation*, *119*(11), 3487–3496. http://dx.doi.org/10.1172/JCI39482. PMID:19805914. PMCID:PMCPMC2769195.

Patterson, G., Day, R. N., & Piston, D. (2001). Fluorescent protein spectra. *Journal of Cell Science*, *114*(Pt. 5), 837–838. PMID:11181166.

Pawley, J. B. (2006). *Handbook of biological confocal microscopy* (3rd ed.). New York: Springer. xxviii, 985 pp.

Piliszek, A., Kwon, G. S., & Hadjantonakis, A. K. (2011). Ex utero culture and live imaging of mouse embryos. *Methods in Molecular Biology*, *770*, 243–257. http://dx.doi.org/10.1007/978-1-61779-210-6_9. PMID:21805267. PMCID:PMCPMC3298811.

Pyrgaki, C., Trainor, P., Hadjantonakis, A. K., & Niswander, L. (2010). Dynamic imaging of mammalian neural tube closure. *Developmental Biology*, *344*(2), 941–947. http://dx.doi.org/10.1016/j.ydbio.2010.06.010. PMID:20558153. PMCID:PMCPMC3873863.

Quinn, K. P., Sridharan, G. V., Hayden, R. S., Kaplan, D. L., Lee, K., & Georgakoudi, I. (2013). Quantitative metabolic imaging using endogenous fluorescence to detect stem cell differentiation. *Scientific Reports*, *3*, 3432. http://dx.doi.org/10.1038/srep03432. PMID:24305550. PMCID:PMCPMC3851884.

Ries, J., Chiantia, S., & Schwille, P. (2009). Accurate determination of membrane dynamics with line-scan FCS. *Biophysical Journal*, *96*(5), 1999–2008. http://dx.doi.org/10.1016/j.bpj.2008.12.3888. PMID:19254560. PMCID:PMCPMC2717290.

Ries, J., & Schwille, P. (2006). Studying slow membrane dynamics with continuous wave scanning fluorescence correlation spectroscopy. *Biophysical Journal*, *91*(5), 1915–1924. http://dx.doi.org/10.1529/biophysj.106.082297. PMID:16782786. PMCID:PMCPMC1544284.

Ries, J., Yu, S. R., Burkhardt, M., Brand, M., & Schwille, P. (2009). Modular scanning FCS quantifies receptor-ligand interactions in living multicellular organisms. *Nature Methods*, *6*(9), 643–645. http://dx.doi.org/10.1038/nmeth.1355. PMID:19648917.

Rivera-Perez, J. A., Jones, V., & Tam, P. P. (2010). Culture of whole mouse embryos at early postimplantation to organogenesis stages: Developmental staging and methods. *Methods in Enzymology*, *476*, 185–203. http://dx.doi.org/10.1016/S0076-6879(10)76011-0. PMID:20691867.

Rocheleau, J. V., Head, W. S., & Piston, D. W. (2004). Quantitative NAD(P)H/flavoprotein autofluorescence imaging reveals metabolic mechanisms of pancreatic islet pyruvate response. *The Journal of Biological Chemistry*, *279*(30), 31780–31787. http://dx.doi.org/10.1074/jbc.M314005200. PMID:15148320.

Rodarmel, C., & Shan, J. J. (2002). Principal component analysis for hyperspectral image classification. *Surveying and Land Information Systems*, *62*(2), 115–122.

Roelfsema, N. M., Grijseels, E. W., Hop, W. C., & Wladimiroff, J. W. (2007). Three-dimensional sonography of prenatal skull base development. *Ultrasound in Obstetrics & Gynecology*, *29*(4), 372–377. http://dx.doi.org/10.1002/uog.3946. PMID:17290416.

Rossant, J., & Tam, P. P. L. (2002). *Mouse development: Patterning, morphogenesis, and organogenesis*. San Diego: Academic Press. xviii, 712 pp.

Roy, R., Hohng, S., & Ha, T. (2008). A practical guide to single-molecule FRET. *Nature Methods*, *5*(6), 507–516. http://dx.doi.org/10.1038/nmeth.1208. PMID:18511918 PMCID:PMCPMC3769523.

Ruf-Zamojski, F., Trivedi, V., Fraser, S. E., & Trinh, L. A. (2015). Spatio-temporal differences in dystrophin dynamics at mRNA and protein levels revealed by a novel FlipTrap line. *PLoS One*, *10*(6), e0128944. http://dx.doi.org/10.1371/journal.pone.0128944. PMID:26083378. PMCID:PMCPMC4471274.

Sauka-Spengler, T., & Bronner-Fraser, M. (2008). A gene regulatory network orchestrates neural crest formation. *Nature Reviews Molecular Cell Biology*, *9*(7), 557–568. http://dx.doi.org/10.1038/nrm2428. PMID:18523435.

Schlosser, G. (2006). Induction and specification of cranial placodes. *Developmental Biology*, *294*(2), 303–351. http://dx.doi.org/10.1016/j.ydbio.2006.03.009. PMID:16677629.

Selleck, M. A., & Bronner-Fraser, M. (1995). Origins of the avian neural crest: The role of neural plate-epidermal interactions. *Development*, *121*(2), 525–538. PMID:7768190.

Serbedzija, G. N., Bronner-Fraser, M., & Fraser, S. E. (1992). Vital dye analysis of cranial neural crest cell migration in the mouse embryo. *Development*, *116*(2), 297–307. PMID:1283734.

Seth, A., Otomo, T., Yin, H. L., & Rosen, M. K. (2003). Rational design of genetically encoded fluorescence resonance energy transfer-based sensors of cellular Cdc42 signaling. *Biochemistry*, *42*(14), 3997–4008. http://dx.doi.org/10.1021/bi026881z. PMID:12680752.

Shaner, N. C., Steinbach, P. A., & Tsien, R. Y. (2005). A guide to choosing fluorescent proteins. *Nature Methods*, *2*(12), 905–909. http://dx.doi.org/10.1038/nmeth819. PMID:16299475.

Simoes-Costa, M., & Bronner, M. E. (2015). Establishing neural crest identity: A gene regulatory recipe. *Development*, *142*(2), 242–257. http://dx.doi.org/10.1242/dev.105445. PMID:25564621. PMCID:PMCPMC4302844.

Speel, E. J. M., Hopman, A. H. N., & Komminoth, P. (1999). Amplification methods to increase the sensitivity of in situ hybridization: Play CARD(S). *The Journal of Histochemistry and Cytochemistry*, *47*(3), 281–288. PMID:ISI:000078701200002.

Stringari, C., Nourse, J. L., Flanagan, L. A., & Gratton, E. (2012). Phasor fluorescence lifetime microscopy of free and protein-bound NADH reveals neural stem cell differentiation potential. *PLoS One*, 7(11), e48014. http://dx.doi.org/10.1371/journal.pone.0048014. PMID:23144844. PMCID:PMCPMC3489895.

Stryer, L. (1978). Fluorescence energy transfer as a spectroscopic ruler. *Annual Review of Biochemistry*, *47*, 819–846. http://dx.doi.org/10.1146/annurev.bi.47.070178.004131. PMID:354506.

Sun, Y., Day, R. N., & Periasamy, A. (2011). Investigating protein-protein interactions in living cells using fluorescence lifetime imaging microscopy. *Nature Protocols*, *6*(9), 1324–1340. http://dx.doi.org/10.1038/nprot.2011.364. PMID:21886099. PMCID: PMCPMC3169422.

Szabo-Rogers, H. L., Smithers, L. E., Yakob, W., & Liu, K. J. (2010). New directions in craniofacial morphogenesis. *Developmental Biology*, *341*(1), 84–94. http://dx.doi.org/10.1016/j.ydbio.2009.11.021. PMID:19941846.

Tam, P. P. (1998). Postimplantation mouse development: Whole embryo culture and micro-manipulation. *The International Journal of Developmental Biology*, *42*(7), 895–902. PMID:9853819.

Trinh, L. A., Hochgreb, T., Graham, M., Wu, D., Ruf-Zamojski, F., Jayasena, C. S., et al. (2011). A versatile gene trap to visualize and interrogate the function of the vertebrate proteome. *Genes & Development*, *25*(21), 2306–2320. http://dx.doi.org/10.1101/gad.174037.111. PMID:22056673. PMCID:PMCPMC3219234.

Truong, T. V., Supatto, W., Koos, D. S., Choi, J. M., & Fraser, S. E. (2011). Deep and fast live imaging with two-photon scanned light-sheet microscopy. *Nature Methods*, *8*(9), 757–760. http://dx.doi.org/10.1038/nmeth.1652. PMID:21765409.

Verveer, P. J., Swoger, J., Pampaloni, F., Greger, K., Marcello, M., & Stelzer, E. H. (2007). High-resolution three-dimensional imaging of large specimens with light sheet-based microscopy. *Nature Methods, 4*(4), 311–313. http://dx.doi.org/10.1038/nmeth1017. PMID:17339847.

Wang, F., Flanagan, J., Su, N., Wang, L. C., Bui, S., Nielson, A., et al. (2012). RNAscope: A novel in situ RNA analysis platform for formalin-fixed, paraffin-embedded tissues. *The Journal of Molecular Diagnostics, 14*(1), 22–29. http://dx.doi.org/10.1016/j.jmoldx.2011.08.002. PMID:22166544. PMCID:PMCPMC3338343.

Wilkie, A. O., & Morriss-Kay, G. M. (2001). Genetics of craniofacial development and malformation. *Nature Reviews Genetics, 2*(6), 458–468. http://dx.doi.org/10.1038/35076601. PMID:11389462.

Yelick, P. C., Driever, W., Neuhauss, S., & Stashenko, P. (1996). Craniofacial cartilage development in zebrafish. *Annals of the New York Academy of Sciences, 785*, 360–361. PMID:8702181.

Yelick, P. C., & Schilling, T. F. (2002). Molecular dissection of craniofacial development using zebrafish. *Critical Reviews in Oral Biology and Medicine, 13*(4), 308–322. PMID:12191958.

Yoshizaki, H., Ohba, Y., Kurokawa, K., Itoh, R. E., Nakamura, T., Mochizuki, N., et al. (2003). Activity of Rho-family GTPases during cell division as visualized with FRET-based probes. *The Journal of Cell Biology, 162*(2), 223–232. http://dx.doi.org/10.1083/jcb.200212049. PMID:12860967. PMCID:PMCPMC2172791.

Young, N. M., Wat, S., Diewert, V. M., Browder, L. W., & Hallgrimsson, B. (2007). Comparative morphometrics of embryonic facial morphogenesis: Implications for cleft-lip etiology. *Anatomical Record (Hoboken), 290*(1), 123–139. http://dx.doi.org/10.1002/ar.20415. PMID:17441205.

Yu, S. R., Burkhardt, M., Nowak, M., Ries, J., Petrasek, Z., Scholpp, S., et al. (2009). Fgf8 morphogen gradient forms by a source-sink mechanism with freely diffusing molecules. *Nature, 461*(7263), 533–536. http://dx.doi.org/10.1038/nature08391. PMID:19741606.

Zipfel, W. R., Williams, R. M., & Webb, W. W. (2003). Nonlinear magic: Multiphoton microscopy in the biosciences. *Nature Biotechnology, 21*(11), 1369–1377. http://dx.doi.org/10.1038/nbt899. PMID:14595365.

Zuzak, K. J., Francis, R. P., Wehner, E. F., Litorja, M., Cadeddu, J. A., & Livingston, E. H. (2011). Active DLP hyperspectral illumination: A noninvasive, in vivo, system characterization visualizing tissue oxygenation at near video rates. *Analytical Chemistry, 83*(19), 7424–7430. http://dx.doi.org/10.1021/ac201467v. PMID:21842837.

INDEX

Note: Page numbers followed by "*f*" indicate figures and "*t*" indicate tables.

D

N

O

Q

R

S

T

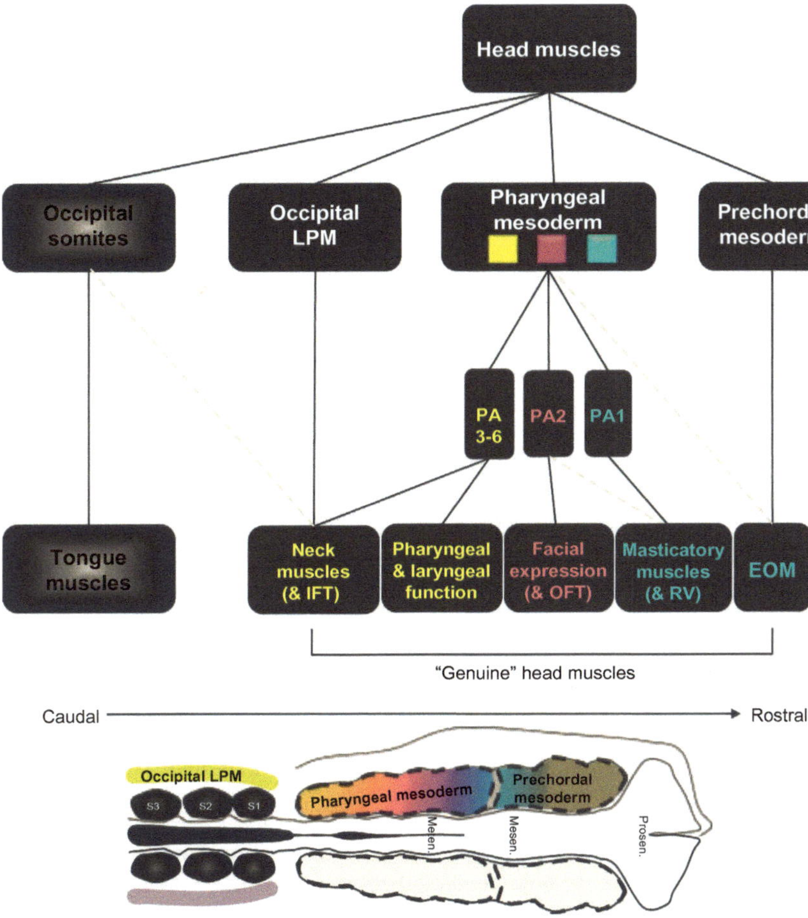

Inbal Michailovici *et al.*, Figure 1 The mesodermal origins of the head muscles. A diagram explaining the contribution of each embryonic tissue to the mature head muscles in caudal to rostral (left to right) order. The summary map shows the mesoderm origin sites of muscles in the chick embryo: prechordal mesoderm, pharyngeal mesoderm, occipital lateral plate mesoderm (LPM), and the occipital somites (S1–S3). Pharyngeal arch (PA), inflow tract (IFT), outflow tract (OFT), right ventricle (RV), and extraocular muscles (EOM).

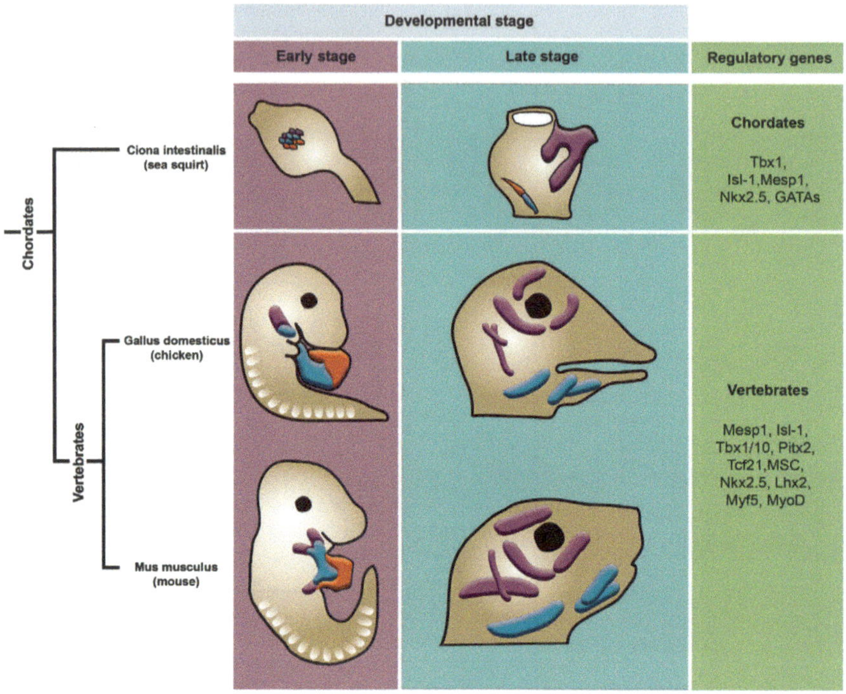

Inbal Michailovici *et al.*, Figure 2 Evolutionarily conserved cardiopharyngeal muscle progenitors. Schematic of early- and late-stage embryonic head muscle development, indicating conservation of cardiopharyngeal precursors. The chordate *Ciona intestinalis* possesses first heart precursors (FHP, red), and second heart precursors (SHP, blue), which contribute to the development of the heart, whereas oral siphon muscles (OSM, purple) are derived from a population of trunk lateral cells. The vertebrate mouse and chick embryos exhibit similar first heart field precursors (FHF, red) and second heart field precursors (SHF, blue), as well as branchiomeric progenitors contributing to craniofacial skeletal muscle (purple). Genes involved in the regulation of craniofacial muscle development and differentiation programs are indicated. *Adapted from Diogo et al. (2015).*

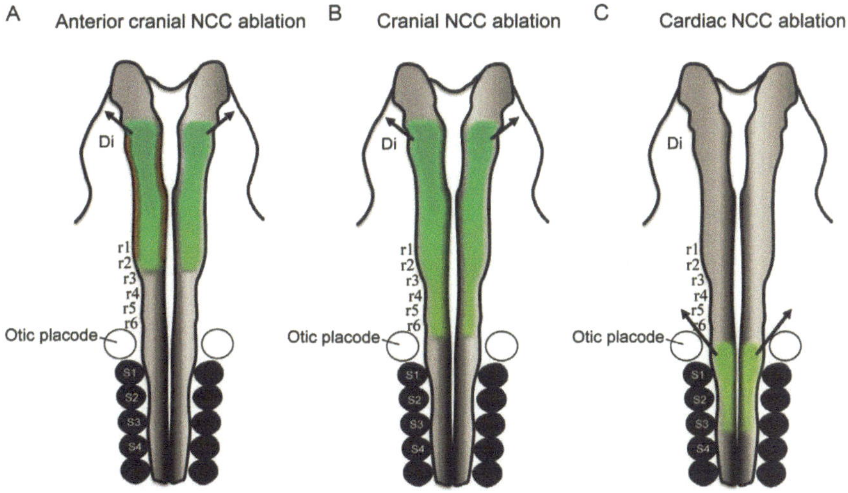

Inbal Michailovici *et al.*, Figure 3 Neural crest cell ablation procedures. (A) Anterior cranial NCC ablation is the ablation of the dorsal half of the neural plate, bilaterally from the mid-diencephalon (Di) down through rhombomere 2 (r2). (B) Cranial NCC ablation is carried out from the diencephalic level to the level of rhombomere 5/6 (anterior to somite 1, S1). (C) Cardiac NCC ablation includes the ablation of the neural folds, bilaterally from the level of the mid-otic placode through the posterior limit of somite 3.

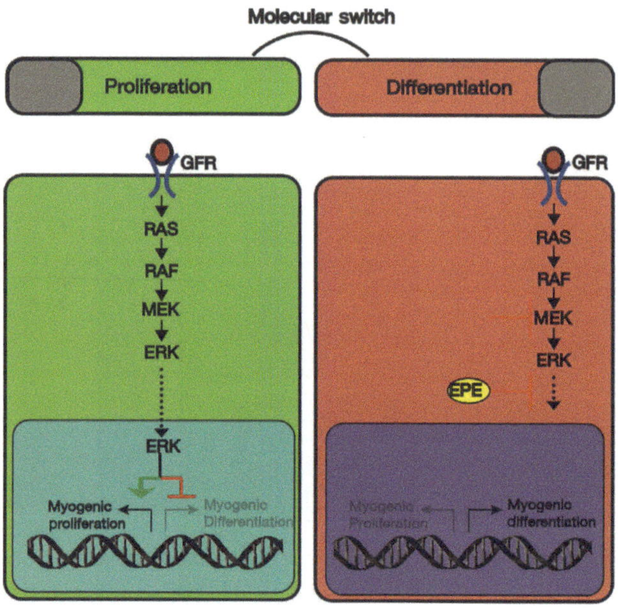

Inbal Michailovici *et al.*, Figure 4 ERK-mediated molecular switch determines myogenic differentiation. As demonstrated by Michailovici et al. (2014), muscle differentiation is determined by the cellular localization of the signaling molecule ERK. Nuclear localization of ERK promotes a progenitor/proliferative phenotype (left side), while cytoplasmic ERK is associated with myogenic differentiation (right side).

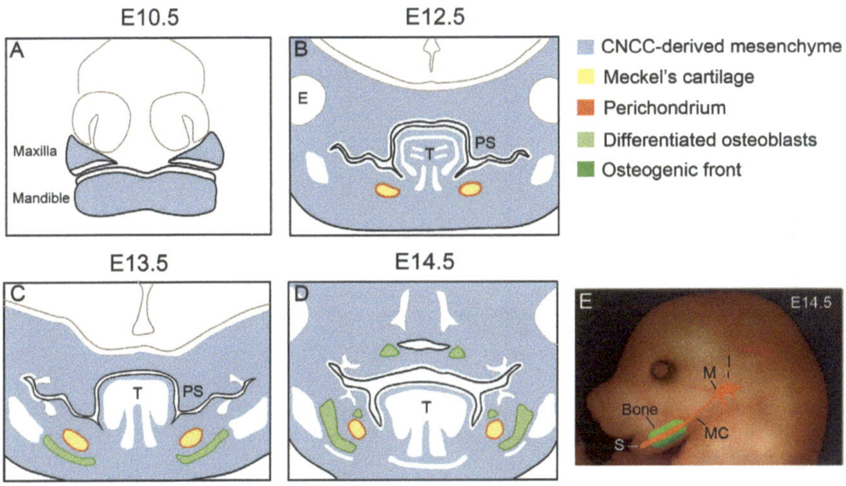

Carolina Parada and Yang Chai, Figure 1 Scheme of mandible development. (A) The first BA subdivides into maxillary and mandibular prominences at E10.5. At this stage, both prominences are solely constituted by CNCC-derived mesenchyme, which is covered by epithelium. (B) At E12.5, the MC develops in the mandibular primordium and mesenchymal cells condense close to it. (C) Mesenchymal cells start differentiating into osteoblasts. An osteogenic front composed of undifferentiated cells surrounds the differentiated osteoblasts. (D) Differentiation progresses in the mandibular primordium at E14.5. (E) The MC gives rise to the symphysis in the distal region and to the malleus and incus in the proximal region of the mandible. *Abbreviations*: E, eye; MC, Meckel's cartilage; I, incus; M, malleus; PS, palatal shelf; S, symphysis; T, tongue.

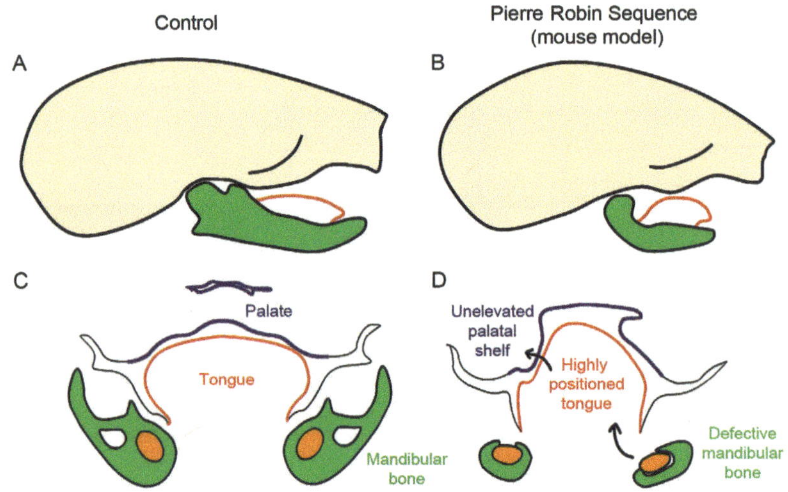

Carolina Parada and Yang Chai, Figure 2 Mechanical relationship between the mandible and tongue. (A) and (B) show schemes of a sagittal view of the skull of control and a PRS mouse model at newborn stage, respectively. (C) and (D) are schemes of corresponding coronal sections.

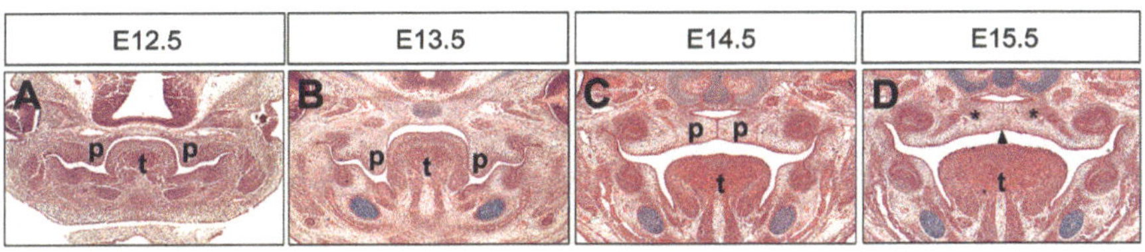

Yu Lan *et al.*, Figure 1 Palate development in mice. (A–D) Representative HE-stained coronal sections through the middle of the anterior–posterior axis of the developing palatal shelves at E12.5 (A), E13.5 (B), E14.5 (C), and E15.5 (D). The palatal shelves grow vertically from the oral side of the maxillary processes from E11.5 to past E13.5 (A and B). By E14.5, the palatal shelves have elevated to a horizontal position above the tongue, adhered to each other at the midline and formed the midline epithelial seam (MES) (C). At around E15.5, the MES is disintegrated to allow confluence of the palatal mesenchyme (D). Arrowhead in (D) points to the disintegrating MES. Asterisks in (D) mark the sites of ossification that will form the palatal bone. p, palatal shelf; t, tongue.

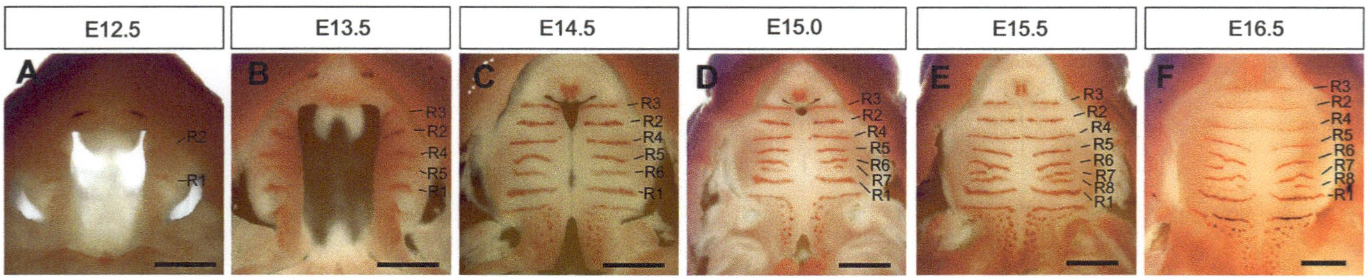

Yu Lan *et al.*, Figure 2 Whole-mount oral view of growth and patterning of the developing palatal shelves in mice. Patterns of *Shh* mRNA expression were detected by whole-mount *in situ* hybridization. R1–R8 mark the Shh-expressing palatal rugae in the sequence of their formation from E12.5 (A) to E16.5 (F). R1 marks the anterior–posterior boundary (all panels are oriented with the tip of the snout pointing up). The patchy *Shh* expression in the posterior palate marks the sensory papilla buds. Scale bar, 500 μm.

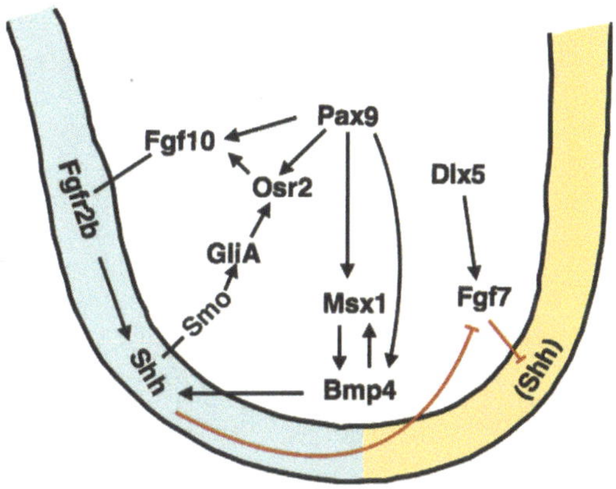

Yu Lan *et al.*, Figure 3 Molecular regulation of palatal shelf growth and oronasal patterning. Schematic diagram of a coronal section of the developing palatal shelf, with the oral side to the left. Black arrows indicate positive regulation whereas red lines indicate repression. Shh is expressed in the palatal epithelium and positively regulates Fgf10 expression through the Osr2 transcription factor in the palatal mesenchyme. Fgf10 signals through the Fgfr2b receptor to maintain Shh expression in the palatal epithelium. Dlx5 activates Fgf7 expression in the palatal mesenchyme. Fgf7 and Shh repress each other's expression such that Fgf7 expression is restricted to the nasal side of the palatal mesenchyme and Shh to the oral side of the palatal epithelium. Msx1 maintains Shh expression through Bmp4 in the anterior palate whereas Pax9 acts upstream of Bmp4, Fgf10, Msx1, and Osr2 in the palatal mesenchyme.

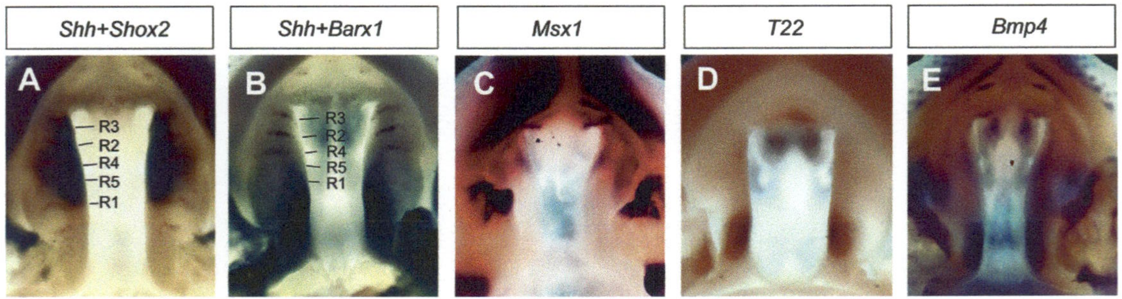

Yu Lan *et al.*, Figure 4 Differential patterns of gene expression along the anterior–posterior axis of the developing palatal shelves at E13.5. (A) Dual color *in situ* hybridization detection of *Shh* (brown) and *Shox2* (blue) mRNA expression. *Shh* is expressed in the palatal rugae (marked as R1–R5 in the sequence of their formation), whereas *Shox2* mRNAs are restricted to the anterior half of the developing palatal shelves. R1 coincides with the posterior boundary of *Shox2* mRNA expression. (B) Dual color *in situ* hybridization detection of *Shh* (brown) and *Barx1* (blue) mRNA expression. *Barx1* is mainly expressed in the posterior halves of the developing palatal shelves. (C) Whole-mount *in situ* hybridization detection of *Msx1* (purple) mRNA expression. *Msx1* mRNAs were detected in the anterior-most quarter of the developing palatal shelves. (D) *Tbx22* mRNA expression is highly restricted in the posterior half of the developing palatal shelves. (E) *Bmp4* mRNA expression was detected in the anterior and posterior subregions of the developing palatal shelves.

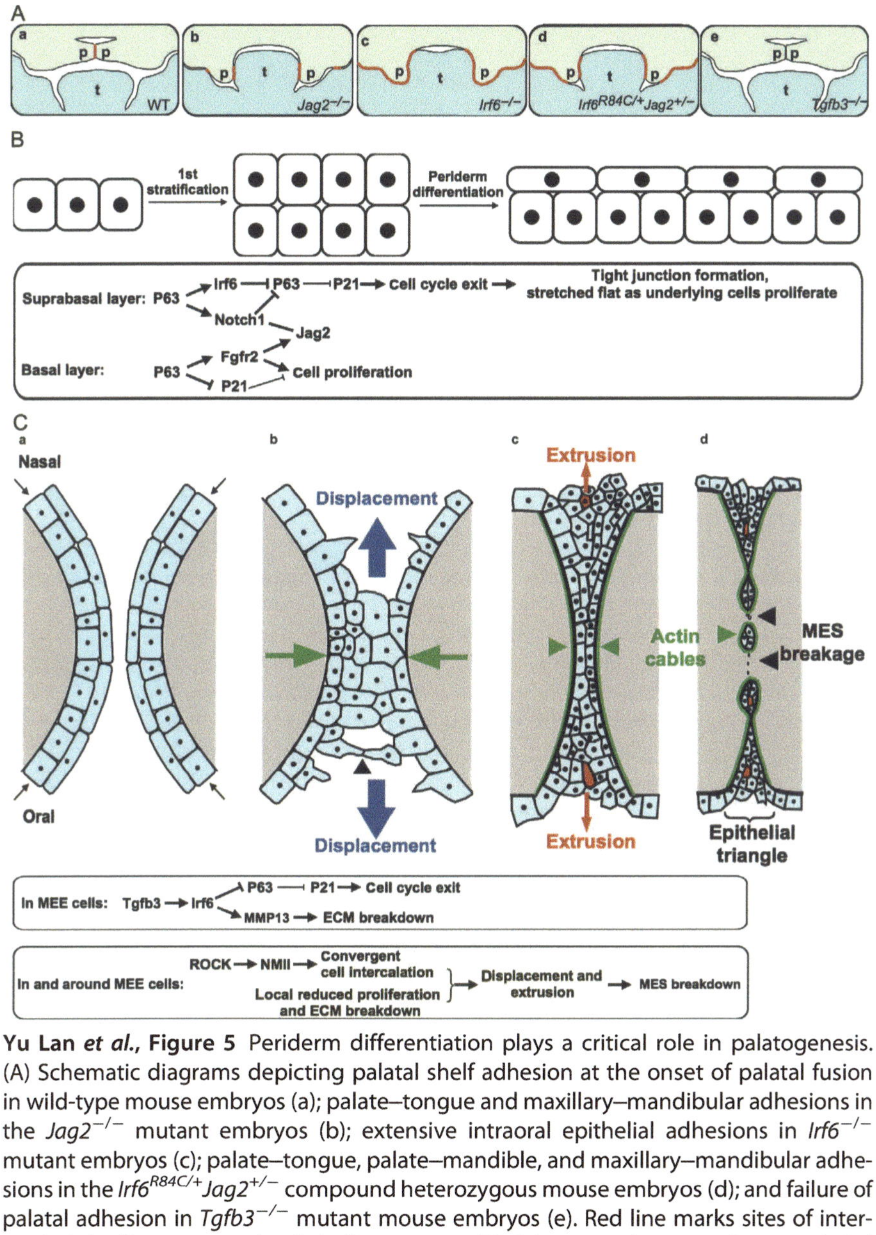

Yu Lan *et al.*, Figure 5 Periderm differentiation plays a critical role in palatogenesis. (A) Schematic diagrams depicting palatal shelf adhesion at the onset of palatal fusion in wild-type mouse embryos (a); palate–tongue and maxillary–mandibular adhesions in the *Jag2*$^{-/-}$ mutant embryos (b); extensive intraoral epithelial adhesions in *Irf6*$^{-/-}$ mutant embryos (c); palate–tongue, palate–mandible, and maxillary–mandibular adhesions in the *Irf6*$^{R84C/+}$*Jag2*$^{+/-}$ compound heterozygous mouse embryos (d); and failure of palatal adhesion in *Tgfb3*$^{-/-}$ mutant mouse embryos (e). Red line marks sites of inter-epithelial adhesion. p, palatal shelf; t, tongue. (B) Schematic diagram of 1st epithelial stratification and periderm differentiation, with the molecular and cellular pathways in the suprabasal and basal cells outlined in the rectangular box below. (C) Schematic diagram of coronal sections of palatal shelves depicting cell convergence and extrusion during palatal adhesion and fusion, with the molecular and

(Continued)

Yu Lan *et al.*, Figure 5—Cont'd cellular processes regulating MES breakdown outlined in the rectangular box below. Palatal epithelial cells are in light blue color and the palatal mesenchyme is in gray. The paired palatal shelves are each covered with a single layer of periderm cells prior to interpalatal adhesion (a). At the onset of palatal fusion, periderm cells and basal MEE cells extend protrusions and converge toward (green arrows) the midline (b), with some cells in the multicellular seam displaced toward the oral or nasal side. As the MES forms, cellular rosettes form and cells at the center of the rosettes are squeezed out by extrusion (marked in red) (c). The actomyosin cables pull the MES apart and surround the epithelial islands, continuing to extrude cells from cellular rosettes (d). Palatal MES breakdown is driven by both the non-myosin II-mediated cell convergence and extrusion as well as by Tgfb3-Irf6 regulated local cell cycle arrest and ECM breakdown. *Panel (C): The cell convergence and extrusion diagrams are modified from Kim et al. (2015).*

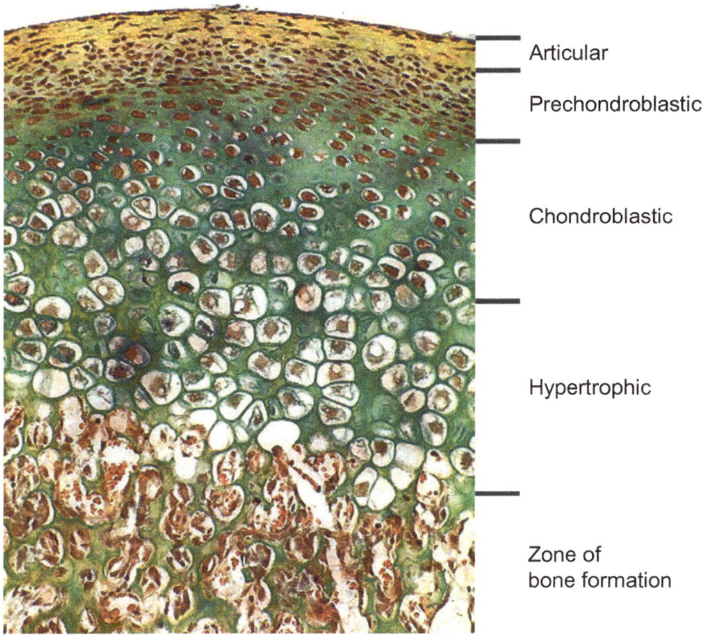

Robert J. Hinton *et al.*, Figure 1 Histology of the growing (postnatal) rat mandibular condylar cartilage. Coronal section, Attwood's stain.

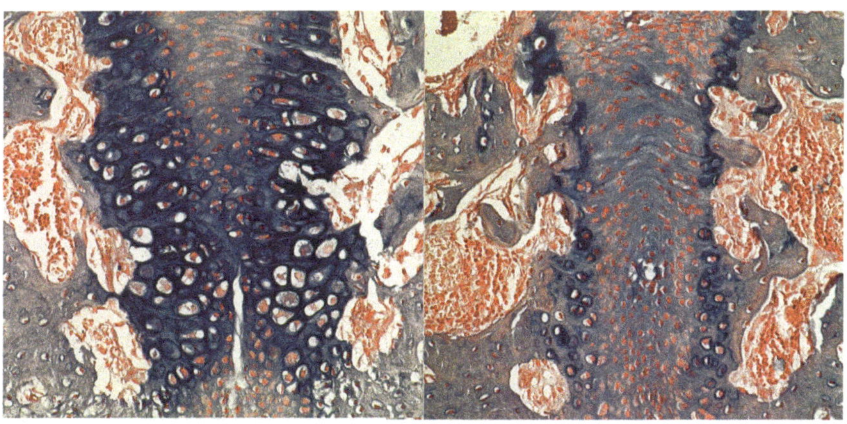

Robert J. Hinton *et al.*, Figure 2 Secondary cartilage at intermaxillary suture in rats fed a normal hard diet (left) or soft, mushy diet (right) for 2.5 weeks. Coronal sections, alcian blue-Kernechtrot aluminum sulfate. Original magnification ×40. Darkest densities correspond to alcian blue staining. Note marked reduction in extent of alcian blue staining and loss of hypertrophic chondrocytes in soft diet animals.

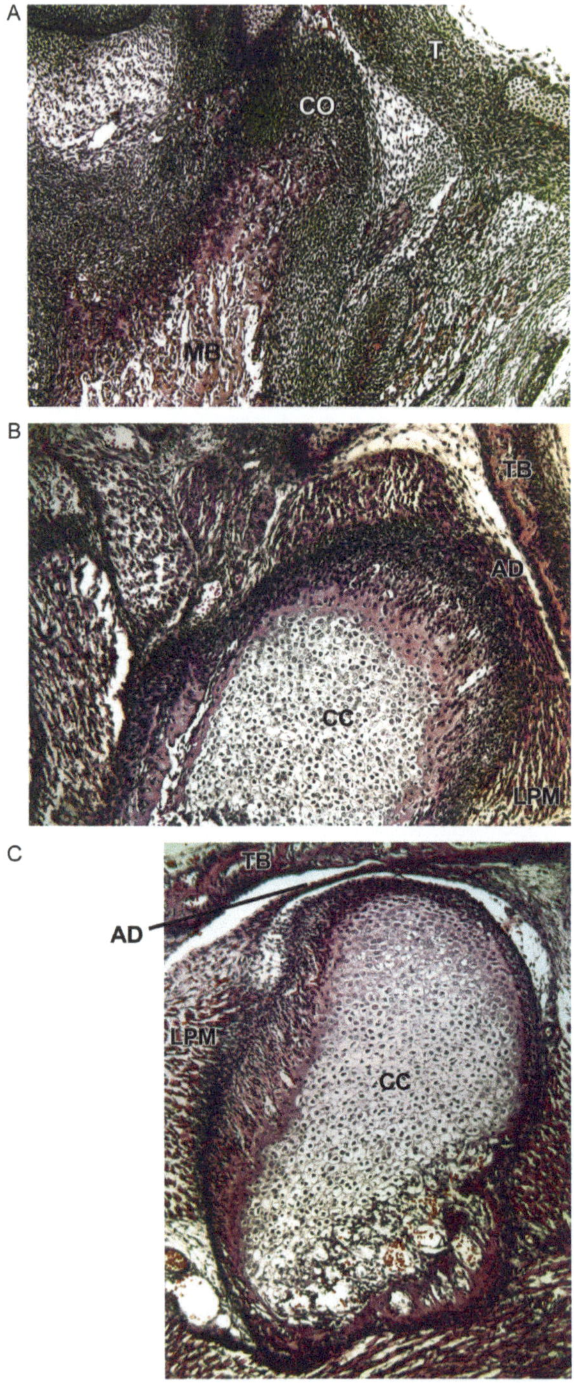

Robert J. Hinton *et al.*, Figure 3 See legend on opposite page.

Robert J. Hinton *et al.*, Figure 3 (A) Temporomandibular joint at *blastematic* stage showing condylar (CO) and temporal (T) blastemas separated by less dense mesenchyme that will form the future disc. Mandibular bone, MB. Mouse, embryonic day 15. Hematoxylin and eosin, sagittal section. (B) Temporomandibular joint at *cavitation* stage illustrating chondro-differentiation of condylar cartilage and formation of joint spaces in progress. Superior joint space is evident (upper right), but inferior joint space is not yet formed. CC, condylar cartilage; AD, articular disc; LPM, lateral pterygoid muscle; TB, temporal bone. Mouse*, embryonic day 17. Hematoxylin and eosin, sagittal section. *The superior joint space forms earlier in a mouse, opposite the human condition. (C) Temporomandibular joint at *maturation* stage. Both inferior and superior joint spaces are fully formed, lateral pterygoid muscle is attached to condylar neck and articular disc begins to resemble postnatal form. CC, condylar cartilage; AD, articular disc; LPM, lateral pterygoid muscle; TB, temporal bone. Mouse, embryonic day 19. Hematoxylin and eosin, sagittal section.

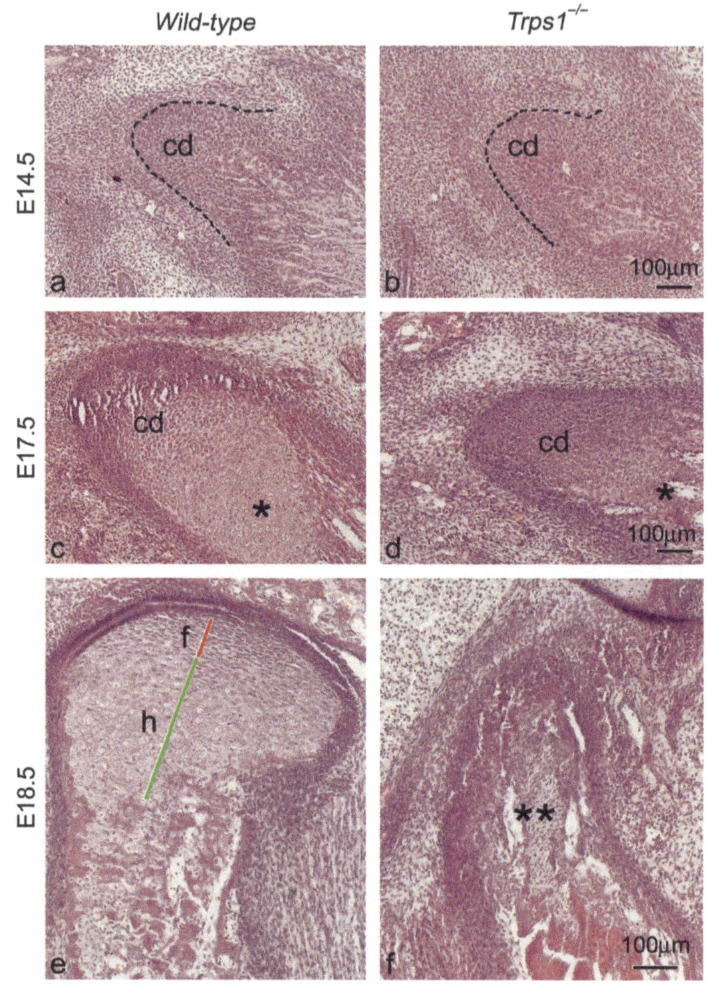

Robert J. Hinton *et al.*, Figure 4 Development of the TMJ in *Trps1* knockout mice at E14.5 (A, B), E17.5 (C, D), and E18.5 (E, F). H–E staining of sagittal sections of the wild-type (A, C, E) and of mutants (B, D, F). Right is anterior in each section. (A, B) The dotted line shows the margin of the developing condyle. (C, D) The star indicates the region of maturating chondrocytes in the condyle. (E, F) Flattened (f) and hypertrophic (h) cell layers are indicated by the red line and green line, respectively (double stars in f indicate abnormally matured chondrocytes in the mutant condyle, cd condyle). *Used with permission of Michikami et al. (2012).*

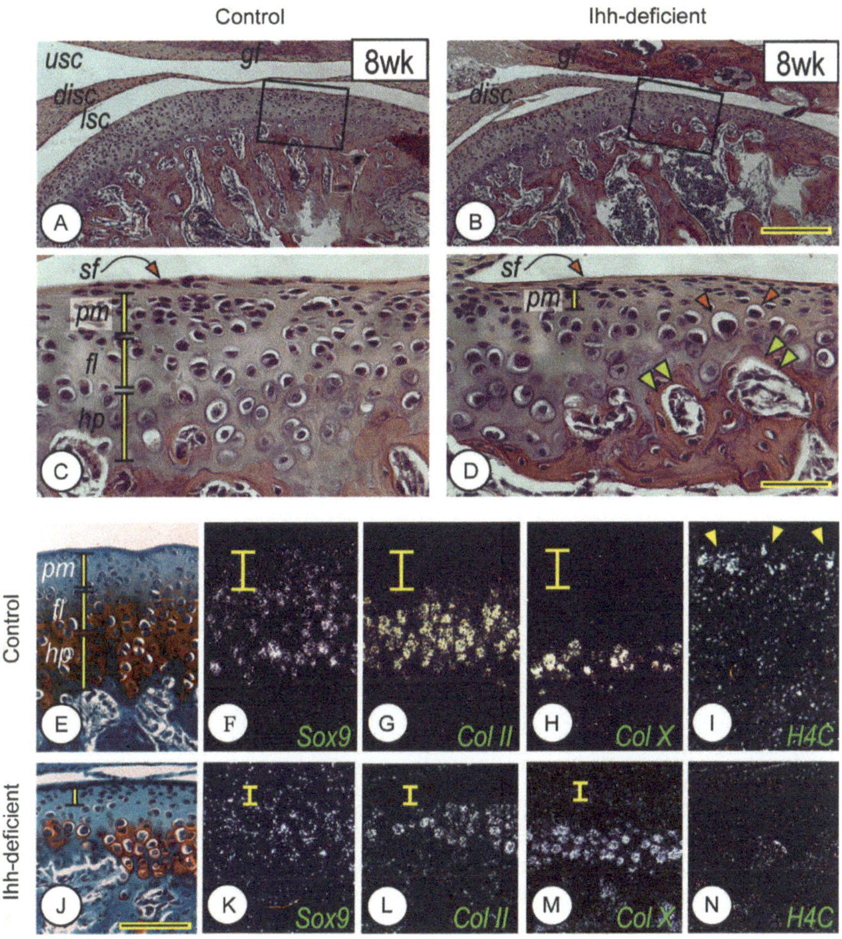

Robert J. Hinton *et al.*, Figure 5 Mandibular condylar cartilage organization is abnormal in *Ihh*-deficient mice. H&E-stained parasagittal sections from 8-week-old control (A, C) and *Ihh*-deficient (B, D) condyles. Control (E) and ihh-deficient (J) condylar cartilage stained with safranin O/fast green show the reduced extent in cKO mice. Serial parasagittal sections from control (F–I) and *Ihh*-deficient condyles (K–N) processed for *in situ* hybridization of: *Sox9* (F, K), *collagen II* (*Col II*) (G, L), collagen X (H, M), and *histone H4C* (*H4C*) (I, N) demonstrate zonal disruption of gene expression and proliferation in cKO mice. Vertical bars in C and E indicate zonal organization in control condyles. *sf*, superficial layer; *pm*, polymorphic zone; *fc*, flattened chondrocyte zone; *hc*, hypertrophic zone. *Used with permission of Ochiai et al. (2010).*

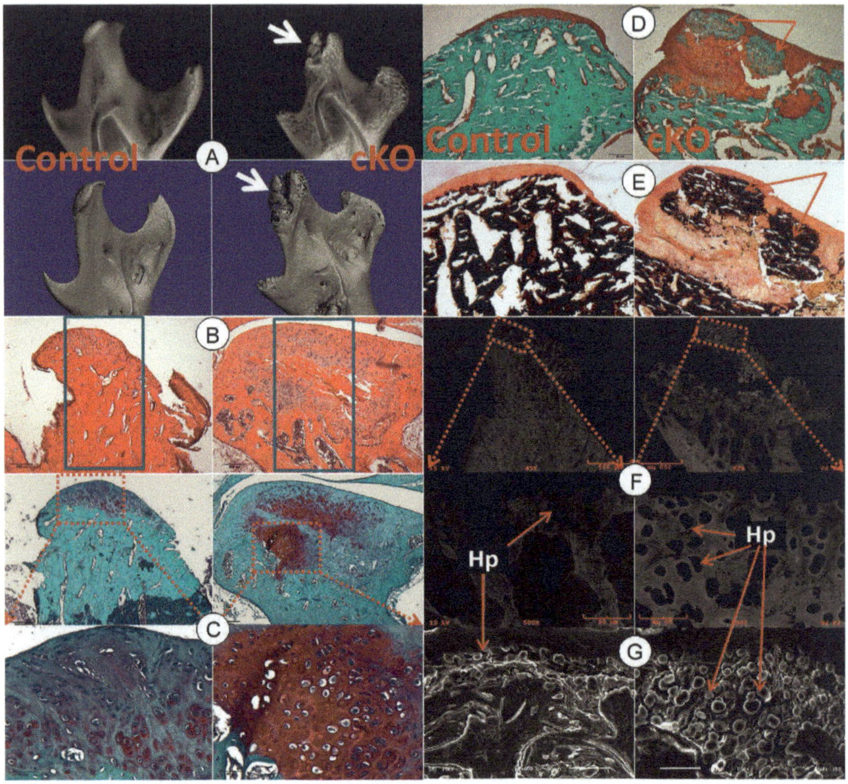

Robert J. Hinton *et al.*, Figure 6 Large calcified cartilage masses are present in condylar cartilage from 5-month-old Osx-cKO mice. (A) Radiographic and μ-CT images showed that the surface of the calcified region in the mandibular condyle of cKO mice was separated from the underlying bone of the condylar neck by a radiolucent gap. (B, C) (upper images) Discrete areas of safranin-O-stained cartilage were present in the MCC of cKO mice, separated by stained regions similar to those of the underlying bone; (lower images) hypertrophic chondrocytes were more numerous in cKO condyles and sometimes clustered. Goldner (D) and von Kossa (E) staining showed large calcified areas within the MCC in cKO mice. (F, G) Back-scattered SEM and acid-etch SEM indicated that the calcified areas in the mutant mice MCC were composed of hypertrophic chondrocytes. Hp, hypertrophic chondrocyte. *Used with permission of Jing, Hinton, Jing, et al. (2014).*

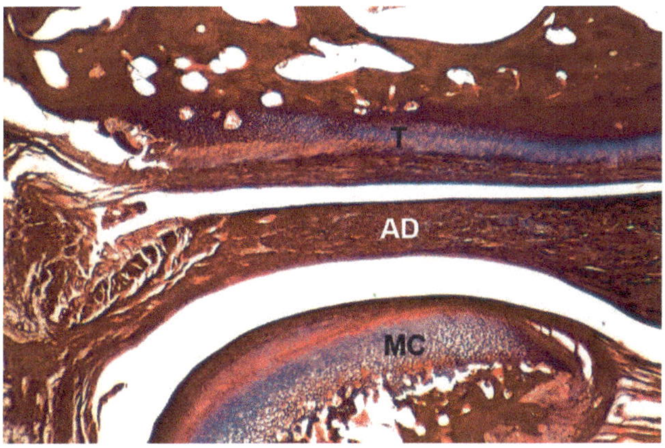

Robert J. Hinton *et al.*, Figure 7 Sagittal section through temporomandibular joint of growing monkey showing mandibular condyle (MC), temporal bone (T), and articular disc (AD). Note that the bluish stain characterizing cartilage is present deep to the perichondrium on the temporal bone as well as the condyle. Masson trichrome.

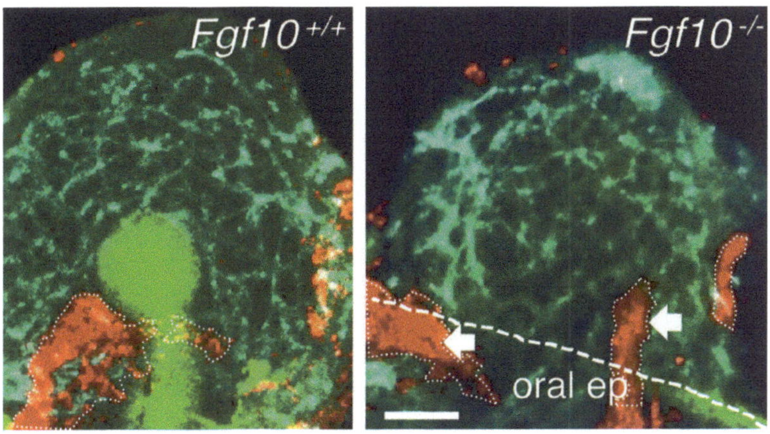

Belinda R. Hauser and Matthew P. Hoffman, Figure 1 Fgf10 is essential for SMG epithelial development. Whole-mount immunostaining of an E12 wild-type (*Fgf10*$^{+/+}$) SMG and *Fgf10*$^{-/-}$ SMG mesenchyme. SMG development involves the formation of an epithelial endbud and primary duct (E-cadherin, green) within a condensed mesenchyme, which is surrounded by an endothelial cell plexus (PCAM, cyan blue) and the formation of a PSG (Tubb3, red). White dotted lines outline PSG, white dashed line is the oral epithelium (oral ep); arrows indicate nerves in the adjacent tongue and oral epithelium. Scale bar, 50 μm. *From Fig. 2B in Knosp et al. (2015).*

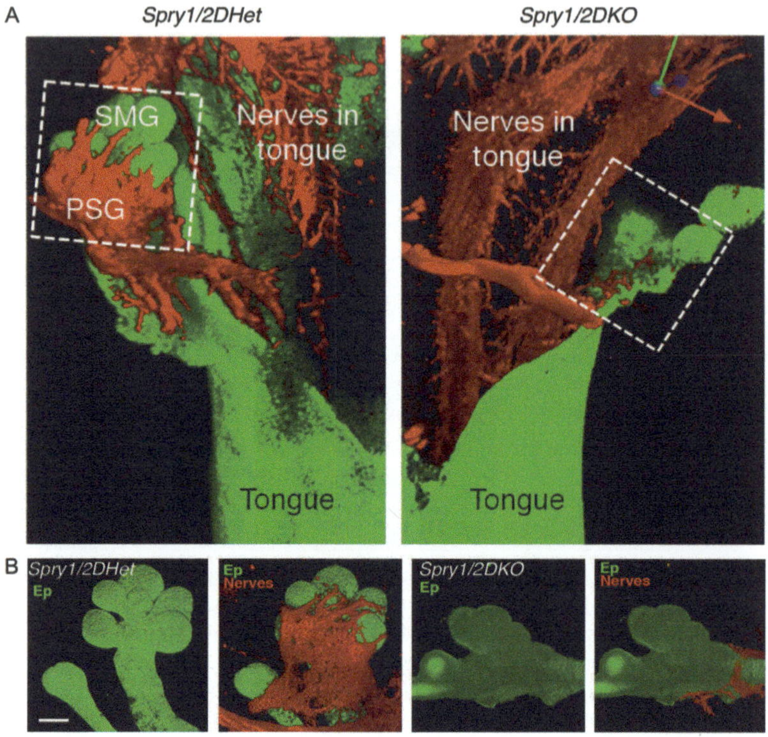

Belinda R. Hauser and Matthew P. Hoffman, Figure 2 There is loss of the PSG and SMG innervation in the *Spry1/2DKO* mouse, although the sensory innervation of the tongue is not affected. (A) Whole-mount immunostaining for the epithelium (Ep, E-cadherin) and nerves (Tubb3) in *Spry1/2DHet* (control) and *Spry1/2DKO* tongues with SMGs attached. (B) The SMGs (green) and PSG (red) that are in white box are shown separately in the lower panels. Scale bar, 100 μm. *From Figure S1A in Knosp et al. (2015).*

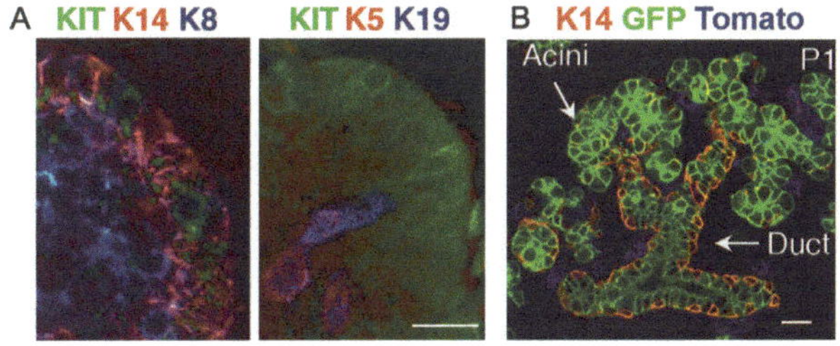

FACS (%) of E13 KIT+K8+ cells:

K14+	14±3
K5+	10±2
K5+K19+	7±2
K14+K5+	7±1
K19+	2±1
K14-K5-K19-	56±4

Belinda R. Hauser and Matthew P. Hoffman, Figure 3 SMG endbuds contain distinct populations of KIT+K14+ and KIT+K5+ progenitors, and the K14+ progenitors are multipotent. (A) Immunostaining of KIT, K14, K8, and K19 in an E13 endbud. Pink cells coexpress K14 and K8 or K5 and K19, 1 μm optical sections. (B) Lineage tracing of K14 in a postnatal day 1 SMG from a K14-Cre x Rosa^{mTmG} mouse. K14 progeny cells (GFP, green) are throughout the gland and label acini, ducts, and myoepithelial cells. The image also shows the endogenous K14 (red and yellow) and non-K14 cells (Tomato, blue), 2 μm sections. Scale bar, 20 μm. (C) FACS analysis to quantitate the subpopulations of epithelial KIT+K8+ cells sorted by K5, K14, and K19 expression from E13 SMGs. Mean ± SEM; $n > 3$ biological samples. *From Figure 4 in Lombaert et al. (2013).*

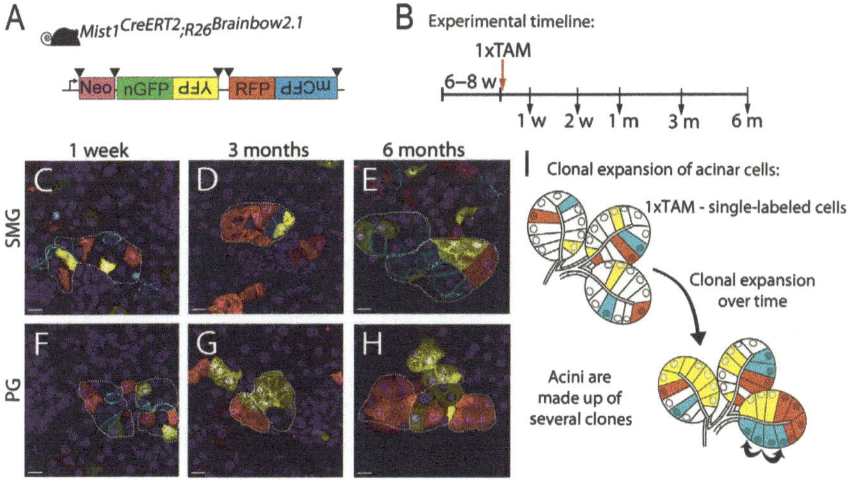

Belinda R. Hauser and Matthew P. Hoffman, Figure 4 Acinar cell renewal occurs by self-duplication, resulting in clonal expansion of acinar cells in the SMG and PG. (A) The inducible Mist1-CreERT2 was used in combination with the R26$^{Brainbow2.1}$ reporter strain (schematic). Black triangles represent LoxP sites. (B) Adult mice (6–8 weeks old) were treated with a single dose of tamoxifen. Tissues were harvested at indicated time points. (C) Labeled cells in the SMG at 1 week following tamoxifen treatment. (D) Clonal expansion of labeled cells in the SMG at 3 months after tamoxifen treatment. (E) Acini comprised of several clones demonstrate tissue turnover 6 months after tamoxifen treatment. (F) Single-labeled cells in the PG 1 week following tamoxifen treatment. (G) Clonal expansion of labeled cells in the PG at 3 months. (H) Clonal analysis after a 6-month chase in the PG. (I) Model of acinar cell proliferation and clonal expansion in adult SMG and PG. *From Figure 2 in Aure et al. (2015).*

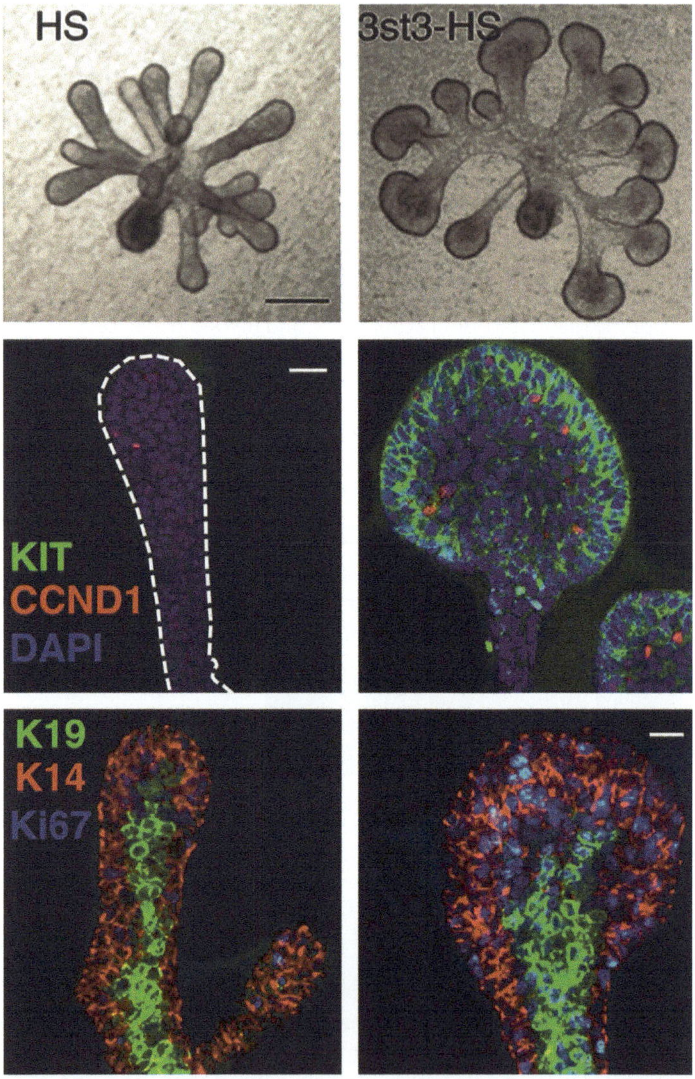

Belinda R. Hauser and Matthew P. Hoffman, Figure 5 3-*O*-sulfated-HS (3st3-HS) increases epithelial endbud morphogenesis and proliferation as well as KIT and K14 expression. Brightfield images of E13 SMG epithelia cultured with Fgf10 and either HS or 3st3-HS for 28 h, top two panels. The epithelia were stained for proliferating cells with CCND1 (red), KIT (green), and nuclei (blue), middle two panels. Lower two panels show K14 (red), K19 (green), and proliferation (Ki67, blue). Images are 2 μm confocal sections. Scale bars, 10 μm. *From Figures 2 and 6 in Patel et al. (2014).*

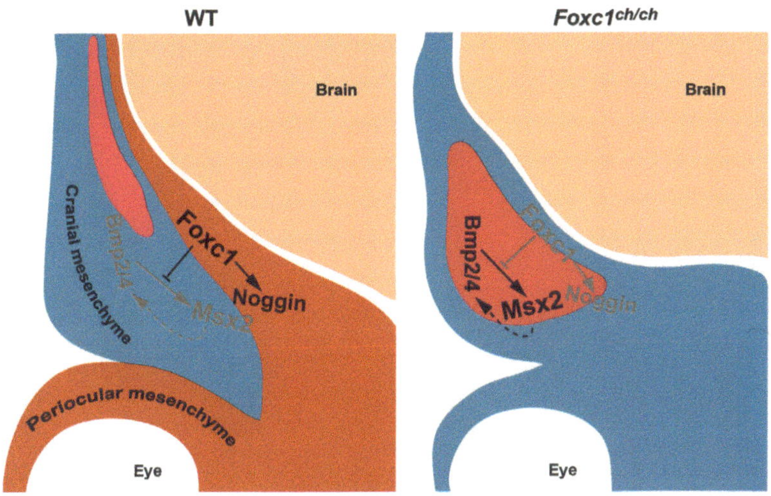

Mamoru Ishii *et al.*, Figure 1 Schematic model showing role of Foxc1 in controlling Bmp-induced *Msx2* expression and specification of osteoprogenitors in the supraorbital ridge. At E12.5, *Foxc1* is expressed in the meninges and adjacent mesenchyme (in red), including the periocular domain. Osteogenic cells are marked in fuchsia, representing the expression of ALP. In wild-type embryos, Foxc1 negatively regulates *Msx2* and *Bmp2/4* and positively regulates *Noggin*, thereby maintaining a level of Bmp signaling appropriate for the specification of the frontal bone primordia as well as the apical migration of undifferentiated progenitor cells. In *Foxc1* mutants, loss of *Foxc1* results in upregulation of *Msx2*, which initiates a positive-feedback loop between *Msx2* and *Bmp2/4*. The result is the propagation of a wave of Bmp–Msx2 pathway from medial to lateral across the supraorbital ridge and the lateral expansion of the ALP-expressing differentiating osteoblast domain. *After Fig. 7, Sun et al. (2013).*

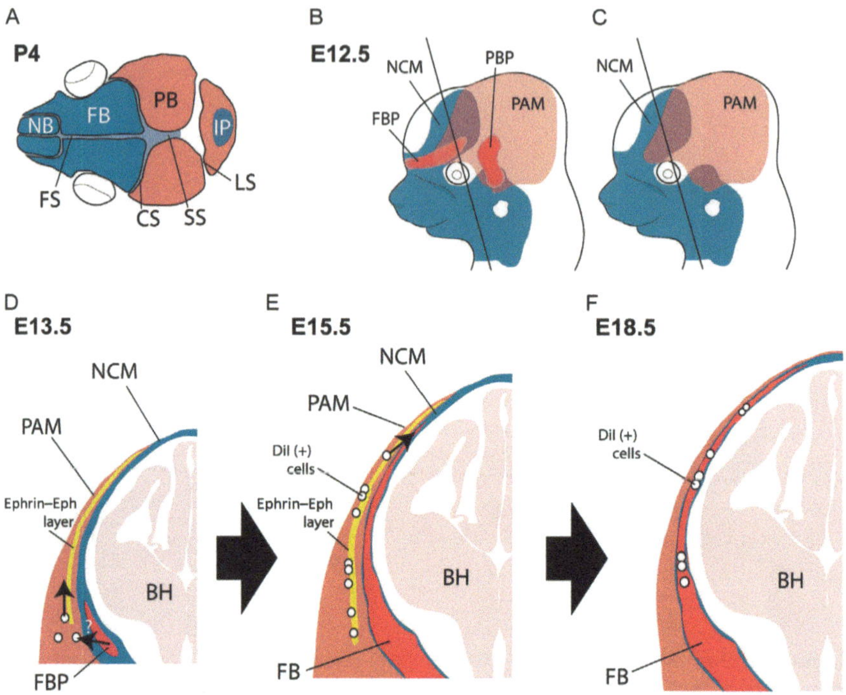

Mamoru Ishii *et al.*, Figure 2 Development of the skull vault. (A) Origins of bones of the skull vault. Blue, neural crest; pink, mesoderm. Note that the paired frontal bones are derived from neural crest and parietal bones from mesoderm. The interparietal bone has a dual origin: the central portion is neural crest-derived; the flanking portions come from mesoderm. (B) and (C) side views of head showing cranial layers. Alp-expressing rudiments are shown in magenta. (D)–(F) Cross-sections at the level of the frontal bone showing cranial layers at different developmental stages. Note that the frontal bone rudiment elongates within layer of neural crest-derived mesenchyme. Flanking this mesenchyme laterally is a layer of ephrin-Eph expressing cells upon which migratory osteogenic cells are found. DiI labeling shows that these migratory cells move from the area of the rudiment apically, ultimately inserting into the growing bone. NB, nasal bone; FB, frontal bone; PB, parietal bone; IP, interparietal bone; FS, frontal suture; CS, coronal suture; SS, sagittal suture; LS, lambdoid suture; NCM, neural crest-derived mesenchyme; PAM, paraxial mesoderm-derived mesenchyme; FBP, frontal bone primordium; PBP, parietal bone primordium; BH, brain hemisphere.

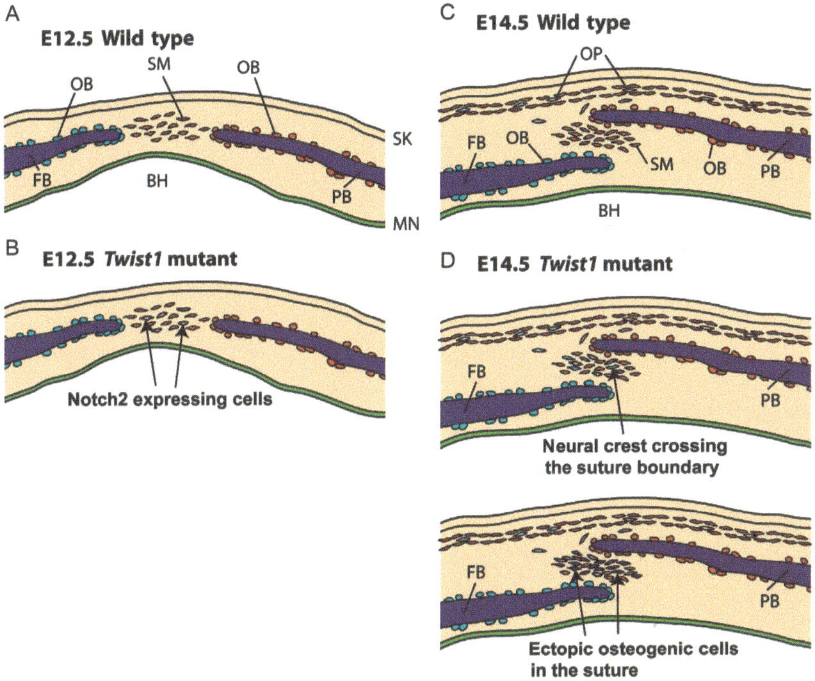

Mamoru Ishii *et al.*, Figure 3 A model of craniosynostosis in *Twist1* mutants; a two-step process leading to ectopic osteogenesis in the suture mesenchyme. (A) and (B) Altered sutural identity. (A) At E12.5, in wild-type mice, nonosteogenic suture mesenchyme cells (SM) exclude expression of Notch2, an early marker of osteogenesis. (B) In the *Twist1* mutants, however, prominent Notch2 expression is evident in this population (shown in white), thus suggesting that the suture assumes a partial osteogenic identity. (C) and (D) Suture boundary defect. (C) At E14.5, in mice, a boundary between neural crest (blue) and mesoderm (red) lineage is formed between the frontal bone (FB, osteogenic) and the SM (nonosteogenic). (D) Top: In *Twist1* mutants, the neural crest–mesoderm boundary is disrupted by intrusion of neural crest-lineage cells (shown in blue) into the SM. Thus, there is a loss of osteogenic-nonosteogenic boundary integrity in *Twist1* mutant mice. Osteogenic precursor cells (OP) migrate through the ectocranial layer (EL), contributing to this loss of boundary integrity. Bottom: The change in sutural cell identity together with the boundary-crossing osteogenic cells induces sutural cells to become fully osteogenic (purple). BH, brain hemisphere; MN, meninges; OB, osteoblast; PB, parietal bone; SK, skin.

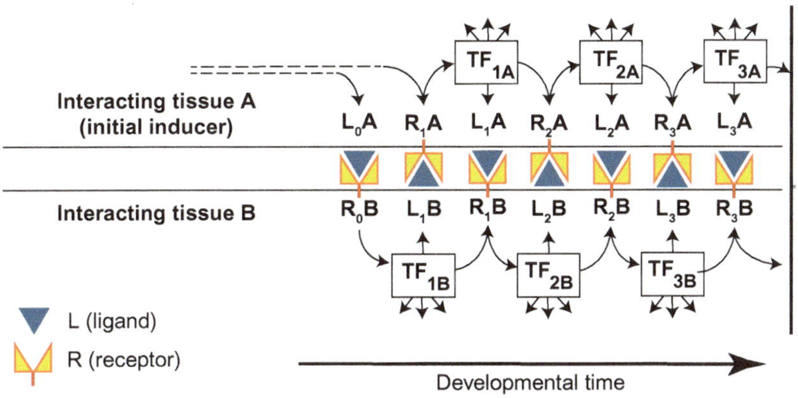

Anamaria Balic and Irma Thesleff, Figure 1 A chain of interactions between adjacent tissues undergoing progressive morphogenetic inductions. Release of the ligand (L) (signaling molecule) from the inductive tissue and its binding to the receptor (R) on the adjacent tissue triggers a sequence of regulatory changes in each tissue (A and B). These changes are controlled by the appearance of new transcription factors (TF) in response to signals from the adjacent tissue. New transcription factors activate a battery of genes (indicated by arrows emanating from the boxes). *Figure adapted from Davidson (1993).*

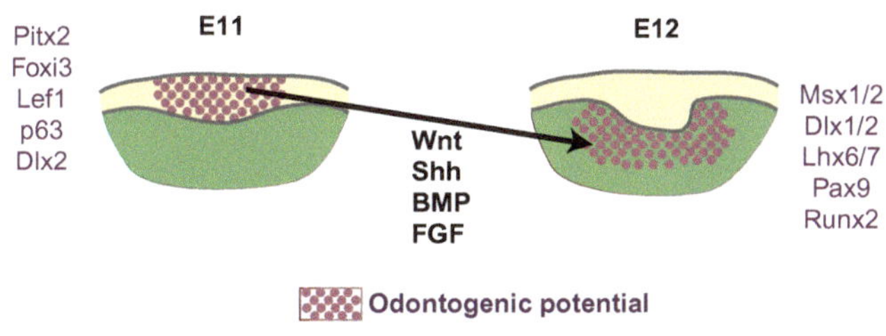

Anamaria Balic and Irma Thesleff, Figure 2 Shift of the odontogenic potential from epithelium to mesenchyme between the dental lamina and placode stages of tooth development. The dental lamina expresses specific transcription factors, which are associated with the dental identity and regulate the expression of signals in four conserved families. These signals mediate the shift of odontogenic potential to the mesenchyme and regulate an array of transcription factors which determine the identity and odontogenic potential of the dental mesenchyme.

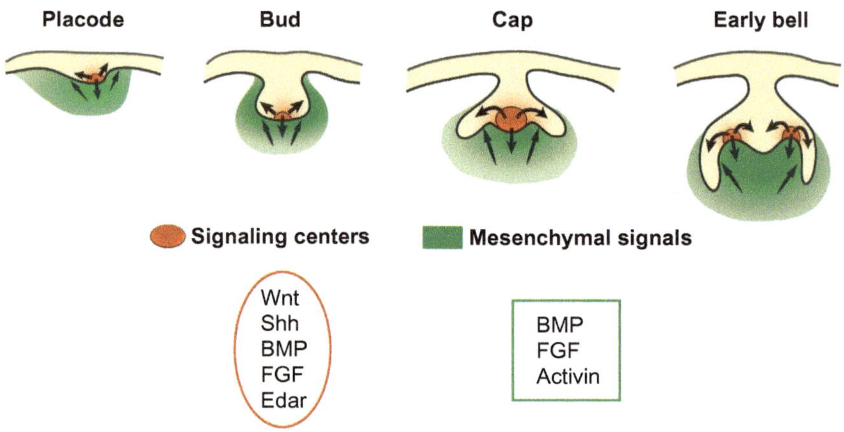

Anamaria Balic and Irma Thesleff, Figure 3 Epithelial–mesenchymal tissue interactions and epithelial signaling centers regulating tooth morphogenesis. Of particular significance for morphogenesis are the epithelial signaling centers which express signals of the four conserved signal families. They also express Edar, the receptor of ectodysplasin (Eda). The early signaling center in the placode regulates epithelial budding, while the primary enamel knot regulates the shift from bud to cap stage as well as the advancement of morphogenesis to bell stage. The secondary enamel knots regulate patterning of the tooth cusps in molar teeth during bell stage. The reciprocal signals expressed in mesenchyme include FGFs, BMPs, and Activin. Arrows indicate signaling across the two tissues and within the epithelium.

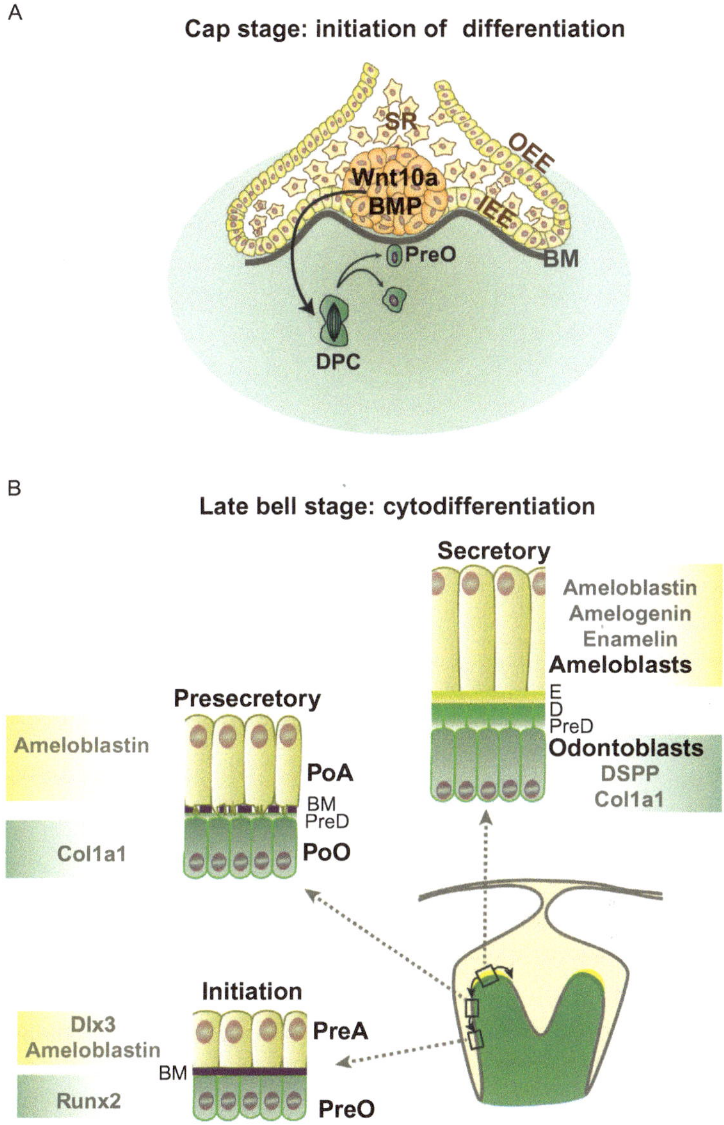

Anamaria Balic and Irma Thesleff, Figure 4 Epithelial–mesenchymal tissue interactions regulating differentiation of odontoblasts and ameloblasts. (A) The enamel organ during cap stage consists of inner and outer enamel epithelium (IEE and OEE, respectively) enclosing the stellate reticulum (SR) compartment. The fully developed primary enamel knot (pink cells in spherical organization) expresses Wnt10A that signals to dental papilla cells in the underlying mesenchyme (green color), which then commence odontoblast differentiation. (B) Ameloblasts and odontoblasts are generated during the bell stage of tooth development. Indicated genes represent the molecular markers that can be used to follow the progression of ameloblast (yellow boxes) and odontoblast (green boxes) lineages. *Abbreviations*: BM, basement membrane; D, dentin; E, enamel; PoA, polarized ameloblasts; PoO, polarized odontoblasts; PreA, preameloblasts; PreD, predentin; PreO, preodontoblasts; DPC, dental papilla cell.

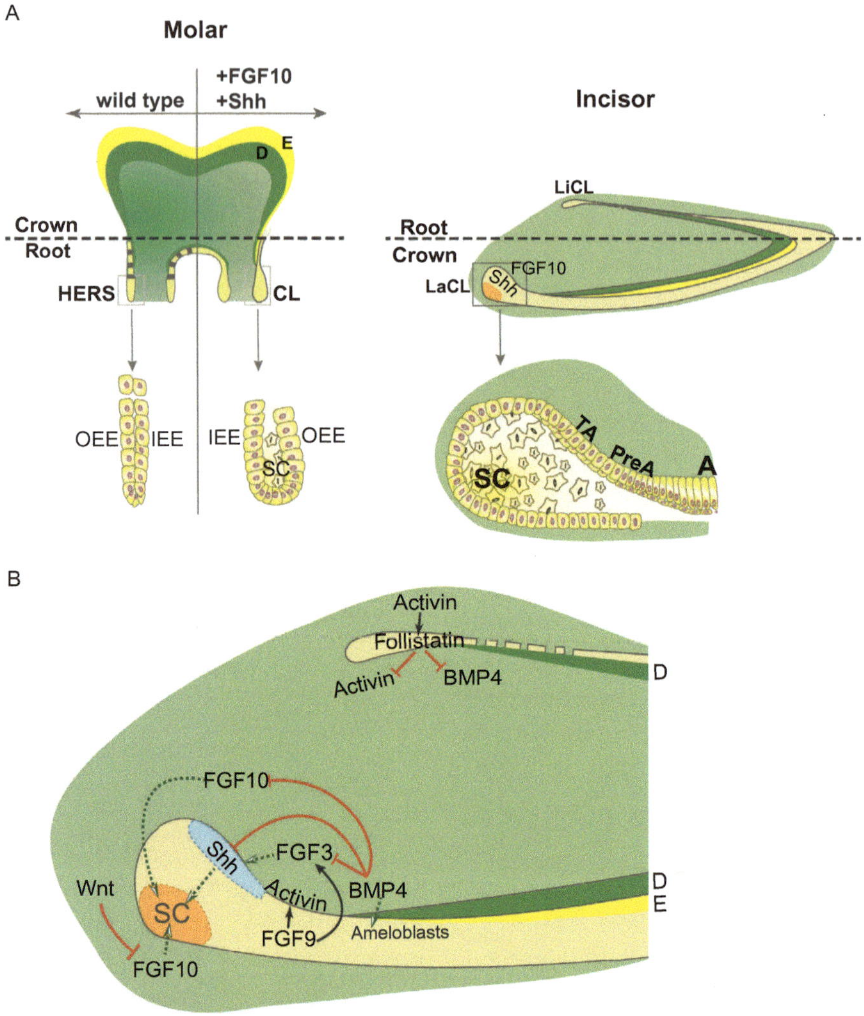

Anamaria Balic and Irma Thesleff, Figure 5 Cervical loops in molar root formation and continuous growth of the incisor, and the gene regulatory network that controls maintenance of epithelial stem cells. (A) In the wild-type molar tooth, disappearance of stellate reticulum (SR) cells in the cervical loop (CL) leads to formation of Hertwig's epithelial root sheath (HERS) and root formation, and is caused by downregulation of Fgf10 and Shh (illustrated in left-side root). Persistence of *Fgf10* and *Shh* expression support maintenance of the SR compartment that houses stem cells in the CL of the molar root (illustrated in right-side root) and in the LaCL of the continuously growing incisor (orange area) (Li et al., 2015; Yokohama-Tamaki et al., 2006). (B) Schematic view of the signals involved in epithelial–mesenchymal tissue interactions regulating the stem cell niche in the continuously growing incisors. Black arrows indicate positive gene regulation, red arrows indicate negative gene regulation, and green dotted arrows indicate positive effects on stem cells (orange area) and TA proliferation (blue area). *Abbreviations*: A, ameloblasts; IEE, inner enamel epithelium; OEE, outer enamel epithelium; PreA, preameloblasts; TA, transit-amplifying cells; SC, stem cells; LaCL, labial cervical loop; LiCL, lingual cervical loop.

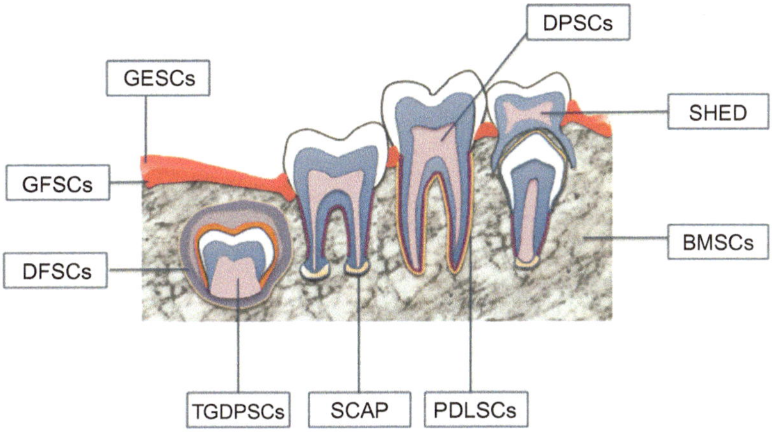

Tian Yu *et al.*, Figure 1 Teeth and supporting tissues are source for different stem cell populations. DPSCs, dental pulp stem Cells; SHED, stem cells from human exfoliated deciduous teeth; BMSCs, bone marrow stem cells; PDLSCs, periodontal ligament stem cells; SCAP, stem cells from apical papilla; TGDPSCs, tooth germ dental pulp stem cells; DFSCs, dental follicle stem cells; GFSCs, gingival fibroblast stem cells; GESCs, gingival epithelial stem cells.

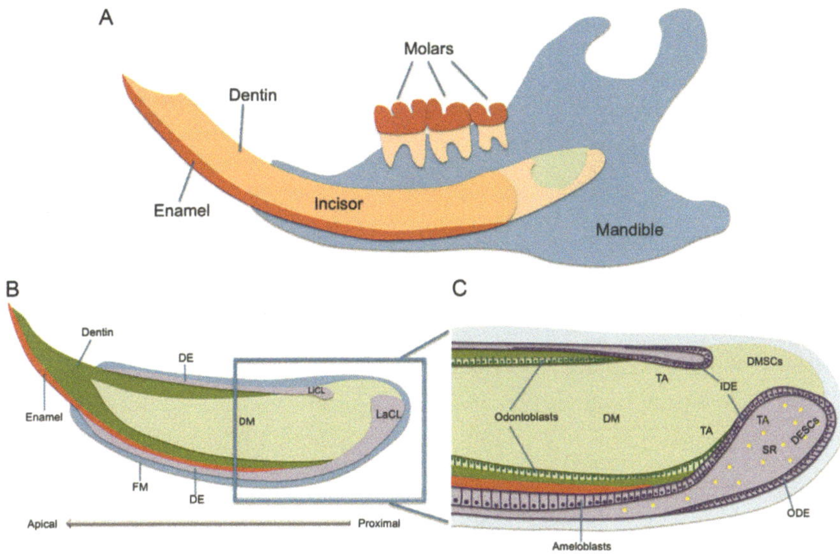

Tian Yu *et al.*, Figure 3 Schematic diagrams of the continuously growing mouse incisor. (A) Basic structure showing the mandible, incisor, and molars in adult mouse. (B) Illustrator of tissue distribution of mouse incisor on a sagittal section, and the enlarged region (C) at the proximal end of incisor showing the individual cells that constitute mouse incisor. (LaCL, labial cervical loop; LiCL, lingual cervical loop; DM, dental mesenchyme; DE, dental epithelium; FM, follicle mesenchyme; TA, transit-amplifying cells; IDE, inner dental epithelium; ODE, outer dental epithelium; SR, stellate reticulum; DESCs, dental epithelial stem cells; DMSCs, dental mesenchymal stem cells).

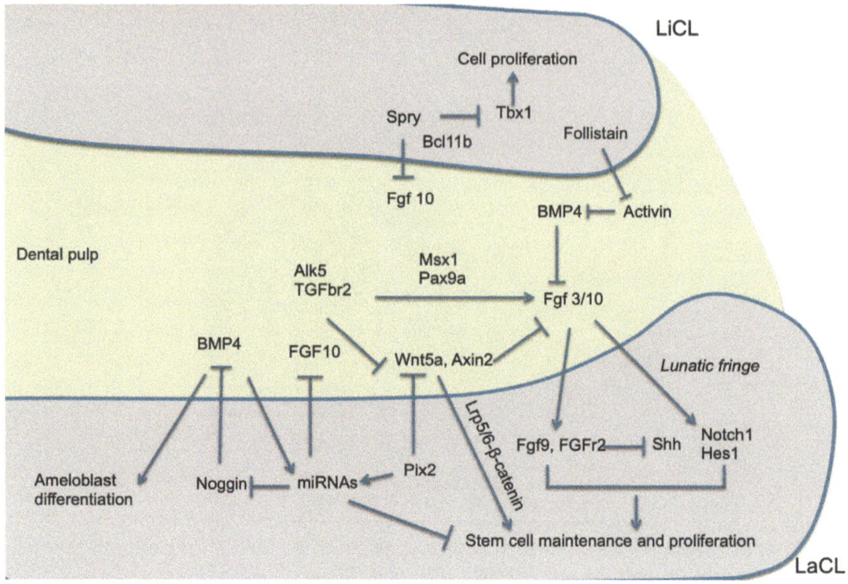

Tian Yu *et al.*, Figure 4 Gene regulatory network in a stem cell niche in the mouse incisors. Signaling molecules, transcription factors, and miRNAs expressed in the dental epithelium and mesenchyme regulate and maintain stem cells proliferation and differentiation.

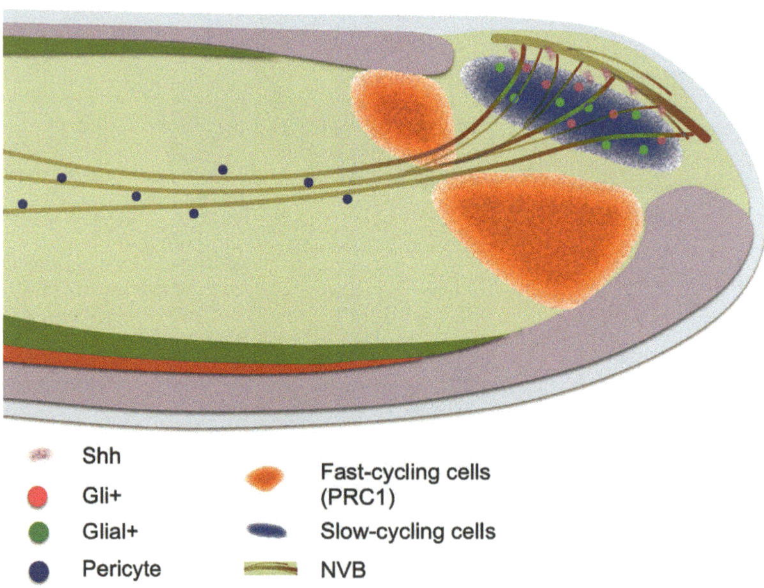

Tian Yu *et al.*, Figure 5 Identification of a mesenchymal stem cell niche in the mouse incisors. Slow-cycling stem cells locate in the mesenchyme between labial and lingual cervical loop region close to the NVB, represented by Gli+ and Glial+ expressing cells. PRC1 expressed in the mesenchyme closed to the labial and lingual cervical loop region, corresponding to the fast-cycling (TA) cells. Pericytes reside on the abluminal surface of endothelial cells in the vascularized connective tissues and contribute to MSC-derived mesenchymal cells responding to tissue damage (NVB, neurovascular bundle).

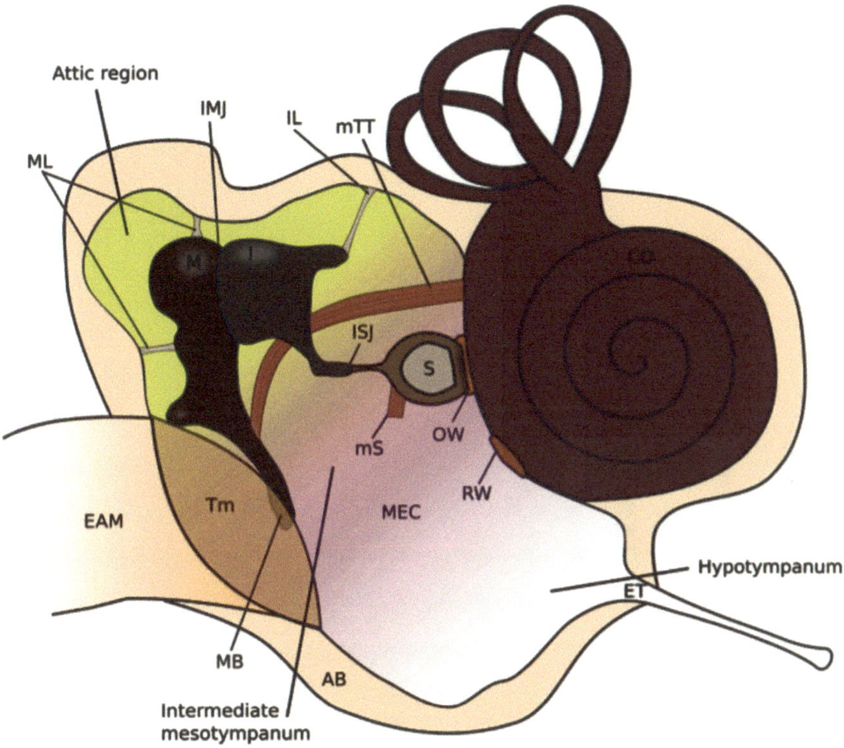

Jennifer C. Fuchs and Abigail S. Tucker, Figure 1 Schematic of the mammalian middle ear. The mammalian middle ear, surrounded by the auditory bulla (AB), is located between the external auditory meatus (EAM) and the spiral-shaped cochlea (CO) of the inner ear. The middle ear cavity (MEC) can be divided into the attic region (yellow), intermediate mesotympanum (purple), and hypotympanum (white). Three middle ear ossicles, the malleus (M), incus (I), and stapes (S) are suspended in the air-filled MEC and form a semirigid bony chain via the synovial incudomalleal joint (IMJ) and the incudostapedial joint (ISJ). The manubrium of the malleus (MB) directly inserts into the tympanic membrane (Tm) to collect sound waves, which are transmitted by the ossicular chain toward the inner ear. The stapes inserts into the membranous oval window (OW), transferring mechanical energy onto the endolymph fluid of the cochlea. The membranous round window (RW), another opening of the medial wall, allows the fluid in the cochlea to move. The ossicular chain is held in place by ligaments, such as the malleolar (ML) and incudal ligaments (IL). Stabilization of the ossicles is ensured by the stapedius muscle (mS) inserting into the crura of the stapes and the tensor tympani muscle (mTT), which is attached to the manubrium and protects the ear against loud sounds. The Eustachian tube (ET) connects the MEC to the nasopharynx and functions in pressure equalization and clearance of effusions.

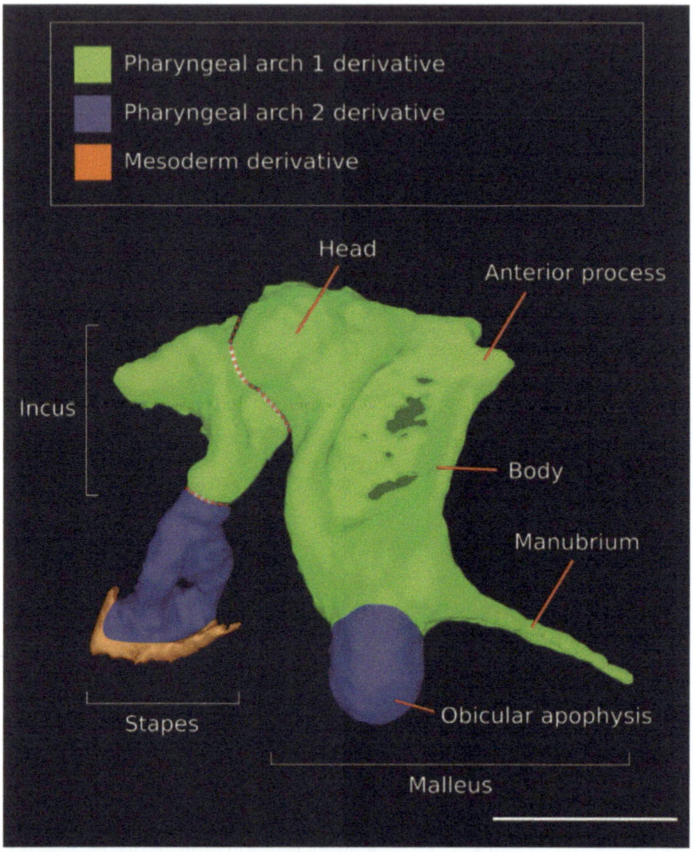

Jennifer C. Fuchs and Abigail S. Tucker, Figure 2 Micro-CT image of the ossicular chain in an adult mouse. Pseudocolored ossicular chain. Structures in green are derived from pharyngeal arch 1 mesenchyme, whereas structures in blue are pharyngeal arch 2 derivatives. The mesoderm-derived part of the stapedial footplate is highlighted in orange. Scale bar: 1 mm.

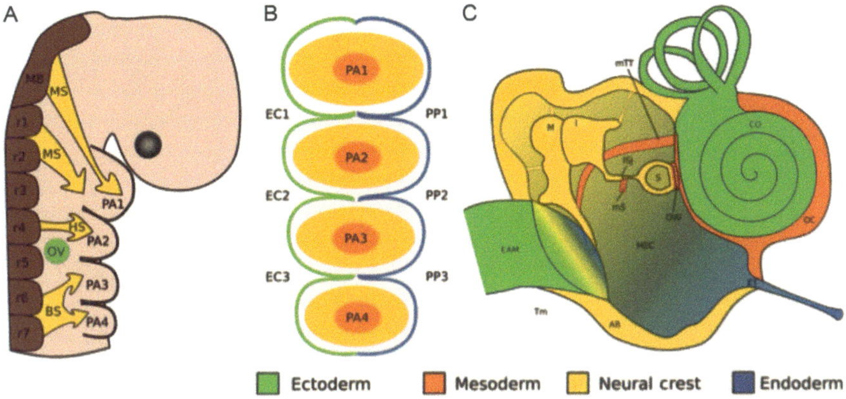

Jennifer C. Fuchs and Abigail S. Tucker, Figure 3 Schematic of mammalian middle ear development. (A) The mammalian pharyngeal apparatus is comprised of pharyngeal arches (PA). During embryonic development, streams of neural crest cells (yellow) from dorsal rhombomeres (r1–r7) and the dorsal edge of the midbrain (MB) migrate into the PA surrounding mesoderm-derived core. The mandibular stream (MS) of NCCs from MB, r1, and r2 migrate into PA1, whereas the hyoid stream (HS) from r4 populates PA2. NCCs from r6 and r7 (branchial stream; BS) migrate into PA3 and PA4. The ectoderm-derived otic vesicle (OV), between PA2 and PA3, gives rise to inner ear structures such as the cochlea (CO). (B) Each PA consists of core mesoderm, derived from paraxial mesoderm (orange), which is surrounded by NCCs (yellow). Both tissues are externally covered by ectoderm (green), which invaginates between adjacent PAs to form ectodermal clefts (ECs) and internally by endoderm (blue) forming pharyngeal pouches (PPs) by directly apposing the overlying ectoderm. (C) The matured ME is derived from all three germ layers and the neural crest. NCCs give rise to ligaments and the auditory bulla (AB) surrounding the MEC as well as the malleus (M) and incus (I) and parts of the stapes (S). The stapes, however, is of dual origin, as parts of the stapedial footplate are mesoderm derived. Other mesoderm-derived structures include both ME muscles, the mTT derived from PP1 and mS derived from PA2, most parts of the otic capsule (OC) and the oval window (OW). The Eustachian tube (ET), as an outpocketing of the PP1, as well as ventral parts of the epithelium lining the MEC are endoderm derived, whereas the mucosa covering the cochlea and attic are of NCC origin. The Tm is multilayered, consisting of the neural crest-derived lamina propia (LP), which is sandwiched in between a lateral ectodermal and a medial endodermal epithelium. The EAM, lined by PA1-derived mesenchyme is of ectoderm origin. *Panels (A and B) adapted from Lindsay (2001).*

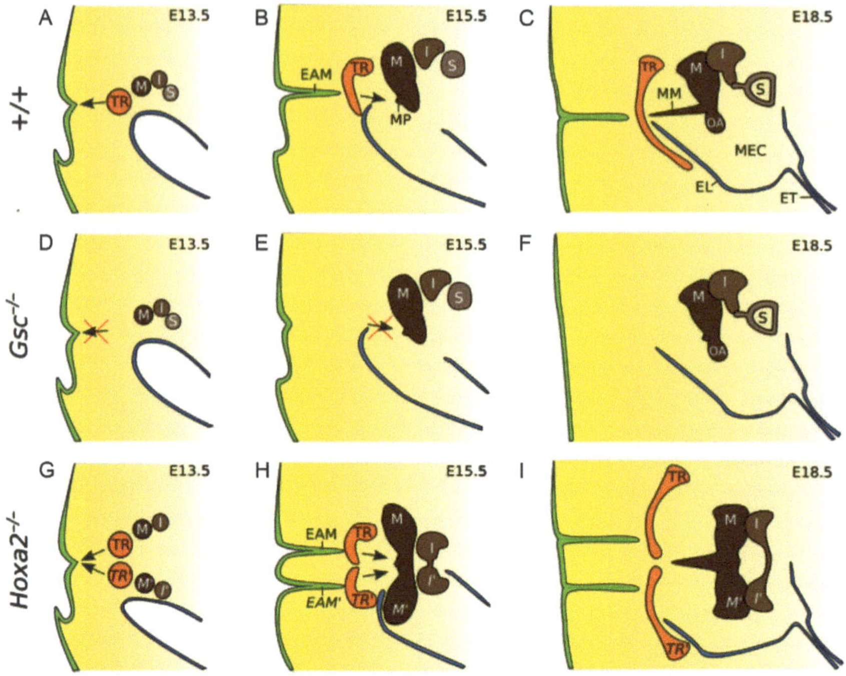

Jennifer C. Fuchs and Abigail S. Tucker, Figure 4 Developmental defects during the integration of the ear. (A–C) Wild-type (+/+). (D–F) *Goosecoid*$^{-/-}$ (*Gsc*$^{-/-}$). (G–I) *Hoxa2*$^{-/-}$. (A) At E13.5, in wild-type embryos, the mesenchymal condensation forming the early tympanic ring (TR) signals (arrow) to the overlying ectoderm (green), leading to invagination of the epithelium. The three ossicles, the malleus (M), incus (I), and stapes (S) form condensations in the surrounding neural crest-derived mesenchyme (yellow). The endoderm-derived tubotympanic recess (blue) extends from the first pharyngeal pouch toward the ossicles and the TR. (B) Two days later, the early external auditory meatus (EAM) has extended toward the TR. The endoderm ruptured, allowing mesenchyme to enter the tubotympanic recess. Signals from the TR (arrow) induce the formation of the manubrium primordium (MP). (C) Shortly before birth, the EAM has collapsed, forming a thin solid ribbon. The malleal manubrium (MM) and orbicular apophysis (OA) have formed. The ossicular chain is still surrounded by mesenchyme. The tubotympanic recess can now be divided into the Eustachian tube (ET) and the epithelial lining (EL) of the middle ear cavity (MEC). (D) The mesenchymal condensation of the early TR is missing in all *Gsc*$^{-/-}$ mice. As a result, mutants lack signals for inducing ectodermal invagination (crossed out arrow). (E) Therefore, at E15.5, the early EAM does not form in mutant mice. In addition, signals from the TR to the MP are missing (crossed out arrow). (F) The lack of inducing signals from the TR has lead to an absent EAM and malleal manubrium. (G) In *Hoxa2*$^{-/-}$ mice, second arch derivatives are transformed into first arch tissues, leading to mirror-image duplication of the tympanic ring (*TR'*), malleus (*M'*), and incus (*I'*), while the stapes is absent. Both TR condensations signal to the overlying surface ectoderm (arrows). (H) As a result of the dual signaling, mutant mice form an additional EAM (*EAM'*), directed to the duplicated TR. Both TRs signal to the fused malleus to induce chondrogenesis of the malleal manubrium (arrows). (I) At E18.5, *Hoxa2* mutants display two collapsed EAMs, directed to their corresponding TRs. The TRs are shortened and closely apposed. In addition, the malleus and the duplicated malleus are fused where the OA normally forms, positioning the fused and shortened manubrium between both EAMs. The incus is fused to the duplicated incus at the position of the incudostapedial joint.

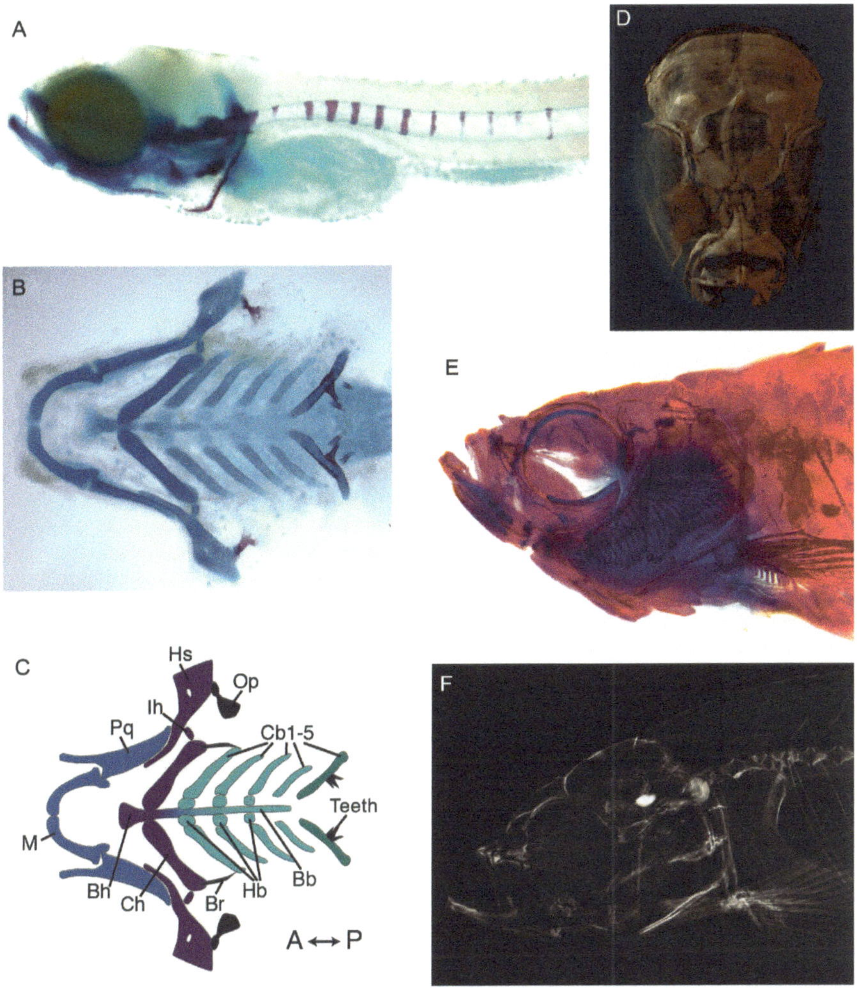

Lindsey Mork and Gage Crump, Figure 1 Zebrafish larval and adult craniofacial skeletons. (A) Larval zebrafish (6 dpf) stained with Alcian blue for cartilage and Alizarin red for mineralized bone reveals the early formation of the craniofacial skeleton. (B) Ventral view of an Alcian blue/Alizarin red-stained larval viscerocranium dissected away from the rest of the head, showing the skeletal derivatives of the seven pharyngeal arches. (C) Schematic of the larval viscerocranium. First-arch structures (blue): Pq, palatoquadrate; M, Meckel's; second-arch structures (purple): Hs, hyosymplectic; Ih, interhyal; Ch, ceratohyal; Bh, basihyal; Op, opercle (bone); Br, branchiostegal ray (bone); arches 3–7 (teal): Cb, ceratobranchial; Hb, hypobranchial; Bb, basibranchial. (D–F) The complex adult zebrafish craniofacial skeleton can be visualized in multiple modalities. (D) Micro-CT projection of an adult zebrafish head (frontal view). (F) Adult zebrafish head stained with Alcian blue and Alizarin red (lateral view). (E) X-ray of an adult zebrafish head (lateral view). Images not to scale. Anterior is to the left in A–C, E, and F. *Panel (D) Image courtesy of S. Ruffins and Panel (E) photo courtesy of S. Paul.*

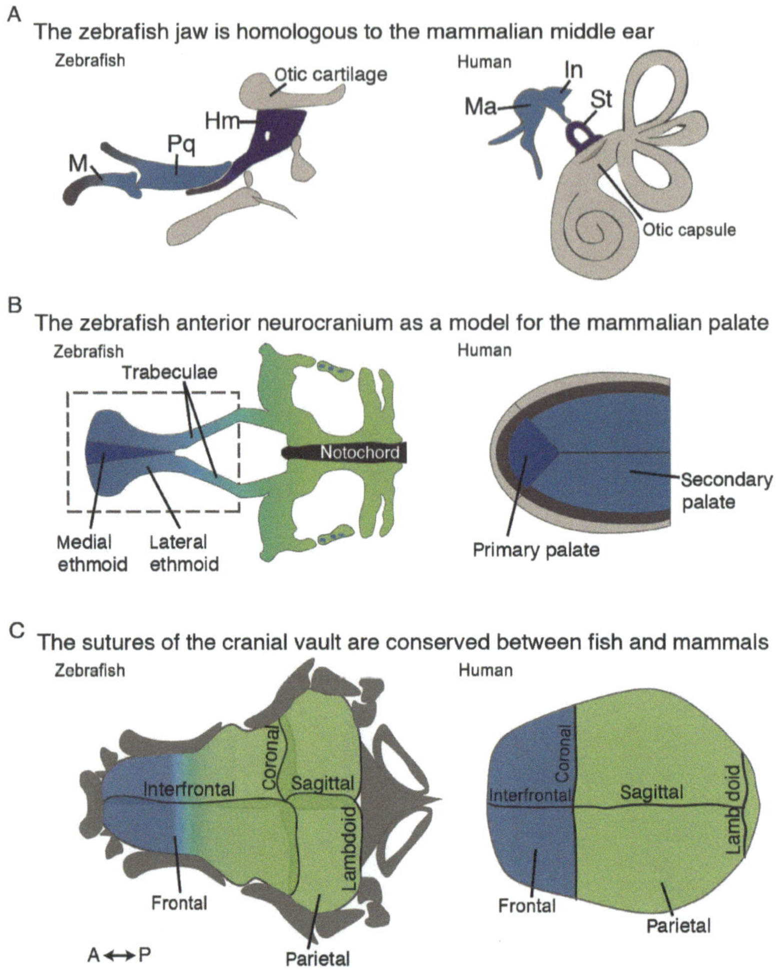

Lindsey Mork and Gage Crump, Figure 2 Zebrafish models of mammalian craniofacial development. (A) The larval zebrafish jaw is evolutionarily homologous to the mammalian middle ear ossicles. The posterior parts of the first arch-derived Meckel's (M) and palatoquadrate (Pq) cartilages (blue), which articulate the jaw in fish, correspond to the malleus (Ma) and incus (In) of mammals. The second arch-derived hyomandibula (purple) (Hm; the dorsal component of the hyosymplectic), which affixes the viscerocranium to the neurocranium via the otic cartilage, is the homolog of the mammalian stapes (St). (B) The anterior end of the larval fish neurocranium is a model for the mammalian hard palate. The medial ethmoid plate (dark blue) derives from frontonasal NCCs, with the lateral ethmoid and trabeculae (light blue) formed by medially migrating anterior maxillary NCCs (left panel), akin to the origins of the primary and secondary hard palate, respectively, in mammals (right panel). The posterior zebrafish neurocranium derives from mesoderm (green). (C) The overlapping calvarial bones of the adult zebrafish form the same suture pattern as observed in mammals. The two frontal bones are separated from each other by the interfrontal (or metopic) suture and from the parietal bones by the bilateral coronal sutures. The parietals are divided along the midline by the sagittal suture and posteriorly bounded by the lambdoid suture. The boundary between neural crest (blue) and mesoderm (green) falls within the frontal bones in fish and along the coronal suture in mammals. Anterior is to the left.

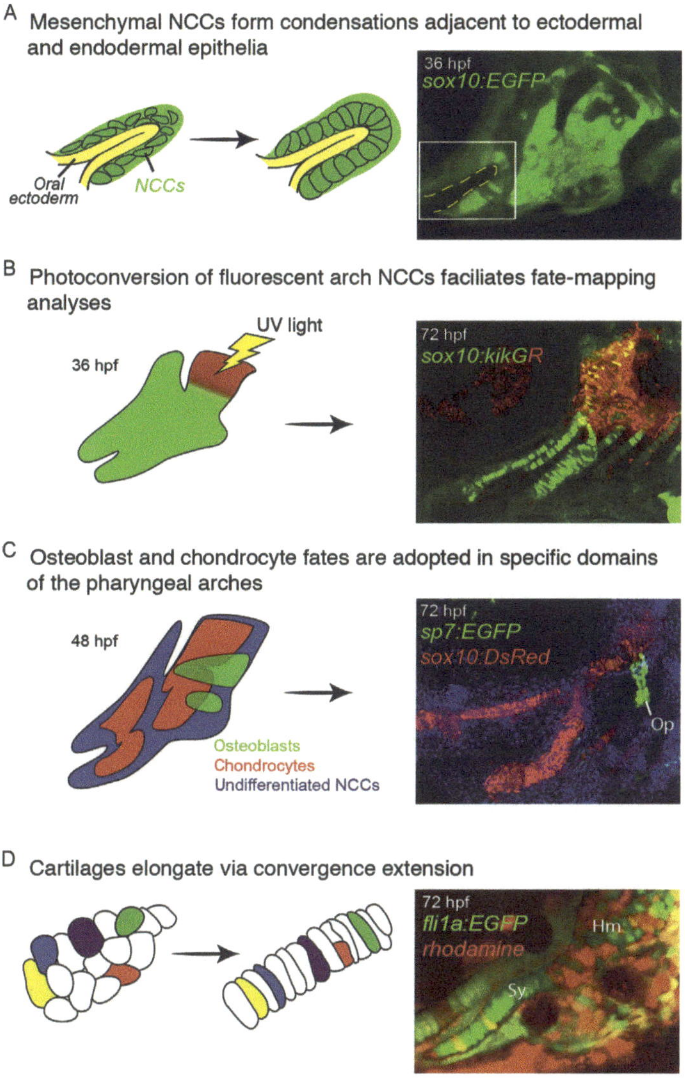

Lindsey Mork and Gage Crump, Figure 3 Morphogenetic processes that shape the facial skeleton. (A) Mesenchymal NCCs form tight condensations next to ectodermal or endodermal epithelia, including the stomodeum (i.e., oral ectoderm). The *sox10:EGFP* transgene (green) is expressed by all NCCs. Boxed area shows NCCs condensed around the oral ectoderm (dashed line). (B) Fate-mapping analyses demonstrated that NCCs at certain locations within the arches reproducibly give rise to specific parts of facial skeletal elements. For example, arch NCCs in the dorsal half of the second arch make the hyomandibular portion of the hyosymplectic. One way to track cells is to use focused applications of ultraviolet light to convert fluorescent proteins such as kikGR from green to red in specific cell populations at early stages. The labeled cells are then reimaged at a

Figure 3—Cont'd later time point. (C) Following mesenchymal condensation, chondrocyte and osteoblast fates are adopted within specific domains of the pharyngeal arches around 48 hpf. At 72 hpf, the *sp7:EGFP* transgene (green) is expressed in osteoblasts, while *sox10:DsRed* (red) is specifically expressed by chondrocytes. Op, opercle. Nuclei are labeled with Hoechst (blue). (D) Convergence-extension movements of differentiating prechondrocytes, rather than cell proliferation, drive the elongation of cartilages from bulky precartilaginous condensations into stacks of disc-shaped chondrocytes. The symplectic rod (Sy) lengthens anteriorly from the plate-like hyomandibula (Hm) in this manner. The right panel shows a *fli1a:EGFP* (green) transgenic embryo with mosaic rhodamine-dextran injection (red), illustrating the intercalation of dye-labeled cells into the elongating Sy.

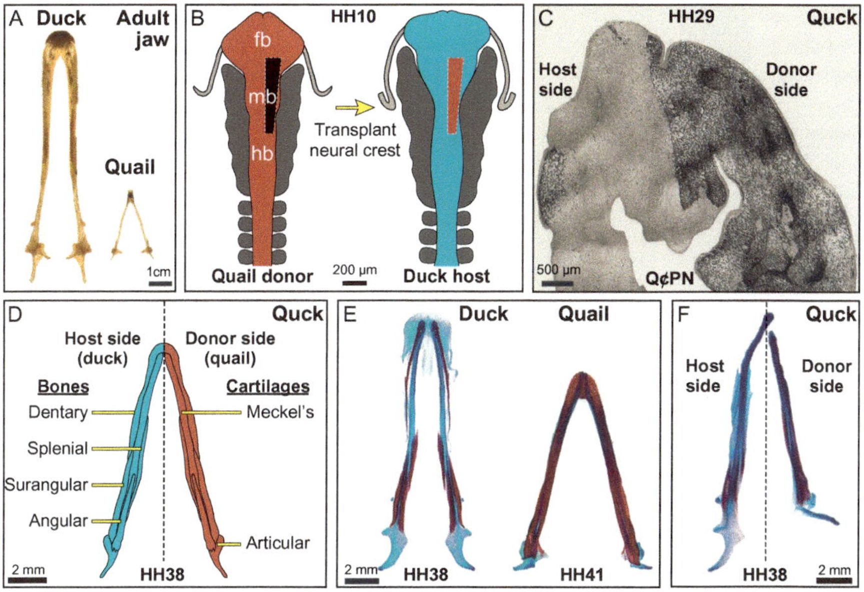

Richard A. Schneider, Figure 1 The quail–duck chimeric system for investigating the origins of species-specific jaw length. (A) Lower jaw skeletons of adult Japanese quail (*Coturnix coturnix japonica*) and white Pekin duck (*Anas platyrhyncos*). (B) Schematic of rostral neural tube at embryonic stage (HH) 9.5, depicting the levels of NCM destined for the jaw primordia and grafted from quail (red) to duck (blue). (C) Horizontal section through the mandibular primordium of a HH29 chimeric quck embryo (rostral at top), which will give rise to the lower jaw skeleton. Quail donor mesenchyme (black), stained with a quail-specific antibody (Q¢PN), is distributed throughout the transplanted side, while only a few quail cells are found on the contralateral duck host side. (D) Schematic of a lower jaw skeleton in a chimeric quck embryo at HH38, showing the contributions of transplanted quail donor NCM (red) to cartilage and bone. (E) The lower jaw skeletons of quail and duck display species-specific differences in size and shape with duck being longer and more curved. Meckel's cartilage is stained with Alcian blue and the bones are stained with Alizarin red. (F) In quck mandibles, the quail donor-derived jaw skeleton is shorter and straighter than that observed for the contralateral duck host-derived jaw skeleton, which is longer and curved. *Panels (A–E) modified from Eames and Schneider (2008) and Panel (F) modified from Fish and Schneider (2014a).*

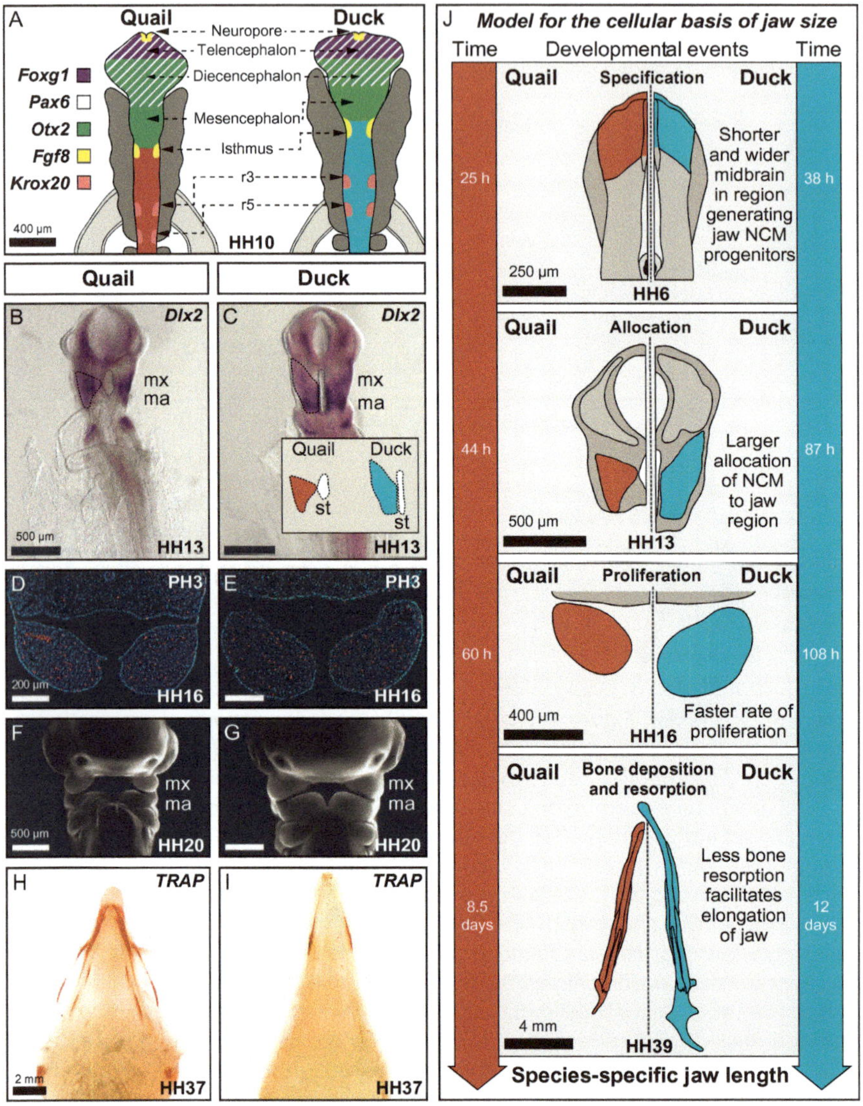

Richard A. Schneider, Figure 2 Molecular and cellular mechanisms regulating jaw length. (A) Quail and duck have distinct head shapes and species-specific regionalization of the neural tube at embryonic stage (HH) 10. Duck have a foreshortened and mediolaterally broader midbrain (mesencephalon). Genes including *Foxg1*, *Pax6*, *Otx2*, *Fgf8*, and *Krox20* are expressed in domains at HH10 with each being shifted more anteriorly in duck versus quail. (B, C) Differences in the allocation of NCM to the maxillary (mx) and mandibular (ma) primordia of the presumptive jaw region can be seen following *in situ* hybridization for *Dlx2* at HH13 in quail versus duck. The *Dlx2*-positive NCM domain and stomodeum (st) demonstrate that duck (blue) have a larger population of

Figure 2—Cont'd NCM relative to quail (red). (D, E) Phosphohistone H3 (PH3) identifies mitotic cells at HH16. While duck develop at a slower rate (taking about 45 h to progress from HH13 to HH20 vs. 32 h in quail), duck NCM completes roughly 3.3 cycles during this developmental window compared to an average of 2.9 cycles for quail. Thus, the rate of proliferation relative to the rate of development is faster in duck. (F, G) As a result by HH20, the jaw of duck is approximately twice the size of that of quail. (H, I) Staining for tartrate-resistant acid phosphatase (TRAP) in the jaw skeleton reveals less bone resorption in duck versus quail at HH37. (J) Multiple developmental events regulate jaw length. Specification at the neural plate establishes a shorter and wider midbrain in duck. This difference, evidenced by distinct *Otx2* expression domains, is evident by HH6 and leads to a larger allocation of NCM to the jaw primordia by HH13. Duck NCM have a higher proliferation rate due to differences in developmental rate (time arrows on side). Finally, lower amounts of bone resorption result in differential growth and elongation of the duck jaw skeleton. *Panels (A–G) and (J) from Fish et al. (2014) and Panels (H) and (I) from Ealba et al. (2015).*

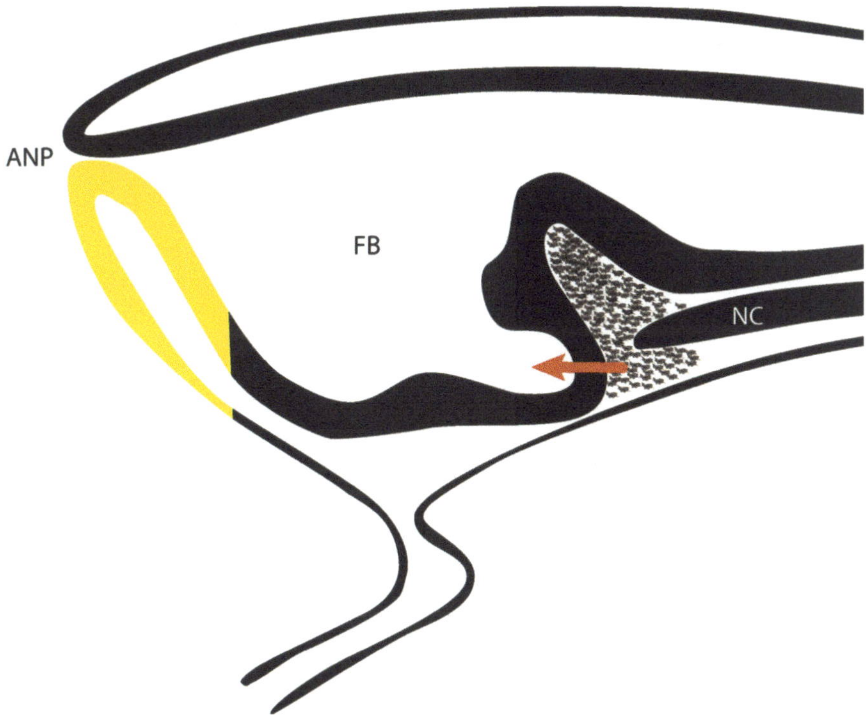

Ralph Marcucio *et al.*, Figure 1 Shared expression domains between the brain and the face. Closure of the anterior neural pore segregates anterior neural and surface ectoderm. At this time, the *Fgf8* domain (yellow) that spans the anterior neural plate and the anterior surface ectoderm becomes two distinct domains. ANP, anterior neuropore; FB, forebrain; red arrow, signaling from mesendoderm to forebrain.

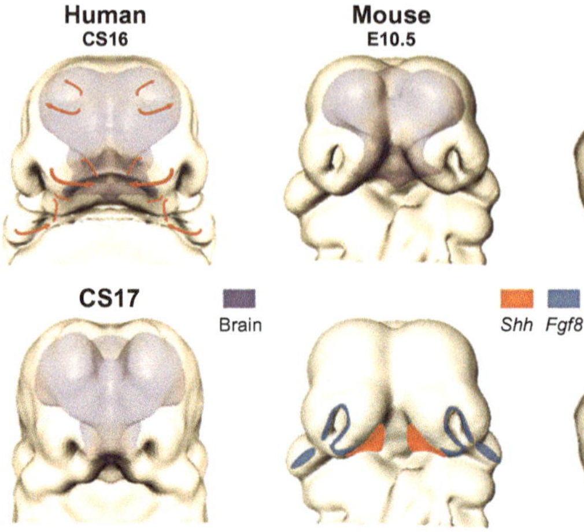

Ralph Marcucio *et al.*, Figure 2 Growth of the brain and face. Relationship of the brain (shaded) to the face during facial morphogenesis shown at comparable stages in human (CS16), mouse (E10.5), and chicken (HH22). Facial prominences grow toward each other (arrows), converging at the midline, while the brain becomes relatively wider and divides into two distinct lobes.

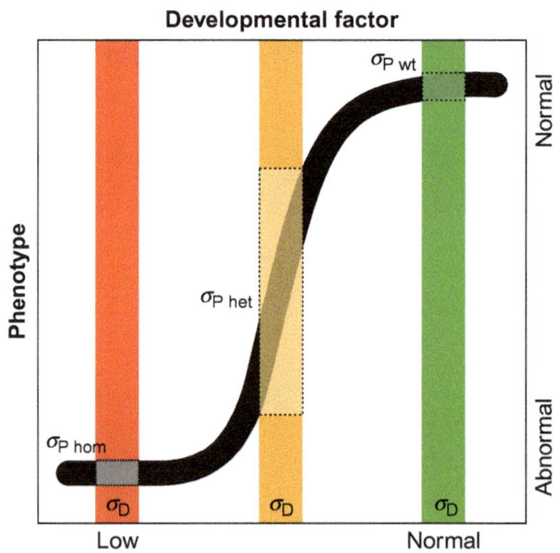

Ralph Marcucio *et al.*, Figure 3 Nonlinearity in the relationship between development and phenotypic outcomes. In a nonlinear relationship, in this case represented as an S-shaped curve, there is a continuous relationship between a genetic or developmental factor and the phenotypic outcome; however, the variance in phenotypic outcomes differs for the same developmental variance. In this example, at normal levels of a developmental factor, phenotypic outcomes tend to be similar. Likewise, when the developmental factor is completely removed, phenotypic outcomes are also similar, such as the case in a highly penetrant genetic knockout. In contrast, when the developmental factor is reduced, such as in a heterozygote animal, the variance in phenotypic outcomes is much greater for the same developmental variance.

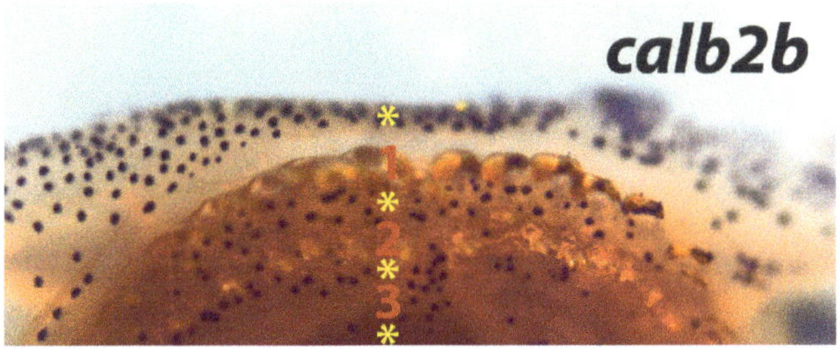

J. Todd Streelman *et al.*, Figure 1 Comingling of teeth and taste buds on the oral jaws of Lake Malawi cichlid fishes. ISH of *calb2* in the cichlid lower jaw, at 4 months of age. Yellow asterisk marks rows of expression in taste buds, while rows of teeth are marked by red numerals. Note that tooth row three has not yet erupted. Dorsal view, labial to bottom of page, scale = 100 μm.

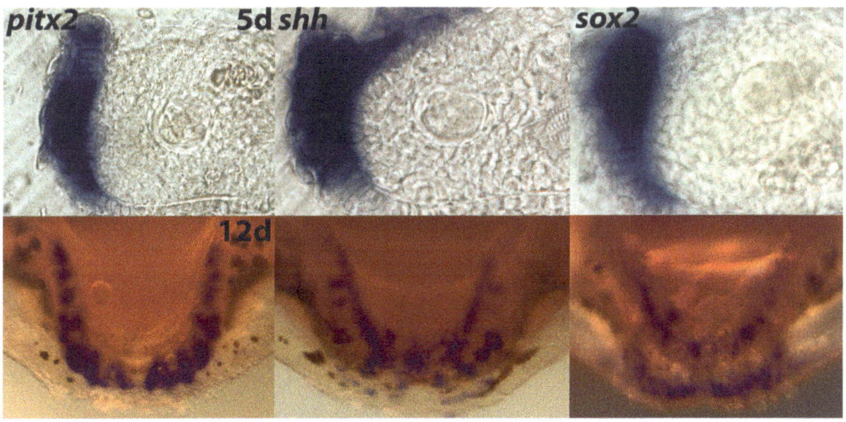

J. Todd Streelman *et al.*, Figure 2 Teeth and taste buds share a bipotent oral lamina. ISH of dental marker *pitx2*, taste/sensory marker *sox2*, and an early marker of both, *shh*, at two representative stages of development. Top panel: sagittal section (18 μm thickness, 40× magnification) at initial stages of jaw formation, 5 dpf. Bottom panel: by 12 dpf, distinct fields of teeth (marked by *pitx2*) and taste buds (*sox2*) have emerged, while *shh* is expressed in both domains. Whole mount dorsal views, labial to bottom of page.

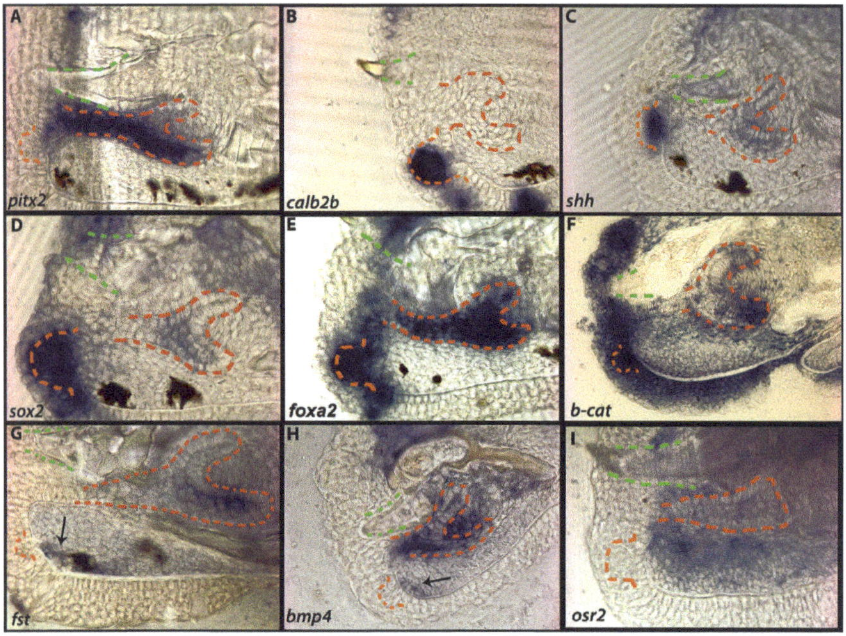

J. Todd Streelman *et al.*, Figure 3 Teeth and taste buds share an epithelial ribbon and molecular marker expression during regeneration stages (~20 dpf). Orange dashes outline taste buds, green outlines the functional tooth, and red outlines replacement teeth. Note expression of *fst* and *bmp4* outside of and juxtaposed to the taste bud unit. These are sagittal sections, 40× magnification; labial is to the bottom of the page, rostral to the left, scale = 10 μm.

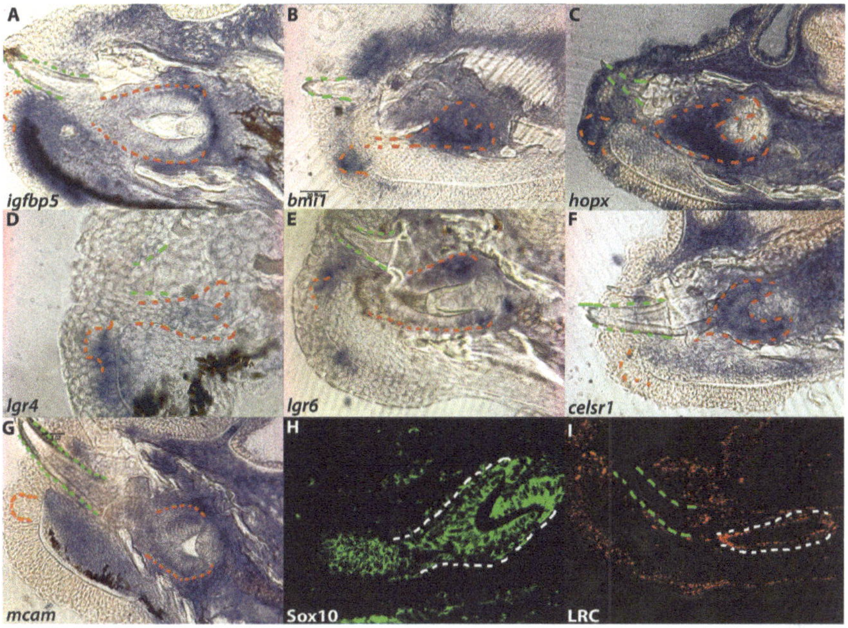

J. Todd Streelman *et al.*, Figure 4 Expression of adult stem cell markers in cichlid tooth and taste bud renewal. In A–G, orange dashes outline taste buds, green outlines the functional tooth, and red outlines replacement teeth. These are sagittal sections, 40 ×; labial is to the bottom of the page, rostral to the left, scale in B = 20 μm. Sox10 IHC (H) in green with replacement tooth epithelium outlined in white. LRC after 100 days chase (I) marked with IHC against BrdU, where LRC clusters are seen at cervical loops, tip of tooth, and in taste bud and surrounding epithelium; replacement tooth outlined in white, 20 × objective.

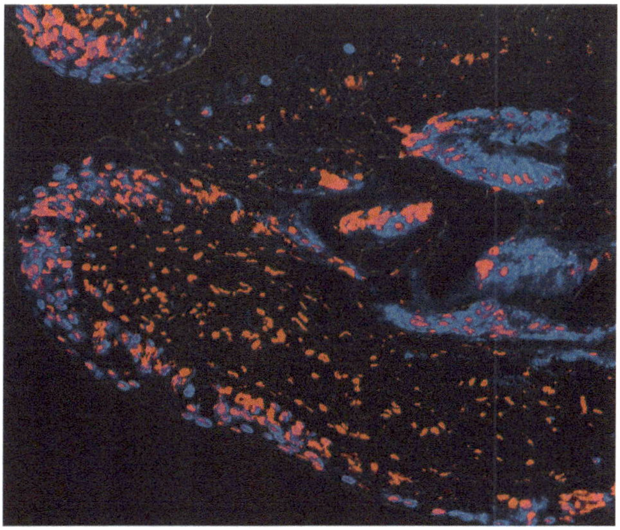

J. Todd Streelman *et al.*, Figure 5 Identification of LRCs/stem cells in cichlid tooth and taste bud regeneration. Double label for LRC after 90 days chase (red) and IHC of adult stem marker Sox2 (blue). This is a sagittal section, 20 × objective; labial is to the bottom of the page, rostral to the left.

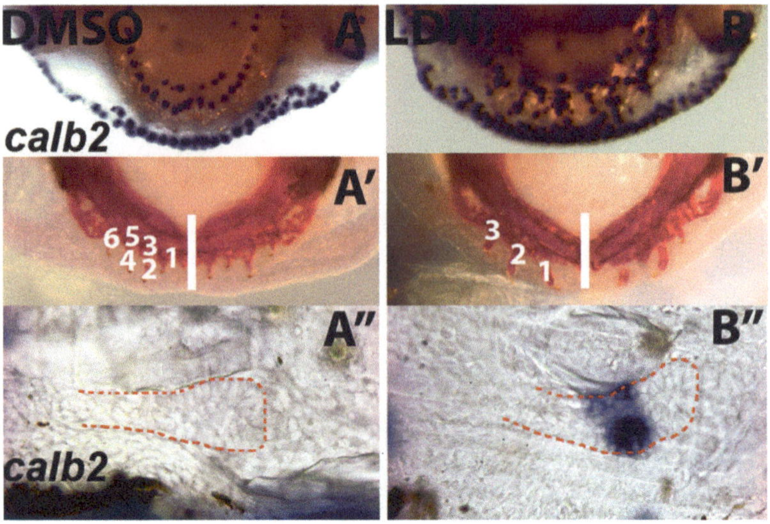

J. Todd Streelman *et al.*, Figure 6 Interorgan plasticity between cichlid teeth and taste buds at patterning and regeneration stages. *calb2* ISH was used to mark taste buds. A–B′ show infiltration of taste buds into the tooth field and reduced tooth density in early (5–6 dpf) LDN-treated animals. In the middle panel, midline of cleared and stained fishes is marked by a white line, and teeth in each half of the dentary are marked by white numerals. A–B′, these are dorsal views of dentaries, labial to bottom of page. At later regeneration stages (20 dpf), LDN treatment results in *calb2* expression in the cervical loops of replacement tooth germs (B″), which is never present in controls (A″). A″–B″ are sagittal sections, 40 × objective. For experiments at patterning stages, animals were treated for 24 h at 5–6 dpf, allowed to recover for a week, and then sacrificed and assayed. For experiments at regeneration stages, animals were treated for 48 h at 20 dpf and then immediately sacrificed and assayed.

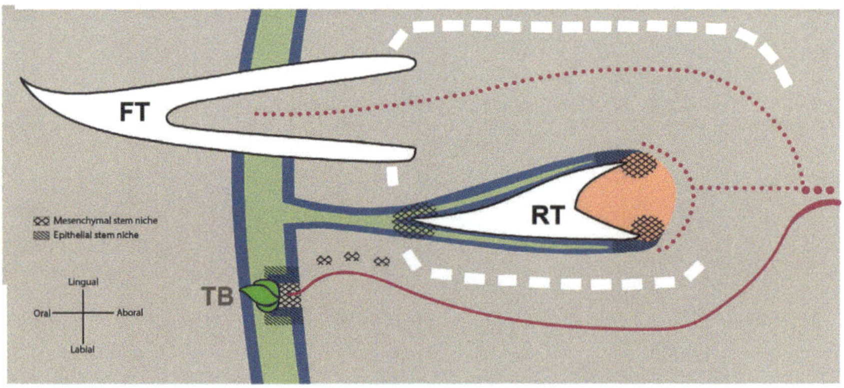

J. Todd Streelman *et al.*, Figure 7 Cichlid teeth and taste buds combine the features of regenerating and replacing organ systems. Here, layers of epithelium are indicated in blue and green. Light pink at the base of the replacement tooth shows dental papilla. Areas of bone are depicted in white, with broken white lines representing remodeling of the bony crypt. Dark pink NVBs are shown penetrating the functional tooth and taste bud, with broken lines in areas of reinnervation. Putative stem cell niches are indicated by crosshatching (mesenchyme) and parallel lines (epithelium). Oral to left of page; labial to bottom.

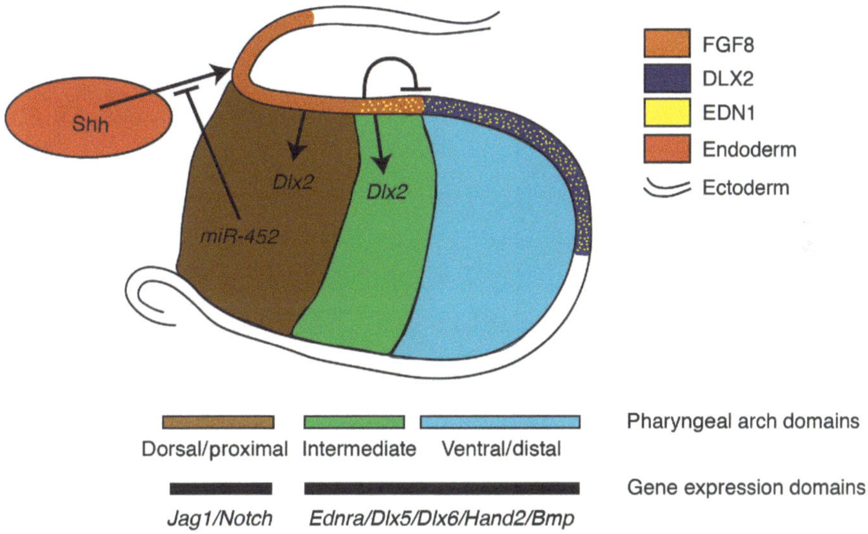

Andre L.P. Tavares *et al.*, Figure 1 Pharyngeal arch organization and miRNA function. A schematized first mandibular arch with dorsal/proximal, intermediate, and ventral/distal domains demarcated. Gene regulatory networks active in these domains are listed below the domain description. The *Ednra/Dlx5/Dlx6/Hand2/Bmp* pathway is induced by ectodermal EDN1. One miRNA-influenced pathway is shown, in which SHH from the endoderm induces ectodermal *Fgf8*, which in turn induces *Dlx2* expression in the NCC-derived mesenchyme. WNT5A negatively regulates this activation of Fgf8 expression and is the target of *miR-452*. Also shown is that FGF8 negatively represses ectodermal *Dlx2* expression.

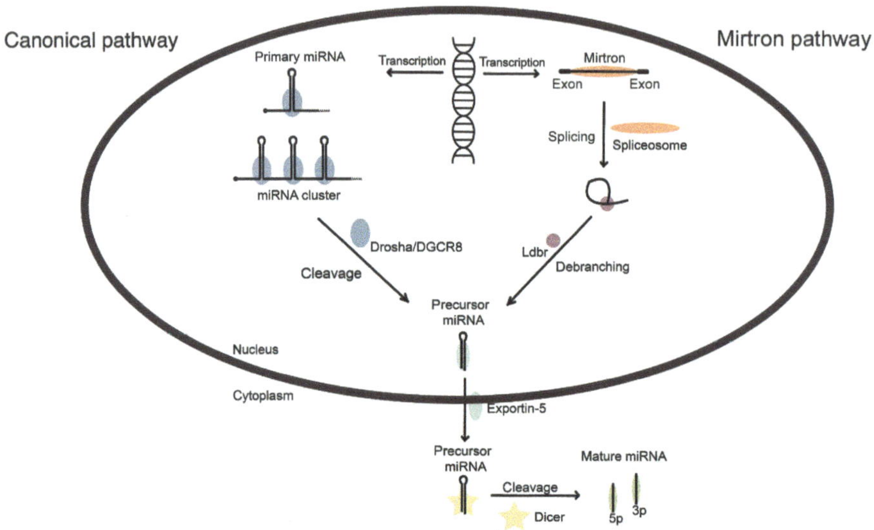

Andre L.P. Tavares *et al.*, Figure 2 miRNA biogenesis pathways. The left side represents the canonical pathway. miRNA genes are transcribed into primary miRNAs (pri-miRNAs) or a miRNA cluster by RNA polymerase II. Drosha/DGCR8 complex cleaves these transcripts into precursor-miRNA (pre-miRNA) hairpins. Export by Exportin-5 transports pre-miRNAs into the cytoplasm, where a second cleavage by DICER1 produces mature miRNAs. Either strand (5p or 3p) can complex with Argonaute (AGO) and produce a functional RISC complex. The right side represents the mirtron pathway. After transcription by RNA polymerase II, splicing of a shorter intronic sequence may generate a lariat that is debranched into a pre-miRNA hairpin by lariat debranching enzyme (LDBR). This process depends on the splicing machinery and is independent of Drosha/DGCR8 function. Export and processing into mature miRNAs are similar to the canonical pathway.

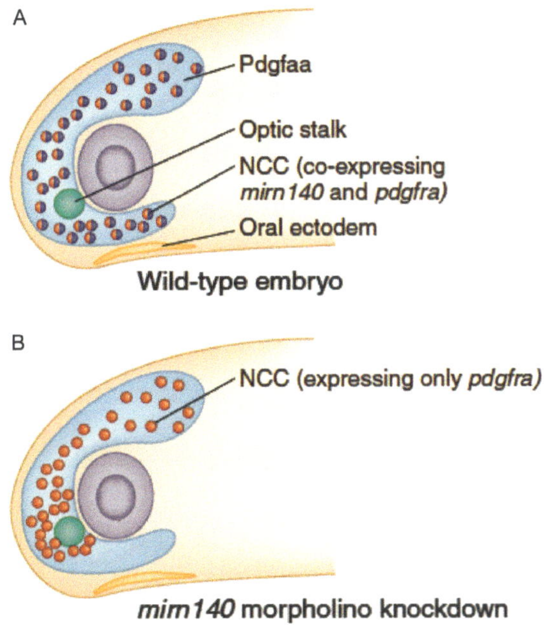

Andre L.P. Tavares *et al.*, Figure 3 *miR-140* attenuates Pdgfra signaling in palatal precursors. (A) *pdgfaa* expression (blue shading) is observed along the migratory route of NCC-derived palatal precursor cells that express both *Pdgfra* and *miR-140* (red/green circles). (B) *miR-140* knockdown results in palatal precursor cells clumping around the optic stalk (yellow circle), rather than continuing to migrate to the oral ectoderm (orange stripe). This subsequently leads to palatal clefting. *From Clouthier (2008).*

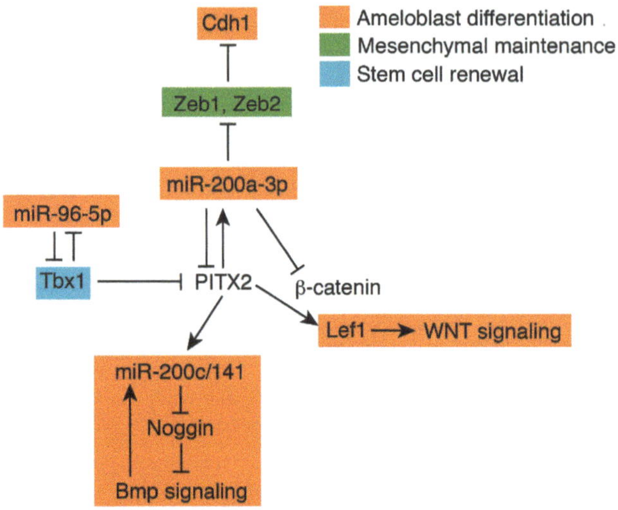

Andre L.P. Tavares *et al.*, Figure 4 Regulation of ameloblast development by PITX2 and miRNAs. *Pitx2* is an essential gene expressed in ameloblastic progenitor cells before differentiation. These cells are in a high proliferative state that is maintained by Tbx1-mediated repression of *Pitx2*. Transition toward differentiation and a nonproliferative state require repression of *Tbx1* by *miR-96-5p*. Pitx2 physically interacts with β-catenin to induce the WNT and BMP pathways, both required for ameloblast differentiation. WNT signaling is activated by PITX2/β-catenin-mediated induction of *Lef1*. Bmp signaling is activated by *miR-200c*. Expression of this miRNA is driven by PITX2/β-catenin. miR-200c directly targets Noggin, allowing Bmp activity. At the same time, Pitx2 drives expression of *miR-200a-3p*, which represses Zeb1 and Zeb2, causing a sequential expression of *Cdh1* (E-cadherin) and transformation of progenitor cells into epithelial-like ameloblastic cells. *miR-200a-3p* also fine-tunes the levels of *Pitx2* and β-catenin, modulating *Lef1* expression, and WNT pathway activity in differentiated cells.

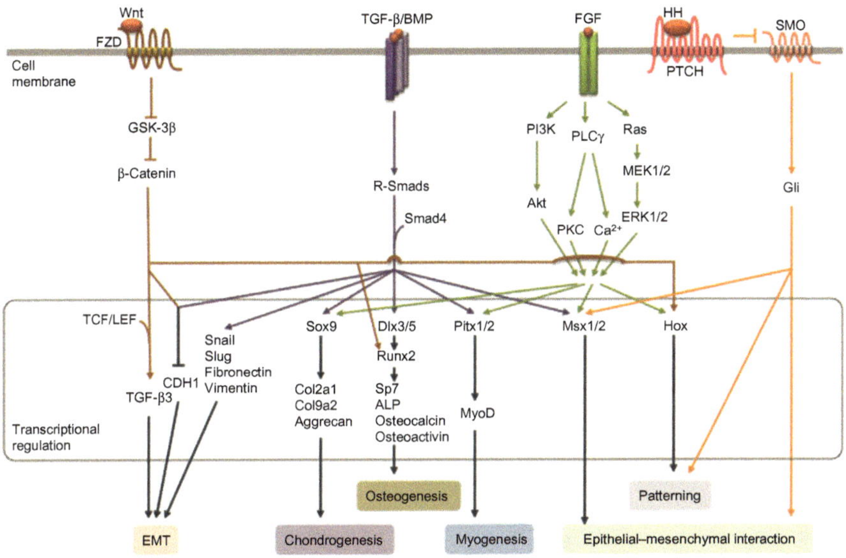

Yongchao Gou *et al.*, Figure 1 Key transcription factors during craniofacial development. TGF-β/BMP signaling induces palatal EMT through Smad-mediated induction of mesenchymal genes and repression of epithelial genes via cooperation with Wnt signaling. Wnt/β-catenin also induces TGF-β3 expression, further enhancing the activation loop. BMP signaling induces Dlx, which in turn induces Runx2 and interact with Runx2 to induce the osteogenic gene program. BMP and FGF signaling induce Sox9, which induce the chondrogenic gene program, and Pitx2, which induces the myogenic gene program in head muscles. BMP, FGF and Hh signaling converge on Msx1/2, which regulate epithelial-mesenchymal interaction. Wnt and FGF signaling induces Hox genes for segment specification.

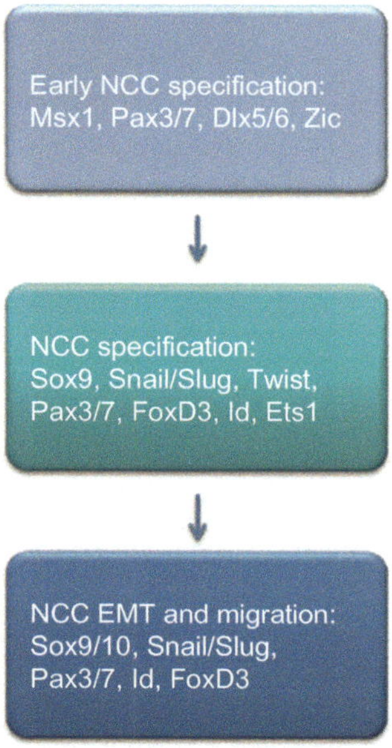

Yongchao Gou *et al.*, Figure 2 Regulatory factors for neural crest cell (NCC) formation. Msx1, Pax3/7, Dlx5/6 and Zic are expressed at the neural plate border prior to NCC specification. They induce Sox9, Snail/Slug, Twist, FoxD3 and Id in the nascent NCC as specifiers, resulting in a discreet cell population. These cells delaminate from the neural tube and migrate dorsolaterally. The expression of Sox9, Snail/Slug, Twist, FoxD3 and Id are maintained to initiate EMT and ensure proper cell dissemination. They further drive the expression of Sox10, which facilitates migration.

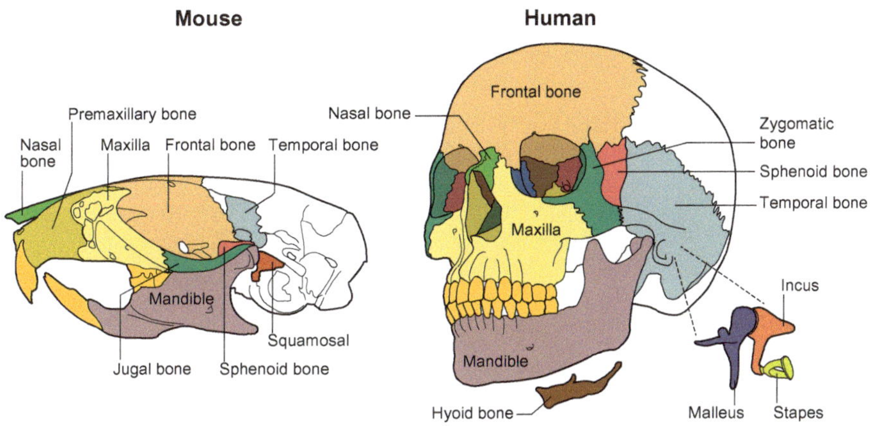

Annita Achilleos and Paul A. Trainor, Figure 1 Contribution of neural crest cells to the craniofacial skeleton. Cranial neural crest cells gives rise to the majority of the craniofacial skeleton. All the colored bones in the mouse (left) and human (right) skull are derived from neural crest cells.

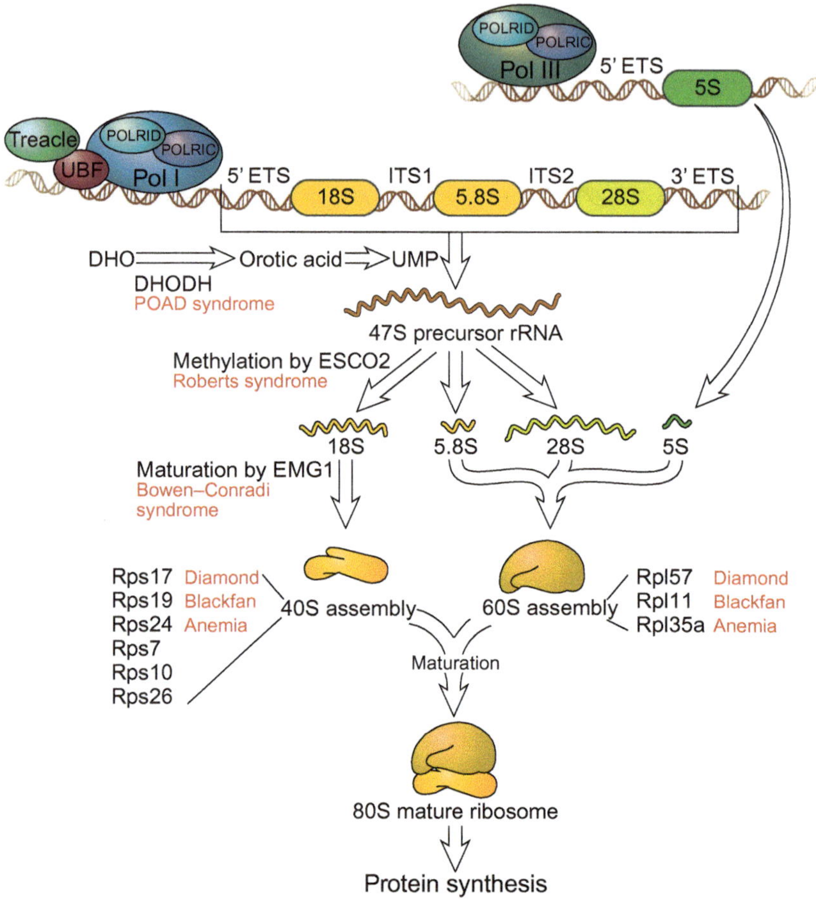

Annita Achilleos and Paul A. Trainor, Figure 2 The ribosome biogenesis pathway and ribosomopathies. Schematic figure summarizing the major stages of ribosome biogenesis. The genes and the corresponding ribosomopathies are depicted. Note that disruption of ribosome biogenesis at different stages leads to different syndromes.

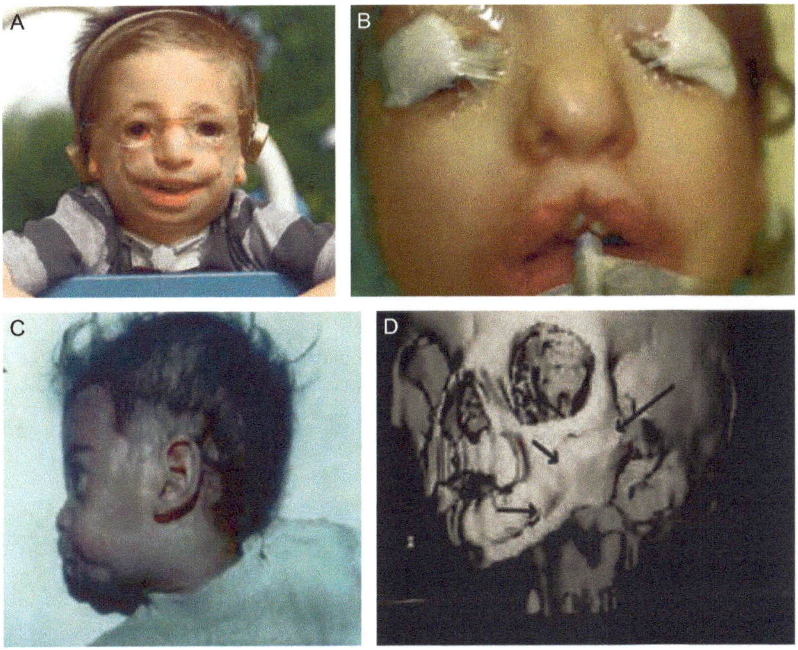

Annita Achilleos and Paul A. Trainor, Figure 3 Examples of rare craniofacial disorders. Images of patients with (A) Treacher Collins syndrome, (B) oral-facial-digital syndrome type 1, and (C) syngnathia. (D) A 3D CT scan image showing a bilateral fusion of the mandible to the maxilla, an underlying feature of syngnathia. *(B) Adapted from Gonzalez, Castro, Nieto, and Bouzan (2014).*

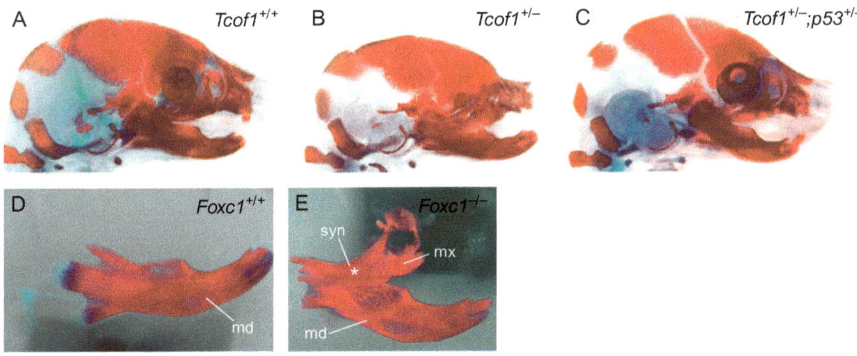

Annita Achilleos and Paul A. Trainor, Figure 4 Mouse models of rare craniofacial disorders. Alizarin red (bone) and alcian blue (cartilage) skeletal staining. (A) Wild-type E18.5 mouse embryo; (B) *Tcof1*$^{+/-}$ littermate that resembles Treacher Collins syndrome (TCS). (C) *Tcof1*$^{+/-}$;*p53*$^{+/-}$ embryo with normal craniofacial morphology. (D) The mandible of a wild-type E18.5 mouse embryo. (E) Mandible of an E18.5 *Foxc1*$^{-/-}$ embryo fused to the maxilla, which is consistent with syngnathia. *Panel (C): Adapted from Jones et al. (2008).*

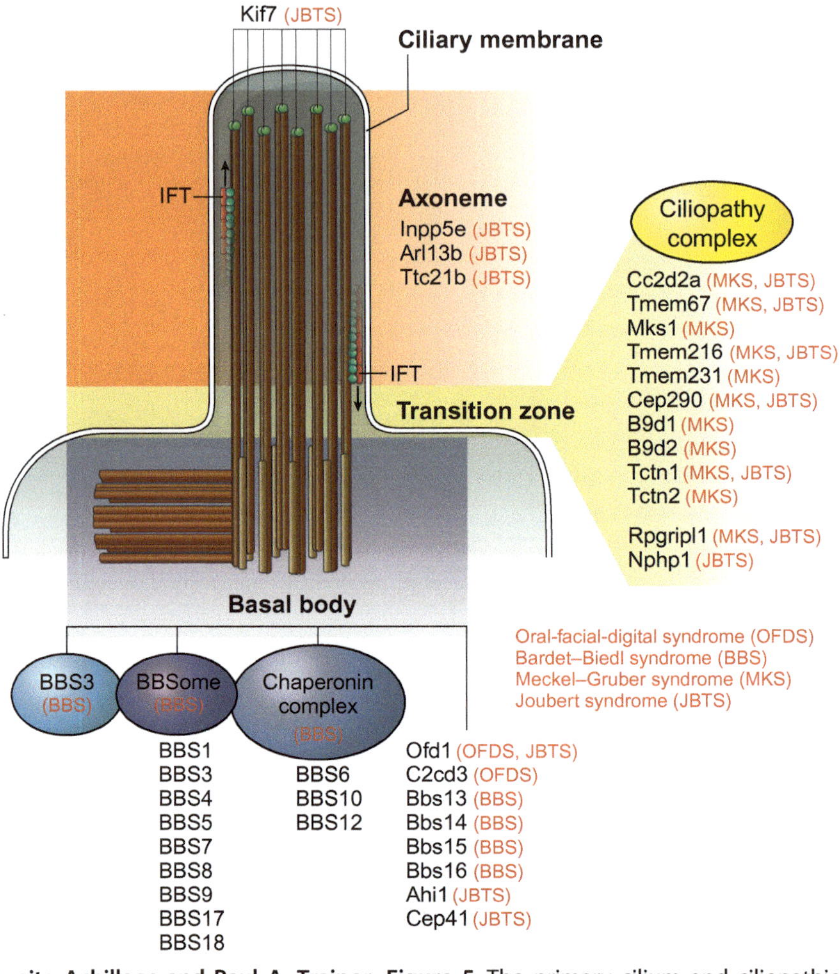

Annita Achilleos and Paul A. Trainor, Figure 5 The primary cilium and ciliopathies. Schematic representation of a primary cilium detailing the main compartments of the cilium as well as the products of the genes and their localization and involvement in specific ciliopathies.

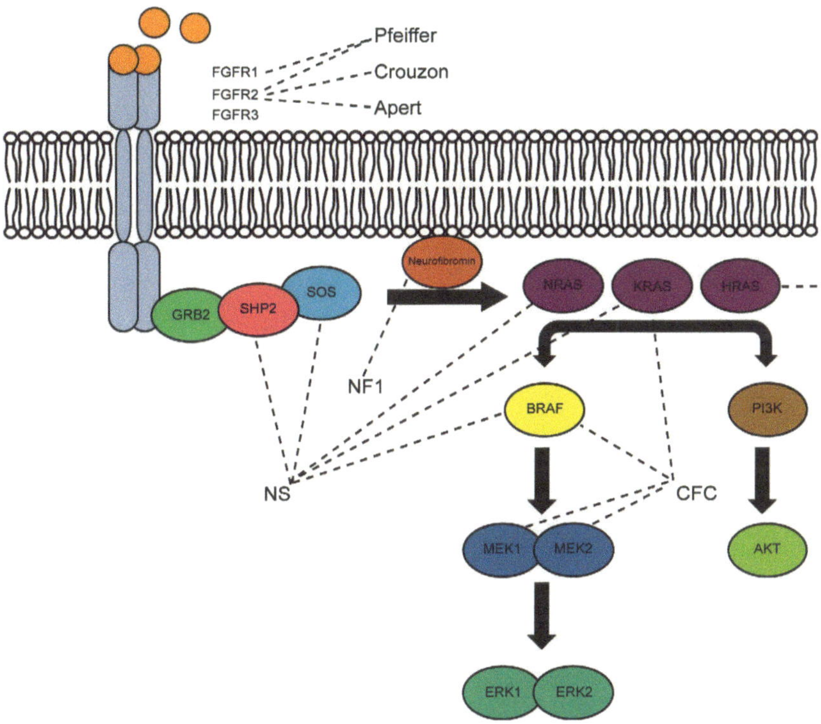

Alice F. Goodwin *et al.*, Figure 1 Syndromes of the RAS pathway. Schematic of the RAS signaling pathway with dashed lines connecting the syndrome with the protein in the pathway encoded by the causative mutated gene. Syndromes described in the text. NS, Noonan syndrome; NF1, neurofibromatosis 1; CFC, cardio-facio-cutaneous syndrome; CS, Costello syndrome.

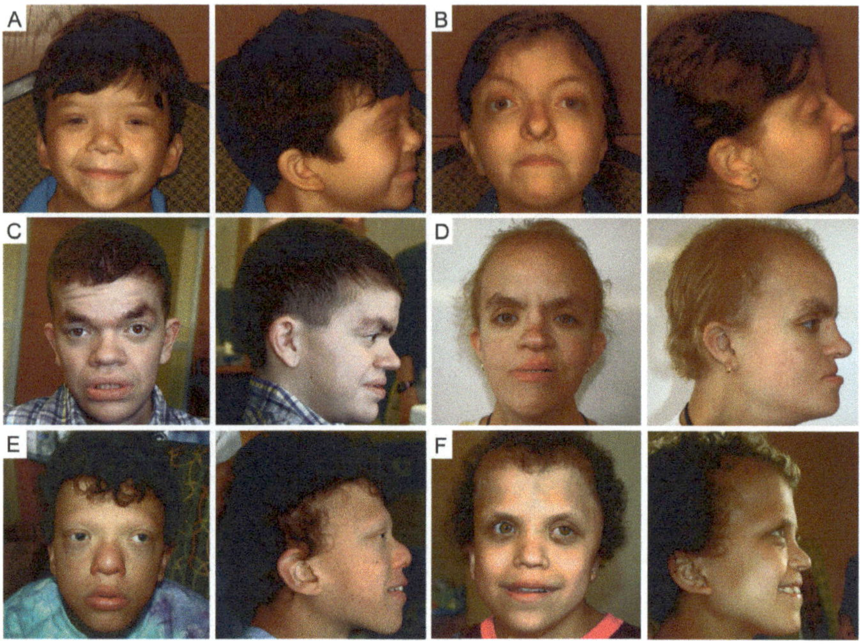

Alice F. Goodwin *et al.*, Figure 2 Craniofacial features of the RASopathies. The RASopathies are caused by mutations in genes encoding different proteins in the RAS pathway and have both unique and overlapping craniofacial characteristics. (A) Frontal and profile photographs of a 7-year-old male and (B) 21-year-old female with Noonan syndrome (NS), (C) a 20-year-old male and (D) 23-year-old female with Costello syndrome (CS), and (E) a 15-year-old male and (F) 15-year-old female with cardio-facio-cutaneous syndrome (CFC).

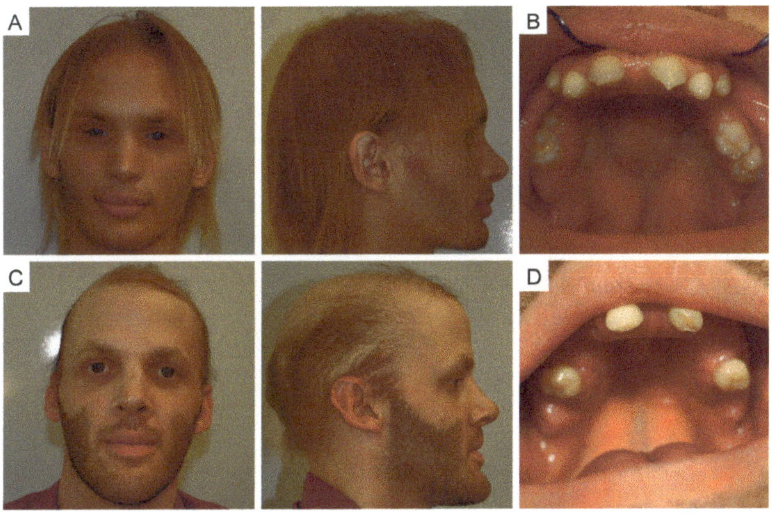

Alice F. Goodwin *et al.*, Figure 3 Craniofacial and dental characteristics of XLHED. (A) 13-year-old male with facial characteristics of XLHED including short face with proportionally longer chin and midface, narrow and pointed nose, narrow mouth with full, rounded lower lip and (B) dental features such as missing teeth and conical-shaped incisors and molars with abnormal cuspal morphology. (C) A 21-year-old male with midface hypoplasia and prognathic mandible and chin and short philtrum, and (D) missing teeth, typical of XLHED.

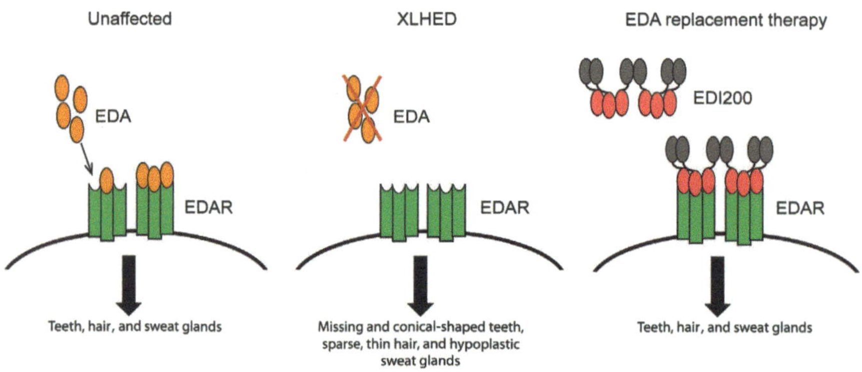

Alice F. Goodwin *et al.*, Figure 4 Schematic of recombinant EDA therapy. In unaffected individuals, EDA binds EDAR, activating downstream signaling that results in tooth, hair, and sweat gland development. XLHED individuals are missing EDA, which results in missing and conical-shaped teeth, sparse, thin hair, and hypoplastic sweat glands. EDI200 therapy replaces missing EDA in XLHED and rescues tooth, hair, and sweat gland development.

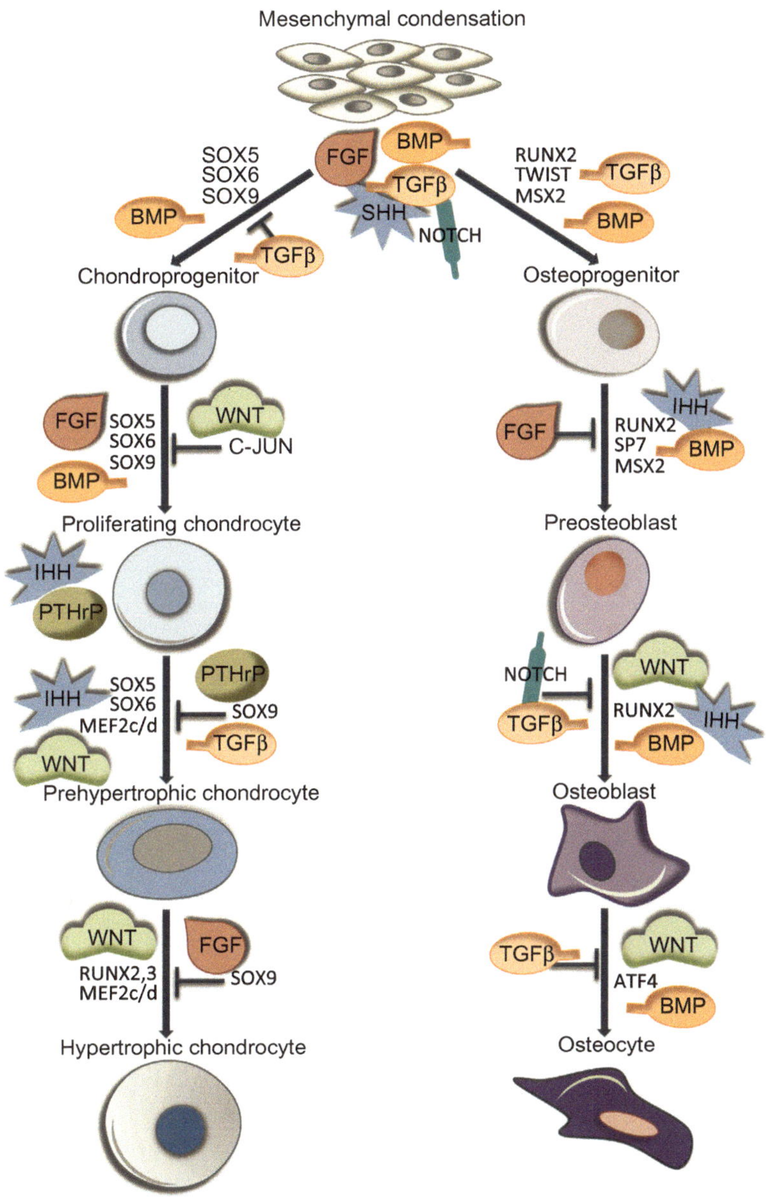

Cynthia L. Neben and Amy E. Merrill, Figure 1 Signaling pathways have combinatorial effects in skeletal development. The key pathways regulating skeletal development include Hedgehog (SHH, IHH), Parathyroid hormone-related protein (PTHrP), WNT, NOTCH, Transforming growth factor β (TGFβ), Bone morphogenetic protein (BMP), and Fibroblast growth factor (FGF). These pathways have combinatorial effects and converge on relatively few transcriptional effectors during the stepwise transition from a self-renewing progenitor cell into a terminally differentiated chondrocyte or osteoblast.

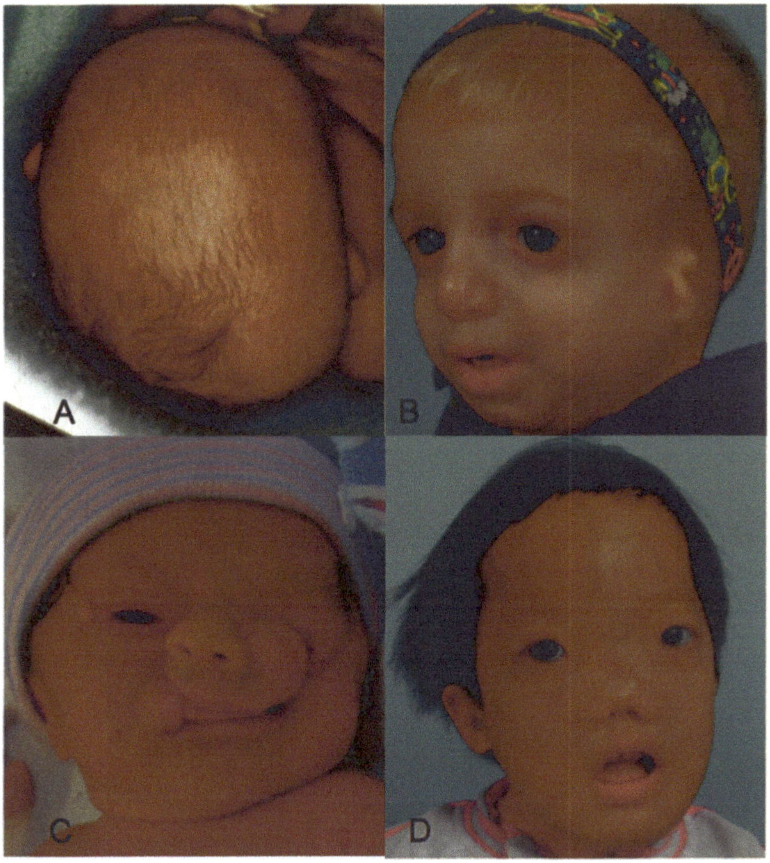

Pedro A. Sanchez-Lara, Figure 1 Examples of a deformation, malformation, disruption, and dysplasia. (A) Deformation example: top view of a child's head with plagiocephaly and left-sided flattening of the occiput (Note the anterior position of the left ear as a secondary finding). (B) Malformation example: child with Treacher Collins syndrome (lower lid colobomas, microtia/anotia, micrognathia, wearing a BAHA for conductive hearing loss). (C) Disruption example: multiple facial defects secondary to amniotic band sequence. (D) Dysplasia example: child with craniometaphyseal dysplasia, a progressive bony dysplasia with thickening of the craniofacial bones.

Father's phenotype	+++	++	+	–	–	–	–
Father's blood test	+++	+/–	+/–	–	–	–	–
Father's Rec. risk	50%	? 1–50%	? 1–50%	? 1–50%	General Pop. risk	General Pop. risk	General Pop. risk
Sons blood test	+++	+++	+++	+++	+++	+/–	–
Sons phenotype	+++	+++	+++	+++	+++	++	+
Sons Rec. risk	50%	50%	50%	50%	50%	? 1–50%	General Pop. risk

Pedro A. Sanchez-Lara, Figure 3 Hypothetical example of an autosomal dominant condition and potential phenotype/genotype spectrum of results when mosaicism is considered. A mutation inherited from the father (A–D); a *de novo* mutation in the male child (E) and post-fertilization somatic mosaicism in the child (F–G). The father's blood DNA test result could be abnormal (+++) (A) potentially abnormal, mosaic, or normal (+/–) (B and C), or normal in (D–G) (–). Figure (D) demonstrates paternal gonadal mosaicism. Figure (E) shows the male child with a *de novo* mutation and (F–G) illustrates decreasing amounts of segmental mosaicism. Affected tissue is illustrated within the colored area (Rec., recurrence; Pop., population).

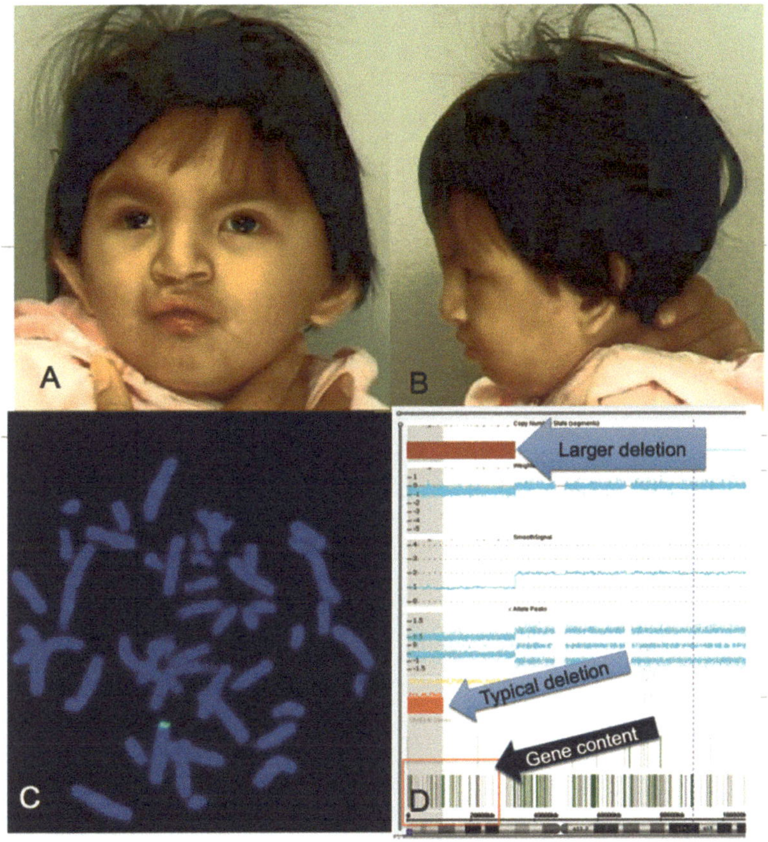

Pedro A. Sanchez-Lara, Figure 4 (A and B) Female patient with cleft lip and palate secondary to Cri Du Chat syndrome (5p minus microdeletion syndrome). (C) Florescence *in situ* hybridization (FISH) green probe for 5p confirmed presence of a single copy. (D) Chromosome microarray detected a 33.7 Mb deletion significantly larger than the typical 5p minus deletion.

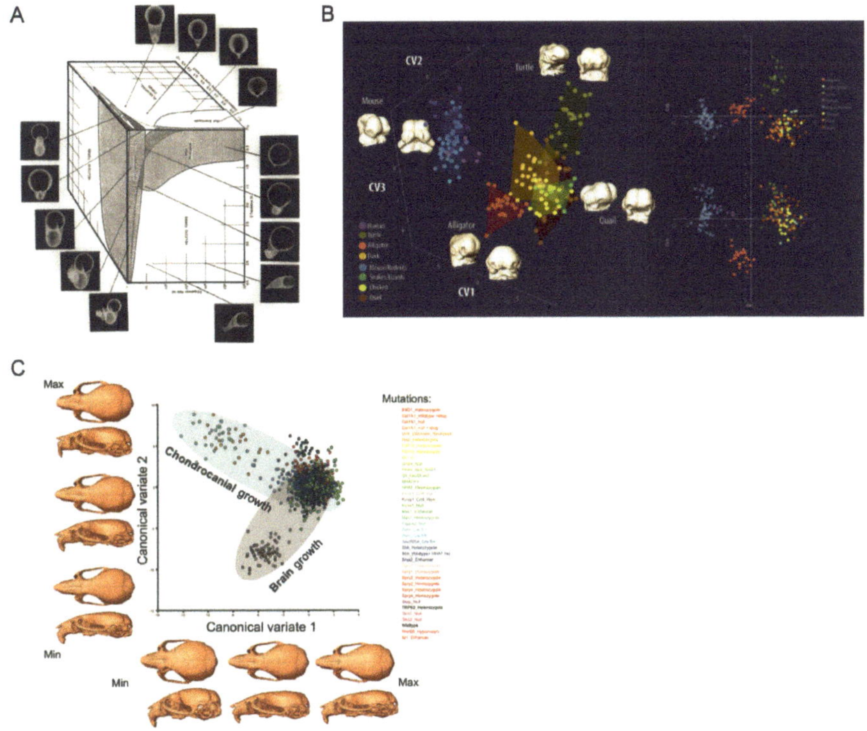

Benedikt Hallgrimsson *et al.*, Figure 1 (A) Raup's morphospace for shell coiling (Raup, 1966). (B) Morphology for amniote embryos, including humans, is shown in a space constructed using principal components analysis from 3D landmark data (Young et al., 2014). (C) Analysis of mouse mutants with in a canonical variates analysis-based space. This plots shows craniofacial effects for mutations affecting chondrocranial growth and brain size forming two distinct common axes of covariation.

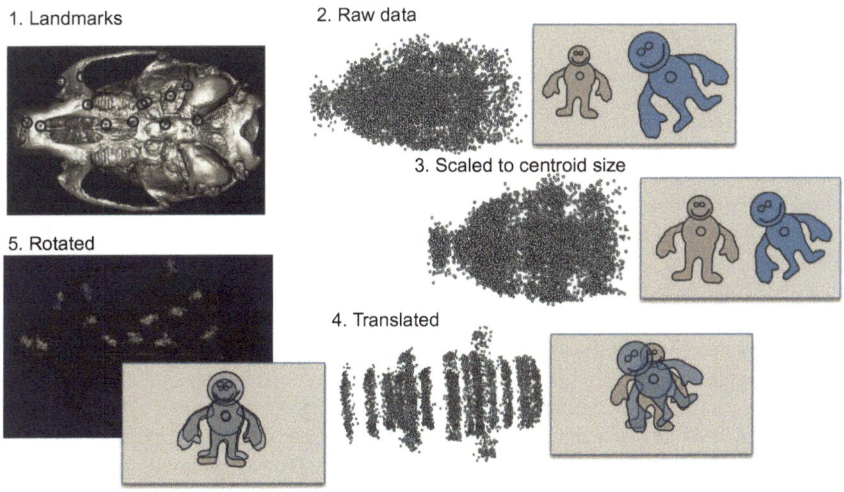

Benedikt Hallgrimsson *et al.*, Figure 2 General procrustes superimposition. (1) A 2D view of a 3D adult mouse landmark set. (2) The full scatter for the raw landmarks in this dataset. In this particular datasets, the two sides of each specimen are treated as separate individuals. (3) The scatter for all individuals (left and right sides) after scaling the LM coordinates for centroid size. (4) The scatter after translation so as to center the centroids of each individual. (5). The distribution of Procrustes residuals after rotation and, where appropriate, reflection superimposed on a ventral view of a mouse.

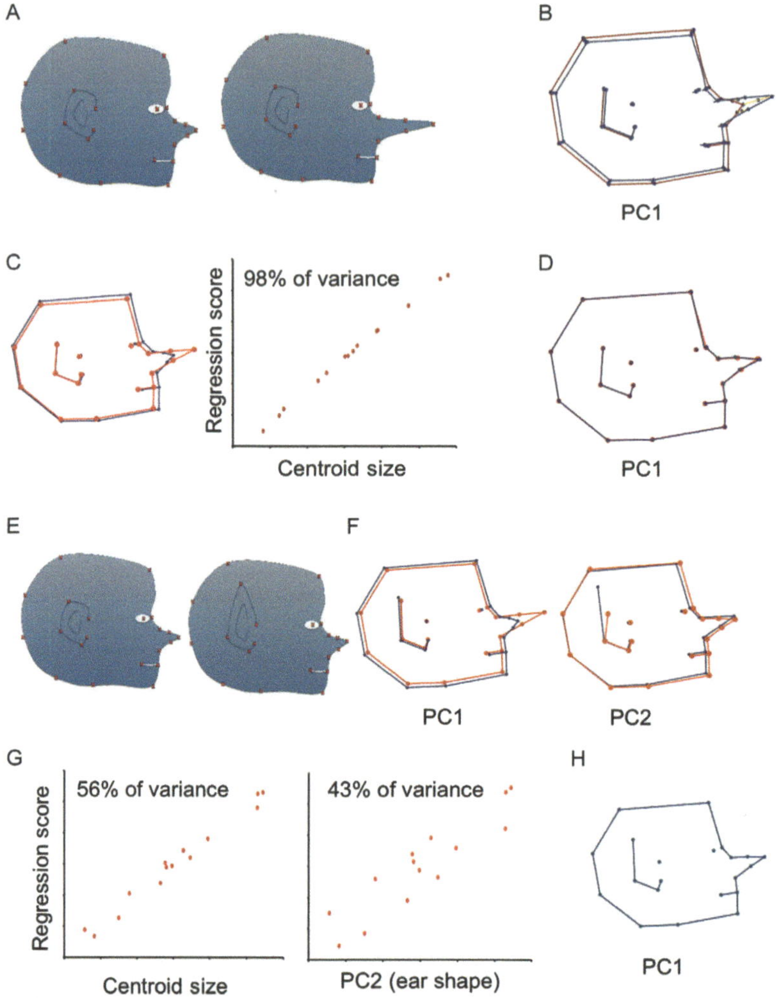

Benedikt Hallgrimsson *et al.*, Figure 4 Illustration of the Pinocchio effect. In (A), a dataset is generated in which a profile of Pinocchio's head varies only in that the nose gets longer. This only displaces the three landmarks on the nose. However, a PCA based on a Procrustes fit of those data shows small displacements of landmarks that were not actually displaced although the largest displacements involve the nose (B). However, the increase in nose length is perfectly correlated with centroid size and so regressing on CS completely removes both the nose length variation and its associated artifacts (C, D). In (D), shape is standardized to the sample average and so shows an intermediate length nose. If we add an additional shape component, ear shape (E) that is uncorrelated with nose length, the Pinocchio effect is complicated slightly. Here, variation is spread across two PCs (F), one which captures nose length and another that captures ear length. These effects and all their associated changes can be removed from the data as shown in (G) and (H), either by regressing out centroid size and PC2 or PC1 and PC2. The artifactual shape changes, however, are preserved in the shape variation associated with these regressions.

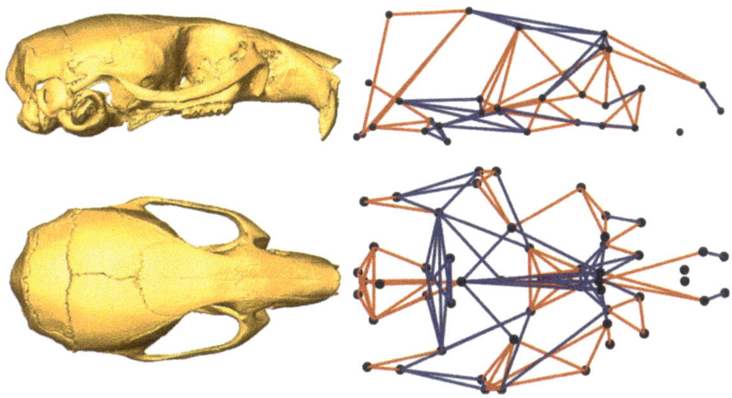

Benedikt Hallgrimsson *et al.*, Figure 5 The strongest (>5% difference in means) significantly ($\alpha = 0.05$) different linear dimensions identified in a EDMA SHAPE comparison of skulls from 129 and C57 mouse strains, viewed from the lateral (top) and superior (bottom) aspects. Measurements were standardized by centroid size before analysis. Red lines are dimensions for which C57 is relatively longer than 129, while blue lines are relatively longer in 129 than C57.

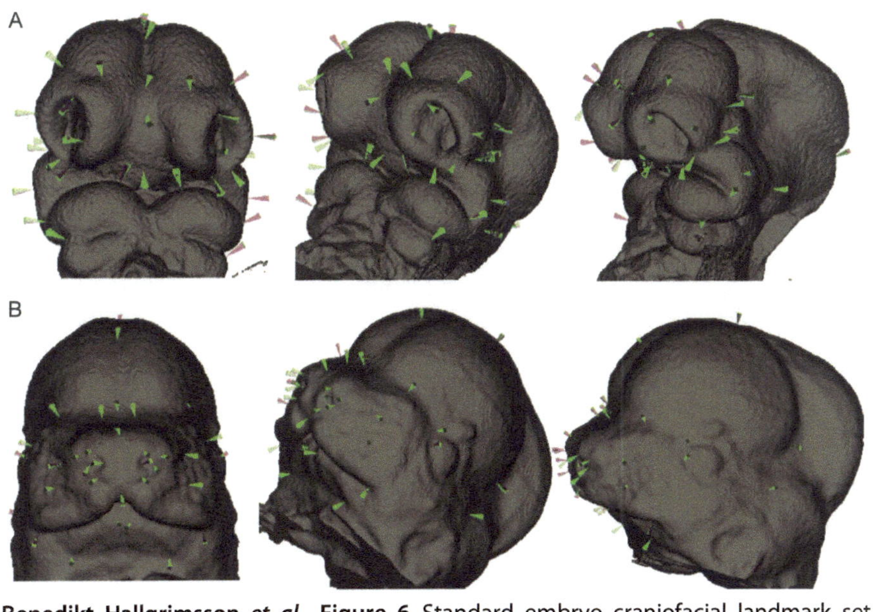

Benedikt Hallgrimsson *et al.*, Figure 6 Standard embryo craniofacial landmark set (Percival et al., 2014).

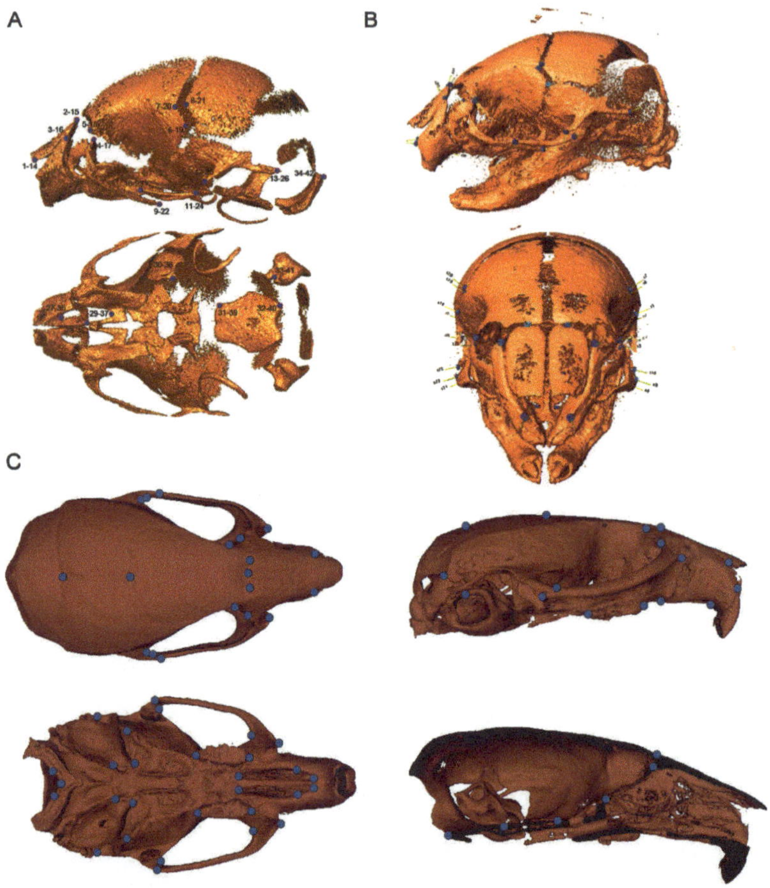

Benedikt Hallgrimsson *et al.*, Figure 8 (A) Late fetal (E17.5) (Gonzalez, Lotto, & Hallgrimsson, 2014), (B) neonatal (Boughner et al., 2008), and (C) adult mouse landmark sets used in recent studies by our group.

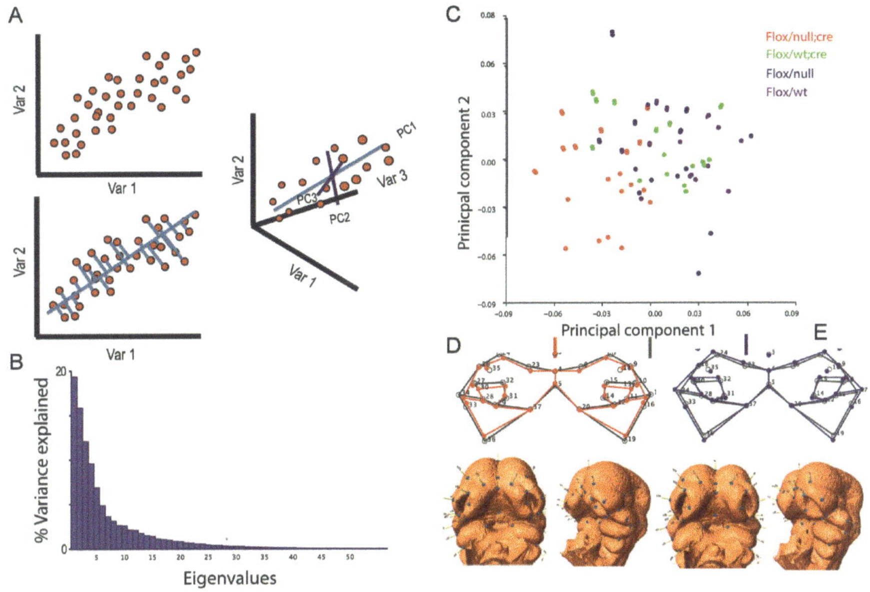

Benedikt Hallgrimsson *et al.*, Figure 9 Principal components analysis in geometric morphometrics. (A) Explanation of PCA (from Zelditch, 2004). The 2D plots illustrate PC1 and the projection of points (individuals) onto PC1 to create PC scores. The 3D plot shows how the scatter among three variables would correspond to principal components. (B–D) Example of the use of principal components analysis to investigate a the effect of a mutation on craniofacial shape in mice (Green et al. 2015).

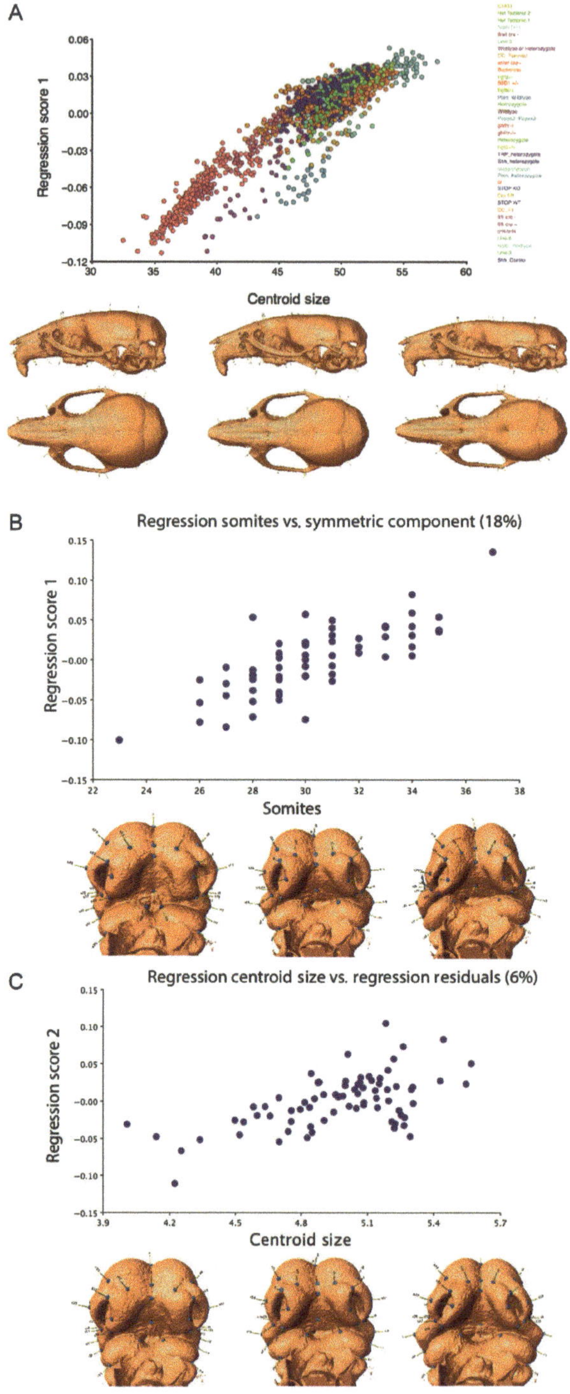

Benedikt Hallgrimsson *et al.*, Figure 10 Allometry in adult mice and in mouse embryos (E1.0–12.5). (A) The allometric component of shape variation in the parental strains and F1 crosses for the collaborative cross mice. (B, C) The results of multiple multivariate regression of shape on both tail somite stage (B) and size (C). (B) shows the component of shape variation that is related to tail somite stage, while (C) shows the static allometry component that is perpendicular to the variation in (C). A challenge in such analyses is the colinearity among the effects of stage and size.

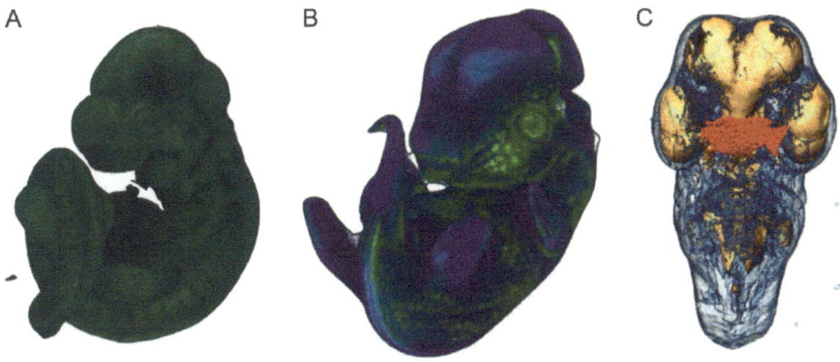

Benedikt Hallgrimsson *et al.*, Figure 12 Examples of optical projection tomography images. (A) E10.5 mouse embryo stained with Sytox Green. (B) E12.5 mouse embryo stained with Sytox Green (blue) and Ser10 phosphohistone H3 (green). (C) Hamburger Hamilton Stage 23 chick embryo showing Shh expression in the frontonasal ectodermal zone (FEZ) highlighted in red.

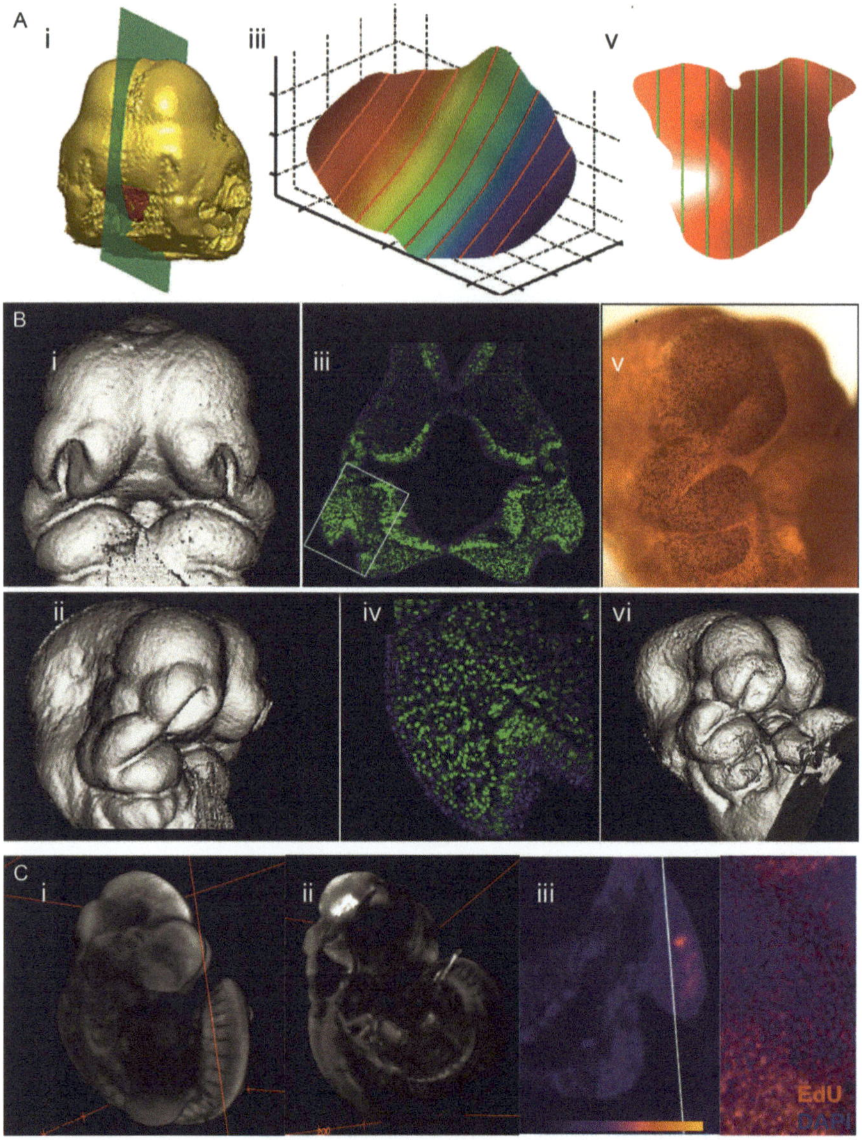

Benedikt Hallgrimsson *et al.*, Figure 13 (A) Analysis of shape of gene expression. (i) OPT scan of whole mount *in situ* showing Shh expression (red). The green plane indicates the slices in (i) and (ii). (ii) Section of FEZ parallel to sagittal plane. (iii) Surface spline extracted from raw FEZ. (B) MicroCT renderings and cell proliferation data from the same specimens. (i and ii) 3D reconstruction of μCT taken after processing but before sectioning. (iii and iv) Hoechst 33342 staining to visualize cell nuclei (blue) with cells in S phase visualized using EdU + Alexa Fluor® 488 labeling (green) at 5 × and 200 ×. (v) Specimen processed whole mount for anti-PHH3 primary antibody to identify M-phase cells. (vi) MicroCT rendering of the same specimen after treatment. (C) Images derived from OPT imaging of EdU-stained embryo counterstained with Sytox Green. (B) Volume view of Sytox Green channel showing regional difference in cell density. (C) Volume view of EdU channel showing regional difference in cell proliferation. (D) Heat map of virtual section of lateral nasal prominence of (C) showing local proliferation differences. (E) EdU incorporation as viewed in a traditional confocal section EdU stain (red) and DAPI (blue).

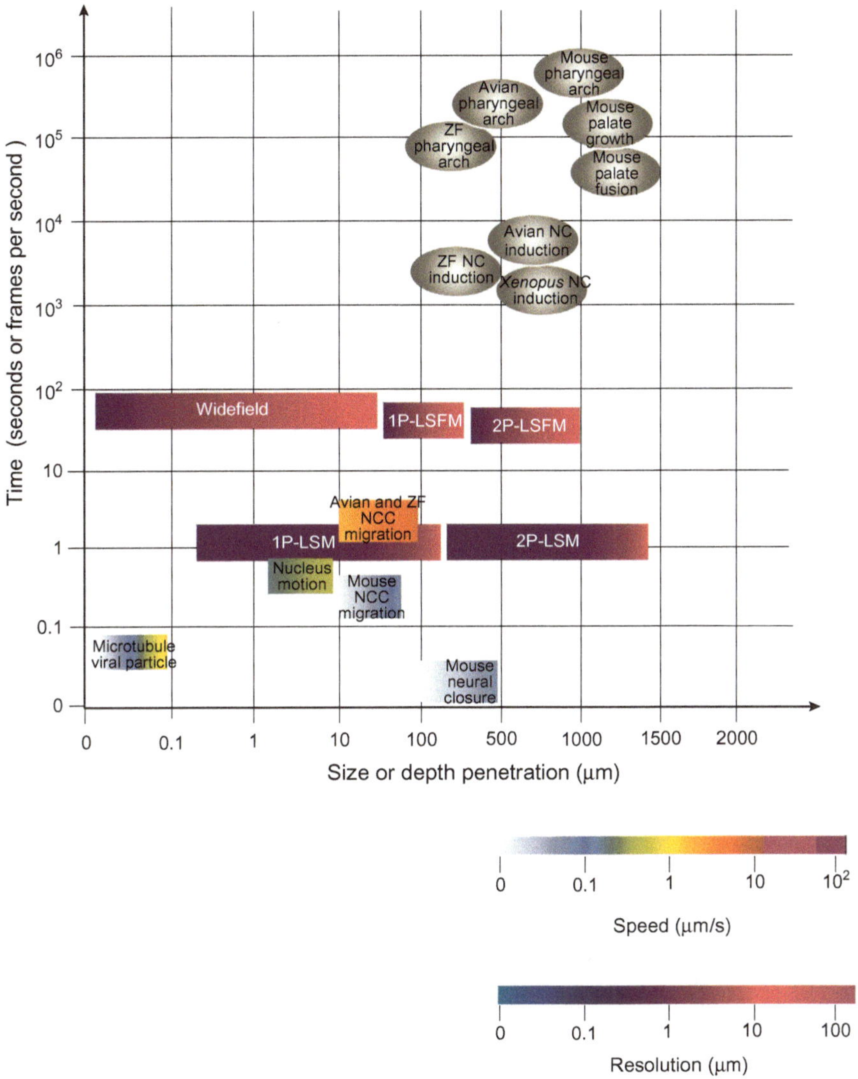

Le A. Trinh and Scott E. Fraser, Figure 1 Varying biological and instrumental spatial and temporal scales encountered in dynamic imaging of development. Plot of biological processes and optical microscopy techniques with *y*-axis representing time or scan rates and *x*-axis representing size or depth penetration. For biological processes, time is mapped based on the length of time a particular process takes to develop (gray ovals) or speed of the delineated biological processes calculated as microns/seconds (rectangular scale bar). For optical techniques, scan rate is calculated for the number of 512×512 pixel frames per second. Scale bar for lateral (*x–y*) resolution microscopy techniques is color-coded in teal to magenta (lower right corner). Scale bar for speed is color-coded in light blue to red (lower right corner). Numbers for biological processes plotted were exacted from the following references; diameter of microtubule (Faisal, White, & Laughlin, 2005), motion of viral particle on microtubule (Arhel et al., 2006), diameter of nucleus and subnuclear motion (Gorisch, Lichter, & Rippe, 2005; Li & Xie, 2011), rate of neural crest migration in mouse, avian, and zebrafish (ZF) (Hochgreb-Hagele & Bronner, 2013; Kawakami, Umeda, Nakagata, Takeo, & Yamamura, 2011; Kulesa & Fraser, 2000), rate of neural tube closure (Pyrgaki, Trainor, Hadjantonakis, & Niswander, 2010), avian, *Xenopus*, and zebrafish neural crest induction (Aybar & Mayor, 2002), avian, mouse, and zebrafish pharyngeal arch formation (Szabo-Rogers et al., 2010; Yelick & Schilling, 2002), and mouse palate growth and fusion (Bush & Jiang, 2012).

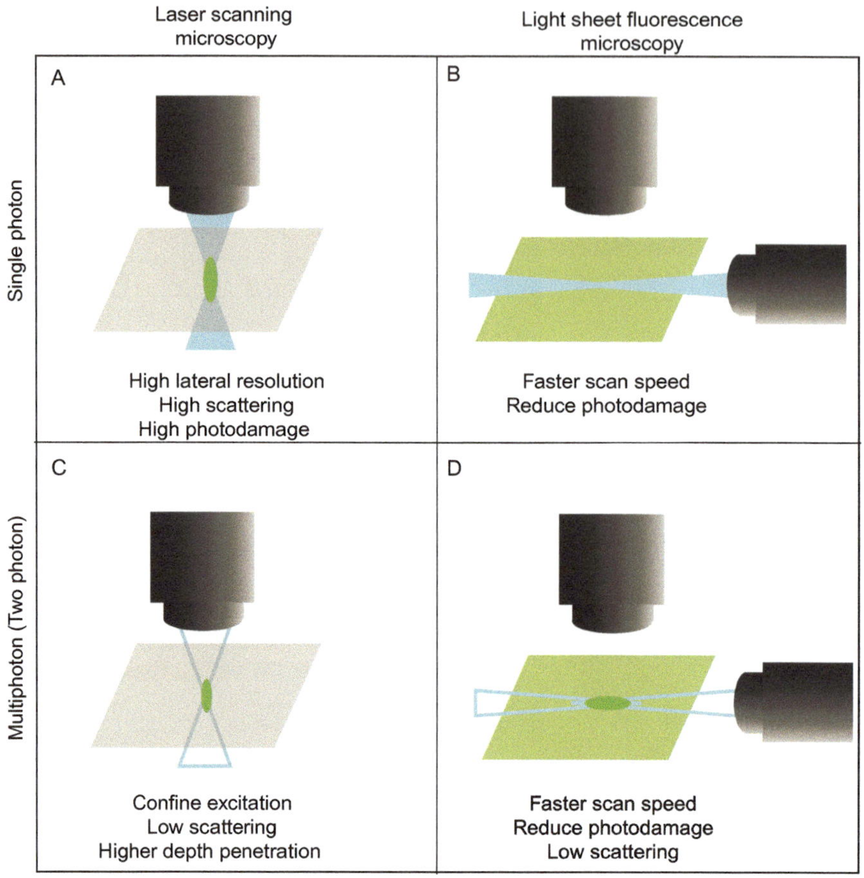

Le A. Trinh and Scott E. Fraser, Figure 2 Approaches in optical microscopy for capturing cell and tissue dynamics. Geometry of laser scanning microscopy (LSM) (A and C) and light-sheet fluorescence microscopy (LSFM) (B and D) with single (A and B) or multiphoton (C and D) excitation. (A) Single-photon LSM provides high lateral (*x–y*) and axial (*z*) resolution through a confocal pinhole that rejects light from above and below the focal plane; however, out-of-focus excitation results in high light scattering and photodamage of the sample. (B) LSFM uses a sheet-shaped beam to selectively excite the plane of interest. Fluorescence is collected with a widefield camera perpendicular to the excitation beam resulting in faster scanning rates and reduce photodamage. (C) Nonlinear excitation of fluorescence in multiphoton LSM confines the excitation to the region of interest. Confined excitation results in reduce photodamage while near-infra-red wavelengths used for excitation allows for lower light scattering and higher depth penetration. (D) Multiphoton LSFM excitation in light-sheet mode provides the same advantages as LSM with an increase in scanning rate.

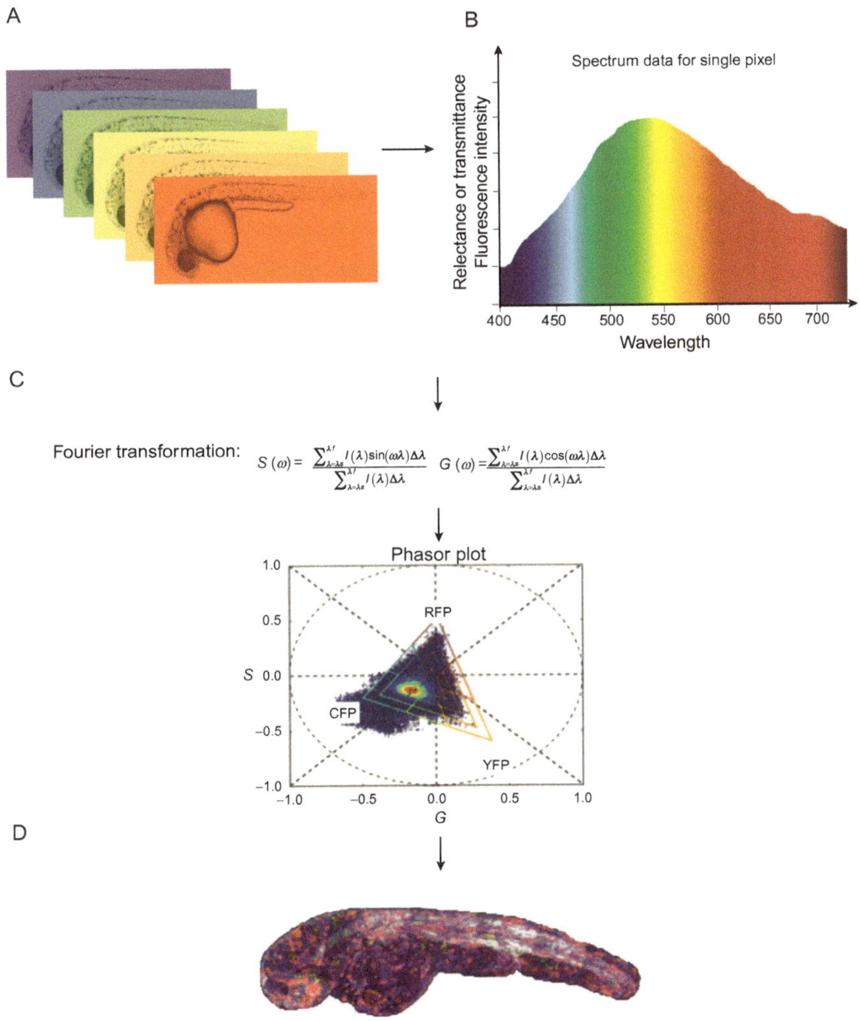

Le A. Trinh and Scott E. Fraser, Figure 3 Hyperspectral imaging workflow of highspectral imaging and data analysis using Spectral Phasor approach. (A) In hyper-spectral imaging, samples are illuminated with a single wavelength and (B) transmitted or reflected fluorescence intensity is collected for the entire visible spectrum at each pixel. (C) Fourier transform of the multidimension data (x, y, z, λ, and/or t) condenses the data into a coordinate S and G on the Phasor plot. S and G coordinates depend on the intensity, $I(\lambda)$, the wavelength, λ, and Harmonic number, ω (Cutrale et al., 2013). (D) Each S and G coordinate on the Phasor plot can be mapped back to corresponding pixels in the image data. *Image of Spectral Phasor plot and corresponding image data courtesy of F. Cutrale.*

FSC
www.fsc.org
MIX
Papier | Fördert gute Waldnutzung
FSC® C083411

Kolibri 360 GmbH
Ferdinand-Jühlke-Straße 7
99095 Erfurt, Deutschland
produktsicherheit@kolibri360.de